第二届先进纺织材料及加工技术国际会议论文集

PROCEEDINGS OF THE SECOND INTERNATIONAL CONFERENCE ON ADVANCED TEXTILE MATERIALS & MANUFACTURING TECHNOLOGY

中国·杭州　HANGZHOU,CHINA 2010.10.20-24

PROCEEDINGS OF THE SECOND INTERNATIONAL CONFERENCE ON ADVANCED TEXTILE MATERIALS & MANUFACTURING TECHNOLOGY

第二届先进纺织材料及加工技术国际会议论文集

中国 浙江理工大学
先进纺织材料与制备技术教育部重点实验室
Zhejiang Sci-Tech University, China
Key Laboratory of ATMMT, Ministry of Education, China

纤维材料和加工技术浙江省重点实验室
生态染整技术教育部工程研究中心
浙江省纺织工程学会
Zhejiang Provincial Key Lab of Fiber Materials and Manufacturing Technology
Engineering Research Center for Eco-Dyeing & Finishing of Textiles, Ministry of Education of China
Textile Engineering Institute of Zhejiang Province

主　编

浙江大学出版社
ZHEJIANG UNIVERSITY PRESS

Conference Academic Committee

Prof. Gao Congjie	State Oceanic Administration of China, China
Prof. Ji Guobiao	National Development and Reform Commission, China
Prof. Shen Zhiquan	Zhejiang University, China
Prof. Sun Jinliang	Shanghai University, China
Prof. Yao Mu	Xian Polytechnic University, China
Prof. Zhou Xiang	Donghua University, China
Prof. Jiang Shicheng	Sinopec Yizheng Chemical Fiber Company Limited, China
Prof. Hogg Paul	The University of Manchester, UK
Prof. Ko Frank	The University of British Columbia, Canada
Dr. Wang Jiping	P & G Company, USA
Prof. Hori Teruo	Fukui University, Japan
Prof. Ni Qingqing	Shinshu University, Japan
Prof. Wang Xungai	Deakin University, Australia
Prof. Tao Xiaoming	Hong Kong Polytechnic University, China
Prof. Zhang Jianchun	Institute of Munitions, General Logistics Department of PLA, China
Prof. Ding Xin	Donghua University, China
Prof. Xiao Changfa	Tianjin Polytechnic University, China
Prof. Chen Guoqiang	Soochow University, China
Prof. Zheng Qiang	Zhejiang University, China
Prof. Wu Dacheng	Sichuan University, China
Prof. Xu Jian	Chinese Academy of Sciences, China
Prof. Zhao Qingzhang	China Textile Academy, China
Prof. Feng Lianfang	Zhejiang University, China
Prof. Chen Jianyong	Zhejiang Sci-Tech University, China
Prof. Chen Wenxing	Zhejiang Sci-Tech University, China
Prof. Liu Guanfeng	Zhejiang Sci-Tech University, China
Prof. Shao Jianzhong	Zhejiang Sci-Tech University, China

Conference Organizing Committee

Prof. Chen Wenxing	Zhejiang Sci-Tech University, China
Prof. Liu Guanfeng	Zhejiang Sci-Tech University, China
Prof. Shao Jianzhong	Zhejiang Sci-Tech University, China
Prof. Wang Lan	Zhejiang Sci-Tech University, China
Prof. Zhu Chengyan	Zhejiang Sci-Tech University, China
Prof. Yao Juming	Zhejiang Sci-Tech University, China
Prof. Zheng Xuming	Zhejiang Sci-Tech University, China
Prof. Wang Xinping	Zhejiang Sci-Tech University, China
Prof. Li Jialin	Zhejiang Sci-Tech University, China
Prof. Zou Fengyuan	Zhejiang Sci-Tech University, China
Prof. Li Chaorong	Zhejiang Sci-Tech University, China
Prof. Xiong Jie	Zhejiang Sci-Tech University, China
Prof. Lu Aihua	Zhejiang Sci-Tech University, China
Associate Prof. Wang Xiaojun	Zhejiang Sci-Tech University, China
Prof. Tao Xiaoming	Hong Kong Polytechnic University, China
Prof. Ni Qingqing	Shinshu University, Japan
Prof. Wang Xungai	Deakin University, Australia
Dr. Wang Jiping	P & G Company, USA
Prof. Ko Frank	The University of British Columbia, Canada
Prof. Chen Xiaogang	The University of Manchester, UK
Prof. Fan Qinguo	University of Massachusetts, USA
Prof. Huh You	Kyung Hee University, Korea

PREFACE

To provide a platform to exchange new ideas, new findings and new technologies for the worldwide textile science and technology, and promote the academic corporation in this area, the Second International Conference on Advanced Textile Materials & Manufacturing Technology, hosted by Zhejiang Sci-Tech University and organized jointly by the Key Laboratory of Advanced Textile Materials & Manufacturing Technology (Ministry of Education of China), Zhejiang Provincial Key Lab of Fiber Materials and Manufacturing Technology, Engineering Research Center for Eco-Dyeing & Finishing of Textiles (Ministry of Education of China) and Textile Engineering Institute of Zhejiang Province, was held on 20-24 October 2010 in Hangzhou, China.

About 110 scholars, specialists and students from universities, research institutes and enterprises in Australia, Canada, China (including Hong Kong), Iran, Japan, R.O. Korea, Singapore, the USA, the UK, and Sweden participated in the conference and presented their papers covering the following topics:

- Novel and high-performance fibers
- Textile composites and industrial textiles
- Advanced manufacturing technology of textiles & clothes
- Functional and smart textiles & clothes
- Eco-dyeing & finishing and pollution control technology
- Environment-friendly textile chemicals and novel synthesis technology
- Basic research and characterization technology in textile area
- Textile machinery, quality control and testing
- Others associated with above themes

The proceedings collect all the papers in the form of oral presentation or poster on the conference, and they can be used as references for the readers in the area of textile materials & manufacturing technology.

Finally, we, the organizers of the conference, would like to express our sincere thanks to all the participants for their supports and contributions to the conference, hoping that the proceedings would bring you nice memory on the conference.

Key Laboratory of Advanced Textile Materials & Manufacturing Technology (Ministry of Education of China), Zhejiang Sci-Tech University, China

目录

1　Structure and Thermal Stability of Nanoclay/flax Nanocomposite ················ 1
Chunhong Wang[1,2], Frank K. Ko[1], Mercedes Alcock[3]
(1. Department of Materials Engineering, the University of British Columbia, Vancouver, Canada
2. Textile College, Tianjin Polytechnic University, Tianjin, China
3. Composites Innovation Centre Manitoba, Inc. Manitoba, Canada)

2　Multifunctional Composite Nanofibers ················ 5
Frank Ko, Masoumeh Bayat, Heejae Yang
(Advanced Fibrous Materials Laboratory, University of British Columbia, Canada)

3　Liquid Crystalline Electrospinning of Carbon Nanotube Reinforced Cellulose Fibers from Bamboo ················ 11
Yuqin Wan, Frank K. Ko
(Advanced Fibrous Materials Laboratory, Department of Material Engineering, University of British Columbia)

4　Structures and Properties of Kapok Fiber ················ 17
Qiuling Cao, Yi Cao, Lin Wang, Xiaowei Sun
(Department of Textiles, Henan Institute of Engineering, Zhengzhou, China)

5　Analysis on Structure of Wool Keratin Film by FT-IR and SEM ················ 21
Liping Chen, Ping Cui
(College of Mechanical and Electric Engineering, Lanzhou University of Technology, Lanzhou, Gansu, China)

6　Preparation and Characterization of CA/CeO$_2$ Composite Nanofibers ················ 24
Rui Chen, Xiaoqiang Zhang, Wenjun Dong, Chaorong Li
(Department of Physics, and Key Laboratory of Advanced Textile Materials and Manufacturing Technology, Ministry of Education of China, Zhejiang Sci-Tech University, Hangzhou, China)

7　Reach on the Structure and Character of a Thermal Regulating Fiber ················ 29
Weilai Chen[1], Ying Gao[1], Jianxiao Zhu[1], Zhebin Tang[1], Yihui Gong[2]
(1. School of Materials and Textiles, Zhejiang Sci-Tech University, Hangzhou, China
2. Zhejiang Sunjazz Garments Co., Ltd., Yiwu, China)

8　Preparation and Characterization of Nanofibrous Bioactive Glass Scaffolds ················ 33
Chunxia Gao, Akira Teramoto, Koji Abe
(Faculty of Textile Science and Technology, Shinshu University, Nagano, Japan)

9 Study on the Preparation and Characterization of SWNTs/Lyocell Composite Fibers ············37
Baohui Guan, Huihui Zhang, Gesheng Yang, Huili Shao, Xuechao Hu
(State Key Laboratory for Modification of Chemical Fibers and Polymer Materials, College of Material Science and Engineering, Donghua University, Shanghai, China)

10 The Effect of Acids on Mechanical Properties of PPS Fibers ·················41
Xiangbing He[1], Bin Yu[1], Jian Han[1,2], Xinbo Ding[1], Mitsuo Matsudaira[3]
(1. School of Materials and Textiles, Zhejiang Sci-Tech University, Hangzhou, China
2. Key Laboratory of Advanced Textile Materials and Manufacturing Technology, Ministry of Education of China, Zhejiang Sci-Tech University, Hangzhou, China
3. Kanazawa Univ., Kakuma-machi, Kanazawa City, Japan)

11 Removal of Indoor Ammonia with Fe (III)-modified PAN Fiber Complexes ···············44
Xing Li[1], Yongchun Dong[1,2], Jiangxing He[1], Zhenbang Han[1,2], Zhichao Wang[1]
(1. Division of Textile Chemistry & Ecology, School of Textiles, Tianjin Polytechnic University, Tianjin, China
2. State Key Laboratory Breeding Base of Photocatalysis, Fuzhou University, Fuzhou, China)

12 Preparation of Catalytic Activated Carbon Fiber and Its Catalytic Oxidation Performance to 4-nitrophenol ·················49
Yanli Li, Chunxia Ma, Qiaosheng Guo, Wenxing Chen
(Key Laboratory of Advanced Textile Materials and Manufacturing Technology, Ministry of Education of China, Zhejiang Sci-Tech University, Hangzhou, China)

13 Influence of SiC Coating on the Oxidation Behavior of PAN Carbon Fiber at Elevated Temperatures ·················54
Ye Li, Lina Wang, Weihui Xie, Jianjun Chen
(Key Laboratory of Advanced Textile Materials and Manufacturing Technology, Ministry of Education of China, Zhejiang Sci-Tech University, Hangzhou, China)

14 Investigation of Osteoblast-like MC3T3-E1 Cells on a Collagen-like Protein and Poly (lactic-co-glycolic acid) Nanofibrous Composite Scaffold ·················59
Yuan Li, A. Teramoto, K. Abe
(Department of Functional Polymer Science, Faculty of Textile Science and Technology, Shinshu University, Ueda, Nagano, Japan)

15 Electrospun Polyvinyl Alcohol/Halloysite Nanotubes Composite Nanofibers ···············63
Ruijuan Liao[1], Baochun Guo[1,2], Jianqing Zhao[1], Wen You Zhou[3]
(1. Department of Polymer Materials and Engineering, South China University of Technology, Guangzhou, China
2. State Key Laboratory of Pulp and Paper Engineering, South China University of Technology, Guangzhou, China
3. Discipline of Orthodontics, Faculty of Dentistry, The University of Hong Kong, Hong Kong, China)

16   Effect of Rheological Properties on Electrospinning of Ultra High Molecular Weight (UHMW) Poly(vinyl alcohol) ··68

Fan Liu[1], Yasushi Murakami[2], Qing-Qing Ni[2]

(1. Interdisciplinary Graduate School of Science and Technology, Shinshu University, Ueda, Japan

2. Faculty of Textile Science and Technology, Shinshu University, Ueda, Japan)

17   Synthesis and Characterization of PSA-PEG Block Copolymer Based on Polysulfonamide and Amine-Terminated Polyethylene Glycol ··72

Li Liu[1,2], Weiwei Gu[2], Jie Zhou[2], Yanan Wu[2], Changfa Xiao[1]

(1. Key Laboratory of Hollow Fiber Membrane Material and Membrane Process of Ministry of Education, Tianjin Polytechnic University, Tianjin, China

2. Dept. of Polymer Science, School of Material Science and Engineering Technology, Shanghai Univ., Shanghai, China)

18   Shape Memory Effect and Actuation Property of Shape Memory Polymer Based Nanocomposites ··76

Qingqing Ni, Li Zhang

(Department of Functional Machinery and Mechanics, Shinshu University, Ueda-shi, Japan)

19   The Relationship between the Structures and Mechanical Properties of *A. pernyi* Silk ········82

Chengjie Fu, Zhengzhong Shao

(Key Laboratory of Molecular Engineering of Polymers of Ministry of Education, Advanced Materials Laboratory, Department of Macromolecular Science, Fudan University, Shanghai, China)

20   Flexible Tactile Sensor Based on PVDF Fibrous Membrane ···84

Yongrong Wang[1], Jianming Zheng[2], Peihua Zhang[1], Wenxing Chen[3], Chunye Xu[2,4]

(1. College of Textiles, Donghua University, Shanghai, China

2. National Laboratory for Physical Sciences at Microscale, University of Science and Technology of China (USTC), Hefei, China

3. Key Laboratory of Advanced Textile Materials and Manufacturing Technology, Ministry of Education of China, Zhejiang Sci-Tech University, Hangzhou, China

4. University of Washington, USA)

21   Modification of Wool Fiber Using Freeze Treatment ··88

Zhengwei Wang, Ruoying Zhu, Jianfei Zhang

(School of Textiles, Tianjin Polytechnic University, Tianjin, China)

22   Preparation and Properties Research of Poly(lactide-co-glycolide)/Silk Blend Nanofibrous Membrane ···93

Hongwei Xiao[1], Jie Xiong[1], Ni Li[1], Hongping Zhang[1], Junjun Xie[2]

(1. Key Laboratory of Advanced Textile Materials and Manufacturing Technology, Ministry of Education of China, Zhejiang Sci-Tech University, Hangzhou, China

2. Clinical Medical College, Hangzhou Normal University, Hangzhou, China)

23　**Effect of Low Temperature Plasma Treatment on Surface Properties of Polysulfonamide Fiber** ·················98

Jianzhong Yang, Lei Ren

(School of Textile and Material, Xi'an Polytechnic Univ., Xi'an, China)

24　**Rheological Behavior and Spinning Performance of Cellulose/[BMIM]Cl Solutions Prepared by Two-Steps Dissolving Process** ·················103

Huihui Zhang, Tao Cai, Huili Shao, Xuechao Hu

(State Key Laboratory for Modification of Chemical Fibers and Polymer Materials, College of Material Science and Engineering, Donghua University, Shanghai, China)

25　**Fabrication and Application of Carbon Nanotube/Magnetite Composites** ·················107

Li Zhang[1], Qing-Qing Ni[2], Yoshio Hashimoto[1]

(1. Faculty of Engineering, Shinshu University, Japan
2. Faculty of Textile Science & Technology, Shinshu University, Japan)

26　**Preparation and Regeneration of Bioplasts for Biomodification of Polyester** ·················111

Weiling Zhang, Jianfei Zhang, Zheng Li, Jixian Gong, Qiujin Li, Cheng Chen, Zhaolong Hu

(Key Laboratory of Advanced Textile Composites, Tianjin Polytechnic University, Ministry of Education, Tianjin, China)

27　**Preparation and Properties Characterization of Butyl-methacrylate Copolymer Absorptive Functional Fiber** ·················115

Zhongjuan Zhang, Changfa Xiao, Naiku Xu

(Tianjin Key Laboratory of Fiber Modification and Function Fiber, Tianjin Polytechnic University, Tianjin, China)

28　**Solubility of Bacterial Cellulose in LiCl/DMAc Solvent System** ·················119

Chuanjie Zhang, Ping Zhu, Liu Wang

(Key Laboratory of Green Processing and Functional Textiles of New Textile Materials of Ministry of Education, Wuhan Textile University, Wuhan, China)

29　**Photocatalytic Properties of TiO_2 Supported on Pd-modified Carbon Fibers** ·················124

Yaofeng Zhu[1], Yaqin Fu[2], Qing-Qing Ni[1]

(1. Graduate School of Science and Technology, Shinshu University, Japan
2. Key Laboratory of Advanced Textile Materials and Manufacturing Technology, Ministry of Education of China, Zhejiang Sci-Tech University, Hangzhou 310018, China)

30　**Impregnation of Metal Complex into Epoxy Insulation Materials Using Supercritical Carbon Dioxide and Its Application for Copper Plating** ·················128

H. Ohnuki[1], S. Sumi[2], T. Hori[1]

(1. Fiber Amenity Engineering Course, Graduate School of Engineering, University of Fukui, Fukui, Japan
2. Guangzhou Meadville Electronics Co., Ltd., Guangzhou, China)

31 Synthesis, Characterization and Application of Polyurethane Modifying Polyether Block Polysiloxane ·········135

Qiufeng An, Fan Yang, Kefeng Wang, Ge Li

(Key Laboratory of Auxiliary Chemistry and Technology for Chemical Industry, Ministry of Education, Shaanxi University of Science and Technology, Xi'an, China)

32 Preparation of Bacterial Cellulose Nanofiber with Silver Nanoparticles by In Situ Method ·········140

Zhijiang Cai[1,2], Guang Yang[1]

(1. School of Textiles, Tianjin Polytechnic University, Tianjin, China
2. Key Laboratory of Advanced Textile Composites, Ministry of Education of China, Tianjin, China)

33 Study and Development of New Motorcyclists' Racing Suit Protector ·········144

Nannan Cao, Jianwei Ma, Shaojuan Chen

(Qingdao University, Qingdao, China)

34 Study on the Morphology and Damping Properties of the Organic Hybrids of Chlorinated Polyethylene and Hindered Phenol ·········149

Xinbo Ding[1], Tao Liu[1], Jian Han[2]

(1. College of Material & Textile Engin., Zhejiang Sci-Tech University, Hangzhou, China
2. Key Laboratory of Advanced Textile Materials and Manufacturing Technology, Ministry of Education of China, Zhejiang Sci-Tech University, Hangzhou, China)

35 Study on Energy Absorption of Epoxy Resin Composite Targets Enforced with SiC Powder and Twaron Staple Subjected to Bullet Impact ·········153

Liping Guan[1,2], Jianchun Zhang[1,3]

(1. College of Textile, Donghua University, Shanghai, China
2. Ningbo Advanced Textile Technology & Fashion CAD Key Laboratory, Zhejiang Textile & Fashion College, Ningbo, China
3. The General Logistics Department of PLA, Beijing, China)

36 Research on Acoustic Performance of Basalt Filament Diaphragm Fabric ·········157

Zongfu Guo, Zhili Zhong, Ruibin Yang

(School of Textile, Tianjin Polytechnic University, Tianjin, China)

37 Preparation and Characterization of Polypropylene Master-batches Containing Electret for Meltblown Nonwoven ·········162

Jian Han, Xiangbing He, Bin Yu, Guoping Xu, Xingbo Ding

(Key Laboratory of Advanced Textile Materials and Manufacturing Technology, Ministry of Education of China, Zhejiang Sci-Tech University, Hangzhou, China)

38 Consolidation of Fragile Silk Fabrics with Fibroin Protein and Ethylene Glycol Diglycidyl Ether by Spaying ·········166

Xiaofang Huang[1], Zhiwen Hu[1], Zhiqin Peng[1], Jing Zhang[1], Xiaoye Cao[1], Yang Zhou[2], Feng Zhao[2]
(1. Key Laboratory of Advanced Textile Materials and Manufacturing Technology, Ministry of Education of China, Zhejiang Sci-Tech University, Hangzhou, China
2. China National Silk Museum, Hangzhou, China)

39 Study on Forming Technology and Mechanical Properties of Biodegradable Hemp Fiber/Polylactic Acid Composites ········170

Lihua Lv, Yongling Yu

(Green Applied Fiber Technology Institute, Dalian Polytechnic University, Dalian, China)

40 Rheological Behavior of Spinning Dope of Polyvinyl Alcohol/attapulgite Nanocomposites ······174

Zhiqin Peng[1], Jinchao Yu[1], Hong Xu[2], Zhiping Mao[2]

(1. College of Materials and Textile, Zhejiang Sci-Tech University, Hangzhou, China
2. College of Materials Science and Engineering, State Key Laboratory for Modification of Chemical Fibers and Polymer Materials, Donghua University, Shanghai, China)

41 Finite Element Calculation on the Quasi-static Impact Performance of 3D Integrated Hollow Woven Structure Composites ········180

Wei Tian, Chengyan Zhu

(College of Materials and Textiles, Key Laboratory of Advanced Textile Materials and Manufacturing Technology, Ministry of Education of China, Zhejiang Sci-Tech University, Hangzhou, China)

42 Effect of Different Fabric Skin Combinations on Predicted Sweating Skin Temperature of a Thermal Manikin ········184

Faming Wang, Kalev Kuklane, Chuansi Gao, Ingvar Holmér

(Thermal Environment Laboratory, Division of Ergonomics and Aerosol Technology, Department of Design Sciences, Faculty of Engineering, Lund University, Lund, Sweden)

43 Preparation of a Novel Anti-bacterial Wool Fabrics ········187

Junhua Wang[1], Zaisheng Cai[2]

(1. Department of Textile and Clothes, Wuyi University, Guangdong, China
2. College of Chemistry, Chemical Engineering and Biotechnology, Donghua University, Shanghai, China)

44 Experimental Study of the High Temperature Resistant Filtration Materials Bonding by of Hydroentanglement Technique ········191

Xiangqin Wang, Lili Wu, Xiangyu Jin, Qinfei Ke

(Key Laboratory of Textile Science & Technology, Ministry of Education, Donghua University, China)

45 Study on the Intensity of Joints for the Textile Geogrids Manufactured with PET High-performance Yarn ········196

Guoping Xu[1], Xinbo Ding[1], Jian Han[2]

1. College of Material & Textile Engin., Zhejiang Sci-Tech University, Hangzhou, China
2. Key Laboratory of Advanced Textile Materials and Manufacturing Technology, Ministry of

Education of China, Zhejiang Sci-Tech University, Hangzhou 310018, China)

46 Study on the Morphology, Structure and Mechanical Performances of Electrospun Aligned PAN/Activated Carbon Nanofiber ·················200

Hua Xue, Ni Li, Jie Xiong, Guan Feng Liu

(Key Laboratory of Advanced Textile Materials and Manufacturing Technology, Ministry of Education of China, Zhejiang Sci-Tech University, Hangzhou, China)

47 Effect of Fabric Structure on the Sound Insulation Property of Honeycomb Weave Fabric/PVC Composites Material ·················205

Tianbing Yang[1], Yaofeng Zhu[2], Bangyong Pang[1], Jin Wang[1], Hao Cen[1], Liyuan Zhang[1], Yaqin Fu[1]

(1. Key Laboratory of Advanced Textile Materials and Manufacturing Technology, Ministry of Education of China, Zhejiang Sci-Tech University, Hangzhou, China
2. Graduate School of Science and Technology, Shinshu University, Japan)

48 Synthesis of Nitrogen/phosphorus/silicon Composite Flame Retardant and Its Charring Properties in Polyester ·················209

Lewei Zhang, Huapeng Zhang, Jianyong Chen

(Key Laboratory for Advanced Textile Materials and Manufacturing Technology, Ministry of Education, College of Materials, Zhejiang Sci-Tech University, Hangzhou, China)

49 Computer Evaluation System of Woven Fabric Smoothness Based on 2D Wavelet Transform ·················214

Yifan Zhang, Ameersing Luximon

(Institute of Textiles and Clothing, The Hong Kong Polytechnic University, Hong Kong, China)

50 Advantages of Basalt Fiber Reinforced Concrete Structures ·················218

Dangfeng Zhao[1], Huawu Liu[1], Zhigang Chen[2], Yinhua Zhang[3], Dangqi Zhao[4]

(1. School of Textiles, Tianjin Polytechnic University, Tianjin, China
2. Tianjin Institute of Electronics and Information, Tianjin, China
3. Yueyang Textile Research Center, Hunan, China
4. China Water Resources and Hydropower 11th Engineering Bureau, Sanmenxia, China)

51 Fabric Drape Prediction and Simulation ·················223

Hua Zhou, Yanfang Shao, Quan Wen

(College of Materials and Textiles, Zhejiang Sci-Tech University, Hangzhou, China)

52 Fabrication and Characterization of Mesoporous Calcium Silicate/silk Fibroin Composite Films ·················228

Hailin Zhu[1,2], Xinxing Feng[1], Huapeng Zhang[1], Jianyong Chen[1]

(1. Key Laboratory of Advanced Textile Materials and Manufacturing Technology, Ministry of Education of China, Zhejiang Sci-Tech University, Hangzhou, China
2. Department of Chemistry, Xiasha Higher Education Zone, Zhejiang Sci-Tech University,

Hangzhou, China)

53 Comparison of Fiber Configurations between Low Torque, Compact and Ring Spun Yarns ···233

Ying Guo[1,2], Xiaoming Tao[2], Bingang Xu[2], Jie Feng[2], Tao Hua[2], Shanyuan Wang[1]

(1. College of Textiles, Donghua University, Shanghai, China
2. Institute of Textiles & Clothing, The Hong Kong Polytechnic University, Hong Kong, China)

54 Hemp Processing with MAE and Alkali-H_2O_2 One-bath Treatment ·················238

Guojun Han, Lijun Qu, Xiaoqing Guo, Xiao Yang, Yuehua Zhao

(Laboratory of New Fiber Materials and Modern Textile, the Growing Base for State Key Laboratory, Qingdao University, Qingdao, Shandong, China)

55 Design of Multicolored Warp Jacquard Fabric Based on Space Color Mixing ···············243

Qizheng Li, Jiu Zhou, Gan Shen, Chenyan Zhu

(College of Materials and Textiles of Zhejiang Sci-Tech University, Hangzhou, China)

56 Design and Development of a Detection System for Recognising Emotions towards Creation of Interactive Fashion ··247

XIA W.J., NG. M.C.F.

(Institute of Textiles and Clothing, The Hong Kong Polytechnic University, Hong Kong, China)

57 Design Creations of Black-and-white Simulative Effect Digital Jacquard Fabric ···············252

Jiu Zhou[1,2], Frankie NG[2], Yejin Jiang[1]

(1. Key Lab of Advanced Textile Materials and Manufacturing Technology of Ministry of Education, Zhejiang Sci-Tech University, Hangzhou, China
2. Institute of Textiles and Clothing, The Hong Kong Polytechnic University, Hong Kong, China)

58 Thickness Variations of Length-distributed Slivers and Draft Conditions in Roller Drafting ···258

Jong S. Kim[1], Jung Ho Lim[2], You Huh[3]

(1. Laboratory of Intelligent Process and Control, College of Engineering, Kyunghee University, Yongin, R. O. Korea
2. Department of Textile Engineering, Graduate School, Kyunghee University, Yongin, R. O. Korea
3. Department of Mechanical Engineering, College of Engineering, Kyunghee University, Yongin, R. O. Korea)

59 The Study on Comprehensive Comfort Property of Garment with Knitted Underwear Fabrics during Exercise ··272

Shan Cong

(Shanghai University of Engineering Science, Shanghai, China)

60 Preparation and Application of a Novel Fluoroalkylpolysiloxane Fabric Finish with Waterproofing and Washing Resistant Properties ··278

Lifen Hao, Qiufeng An, Wei Xu, Qianjin Wang

(Key Laboratory of Auxiliary Chemistry and Technology for Chemical Industry, Ministry of Education, Shaanxi University of Science and Technology, Xi'an, China)

61 Research on Pattern Grading of Apparel Shoulder Slope ·············283

Canyi Huang[1], Lina Cui[2]

(1. Department of Business and Information, Quanzhou Normal University, Quanzhou, Fujian, China
2. Department of Arts and Design, Quanzhou Normal University, Quanzhou, Fujian, China)

62 Study on the Elastic Recovery of PLA/Cotton Fabric ·············288

Peng Liu, Wei Tian, Yanqing Li, Zhaohang Feng, Chengyan Zhu

(College of Materials and Textiles, Key Laboratory of Advanced Textile Materials and Manufacturing Technology, Ministry of Education of China, Zhejiang Sci-Tech University, Hangzhou, China)

63 Functional Properties of Hemp Union Fabrics for Home Textiles ·············293

Lin Lou, Xiaohang Zhu, Hongyan Xu, Jianfang Wang, Xinghai Pei, Jianliang Li

(Zhejiang Sci-Tech University, Hangzhou, China)

64 Effects of Annealing Atmosphere on the Structures and Photocatalytic Properties of Gd-Doped TiO_2/CF Photocatalysts ·············298

Bangyong Pang[1], Yaofeng Zhu[2], Tianbing Yang[1], Yaqin Fu[1], Hao Chen[1], Liyuan Zhang[1], Jin Wang[1]

(1. Key Laboratory of Advanced Textile Materials and Manufacturing Technology, Ministry of Education of China, Zhejiang Sci-Tech University, Hangzhou, China
2. Graduate School of Science and Technology, Shinshu University, Ueda, Japan)

65 Study on Preparation Techniques and Function of Thermochromic Microcapsule ·············302

Zanmin Wu, Wenzhao Feng, Xiaozhu Sun

(Institute of Textile, Tianjin Polytechnic University, Key Laboratory of Advanced Textile Composites, Tianjin Polytechnic University, Ministry of Education, Tianjin, China)

66 The Influence of Fabric Structures on the Property of Anti-electromagnetic Radiation of Fabrics with Embedded Silver-plated Fibers ·············307

Hongxia Zhang, Zhilei Chen, Lijia Shi, Yanqing Li

(Key Laboratory of Advanced Textile Materials and Manufacturing Technology, Ministry of Education of China, Zhejiang Sci-Tech University, Hangzhou, China)

67 Study on the Patterns of Small Black-and-white Grid in Costumes ·············312

Jian Zhao, Cai Qian Zhang

(Textile and Apparel Institute, Shaoxing University, Shaoxing, Zhejiang, China)

68 Thermodynamics Study of Monomer Adsorption Process for Fabrication of Conductive Textiles ·············316

Yaping Zhao[1], Zhaoyi Zhou[2], Xiaolan Fu[1], Zaisheng Cai[1]

(1. College of Chemistry, Chemical Engineering & Biotechnology, Donghua University,

Shanghai, China
2. Shanghai Institute of Fibre Inspection, Shanghai, China)

69 An Analytical Model for Ballistic Impact on Textile Based Body Armour ·············321
Fuyou Zhu[1], Xiaogang Chen[1], Garry Wells[2]
(1. Textiles and Paper, School of Materials, University of Manchester, Manchester, UK
2. Physical Science Department, Dstl, Porton Down, Salisbury, Wiltshire, UK)

70 Study on Automatic Classification of Size Designation in Clothing MC Based on Improved LBG Algorithm ·············326
Fengyuan Zou, Li Dong, Lifeng Pan, Xiaojun Ding, Minzhi Chen
(Fashion College, Zhejiang Sci-Tech University;
Key Laboratory of Advanced Textile Materials and Manufacturing Technology, Ministry of Education of China, Zhejiang Sci-Tech University, Hangzhou, China)

71 Thermodynamic Behavior on the Binding of the Polymers and Acid Dyes in Inkjet Ink for Textiles ·············331
Juyoung Park, Yuichi Hirata, Kunihiro Hamada
(Faculty of Textile Science and Technology, Shinshu University, Japan)

72 Applications of UV Curing on Textiles ·············336
Shiqi Li, Henry Boyter Jr.
(Institute of Textile Technology, North Carolina State University, Raleigh, NC 27695, USA)

73 A Study on the Kinetics and Thermodynamics of UV-absorber Taken Up to Polyester ·········344
Weiguo Chen[1,2], Xiaofang Wang[2], Son Fang[2], Qingqing Hu[2], Yining Cao[2]
(1. College of Chemistry and Chemical Engineering, Donghua University, Shanghai, China
2. Key Laboratory of Advanced Textile Materials and Manufacturing Technology, Ministry of Education of China, Zhejiang Sci-Tech University, Hangzhou, China)

74 Improving Dyeing Behavior of PLA Fiber with Plasticizing and Solubilizing System ·········347
Xiaonan Dang[1], Jinhuan Zheng[1,2], Jianjian Fu[1]
(1. Key Laboratory of Advanced Textile Materials and Manufacturing Technology, Ministry of Education of China, Zhejiang Sci-Tech University, Hangzhou, China
2. Engineering Research Center for Eco-Dyeing & Finishing of Textiles, Ministry of Education, Hangzhou, China)

75 Surface Resistivity of PET/COT Fabrics Treated with 3,4-ethylenedioxythiophene via Vapor Phase Polymerization ·············351
Qinguo Fan[1,2], Okan Ala[1], Jianzhong Shao[2], Jinqiang Liu[3]
(1. Department of Materials & Textiles, University of Massachusetts, Dartmouth, USA
2. Key Laboratory of Advanced Textile Materials and Manufacturing Technology, Ministry of Education of China, Zhejiang Sci-Tech University, Hangzhou, China
3. Engineering Research Center for Eco-Dyeing & Finishing of Textiles, Ministry of Education,

Zhejiang Sci-Tech University, Hangzhou, China)

76 Synthesis and Application of MDI Water-borne Polyurethane Fixing Agent ················356

Baozhou Li, Shuling Cui, Rui Li

(Hebei University of Science and Technology, Shijiazhuang, China)

77 A Novel Approach for Evaluating Cotton Fabric Strength Damage Caused by Bleaches during Laundering ················362

Yongqiang Li[1,2], Jinqiang Liu[1,2], Chunjie Qian[1,2], Liming Deng[1,2]

(1. Key Laboratory of Advanced Textile Materials and Manufacturing Technology, Ministry of Education of China, Zhejiang Sci-Tech University, Hangzhou 310018, China

2. Engineering Research Center for Eco-Dyeing & Finishing of Textiles, Ministry of Education, Zhejiang Sci-Tech University, Hangzhou, China)

78 Dyeing of Polyurethane Fiber and Polyurethane/Nylon Blend with Temporarily Solubilised Disperse Dyes ················366

Hongfei Qian[1], Xinyuan Song[2]

(1. College of Textiles and Apparel, Shaoxing University, Shaoxing, China

2. College of Chemistry and Chemical Engineering, Donghua University, Shanghai, China)

79 Dyeing Kinetics of Carrier Cindye Dnk in Dyeing Process of Aramid Fiber with Disperse Dyes ················372

Lan Wang[1], Ping Liang[2], Junxiong Lin[3], Duan Ni[4]

(1. Key Laboratory of Advanced Textile Materials and Manufacturing Technology, Ministry of Education of China, Zhejiang Sci-Tech University, Hangzhou, China

2. Zhejiang Huatai Silk Co., Ltd, Hangzhou, China

3. Engineering Research Center for Dyeing and Finishing of Textiles, Ministry of Education, Zhejiang Sci-Tech University, Hangzhou, China

4. Keyi College of Zhejiang Sci-Tech University, Hangzhou, China)

80 Energy Savings via Fast Drying in Textile Industry Part I: Theoretical Discussion and Technical Strategies ················377

Jiping Wang[1], Jinqiang Liu[2]

(1. The Procter & Gamble Company, Cincinnati, Ohio, USA

2. Zhejiang Sci-Tech University, Hangzhou, China)

81 Utilization of Enzyme in the Degumming Process of Jute Fibers ················382

Weiming Wang[1], Zaisheng Cai[2,3], Jianyong Yu[4]

(1. College of Textile & Apparel Engineering, Shaoxing University, Shaoxing, China

2. College of Chemistry, Chemical Engineering & Biotechnology, Donghua University, Shanghai, China

3. Key Laboratory of Science & Technology of Eco-Textile, Ministry of Education, Donghua University, Shanghai, China

4. Modern Textile Institute, Donghua University, Shanghai, China)

82 Kinetics of Ultrasonic Dyeing of PTT Fiber with Emodin ·················387
Xueni Hou, Xiangrong Wang
(National Engineering Laboratory for Modern Silk, Soochow University, Suzhou, China)

83 Study on the Effect of Cellulase on Cotton Properties ·················392
Yingzhe Wu[1], Jianzhong Shao[1], Lan Zhou[2]
(1. Key Laboratory of Advanced Textile Materials and Manufacturing Technology, Ministry of Education of China, Zhejiang Sci-Tech University, Hangzhou, China
2. Engineering Research Center for Eco-Dyeing & Finishing of Textiles, Ministry of Education, Zhejiang Sci-Tech University, Hangzhou, China)

84 Double-temperature Cationic Modification Process for Uniform Reactive Dyeing of Cotton Fabrics ·················397
Min Xu[1], Jianzhong Shao[1], Liqin Chai[1], Lan Zhou[1], Qinguo Fan[2]
(1. Engineering Research Center for Eco-Dyeing and Finishing of Textiles, Ministry of Education, Zhejiang Sci-Tech University, Hangzhou, China
2. Department of Materials & Textiles, University of Massachusetts, Dartmouth, USA)

85 Modification of *Mulberry Silk* by Calcium Salt Treatment and Epoxy Crosslinking ·················402
Wei Zhang, Shizhong Cui
(College of Textiles, Zhongyuan University of Technology, Zhengzhou, China)

86 Prospects of Cellulose Fiber Industrial Production in Solvent Process ·················407
Qingzhang Zhao
(China Textile Academy, China)

87 Surface Modification of Polypropylene Nonwoven Fabric with Plasma Activation and Grafting ·················410
Ahmad Mousavi Shoushtari[1], Aminoddin Haji[2], Azadeh Jafari[1]
(1. Textile Engineering Department, Amirkabir University of Technology, Tehran, Iran
2. Islamic Azad University, Birjand Branch, Birjand, Iran)

88 Characterization and Mechanical Behavior of a Tri-layer Polymer Actuator ·················415
Akif Kaynak, Chunhui Yang, Yang C. Lim, Abbas Kouzani
(School of Engineering, Deakin University, Geelong, Australia)

89 Improvement of Dyeability of Cotton with Natural Cationic Dye by Plasma Grafting ·················420
Aminoddin Haji[1], Ahmad Mousavi Shoushtari[2]
(1. Islamic Azad University, Birjand Branch, Birjand, Iran
2. Textile Engineering Department, Amirkabir University of Technology, Tehran, Iran)

90 The Impact of Textile Fibres on the Environment ·················425
Chris Hurren, Qing Li, Xungai Wang

(Centre for Material and Fibre Innovation, Deakin University, Geelong, Australia)

91 Thermal and Chemical Properties of Wool Powders ··············429
G. Wen, X. Liu, X.G. Wang
(Centre for Material and Fibre Innovation, Institute for Technology Research and Innovation, Deakin University, Geelong, Australia)

92 Influence of Spacer on Spun Yarn Quality in Cotton Spinning ··············438
Sayyed Sadroddin Qavamnia, Amir Hossein Raei, Arsham Zibaee
(Islamic Azad University, Birjand Branch, Birjand, Iran)

93 Thermobonding of Wool Nonwovens ··············445
R. H. Gong, K. M. Nassar, I. Porat
(Textiles and Paper, School of Materials, University of Manchester, UK)

94 PEG-[Si(OEt)$_3$]$_2$ Sol Agent Applied to PP Fabrics for Moisture Management ··············449
Yi Hu[1,2], Jinghong Yuan[3], Wenjie Chen[1,2], Jinqiang Liu[1,2]
(1. Key Laboratory of Advanced Textile Materials and Manufacturing Technology, Ministry of Education of China, Zhejiang Sci-Tech University, Hangzhou, China
2. Engineering Research Center for Eco-Dyeing & Finishing of Textiles, Ministry of Education, Zhejiang Sci-Tech University, Hangzhou, China
3. Institute of Costume, Zhejiang Textile & Fashion College, Ningbo, China)

95 Synthesis of Pyrazolone-Containing Carboxyester Dyes and Their Application to Poly(lactic acid) Fabric ··············453
Kai Liu[1], Zhihua Cui[1,2], Weiguo Chen[1,2]
(1. Key Laboratory of Advanced Textile Materials and Manufacturing Technology, Ministry of Education of China, Zhejiang Sci-Tech University, Hangzhou, China
2. Engineering Research Center for Eco-Dyeing & Finishing of Textiles, Ministry of Education, Hangzhou, China)

96 Preparation of Baicalin-Al(III) Complex Dye and Application in Dyeing of Silk Fabric ··············458
Zhengming Liu[1,2], Qibing Wang[1,2], Zhicheng Yu[1,2]
(1. Engineering Research Center for Eco-Dyeing & Finishing of Textiles, Ministry of Education, Zhejiang Sci-Tech University, Hangzhou, China
2. Key Laboratory of Advanced Textile Materials and Manufacturing Technology, Ministry of Education of China, Zhejiang Sci-Tech University, Hangzhou, China)

97 Influence of 1:1 Acid Metal Complex Dyes on Extractable Chromium of Wool Fabric ··············464
Baihua Wang[1,2]
(1. School of Materials Science and Engineering, Beijing Institute of Fashion Technology, Beijing, China
2. Beijing Key Laboratory of Clothing Materials R&D and Assessment, Beijing, China)

98 Preparation of Hydrostable Fiber Crosslinking Agent Containing Glycidyl Groups ··············468

Xiumei Zhang, Hualiang Wen, Jinlong Chen
(College of Materials and Textile, Zhejiang Sci-Tech University, Hangzhou, China)

99 Silver Modified Silk Fibroin Composite Film and Its Antibacterial Property ········· 472
Lan Zhou, Jianzhong Shao, Xinxing Feng, Bin Sun
(Engineering Research Center for Eco-Dyeing & Finishing of Textiles, Key Laboratory of Advanced Textile Materials and Manufacturing Technology, Ministry of Education of China, Zhejiang Sci-Tech University, Hangzhou, China)

100 Nondestructive Observation of Textile Reinforced Composites by Low Energy X-ray Beam ········· 477
Hyungbum Kim[1], Jung H. Lim[1], You Huh[2]
*(1. Dept. of Textile Eng., Graduate School, KyungHee Univ., Youngin, R. O. Korea
2. Dept. of Mechanical Engineering, KyungHee Univ., Youngin, R. O. Korea)*

101 Structural Differences of Wild Silks and B. Mori Silk Characterized by FTIR Spectroscopy, XRD Diffraction and TG Analysis ········· 482
Jianxin He[1,2], Yan Wang[1], Kejing Li[1], Shizhong Cui[1]
*(1. College of Textiles, Zhongyuan University of Technology, Zhengzhou, China
2. Key Laboratory of Textile Science & Technology, Ministry of Education, Donghua University, Shanghai, China)*

102 The Tearing Strength of Plain Woven Fabric with the High Strength ········· 488
Xinling Li, Zhiyu Zheng, Xiaohong Zhou
(Key Laboratory of Advanced Textile Materials and Manufacturing Technology, Ministry of Education of China, Zhejiang Sci-Tech University, Hangzhou, China)

103 The Preparation of Mesoporous Silica Inorganic Antibacterial Material and Its Application on Cotton Fabrics ········· 492
Bin Sun, Xinxing Feng, Lan Zhou, Na Liu, Zhangwei Wu, Jianyong Chen
(Key Laboratory of Advanced Textile Materials and Manufacturing Technology, Ministry of Education of China, Zhejiang Sci-Tech University, Hangzhou, China)

104 Observation and Modeling of the Microstructure of Heavyweight Hydroentangling Nonwoven Fabrics ········· 497
Hong Wang, Xiangqin Wang, Xiangyu Jin, Haibo Wu, Baopu Yin
(Key Laboratory of Textile Science & Technology, Ministry of Education, Donghua University, China)

105 Frequency Features of Acoustic Emission on Failure Mechanisms in PE Self-reinforced Laminates ········· 501
Xu Wang, Huipin Zhang, Xiong Yan
(Key Lab of Textile Science and Technology, Ministry of Education, College of Textiles, Donghua University, Shanghai, China)

106 Research on Performance of Filaments/Short Fibers Composite Yarn ·········506
Yu Xie, Ruicai Jing, Min Guo, Shiqin Song, Bin Jiang, Tonghua Zhang
(College of Textile & Garments, Southwest University, Chongqing, China)

107 Research on the Constant Tension Yarn Feeder of Weft Knitting Machine ·········510
Xiaochuan Bian, Yuan Fang
(School of Materials and Textiles, Zhejiang Sci-Tech University, Hangzhou, China)

108 Optimization of Prediction Performance for Worsted Yarn based on Neural Network ·········516
Xiang Li[1], Zhiqin Peng[1], Zongdong Gu[2], Yuan Xue[3], Guoliang Hu[1]
(1. College of Materials and Textiles, Zhejiang Sci-Tech University, Hangzhou, China
2. Zhejiang Linglong Textile Co., Ltd., Jiashan, Zhejiang, China
3. College of Garment and Art Design, Jiaxing University, Jiaxing, Zhejiang, China)

109 Effect of Bending Stiffness on Drape of Soft Weft-Knitted Fabric ·········521
Chengxia Liu[1], Caiqian Zhang[2], Weilai Chen[1]
(1. Zhejiang Sci-Tech University, Hangzhou, China)
(2. Shaoxing University, Shaoxing, Zhejiang, China)

110 Regression Analysis of the Influences of Cylinder Speed on Drawn Sliver in the Irregularity ·········526
Kui Mu, Chongqi Ma
(Tianjin Polytechnic University, Tianjin, China)

111 The Effect of Ring-spinning Parameters: Result from Theoretical Model ·········530
Rong Yin[1], Yang Liu[2], Hongbo Gu[1]
(1. College of Mechanical Engineering, Donghua University, Shanghai, China
2. Textile Materials and Technology Laboratory, Donghua University, Shanghai, China)

112 Repeatability Research of New Fabric Anti-crease Evaluation Method ·········535
Caiqian Zhang[1], Chengxia Liu[2]
(1. Shaoxing University, Shaoxing, Zhejiang, China)
(2. Zhejiang Sci-Tech University, Hangzhou, China)

113 Embedded Microcontroller Based Flat Knitting Machine Controller Design ·········539
Zhang Hua, Xudong Hu, Yanhong Yuan, Xianmei Wang, Laihu Peng, Jianyi Zhang
(Faculty of Mechanical Engineering & Automation, Zhejiang Sci-Tech University, Hangzhou, China)

114 A Durable Flame-retardant Finish for Cotton ·········544
Bin Fei, Zongyue Yang, John H. Xin
(Institute of Textiles & Clothing, The Hong Kong Polytechnic University, Hong Kong, China)

115 Influence of Extension on the Side-glowing Properties of POF ·········548
Rui Wu, Jinchun Wang, Bin Yang

(Key Laboratory of Advanced Textile Materials and Manufacturing Technology, Ministry of Education of China, Zhejiang Sci-Tech University, Hangzhou, China)

116 A Rechargeable Antibacterial Poly(ethylene-co-methacrylic acid) (PE-co-MAA) Nanofibrous Membrane: Fabrication and Evaluation ···················553

Jing Zhu, Dong Wang, Gang Sun

(Polymer and Fiber Science, University of California, Davis, California, USA)

117 Recent Developments in Electrospinning of Nanofibers and Nanofiber Yarns ···················560

Tong Lin, Xungai Wang

(Centre for Material and Fiber Innovation, Deakin University, Geelong, VIC 3217, Australia)

118 Fabric Wrinkle Characterization and Classification Using Modified Wavelet Coefficients and Support-Vector-Machine Classifiers ···················564

Jingjing Sun[1], Ming Yao[1], Patricia Bel[2], Bugao Xu[1]

(1. School of Human Ecology, University of Texas at Austin, Austin, USA
2. USDA Southern Regional Research Center, New Orleans, USA)

1 Structure and Thermal Stability of Nanoclay/flax Nanocomposite

Chunhong Wang[1,2], Frank K. Ko[1*], Mercedes Alcock[3]

(1. Department of Materials Engineering, the University of British Columbia,
111 AMPEL, 2355 East Mall, Vancouver, BC, Canada V6T 1Z4
2. Textile College, Tianjin Polytechnic University, 63 Chenglin Road, Hedong District, Tianjin, China, 300160
3. Composites Innovation Centre Manitoba, Inc. Manitoba, Canada)

Abstract: Nanocomposite of nanoclay and natural flax fibers were generated by the electrospinning process. The structure of the nanocomposite was evaluated by the X-ray diffraction (XRD), Fourier transform infrared spectroscopy (FTIR) and scanning electron microscopy (SEM). The thermal stability of the nanocomposites was examined by thermogravimetry (TG). The X-ray diffraction and FTIR spectra suggested the inclusion of nanoclay in the nanocomposite fibers. The SEM images showed that submicron scale nanoclay/flax nanocomposite fibers can be obtained and the addition of nanoclay reduced the diameter of the nanofibers and made the fibers more uniform. TG analyses revealed that the thermal property of the flax fibers was notably improved by the introduction of nanoclays.

Keywords: Nanoclay/flax nanocomposite; Structure; Thermal stability

1 Introduction

Natural fibers are low in cost and density, high in toughness and specific strength properties, biodegradable and CO_2 neutral when burned. Among nearly all the natural fibers, flax fibers are equipped with excellent mechanical properties because of the high cellulose content, for which they are more widely used as the reinforcement in polymer composites[1]. But the low thermal stability and nonuniformity nature of flax fibers are the limiting factors for broader application of this kind of composites, such as automotive interior and aircraft materials.

In order to make high thermal resistant and more uniform flax fibers, we plan to transform flax fibers into nanocomposite fibers by electrospinning of flax fibers solution with nanoclay. Electrospinning is an efficient process for producing fibers with submicron scale diameters through the action of electrostatic forces[2]. The addition of very low percent of nanoclays can lead to an enhancement in mechanical and thermal properties of the nanocomposite because the large surface area the nanoclays poses[3].

Recently although electrospinning process has been applied to the generation of several cellulose nanofibers[3-5], little work has been done on flax and the generation of nanoclay/cellulose nanocomposite fibers by electrospinning.

2 Experiments

2.1 Solution Preparation

The water in the 50 wt% aqueous NMMO solution (provided by Sigma–Aldrich) was removed by the HWS 6210A BUCHI Rotavapor R-200 and BUCHI Heating bath B-490. The weight ratio of NMMO to water was about 85/15 in the final solution. The flax fibers (provided by the Composites Innovation Center) were cut into the length of 1-2 mm.

In preparing the solution without nanoclay, first 1 wt% flax fibers and propyl gallate (1.5 wt% of that of flax fibers) were placed in the vial, then 85% NMMO/water solution was added to reach the desired solution composition. Then the mixtures were stirred by a stirrer(Corning PC-4200) at the temperature and speed of 85℃ and 1000rpm until the flax fibers were completely dissolved.

In preparing the solutions with nanoclay, first the nanoclay with two different weight percentages (0.25 and 1) was added into the 85% NMMO/water solution. Then the solutions were stirred for 2 hours so as to disperse the nanoclay. Then the propyl gallate and flax fibers were placed inside the vials and the mixture was stirred until the complete dissolution of flax fibers.

2.2 Electrospinning

Before electrospinning, the solution was loaded in a syringe and placed in the micropump (Kd Scientific) at a spinning angle of about 45°. A swirling water bath was used to collect the electrospun fibers so as to

---

* corresponding author

prevent the fibers from agglomerating. The horizontal and vertical gaps between the tip of the needle and the surface of the water were about 5 and 10 cm respectively.

During electrospinning, the temperature of the syringe and the needle was maintained at around 85℃. A voltage of 10 kV was applied to the tip of the needle by using a high-voltage regulated direct current power supply (Gamma High Voltage Research Inc.). The polymer solution was pumped at a rate of 0.9 ml/h continuously.

3 Results
3.1 X-Ray diffraction

X-ray diffraction photographs of electrospun flax, Cloisite 25A nanoclay and 0.25% nanoclay/flax nanocomposite fibers were characterized by XRD (Rigaku X-Ray diffractometer operated at 40 kV and 20 mA with Cu-Kα radiation, scan rate of 5°/min) are shown in Figure 1(a). The peak of $2\theta= 18.6°$ suggests the Cloisite 25A is inside the nanoclay/flax nanocomposite fibers.

3.2 Fourier transform infrared (FTIR) spectroscopy analysis

As shown in Figure 1(b), the electrospun flax, Cloisite 25A nanoclay and 1% nanoclay/flax nanocomposite fibers were characterized by FTIR (Spectrum One Q500 FTIR Spectrophotometer, Perkin-Elmer). The appearance of two characteristic bands at 2,849 cm^{-1} and 3,632 cm^{-1} indicate the nanoclay is inside the nanoclay/flax nanocomposite fibers.

3.3 Scanning electron microscopy (SEM)

As shown in Figure 2 (a) and (b), the diameter of the electrospun flax fibers was dramatically reduced with the addition of 0.25wt% nanoclay. The average diameter of the electrospun fibers observed by SEM was calculated based on about 15 observations (by using the Quartz Xone software), as shown in Table 1. From this table we can see that the electrospun flax fibers are an order of magnitude finer than that of the raw flax fibers, which is 56.024±27.717 μm. It is interest to note that the introduction of 1 wt% nanoclay produces not only finer fiber but also significantly improves the uniformity of the fiber diameter with a coefficient of variation of less than 18% as compared to 50% for the raw flax fibers, as shown in Figure 2 (c) and (d). This may be due to the increase of conductivity of electrospining solution by the introduction of nanoclay[6].

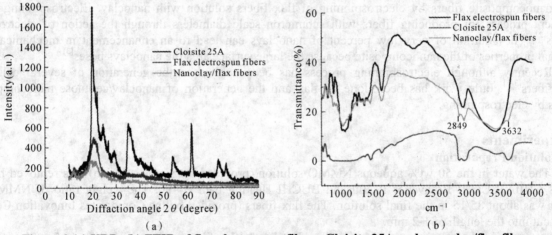

Figure 1 (a) XRD, (b) FTIR of flax electrospun fibers, Cloisite 25A and nanoclay/flax fibers

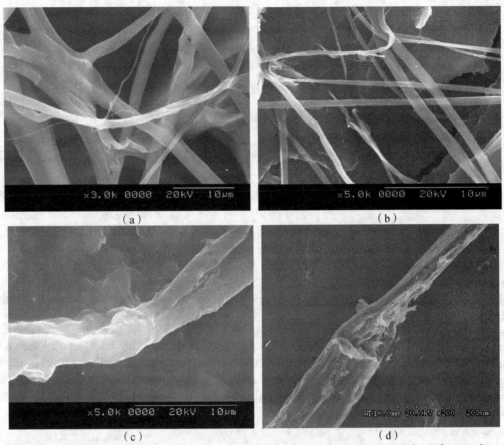

Figure 2 SEM image of electrospun 1 wt% flax fibers at the concentration of nanoclay
(a) 0 wt% (b) 0.25 wt% (c) 1 wt% (d) natural flax fibers

Table 1 Electrospinning parameters and fiber diameters for electrospun flax and nanoclay/flax nanocomposite fibers

Flax (wt%)	Nanoclay (wt%)	Working distance (cm)	Applied voltage (kV)	Pump rate (ml/h)	Mean fiber Diameter (μm)	Standard deviation (μm)
1	0	10	10	0.9	2.188	2.363
1	0.25	10	10	0.9	0.313	0.151
1	1	10	10	0.9	1.749	0.312

3.4 Thermogravimetric Analysis (TGA)

The thermal stability of raw flax fibers, flax and nanoclay/flax electrospun nanocomposite fibers was evaluated by thermogravimetry (TA instruments Q500 thermogravimetric balance, heating rate of 20℃/min). It can be observed that decreases in thermal stability by about 45℃ and 20℃ at 5% and 50% weight losses for the electrospiun flax fibers compared to raw flax fibers, as shown in Figure 3 (a). However from Figure 3 (b) we can see that the presence of nanoclay greatly improves the thermal stability of the flax fibers. The addition of 1wt% nanoclay to the electrospun flax resulted in close to 100℃ improvement in thermal stability, from 350℃ to 450℃ at 50% weight loss comparing to the electrospun flax and raw flax fibers.

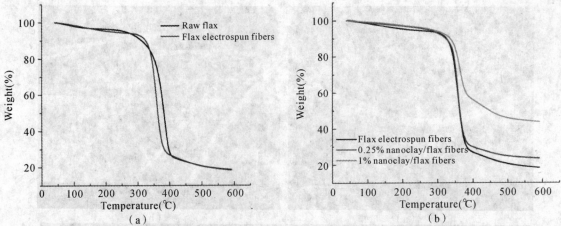

Figure 3 TGA curves for (a) raw flax and electrospun flax fibers
(b) electrospun flax fibers with 0, 0.25, 1% nanoclay

4 Conclusions

In this study the flax and nanoclay/flax nanocomposite fibers were generated by electrospinning process. X-ray diffraction and FTIR spectra suggested that the nanoclay is inside the nanoclay/flax electrospun nanocomposite fibers assembly. The SEM images revealed that the addition of appropriate level of nanoclay facilitates the production of more uniform and finer electrospun fibers. The TGA results showed that the thermal stability of flax nanofibers could be improved by the presence of an appropriate level of nanoclay.

Using electrospun composite nanofibers as a model this study provides the basis for further scale up study of the role of nanoclay in improving the fiber uniformity, thermal stability and mechanical properties of natural fiber composites.

Acknowledgement

The authors gratefully acknowledge the financial support provided by the China Scholarship Council-University of British Columbia Joint Program and the Composites Innovation Centre under a collaborative research program.

References

[1] Williams G.I. and Wool R.P. Composites from natural fibers and soy oil resins. Applied Composite Materials, 2000, (7): 421-432.
[2] Yamashita Yoshihiro, Ko Frank, Miyake Hajime, et al. Establishment of nanofiber preparation technique by electrospinning. Sen'i Gakkaishi, 2008, 64: 24-28.
[3] Cerruti Pierfrancesco, Ambrogi Veronica, Postiglione Alessandro, etal. Morphological and Thermal Properties of Cellulose-Montmorillonite Nanocomposites. Biomacromolecules, 2008, (9): 3004-3013.
[4] Kim Choo-Won, Kim Dae-Sik, Kang Seung-Yeon. Structural studies of electrospun cellulose nanofibers. Polymer, 2006, (47): 5097-5107.
[5] Graham K, Gogins M, Schreuder-Gibson H. Incorporation of Electrospun. Nanofibers into Functional Structures. International Nonwovens Journal, 2004, (13): 21-27.
[6] Lei Yu, Peggy Cebe. Crystal polymorphism in electrospun composite nanofibers of poly(vinylidene fluoride) with nanoclay. Polymer, 2009, 50: 2133-2141.

2 Multifunctional Composite Nanofibers

Frank Ko, Masoumeh Bayat, Heejae Yang

(Advanced Fibrous Materials Laboratory, University of British Columbia, Canada)

Abstract: Recent advances in nanotechnology have greatly expedited the development of the new generation of multifunctional nanomaterials. The physical and chemical properties of these nanomaterials, which usually exist in particle form, are extremely sensitive to the change in the environment such as temperature, pressure, electric field, magnetic field, optical wavelength, and adsorbed gas molecules. To facilitate the translation of these functions to higher order structures the nanoparticles are mixed with a polymer solution and converted to composite nanofibers. In this presentation the approach of translating functions from nanoparticles to advanced fibrous structures by the co-electrospinning process will be presented. As an example of multifunctional material the electrical conductivity and magnetic properties of Fe_3O_4/C electrospun composite nanofiber will be demonstrated. Specifically polyacrylonitrile (PAN) solutions containing (10 wt% Fe_3O_4) magnetite nanoparticles were electrospun and pyrolized at two different temperatures. The electrical conductivity of the composite nanofibrous structures were measured using four-point probe method and the magnetic properties were also obtained using Superconducting Quantum Interference Devices (SQUID). Electromagnetic composite nanofiber with the electrical conductivity of 9.2S/cm and saturation magnetization equal to 16emu/g was acquired. Raman, XRD, SEM and TEM were used for investigating the morphological characteristic of these electrospun composite nanofibers.

Keywords: Multifunction; Composite nanofibers; Fe_3O_4; PAN; Carbonization; Superparamagnetic; SQUID; Electrical conductivity; Raman; XRD; SEM; TEM

1 Introduction

Recent advances in nanotechnology have greatly expedited the development of the new generation of multifunctional materials. Multifunctional materials are materials that possess more than one physical, chemical and biological property that can be changed significantly in a controlled fashion by external stimuli such as temperature, pressure/stress, electric field, magnetic field, optical wavelength, adsorbed gas molecules and the pH value. Many of the functional materials are in the form of 0-D, 1-D and 2-D nanoparticles. The discrete nature of the nanoparticles makes them difficult to create structural forms. In order to bridge the gap between nano and macro length scales and connect the nano effect to macro-performance we propose to create composite fibers containing the functional nanoparticles. We specifically produce the composites in nanofiber form in order to preserve the nano effect. Therefore a major challenge for this composite nanofiber concept is to demonstrate that the unique functional properties of the nanoparticle can be translated to the composite nanofibrous structure. In this study we illustrate the creation of electromagnetic composite nanofibers using Fe_3O_4 nanoparticles in a carbon matrix thus combining the superparamagnetic properties of Fe_3O_4 with the electrical conductivity derived from the carbon nanofiber matrix.

Carbon nanofibers have attracted great interest in the last few decades due to their widespread applications such as gas sensors [1], hydrogen storage media [2], electromagnetic interference shielding [3], catalysts and catalytic supports [4] and field-emission microelectronic devices [5]. Among different methods of synthesizing carbon nanofibers, electrospinning is an easy and cost-effective technique. Composite materials containing both electrical conductivity and magnetic properties are promising materials with potential applications as lithium ion batteries [6], microwave absorbers [7] and membrane-less bio-fuels [8].

Magnetite (Fe_3O_4) nanoparticles due to their specific magnetic properties are one of the preferred and well characterized filler materials which are used in combination with polymers for different purposes such as recording media, medical applications and for absorption of radiation [9,10]. In the last few decades, the magnetic nanoparticle/C composites have been recognized as promising materials in numerous fields of applications. For example, Reneker and Hou [11] have prepared a catalyst for Carbon nanotube growth by heat treating the PAN/iron salt and reducing the iron salt into the iron nanoparticles in an appropriate atmosphere. Wang et al. [6] synthesized the Fe_3O_4/carbon electrospun nanofiber composite by using iron oxide precursor and investigated their application as promising anode materials for high performance lithium-ion batteries without reporting its magnetic properties. Panel et al. [12] also synthesized carbon /iron oxide nanofiber composites with hierarchical structure and reported both electrical and magnetic properties. All these previous studies on iron oxide/carbon nanofiber composites, there is a lack of studying how to control the effect of filler content and heat treatment temperature on the electromagnetic properties as well as investigating the properties of non-carbonized fibers in comparison with Fe_3O_4 nanoparticles.

In the present work, pristine carbon nanofibers and Fe_3O_4/carbon nanofiber composites were prepared using electrospinning and subsequent pyrolysis to make the electromagnetic nanofiber composites. Fe_3O_4 nanoparticles were used as magnetic fillers into the PAN matrix due to its interesting magnetic properties. On the other hand, PAN was selected as a polymer matrix of this composite, since it is an electrospinnable polymer with the ability to be pyrolyzed at high temperatures to make electrically conductive carbon nanofibers with their high Thermal and chemical resistant properties. The main objective of the present study is to demonstrate the feasibility of preparing electromagnetic nanofiber composites and characterization of their microstructural and physical properties.

2 Experimentation

The materials used in this study include the following: Polyacrylonitrile (PAN) (150,000 average molecular weight) was purchased from Scientific Polymer Products, Inc. N,N-Dimethylformamide (DMF, 99.9%) was purchased from Fisher Scientific. Fe_3O_4 magnetite nanoparticles (~20-30 nm) were purchased from Nanostructured and Amorphous Materials Inc. Polyvinylpyrollidone (PVP) was purchased from Sigma-Aldrich.

The production of the nanofibers begins with the preparation of pure 10wt%PAN solution using DMF as solvent. In another spinning dope, nanoparticles were dispersed in DMF and sonicated for 9 hrs. Subsequently PVP (surfactant) was added into the previously prepared solutions and sonicated for 9hrs. Finally, the PAN (polymer powder) was added and the solutions were sonicated for another 9hrs to make composite solutions.

The prepared composite solutions were loaded into 10ml plastic syringe with 1mm stainless steel 18-gauge needle. A high potential of 11kV was applied to the needle. A constant supply of the Fe_3O_4/PAN solutions was delivered to the needle using a syringe pump at a flow rate of 0.05mm/min. A fiber collector at 17cm distance to the needle tip covered with aluminum foil was used as a grounded counter electrode. The non-woven fibers were collected on the aluminum foil and peeled off for carbonization process.

The as-spun nanofiber composites were then carbonized according to the follow steps: 1) heating up to 250℃ with ramping rate of 5℃/min, 2) stabilization at 250℃ for 2hrs, (1&2 under the oxygen atmosphere) 3) heating up to 700℃ or 900℃ with ramping rate of 5℃/min and 4) carbonization at 700℃ or 900℃ for 1hr (steps 3 &4 were carried out under nitrogen atmosphere). The samples were then cooled down to the room temperature in the furnace under nitrogen atmosphere. A summary of sample designation code for different samples used in this study is shown in Table 1.

Table 1 Summary of sample codes

Sample	As-spun nanofibers	Carbonized at 700℃	Carbonized at 900℃
Pure PAN	APAN	APAN700	APAN900
10wt% Fe_3O_4	A10F	A10F700	A10F900

3 Results and discussion
3.1 Microstructural analysis
3.1.1 SEM

Figure 1 shows the morphology of nanofibrous mat of APAN, A10F, APAN900 and A10F900 samples and fiber diameters were measured using Image J analysis software. SEM images show a relatively uniform fibrous mat with random distribution was obtained for all samples. Results show that fiber diameter decreases by carbonization and decreases more for samples carbonized at higher temperature. Therefore, the fiber diameter is lower for samples carbonized at 900℃ rather than 700℃. As an example, fiber diameter increases from 540nm to 430nm and 370nm for A10F, A10F700 and A10F900 samples, respectively. Also, fiber diameters are higher for samples containing Fe_3O_4 nanoparticles rather than pure carbon nanofibers.

It can be seen from the SEM results that fiber diameter decreases after carbonization process. During the carbonization of PAN nanofibers due to the residual solvent evaporation and removal of unwanted elements like nitrogen the fiber diameter decreases. The same phenomenon was observed for fibers carbonized at higher temperature, i.e. 900℃[13].

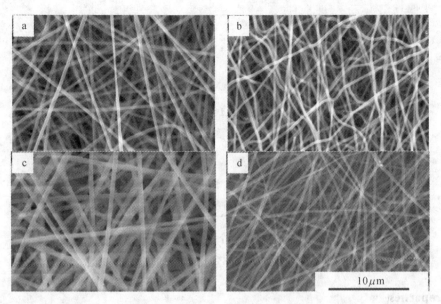

Figure 1 SEM micrographs of (a) APAN, (b) APAN900, (c) A10F and (d) A10F900

3.1.2 TEM

Nanoparticles distribution in the fibers before and after carbonization was observed using TEM. Figure 2 shows the TEM micrographs of A10F and A10F900 samples. Relatively uniform particles distribution has been obtained for A10F sample and after carbonization, it seems that nanoparticles agglomerate together and form relatively large clusters, as shown in Figure 2. The TEM images show that larger clusters of magnetite were formed in carbonized samples. This can be attributed to the effect of high carbonization temperature. During the carbonization process, nanoparticles became agglomerated together and made large clusters as shown in TEM pictures.

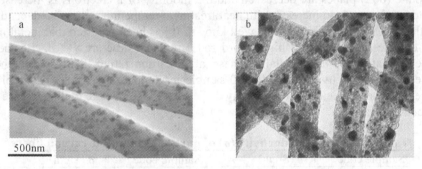

Figure 2 TEM micrographs of (a) A10F and (b) A10F900 samples

3.1.3 XRD

Phase identification has been studied using XRD and the results have been presented in Figure 3 for APAN900 and A10F900 samples. General observations show two characteristic peaks of graphite at $2\theta=24°$ and 44°. Besides, the Fe_3O_4 peaks have been identified using MDI Jade software for A10F900 sample. In addition to the two peaks, the formation of two other phases, Fe_3C and α–Fe, was also observed. Similar results were obtained for samples carbonized at 700℃. The XRD results showed the formation of graphite, Fe_3O_4, α-Fe and Fe_3C compounds. This agrees with the report by Kaburagi et al.[14] who showed that Fe_3C can be formed by the reaction of Fe and C during carbonization in the carbonizing temperature range between 700℃ to 1000℃. The formation of α-Fe has also been shown by Park et al.[15] and Xu et al.[16].

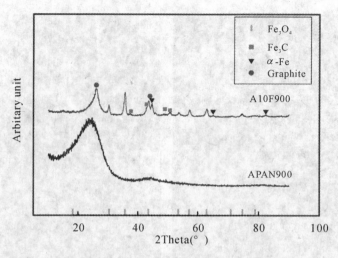

Figure 3 XRD pattern of APAN900 and A10F900 samples

3.2 Physical properties
3.2.1 Electrical conductivity

Electrical conductivity of the composite nanofibers was measured using the four-point probe measurement technique and the results are summarized in Table 2. It can be seen that electrical conductivity increases with carbonization temperature from 700℃ to 900℃. Electrical conductivity of A10F700 and A10F900 samples increases from 2.3 S/cm to 9.2 S/cm, respectively. Similar increasing trend is shown for Fe_3O_4 concentration in the matrix. The heat treatment temperature dependence of the carbon nanofibre mats was also reported by Wang et al. [13]. During the carbonization graphitic structure will be formed and depending on the temperature, the degree of graphitization will be different. Previous results have revealed that the higher order graphitic structure can be formed as the temperature increases. As a result, the planar d-spacing of graphite (002) planes are decreased and the mobility of π electrons is increased. Consequently, the electrical conductivity tends to increase with heat treatment temperature [17]. The dependence of electrical conductivity on the Fe_3O_4 content in matrix is well known. It has been reported that metal/metal oxides have catalytic effect on the formation of highly ordered graphite structure [18,19]. Besides, XRD results demonstrated the formation of α-Fe metallic phase after carbonization. This is yet another reason for increasing the electrical conductivity for A10F900 sample in compare to APAN900 sample. Chen et al. [20] also attributed the increase in electrical conductivity of Fe_3O_4-PPy composite to the increase in the Fe content during the composite preparation.

Table 2 Electrical conductivity (σ) of pure and composite carbon fibers

Sample code	σ (S/cm)	Sample code	σ (S/cm)
APAN700	0.055±0.004	APAN900	2.6±0.9
A10F700	2.3±0.6	A10F900	9.2±0.5

3.2.2 Magnetic properties

The magnetic properties of Fe_3O_4 nanoparticles and nanofibre composites were characterized using SQUID. Magnetization of nanofibre composites in response to the applied magnetic field has been demonstrated as M-H curves for the A10F and A10F900 samples (Figure 4). Saturation magnetization (Ms), remanence magnetization (Mr), and coercivity (Hc) have been summarized in Table 3 for all magnetic samples. An interesting result obtained from SQUID measurements is that the superparamagnetic properties of Fe_3O_4 nanoparticles with a very small coercivity and remanence has also been observed for as-spun composite fibre (A10F). Therefore, it proves the fact that we were able to transfer superparamagnetic properties of the nanoparticles into the fibrous structure.

However, magnetic hysteresis has been observed for carbonized samples compared to non-carbonized one especially at the higher carbonization temperature, i.e., 900℃. A possible explanation for this observation is the increasing the possibility of particle growth and agglomeration after carbonization. Since particle growth is more pronounced in samples carbonized at 900℃ rather than 700℃, it can be seen that hysteresis in the A10NF900 sample is larger than the A10F700 sample. Larger particles have higher magnetocrystalline anisotropy [22]. The resulting hysteresis is higher for the A10F900 sample, which can be due to the larger particle formation at higher temperatures in comparison with the A10F700 sample.

Besides, the Ms level can be controlled by the rule of mixtures for as-spun nanofibre, meaning that the A10F samples have 10% of the Ms of Fe_3O_4 nanoparticles (55emu/g vs. 5.5emu/g). This is an important result which shows the capability of this work to control the level of magnetization by making magnetic nanofibres.

Moreover, Ms increases by carbonizing the samples and by increasing the carbonization temperature from 5.2emu/g to 12emu/g and 16emu/g for A10F, A10F700 and A10F900 samples, respectively. It has also been observed that saturation magnetization increases by carbonizing the samples especially at higher temperature (900℃). The increase in the saturation magnetization (Ms) value for the carbonized samples in comparison to the non-carbonized samples can be partly attributed to the PAN weight loss (~50wt%) due to the carbonization. As a result, Fe_3O_4 portion in the matrix rises up. Since PAN weight loss is higher at higher temperatures, the Ms value obtained for the A10F900 sample would be higher than for the A10F700 sample. Another important reason can be related to particle growth because of carbonization. The increasing size of the particles results in an increase in Ms value [15,22].

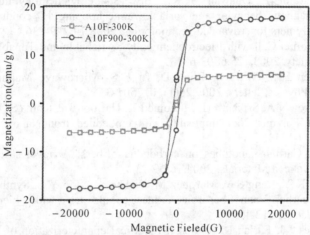

Figure 4 M-H Curves of A10F and A10F900 Samples

Table 3 Ms, Mr and Hc of Fe_3O_4 and Fe_3O_4/Carbon Nanofibre Composites at 300K

Sample code	Ms(emu/g)	Mr(emu/g)	Hc(Oe)
A10F	5.2	0.15	105
A10F700	12.0	2.5	500
A10F900	16.0	5.2	600
Fe_3O_4 NPs	55	2	100

4 Summary

In this study, as an illustration of the concept of multifunctional composite nanofiber, Fe_3O_4 nanoparticles were embedded in a carbon nanofiber matrix to produce nanofibers that have a combination of electrical conductivity and magnetic properties. Experimental results showed that superparamagnetism of Fe_3O_4 nanoparticles were transferred to the nanofibrous structure. Electrical conductivity and magnetic properties were enhanced by increasing carbonization temperature and by the addition of magnetite nanoparticles. Additionally, as evident in the hysteresis, it was shown that the magnetic properties of the nanofibrous structures could be tailored by varying the heat treatment temperature for the Fe_3O_4/PAN composite. The resultant electromagnetic composite nanofibers have the potential to be used as drug carriers and drug delivery scaffold; electromagnetic sensors; and as shielding materials for electromagnetic radiation.

Acknowledgement

This study was supported by AOARD/AFOSR and NSERC through a Discovery Grant. The equipment used in this study is funded by CFI.

References

[1] Jang J, Bae J. Carbon nanofiber/polypyrrole nanocable as toxic gas sensor. Sensors and Actuators B, 2007,122:7–13.

[2] Zhu H W, Li X S, Ci L J, Xu C L, Wu D H, Mao Z Q. Hydrogen storage in heat-treated carbon nano-fibers prepared by the vertical floating catalyst method. Materials Chemistry and Physics, 2003,78:670–675.

[3] Yang S, Lozano K, Lomeli A, Foltz H D, Jones R. Electromagnetic interference shielding effectiveness of carbon nanofiber/LCP composites. Composites: Part A, 2005, 36:691–697.

[4] Endo M, Kim Y A, Ezaka M, Osada K, Yanagisawa T, Hayashi T, Terrones M, Dresselhaus M S. Selective and efficient impregnation of metal nanoparticles on cup-stacked-type carbon nanofibers. Nanoletters, 2003,3: 723–727.

[5] Teo K B K, Chhowalla M, Amareatunga G A J, Milne W I, Pirio G, Legagneux P, Wyczisk F, Pribat D, Hasko D G. Field Emission from dense, sparse and patterned arrays of carbon nanofibers. Applied Physics Letters, 2002,80:2011–2013.

[6] Wang L, Yu Y, Chen P C, Zhang D W, Chen C H. Electrospinning synthesis of C/ Fe_3O_4 composite nanofibers and their application for high performance lithium-ion batteries. Journal of Power Sources, 2008,183: 717–23.

[7] Singh K, Ohlan A, Saini P, Dhawan S K, Poly (3,4 ethylenedioxythiophene) g-Fe2O3 polymer composite–super paramagnetic behavior and variable range hopping 1D conduction mechanism–synthesis and characterization. Polymers for Advanced Technologies, 2008,19:229-236.

[8] Katz E, Willner I. A Biofuel Cell with Electrochemically Switchable and Tunable Power Output. Journal of American Chemical Society, 2003,125:6803-6813.

[9] Zheng H, Yang Y, Wen F S, Yi H B, Zhou D, Li F S. Microwave Magnetic Permeability of Fe_3O_4 Nanoparticles. Chinese Physics Letters, 2009,26(1): 017501-3.

[10] Wilson J L, Poddar P, Frey N A, Srikanth H, Mohomed K, Harmon J P, Kotha S, Wachsmuth J. Synthesis and Magnetic properties of Polymer Nancomposites with embedded Iron Nanoparticles. Journal of Applied Physics, 2004,95(3):1439-1443.

[11] Hou H, Reneker D H. Carbon nanotubes on carbon nanofibers: A novel structure based on electrospun polymer nanofibers. Advanced Materials, 2004,16:69-73.

[12] Panels J E, Lee J, Park K Y, Kang S Y, Marquez M, Wiesner U, Joo Y L. Synthesis and characterization of magnetically active carbon nanofiber/iron oxide composites with hierarchical pore structures. Nanotechnology, 2008,19:455612-19.

[13] Wang Y, Santiago-Aviles J J, Furlan R, Ramos I. Electrical characterization of a single electrospun porous nanostructured tin oxide ribbon. IEEE Transactions on Nanotechnology, 2003,2(1):39-43.

[14] Kaburagi Y, Hishiyama Y, Oka H, Inagaki M. Growth of iron clusters and change of magnetic property with carbonization of aromatic polyimide film containing iron complex. Carbon, 2001,39:593–603.

[15] Park S H, Jo S M, Kim D Y, Lee W S, Kim B C. Effects of iron catalyst on the formation of crystalline domain during carbonization of electrospun acrylic nanofiber. Synthetic Metals, 2005,150: 265–270.

[16] Xu J, Yang H, Fu W, Sui Y, Zhu H, Li M, Zou G. Preparation and characterization of carbon fibers coated by Fe_3O_4 nanoparticles. Materials Science and Engineering B, 2006,132: 307–310.

[17] Panapoy M, Dankeaw A, Ksapabutr B. Electrical Conductivity of PAN-based Carbon Nanofibers Prepared by Electrospinning Method. Thammasat International Journal of Science and Technology, 2008,13:11-17.

[18] Kim P, Joo J B, Kim J, Kim W, Song I K, Yi J. Sucrose-derived graphitic porous carbon replicated by mesoporous silica. Korean Journal of Chemical Engineering, 2006,23(6):1063-1066.

[19] Mochida I, Yoon S H, Qiao W. Catalysts in Syntheses of Carbon and Carbon Precursors. Journal of the Brazilian Chemical Society, 2006,17(6):1059-1073.

[20] Chen A, Wang H, Li X. Influence of concentration of FeCl3 solution on properties of polypyrrole–Fe_3O_4 composites prepared by common ion absorption effect. Synthetic Metals, 2004,145:153–157.

[21] Lin C R, Chu Y M, Wang S C. Magnetic properties of magnetite nanoparticles prepared by mechanochemical reaction. Materials Letters, 2006,60:447–450.

[22] Liu C, Zhang Z J. Size-Dependent Superparamagnetic Properties of Mn Spinel Ferrite Nanoparticles Synthesized from Reverse Micelles. Chemistry of Materials, 2001, 13:2092-2096.

3 Liquid Crystalline Electrospinning of Carbon Nanotube Reinforced Cellulose Fibers from Bamboo

Yuqin Wan, Frank K. Ko[*]
(Advanced Fibrous Materials Laboratory, Department of Material Engineering, University of British Columbia)

Abstract: Multiwalled carbon nanotubes (MWNTs) were introduced into the bamboo cellulose matrix to form composite nanofibres by the liquid crystal electrospinning process. Aligned MWNTs in the fibres were observed through AFM and SEM. TGA testing showed 23wt% of MWNTs were present in the composite nanofibres. The average strength and breaking elongation 300 MPa and 35% respectively.

Keywords: Carbon nanotube; Bamboo; Liquid crystal; Nanocomposite; Nanofiber

1 Introduction

The increasing awareness of energy shortage and environmental concern has prompted the resurgent of interest in non-petroleum based materials. Of particular interest are the natural and sustainable resources such as cellulose [12,22]. With its abundant availability, fast maturity cycle, and rich content of cellulose, bamboo is one of the most favorite sources of cellulose[4]. Many researches about developing and exploring new applications of bamboo have been carried up[4,6]. In search of non-petroleum based multifunctional materials, we explored the feasibility of regenerating cellulosic nanofibres as a carrier for functional nanoparticles by electrospinning.

Possessing superior thermal, electrical and mechanical properties, carbon nanotube is one of the most interest functional materials in the nanoscale[21,5274,5407,6584,6624]. It has been shown by Ko et al.[1,10,14] that carbon nanotube reinforced nanofibrous yarns, can be produced by the electrospinning proccess. Accordingly we hypothesize that the combination of carbon nanotube and bamboo based cellulosic into composite nanofibres will open up new opportunities for the tailoring of new functions for engineering materials. We specifically coelectrospin high concentration of Multiwalled carbon nanotubes (MWNTs) with bamboo cellulose matrix to form composite nanofibres by liquid crystal electrospinning process.

2 Experiment

Solution preparation 0.7wt% MWNTs, 0.05wt% PVP and 0.01% Propyl Gallate were mixed with 85wt% NMMO solution while heated to 80℃. The mixture was then sonicated until MWNTs were well dispersed. 3wt% bamboo cellulose fibre and 0.01% Propyl Gallate were dissolved in 85wt% NMMO solution at 80℃. The two solutions at 1:1 weight ratio were mixed, heated and sonicated.

Electrospinning process The schematic of electrospinning setup is shown in Figure 1. The solution temperature was maintained at 90℃. The voltage was 10kV and the spinning distance between the tip of needle and the surface of aqueous coagulant was 7 cm.

Characterization The morphology of the collected fibre was observed using optical microscope, scanning electron microscope (SEM) and atomic force microscope (AFM). The thermo-degradation property was characterized by thermo-gravimetric analysis (TGA). Mechanical properties of the bamboo/MWNT fiber assemblies were tested using KES-G1 tensile tester with a gauge length of 1 cm; the results were compared with the data of commercial regenerated bamboo fibers that was measured in our lab.

---

* Corresponding author's email: yuqin@interchange.ubc.ca, frank.ko@ubc.ca

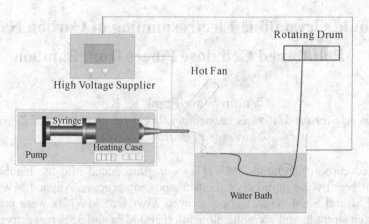

Figure 1 Schematic diagram of dry jet wet co-electrospinning and collection setup

3 Results

The continuous nanotube cellulosic yarns were produced by dry jet wet co-electrospinning. Adjusting the flow rate and applied voltage nanofibers are achievable, but to ensure the fibers strong enough to overcome the surface tension of water while being collected, the fiber diameter is kept in micro scale. The collected continuous MWNTs reinforced bamboo fiber yarn was shown in Figure 2.

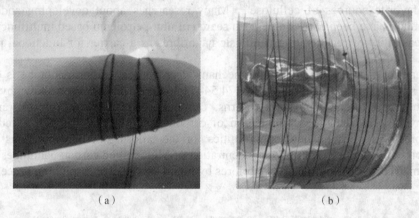

Figure 2 Co-electrospun MWNTs reinforced continuous bamboo yarn (a) MWNTs reinforced bamboo yarn collected on a drum. (b) MWNTs reinforced bamboo yarn wrapped on fingers shows flexibility of the yarn

As the whipping wet jet driven by the electric force reached the surface of the water bath while whipping, alignment of the fiber segments was formed. The segments then stuck to each other when collected to a drum, Figure 3. Trough controlling collection speed, continuous yarns with different diameters can be achieved.

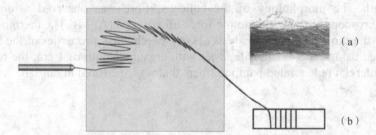

Figure 3 Mechanism of collection of continuous wet spinning fiber yarn (a) a bunch of aligned MWNTs reinforced bamboo fibers in water; (b) collection mechanism

The inclusion distribution and orientation of MWNT in bamboo fiber was conformed using optical microscopy, atomic force microscopy and scanning electron microscopy, as shown in Figure 4. MWNTs were well aligned along the axis of bamboo fibers. The distribution is almost homogeneous. Only a few MWNT aggregates were observed.

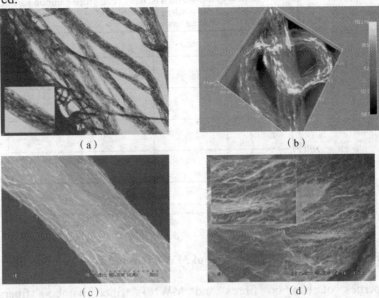

Figure 4 Aligned MWNTs in bamboo fibers (a) microscopy; (b) AFM; (c) SEM of fiber surface; (d) SEM of fiber inner structure

Fusing, which was reported by other researches on electrospun cellulosic fibers[4,14] before, was observed, see Figure 5. The reason was explained as: when the electrospun jets reach the coagulant, they are not solidified into fibers instantaneously due to it takes time for the coagulant to absorb NMMO from fibers; on the other hand, to prevent these micro fibers from entanglement and obtain a well aligned fiber yarn, the fibers must be collected soon after they formed aligned segments. When collected, those fibers are in semi-gel conditions and once contacted, they stick together and slowly fuse together.

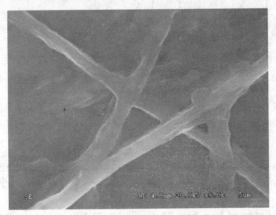

Figure 5 Fusing structure of collected bamboo fibers

Mechanical properties testing results of MWNTs reinforced bamboo yarns and pure bamboo fiber yarns were shown in Figure 6. Because of the existence of weak points, perhaps introduced while colleting these nanofibres into bundles, the testing results show wide The strengths and modulus are not as high as we expected which can be explained by three aspects: 1) the uncompleted post process such as post draw and washing which can be seen in Figure 6; 2) the imperfection structure of yarn caused by continuous collection, e.g. the non-uniform diameter; 3) the non-accurate calculation of the cross sectional area. Because the yarns were composed of numerous segmented fibers, interspaces are inevitable. However, the reinforcement of MWNTs is evident and significant. Comparing with the non-washed pure bamboo yarn which has experienced the same processes with the MWNTs reinforced bamboo yarn, the average strength increased more than 19 times from 16MPa to 311.7MPa, and the strain ranges from 22% to 42%. The modulus was more 30 times enhanced and reached 2.39 GPa.

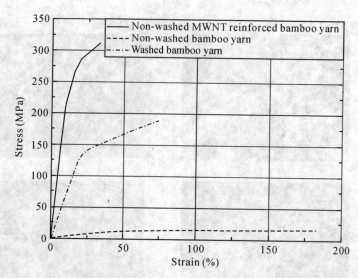

Figure 6 Mechanical properties of MWNTs reinforced bamboo fiber yarn

Thermo properties of bamboo fibres and MWNTs filled bamboo fibers were examined by thermogravimetry analysis, as shown in Figure 7. Bamboo fibres and the MWNT filled bamboo fibres have similar onset thermo decomposition temperature at 324°C. As the temperature increases the MWNT filled bamboo fibre show significantly higher thermal stability. At 50% weight loss the degradation temperature is 450°C showing 125°C higher degradation temperature comparing to that of the unreinforced bamboo fibres. The thermal analysis results also verified the incorporation of 23 wt% MWNTs in fibers.

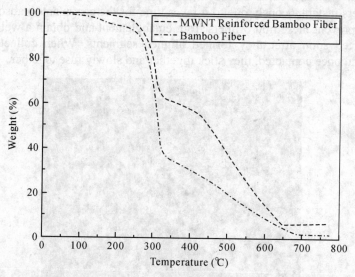

Figure 7 TGA curves of bamboo fibres and MWNTs reinforced bamboo fibres

To investigate the effect of nanotubes applied to bamboo fibers, electric properties of collected MWNTs reinforced continuous bamboo yarn were measured through two-point probes method. The results were shown in Figure 8. The conductivity of untwisted yarns is 0.19 S.cm^{-1} and for twisted yarn is about 0.14 S.cm^{-1} which demonstrated that embodiment of MWNTs in bamboo fibers converted the insulate substance into a conductor. The conductivity nearly two or three orders magnitude higher than many of the reported nanotubes reinforced nanocomposites[1-2,9,14,24] It is also can be assured that the MWNTs reinforced bamboo fiber yarn will have a higher conductivity after carbonization, since carbon fiber is a more conductive material than cellulose.

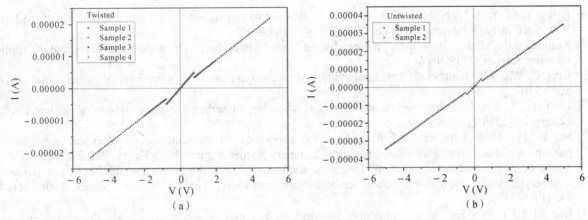

Figure 8 Voltage-current

4 Conclusions and future work

MWNTs were successfully introduced into regenerated bamboo cellulosic fibres with a high weight ratio by the liquid crystal electrospinning process. MWNTs were well aligned both in the fibres and on the surface of fibres. The thermal stability and modulus of the bamboo fibres were significantly improved through the introduction of MWNTs.

With further improvement of the interface between MWNTs and bamboo fibres and further improvement of the uniformity of the performance of the electropun yarns are expected to be further improved thus serving as a nonpetroleum source of renewable advanced multifunctional materials.

Acknowledgement

This work is supported by an NSERC Discovery Grant and by AOARD using equipment funded by a CFI Grant.

References

[1] Kim, J., S. Yun, et al. Discovery of cellulose as a smart material. Macromolecules, 2006,39(12): 4202-4206.
[2] Klemm, D., B. Heublein, et al. Cellulose: fascinating biopolymer and sustainable raw material. Angewandte Chemie International Edition, 2005,44(22): 3358.
[3] Tono, T. Studies on Bamboo Culm and its Fiber as Cellulose Material: Part 2. Microscopical Studies on Beating of Bamboo Fiber. Bulletin of the University of Osaka Prefecture. Ser. B, Agriculture and biology, 1961,11: 155-160.
[4] Parameswaran, N. and W. Liese. On the fine structure of bamboo fibres. Wood Science and Technology, 1976,10(4): 231-246.
[5] Wan, Y., L. Wu, et al. Research on hydroscopic property of original bamboo fibers. Journal of Textile Research, 2004,25(4).
[6] Wan, Y., Y. Cui, et al. Development and technical application of bamboo fibers. Journal of Textile Research, 2004,25(6).
[7] M. Motta, A. M., I. A. Kinloch, Alan H. Windle. High Performance Fibres from "Dog Bone" Carbon Nanotubes. Advanced Materials, 2007,19(21): 3721-3726.
[8] Thess, A., R. Lee, et al. Crystalline Ropes of Metallic Carbon Nanotubes. Science, 1996, 273(5274): 483.
[9] Poncharal, P., Z. L. Wang, et al. Electrostatic Deflections and Electromechanical Resonances of Carbon Nanotubes. Science, 1999, 283(5407): 1513.
[10] Treacy, M. M. J., T. W. Ebbesen, et al. Exceptionally high Young's modulus observed for individual carbon nanotubes. Nature, 1996,381(6584): 678-680.
[11] Tans, S. J., M. H. Devoret, et al. Individual single-wall carbon nanotubes as quantum wires. Nature, 1997, 386(6624): 474-477.
[12] Ko, F. K., H. Lam, et al. Coelectrospinning of carbon nanotube reinforced nanocomposite fibrils. A. C. S. symposium series, 2006, 918: 231-245.
[13] Ko, F. K., S. Khan, et al. Structure and properties of carbon nanotube reinforced nanocomposites. 2002.
[14] Ayutsede, J., M. Gandhi, et al. Carbon Nanotube Reinforced Bombyx mori Silk Nanofibers by the Electrospinning Process. Biomacromolécules, 2006,7(1): 208-214.
[15] Ye, H., H. Lam, et al. Reinforcement and rupture behavior of carbon nanotubes—polymer nanofibers. Applied Physics Letters, 2004, 85(10): 1775-1777.

[16] F. Ko, Y. G. A. A. N. N. H. Y. G. L. Y. C. L. P. W. Electrospinning of Continuous Carbon Nanotube-Filled Nanofiber Yarns. Advanced Materials, 2003,15(14): 1161-1165.

[17] Kulpinski, P. Cellulose nanofibers prepared by the N-methylmorpholine-N-oxide method. Journal of Applied Polymer Science, 2005, 98(4).

[18] Kim, C. W., D. S. Kim, et al. Structural studies of electrospun cellulose nanofibers. Polymer, 2006,47(14): 5097-5107.

[19] Li, Y.-H., J. Wei, et al. Mechanical and electrical properties of carbon nanotube ribbons. Chemical Physics Letters, 2002, 365(1-2): 95-100.

[20] So, H. H., J. W. Cho, et al. Effect of carbon nanotubes on mechanical and electrical properties of polyimide/carbon nanotubes nanocomposites. European Polymer Journal, 2007,43(9): 3750-3756.

[21] Sundaray, B., V. Subramanian, et al. Electrical conductivity of a single electrospun fiber of poly(methyl methacrylate) and multiwalled carbon nanotube nanocomposite. Applied Physics Letters, 2006, 88(14): 143114-143113.

[22] Du, F., R. C. Scogna, et al. Nanotube networks in polymer nanocomposites: rheology and electrical conductivity. Macromolecules, 2004, 37(24): 9048-9055.

4 Structures and Properties of Kapok Fiber

Qiuling Cao[*], Yi Cao, Lin Wang, Xiaowei Sun

(Department of Textiles, Henan Institute of Engineering, Zhengzhou, Henan 450007, China)

Abstract: The structures and properties of kapok fibers were characterized and compared with cotton fiber (G.herbaceum and G. hirsutum). The results show that kapok fiber has an average length of 19.5 mm, close to G.herbaceum fiber, the fineness of kapok fiber is finer than that of cotton fiber, and the width is close to G. hirsutum fiber's. Cotton fiber look like ribbons, rolled in a helicoidal manner around the axis while Kapok fiber without convolutes. The feature of infrared spectrogram of kapok fiber is similar to cotton fiber's, but it indicates that lignin existed in kapok fiber. The XRD intensity profile of kapok fiber shows the well-resolved spectrum of Cellulose I and the crystallinity of kapok fiber is 46.4%, below cotton fiber's 70.3%. The strength and elongation of kapok fiber is lower than cotton fiber's significantly because its thinner cell wall and lower crystallinity.

Keywords: Kapok fiber; Cotton fiber; Structure; Property

1 Introduction

Kapok fiber is the fruit fiber that draws from parts of Bombacaceae plants. The kapok fiber of China mainly belongs to Bombax malabaricum, and has the long spinning application history [1]. In recent years, kapok fiber has a great application in wadding materials and developing the shell fabrics [2-4]. So the structures and properties of kapok fiber were characterized and compared with cotton fiber.

2 Experimental

Samples of kapok fibers (belong to Bombax malabaricum) came from Hainan province. Samples of cotton fibers (belong to G. herbaceum and G. hirsutum) were provided by the Chinese Cotton Germplasm Resource Center, Anyang, China.

2.1 Length, fineness, width and convolutions measurements

The length of fibers was measured by manual method. The fineness of fibers was measured by middle cut and weigh method. Width and convolution were then determined using an Olympus CH-2 video microscope.

2.2 Scanning electron microscopy tests

The shapes of kapok fibers and G. herbaceum fibers were investigated in more detail using a Hitachi JSM-5600LV scanning electron microscope with an accelerating voltage of 15 kV. Specimens were coated with platinum to avoid charging effects.

2.3 Infrared spectra

The IR spectra were obtained with a Nicolet 5700 spectrophotometer using the attenuated total reflection technique, and recorded at 4 cm^{-1} resolution. Sixty-four scans were taken per sample.

2.4 X-Ray diffraction tests

Wide-angle X-ray diffraction measurements of samples were carried out with a D/max-2550PC diffractometer of CuKα radiation operated at 40 kV, 300 mA with the wavelength λ =1.54056 Å. The diffraction patterns of all samples were recorded over the 5°~50° 2θ range continuously at a scan rate of 10°/min.

2.5 Tensile properties tests

Based on the ASTM D3822 Standard Test Method for Tensile Properties of Single Textile Fibers, the tensile properties of the samples were tested using a YG001N Tensile Tester at a gauge of 10 mm and drawing speed of 10 mm/min.

3 Results and discussion

3.1 Comparison of fiber characteristics

The measured data of fiber length, fineness, width and convolutions are shown in Table 1.

---

* Corresponding author's email: cqlwwd@163.com

Table 1 The length, fineness, width and convolutions of fibers

Sample	Length/mm	Fineness/dtex	Width/μm	Convolutions/cm^{-1}
Kapok fibers	19.5	0.67	16.4	0
G. herbaceum	18.7	2.3	21.4	59
G. hirsutum	28.9	1.6	17.3	77

Those data in Table 1 reveal that the length of kapok fibers is 19.5 mm, similar to the 18.7 of G. herbaceum fiber used widely in ancient China, while shorter significantly than that of G. hirsutum used in modern spinning now; the fineness of kapok fibers, 0.67dtex, is much finer than that of cotton fibers; the width of kapok fibers is about 16.4 μm, narrower than that of G.herbaceum fiber, while closed to G. hirsutum fiber's. The convolutions are characterized by the number of 180° reverses per centimeter in length. The G.herbaceum fibers convolute about 59 times per centimeter, while kapok fiber without convolutes, so that the cohesion and spinnablity of kapok fibers are poor.

3.2 Morphology analysis by SEM

Figure 1 to Figure 4 show the SEM images of both kapok fibers and G.herbaceum.

Figure 1 Longitudinal morphology of kapok fibers

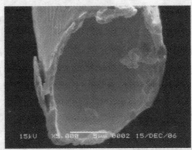
Figure 2 Cross-sectional morphology of kapok fibers

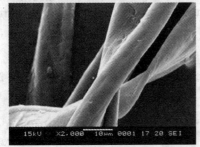
Figure 3 Longitudinal morphology of G.herbaceum

Figure 4 Cross-sectional morphology of G.herbaceum

The SEM analysis reveals that cotton fibers have the ribbon shape rolled in a helicoidal manner around the axis, and bean-like cross sections, while the shape of kapok fiber is different from cotton fibers'. It appears smooth with the thickness of 1 μm and uneven cell wall. The diameter of the lumen is about 15 μm, the degree of hollowness reaches up to 90%, in which there is full of air. So the volume mass of kapok fiber is lower and heat retention is better than other fibers.

3.3 Infrared spectra analysis

The IR spectra of both the kapok fibers and the G.herbaceum fibers are provided in Figure 5.

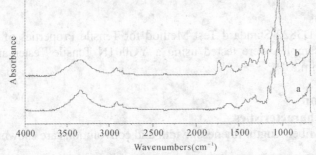

Figure 5 IR spectra of samples: (a) G. herbaceum, (b) kapok fiber

The IR spectra of the kapok fibers are similar to the spectra of the cotton fibers. The peak due to –OH stretching vibration in the absorption range of 3200 to 3400 cm^{-1} is the major characteristic absorption peak for cellulose fibers. Nearly identical tiny peaks at 2895 cm^{-1} associated with C–H stretching, and 2849 cm^{-1} associated with –CH_2 symmetrical stretching vibration are seen on both curves; the physically absorbed water molecule shows a characteristic peak at 1630 cm^{-1}[5]. Peaks at 1357 are due to C–H bending, 1335 to O–H in-plane bending, and 1314 cm^{-1} to –CH_2 wagging; the band at 1160 cm^{-1} is associated with C–O stretching or O–H bending of the C–OH group[6]; peaks near 1100 cm^{-1} are attributed to the hydrogen bonding on the skeletal vibrations which involve stretching of the C–O bond; the band at 895 cm^{-1}, assigned to motions of atoms attached to C_1, reflects the changes in molecular conformation due to rotation about the C_1–O–C_4[7].

From Figure 5 it can be seen that there are some difference between the kapok fibers and cotton fibers. The main composition of kapok fiber are cellulose of 63.5%, lignin of 13% and something others [8]. Peaks at 1740 cm^{-1} are due to C-O stretching vibration of non-conjugated ketone and carboxyl group, 1510 cm^{-1} to vibration of aromatic ring, and 1240 cm^{-1} to phenolic hydroxyl group. All of them indicates that lignin existed in kapok fibers [9].

3.4 X-Ray diffraction analysis

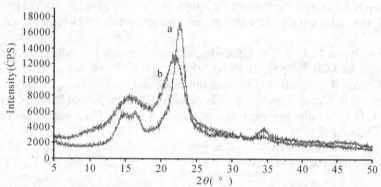

Figure 6 XRD spectra of samples: (a) G. herbaceum, (b) kapok fiber

The XRD spectra of the samples are shown in Figure 6. Both the XRD intensity profiles of the fibers show a well resolved spectrum of Cellulose I. The crystallinity of kapok fiber is 46.4%, below cotton fiber's 70.3%. The degree of crystal orientation of kapok fiber is 82.4%, comparable to 72.4% of cotton fiber. It is noted that the degree of crystal orientation is different from the comprehensive orientation measured by Senarmont's method.

There is more amorphous region in kapok fiber, which is in favour of the entrance of water. The kapok fiber's water retaining capacity is also higher than that of cotton fiber once absorption occurs.

3.5 Strength and elongation of fibers

Table 2 shows the strength and elongation of kapok fibers and cotton fibers.

Table 2 The strength and elongation of fibers

Sample	Strength /cN	Elongation /%
kapok fiber	1.05	2.89
G.herbaceum	5.09	5.32
G.hirsutum	3.58	5.92

The strength and elongation of kapok fibers is less than those of cotton fibers significantly because its thinner cell wall and lower crystallinity.

4 Conclusion

Kapok fiber is the fruit fiber that draws from parts of Bombacaceae plants. The length of kapok fiber is shorter than that of G.hirsutum while closed to G.herbaceum fiber, the fineness of kapok fiber is finer than that of cotton fiber, and the width is close to G. hirsutum fiber's. Cotton fiber look like ribbons, rolled in a helicoidal manner around the axis while kapok fiber without convolutes, the spinnability of kapok fiber is poor. The feature of infrared spectrogram of kapok fiber is similar to cotton fiber's, but it indicates that lignin existed in kapok fiber. The XRD intensity profile of kapok fiber shows the well-resolved spectrum of Cellulose I and the crystallinity of kapok fiber is 46.4%, below cotton fiber's 70.3%. The strength and elongation of kapok fiber is lower than cotton fiber's significantly because its thinner cell wall and lower crystallinity.

Acknowledgement

The research was supported by the Doctor Foundation of Henan Institute of Engineering and Henan Province Education Department. The authors wish to thank Professor Yuping Zhu of Donghua University for his help and efforts completing this research.

References

[1] Cao Qiuling, Tu Hengxian, Zhu Sukang. Study on the Names of Cotton and Kapok in Ancient China. Agricultural Archeology, 2007, (3):20-22.
[2] Lou Ying, Wang Fumei, Liu Wei. Compressibility of Kapok Wadding. Journal of Textile Research, 2007, 28(1):10-13.
[3] Zhao Kongwei, Wang Fumei. Test Analyses of Twist and Elongation & Strength of Java Cotton and Cotton Blended Rotor Yarn. Cotton Textile Technology, 2007, 35(3):154-156.
[4] Yang Xiaoxia. Kapok Fibers and the Product Development. Shanxi Textile, 2007, (3):46-47.
[5] M S Parmar, C C Giri. Spectral Characterisation and Thermal Study on Natural Coloured Cotton. Colourage, 2001(9):21-26.
[6] R T O'conner, E F Dupre, McCall E R. Infrared Spectrophotometric Procedure for Analysis of Cellulose and Modified Cellulose. Anal Chem, 1957, 29:998-1005.
[7] Nelson M L, O'Conner R T. Relation of Certain Infrared Bands to Cellulose Crystallinity and Crystal Lattice Type. Part Ⅰ, Spectra of Lattice Type Ⅰ, Ⅱ, Ⅲ and of Amorphous. J Appl Poly Sci, 1964, 8(3):1311-1324.
[8] B M D Dauda, E G Kolawole. Processibility of Nigerian Kapok fibre. Indian Journal of Fibre & Textile Research, 2003, 28:147-149.
[9] Chen Jiaxiang, Yu Jialuan. Research Method of Plant Fibers' Chemical Structures. Guangzhou: South China University of Technology Press, 1989:142-147.

5 Analysis on Structure of Wool Keratin Film by FT-IR and SEM

Liping Chen, Ping Cui

(College of Mechanical and Electric Engineering, Lanzhou University of Technology, Lanzhou 730050 Gansu, China)

Abstract: The preparing methods of wool's keratin film were introduced in this paper. The chemical structure of wool's keratin film was studied by FTIR. Through the analysis of characteristic IR bands of wool and the keratin film, the changes of the macromolecule structure have been proved and these reasons were been introduced. The surface feature and the stretched section of the keratin film were observed by SEM.

Keywords: Keratin film; Structure; Morphology; FT-IR; SEM
CLC Number: TS 102

0 Introduction

Wool fiber is an excellent natural polymer material with many kinds of amino acids, which is very good affinity to human skin[1]. In 1964, O'Donnell and Thompson prepared soluble keratin from wool by chemical reduction method[2]. Since then many researchers utilized the classical method directly or modified it to expand the applications of wool keratin and to make full use of regenerated wool[3][4]. However, people have not paid enough attention to the wool keratin film. Several methods have been reported to describe the keratin film. Nevertheless, the chemical structure of keratin film extracted from wool is seldom stated.

1 Extraction of keratin and preparation of keratin film

The soluble keratin from 64s wool was extracted by the following compounds, urea (H_2NCONH_2), 2-mercaptoethanol ($HSCH_2CH_2OH$), sodium dodecyl sulfate (SDS) etc. Dialysis was performed using a cellulose bag, 40 mm diameter with molecular weight cut off 8000-12000 Daltons. First, the cleaned wool was mixed with above mentioned chemicals in a flask to stir at 40-80℃ for 6-12 h. Second, the resulting mixture was filtered through a mesh and third, the filtrate was dialyzed using deionised water to afford a clear solution of the reduced keratin.

An aqueous solution of the reduced keratin was spread on a surface plate in heat atmosphere with 50℃. Then, the keratin film was obtained.

2 Chemical structure of keratin film by FT-IR spectra analysis

2.1 Test instrument

Nicolet Nexus-670 FTIR was used with transmission type polarizer, 200 times of scanning and 8 cm^{-1} of resolution. The size of aperture is 20 $\mu m \times 60$ μm.

2.2 FT-IR spectrogram of keratin film

Figure 1 is the FT-IR spectrogram of wool keratin and Figure 2 is the FT-IR spectrogram of keratin film. By comparing the spectra in Figure 1 and Figure 2, we could find the FT-IR spectrograms of wool and keratin film are very similar. They are all belong to the FT-IR spectrograms of classical proteins, that is to say, classical characteristics of absorbance bands of hydrogen bond, amide I, amide II and amide III are existing.

The classical absorbance bands of protein are Hydrogen bond (around 3300 cm^{-1}), Amide I (1690-1600 cm^{-1}), Amide II (1575-1480 cm^{-1}) and Amide III (1301-1229 cm^{-1}) [5].

The following Table 1 is the summarization of data based on Figure 1 and Figure 2.

Table 1 Absorbance peaks of wool and keratin film

	amide I ratio (cm^{-1})	amide II ratio (cm^{-1})	amide III ratio (cm^{-1})	Hydrogen bond ratio (cm^{-1})	unsymmetrical flex of $CH_3(CH_2)$ ratio (cm^{-1})	symmetrical flex of $CH_3(CH_2)$ Ratio (cm^{-1})
Tested wool	1658	1551	1243	3306	2963	2870
Tested keratin film	1656~1670	1541	1240	3292	2922	2850

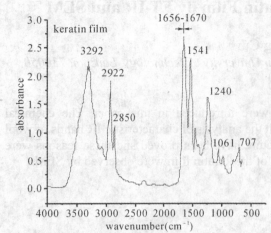
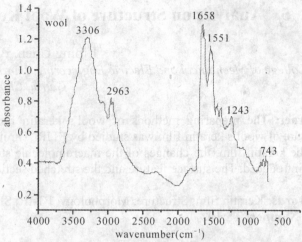

Figure 1 FT-IR spectrogram of wool keratin   Figure 2 FT-IR spectrogram of wool (64s Merino wool)

2.3 Amide I analysis of keratin film

The absorbance peak of Amide I is caused by C=O flex vibration of amide, which can be used as a characterization of protein structure[6]. The region has some peaks and each peak implies a conformation, such as α-helix, β-pleated, β-turns and disordered region. It is testified that the peak of 1700~1660 cm^{-1} is belong to β-turns vibration, the peak of 1658~1650 cm^{-1} is belong to α-helix vibration, the peak of 1650~1640 cm^{-1} is belong to the disordered region vibration and the peak of 1640~1610 cm^{-1} is β-pleated vibration[7].

From the above Table 1, we can see the changes from wool to keratin film by chemical extracting are that both of molecule perssad and kinds of vibration are same and the molecular conformation of keratin film was changed. We know that there are α-helix (1658 cm^{-1}), β-pleated (1640~1610 cm^{-1}) and the disordered region (1650~1640 cm^{-1}) in wool structure. The α-helix is main and this is tested through our test (Table 1 and Figure 2). However, the absorbance peak of amide I of keratin film was widened (1656~1670 cm^{-1}). In this wide region, the band of 1656~1658 cm^{-1} is belong to α-helix vibrant region and others are belong to β-turns. It exhibits that the molecular chain of keratin in wool fiber has not been broken and that its macromolecule conformation transforms from α-helix to β-turns during solving wool fiber.

2.4 Amide II and amide III analysis of keratin film

The absorbance peaks of Amide II and amide III are all caused by C-N flex and N-H curve in amide vibration. Both of them all moved to low band slightly, combining change of the absorbance peak of hydrogen bond, it exhibits that the hydrogen bond has been broken during solving wool to keratin film.

2.5 Disulfide bond (-S-S) vibrant analysis

Disulfide bond (-S-S-) vibrant region in IR microscope spectra of keratin film have two region. One is 2922 cm^{-1}, the other is 2850 cm^{-1} (Table 1). Both of peaks are all caused by symmetrical or unsymmetrical flex of CH_3 (CH_2) and are all from high to low. It is attributed to CH_2–S-S-CH_2 has been broken.

2.6 SEM of wool keratin film

The Scan Electronic Microscopy (JSM—5600LV SEM) was utilized to observe the surface of the keratin, as seen in Figure 3 and Figure 4. It can be conclude that the keratin surface is smooth with some tiny granule material (Figure 3). However, the profile of stretched keratin film arranged disorderly, which was similar to something was pulled out.

3 Conclusions

Keratin is extracted by wool. During solving wool, its macromolecule conformation transforms from α-helix to β-turns because of the broken of hydrogen bond and disulfide bond. That is to say, there are more β-turns and more disordered regions in the structure of keratin film comparing wool and the molecular chain of keratin in wool fiber is remained. It is very important for keratin film development.

 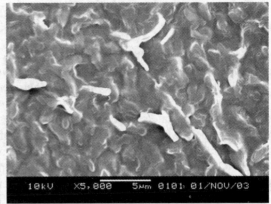

Figure 3 Surface of keratin   **Figure 4 Profile of stretched keratin film**

By SEM, we know the keratin surface is smooth with some tiny granule material and the profile of stretched keratin film arranged disorderly, which was similar to something was pulled out.

References
[1] Zhang D S, Zhang H W. Wool[M]. Xian: Shanxi Science and Technology Publication.1986.12: 249.
[2] O'Donnell I J, Thompson E O P. Reduced Wool:IV. The Isolation of a Major Component[J]. Australian Journal of Biological Science, 1964, 17(4):.973-979.
[3] Hiroaki Y, Hiroyuke M, Isao A, 13C CP/MAS NMR Study of the Conformation of stretched or Heated Low-sulfur Keratin Protein Films[J]. Macromolecules, 1991, 24(5): 862-866.
[4] 柴山幹生, 加藤一德, ウールハラチソ及びそ複合物の纖維化技術, 纖維加工. 2000, 52 (8): 351-361.
[5] WENG S F. Fourier Transform infrared spectrometry[M]. Beijing: Chemical Industry Publication, 2005: 294-295.
[6] Singh B R. Infrared Analysis of Peptides and Proteins: Principles and Applications[J]. Washington, D C: American Chemical Society, 2003, 3, 122, 60.
[7] Xie M.X. and Liu Y. Studies on the Hydrogen Bonding of Aniline's Derivatives by FT-IR[J] Spectrochimica Acta, Part A, 2002, 58(13): 2817-2826.

6 Preparation and Characterization of CA/CeO$_2$ Composite Nanofibers

Rui Chen, Xiaoqiang Zhang, Wenjun Dong, Chaorong Li[*]

(Department of Physics, and Key Laboratory of Advanced Textile Materials and Manufacturing Technology, Ministry of Education of China, Zhejiang Sci-Tech University, Hangzhou 310018, China)

Abstract: Cellulose acetate/ceria composite nanofibers were prepared via the electrospinning technique. The composite nanofibers were characterized by means of scanning electron microscopy, transmission electron microscopy, X-ray diffraction and UV–vis absorption. The X-ray diffraction patterns revealed that the particles on Cellulose acetate fibers were face-centered cubic fluorite-type CeO$_2$. Scanning electron microscopy and transmission electron microscopy demonstrated that CeO$_2$ nanoparticles was successfully doped into CA polymer and well dispersed along the fibers overall. Moreover, due to the existence of CeO$_2$, the composite nanofibers showed strong UV absorption properties.

Keywords: Electrospinning; Composite nanofibers; Cellulose acetate; CeO$_2$ nanopaticles

1 Introduction

Electrospinning has been attracted considerable attention recently since it is the most convenient technique for preparing ultra-fine fibers with diameters ranging from nanometer to micrometer scale [1-2]. Electrospun nanofibers have a number of important characteristics, such as high surface/volume value and small fiber diameter [3-5]. Particularly, it is fairly easy to improve the functionality of nanofibers by incorporating additives during the electrospinning process [6].

In recent years, incorporating inorganic nanoparticles into polymer fibers thus preparing organic-inorganic composite nanofibers has been widely investigated [7-10]. These composite nanofibers combine both the advantages of organic materials, such as light weight, flexibility and good moldability, and inorganic materials, such as high strength, heat stability and chemical resistance [11-13]. Cellulose acetate (CA), an acetylated derivative of cellulose, is widely used, due to its nontoxicity, biodegradability, and good processability [14-18].

Ceria (CeO$_2$), as a fascinating rare earth material, exhibiting strong UV absorption properties, are good candidates in the application of UV blocking or shielding materials [19-20]. The functionalized CA fiber with CeO$_2$ will have wide potential applications in textile industry, environment protection, catalysis etc. In this paper, we investigated the fabrication of composite nanofibers of CA/CeO$_2$ via electrospinning technique and their optical properties. The cellulose acetate was selected as the filament-forming matrix, while the inorganic nanopaticles serve as functional component.

2 Experimental

2.1 Reagents

Cellulose acetate (CA; white powder; Mw=30,000; acetyl content 40wt%) was supplied by Aladdin Co. Acetone and N,N-Dimethylacetamide (DMAC) were purchased from Shanghai Chemical Reagents. Ce(NO$_3$)$_3$·6H$_2$O was from Shanghai Yaojing Chemical Industry Co., Ltd. Ethylenediamine (EDA) was from Tianjin Bodi Chemical Industry Co., Ltd. All chemicals were used directly without further purification.

2.2 Preparation of spinning solutions

CeO$_2$ nanoparticles were synthesized by the hydrothermal method. In a typical experiment, 1g Ce(NO$_3$)$_3$·6H$_2$O were dissolved in 45 ml deionized water. A total of 0.5 ml EDA was added dropwise to the solution. Then the obtained solution was transferred into Teflon-lined steel autoclaves and heated for 9h at 180°C in a thermostatted oven. After cooling to room temperature naturally, the precipitated powder was filtered off, washed with deionized water and anhydrous alcohol for three times. A straw yellow powder of CeO$_2$ nanoparticles was finally obtained after drying for 12h at 60°C in vacuum.

The process of preparing composite solutions for electrospinning as following: 0.4g of pre-prepared CeO$_2$ nanoparticles were firstly dispersed in 10ml of DMAC/acetone (1:2, V/V) solvent with 0.5h sonication, then 1.5 g CA powder was directly added to the above solution. To improve homogeneous distribution of CeO$_2$ particles in the polymer pre-spinning solution, the resulting milky slurry was ultrasonicated for 1h prior

---

[*] Corresponding author's email: crli@zstu.edu.cn

to electrospinning.
2.3 Electrospinning
Electrospinning was done by a self-made apparatus that consists of a high voltage supply (Glassman high voltage, FC60P2), a peristaltic pump (USA KDS Scientific Inc, KDS100), a syringe with metal needle (ID=0.84mm) and a collector. In the typical electrospinning process, the precursor was electrospun at 18 kV voltage, 15 cm working distance (the distance between the needle tip and the collector), and 0.6ml/h flow rate at room temperature. The as-collected CA/CeO_2 nanofibers were dried for 24h at 80°C under vacuum prior to the subsequent characterizations. In contrast, 15wt% of CA in DMAC/acetone (1:2, V/V) was used for electrospinning to prepare CA fibers.

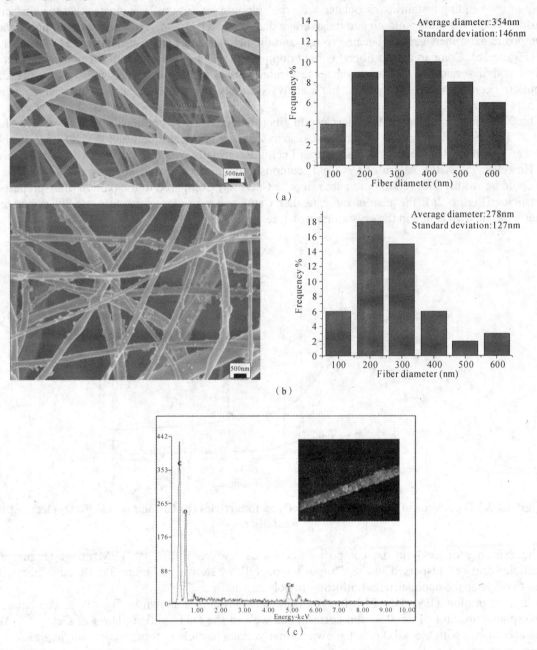

Figure 1 SEM images and diameter distribution of (a) CA and (b) CA/CeO_2 electrospun nanofibers (c) EDS spectrum of CA/CeO_2 composite nanofibers

2.4 Characterization
The morphologies and composition of electrospun nanofibers were studied by field-emission scanning electron microscope (FESEM, HitachiS-4800) equipped with an Energy-dispersive X-ray spectroscopy (EDS),

and transmission electron microscope (JEM2100, 200kV). The structural property of composite nanofibers was characterized by X-ray diffraction (XRD, Bruker AXS D8-discover) using CuKα1 radiation (λ=0.15405nm at 40kV and 40mA). UV-Vis absorption spectra of the fiber were obtained on a Lambda 900 UV spectrometer. In order to analyze fiber diameter and its distribution, 50 electrospun fibers from each sample were randomly selected from SEM image and their diameters were measured using Image-Pro®Plus software.

3 Results and discussion

The SEM micrographs and the corresponding diameter distribution chart of CA and CA/CeO_2 electrospun nanofibers are shown in Figure 1. It can be seen that the orientation of all fibers are random because of the bending instability associated with the spinning jet. Figure 1a indicates that the surface of CA electrospun fibers was quite smooth and the average diameter was 354 nm. However, after dopping the CeO_2, the composite nanofibers surface became rough, and the nanoparticles were well-dispersed on the surface of fibers (Figure 1b). Compared with fibers without dopping, the average diameter of the composites reduces to 278 nm, and the standard deviation decreases from 146nm to 127nm which means that the diameter distribution becomes even. Moreover, EDS analysis shows the presence of Ce element in composite nanofibers (Figure 1c).

The XRD patterns of pure CA electrospun fibers and CA/CeO_2 composite nanofibers are shown in Figure 2. The diffraction pattern of pure CA displays two broad peaks between 5° and 25°, as shown in Figure 2b. This indicates the amorphous nature of cellulose acetate, which is similar to those previous works [21-24]. However, the XRD pattern of CA/CeO_2 composite nanofibers reveals some sharp and intense peaks, which could be readily indexed to pure cubic fluorite CeO_2 [25]. Compared to the pattern of as-prepared CeO_2 nanoparticles (Figure 2a), the results indicate that CeO_2 nanoparticles were successfully doped into CA polymer without any change in the crystalline structure.

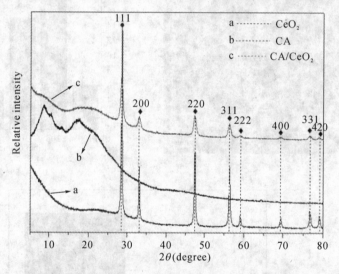

Figure 2 XRD patterns of (a) as-prepared CeO_2 nanoparticles (b) CA and (c) CA/CeO_2 electrospun nanofibers

The existence of CeO_2 in the composite fibers was also confirmed by TEM image (Figure 3). The nanoparticles are well dispersed in the composites overall, as shown in Figure 3a, although normally it is difficult to disperse the nanoparticles uniformly in polymer matrix.

A high resolution TEM (HRTEM) image of the nanoparticles is shown in Figure 3b. The lattice fringe with interplanar spacing of ca. 0.32 nm corresponds well to the (111) lattice planes of CeO_2 [26]. And this result is accordance with the XRD result shown formerly. Each particle is composed of single grain.

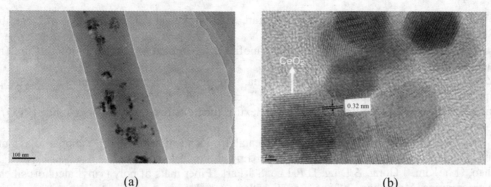

(a) (b)
Figure 3 (a) TEM and (b) HRTEM images of CA/CeO$_2$ composite nanofibers

The optical properties of electrospun CA fibers and CA/CeO$_2$ composite nanofibers are characterized using UV-absorption spectroscopy (Figure 4). From Figure 4a, it can be seen that the electrospun CA fibers have hardly any absorption band above 250 nm. But the CA/CeO$_2$ nanocomposites appeared strong absorption at 340 nm, as shown in Figure 4b, corresponding well with the known spectral behavior of CeO$_2$ nanoparticles [19].

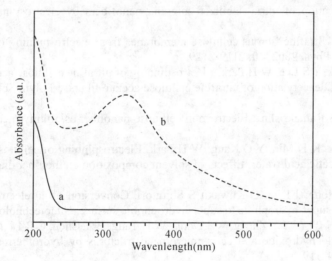

Figure 4 UV-vis spectra of (a) CA and (b) CA/CeO$_2$ electrospun nanofibers

4 Conclusions

In conclusion, the composite nanofibers of CA/ CeO$_2$ with an average diameter of 278 nm were simply prepared via the electrospinning of CA spinning solutions with CeO$_2$ nanoparticles. The effects of adding nanoparticles on the morphology and structure of the electrospun nanofibers were studied in this work. When the CeO$_2$ particles were added into the solution, the composite nanofibers became coarse along with diameter decreasing. Meanwhile, the CeO$_2$ nanoparticles was successfully doped into CA polymer without any change in the crystalline structure and well dispersed along the fibers overall. Furthermore, the composites shows strong absorption property in the UV range, which is expected to have a potential as shielding material in future.

Acknowledgement

This work was supported by the National Natural Science Foundation of China (Nos 10874153, 50772100, 50972130, 20701033), Zhejiang National Natural Science Foundation (No. Y407188), Science and Technology Department of Zhejiang Province Foundation (No. 2008C23011).

References

[1] D Li, Y Xia. Electrospinning of Nanofibers: Reinventing the Wheel? Adv Mater, 2004, 16: 1151-1170.
[2] E Zussman, A Theron, A L Yarin. Formation of nanofiber cross bars in electrospinning. Appl Phys Lett, 2003, 82: 973-975.

[3] T Subbiah, G S Bhat, R W Tock, S Parameswaran, S S Ramkumar. Electrospinning of Nanofibers. J Appl Poly Sci, 2005, 96: 557-569.

[4] A Frenot, I S Chronakis. Curr Opin. Polymer nanofibers assembled by electrospinning. Curr Opin Colloid Interface Sci, 2003, 8: 64-75.

[5] M H Zheng, Y Z Zhang, M Kotaki, S Ramakrishna. A review on polymer nanofibers by electrospinning and their applications in nanocomposites. Compos Sci Technol, 2003, 63: 2223-2253.

[6] T Uyar, F Besenbacher. Electrospinning of cyclodextrin functionalized polyethylene oxide (PEO) nanofibers. Eur Polym J, 2009, 45: 1032-1037.

[7] S K Lim, S K Lee, S H Hwang, H Kim. Photocatalytic Deposition of Silver Nanoparticles onto Organic/Inorganic Composite Nanofibers. Macromol Mater Eng, 2006, 291: 1265-1270.

[8] C L Shao, H Y Kim, J Gong, B Ding, D R Lee, S J Park. Fiber mats of poly (vinyl alcohol)/silica composite via electrospinning. Mater Lett, 2003, 57: 1579-1584.

[9] G M Kim, A Wutzler, H J Radusch, G H Michler, P Simon, R A Sperling, W J Parak. One- dimensional arrangement of gold nanoparticles by electrospinning. Chem Mater, 2005, 17 (20): 4949-4957.

[10] J S Kim, D S Lee. Thermal Properties of Electrospun Polyesters. Polym J, 2000, 32: 616-618.

[11] X M Sui, C L Shao, Y C Liu. White-light emission of polyvinyl alcohol/ZnO hybrid nanofibers prepared by electrospinning. Appl Phys Lett. 2005, 87: 1135.

[12] Y L Hong, D M Li, J Zheng, G T Zou. Sol-gel growth of titania from electrospun polyacrylonitrile nanofibres. Nanotechnology, 2006, 17: 1986-1993.

[13] A M Azad. Fabrication of yttria-stabilized zirconia nanofibers by electrospinning. Mater Lett, 2006, 60: 67-72

[14] H Liu, YL Hsieh. Ultrafine fibrous cellulose membranes from electrospinning of cellulose acetate. J Polym Sci. Part B: Polym Phys, 2002, 40: 2119-2129.

[15] W K Son, J H Youk, T S Lee, W H Park. Electrospinning of ultrafine cellulose acetate fibers: Studies of a new solvent system and deacetylation of ultrafine cellulose acetate fibers. J Polym Sci. Part B: Polym Phys, 2004, 42: 5-11.

[16] Z Ma, M Kotaki, S Ramakrishna. Electrospun cellulose nanofiber as affinity membrane. J Membr Sci, 2005, 265: 115-123.

[17] S O Han, J H Youk, K D Min, Y O Kang, W H Park. Electrospinning of cellulose acetate nanofibers using a mixed solvent of acetic acid/water: Effects of solvent composition on the fiber diameter. Mater Lett, 2008, 62: 759-762.

[18] B Ding, C Li, Y Hotta, J Kim, O Kuwaki, S Shiratori. Conversion of an electrospun nanofibrous cellulose acetate mat from a super-hydrophilic to super-hydrophobic surface. Nanotechnology, 2006, 17: 4332-4339.

[19] M Y Cui, J X He, N P Lu, Y Y Zheng, W J Dong, W H Tang, B Y Chen, C R Li. Morphology and size control of cerium carbonate hydroxide and ceria micro/nanostructures by hydrothermal technology. Mater Chem Phys, 2010, 12: 314-319.

[20] M Y Cui, X Q Yao, W J Dong, K Tsukamoto, C R Li. Template-free synthesis of CuO–CeO2 nanowires by hydrothermal technology. J. Cryst. Growth, 2001, 312: 287-293.

[21] Q Zhou, L Zhang, M Zhang, B Wang, S J Wang. Miscibility, free volume behavior and properties of blends from cellulose acetate and castor oil-based polyurethane. Polymer, 2003, 44: 1733-1739.

[22] H Wu, F Xin, X Zhang. Cellulose acetate-poly (N-vinyl-2-pyrrolidone) blend membrane for pervaporation separation of methanol/MTBE mixtures. Sep Purif Technol, 2008, 64: 183-191.

[23] W K Son, J H Youk, W H Park. Preparation of ultrafine oxidized cellulose mats via electrospinning. Biomacromolecules, 2004, 5: 197-201.

[24] B Ding, E Kimura, T Sato, S Fujita, S Shiratori. Fabrication of blend biodegradable nanofibrous nonwoven mats via multi-jet electrospinning. Polymer, 2004, 45: 1895-1902.

[25] C Sun, J Sun, G Xiao, H Zhang, X Qiu, H Li, L Chen. Mesoscale Organization of Nearly Monodisperse Flowerlike Ceria Microspheres. J Phys Chem B, 2006, 110: 13445-13452.

7　Reach on the Structure and Character of a Thermal Regulating Fiber

Weilai Chen[1*], Ying Gao[1], Jianxiao Zhu[1], Zhebin Tang[1], Yihui Gong[2]

(1. School of Materials and Textiles, Zhejiang Sci-Tech University, Hangzhou 310018, China
2. Zhejiang Sunjazz Garments Co., Ltd., Yiwu 322000, China)

Abstract: The thermal regulating fibers based on acrylic fiber were characterized by Scanning electron microscopy (SEM), Differential scanning calorimetry (DSC), thermogravimetry (TG) and tensile tester. The results showed that the fiber possessed the morphological structure similar to acrylic fiber, and the outlast fibers had thermal properties were: the latent heat of the outlast fibers were 6.5J/g in the heating process and 7.5J/g in the freezing process; the tensile strength was 6.5cN, the elongation at break was 47.1%.

Keywords: Thermal regulating fiber; Phase change material; Acrylic fiber

1 Introduction

As technology and people's living standards advances, the traditional clothing does not satisfy the demand for a simple dress. Utilizing phase change materials (PCMs) to achieve energy storage is very attractive because of its high storage density with small temperature swing [1]. This technology was initially developed by outlast in the early 1980s under NASA research programmer [2]. There are several methods to prepare thermal regulating fibers, for instance, hollow fibers filling, micro encapsulation spinning, composite spinning, et al. Vigo and his co-worker [3, 4] produced fibbers with energy storage function by immersing hollow fibbers into polyethylene glycol (PEG) solution or melt. Jiang [5] chose polyvinyl alcohol (PVA) as fibers matrix, ethyl orthosilicate (TEOS) as membrane-forming reagent and paraffin as PCM to produce thermal regulating fibers through wet spinning and the paraffin was microencapsulated by in-situ method when the as-spun fibers was treated in acidic or basic aqueous Na_2SO_4 solution. The fibers treated in acidic condition formed better membrane structure of microcapsule. A kind of thermal regulating fibers based on polypropylene (PP) and mixture of methyl separate (MS) and methyl palmitate (MP) was produced by melting spinning by XU [6]. In this paper, the thermal regulating fibers based on acrylic fiber was prepared by melting spinning. The fibers were characterized by DSC, SEM, TGA and tensile tester.

2 Experimental
2.1 Materials and Tests
2.1.1 Materials

Table 1 The character of the outlast fiber and acrylic fiber

Samples	Length/mm	Linear density/ dtex
Outlast fiber	38	2.2
Acrylic fiber	38	1.67

Outlast fibers and acrylic fibers were provided by Hua Xin Industry Business Company. Phase change materials were prepared via interracial polymerization with low melt-point paraffin wax as the core. The characters of the outlast fiber were listed in the Table 1.

2.2 Measurement
2.2.1 Morphological examinations

Morphology of outlast fiber and acrylic fiber was observed by digital vacuum scanning electron microscope (JSM-5600LV, Japan Electron Optical Laboratory) at the accelerating voltage of 5 kV. Samples for SEM observations were sputter coated with gold before observation.

2.2.2 DSC Test

Thermal property of outlast fiber and acrylic fiber was investigated by means of DSC (Pyris Diamond) made by Perkin Elmer Corporation, America. The samples were rapidly heated from $-30℃$ to $100℃$, then were decreased to $-30℃$ after 3 minutes at the rate of $10℃/min$. Dry nitrogen was used as purging gas. Samples were cut into powders, and the weights of them varied from 5 to 8 mg.

---

* Corresponding author's email: wlchen193@163.com

2.2.3 TG Test

Thermo gravimetric (TG) analyses were carried out using a Pyris1 (Perkin Elmer) under N_2 atmosphere. Both of samples were heated from room temperature to 600℃ at the rate of 20°C/min.

2.2.4 Mechanical properties

The mechanical properties of fibers including tensile strength, elongation at break and young's modulus are tested by fiber tensile testing machine (XQ-2). Test conditions were as follows: temperature, 20±1℃; gauge length, 20mm; and crosshead speed, 10mm/min.

3 Results and discussion
3.1 Morphology

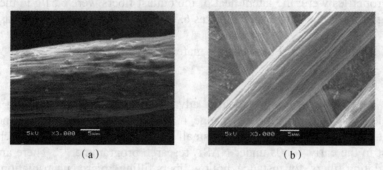

Figure 1 SEM photographs of outlast fiber (a) and acrylic fiber (b)

SEM photographs of the surface of the outlast fiber and acrylic fiber are shown in Figure 1. We can clearly see many grains that were called capsule on the surface of the outlast fiber, which cannot be seen on the surface of the acrylic fiber, so acrylic fiber looks relatively smooth compared with the outlast fiber. In addition, we can see stripes that are similar to acrylic fiber in the longitudinal part as seen in the longwise of the outlast fiber. The reason was that the outlast fiber was manufactured based on acrylic fiber.

3.2 Thermal property

Figure 2 is the DSC curves of the outlast fiber and acrylic fiber, and the related thermodynamic parameters for the outlast fiber were shown in Table 2. In Figure 2, there were obvious fusion and freezing peaks for the outlast fiber, which were not seen in the DSC curve of acrylic fiber. As we can see from Table 2, the phase change temperature started at 26.7℃ and the end at 33.7℃ in the process of the heating, and the phase change temperature started at 17.8℃ and end at11.6℃ in the process of the freezing. Latent heat of outlast fiber had melting peaks at about 30.1℃ and 15.6℃. From the Figure 2, we also obtained two heat fusion for outlast fiber about 6.5 J/g in the heating process and 7.5 J/g in the freezing process after calculated the area under the peak of the DSC curve. So, the ability of the thermal regulation in the freezing process was better than that in the heating process.

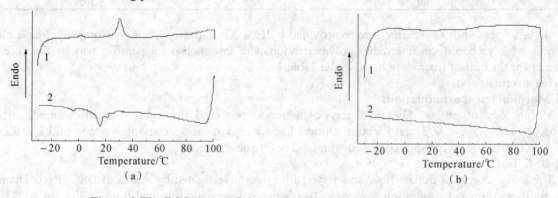

Figure 2 The DSC curve of the outlast fiber (a) and acrylic fiber (b)
Note: 1-heating curve; 2-freezing curve

Table 2 Latent heat of fusion and melt temperature of the outlast fiber

Phase change process	Start temperature/°C	End temperature/°C	Peak temperature/°C	Heat of fusion (J/g)	Phase change temperature range/°C
heating	26.7	33.7	30.1	6.516	5.6
freezing	17.8	11.6	15.6	7.501	6.1

3.3 Thermal stabilities

Thermo gravimetric (TG) analysis of outlast fiber and acrylic fiber are shown in Figure 3. Both of the fibers had two obvious weight loss band, and a major weight loss proceeded from 310℃ to 480℃ for outlast fiber and acrylic fiber. There is the first significant loss of weight for the outlast fiber with 20% weight loss at 310℃-340℃, which was lower than that of acrylic fiber. The second major decomposition temperature for outlast fiber was from 360℃, which was similar to that of the acrylic fiber. When the weight loss was about 5%, the outlast fiber was at 307.3℃, and the acrylic fiber was at 325.6℃. Thermal stability of outlast fiber was worse than that of acrylic fiber and it can be seen that the shell of the capsules inner the outlast fiber kept intact when the temperature was low.

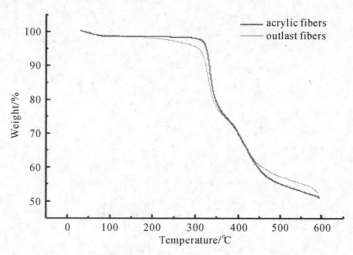

Figure 3 TG curve of the outlast fiber and acrylic fiber

3.4 Mechanical properties

The mechanical properties of the outlast fiber and acrylic fiber were listed in Table 3. As we can see from Table 3, the tenacity of outlast fibers was lower than that of acrylic fiber, while elongations at break of the outlast fiber were higher than that of acrylic fiber. This was attributed to the weak link increasing inner the outlast fibers, which were filled with the PCM.

Table 3 Mechanical properties of outlast fiber and acrylic fiber

Sample	Outlast fiber	Acrylic fiber
Tenacity(cN/dtex)	2.95	4.25
Elongation at break/%	47.1	44.1

4 Conclusions

This paper investigated morphology; thermal properties and mechanical properties of outlast fiber and acrylic fiber. The results showed that there were many capsules on the surface of the outlast fiber, the Phase change temperature range was about 5.6℃ in the heating process and 6.1℃ in the freezing process, and the tenacity is lower (2.95 cN/dtex) than that of the acrylic fiber, but elongations at break of the outlast fiber was high (47.1%).

Acknowledgement

This work was financially supported by the major textile foundation of Zhejiang Province (2008C11079), and we also thank Zhejiang Sunjazz Garments Co.Ltd for offering the materials of outlast fibers and acrylic fibers.

References
[1] A. Pasupathy, R. Velraj, R. V. Seeniraj, Renew. Sust. Energ. Rev., 2008, 12: 39-64.
[2] S. Mondal, Appl. Therm. Eng., 2008, 28, 1536-1550.
[3] T. L. Vigo, US Pat., 5 885 475, 1999.
[4] T. L. Vigo, US Pat., 4 908 238, 1990.
[5] M. Jiang, X, Song, G. Ye, et al., COMPOS. SCI. TECHNOL., 2008, 68: 2231-2237.
[6] S.XU, Liming ZOU. Proceedings of 2009 International Conference on Advanced Fibers and Polymer Materials: Volume II. Beijing: Chemical Industry Press, 2009,707-709.

8 Preparation and Characterization of Nanofibrous Bioactive Glass Scaffolds

Chunxia Gao, Akira Teramoto, Koji Abe*

(Faculty of Textile Science and Technology, Shinshu University, 3-15-1 Tokita, Ueda City, Nagano, 386-8567, Japan)

Abstract: Sol–gel derived glass nanofibers with high bioactivity were prepared by electrospinning technique and combining poly (vinyl butyral) (PVB) as chain entanglements. The hierarchical bioactive glass scaffolds with interconnected macropores and mesopores on the surface of the fibers were obtained. This study concentrated on the effect of PVB concentration on the diameter and morphology of the composite fibers and the process of in vitro biomineralization. The early stage of in vitro biomineralization of the glass nanofibers in the simulated body fluid (SBF) was studied. The in vitro biomineralization behavior of bioactive glass nanofibers was different with the conventional ones, and the structure of nanofibers contributed to accelerate the formation of hydroxyapatite.

Keywords: Bioactive glass; Hydroxyapatite; Nanofibers; Bone tissue scaffold

1 Introduction

Bioactive glasses, especially silica-based materials, have attracted much attention as artificial biomaterials in bone tissue engineering due to their excellent bioactivity, biocompatibility, osteoconductivity and even osteoinductivity [1, 2]. For practical applications, bioactive glasses have been synthesized in many forms, e.g. powders, foams, monoliths and films [3]. Recently, nanofibrous materials are gaining considerable interest in bone tissue engineering, due to the structural similarity to the bone extracellular matrix (ECM). Electrospinning has been widely recognized as an effective technique for fabricating polymer nanofibrous biomaterials. Because electrospun submicron fibers which mimic the features of native ECM. Moreover, nanofibrous scaffolds have significant effects on various cell behaviors including cell migration, orientation, adhesion and proliferation. Compared with the other shapes, the nanofibrous bioactive glass has great potential as a bone regeneration material owing to their special structure. However, there are a few reports on the study of electrospun inorganic biomaterials with nanofibrous 3D structure [4].

In this study, nanofibrous bioactive glass ($70SiO_2$-$25CaO$-$5 P_2O_5$ (mol%)) was fabricated by electrospinning technology. The effect of the PVB concentration on the morphology of the fibers was investigated. Meanwhile, the in vitro biomineralization of nanofibrous silicate glass was evaluated by the formation rate of hydroxyapatite crystals in SBF solution for different immersed time.

2 Experimental

2.1 Raw Materials

Tetraethyl orthosilicate(TEOS), calcium nitrate tetrahydrate ($Ca(NO_3)_2 \cdot 4H_2O$, 99%), triethyl phosphate (TEP) and poly (vinyl butyral) (PVB, Mw=144000). Acetic acid was used to act as a catalyst. All chemicals were analytical grade reagents and were used as received, without further purification.

2.2 Preparation of bioactive silicate glass solution

The composition of silicate glass used in this study is: $70SiO_2$-$25\%CaO$-$5\%P_2O_5$ (mol%) which was chosen based on these previously developed bulk glasses [2, 4]. Initially, 5ml TEOS was added to 80 g water/ethanol mixture (1:1, molar ratio) containing 6.5mol/L acetic acid. After stirring for 6 h to allow complete hydrolysis of TEOS, 0.545ml TEP and 1.89g $Ca(NO_3)_2 \cdot 4H_2O$, were added in sequence into the TEOS solution with a interval of 6 h. After fully mixing the chemicals, the mixture solution was stirred for 24 h and aged without stirring at 40℃ for 24 h followed by a further 12 h at 60℃.

2.3 Electrospinning

For the electrospinning procedure, the aged solution was mixed with PVB at appropriate ratio, and the amount of PVB in the mixed solution was 10wt%. After fully stirring for 12 h, the solution was loaded in a syringe coupled with nozzles(1.2mm×38mm) and followed by electrospinning under controlled conditions (electric voltage: 7 kV, distance: 8 cm and injection speed: 3.75ml/h). To get the bioactive glass nanofibers,

---

* Corresponding author's phone: +81-26-821-5488, email: kojiabe@shinshu-u.ac.jp

the composite fibers were heat-treated at 700℃ for 3 h in air at a heating rate of 5℃/min.

2.4 Characterization and biomineralization test

The morphology of electrospun nanofibers were observed with scanning electron microscopy (SEM, S-5000, Hitachi. Co., Japan). An energy disperse spectroscopy (EDS) analysis was carried out to get the atomic proportion of Si and Ca elements in the scaffold. The structure of the nanofibers was conducted with Fourier translate infrared (FT-IR) spectrometer. The wide angle X-ray diffraction (WAXD) measurements of bioactive glass nanofibers were carried out with a fixed anode X-ray generator (Rotaflex RTP300, Rigaku.Co., Japan; 40 kV, 150mA, Japan) with Cu (Kα) radiation (λ =1.5402 Å) at a scanning speed of 2θ=4°/min.

The biomineralization in vitro of nanofibrous bioactive glass were assessed by incubating samples in a simulating body fluid (SBF) which contains similar ion concentration of human body plasma, described by Kokubo et.al [5]. In this study, 50 mg glass nanofibers immersed in 50ml SBF at 37℃ for up to 3 days without refreshing. The SBF-immersed samples were measured using SEM, EDS, FT-IR spectroscopy and XRD.

3 Results and discussion

3.1 Effect of the PVB concentration

The effect of the concentration of PVB on the morphology and diameter of the as-spun fibers were investigated. The Figure 1 showed that electrospun composite fibers with the concentration of PVB between the 9wt% and 11wt% have good morphology without beads and the average diameter of the fibers ranged from 200nm to 1.5um. According to the resultant morphology and diameter at different concentration of PVB, 10 wt% was chosen for the amount of PVB in the following experiments.

Figure 1 SEM images of the composite fibers at different concentration of PVB ((a)~(e): 5, 7, 9, 11, 13 wt%)

3.2 Morphology and structure of bioactive glass nanofibers

The as-spun composite fibers were heat-treated at 700℃ for 3h with a heating rate of 5℃/min to prepare a nanofibrous mesh of bioactive glass. Figure 2 showed the morphology of glass precursor/PVB composite and glass nanofibers after heat-treatment. After the calcination, the average diameter decreased from 300nm with a smooth surface to 150nm with a mesoporous surface, due to the burn-out of residual and the consolidation of the glass network. The EDS result confirmed that the composition of the fibers after calcination accorded with that of the silicate glass precursor ($75SiO_2 \cdot 20CaO \cdot 5P_2O_5$).

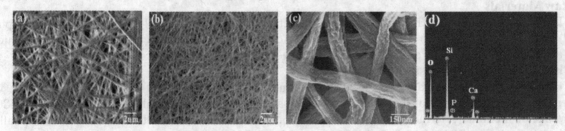

Figure 2 SEM images of electrospun bioactive glass fibers before and after calcined at 700℃
((a) Composite fiber, (b) and (c) bioactive glass fibers, (d) EDS of bioactive glass fibers)

FT-IR spectra of nanofibers before and after calcination were shown in Figure 3. According to the spectra, the organic and other impurities were removed completely after heat-treated at 700℃ for 3 h. In the spectrum of fibers after calcination, the observed bands at about 821cm^{-1} and 1050cm^{-1} can be attributed to the Si-O-Si symmetric and asymmetric stretching vibration respectively. The band at about 2322cm^{-1} may be assigned to C=O of the CO_2 in the air. The XRD spectrum shown in Figure 3 illustrated that the nanofibrous silicate glass calcined at 700℃ for 3h were amorphous.

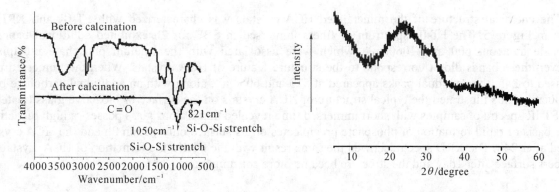

Figure 3 FT-IR and XRD spectra of electrospun glass fibers before and calcined at 700℃

3.3 In vitro biomineralization test

It is commonly accepted that the HCA crystals formation rate on the glass surface is related to the in vivo bone-forming bioactivity. The in vivo bone-forming bioactivity of glass fibers was tested by immersing samples in SBF solution and monitoring the formation rate of HCA crystals on the surface of samples over time. Figure 4 showed SEM images of glass fibers immersed in SBF for 1 and 3 days.

According to Figure 4, many small particles with the diameter of 20-30nm formed on the surface of glass nanofibers after incubating for 1 day. When the immersed time reached to 3 days, the surface of the fibers became more compact, with numerous needle-like crystals covering almost entire surface of the fibers. High magnification images of the fibers incubated for 3 days showed the needle-like crystals grew gradually into plate-like in a higher density and some plate-like crystals aggregated to form some flower-like clusters crystals which are a typical morphology feature of HCA crystal. More important, it was found that fibrous structure remained after immersing for 3 days, which was important feature for potential applications as scaffolds in bone tissue engineering.

Of special note, when compared with the conventional bulk silicate glasses which usually need several days to weeks for nucleation and growth of HCA crystals, the nanofibrous glass exhibited a more rapid speed for nucleation and growth of HCA crystals. This was mainly attributed to the large surface area afforded by nanofibers which resulted in faster dissolution and supersaturation of the medium with respect to the HCA crystals nucleation. Meanwhile, the rough surface morphology provided the nucleation sites which are important for rapid speed of HCA crystals formation.

The results of EDS analysis of the crystal area showed a high concentration of Ca and P without Si. The result revealed the crystals formed on the surface of fibers were calcium phosphate and the molar ratio of Ca/P was close to 1.67, which further confirmed the precipitate on the surface of the fibers were HCA crystals. With the time increasing, the quantity of calcium phosphate increased and the Ca/P molar ratio was still constant.

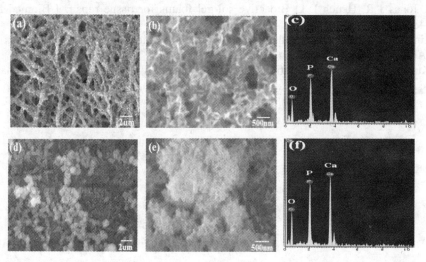

Figure 4 SEM images and EDS analysis of electrospun bioactive glass fibers after immersing in SBF solution for 1 and 3 days ((a) (b) SEM and (c) EDS for 1 day; (d) (e) SEM and (f) EDS for 3 days)

The chemical structure of the mineralized HCA crystals was characterized with FT-IR and XRD as shown in Figure 5. The FT-IR spectrum of fibers immersed in SBF for 2h exhibited additional bands of phosphate at about 600 and 1060cm^{-1}, which were associated with the calcium phosphate precipitate. However, these bands didn't correspond to the structure feature of HCA crystals. When the immersed time increased to 24h, the new dual peaks appeared at 570 and 605cm^{-1}, along with an additional carbonate band at 1400cm^{-1}. This illustrated the typical structure of HCA crystals on the surface of bioactive glass treated in SBF. FT-IR spectra of samples with short immersed time revealed that fibrous glass possess a high bioactivity, on the basis of rapid formation of phosphate precipicates on its surface only within 2h and the HCA crystals formed after 24h. The XRD spectra showed the same result with the FT-IR, the formation of HCA crystals on the fibers surface after 1day and the structure became more consummate.

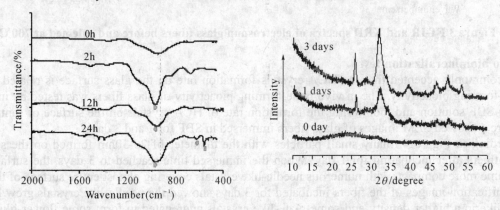

Figure 5 FT-IR and XRD spectra of silicate bioactive glass nanofibers before and after immersing in SBF

4 Conclusion

Bioactive glass nanofibers were fabricated by the electrospinning technique. The results revealed the bioactive glass scaffold in the form of nanofibers possessed large surface area, mesoporous surface morphology and interconnected pore network structure. Because of these special morphology and structure, electrospun silicate glass nanofibers have excellent bioactivity by rapid formation of bone like HCA crystals in SBF solution. The results implied that the electrospun bioactive glass with excellent vitro bioactivity is a potential candidate for bone tissue engineering.

References
[1] Hench L L. Bioceramics. J Am Ceram Soc, 1998, 81(7): 1705-1728.
[2] Zhong J, Greenspan DC. Processing and properties of sol-gel bioactive glasses. J Biomed Mater Res. 2000, 53(6): 694-701
[3] Sepulveda P, Jones J R, Hench L L. Bioactive sol-gel foams for tissue repair. J Biomed Mater Res, 2002, 59(2): 340-348.
[4] Kim H W, Kim H E, Knowles J C. Production and potential of bioactive glass nanofiber as a next generation biomaterial. Adv Funct Mater, 2006, 16 (12): 1529-1535
[5] Kokubo T, Kushitani H, Sakka S. Solutions able to reproduce in vivo surface-structure changes in bioactive glass-ceramic A-W. J Biomed Mater Res, 1990, 24 (6): 721-734.

9 Study on the Preparation and Characterization of SWNTs/Lyocell Composite Fibers

Baohui Guan, Huihui Zhang*, Gesheng Yang, Huili Shao, Xuechao Hu

(State Key Laboratory for Modification of Chemical Fibers and Polymer Materials, College of Material Science and Engineering, Donghua University, Shanghai 201620, China)

Abstract: In this paper, single-walled carbon nanotubes (SWNTs) were used as the additives to prepare SWNTs/Lyocell composite fibers. To realize the homogeneous dispersion of SWNTs, the purification and functionalization of SWNTs were carried out, then SWNTs/Lyocell composite fibers were spun by the dry-jet wet spinning process and the properties of the composite fibers were investigated. The results showed that SWNTs treated with the purification and functionalization could be dispersed homogeneously in NMMO aqueous solution. The addition of appropriate amount of SWNTs could improve the mechanical properties of composite fibers. Compared with the Lyocell fiber, the tensile strength and initial modulus of the SWNTs/Lyocell composite fibers containing 1wt% SWNTs were increased by 24.2% and 55.8%, respectively. Moreover, the thermal stability of composite fibers was improved by the addition of SWNTs.

Keywords: Single-walled carbon nanotubes; Lyocell; Composite fibers; Properties

1 Introduction

Lyocell fiber is a new kind of regenerated cellulose fiber, which is spun from cellulose solutions in the organic solvent N-methyl morpholine-N-oxide (NMMO) hydrate by the dry-jet wet spinning process. Now, Lyocell fibers have been proven commercially successful in textile products because of their excellent mechanical properties and high moisture absorption characteristics. In addition, Lyocell fibers modified with certain additives can expectedly be used as the industrial fibers such as the tire cord, the precursor for carbon fibers.

Since carbon nanotubes (CNTs) were discovered by Iijima in 1991, they have received much attention due to their remarkable properties such as high tensile strength, superior modulus and exceptional electrical conductivity, so they are able to enhance the strength of the composites [1]. Up to date, CNTs have been used to reinforce many polymer fibers. In 2007, Lu et al [2] have prepared the MWNTs/Lyocell composite fibers, and they found that the addition of an appropriate amount of MWNTs could improve the mechanical properties of Lyocell fibers.

The reason of reinforcement is that CNTs and the fiber substrate will form more connected interface, which can absorb more force when the CNTs were dispersed uniformly in the fiber, therefore the loads can be effectively transferred to the CNTs from the matrix, so that the mechanical properties can be improved. Therefore, if an effective load transfer from the surrounding matrix into CNTs is important to estimate the reinforcement potential of CNTs. However, stress transfer occurs only via the outermost layer of nanotubes. For multi-walled carbon nanotubes (MWNTs), the stress transfer between the concentric layers has to occur via interlayer shearing to be transferred by van der Waals forces, which are relatively weak, therefore, MWNTs can considered to be the less efficient concerning a mechanical reinforcement.[3] In contrast, single-layer structure of single-walled carbon nanotubes (SWNTs) can improve the mechanical properties in the maximal extent.

In this paper, the SWNTs were added to prepare the SWNTs/Lyocell composite fibers. The functionalization of SWNTs with sodium dodecylbenzene sulfonate (SDBS) and its dispersion in the solvent system (i.e. N-methylmorpholine-N-oxide aqueous solution) of Lyocell process were studied. In addition, the morphology and properties of SWNTs/Lyocell composite fibers were also investigated.

2 Experimental
2.1 Materials

Cellulose pulp (DP=488, α-cellulose content 98%) used in this work was obtained from Baoding Chemical Fiber Co., China.

N-methylmorpholine-N-oxide (NMMO) aqueous solution (50 wt%) was purchased from BASF AG,

---

* Corresponding author's email: hhzhang@dhu.edu.cn

Germany. n-propyl gallate and SDBS were reagent grade. They were purchased from Shanghai Chemical Co. and Farco chemical supplies, respectively.

SWNTs with purity higher than 90% were prepared by chemical vapor deposition process and provided by Chengdu Organic Chemistricals Co., Ltd. The SWNTs exhibit an average outer diameter of 1-2nm and a length of 5-30 μm.

2.2 Purification and functionalization of SWNTs

In this work, the raw SWNTs were purified by being refluxed in 2.6 m HNO_3 for 48h. On cooling, the refluxing mixture was washed with deionized water on a sintered glass filter until the washing showed no acidity. To modify the surface of SWNTs, the purified SWNTs were homogeneously dispersed in 1 wt% SDBS aqueous solution by ultrasonication for 30 min and the obtained suspension was filtered with membrane with a large pore size of 5 μm. The functionalized SWNTs treated with SDBS were obtained by further rinsing and drying.

2.3 Preparation of SWNTs/Lyocell composite fibers

NMMO aqueous solution distilled to 74 wt% and a certain amount of functionalized SWNTs were mixed by magnetic stirring and ultrasonic treatment, then the mixture was distilled under vacuum to remove extra water to obtain NMMO·H_2O/SWNTs solution. Whereafter, the NMMO·H_2O/SWNTs solution, cellulose pulp and n-propyl gallate (antioxidant) were mixed and stirred continually for 3 h at 100°C. The mixture gradually turned into a homogeneous solution, and the solutions with 10wt% cellulose in NMMO·H_2O with different SWNTs content can be obtained.

The spinning solution was extruded through a spinneret with 0.145mm in diameter, and then passed through an air gap of 50 mm in length and immersed in a coagulation bath of water to precipitate cellulose in filament form. The filaments were washed with water, wound and dried. Then SWNTs / Lyocell composite fibers containing various amounts of SWNTs were prepared.

2.4 Characterization

TGA analysis of SWNTs/Lyocell fibers was performed on a TG 209 F1 Iris Thermogravimetric Analyzer (Netzsch, Germany) in the range of temperature from 50 to 650°C under a nitrogen atmosphere. The scan rate was maintained at 10°C/min. The surfaces and cross-sections morphology of SWNTs/Lyocell composite fibers were investigated using JSM-5600LV SEM (JEOL Co., Japan). Mechanical test for SWNTs/Lyocell fibers was performed on XQ-1 Tensile Tester (China Textile University, China).

3 Results and discussion

3.1 The dispersion of SWNTs in NMMO aqueous solution

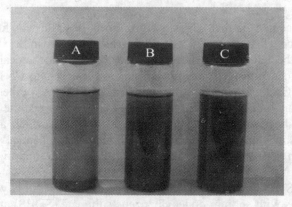

Figure 1 The photographs of the dispersion of SWNTs with different treatment in 74wt% NMMO solution stored at the room temperature after two months (A-raw SWNTs, B-purified SWNTs, C- functionalized SWNTs)

To realize the homogeneous dispersion of SWNTs, the purification and functionalization of SWNTs were carried out in this paper. Figure 1 shows the photographs of the dispersion of SWNTs with different treatment in 74wt% NMMO solution stored at the room temperature after two months. It was found that compared to the raw SWNTs, the dispersion of purified SWNTs in NMMO aqueous solution was improved although the dispersion uniformity of the solution was not very well. In order to realize the homogeneous dispersion of SWNTs in NMMO aqueous solution, a kind of anionic surfactant SDBS, which was considered as a good dispersant for carbon nanotubes, was used to functionalize SWNTs. It can be seen from Figure 1 that functionalized SWNTs could be dispersed more homogeneously in NMMO aqueous solution, which is helpful to the homogeneous dispersion of SWNTs in Lyocell matrix.

3.2 Morphology of SWNTs/Lyocell composite fibers

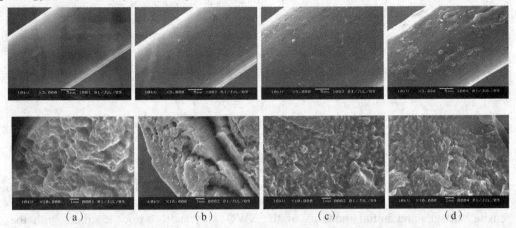

Figure 2 SEM photographs of the surfaces and cross-sections of SWNTs/Lyocell composite fibers with 0.5wt% (a), 1 wt% (b), 3 wt% (c) and 5 wt% (d) SWNTs

Figure 2 shows the SEM photographs of SWNTs/Lyocell composite fibers with 0.5, 1, 3 and 5wt% SWNTs, respectively. It can be seen that with the increasing of SWNTs content, the surfaces of SWNTs/Lyocell composite fibers becomes more and more rough, which maybe resulted from the aggregation of SWNTs.

In addition, it is apparent from Figure 2 that there were more bright dots and lines, which were attributed to the SWNTs, in the SWNTs/Lyocell composite fibers with higher SWNTs content. It can be found that SWNTs were dispersed homogeneously in Lyocell matrix in the case of lower SWNTs content (Figure 2 (a)-(c)). However, when the amount of SWNTs filled in the fiber is higher the aggregation of SWNTs can be obviously observed (Figure 2 (d)).

3.3 Thermal stability of SWNTs/Lyocell composite fibers

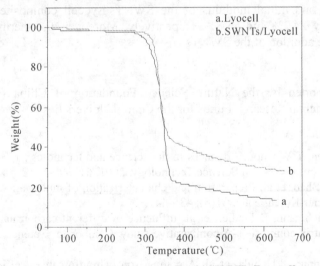

Figure 3 TGA curves of Lyocell fiber and 5wt%SWNTs/Lyocell composite fiber

Figure 3 shows the TGA curves of Lyocell fiber and 5wt%SWNTs/Lyocell composite fiber. It can be seen that the thermal decomposition temperature and the residue of SWNTs/Lyocell composite fiber were higher than those of Lyocell fiber without SWNTs. That is to say, the thermal stability of SWNTs/Lyocell composite fibers was improved by the addition of SWNTs because of the excellent thermal stability of SWNTs and the interaction between SWNTs and cellulose matrix [4].

3.4 Mechanical properties of SWNTs/Lyocell composite fibers

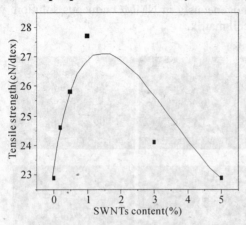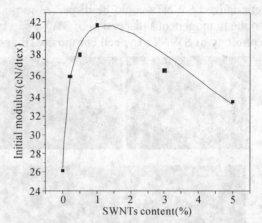

Figure 4 The tensile strength (left) and initial modulus (right) of SWNTs/Lyocell composite fibers

The tensile strength and initial modulus of the SWNTs/Lyocell composite fibers with the different content of SWNTs are shown in Figure 4. It can be found that the tensile strength and initial modulus of the SWNTs/Lyocell composite fibers were improved with the increasing of the SWNTs content, and reach the maximum when the SWNTs content is 1wt%. Compared with the Lyocell fiber, the tensile strength and initial modulus of the SWNTs/Lyocell composite fibers containing 1wt% SWNTs were increased by 24.2% and 55.8%, respectively. Then the strength and modulus of the composite fibers show downward trend with the increasing of SWNTs when the content of SWNTs is further increased, suggesting that the localized clusters or aggregations of SWNTs might occur [5].

4 Conclusions

SWNTs treated with the purification and functionalization could be dispersed homogeneously in NMMO aqueous solution, which is helpful to prepare a spinnable cellulose spinning dope. The mechanical properties of composite fibers was increased firstly with the addition of SWNTs and reached the maximum with adding 1wt% SWNTs, then decreased with the further addition of SWNTs. Compared with the Lyocell fiber, the tensile strength and initial modulus of the SWNTs/Lyocell composite fibers containing 1wt% SWNTs were increased by 24.2% and 55.8%, respectively. Moreover, the thermal stability of composite fibers was improved by the addition of the SWNTs.

Acknowledgement

This work was supported by the Nature Science Foundation of China (Grant No.50873024 and 50673016) and the Fundamental Research Funds for the Central Universities.

References

[1] E T Thorstenson, Z. Ren, T W Chou. Advances in the science and technology of carbon nanotubes and their composites: a review. Composites and Science Technology, 2001, 61: 899-1912.
[2] J Lu, H H Zhang, H L Shao, et al. Preparation and characterization of multiwalled carbon nanotubes/Lyocell composite fibers. Polymer-Korea, 2007, 31(5): 436-441.
[3] F H Gojny, M H G Wichmann, B Fiedler, et al. Influence of different carbon nanotubes on the mechanical properties of epoxy matrix composites-A comparative study. Composites Science and Technology, 2005, 65: 2300-2313.
[4] H G Chae, T V Sreekumar, T Uehida, et al. A comparison of reinforcement efficiency of various types of carbon nanotubes in polyacrylonitrile fiber. Polymer, 2005, 46(24): 10925-10935.
[5] J W Ning, J J Zhang, Y B Pan, J K Guo. Fabrication and mechanical properties of SiO_2 matrix composites reinforced by carbon nanotube. Materials Science and Engineering A, 2003, 357: 392-396.

10 The Effect of Acids on Mechanical Properties of PPS Fibers

Xiangbing He[1], Bin Yu[1], Jian Han[1,2*], Xinbo Ding[1], Mitsuo Matsudaira[3]

(1. School of Materials and Textiles, Zhejiang Sci-Tech University, Hangzhou, China
2. Key Laboratory of Advanced Textile Materials and Manufacturing Technology, Ministry of Education of China, Zhejiang Sci-Tech University, Hangzhou 310018, China
3. Kanazawa Univ., Kakuma-machi, Kanazawa City, 920-1192, Japan)

Abstract: In this article, the effect of HCl, H_2SO_4 and HNO_3 on the mechanical properties of PPS fibers were studied by fiber tensile tester, FESEM and DSC, respectively. It was found that the retention of strength of PPS fibers declined linearly with increasing concentration of acids, and HNO_3 solution showed the greatest impact on the mechanical properties of PPS fibers. The retention of strength of PPS fibers treated with 4mol/L HNO_3 was 60%, and that of PPS fibers treated with 4mol/L HCl and H_2SO_4 were over 77%. The FESEM results showed that many holes appeared on the surface of PPS fibers after HNO_3 treated. The degree of crystallinity of PPS treated with the acids fibers decreased compared with that of untreated PPS fibers.

Keywords: PPS fiber; Mechanical property; Acid

1 Introduction

Polyphenylene sulfide (PPS) was often regarded to be durable at high temperature (160-240℃) and resistant against several chemicals. PPS non-woven fabrics, often used as bag filter materials in incinerating plants, were exposed to acidic gases such as hydrochloride, SO_x (SO_2 and SO_3) and NO_x (NO and NO_2), which could form a solution of HCl, H_2SO_4 and HNO_3, when the operational temperature was under the acid dew point. These acid solutions in condensed droplets could stay long enough to severely degrade the PPS fabric. Durability of PPS fibers against these acids was therefore significant to such industries as coal-fired power plant [1-6].

In this work, the effect of the acids on the mechanical properties of PPS fibers will be investigated and the reason of mechanical loss of acid-treated samples will also be presented. Thus, the surface morphology and thermal properties of PPS fibers treated with the acids are studied and the PPS fibers are expected to have important applications as high temperature resistance filtration materials.

2 Experimental

2.1 Materials and Samples preparation

The specifications of the PPS fibers were listed as Tables 1. The treated time and temperature was set at 24 h and 92℃, and the acid concentrations for HCl, H_2SO_4 and HNO_3 were varied as 1, 2.5 and 4 mol/L. After the acid exposure, the samples were rinsed with distilled water for five times, and then dried in the incubator (DHJ-9146A) at 80℃ for 2 hours.

2.2 Characterization

Tensile tests were measured by the fiber tensile tester (XQ-2) with 20 mm/min. The surface of the sample was fractured in liquid nitrogen, sputter-coated with gold and observed with the Field Emission Scanning Electron Microscopy (FETX3 7426, England) at a voltage of 2 kV. Thermal analysis was performed with a differential scanning calorimeter (Pyris Diamond, Perkin-Elmer). All the samples were heated from 50℃ to 300℃ at 10℃/min under the nitrogen atmosphere, and the weights of them varied from 4 to 7mg.

Table 1 Materials and their specifications

Material type	Fiber number/dtex	Length/mm	T_g/℃	T_m/℃	Strength/cN
PPS fibers	2.4	51	190	280.5	10.73

3 Results and discussion

Figure 1 and Figure 2 shows the retention of strength and breaking elongation of PPS fibers treated by the acids with different concentrations, respectively. It is clear shown that the retention of strength of PPS

---

* Corresponding author's email: hanjian8@zstu.edu.cn

fibers declined linearly (all of the linear correlation were over 0.94) with increasing concentration of acids, as seen in Figure 1. With increasing acids' concentration, the retention of strength of PPS fibers treated with HNO_3, HCl and H_2SO_4 is decreased with the straight slope of -9.63, -4.55 and -5.37, respectively. The retention of strength of PPS fibers treated with 1mol/L HNO_3 is 84%, while that of PPS fibers treated with 1mol/L HCl and H_2SO_4 is still more than 91%. The retentions of strength of PPS fibers treated with 4mol/L HNO_3, HCl and H_2SO_4 were 60%, 82% and 77%, respectively. Moreover, breaking elongation of PPS fibers are affected notably by the acids, as shown in Figure 2. Breaking elongation of PPS fibers decrease as increasing the acids' concentration. Namely, HNO_3 solution shows the greatest impact on the mechanical properties of PPS fibers.

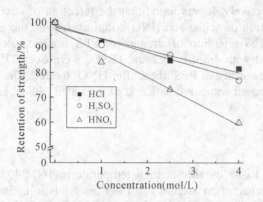
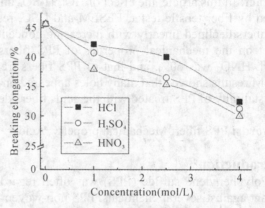

Figure 1 The retention of strength of PPS fibers treated with HCl, H_2SO_4 and HNO_3

Figure 2 The Breaking elongation of PPS fibers treated with HCl, H_2SO_4 and HNO_3

To examine the effect of HNO_3 on the mechanical properties of the PPS fiber, the FESEM measurements are carried out and the surface morphology of PPS fibers treated by HNO_3 with different concentration are shown in Figure 3. The FESEM results show that there are more and more holes and cracks in the surface of PPS fibers with increasing the acid concentration. Apparently the fibers have a rougher surface than that treated by acids with lower concentrations. These can be explained that the surface of PPS fibers is corroded due to the oxidation effect of HNO_3, resulting in the decrease of the mechanical properties of the PPS fiber treated with the HNO_3.

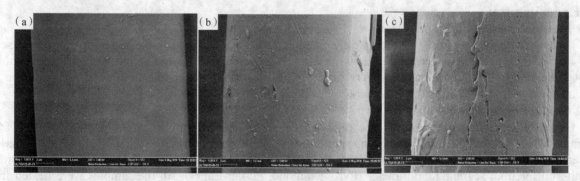

Figure 3 FESEM images of samples treated with different concentration of HNO_3 (92℃, 24h)
(a)1 mol/L 5000×   (b)2.5 mol/L 5000×   (c)4 mol/L 5000×

For further understand and investigating of the reason of mechanical loss of the acid-treated PPS fibers, the DSC measurements are also carried out and the DSC curves of the PPS fibers are depicted in Figure 4 and the thermal properties data are listed in Table 2. As can be seen from Figure 4, the melting temperature of PPS fibers treated with 4mol/L HCl and H_2SO_4 show no significant changes, while those treated with 2.5mol/L and 4mol/L HNO_3 decreases notably. As shown in Table 2, it could be seen that the melting temperature and melting enthalpy of the fibers treated with HCl, H_2SO_4 and HNO_3 are lower than that of the untreated PPS. The melting temperature of PPS treated with 4mol/L HCl and H_2SO_4 were about 275℃, while that of PPS fibers treated with 4mol/L HNO_3 decreased from 280.5 to 264.5℃. The melting enthalpy of the PPS fibers treated with 4mol/L HCl, H_2SO_4 and HNO_3 decreased from 54.5 J/g to 41.8 J/g, 32.5 J/g and 18.0

J/g, respectively. These results display that the crystalline structure of PPS fibers is destroyed by the acids and the HNO_3 plays a great role on the crystalline structure of PPS fibers among the three acids. High degree of crystallinity reflected high tensile strength of the polymer [7]. Thus, the decrease of the degree of crystallinity of the treated PPS with acids may be another one of the reasons resulting in the mechanical loss of acid-treated PPS fibers.

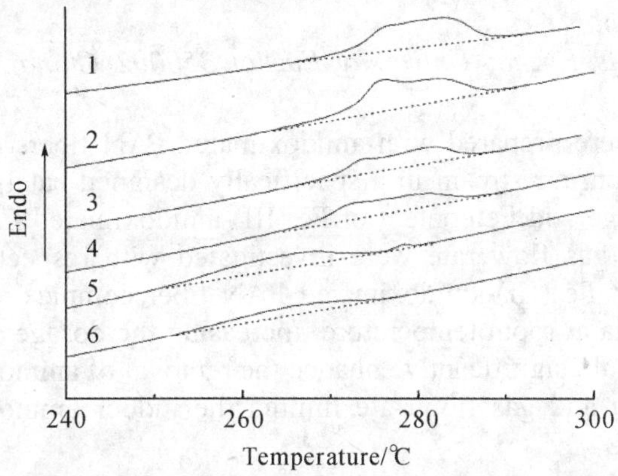

Figure 4 DSC curves of the acid-treated PPS fibers

Table 2 The thermal properties data of PPS fibers

Sample	Sample preparation	T_m/℃	△H (J/g)
1	untreated	280.5	54.5
2	4mol/L HCl	275.7	41.8
3	4mol/L H_2SO_4	275.4	32.5
4	1mol/L HNO_3	272.8	32.8
5	2.5mol/L HNO_3	265.9	22.5
6	4mol/L HNO_3	264.5	18.0

4 Conclusions

In this article, the effect of HCl, H_2SO_4 and HNO_3 on the mechanical properties of PPS fibers were studied. The results showed that the retention of strength of PPS fibers treated with HNO_3 was decreased quicker than these treated with HCl and H_2SO_4 as increasing concentration of acids. The retentions of strength of PPS fibers treated with 4mol/L HNO_3 was 60%, while that of PPS fibers treated with 4mol/L HCl and H_2SO_4 were over 77%. The FESEM results showed that many holes and creacks were presented in the surface of the acid-treated PPS fibers. For further investigating, the crystalline structure of PPS fibers is destroyed by the acids and the HNO_3 plays a great role on the crystalline structure of PPS fibers. These results might be the reasons resulting in the mechanical loss of acid-treated PPS fibers.

Acknowledgement

This work was supported by Science and technology department plan project of Zhejiang province (2008C21001), by program for Changjiang scholars and innovative research team in university (IRT0654) and by The key laboratory of advanced textile materials and manufacturing technology tender project of Zhejiang province (S2010004).

References

[1] W. Tanthapanichakoon, M. Furuuchi, K. Nitta, et al. Degradation of semi-crystalline PPS bag-filter materials by NO and O_2 at high temperature. Polymer Degradation and Stability, 2006, 91(8): 1637-1644.
[2] Winyu Tanthapanichakoon, Masami Furuuchi, Koh-Hei Nitta, et al. Degradation of bag-filter non-woven fabrics by nitric oxide at high temperatures. Advanced Powder Technology, 2007, 18(3): 349-354.
[3] WANG Hua, QIN Junfeng, HE Yong, et al. Effect of Conditions in Application on PPS Filter Media. Synthetic Fiber in China, 2009, (9): 43-46.
[4] ZHUANG Yuling. Research on anticorrosion of PPS f ilter medium in SO_2. Journal of Safety and Environment, 2008, 8(1): 51-55.
[5] Vives VC, Dix JS, Brady DG. Polyphenylene sulfide (PPS) in harsh environments. *ACS Symposium Series* 1983, 229(6): 65-85.
[6] Winyu Tanthapanichakoon, Mitsuhiko Hata, Koh-hei Nitta, et al. Mechanical degradation of filter polymer materials: Polyphenylene sulfide. Polymer Degradation and Stability. 2006, 91(11). 2614-2621.
[7] YU Weidong, Textile material. Beijing: China Textile & Apparel Press, 2005: 104.

11 Removal of Indoor Ammonia with Fe (III)-modified PAN Fiber Complexes

Xing Li[1], Yongchun Dong[1,2], Jiangxing He[1], Zhenbang Han[1,2], Zhichao Wang[1]

(1. Division of Textile Chemistry & Ecology, School of Textiles, Tianjin Polytechnic University, Tianjin 300160, China
2. State Key Laboratory Breeding Base of Photocatalysis, Fuzhou University, Fuzhou 350002, China)

Abstract: Fe (III)-amidoximated PAN fiber complexes were prepared with amidoximated PAN fibers and ferric chloride and used for removing indoor ammonia from air stream in a specifically designed catalytic system. Some important effecting factors such as the dosage and Fe content of Fe (III)-amidoximated PAN fiber complex, the initial concentration of ammonia and gas flow rate were investigated with respect to removal efficiency of ammonia. The results indicated that Fe (III)-amidoximated PAN fiber complex was used as a catalyst for the decomposition of indoor ammonia at room temperature. Increasing the dosage and Fe content of Fe (III)-amidoximated PAN fiber complex could significantly enhance the removal of ammonia from air stream. Higher initial concentration of ammonia and gas flow rate limited the indoor ammonia decomposition.

Keywords: Ammonia removal; Modified polyacrylonitrile fiber; Fe (III) ions; Complex

1 Introduction

The gaseous ammonia is generated by a continuous decomposition of antifreeze admixtures based on urea compounds in the concrete wall under alkaline and warm condition, and then release to indoor environment through slow diffusion and have led to the increasing indoor air pollution[1].Therefore, how to reduce the risk caused by ammonia in indoor air becomes a big issue in some countries particularly China. In recent years, several studies involved in the decomposition of indoor ammonia with nano-TiO_2 loaded woven fabrics as the photocatalysts have been reported[2-3]. However, it is known that some disadvantages such as higher cost hindered the nano-TiO_2 particles from using as the photocatalyst in an industrial scale. Hence, it is necessary to explore new catalysts for the decomposition of indoor ammonia by using lower-cost polymer materials. Fe (III)-modified polyacrylonitrile (PAN) fiber complexes have been used as a low-cost and effective heterogeneous Fenton catalyst in the decomposition of textile dyes in wastewater since they could enhance the decomposition of H_2O_2 into hydroxyl radicals with high oxidative power[4-5]. In this work, the Fe-(III)-amidoximated PAN fiber complexes (Fe-AO-PAN) were firstly prepared and expected to serve as the catalyst for the oxidation of ammonia in indoor air. And some important effecting factors such as catalyst dosage, Fe content of the catalyst, initial ammonia concentration and gas flow rate were investigated and discussed.

2 Experimental methods
2.1 Materials and reagents

Acrylic knitting yarns consisted of twisted acrylic fibers containing 87.07% acrylonitrile monomer are purchased from Kunshan Shilin Woolen Spinning Company (Shanghai, China). Hydroxylamine hydrochloride, sodium hydroxide, ammonia water and ferric chloride were of agent grade.

2.2 Preparation of Fe-AO-PAN

In order to prepare Fe-AO-PAN, the following two steps were applied: (1) amidoximation of PAN fiber with NH_2OH in aqueous NaOH solution for 2 h with stirring at 68℃, and (2) impregnation of the amidoximated PAN fiber (AO-PAN) by 0.10 mol/L $FeCl_3$ aqueous solution at room temperature. $C_P\%$ (the degree of conversion from cyano group to amidoxime group) of the amidoximated PAN fiber was calculated to be 67.86 by the method presented in literature[5]. Fe-AO-PAN was treated in an aqueous 2.0 mol/L H_2SO_4 solution for 24 h to remove the total Fe ions from the surface of Fe-AO-PAN. The color transformation of Fe-AO-PAN from brown to white indicated that the Fe ions were removed absolutely from the catalyst. And then the Fe ion concentration in acidic solution was determined by using atomic absorption method. and Fe content (C_{Fe-PAN}) of Fe-AO-PAN was calculated as a mass.

2.3 Decomposition of indoor ammonia

The degradation system used in this research consists mainly of a small stainless steel environmental condition simulated chamber (121 L) and a glass catalytic reactor. The schematic diagram of degradation system is shown in our previous study[3].

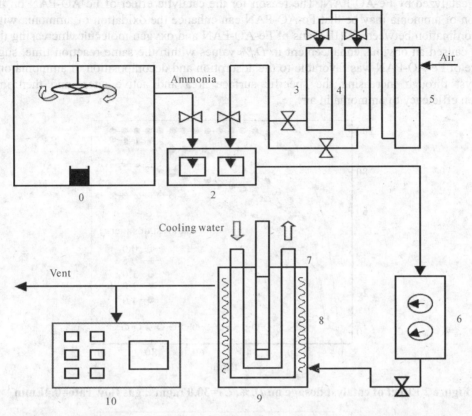

Figure 1 The schematic diagram of catalytic and testing system for ammonia in air. 0: source of ammonia; 1: environmental chamber; 2: gas mixer; 3: dewater reagent; 4: water bottle; 5: activated carbon filter; 6: thermo/humidity meter; 7: quartz shelter (inside cooling water); 8: catalysts; 9: reactor; 10: ammonia detector

The catalytic reactor was a 38 cm long glass plug flow tubular reactor with an effective volume 2.5 L. Environmental chamber and reactor were thoroughly cleaned with a solution containing 2.0 gL^{-1} sodium carbonate and deionized water, respectively before the experiments. An aqueous solution containing about 0.1 gL^{-1} ammonia in a 100 ml beaker was placed into the environmental chamber for simulating the indoor ammonia as a stable ammonia emitting source. Gaseous ammonia emitted environmental chamber and clean air were mixed in a gas mixer and maintained constant for providing ammonia gas with a steady state concentration. The temperature and humidity of the mixed gas flow in the reactor were kept at 20±1 ℃ and 50±2%, respectively. A NH_3 gas detector Z-800XP (Environmental Sensors Company, USA) was used to determine the concentration of gaseous ammonia in the air stream from the reactor. The reaction was not considered to reach balance until the concentration of gaseous ammonia in the reactor was kept steady for 5 min. The ammonia concentration in the reactor at this time was referring to the balance concentration of ammonia. Catalytic decomposition percentage of ammonia ($D_p\%$) at constant humidity and temperature was calculated as follows: $D_p\%=(1-C/C_0)\times 100\%$, where $D_p\%$ is the catalytic decomposition percentage of ammonia, C_0 the initial concentration of ammonia (mgm^{-3}) and C is the residual concentration of ammonia (mgm^{-3}).

3 Results and discussion
3.1 Catalyst dosage

The decomposition of gaseous ammonia in the presence of different dosage of Fe-AO-PAN ($C_{Fe\text{-}PAN}$ =112.1 mgg^{-1}) was carried out, and the results are shown in Figure 2.

It can be observed in Figure 2 that the decomposition of ammonia without the catalysts was found to be negligible within the reaction time. It is worth to notice that after the addition of Fe-AO-PAN to the reactor,

$D_p\%$ values increased significantly in the first 20 min because of adsorption of ammonia on the catalysts, and further decomposition of ammonia after 20 min may be the results from the oxidative decomposition of ammonia by catalyzed by Fe-AO-PAN. The reason for the catalytic effect of Fe-AO-PAN on the oxidative decomposition of ammonia may be that Fe-AO-PAN can enhance the oxidation of ammonia with oxygen in air by the coordination between Fe (III) ions of Fe-AO-PAN and oxygen molecule. Increasing the dosage of Fe-AO-PAN caused an obvious enhancement in $D_p\%$ values within the same reaction time, suggesting that higher dosage of Fe-AO-PAN was favorable to the adsorption and decomposition of ammonia on the surface of the catalyst through increasing the specific surface area and active sites, and then improved the decomposition efficiency of ammonia in air.

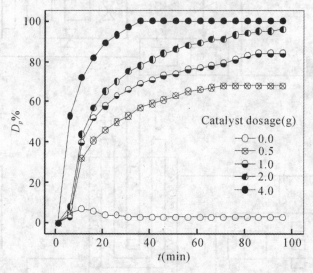

Figure 2 Effect of catalyst dosage on $D_p\%$, C_0=30.0 mgm^{-3}, gas flow rate=0.4Lmin^{-1}

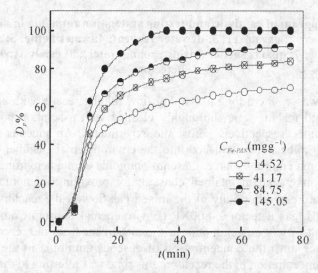

Figure 3 Relationship between C_{Fe-PAN} of Fe-AO-PAN (2.0 g) and $D_p\%$, C_0=28.8 mgm^{-3}, gas flow rate=0.4Lmin^{-1}

3.2 Fe content of catalyst

As can be seen in Figure 3 that $D_p\%$ values gradually increased with increasing C_{Fe-PAN} of Fe-AO-PAN, indicating that increasing the amount of Fe (III) ions on the fiber could accelerate the ammonia decomposition in the reactor. This is because that higher C_{Fe-PAN} of Fe-AO-PAN may enhance the coordination between Fe (III) ions of Fe-AO-PAN and oxygen molecule and increase the number of the activated oxygen molecules, thus promoting the ammonia decomposition.

3.3 Gas flow rate

It is found from Figure 4 that $D_p\%$ values gradually decreased with increasing gas flow rates from 0.20

to 0.80 Lmin^{-1}, implying that the decomposition of ammonia in air is dependent on the level of gas flow rates. Wang and coworkers[6] reported that mass transfer and surface reaction are important factors controlling overall reaction rates in the heterogeneous catalytic reaction. Therefore, in lower flow rates, decomposition reaction rates of ammonia may be significantly affected by mass transfer. However, when gas flow rate was higher, surface reaction may control the progress of reaction. Thus increasing flow rate may shorten the residence time of ammonia on the surface of the catalyst and an insufficient contact of gas stream with the catalyst may occur, thus lowing decomposition of ammonia.

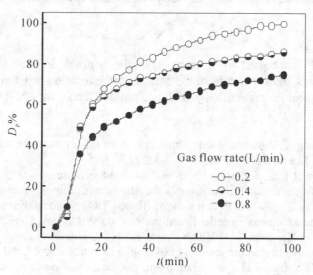

Figure 4 Effect of gas flow rate on $D_p\%$, Fe-AO-PAN ($C_{Fe\text{-}PAN}$ =115.3 mgg^{-1}) =2.0 g, C_0=28.8 mgm^{-3}

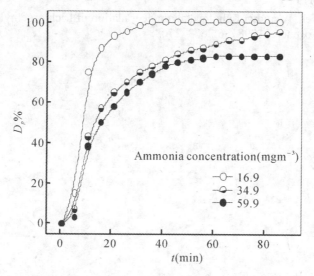

Figure 5 Concentration change of ammonia under different initial concentrations, Fe-AO-PAN ($C_{Fe\text{-}PAN}$ =102.6 mgg^{-1}) =2.0 g, gas flow rate=0.4Lmin^{-1}

3.4 Initial concentration of ammonia

It is clear that increasing ammonia concentration caused a significant decrease in $D_p\%$ values. For experiment conducted at low initial ammonia concentration (16.9 mgm^{-3}), $D_p\%$ value reaches nearly 100% within 40 min. On the contrary, for experiment conducted for high initial ammonia concentration, less than 80% of ammonia was removed during reaction of 70 min. The similar result was obtained for the decomposition of gaseous ammonia using TiO$_2$-loaded cotton fabric[2]. This is probably due to the fixed amount of active sites on the surface of the catalyst available for the adsorption of ammonia molecules before the decomposition of ammonia.

4 Conclusion

Gaseous ammonia can be effectively removed from air stream using the Fe (III)-amidoximated PAN fiber complex through its combined effect of absorption and catalysis in a catalytic reactor. The decomposition efficiency of ammonia depended greatly upon the dosage and Fe content of Fe (III)-amidoximated PAN fiber complex, the initial concentration of ammonia and gas flow rate. Increasing the dosage and Fe content of Fe (III)-amidoximated PAN fiber complex can significantly enhance the degradation of ammonia. Gaseous ammonia was easily removed at low initial concentration of ammonia and slow gas flow rate.

Acknowledgement

This work was kindly supported by a grant from the Natural Science Foundation of China (No. 20773093). The authors thank the support from the Ministry of Education of China through a grant from the Research Fund for the Doctoral Program of Higher Education of China (No. 20070058005).

References

[1] Z. Bai, Y. Dong, Z. Wang, T. Zhu. Emission of ammonia from indoor concrete wall and assessment of human exposure. Environment International, 2006,32: 303-311
[2] Y C Dong, Z P Bai, R H Liu, T Zhu. Decomposition of indoor ammonia with TiO_2-loaded cotton woven fabrics prepared by different textile finishing methods. Atmospheric Environment, 2007,41: 3182-3192
[3] Y C Dong, Z P Bai, R H Liu, T Zhu. Preparation of fibrous TiO_2 photocatalyst and its optimization towards the decomposition of indoor ammonia under illumination. Catalysis Today, 2007,126: 320–327
[4] V V Ishtchenko, K D Huddersman, K D Vitkovskaya. Part 1. Production of a modified PAN fibrous catalyst and its optimisation towards the decomposition of hydrogen peroxide. Appl Catal A, 2003, 242: 123-137.
[5] Y C Dong, Z B Han, C Y Liu, F Du. Preparation and photocatalytic performance of Fe (III)-amidoximated PAN fiber complex for oxidative degradation of azo dye under visible light irradiation. Sci. Total Environ., 2010, 408(10): 2245–2253.
[6] K Wang, H Tsai, Y Hsieh. A study of photocatalytic degradation of trichloroethylene in vapor phase on TiO_2 photocatalyst [J]. Chemosphere, 1998, 36:2673–2763.

12 Preparation of Catalytic Activated Carbon Fiber and Its Catalytic Oxidation Performance to 4-nitrophenol

Yanli Li, Chunxia Ma, Qiaosheng Guo, Wenxing Chen

(Key Laboratory of Advanced Textile Materials and Manufacturing Technology, Ministry of Education of China, Zhejiang Sci-Tech University, Hangzhou 310018, China)

Abstract: A novel catalytic activated carbon fiber (ACF-FePcS) was prepared by immobilizing highly active iron tetrasulfophthalocyanine (FePcS) onto the ethylenediamine-modified ACF. This new catalytic fiber was investigated in the degradation of 4- nitrophenol (4-NP) in the presence of H_2O_2. The results showed that the removal efficiency of 4-NP reached more than 90% in 4 h at 25°C, and it still possessed a good catalytic behavior over a wide range of pH. The addition of anionic surfactant could reduce the treatment effect by retarding the absorption of 4-NP onto ACF-FePcS. Thus, the performance of ACF-FePcS was related with its adsorption rate, exhibiting the process of "enrichment and in-situ catalytic degradation". Furthermore, experiment of adding isopropanol as radical scavenging agent suggested that ·OH played a key role in the degradation of 4-NP by ACF-FePcS/H_2O_2 system.

Keywords: Active carbon fiber; Iron tetrasulfophthalocyanine; 4-nitrophenol; Adsorption; Catalytic oxidation

1 Introduction

Active carbon fiber (ACF) exhibits very interesting properties, in terms of texture and surface chemistry, which has been widely used in environmental protection, electronic, medicine sanitation, chemical industry and so on[1-2]. This material can be obtained from a variety of precursors, such as polyacrylonitrile(PAN), phenolic resins, pitch or from hydrocarbons. The main advantage of it possesses a very high apparent surface area (600-2000m^2/g) and small diameters (10-20 μm), which lead to excellent physical adsorption capability [3]. The functional groups such as hydroxyl, carbonyl, carboxyl and lactones could be gained on the surface of ACF after the oxidization treatments, providing it redox characteristic and making it an excellent catalyst support. For this reason, we can anchor catalyst on it to prepare the catalytic activated carbon fiber that can continuously oxidize organic pollutants, thereby it extend the application of ACF in the treatment of organic pollution.

Phthalocyanine complexes of transition metals are attractive as potential oxidation catalysis because of their cheap and facile preparation in a large scale and their chemical and thermal stability [4]. Iron tetrasulfophthalocyanine (FePcS) containing a central Fe ion surrounded by a macrocyclic system has been investigated and applied plentifully as catalyst [5-7]. Sorokin et al [6] have reported their research on H_2O_2 oxidative dechlorination and aromatic ring cleavage of chlorinated phenols catalyzed by FePcS, discovering that the HOOFePcS was the main functionary species during reaction process. In light of its high activity, we chose FePcS as the catalyst anchored on the ethylenediamine-modified ACF and a novel catalytic active carbon fiber (ACF-FePcS) was obtained. The objective of the present work is to study the catalytic properties of this material in the oxidation of 4-nitrophenol (4-NP). The results showed it had a good catalytic behavior over a wide range of pH. Adding isopropanol as ·OH trapper indicates that ·OH plays a key role in the degration of 4-NP by ACF-FePcS/H_2O_2 system. Therefore, the ACF-FePcS/H_2O_2 system presents promising prospect in the field of organic wastewater treatment.

2 Experiments
2.1 Materials

ACF (Jiangsu Sutong Carbon Fiber Co., Ltd., China), 4-Sulfophthalic Acid Triammonium Salt (Tokyo Kasei Kogyo Co., Japan); Iron (Ⅱ) chloride tetrahydrate, urea and ammonium were purchased from Shanghai Pharmaceutical Co., ltd. (Shanghai, China); 4-NP (East China Normal University Chemical Plant, China); hydrogen peroxide and Ethylene diamine (EDA) (AR, Hangzhou Gaojing Fine Chemical Industry Co., Ltd., China); N,N-Dimethylformamide(DMF), sulfinyl chloride(AR, Tianjin Bodi Chemical Holding Co., ltd.). All the other solvents and reagents were of analytical grade and used without further purification.

2.2 Preparation of catalytic active carbon fiber ACF-FePcS
2.2.1 Preparation and modification of FePcS

Synthesize of FePcS refer to references [8, 6], following the solidoid benzoic anhydride-urea route. The molecular weight of FePcS was 888.695, measured by mass spectrometer (Qtrap LC/MS/MS System, Applied Biosystems Inc., United States), consistent with the theoretical molecular weight ($MW_{C_{12}H_{16}FeN_8O_{12}S_4} = 888.655$).

FePcS was modified according to reference [8]. 0.1 g FePcS was dissolved in 100 ml DMF, and 5 ml $SOCl_2$ was added. The reaction was kept at 25°C for 24 h, then, raising the temperature to 40°C and vacuum distilled the residual $SOCl_2$ later, finally, the modified iron tetrachlorosulfonylphthalocyanine($FePc(SO_2Cl)_4$) was attained.

2.2.2 Grafting of ACF with FePcS

Firstly, 10 g ACF was immersed in nitric acid ($V_{HNO_3}/V_{H_2O} = 1/2$) for 24 h, washed by distilled water to neutrality, then, diverted to vacuum oven at 105 °C to remove the water, and the acid treated ACF were gained(ACF-T). Then, EDA was grafted to ACF-T as described[9], denoted by ACF-E.

10 g anhydrous ACF-E were added into the $FePc(SO_2Cl)_4$ solution at 25°C. 12 h later, took them out and washed to colorless and tasteless, dried in the oven at 60°C, so were the ACF- FePcS obtained. Atomic absorption spectrometer (Sollaar M6, Thermo Electron Co., United States) was used to measure the content of Fe in FePcS-ACF: the content of FePcS in ACF-FePcS was 0.55%.

2.3 Analytical methods

Quantitative analysis of 4-NP was carried out on a high performance liquid chromatography (HPLC) equipped with a C18 column (1.7 μm, 2.1 mm × 50 mm). The column temperature was 30°C; acetonitrile and H_3PO_4 aqueous solution (volume ratio was 0.1%) as mobile phase (60/40, V/V); current velocity was 0.3 ml/min and the injection capacity was 2 μL. The catalytic performance of ACF-FePcS were characterized by the residual ratio of 4-NP.

3 Results and discussions

3.1 Catalytic oxidation performance of FePcS-ACF to 4-NP

Results of Figure 1 showed the concentration changes of 4-NP under different conditions. Catalytic oxidation experiments were performed by comparing the behaviors of $FePcS/H_2O_2$, ACF, $(FePcS+ACF)/H_2O_2$, ACF-FePcS (FePcS anchored to ACF by a formation of covalent bond) and ACF-FePcS/H_2O_2 to investigate the catalytic ability of ACF-FePcS/H_2O_2. It was found that homogeneous FePcS performs no catalytic activity, which had been supported by the previous works of Hadasch et al[8]. As shown in Figure 1, the residual of 4-NP in the FePcS-ACF/H_2O_2 system accounted for 10.8% while the system without H_2O_2 was 33.1% in 240 min. When 4-NP exposed to bare ACF and $(FePcS+ACF)/H_2O_2$, the surplus achieved at 36.8% and 34.5% respectively, both exhibiting the ACF adsorption behavior, which were similar to the system of FePcS-ACF. The results demonstrate that FePcS had no catalytic activity whether in the system FePcS+ACF or only with homogeneous FePcS. However, when ACF-FePcS exposed into 4-NP aqueous solution, under the action of H_2O_2, the concentration of 4-NP decreased most. From these suggest ACF-FePcS/H_2O_2 present a good catalytic oxidative activity.

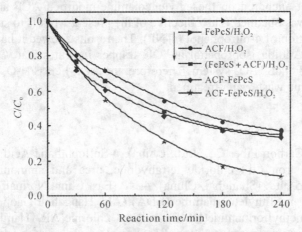

Figure 1 Catalytic degradation performances of 4-NP by ACF-FePcS. (pH=7, T=25 °C, FePcS (3.74 μmol/L), ACF (1.33 g/L), FePcS-ACF (1.33 g/L, containing 3.74 μmol/L FePcS), H_2O_2 (50 mmol/L)

3.2 Influence of catalytic oxidation performance of FePcS-ACF to 4-NP
3.2.1 Effect of initial pH

It is of great importance to study the influence of pH on the oxidative removal of 4-NP under different pH values. As shown in Figure 2, ACF-FePcS still had catalytically activity under neutral even alkaline condition. In basic condition, the adsorption of 4-NP from aqueous phase to ACF-FePcS was lower compared with that in acidic environments. This phenomena can be explained by means of the dissociation of 4-NP[10]. As pH increased, more and more of the soluble 4-NP was in its dissociated form, which reduced the adsorption of 4-NP because of the dominant electrostatic repulsion between the dissociated 4-NP and the surface of ACF. Unlike the traditional Fenton reactions with ferrous ions whose optimal working pH value is limited by its acidic requirement, admittedly, the ACF-FePcS/H_2O_2 system could efficiently eliminated 4-NP in neutral and alkaline aqueous solutions. Therefore, our results confirm this catalytic oxidation system is more advantageous than Fenton systems.

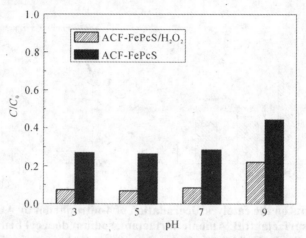

Figure 2 Effect of pH on the catalytic degradation of 4-nitrophenol by ACF-FePcS
(pH 7, t =25°C, C(4-NP)=5×10^{-4}mol/L, C(ACF-FePcS) =1.33g/L (containing 8.23×10^{-6}mol FePcS))

3.2.2 Influence of applied H_2O_2 on the removal of 4-NP

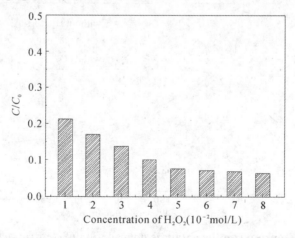

Figure 3 Effect of applied H_2O_2 on the removal of 4-NP
(pH 7, t =25°C, c(4-NP)=5×10^{-4}mol/L, c(ACF-FePcS) =1.33g/L (containing 8.23×10^{-6}mol FePcS))

The amount of oxidant also have an effect on the catalytic degradation, the effect of applied H_2O_2 on the catalytically degrade 4-NP was shown in Figure 3. From the figure, we observed the removal efficiency of 4-NP increased with a higher content of H_2O_2 when the concentration of H_2O_2 was less than 5.0×10^{-2}mol/L. However, when the dosage of H_2O_2 is higher than that of it, the residual of 4-NP nearly unchanged. From this, we could conclude 5.0×10^{-2}mol/L is the best dosage of oxidant.

3.2.3 Effect of surfactants on the catalytic degradation of 4-nitrophenol by ACF-FePcS

The interaction between surfactants and the surface of ACF-FePcS are important factors affecting the adsorption rates of 4-NP uptake on the ACF-FePcS. Thus, we studied the relationship between the adsorption rates and degradation efficiency. Figure 4 (a) was the adsorption rates of 4-NP by only ACF-FePcS under different kinds of surfactants after 4h. The removal ratios were 69.9%, 30.4%, 65.5% and 68.7% corresponding to the condition without surfactant, anionic surfactant, nonionic surfactant, cationic surfactant respectively. Graph (b) depicted the residual ratios of 4-NP by ACF-FePcS/H_2O_2 system. The degradation tendency was similar to the adsorption rule by ACF-FePcS: the addition of anionic surfactant could retard the removal rates of 4-NP and the presence of other surfactants almost had no effect on the degradation. This was because the surface properties of ACF-FePcS changed after adding anionic surfactant, leading to the negative charge increased and to enhance the electrostatic repulsion between the surface of ACF-FePcS and 4-NP. As a result, anionic surfactant hinder the adsorption of 4-NP. We can get the conclusion that the addition of anionic surfactant could reduce the treatment effect by lowering the absorption of 4-NP onto ACF-FePcS. The performance of ACF-FePcS was related with its adsorption rate, exhibiting the process of "enrichment and in-situ catalytic degradation".

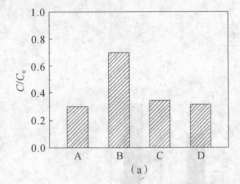 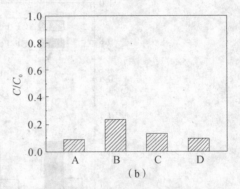

Figure 4 Effect of surfactants on the catalytic degradation of 4-nitrophenol by ACF-FePcS; (a) without H_2O_2, (b) with H_2O_2(A: Without surfactant; B: Anionic surfactant, sodium dodecyl benzene sulfonate; C: Nonionic surfactant, polysorbate 80; D: Cationic surfactant, cetyltrimethylammonium bromide, the amount of surfactant are equal: 0.6g/L)

3.3 Influence of ·OH scavenging agent on the catalytic activity of ACF-FePcS

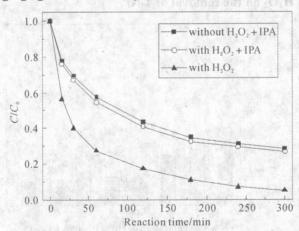

Figure 5 Effect of isopropyl on the catalytic degradation of 4-nitrophenol by ACF-FePcS
(pH 7, t =25°C, C(4-NP)=5×10^{-4}mol/L, C(ACF-FePcS) =1.33g/L (containing 8.23×10^{-6}mol FePcS), $C(H_2O_2)$ = 5×10^{-2}mol/L, C(isopropanol) = 5×10^{-2}mol/L)

While ·OH has an extreme short half-life and can not be measured directly, substantial evidence of ·OH formation in the ACF-FePcS system can be obtained by using isopropanol as an ·OH trap, which could produce acetone [11]. Figure 5 showed the effect of isopropanol on the catalytic oxidation activity of 4-NP by ACF-FePcS. As shown, the residual of 4-NP was 26.6% in the presence of isopropanol and H_2O_2 which was similar to the condition of adding isopropanol, however, it reached 5.06% on the ACF-FePcS/H_2O_2 system. The results indicated that ·OH were probably produced in the ACF-FePcS/H_2O_2 system and these active free radicals can decompose organic material 4-NP. The reaction in the system may agree to ·OH mechanism and the catalytic oxidation mechanism in this process is still under investigation further.

4 Conclusions

In this paper, a novel catalytic activated carbon fiber (ACF-FePcS) was prepared. Experiments were carried out on the catalytic degradation of aromatic compound 4-NP by this catalytic functional materials in the neutral aqueous solution. We observed that FePcS monomer had no activity on the removal of 4-NP in neutral condition, however, the catalytic activity and stability of FePcS have been greatly enhanced after anchored on the ACF. Experimental results showed that the removal efficiency of 4-NP reached more than 90% in 4 h at 25°C, and it still had a good catalytic behavior over a wide range of pH. The addition of anionic surfactant could reduce the treatment effect by lowering the absorption of 4-NP onto ACF-FePcS. Thus, the performance of ACF-FePcS was related with its adsorption rate, exhibiting the process of "enrichment and in-situ catalytic degradation". Experiment of adding isopropanol as radical scavenging agent suggests that ·OH plays a key role in the degradation of 4-NP by ACF-FePcS/H_2O_2 system. Furthermore, this catalytic functional material could avoid the environment problem such as secondary pollution caused by regeneration of ACF and it shows promising prospect in the field of wastewater treatment.

References

[1] Shen Z M, Zhang W H, Zhang X J, et al. Preparation and Application of Active Carbon Materials. Beijing: Chem Ind Press [M]. 2006. 295, 287.
[2] Qu X F, Zheng J T, Zhang Y Z. J Colloid Interf Sci [J]. 2007, 309: 429.
[3] Manuel F R, José J.M.Órfão, josé L.F. Oxidative dehydrogenation of ethylbenzene on actived carbon fibers.carbon [J]. 2002,40:2393-2401.
[4] Debasish D, Vivekanand G, Nishith V. Carbon [J]. 2004, 42: 2949.
[5] Fukushima M, Tatsumia K. Bioresour Technol [J]. 2006, 97(14): 1605.
[6] Sorokin A, Seris J, Meunier B. Science [J].1995, 268: 1163.
[7] Mahtab P, Mostafa M A, Nasser S. J Colloid Interf Sci [J]. 2008, 319: 199.
[8] Sanchez M, Hadasch A, Rabion A, Meunier B. Surf Chem and Catal, series [J]. 1999: 241.
[9] Nemykin V N, Polshyna A E, Borisenkova S A, Strelko V V. J Mol Catal A-Chem [J].2007, 264: 103
[10] Lv W Y, Chen W X, Li N. Oxidative removal of 4-nitrophenol using activated carbon fiber and hydrogen peroxide to enhance reactivity of metallophthalocyanine, Appl Cata B-Env [J].2009,87:146-151.
[11] Zhu B., Kitrossky N., Chevion M. Evidence for production of hydroxyl radicals by pentachlorophenol metabolites and hydrogen peroxide: A metal-independent organic Fenton reaction [J]. Biochem. Biophys. Res. Commun., 2000, 270(3): 942-946.

13 Influence of SiC Coating on the Oxidation Behavior of PAN Carbon Fiber at Elevated Temperatures

Ye Li, Lina Wang, Weihui Xie, Jianjun Chen*

(Key Laboratory of Advanced Textile Materials and Manufacturing Technology, Ministry of Education of China, Zhejiang Sci-Tech University, Hangzhou 310018, China)

Abstract: SiC was coated on the surface of carbon fibers substrate by a reaction between thermally evaporated silicon and carbon at 1600℃. SEM and XRD analyses show that the coatings obtained are composed of SiC grains and micro-crystals. The oxidation behavior of coated samples was investigated, and then their oxidation mechanisms were studied. The weight results show that carbon fibers loss weight after oxidation. However, carbon fibers with SiC coating gained weight after oxidation and the weight gain increased with the treatment temperature. The oxidation steps included oxidation of substrate by diffusion of oxygen through coating cracks, the reaction of oxygen and SiC coating and oxidation of coating by diffusion of oxygen through the oxide film. It can be concluded that SiC coatings which were prepared on carbon fibers can improve the oxidation resistance of carbon at elevated temperatures.

Keywords: SiC; PAN carbon fibers; Coating; Oxidation behavior

1 Introduction

Carbon fibers and carbon fiber reinforced composites are being considered for the defense and aerospace industries due to low density and excellent high temperature mechanical properties [1]. But a high reactivity to oxygen or hydrogen at high temperature has been known as a major drawback of the carbon materials [2-4]. One of the ways to prevent these materials from eroding through formation of hydrogen or carbon oxide is to cover the surface with a refractory ceramics such as SiC, Si_3N_4, and oxide glasses protective layer. Among them, a SiC coating is considered to be most effective in protecting the carbon from oxidation at high temperatures up to 1700℃ [5].

Various methods such as chemical vapor deposition(CVD), sol-gel process and calcinations of polycarbosilane have been tested for coating carbon fibers with SiC films. CVD is more popular than the other two due to its convenience and continuous reaction process. But the optimization of CVD reactions is rather difficult. The morphology of SiC films is strongly affected by small changes in preparation conditions, and the tensile strength of SiC-coated fibers depends on the content of carbon in the SiC films [6]. Here, we report a simple thermal evaporation approach to prepare SiC coating on the substrate of PAN carbon fibers. In this approach, silicon vapor due to silicon melt evaporation migrated to the surface of carbon fibers, and reacted with carbon to form SiC coating.

2 Experimental

The SiC coating on the substrate of PAN carbon fiber was prepared via the following procedure: 10g silicon (99.99%purity, Aldrich) was placed inside an open graphite crucible(50mm in length and 80mm in diameter). Then, a bunch of PAN carbon fibers (Toray, Japan) were laid to bridge the top of the crucible. Other graphite crucible with the same size was put upside down to cover the carbon fibers as well as the crucible containing silicon. The two-crucible set was then moved to the center of a vertical graphite furnace (VSF-120/150, Institute of Vacuum Technologies, Shenyang, China). The furnace was initially evacuated to $\sim 5\times 10^{-3}$ Pa by a rotary and diffusion pumps. High purity argon (>99.999%) gas was then introduced at a pressure of 0.11MPa and kept forever until the experiment ceased so that any oxidation effect was avoided. Then the furnace was heated to 1600℃ at 25℃/min and maintained at this temperature for 6 h, followed by cooling to 1200℃ in 60 min. Finally, the furnace was naturally cooled to room temperature by simply turning the power off. The reacted carbon fibers were taken out and greenish product was found on the surface of the carbon fibers.

The morphology and crystalline structure of the samples were analyzed by scanning electron microscopy

---

* Corresponding author's email: chen@zstu.edu.cn

(JEOL JSM-5610LV) and X-ray powder diffraction which was conducted on a Rigaku, Geigerflex/D using Cu K_ (λ = 1.5406 nm) radiation with a step-scanning technique in the 2θ range from 10° to 80°. PAN carbon fibers and fibers with SiC coating were treated at 400℃, 600℃, 800℃, 1000℃ separately in the muffle to inspect the oxidation behavior. The electronic balance with the precision of 1/10000 wa used to weight the samples. The evolution of the sample surface morphology with the oxidation temperature was systematically examined by SEM.

3 Results and discussion

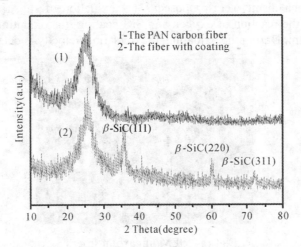

Figure 1 XRD patterns of the PAN carbon fiber (1) and the fiber with coating (2)

The XRD patterns of the PAN carbon fiber and the fiber with coating are shown in Figure 1. The shape of the XRD pattern from the original PAN carbon fiber illustrates that it is composed of crystallized graphite and amorphous carbon. After coating however, main peaks, which can be indexed to β-SiC zinc-blende SiC phase(3C-SiC), are found besides the graphite and amorphous carbon still existing. XRD results explicitly show that SiC was produced during heating the carbon fibers at 1600℃ for 6h in the presence of silicon melt. The reaction of the PAN carbon fibers with silicon vapor instead of silicon melt is responsible for the formation of SiC since the silicon melt was separated to the carbon fiber in the experiment.

Then the oxidation behavior of carbon fibers with coating or without coating was investigated. From Table 1, the carbon fiber started to oxide at 400℃, even oxided completely at 600℃ for 2 hr. From equation (1), the weight of coating fibers gained after high temperature oxidation because the mole mass of SiO_2 is bigger than that of SiC. And the weight gain increased with increasing treatment temperature.

$$SiC(s) + 2O_2 \rightarrow SiO_2(s) + CO_2 \uparrow \qquad (1)$$

Table 1 The oxidation weight data (g)

	Fiber weight (g)	Fiber weight after oxidation (g)	Weight loss (g)	Loss rate (%)
fibers after 400℃ oxidation for 2hr	0.0024	0.0017	0.0007	29.17
fibers after 600℃ oxidation for 2hr	0.0013	0	00013	100
coating fibers after 400℃ oxidation for 2hr	0.0017	0.0020	-0.0003	-17.65
coating fibers after 600℃ oxidation for 2hr	0.0010	0.0022	-0.0012	-120
coating fibers after 800℃ oxidation for 2hr	0.0004	0.0021	-0.0017	-425
coating fibers after 1000℃ oxidation for 2hr	0.0014	0.0053	-0.0039	-279

Figure 2 shows the carbon fibers were stained with dirty impurity. After the 400℃ oxidation for 2 hours (Figure 3), the fibers surface become cleaner and harsher which shows that the surface already start to oxide. And this is consistent to the result of fibers weight loss. Due to the mismatch of coefficient of thermal expansion between carbon fiber and sic coating, it can be hardly coated to obtain a continuous coating. For the reasons above, it is reasonable to believe that the formation of defects in the single-layer sic coating is inevitable. The fibers surface appearance with SiC coating show that the coating thickness is approximately 0.5~1.5 μm and the coating cracks still exist after the high temperature oxidation. The cracks healing which is mentioned in other article[5] occurs obviously at elevated temperature (Figure 4-Figure 7).

Different oxidation mechanisms were proposed[7]. (1) oxidation of substrates by diffusion of oxygen through coating cracks; (2) the reaction of oxygen and SiC coating; (3) oxidation of coating by diffusion of oxygen through the oxide film. During the oxidation, the oxidation velocity is determined by the slower step.

Figure 2 SEM of carbon fibers

(a)

(b)

Figure 3 SEM of carbon fibers after the oxidation of 400℃ for 2 hours

(a)

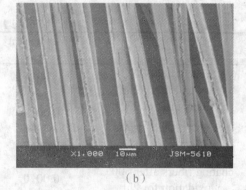
(b)

Figure 4 SEM of coating fibers after the oxidation of 400℃ for 2 hours

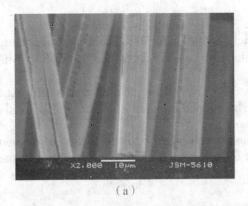 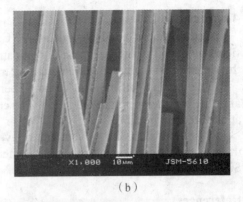

(a) (b)

Figure 5 SEM of coating fibers after the oxidation of 600℃ for 2 hours

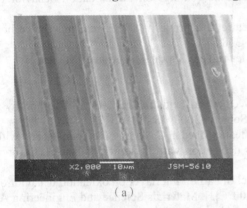 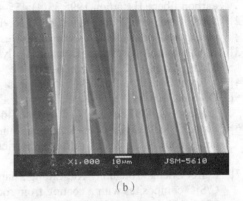

(a) (b)

Figure 6 SEM of coating fibers after the oxidation of 800℃ for 2 hours

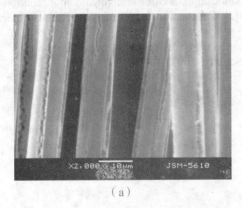 

(a) (b)

Figure 7 SEM of coating fibers after the oxidation of 1000℃ for 2 hours

As concerned with SiC coated carbon fibers, oxygen diffusing through these defects is much easy because the penetrating defects such as cracks are formed in the coating. Oxidation occurred between inner carbon fibers substrate and oxygen, and the cracks are filled with silica. Meanwhile, when the carbon fiber surface is covered with SiC coating, the amount of active carbon atom exposed to air is small, thus the chemical reaction rate is very low[8]. For this reason, it is believed that the oxidation rate of the SiC coated carbon fibers at different temperatures is mainly concerned with reaction rate. According to Arrhenius equation: $k=A\exp(-E/RT)$, the oxidation rate at elevated temperature is higher than that at low temperature, thus the fiber surface weight gain increases too. When the surface is coved with a layer of silica, the oxidation is controlled by the diffusion of oxygen through the oxide film and oxidation is then retarded because the diffusion coefficient of oxygen through the SiO_2 layer is very low[9]. The further experiment and research is still needed to understand the oxidation mechanism of carbon fibers with sic coating.

4 Conclusion

Carbon fiber started to oxide at 400℃, even oxided completely at 600℃ for 2hr. The weight of coating fibers gained after high temperature oxidation because the mole mass of SiO_2 is bigger than that of SiC. And the weight gain increased with increasing treatment temperature.

The coating thickness of carbon fiber is approximately 0.5~1.5 μm and the coating cracks still exist after the high temperature oxidation. The cracks healing which is mentioned in other article occurs obviously at elevated temperature.

When the surface is coved with a layer of silica, the oxidation is controlled by the diffusion of oxygen through the oxide film and oxidation is then retarded because the diffusion coefficient of oxygen through the SiO_2 layer is very low.

References

[1] T. Piquero, H. Vincent, C. Vincent, J. Bouix. Influence of carbide coatings on the oxidation behavior of carbon fibers[J].Carbon,1995,33(4):455-467.
[2] Juan Zhao, Gui Wang, Quangui Guo, Lang Liu. Microstructure and property of SiC coating for carbon materials[J]. Fusion Engineering and Design, 2007, 82(4): 363-368.
[3] E. Gauthier, A. Grosman, J. Valter. Analysis of long term samples in Tore Supra[J]. Journal of Nuclear Materials, 1995, 220–222: 506–510.
[4] J. Roth, C. Garcia-Rosales[J]. Nucl. Fus, 1996, 36:1647.
[5] Oh-Sang Kwon, Seong-Hyeon Hong*, Hwan Kim.The improvement in oxidation resistance of carbon by a graded SiC/SiO2 coating[J]. Journal of the European Ceramic Society, 2003, 23:3119–3124.
[6] Katsuki Kusakabe, Bong-Kuk Sea, Jun-Ichiro Hayashi, Hideaki Maeda. Coating of carbon fibers with amorphous SiC films as diffusion barriers by chemical vapor deposition with triisopropylsilane[J]. Carbon, 1996, 34(2): 179-185.
[7] Laifei Cheng, Yongdong Xu, Litong Zhang, Xiaowei Yin. Effect of carbon interlayer on oxidation behavior of C/SiC composites with a coating from room temperature to 1500°C[J]. Materials Science and Engineering A, 2001, 300(1-2): 219-225.
[8] Xin Yang, Qi-zhong Huang, Yan-hong Zou, Xin Chang. Anti-oxidation behavior of chemical vapor reaction SiC coatings on different carbon materials at high temperatures[J]. Transactions of Nonferrous Metals Society of China, 2009, 19(5): 1044-1050.
[9] Oh-Sang Kwon, Seong-Hyeon Hong, Hwan Kim. The improvement in oxidation resistance of carbon by a graded SiC/SiO2 coating[J]. Journal of the European Ceramic Society, 2003, 23(16): 3119-3124.

14 Investigation of Osteoblast-like MC3T3-E1 Cells on a Collagen-like Protein and Poly (lactic-co-glycolic acid) Nanofibrous Composite Scaffold

Yuan Li, A. Teramoto, K. Abe

(Department of Functional Polymer Science, Faculty of Textile Science and Technology, Shinshu University, Ueda, Nagano, Japan)

Abstract: We present an effective emulsion eletrospinning method in obtaining collagen-like protein (denoted as Fol-8Col) and poly(lactic-co-glycolic acid) (PLGA) nanofibrous composite scaffolds. The reservation of Fol-8Col's chemical structure has been confirmed by FT-IR spectra on electrospun composite fibers. Importantly, compared to fibrous mat without Fol-8Col, the osteoblastic behavior on the nanofibrous composite scaffolds was enhanced. The novelty of the process was that Fol-8Col in the scaffold had a positive modulation on early osteoblast adhesion and extension (within 24 hours) compared to neat PLGA. We concluded that this effective emulsion electrospinning is able to fabricate novel Fol-8Col/PLGA nanofibrous composite scaffolds which is promising for bone tissue engineering application.

Keywords: Nanofibrous Composite; Electrospun composite fibers; Collagen-like protein

1 Introduction

In previous study, the initial investigation of Fol-8Col with triple helix structure showed this collagen-like protein was promising for tissue engineering [1]. In order to develop three dimensional scaffold of Fol-8Col for bone regeneration application, electrospinning was used to fabricate nanofibrous scaffold [2]. Since the low yield of Fol-8Col, we chose to blend it with PLGA which is widely used as an interesting biomaterial [3]. We employed emulsion electrospinning for Fol-8Col/PLGA nanofiber fabrication in this study. Traditional electrospinning as a promising technique to prepare polymer fibers is mainly confined to homogeneous polymer solutions. One of the most challageneing issues when preparing hybrid scaffolds is how to uniformly distribute the dispersed phase in continuous phase. Unlike coaxial electrospinning which need a special apparatus, emulsion electrospining is recently used to prepare core-shell nanofibers [4, 5]. These nanofibers with drugs and protein encapsulated would be developed into bioactive tissue engineering scaffolds [6].

2 Experimental details
2.1 Materials

PLGA (poly (D, L-lactic-co-glycolic acid)) with a 75:25 monomer ratio, ester-terminated, and viscosity of 0.55-0.75dl/g was purchased from Durect Corporation (Pelham,AL). All other solvents and reagents used were analytical grade.

2.2 Preparation of Fol-8Col/PLGA fibrous composite scaffolds

All concentrations measurements were done in weight by weight (w/w). PLGA (10%) was fully dissolved in chloroform/toluene (75:25) mixed solvent under gentle stirring with the 1% of surfactant SPAN80. Then, Fol-8Col water solution (5%) was added dropwise into the chloroform/toluene phase. The mixture was sonicated over an ice bath to emulsify for about 30 min. Katotech Electrospinning Unit was used to perform the emulsion electrospinning. The stable W/O emulsion was transferred to a 10 ml syringe fitted to a 18-G needle. The tip-to-collection plate (covered with aluminum foil) distance was fixed at 15cm in front of the needle tip. Voltage applied between the needle (anode) and the collector (cathode) is 17kV. Feed rate of the emulsion spinning dope was 0.012 ml/min. The fibrous constructs collected on the plate were placed under vacuum for overnight for the removal of residual solvent. The morphology of the gold sputtered (DENTON VACUUM DESK II) electrospun fibers was observed with SEM (HITACHI S-2300, Japan) at an accelerating voltage of 20kV.

2.3 Cell culture

MC3T3-E1 cells (2.5×10^4 cells/well) seeded on dishes setting with Fol-8Col/PLGA fibrous composite scaffolds were incubated at 37°C, in a 5% CO_2 atmosphere incubator, using α-modified minimal essential medium (α-MEM; GIBCO). The medium comprised of 10% heat-inactivated fetal bovine serum (FBS), 100 U/ml penicillin and 100U/ml streptomycin. For all cell investigations, MC3T3-E1 cells cultured on collagen/PLGA fibrous composite and TCDs were evaluated as controls. TCDs are composed of optically clear, high-grade polystyrene Nunc™ Dishes, prepared using a vacuum-gas plasma sterilization treatment. The incorporation of nitrogen-containing cations on the surface of the dishes has been correlated with the attachment and spreading of primary endothelial cells in a clonal cell-growth assay [7].

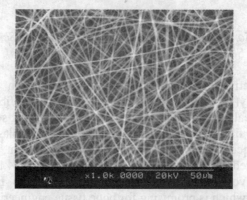

Figure 1 SEM image of nanofibers with Fol-8Col/PLGA=0.25%

3 Results and discussion

Figure 1 shows that smooth, uniform emulsion electrospun fibers were prepared. Average diameter is 635±20nm(data not shown). The FT-IR scans (Figure 2) of the emulsion electrospun fibrous scaffold were analyzed to evaluate the structure of the neat PLGA and the composite fibrous nonwoven. Typical bands for PLGA are ester carbonyl stretch(C=O) at 1747 cm^{-1}, C-O stretch at 1128 cm^{-1} and C-O-C group at 1083 cm^{-1}.

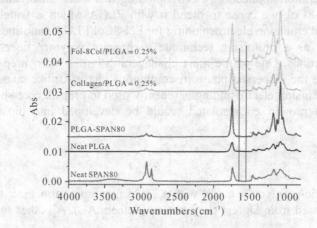

Figure 2 FTIR spectra of nanofibers with Fol-8Col/PLGA=0.25%

These three typical bands were found for neat PLGA[3], Fol-8Col/PLGA=0.25% and Collagen/PLGA=0.25% emulsion electrospun fibers respectively. The amide I and amide II bands at 1645 and 1552, respectively, were found for emulsion electrospun composite fibers Fol-8Col/PLGA=0.25% and Collagen/PLGA=0.25%.

Cytotoxicity assays (Figure 3 (a) (b)) indicate that MC3T3-E1 cells have good viability on Fol-8Col/PLGA nanofibrous scaffold. From the live/dead fluorescence micrographs, the majority of cells incubated for 24 hours on were alive and revealed polygonal-shaped morphology. This figure also semi-qualitatively shows nearly 90% viable MC3T3-E1 cells attached to Fol-8Col/PLGA fibers after 24 hours cultivation. This results was consistent with the LDH (lactate dehydrogenase) measurements by determining the concentration of LDH liberated from cells after chemically lysing the cell membrane, using a LDH-Cytotoxicity Kit (Wako Pure Chemical Industries Ltd.). LDH release from all the composite nanofibers are around 10% which is comparable to cells on TCD.

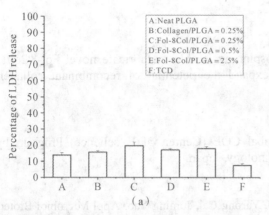 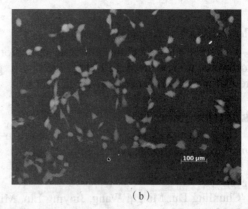

Figure 3 (a) LDH release ratio of MC3T3-E1 cells on nanofibers with Fol-8Col/PLGA=0.25% for 24 hours; (b) live/dead cells staining of MC3T3-E1 cells on nanofibers with Fol-8Col/PLGA=0.25% for 24 hours

The results of the adhesion ratio and immunolabeling analysis (Figure 4 (a), (b)) show that Fol-8Col/PLGA nanofiber scaffold can support the attachment and stretching of MC3T3-E1 cells. MC3T3-E1 cells on Fol-8Col/PLGA composite fibers show intensive vinculin signals along extended actin microfilaments, with the highest concentration seen at the extremities of cellular extensions, resulting in focal adhesion patches.

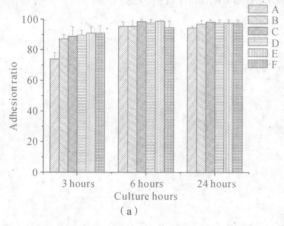 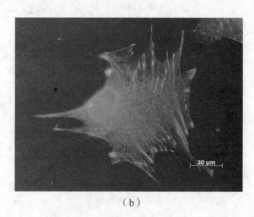

Figure 4 (a) Adhesion ratios of MC3T3-E1 cells on nanofibers with Fol-8Col/PLGA=0.25% for 24 hours; (b) fluorescent staining of F-actin (green), vinculin (red) and cell nuclei (blue) for MC3T3-E1 cells cultured on nanofibers with Fol-8Col/PLGA=0.25% for 24 hours

Figure 5 shows the morphology of MC3T3-E1 cells cultured on Fol-8Col/PLGA nanofiber scaffold is conductive for cell differentiation after 2 weeks cultivation. Cells show the morphology of fully extension in sheet form.

Figure 5 SEM image of MC3T3-E1 cells on nanofibers with Fol-8Col/PLGA=0.25% for 14 days

4 Conclusion

We concluded that this effective emulsion electrospinning is able to fabricate novel Fol-8Col/PLGA nanofibrous composite scaffolds which will greatly expand the potential of recombinant collagen-like proteins for tissue regeneration applications.

Acknowledgement

This work was supported by a Grant-in-Aid Global COE (Center Of Excellence) Program by the Ministry of Education, Culture, Sports, Science and Technology, Japan.

References

[1] Chunling Du, Mingqi Wang, Jinying Liu, Mingli Pan, Yurong Cai, Juming Yao, Appl Microbiol Biotechnol. 2008, 79:195.
[2] Wan-Ju Li, Cato T. Laurencin, Edward J. Caterson, Rocky S. Tuan, Frank K. Ko,Journal of biomedical materials research. 2002, 60:613.
[3] Moncy V. Jose, Vinoy Thomas, Derrick R. Dean, Elijah Nyairo,Polymer. 2009, 50: 3778.
[4] Xiuling Xu, Xiuli Zhuang, Xuesi Chen, Xinri Wang, Lixin Yang, Xiabin Jing, Macromolecular Rapid Communications. 2006, 27: 1637.
[5] Y. Z. Zhang,X. Wang, Y. Feng, J. Li, C. T. Lim, and S. Ramakrishna, Biomacromolecules. 7(2006) 1049.
[6] Yiliang Liao, Lifeng Zhang, Yi Gao, Zheng-Tao Zhu, Hao Fong, Polymer. 49 (2008) 5294.
[7] Chilkoti A, Schmierer AE, Pérez-Luna VH, Ratner BD,Anal Chem. 67 (1995)2883.

15 Electrospun Polyvinyl Alcohol/Halloysite Nanotubes Composite Nanofibers

Ruijuan Liao[1], Baochun Guo[1,2]*, Jianqing Zhao[1], Wen You Zhou[3]

(1. Department of Polymer Materials and Engineering, South China University of Technology, Guangzhou 510640, China
2. State Key Laboratory of Pulp and Paper Engineering, South China University of Technology, Guangzhou 510640, China
3. Discipline of Orthodontics, Faculty of Dentistry, The University of Hong Kong, 34 Hospital Road, Hong Kong, China)

Abstract: Polyvinyl alcohol (PVA)/halloysite nanotubes (HNTs) composite fibers have been produced by electrospinning. The electrospun composite fibers were further crosslinked by glutaraldehyde (GA). The effects of voltage, HNTs contents and GA contents on the electrospinnability were investigated. The crystallization and interaction between PVA and HNT were discussed. The results revealed that the HNTs/PVA composite nanofibers have diameters in the range of 100-400 nm. The hydrogen bonding between PVA and HNTs was substantiated and HNTs was proposed to locate close to the fiber surface. With the increasing of HNTs loading, the PVA crystallinity was firstly increased and then decreased, which was attributed to the nucleating and confinement effects of the aligned HNTs in the composite fibers. The crosslinking did not change the fiber morphology significantly and the porous characteristics of the mat were retained.

Keywords: Polyvinyl alcohol; Halloysite nanotubes; Electrospinning; Crosslinking

1 Introduction

Electrospinning is a highly versatile method to process solutions or melts, mainly of polymers, into continuous fibers with diameters ranging from a few micrometers to nanometers [1,2].

PVA is a polyhydroxy polymer that has been studied intensively because of its good film forming and physical properties, high hydrophilicity, biocompatibility [3]. As a biocompatible polymer, PVA has been widely used in biomedical applications.

Nanofiller such as nanotubes have been incorporated into electrospun polymer fiber [4] to increase the strength, biocompatibility etc. HNTs, a type of tubular nanoclay, have been recognized as the additive of polymers for better mechanical properties or thermal properties [5]. Recently, we found [6] that HNTs may effectively increase the strength and cell viability of PVA film. HNTs were found exposed on the surface of the PVA/HNTs composites film which is benefit for cell attachment. Actually, Lvov also reported the cytotoxicity of halloysite. They added halloysite to two different cell lines, demonstrated that it is nontoxic up to concentrations of 75 ug/ml, which improved that halloysite is not toxic for cells [7].

In this study, PVA nanofibers and HNTs/PVA composite nanofibers have been produced by electrospinning. The electrospun composite fibers were further crosslinked by glutaraldehyde (GA). We studied the electrospinnability of PVA/HNTs composites at different voltages, HNTs and GA contents. The structure and properties of these composites were preliminarily evaluated and discussed.

2 Materials and methods

2.1 Raw materials

PVA (2488, with the polymerization degree of ~2400 and the degree of hydrolysis of ~88%) was purchased from ChuanWei Group, China. The HNTs were mined from Yichang, Hubei, China. GA used as a chemical crosslinking reagent, was purchased in 25 wt% aqueous solution.

2.2 Preparation of PVA/HNT composite suspensions

The stock suspensions of PVA and HNTs suspension were prepared separately. The purified HNTs were dispersed in deionized water for 2 hours by ultrasonic treatment. Aqueous PVA solutions were prepared by dissolving PVA powder in deionized water at 90°C with constant stirring for at least 20 hours. When the solution was cooled to room temperature, HNTs suspension and GA was added. The mixture was stirred further for 30 min before electrospinning.

---

* Corresponding author's email: psbcguo@scut.edu.cn

2.3 Electrospinning of PVA/HNT composite suspensions

Approximately 5 ml of the uniform mixture of PVA/HNTs was sucked into a 10 ml plastic syringe. A syringe pump was used to control the feeding rate. The positive electrode of the high-voltage power supply was connected to the needle, and the grounding electrode was connected to a plastic collector plate wrapped with aluminum foil located at a distance of 15 cm from the needle tip. To get crosslink nanofiber mats, the collected mats were exposed to concentrated HCl vapour for about 1min and then dried at 60°C for 15 min to catalyze the crosslinking between PVA and GA.

2.4 Characterizations

The diameters of collected fibers were measured by SEM. The fiber diameter was determined from SEM images with the aid of image analysis software (Image Pro + 6.0). DSC and TGA data of all samples were measured by a TA Q2000. Using the enthalpy of 155 J/g for a theoretical 100% crystalline PVA [8]. XRD data were taken by D8 ADVANCE from Bruker company. FTIR spectrum of PVA and PVA/HNTs composites nanofibers were measured using attenuated total reflectance model.

3 Results and discussion
3.1 Electrospinability of PVA/halloysite composites

For pure PVA nanofiber mat, the voltage was varied from 12 kV to 20 kV to find the relationship between the fiber morphology and the applied voltage. From Figure 1, one can see that with the increase in voltage, the mean diameter of the fibers increases from 333 nm to 433 nm. In addition, at higher voltage, the distribution of fiber diameter become wider. This observation was similar to other reported result [9]. With increasing the applied voltage, the electric field strength increases the electrostatic repulsive force on the fluid jet which favors the thinner fiber formation. On the other hand, at higher voltage, the solvent will be evaporated from the capillary tip more quickly as the suspension is ejected from Taylor cone. In the present situation, the latter effect dominates. Consequently the fiber diameter was increased with increasing voltage.

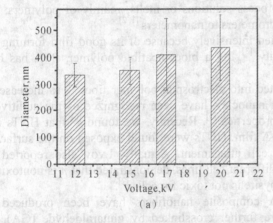
(a)

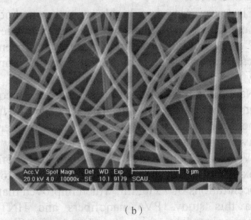
(b)

Figure 1 (a) Effect of voltage on morphology and fiber diameter distribution from a 11.11wt% PVA solution; (b) PVA nanofibers by 15kV

In order to investigate the effect of HNTs content on the morphology of the PVA/HNTs composite fibers, six different HNTs contents (relative to PVA weight) such as 0, 6, 10, 20, 25 and 30 wt% were examined. Figure 2 showed that with the adding of HNTs, the fiber diameter decreases sharply. This may be due to the change suspensions viscosity by the incorporation of HNTs. Li etc. [10] also suggested this. The TGA results of PVA/HNTs composites fibers indicated that HNTs was successfully loaded in the fiber during electrospinning and the thermal stability is increased with the loading of HNTs.

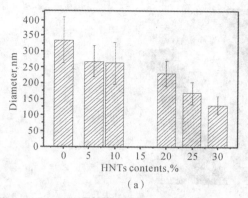
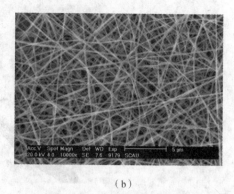

Figure 2 (a) Effect of HNTs contents on the diameter of the PVA/HNTs nanocomposite fibers;
(b) 30% HNTs contents of PVA/HNTs nanocomposite fibers

3.2 Crystallization of composite fibers and interactions between PVA and HNTs

Table 1 shows the crystallization and glass transition temperatures of pure PVA and PVA/HNTs composite fibers. As can see from Table 1, the pure PVA nanofiber mat exhibited a relatively low crystallinity of 15.3%. With the adding of 3% and 6% HNTs, the crystallinity increased to 16.4% and 19.2%, respectively. However if the loading of HNTs is further increased, the crystallinity tends to decline. When HNTs are incorporated, they may act as the nucleating agent for PVA and thus facilitating the crystallization. However, overloading of HNTs may effectively suppressed the crystal growth due to the confinement by the aligned HNTs.

Table 1 Effects of HNTs on the crystallization and transition behavior of the composite fibers

HNTs content/%	T_g/°C	T_m/°C	Melting enthalpy/J.g^{-1}	X_c/%
Neat PVA	82.5	194.9	23.7	15.3
3	82.7	195.8	26.2	16.4
6	83.3	195.6	31.5	19.2
10	82.6	196.2	21.3	12.5
15	78.1	197.9	22.0	12.3

WAXD was used to further study the PVA crystallization in both the pure PVA and HNTs/PVA composite nanofibers. The typical diffraction peak for the HNTs and PVA appeared at $2\theta = 11.90°$ (001) and 19.44° (101) respectively. The (101) phase was the main crystalline due to the high integrated area of PVA [8]. The XRD curves of PVA/HNTs composites fibers revealed that with different HNTs contents, the crystal structures of the fibers were largely changed. Interestingly, as seen in Figure 3, when the HNTs contents was only about 10% to PVA content, the diffraction peak for the HNTs (001) was even stronger than the PVA diffraction (101), indicating that the HNTs mostly likely distributed closed to the surface of the PVA fiber. This can be further demonstrated by Figure 4.

As seen in Figure 5, with the increasing of HNTs contents, the –OH stretching vibration peak for PVA was red shifted consistently from 3280 cm^{-1} to 3320 cm^{-1}. The spectral evolution is indicative to the hydrogen bonding between PVA hydroxyls and siloxane of HNTs. The authors have also demonstrated the hydrogen bonding between HNTs and PVA in PVA/HNTs composite films [6].

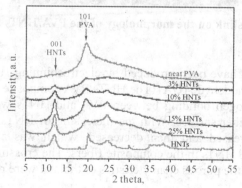

Figure 3 The WAXD curves of PVA/HNTs composites fibers with different HNTs contents

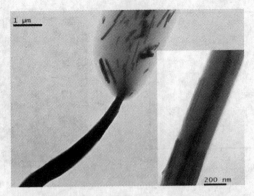

Figure 4 TEM image of fiber morphology taken from a sinlge electrospun PVA/HNTs nanocomposite fiber

3.3 Effect of crosslinking on PVA/HNTs composite fiber mats

As seen in Figure 6, the morphology of the fiber mats has not been changed obviously except some conglutination compared with above SEM photos. Importantly, the porous morphology of mat, which is crucial for many biomedical applications, was retained.

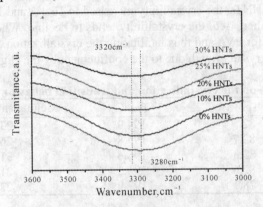

Figure 5 The FTIR curves of PVA/HNTs composites fibers with different HNTs contents

Figure 6 Effect of crosslink on the morphology of the PVA/HNTs nanocomposite fibers

4. Conclusions

PVA/HNTs composite fibers have been produced by electrospinning. The electrospun composite fibers were further crosslinked by glutaraldehyde (GA). The HNTs/PVA composite nanofibers have diameters in the range of 100-400 nm. The hydrogen bonding between PVA and HNTs was substantiated and HNTs was proposed to locate close to the fiber surface. With the increasing of HNTs loading, the crystallinity of the composite fiber was firstly increased and then decreased, which was attributed to the nucleating and confinement effects of the aligned HNTs in the composite fibers. The crosslinking of the fiber did not change the morphology significantly and the porous characteristics of the mat were retained.

Acknowledgement

We authors appreciate the financial supports from the National Natural Science Foundation of China (50603005).

References

[1] Reneker, D.H. and A.L. Yarin, Electrospinning jets and polymer nanofibers. Polymer, 2008. 49(10): p. 2387-2425.
[2] Greiner, A. and J.H. Wendorff, Electrospinning: A fascinating method for the preparation of ultrathin fibres. Angewandte Chemie-International Edition, 2007. 46(30): p. 5670-5703.
[3] Yang, E.L., X.H. Qin, and S.Y. Wang, Electrospun crosslinked polyvinyl alcohol membrane. Materials Letters, 2008. 62(20): p. 3555-3557.
[4] Kim, G.M., et al., Electrospun PVA/HAp nanocomposite nanofibers: biomimetics of mineralized hard tissues at a lower level of complexity. Bioinspiration & Biomimetics, 2008. 3(4): 046003.
[5] Guo, B.C., et al., Crystallization behavior of polyamide 6/halloysite nanotubes nanocomposites. Thermochimica Acta, 2009. 484(1-2): p. 48-56.
[6] Zhou, W.Y., et al., Poly(vinyl alcohol)/Halloysite nanotubes bionanocomposite films: Properties and in vitro osteoblasts and fibroblasts response. Journal of Biomedical Materials Research Part A, 2010. 93A(4): p. 1574-1587.
[7] Vergaro, V., et al., Cytocompatibility and Uptake of Halloysite Clay Nanotubes. Biomacromolecules, 2010. 11(3): p. 820-826.
[8] Naebe, M., et al., Electrospun single-walled carbon nanotube/polyvinyl alcohol composite nanofibers: structure-property relationships. Nanotechnology, 2008. 19(30): 305702.
[9] Zhang, C.X., et al., Study on morphology of electrospun poly(vinyl alcohol) mats. European Polymer Journal, 2005. 41(3): p. 423-432.
[10] Li, L., et al., Formation and properties of nylon-6 and nylon-6/montmorillonite composite nanofibers. Polymer, 2006. 47(17): p.6208-6217.

16 Effect of Rheological Properties on Electrospinning of Ultra High Molecular Weight (UHMW) Poly(vinyl alcohol)

Fan Liu[1], Yasushi Murakami[2], Qing-Qing Ni[2]

(1. Interdisciplinary Graduate School of Science and Technology, Shinshu University, Ueda, Japan
2. Faculty of Textile Science and Technology, Shinshu University, Ueda, Japan)

Abstract: The effects of rheological properties on the electrospun fiber structure of polyvinyl alcohol (PVA) with ultra high molecular weight (UHMW) and high hydrolysis degree have been studied in this article. A hydrazine salt was used to modify the rheological properties of PVA aqueous solution in order to facilitate electrospinning. Rheological properties of the solution and fiber morphology were investigated by a rheometer and a scanning electron microscope respectively. The results show that it is impossible for the pure PVA solution to electro spin. But at the presence of hydrazine salt, the apparent viscosity of PVA solution decreased with storage time increasing and nano fibers can be fabricated by electrospinning. Furthermore, the diameters of electrospun fibers also reduced with viscosity decreased.

Keywords: Rheological properties; Electrospun fiber; Polyvinyl alcohol; Electrospinning

1 Introduction

PVA is a semi-crystalline, hydrophilic polymer with good chemical and thermal stability and these prominent properties led PVA to find its wide applications[1-3]. Ultrafine PVA fibers, which may have potential applications in filtration and biomedical engineering but cannot be produced by conventional spinning techniques. However, it is difficult to prudence fibers at submicron scales by traditional spinning method.

Electrospinning is a straightforward, convenient, and inexpensive method of preparing polymer fibers at submicron scales.[4-6] In electrospinning, a high electric potential is applied to a polymer solution or melt. This technique has received a lot of attention in recent years because of the relative simplicity with which a wide range of porous structures can be produced.

The fibrous of PVA in electrospinning is greatly affected by molecular weight, degree of hydrolysis degree and concentration of PVA solution[7]. It is difficult for PVA with high molecular weight and high degree of hydrolysis to be used in electrospinning due to the strong interaction between PVA molecules. In our research, we found a kind of hydrazine salt, N_2H_4HCl, which can modify the rheological properties of PAV solution (polymerization degree, Pn 8000; hydrolysis degree >99%) and make it possible to be used in electrospinning.

2 Experimental

2.1 Materials

PVA with polymerization degree of 8000 and hydrolysis degree of +99% were obtained from Kurary company (Japan). Hydrazine monochloride (N_2H_4HCl) (GR grade, Tokyo Chemical Industry Co., Ltd.) and isopropanol (GR grade, Wako Pure Chemical Industries, Ltd.) were used without any treatment.

2.2 Preparation of PVA solution

PVA was dissolved in distilled water at 95°C for 3h with high speed stirring to get 5wt% solution and the solution was kept for 2 h to ensure homogenization. Different amount of N_2H_4HCl was added into the solution to get PVA solutions with N_2H_4HCl concentration of 0.05mol/L, 0.1mol/L and 0.3mol/L. These solutions were stirred for 1h in order to make the N_2H_4HCl disperse absolutely and then kept at 40°C.

2.3 Electrospinning conditions

The electrospinning experiments were performed at 25°C. The polymer solution was placed into a 12 ml syringe with a needle tip having an inner diameter of 0.8 mm. A copper wire connected to the positive electrode was clamped on the needle. A rotating cylinder wrapped with aluminum foil was used as the collector and the collector was connected to the ground. The applied voltage and the tip-to-collector distance (TCD) were fixed at 15 kV and 15 mm, respectively and the flow rate was approximately 0.2 ml/h, measured with syringe pump.

2.4 Rheological properties measurements and fiber morphology

The rheological properties of the PVA solutions were measured by means of an controlled stress rheometer (Elquest, Rheologia-A300). A double-concentric cylinder with a gap of 2.1mm was used to measure dynamic viscoelastic parameters. The morphology of electro-spun fibers was observed on a scanning electron microscope (SEM) (Hitachi S-3000N) after platinum coating.

3 Results and discussion
3.1 Effect of N_2H_4HCl on the rheological properties of PVA

A lot of organic and inorganic chemicals have an effect on the viscosity of PVA solutions and most of them will lead to increase of the viscosity[8]. But in our research, we found that hydrazine monochloride (N_2H_4HCl) had an opposite effect on the viscosity of PVA solution. As shown in Figure 1, the apparent viscosity of pure PVA solution remains almost unchanged over 168h. But at the presence of N_2H_4HCl, solution viscosities decreased gradually with storage time. It is indicate that the addition of N_2H_4HCl could disrupt the inter and intra chain hydrogen bonding of the PVA chains. As a consequence, without of the physical cross-linking formed by hydrogen bonding, the PVA molecules can move more freely. On the other hand, the N_2H_4HCl could form hydrogen bonds with hydroxyl groups on PVA chains, which will also weaken the interaction between PVA molecules.

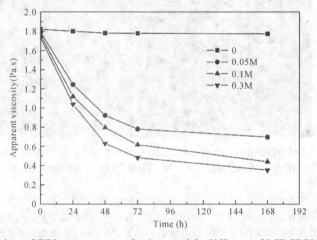

Figure 1 Apparent viscosity of PVA aqueous solutions with different N_2H_4HCl concentration after different storage time. (T: 25°C; shear rate: 100 1/sec)

Figure 2 shows the interrelationship between apparent viscosity and shear rate for PVA solutions with different N_2H_4HCl concentrations after 72 h. It is seen from the figure that the apparent viscosities decrease significantly when the N_2H_4HCl concentration increases from 0 to 0.3mol/L. For pure PVA solution, the apparent viscosity is extremely high and a pronounced shear thinning behaviour is observed. For the solution with addition of N_2H_4HCl, the apparent viscosities parheological properties changed little over the same shear rate range. The shear thinning behaviour indicates the presence of stronger polymer chain associations because of inter and intra chain hydrogen bonding and chain entanglement. However, this associations could be weakened by N2H4HCl result in the disappearance of shear thinning.

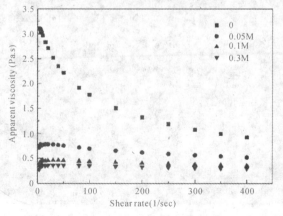

Figure 2 Apparent viscosity and shear rate interrelationship for PVA aqueous solutions with different N_2H_4HCl concentration at 25°C (after 72h)

3.2 Electrospinnability of PVA solution

As we know, in electrospinning only when the electric field force overcomes the interaction force between polymer molecules, it is possible to form a ultrafine fibers. In PVA solutions, the interaction between molecules chains is more complex and stronger than other common polymers due to the hydrogen bonding inter and intra the molecules. It more difficult to produce ultrafine fiber by electrospinning for PVA with high hydrolysis degree and ultra high molecular weight. However, in our research, the interaction between PVA molecules was weakened by addition of N_2H_4HCl and it became possible for electrospinning.

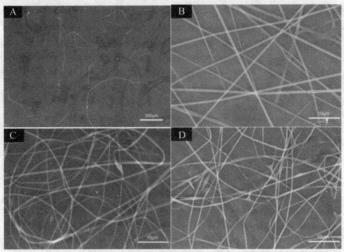

Figure 3 SEM images of electrospun PVA nanofiber containing different amount of N_2H_4HCl. (A, PVA only; B, 0.05M; C, 0.1M; D, 0.3M) (Storage time: 72h)

As shown in Figure 3, for pure PVA solution, there was no ultra fine fibers formed on because the strong interaction between PVA chains resists the breakup of the viscoelastic jet, leading to the formation of some long threads and irregular blocks of PVA (Figure 3-A). At the presence of N_2H_4HCl, as the viscosity of spinning solution decreased, nanofibers can be fabricated. In addition, the fiber diameter reduced and the fibers shape became irregular with the amount of N_2H_4HCl increasing.

The viscosity of PVA solution changed with storage time as shown in Figure 1, so we also investigated the morphology change of fibers produced by solutions after different storage time. Although N_2H_4HCl was added in the solution, the viscosity of PVA solution was still high at the beginning, so in Figure 4-A, only some thick and irregular fibers can be investigated. As the storage time increasing, the viscosity of the solution decreased gradually and the fiber diameter decreased correspondingly the shape of fibers also became more regular.

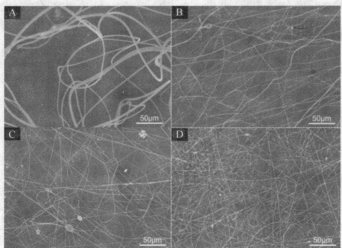

Figure 4 SEM images of electrospun PVA nanofiber after storage time: A, 5min; B, 24h; C, 48h; D, 72h (N_2H_4HCl, 0.05M)

4 Conclusions

The effects of rheological properties on the electrospun fiber structure of PVA (Pn 8000; hydrolysis degree>99%) have been studied. At the presence of N_2H_4HCl, apparent viscosity of PVA solutions decreased gradually to certain degree depending on the concentration of N_2H_4HCl and storage time. As corresponding results, the diameters of electrospun fibers also become smaller and the fiber morphology became more regular. This method makes it possible to fabricate nanofibers by electrospinning technology utilizing PVA with UHMW and high hydrolysis degree.

References

[1] Shao C, Kim H, Gong J, Ding B,. Lee D, Park S. Materials Letters. 57 (2003) 1579-1584.
[2] Krumova M, Lopez D, Benavente R, Mijangos C, Peren J.M, Polymer 41(2000) 9265-9272.
[3] Hodge R M, Edward G H, Simon G P, Polymer 37 (1996)1371-1376.
[4] Shin Y M, Hohman M M, Brenner M P, Rutledge G C. Polymer. 42(2001) 9955.
[5] Yarin A L, Koombhongse S, Reneker D H. J. Appl. Phys. 90 (9)(2001) 4836-4836.
[6] Yarin A L, Koombhongse S, Reneker D H. J. Appl. Phys. 89 (5)(2001) 3018-3026.
[7] Koski, A, Yim K, Shivkumar S. Materials Letters 58 (2004) 493–497
[8] Finch C. Poly(vinyl alcohol), properties and applications, New York:Wiley, 1973.
[9] sLyoo W S, Yeum J H, Ghim H D, Ji B C, Yoon N S, Ha J B, Lee J. J Appl Polym Sci 90(2003) 227.

17 Synthesis and Characterization of PSA-PEG Block Copolymer Based on Polysulfonamide and Amine-Terminated Polyethylene Glycol

Li Liu[1,2*], Weiwei Gu[2], Jie Zhou[2], Yanan Wu[2], Changfa Xiao[1]

(1. Key Laboratory of Hollow Fiber Membrane Material and Membrane Process of Ministry of Education, Tianjin Poly-technic University, Tianjin 300160, China
2. Dept. of Polymer Science, School of Material Science and Engineering Technology, Shanghai Univ., Chengzhong Road 20, Jiading District, Shanghai 201800, China)

Abstract: A novel PSA-PEG block copolymer was synthesized by polycondensation of aromatic polysulfonamide and amine-terminated polyethylene glycol. The structure of the copolymer was characterized by IR spectra. The solubility test proved that the solubility of copolymer was changed in 1-methyl-2-pyrrolidinone(NMP), N,N- dimethylformamide (DMF) and N,N- dimethylacetamide (DMAc) with the addition of PEG. And the most suitable spinning solution concentration of copolymer/DMAc was 14%~16%. TGA test indicates that copolymers still maintain excellent thermal stability.

Keywords: Polysulfonamide; Polyethylene Glycol; Block Copolymer

1 Introduction

Polysulfonamide (PSA) is a kind of semi-aromatic polyamide material containing aryl sulfone linkages between amide groups and often used as heat resistance fiber[1,2]. The main processing method of PSA fiber is wet spinning. As its main chain contains sulfonyl(SO_2), PSA fiber has exhibited excellent properties of heat resistance, thermal stability and inoxidizability. However, there are also some obvious disadvantages, such as low degree of crystallinity and poor mechanical property, so how to improve the combination property of PSA is still one of the most important tasks of today's materials science[2,3]. The structural modification is one of the most effective methods to improve the property of polymers.

In this paper, we chose amine-terminated polyethylene glycol(PEG) as soft segment copolymerization with polysulfonamide. PEG was a kind of extremely extensive polyether polymer, which was used in the fields of medicine, health, food and chemical industry[4,5].PEG was also applied in the preparation of thermoplastic polyamide elastomer(TPAEs), and the TPAEs presented better toughness and forming performance than pure polyamide[6].Moreover, PEG has a good compatibility with many other polymers, and its excellent property will transfer to the copolymer when copolymerization with other polymers[7].Therefore, it would be a good way to ameliorate the flexibility and aggregation structure of PSA, enhance its combination property by means of appending PEG soft segment.

2 Experimental

2.1 Synthesis of block copolymer

a. Synthesis of polysulfonamide(PSA) prepolymer

With the protection of N_2, certain 4, 4'- diaminodiphenylsulfone was added into DMAc and kept stirring the mixture in the four-necked flask. After the 4, 4'- diaminodiphenylsulfone was dissolved, the terephthaloyl chloride with excessive was added at low temperature (below 0℃). The reaction was continued for 5~6 hours at 1500rpm at room temperature. The calcium oxide used as antacid was added at the end, the pH of resulting polymer was adjusted to approach neutral.

b. Synthesis of amine-terminated polyethylene glycol(PEG)

Amino-terminated polyethylene glycol was synthesized from PEG by esterification with p-toluenesulfonyl chloride, and then reacting with $NH_3·H_2O$ at high temperature and pressure followed by basification with NaOH[8].

c. Synthesis of PSA-PEG block copolymer

With the protection of N_2, PSA prepolymer and amine-terminated PEG prepared previous were added into DMAc and kept stirring the mixture in the water-ice bath for about 30 minutes. Then added the calcium oxide into the mixture and kept the reaction for 12 hours at room temperature without light. At last, the solution was deposited with deionized, and the deposite was washed to neutral, dried to product. The chemical structure of a repeat unit of the block copolymer is shown in Figure 1.

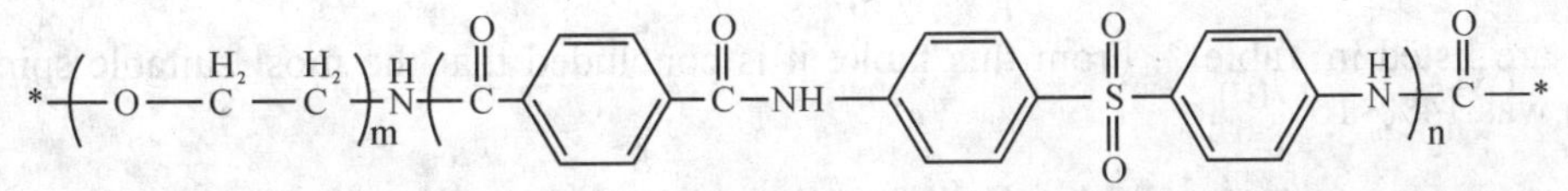

Figure 1 Chemical structure of a repeating unit in the PSA-PEG block copolymer

2.2 Measurements

Infrared (IR) spectra were recorded by a reflection method on Thermo Nicolet Nexus 470 Fourier transform infrared (FTIR) spectrometer. The solubility test for block copolymer and traditional PSA was carried out in several kinds of solvents. The viscosity of the copolymer solution was studied by falling ball viscosity experiment. Thermal behavior of copolymers was evaluated by thermo gravimetric analysis (TGA), which was carried on a Netzsch STA 409 at a heating rate of 20℃/min and the testing temperature ranged from 30℃ to 800℃.

3 Result and discussion
3.1 Infrared measurement

Structure of PSA-PEG block copolymer was characterized by IR spectra, as shown in Figure 2. We can observe the characteristic peaks of 3310 cm^{-1}(-NH-stretching), 1640 cm^{-1} (-CO-stretching), 1540 cm^{-1} (-NH-bending)[9], 1154 cm^{-1} and 1322 cm^{-1} (-SO_2- symmetric and asymmetric stretching)[10], 1108 cm^{-1} (-C-O-C- stretching) and 726 cm^{-1} (-CH_2-stretching). According to the expected structure, it is proved that PSA-PEG block copolymer was synthesized successfully.

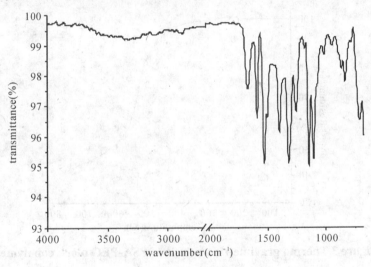

Figure 2 IR spectrum of PSA-PEG block copolymer

3.2 Solubility test

Polymer solubility in solvents is a very important factor for getting high strength fibers by the wet spinning method. Table.1 listed the solubility of traditional PSA and PSA-PEG block copolymer in some kinds of organic solvents. As we have seen, solubility of the copolymer with PEG monomers' addition was changed obviously. It is because that the flexibility of the segment increased with the addition of PEG.

Table 1 Solubility of PSA and PSA-PEG in organic solvents

	NMP	DMF	DMAc
PSA	0.03g/ml	0.07g/ml	0.09g/ml
PSA-PEG	0.1g/ml	0.2g/ml	0.5g/ml

3.3 Falling ball viscosity test

In order to further study spinnability of the block copolymer prepared, the viscosity of various concentration of copolymer/N,N-dimethylacetamide was researched. Here we used falling ball viscosity to study the relations between concentration and viscosity of PSA-PEG block copolymer. Results of falling ball viscosity test are listed in Table 2. From this table, it is concluded that the most suitable spinning solution concentration was 14%~16%[11].

Table 2 Falling ball viscosity test results

	Concentration/wt%	Falling ball time/s	Solution condition
	10.0	1	S
	14.0	23	S
	14.5	28	S
PSA-PEG	15.0	32	S
	15.5	69	S
	16.0	87	S
	17.0	147	S
	20.0	384	S

Note: S means the solution flowing well.

3.4 Thermal analysis

As mentioned in introduction, PSA is a kind of heat resistance fibers, so it is necessary to characterize if the heat resistance of copolymer would decrease when PEG addition. Thermal stability of polymers was evaluated by thermo gravimetric analysis (TGA), as shown in Figure 3. And the results were compared with traditional PSA[1], as listed in Table 3. It is shown that the decomposition temperature of copolymer was a bit lower than pure PSA, while it still maintains excellent thermal stability.

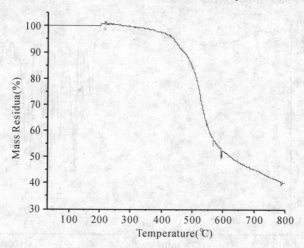

Figure 3 Thermo gravimetric analysis of PSA-PEG block copolymer

Table 3 Thermal analysis of PSA-PEG and traditional PSA

	PSA-PEG	Traditional PSA
Decomposition temperature (t_e)/℃	390.0	422.0
Heat loss rate at t_e /%	8.0	4.0
Heat loss rate at 500℃/%	28.0	25.0
Heat loss rate at 600℃/%	50.0	47.0

4 Conclusion

In this study, we have prepared successfully a novel PSA-PEG block copolymer based on polysulfonamide and amine-terminated polyethylene glycol. Compared with traditional PSA, as the flexibility of the segment increased, the solubility of copolymer in some kinds of organic solvents was improved. From falling ball viscosity experiment, it is concluded that the most suitable spinning solution concentration was 14%~16%, taking DMAc as solvent. TGA test indicates that the decomposition temperature of copolymer was a bit lower than PSA, while still maintain excellent thermal stability. The study of the crystallization and mechanical properties of the wet spinning fiber will be reported in the later work.

Acknowledgement

Project supported by the National Natural Science Foundation of China (NSFC No.50973059) and Science & Technology Commission of Shanghai Municipality (STCSM No.10JC1405200)

References

[1] L Zhang, Y B Li, J Z Liu. Research on heat resistance of polysulfonamide fibers. Journal of Tianjin Polytechnic University, 2006, 25(6): 17-19
[2] L Liu, W T Wang, C F Xiao. Preparation and Characterization of Polysulfonamide Statistical Copolymers from Three Aromatic Diamines. Journal of Shanghai Jiaotong Univ. 2010, 15(1): 114-118
[3] Y Ban, Y H Zhang, J R Ren. The Application of Polysulfonamide Fiber in High Temperature Resistant Filtration Materials. Technical Textiles. 2006, 12(11): 33-36
[4] J Chen, S K Spear, J G Huddleston, R D Rogers. Polyethylene glycol and solutions of polyethylene glycol as green reaction media. Green Chem., 2005, 7: 64 - 82
[5] M J Mahoney, K S Anseth. Three-dimensional growth and function of neural tissue in degradable polyethylene glycol hydrogels. Biomaterials, 2006, 27(10): 2265-2274
[6] S Mallakpour, F Rafiemanzelat. New optically active poly(amide-imide-urethane) thermoplastic elastomers derived from poly(ethylene glycol diols), 4,4-methylene-bis(4-phenylisocyanate), and a diacid based on an amino acid by a two-step method under microwave irradiation. Journal of Applied Polymer Science.2005, 98(4): 1781-1792
[7] J M Harris. Poly(ethylene glycol) Chemistry: Biotechnical and Biomedical Application. New York: Plenum Press, 1992: 1-10
[8] Q M Wang, S R Pan, J X Zhang. Preparation and Identification of Amino-terminated Polyethylene Glycol. Chinese Journal of Pharmaceuticals. 2003, 34(10): 490-492
[9] W H Chan, Y L Suei. Synthesis and characterization of random poly (amide-sulfonamide)s: 3 copolymers from two diamino monomers. Polymer.1995, 36(23): 4503-4508
[10] G Fan, J Zhao, Y Zhang. Grafting modification of Kevlar fiber using horseradish peroxidase. Polym Bull. 2006, 56: 507-515
[11] D R Christiaan, M Eduardo, H Boerstoel. Orientational order and mechanical properties of poly (amide-block-aramid) alternating block copolymer films and fibers. Polymer. 2006, 47(26): 8517-8526

18 Shape Memory Effect and Actuation Property of Shape Memory Polymer Based Nanocomposites

Qingqing Ni, Li Zhang

(Department of Functional Machinery and Mechanics, Shinshu University, Ueda-shi, 386-8567, Japan)

Abstract: In this paper, carbon nanotube and fullerenol nanocomposites with the matrix of shape memory polymer (SMP) were developed and their shape memory properties, EMI SE and electric-field response at different electrical voltages were examined. The strain recovery ratio after the second cycle is more than 90% and tended to be constant 95% after more cycle numbers. This phenomenon is found and called as a training effect. It is clear that the recovery stress of CNT/SMP or fullerenol/SMP nanocomposites were much larger than that of SMP bulk, especially for the nanocomposites of 3.3wt% CNT and 0.3% fullerenol weight fraction. The electrical-field response under different DC voltages was also investigated and their actuation property was partly made clear. This will be helpful for the development of innovative actuators of polymer composites.

Keywords: Shape memory polymer; Shape memory effect; Electrical-field response; Actuator; Nanocomposites

1 Introduction

Carbon nanotubes (CNTs) are receiving steadily increasing attention because of their unique structural and electrical characteristics. The tensile modulus and strength of CNTs reach 270 GPa to 1 TPa and 11-200 GPa, respectively [1-2]. The unusual electrical and mechanical properties of carbon nanotubes have motivated a flurry of interests to exploit their applications in advanced composite materials, particularly polymer based composites, to improve the performance of a matrix or to achieve new properties [3]. Recently, because various composition technologies were established and the cost of CNT decreased due to a large quantity production, the more and more applications for CNTs have been developed. Many researches on mechanical, thermal and electric characterizations of CNT reinforced polymers are reported, such as CNT/polystyrene, CNT/PVA, CNT/PVDF, CNT/PP, CNT/nylon, and CNT/epoxy etc. Some researches on the surface modification were also reported in order to improve interfacial property.

On the other hand, shape memory polymers (SMPs) have the characteristics such as large recoverability, lightweight, superior molding property and lower cost. These advantages have resulted in that the SMPs become one of functional materials with much attention from many fields. For shape memory polymer (SMP) of polyurethane series, its glass transfer temperature (T_g) may be set up around the room temperature, and its characterizations such as shape recovery and/or shape fixation may appear to be quite different at the temperature above and below T_g [4-7]. Thus the polyurethane SMP is respected to have wide applications in the field of industry as an actuation material.

In this paper, SMP nanocomposites with carbon nanotubes (CNTs) or fullerenols are developed and their unique characteristics, such as shape memory effect, and electrical-field response under different DC voltages [8], are investigated for the wider applications in mechanical and electrical/electronic fields.

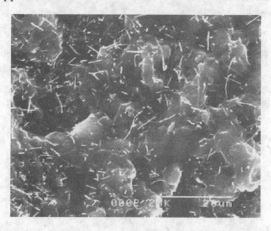

Figure 1 Scanning electron micrograph of 5.0wt% VGCF/SMP nanocomposite

2. Experimental
2.1. Materials

The carbon nanotubes of VGCFs provided from Showa Denko K.K. are used as fillers. The diameter of VGCF is about 150 nm, and the length is 10~20 μm. The polyesterpolyol series of polyurethane shape memory polymer (SMP) is used as matrix and its glass transition temperature T_g is about 45°C. The raw material is liquid. The weight ratio of polymer to solvent was set to be 3:7. SEM micrograph (see Figure 1) for CNT/SMP nanocomposite developed shows that VGCFs exhibit relatively good dispersion in SMP and are of 1~2 hundred nanometers in thickness and several microns in length. An interconnected conductive network has been formed in the materials. The fullerenol/SMP nanocomposites with fullerenol of 0.1, 0.2, 0.3wt% were also prepared.

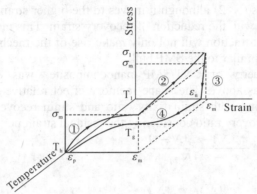

Figure 2 Three-dimensional stress-strain-temperature schematic in a thermo-mechanical cycle test (1) memorized shape after molding and cooling; (2) free deformation due to the rubber elasticity of the amorphous portion by heating over T_g under an applied force; (3) shape fixity by cooling below T_g; (4) shape recovery by heating over T_g under free load condition

2.2. Thermo-mechanical cycle test

In order to investigate the shape recovery property of the developed nanocomposites, the thermo-mechanical cycle test was conducted (see Figure 2). The specimen was pulled up to maximum strain ε_m at a constant tensile speed at the high temperature T_h above T_g (at the rubbery state) (Process 1).

Maintaining the strain at ε_m, the specimen was cooled to the low temperature T_l below T_g (at the glassy state) and kept for 20 minutes (Process 2). The specimen was unloaded at the temperature T_l (Process 3), where small unloading strain ε_u occurred. Then the specimen was heated from T_l to T_h under no-load and kept for 5 minutes (Process 4), where the strain of the specimen was recovered. This is one completed thermo-mechanical cycle.

2.3. Recovery stress test

To measure the recovery stress, the specimen was kept for 10 minutes at low temperature T_l under no-load in the process 4, and then the test distance of the specimen was adjusted to have a constant strain. After that the specimen was heated from T_l to T_h and kept for 5 minutes at T_h under a constraint strain condition and the recovery stress was measured.

2.4. Measurement of electrostrictive response

For the fullerenol/SMP nanocomposites, the electrostrictive response under different electric fields of 0~1000V was investigated by using the testing system as shown in Figure 3.

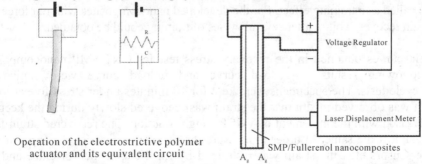

Figure 3 The testing system to measure electrostrictive response of fullerenol/SMP nanocomposites

3 Result and discussion
3.1 Shape memory property

The stress-strain curves obtained in the thermo-mechanical cycle test of the maximum strain ε_m=100% for CNT/SMP nanocomposites and SMP bulk were shown in Figure 4. Observing the same cycle number, when the VGCF weight fraction in CNT/SMP nanocomposites became high, the stress at the maximum strain became large, and the residual strain increased. The stress corresponding to any strain value during cycle loading also increased with the VGCF weight fraction, but the tendency of stress-strain curve on cycle number in CNT/SMP nanocomposites with 1.7, 3.3 and 5.0wt% of VGCFs weight fraction, respectively, is similar to that in SMP bulk. When cycle number increased the variation of stress-strain loop was large at an early stage, and it became small pronouncedly after cycle number N=2. The shape of stress-strain loop was almost the same after two cycles (N>2) although it moves to the higher strain side. The residual strain $\varepsilon_p(N)$ increased slowly accompanying with the reduction of recovery strain. This means that the addition of carbon nanotubes up to 5.0 wt% weight fraction will not only make rise of the mechanical property but also almost keep the shape fixture property similar to the SMP bulk.

The shape recovery property of CNT/SMP nanocomposites was examined from the result of thermo-mechanical cycle tests as above. The shape memory effect relative to both strain fixity and strain recovery were evaluated by using both strain fixity ratio and strain recovery ratio at cycle number N as defined by Eq.(1). The strain recovery ratio considering the residual strain per each cycle was used.

$$R_f(N) = \frac{\varepsilon_u(N) - \varepsilon_p(N-1)}{\varepsilon_m - \varepsilon_p(N-1)} \tag{1}$$

$$R_r(N) = \frac{\varepsilon_u(N) - \varepsilon_p(N)}{\varepsilon_u(N) - \varepsilon_p(N-1)} \tag{2}$$

where $R_f(N)$ and $R_r(N)$ is the strain fixity ratio and the strain recovery ratio in the cycle number N, $\varepsilon_u(N)$ is the unloading strain in the process 3 at T_l in the cycle number N, $\varepsilon_p(N)$ and $\varepsilon_p(N-1)$ were residual strain in the cycle number N and $N-1$.

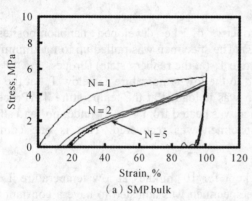
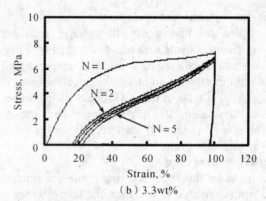

**Figure 4 Stress-strain curves in the thermo-mechanical cycle tests at for %100 m
(a) SMP bulk and (b) 3.3wt%**

The relationship between the strain recovery ratio and the cycle number is shown in Figure 5. The strain recovery ratio after the second cycle is more than 90% and tended to be constant 95% after more cycle numbers. This phenomenon is called as a training effect. For the developed nanocomposites with reinforcement of carbon nanotubes, stable strain recovery ability after several cycles of training could be obtained.

3.2 Recovery stress

The stress-strain curves obtained in the recovery stress test for CNT/SMP nanocomposites and SMP bulk shows the following results. The load curve and unload curve were similar to those in thermo-mechanical cycle tests. The specimens were kept for 10 minutes under stress free at low temperature T_l after unloading. It was observed in situ that the strain was recovered slowly during the keeping time. This recovered strain decreased with the increment of VGCF weight fraction. The recovered strain for 3.3wt% and 5.0wt% VGCF was almost the same. In order to measure the recovery stress for the application of actuator, such as temperature sensors etc., the strain was restrained by fix the specimen distance and then recovery stress was measured.

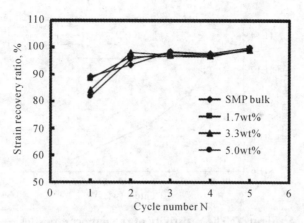

Figure 5 Relationship between strain recovery ratio and cycle number

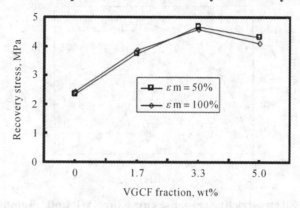

Figure 6 Relationships between recovery stress and VGCF weight fraction for CNT/SMP nanocomposites

The relationship between the recovery stress and the VGCF fraction was shown in Figure 6 and their typical values are listed in Table 1. It is clear that the recovery stress of CNT/SMP nanocomposites is larger than SMP bulk. Especially, for the nanocomposites of 3.3wt% VGCF weight fraction, the recovery stress increased about 2 times as large as that in SMP bulk. The reason for this result could be considered as that when a specimen was loaded under both constant strain and high temperature and then cooled to low temperature and unloaded, the carbon nanotubes may store elastic strain energy. When reheating the specimen, this stored elastic strain energy will release and CNT/SMP nanocomposites obtain larger recovery stress.

3.3 Resistivity

Surface resistivity can be expressed as follows.

$$\rho s = R * RCF(S) \qquad (3)$$

where, RCF is resistivity correction factor. In this study, RCF is derived from the shape of the electrode. We used a standard electrode URSS, which RCF(S) is 9.065. The surface resistance of each nanocomposite is shown in Figure 7. As CNT is quite a good conductor, the more it is added the lower the resistivity of the nanocomposites decreases. When the CNT content reaches to 2 wt%, the resistivity of the nanocomposite decreases to 1E6Ω, which is close to semiconductors. Comparing to the CNT, Fullerene is almost an insulator. For this reason, the samples with fullerene 0.1 wt%, 0.5 wt%, 1 wt%, 2 wt% show no change in their resistivity.

3.4 Electrostrictive property

Electrostrictive response were observed at 0~1000V electric field for 150 μm thickness Monomorph type electrostrictive actuator. Bending displacement of SMP and SMP/fullrenol nanocomposites at 800V were shown in Figure 8. For the normal SMP bulk sample, electrostrictive response is slow against electric field and it takes several seconds to achieve maximum displacement.

For the fullerenol/SMP nanocomposite (see Figure 8), the electrostrictive response is much quicker than that in SMP bulk although the maximum displacement reduced. At the same time, the vibration will tend to be constant at less than 2 seconds.

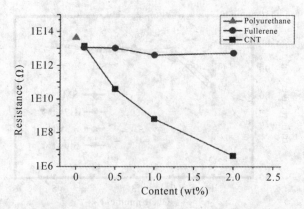

Figure 7 The resistivity of the nanocomposites

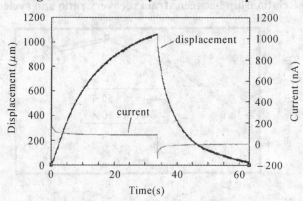

Figure 8 Electrostrictive response curve for SMP bulk sample at 800V

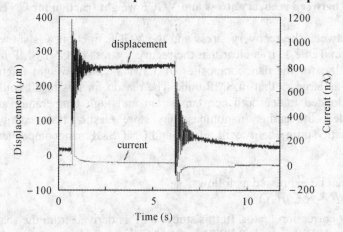

Figure 9 Electrostrictive response curve for 0.1wt% SMP/fullerenol nanocomposite at 900V

4 Conclusions

The carbon nanotube and fullerenol based shape memory polymer (SMP) nanocomposites were developed and their shape memory properties and electrostrictive property were examined. The results obtained are concluded as follows:

1. The developed nanocomposites have excellent shape recoverability. They can keep high strain recovery ability more than 90% after several cycles of training. This will lead that the developed materials can be utilized for the cycle application with any shape in daily life.

2. The recovery stress of CNT/SMP nanocomposites is much larger than SMP bulk. For the nanocomposite with 3.3wt% VGCF weight fraction, the recovery stress is about twice as large as that in SMP bulk. This characterization may be expected to the application of temperature sensor materials for CNT/SMP nanocomposites.

3. For the nanocomposite with fullerenol filler and SMP, its mechanical strength increased and the

electrostrictive response becomes much quicker than the bulk sample.

Acknowledgement

The authors gratefully acknowledge the support of the CLUSTER of Ministry of Education, Culture, Sports, Science and Technology (MEXT), Japan.

References

[1] Iijima S. Helical microtubules of graphitic carbon. Nature 1991; 354: 56-8.
[2] Subramoney S. Novel nanocarbons—structure, properties, and potential applications. Adv Mater 1998; 15: 1157-71.
[3] Baughman RH, Zakhidov AA, Heer WAD. Carbon nanotubes—the route toward applications. Science 2002; 297: 787-92.
[4] Otsuka K, Wayman C. Shape Memory Materials. Cambridge University Press; 1998 pp. 203-19.
[5] Tobushi H, Hayahi S. Properties and application of shape memory polymer polyurethane series. Res Mach 1994; 46(6): 646-52.
[6] Ohki T, Ni Q-Q, Ohsako N, Iwamoto M. Mechanical and shape memory behavior of composites with shape memory polymer. Composites Part A 2004; 35: 1065-73.
[7] Ni Q-Q, Ohsako N, Ohki T, Wang W, Iwamoto M. Development of smart composites based on shape memory polymer. International Symposium on Smart Structure & Microsystems (IS3M) 2000; 1-6(C2-1).
[8] Yiping Liu, Ken Gall, Martin L. Dunn, Patrick McCluskey. Mechanics and Materials. 2003; 36: 929-940.

19 The Relationship between the Structures and Mechanical Properties of *A. pernyi* Silk

Chengjie Fu, Zhengzhong Shao

(Key Laboratory of Molecular Engineering of Polymers of Ministry of Education, Advanced Materials Laboratory, Department of Macromolecular Science, Fudan University, Shanghai 200433, China)

Animal silks have drawn the attention of biologists, chemists, and material scientists for more than a century. The recent resurgence in scientific interest in such materials, which have evolved independently in spiders as well as a wide range of insects, is due to the silk's marvelous combination of modulus, strength and extensibility. After intensive study, the scientists have gained considerable knowledge about the primary sequences (although often only partial) of a wide range of natural silk proteins, and about the secondary structure of certain motifs. However, it is still less understood about the details of how the secondary structures arrange and organize into the condensed state structure at a higher level in the natural silks, and how these secondary structures contribute to the remarkable mechanical properties of the natural silks. It seems to still be a long way from producing artificial silk fibers comparable to those masterpieces produced by the spiders or silkworms themselves, whether the regenerated silk protein solutions or recombinant ones are used.

Compared to major ampullate silks of spider and mulberry silk, which were widely studied, *A. pernyi* silk is quite special in the family of animal silks. Like mulberry (*B. mori*) silk, *A. pernyi* silk is used to construct cocoon. However, examination of the primary structure of *A. pernyi* fibroin reveals that it is more like major ampullate spidroins than *B. mori* fibroin. Therefore, obtaining reliable mechanical properties data of *A. pernyi* silk and figuring out its relationship with the structures of *A. pernyi* silk will provide important insight into the structure-property relationship of the whole family of animal silks.

At the present, *A. pernyi* silk fibers with few macro defects were obtained by forcibly reeling from the silkworm. The mechanical properties of such silk were studied in normal and moist environment as well as at low temperatures. Also, the effects of water on the structure and properties of forcibly reeled *A. pernyi* silk were discussed. Furthermore, the failure behavior of it was investigated at both ambient temperature and cryogenic temperature.

It was found that the forcibly reeled *A. pernyi* silk fibers displayed breaking stress and toughness of the same magnitude as spider major ampullate silks and forcibly reeled mulberry silk. The other mechanical properties, such as elasticity, supercontraction, and the effect of water on modulus were between those of spider major ampullate silks and mulberry silk (Figure 1, left). Therefore, an interpretation of the connection between the primary structures of silk proteins and the mechanical properties of silks is proposed here based on the ordered fraction, which is determined by both the protein sequence and spinning process of the silk (Figure 1, right).

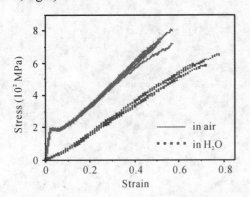 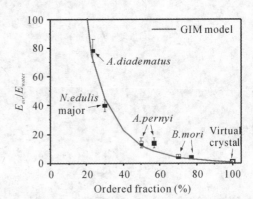

Figure 1 Left: Typical stress-strain curves of water-contracted *A. pernyi* silk fibers at wet and dry state. Right: Correlation between the ordered fraction of silks and the ratio of dry modulus to wet modulus. Half-filled symbols adopt the ordered fraction excluding serine residue, while full-filled symbols adopt the ordered fraction including serine residue.

On the other hand, dynamic mechanical thermal analysis (DMTA), tensile test in liquid nitrogen and fractography were used to explore the mechanism, by which forcibly reeled *A. pernyi* silk fiber remains tough at low temperatures. The silk breaks in a ductile way even at the temperature of liquid nitrogen, i.e. -196℃, and its breaking elongation doesn't differ from that at room temperature. The result implies that previous attribution of animal silk's low temperature toughness to the relaxation at -70℃ is unsatisfactory. Furthermore, the failure mode of forcibly reeled *A. pernyi* silk fiber in liquid nitrogen was found to depend on sericin coating, moisture content, artificial notch and so on. For example, the silk fiber breaks in semi-ductile (or brittle) way in liquid nitrogen if it is undegummed and wet or it is artificially notched. According to fractography of forcibly reeled *A. pernyi* silk fiber and the previous observation on high tenacity nylon and PET fibers, we speculate that the well-defined nano-fibrillar structure in natural animal silks is the origin of their low temperature toughness.

Acknowledgement

This work was supported by the National Natural Science Foundation of China (NSFC 20974024) and the Programme for Changjiang Scholars and Innovative Research Team in the Fudan University.

20 Flexible Tactile Sensor Based on PVDF Fibrous Membrane
==

Yongrong Wang[1], Jianming Zheng[2], Peihua Zhang[1], Wenxing Chen[3], Chunye Xu[2,4*]

(1. College of Textiles, Donghua University, Shanghai 201620, China
2. National Laboratory for Physical Sciences at Microscale, University of Science and Technology of China (USTC), Hefei 230026, China
3. Key Laboratory of Advanced Textile Materials and Manufacturing Technology, Ministry of Education of China, Zhejiang Sci-Tech University, Hangzhou, China
4. University of Washington, USA)

Abstract: Piezoelectric polymer, polyvinylidene fluoride (PVDF) film, has been widely investigated as sensor and transducer material due to its high piezo-, pyro-, and ferroelectric properties. However, there are many limitations for PVDF film as human-related tactile sensor, such as non-breathability, non-stretching, requirement of additional processes like poling, etc. In this paper, PVDF nano-fibrous membrane which was flexible, porous, wearable, was prepared by electrospinning technique. The morphology of PVDF nanofiber was determined by scanning electron microscopy (SEM), crystal structure was evaluated by FTIR spectroscopy. The PVDF fibrous membranes were well designed and fabricated for tactile sensor. We found the feasibility of applying PVDF fibrous membranes prepared through electrospinning technology, to be flexible human-related tactile sensors.

Keywords: Tactile sensor; PVDF; Electrospinning; Electroactive polymer

1 Introduction

Poly(vinylidene fluoride) (PVDF) is a chemically stable piezoelectric polymer with high piezo-, pyro- and ferroelectric properties [1]. PVDF film under poling is most widely used in transducers and sensing mechanisms for its inexpensive, biologically compatible and excellent sensitivity particularly in harsh and biological environment [2,3].

In recent years, varied technologies have been developed to process nanoscale PVDF fibrous membranes. Electrospinning is a simple, efficient technique of producing PVDF fibrous membranes [4]. PVDF nano-fibrous membranes are reported for various applications [5-7], such as biomedical materials, polymer Electrolytes, filtration membranes, etc. Little information reported about the application for tactile sensor [8].

The objective of this study is to investigate piezoelectric response of PVDF nano-fibrous membranes, and develop human-related flexible tactile sensor using PVDF nano-fibrous membranes which are light weight, flexible, breathability, and wearable, etc. The optimized process parameters during electrospinning were obtained, the morphology and distribution of PVDF nanofiber were discussed, and the crystalline structure of PVDF nanofiber was evaluated. The fabrication of tactile sensor is in the preliminary phase where the feasibility of the approach is evaluated by the piezoelectric characterization on the home-made system.

2 Experiment

2.1 Materials

Poly(vinylidene fluoride) (PVDF) with molecular of 534,000 (g/mol), N,N-Dimethylformamide (DMF), and acetone were purchased from Sigma Aldrich (U.S.A.).

2.2 Solution preparation

Measured amount of PVDF mixed with DMF/Acetone (weight ratio is 2:3) co-solvent was continuously ultrasonic stirred at 40°C until a transparent and homogeneous polymer solution is formed, and then cooled to ambient temperature.

2.3 Electrospinning process

The Nanofiber Electrospinning Unit was purchased from KESKATOTECH CO., LTD. 12% PVDF was electrospun with a voltage range of 9-18kV and a flow rate range of 0.030-0.135mm/min. A grounded copper flat plate collector located in front of the needle-tip at a constant distance of 15cm, and the electrospinning process was continued for around 4h.

---

* Corresponding author's email: chunye@ustc.edu.cn

2.4 Characterization

The morphologies of electrospun PVDF fibers were observed using scanning electron microscope (FEI Sirion SEM). Each sample was sputter-coated with gold for analysis.

Unpolarized Fourier transform infrared-transmission spectroscopy (FTIR-TS, 100 scans, 4cm^{-1} resolution in the range of 1600-400cm^{-1}) was collected using Bruker-IFS66Vspectrometer.

2.5 Tactile sensor measurement

Figure 1 shows the experimental setup for the characterization of the PVDF tactile sensor. It consists of two components, hammer and signal collection. A home made hammer, which was managed by a controller, hit the tactile sensor at a certain frequency. The probe of load cell, having a sensitivity of 22.5mV/N, was attached to the tip of hammer to measure the force input to the tactile sensor, and the force from the load cell was from 3N to 4.5N. The peak-to-peak voltage signals output from the load cell and tactile sensor were measured and recorded using the oscilloscope (Tektronix TDS 540A).

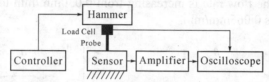

Figure 1 The experimental setup for the characterization of the PVDF tactile sensor

3 Results and discussion

3.1 Morphology and diameter distribution of PVDF nanofiber

In this study, 12% PVDF solution was electrospun with a voltage range of 9-18 kV and a flow rate range of 0.030-0.135mm/min, perfect fibers formed. In Figure 2 and Figure 3 A, B, C and D show the SEM images of electrospun PVDF membranes. It is found that the diameter of the PVDF fibers did not change significantly with the varying voltage and flow rate, the diameters of the PVDF fibers were in the range of 20nm to 800nm. However, the distribution of diameters was affected by these processing conditions. Figure 2 shows how the diameter distribution was effected by the applied voltage, a narrower diameter distribution occurs at the voltage of 12kV. Other changes in diameter distribution, due to the charge interactions when the flow rate was varied, are shown in Figure 3. It can be concluded that there is a narrower diameter distribution at flow rate of 0.065mm/min while the applied voltage is 12kV.

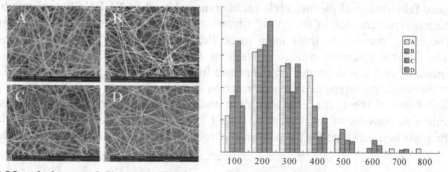

Figure 2 Morphology and diameter distribution of PVDF nanofiber at flow rate 0.065mm/min, traveling distance of 15cm, and voltage of A) 9kV, B) 12kV, C) 15kV and D) 18kV

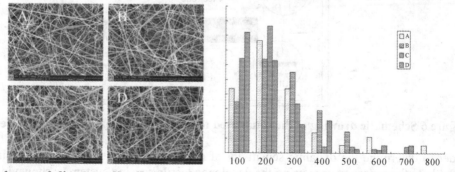

Figure 3 Morphology and diameter distribution of PVDF nanofiber at voltage of 12kV, traveling distance of 15cm, and flow rate of A) 0.030mm/min, B) 0.065mm/min, C) 0.100 mm/min and D) 0.135 mm/min

3.2 Crystalline structure of PVDF nanofibers

Although there are five different crystalline structures in PVDF, only the β-phase structure has the strong piezoelectricity and ferroelectricity. Therefore it is crucial to understand the crystalline structure of electrospun nanofibers of PVDF.

FTIR spectroscopy was used to characterize the crystal structures of the electrospun PVDF nanofiber because different crystal phases show different characteristic aborpsions in IR spectroscopy, Figure 4 shows FTIR spectra of electrospun PVDF fibrous membrane. The characteristic bands were observed at 613, 763cm^{-1} for α-phase, at 474, 510, 1276cm^{-1} for β-phase.

FTIR spectra confirm that electrospun PVDF nanofibers manufactured by electrospining have a predominantly β-phase structure. It means that the α-phase of the raw PVDF is converted into β-phase during electrospinning. For the same collector distance, it was observed that as the applied voltage increases, the most β-phase formation in the electrospinning fibers when the voltage is 12kV. As the Figure 4 suggested, when the voltage is 12kV, the flow rate is increasing from 0.030mm/min to 0.135mm/min, the β-phase peak is most when the flow rate is 0.065mm/min.

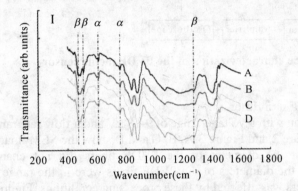
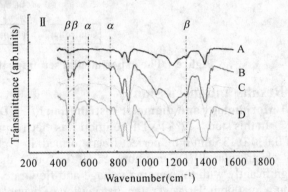

Figure 4 FTIR spectra of electrospun PVDF fibrous membrane prepared under different electrospinning conditions: (Ⅰ) at different voltage: (A) 9kV, (B) 12kV, (C) 15kV and (D) 18kV; (Ⅱ) at different flow rate: (A) 0.030mm/min, (B) 0.065mm/min, (C) 0.100mm/min and (D) 0.135mm/min

3.3 Design and fabrication of piezoelectric tactile sensor based on PVDF fibrous membrane

A schematic representation of the PVDF fibrous membrane setup used as a tactile-sensor was shown in Figure 6. The sensor consists of three main parts: flexible electrode, PVDF fibrous membrane, and fixed electrode. The flexible electrode consists of Indium Tin Oxide (ITO) layer and plastic layer, ITO as the conducting material first was deposited on the plastic film as the top electrode, the plastic film as an insulator film protests the whole prototype avoiding destruction when the hammer hits. PVDF nano-fibrous membrane as the tactile-sensing element, connected with top and bottom electrodes by conductive silver painter. The rigid electrode also consists of two layers, ITO and glass. ITO was deposited on the glass as the bottom electrode, the glass as a substrate for the bottom electrode and a platform supports the whole prototype.

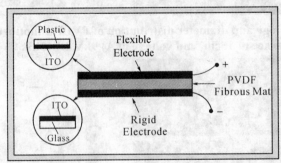

Figure 6 Schematic drawing of the setup used to measure the piezoelectric response of the sensors

In order to determine the dynamic sensitivity of each sensor, input forces at a certain frequency were applied directly to the sensor surface using the developed tactile sensing system as shown in Figure 1.

Table 1 and Table 2 show the testing results for the PVDF tactile sensors prepared under different electrospinning conditions. Piezoelectric response for the electrospun samples showed a correspondingly varied trend with β-phase content, also with the applied voltage and flow rate. The sample prepared under the electrospinning condition of 12kV applied voltage and 0.065mm/min flow rate has the highest sensitivity of 42.00mV/N, which is in good agreement with the results from Figure 4.

Table 1 Piezoelectric characteristic of the PVDF tactile sensors prepared at different electrospinning voltage

Sample #	Input Load (N)	Voltage (mV)	Sensitivity (mV/N)
A	4.44	130	29.25
B	3.33	140	42.00
C	3.56	135	37.97
D	4.44	107	24.08

Note: A) 9 kV, B) 12 kV, C) 15 kV, D) 18 kV.

Table 2 Piezoelectric characteristics of the PVDF tactile sensor prepared under different electrospinning flow rate

Sample #	Input Load (N)	Voltage (mV)	Sensitivity (mV/N)
A	3.56	135	37.97
B	3.33	140	42.00
C	4.44	105	23.63
D	4.44	50	11.25

Note (unit: mm/min): A) 0.030, B) 0.065, C) 0.100, D) 0.135.

4 Conclusions

In this study, we designed and fabricated tactile sensor based on piezoelectric PVDF nano-fibrous membrane. The morphology of the PVDF nanofiber, the crystal structure and piezoelectricity of the PVDF tactile sensor were investigated and reported. The major findings are summarized below:

Electrospinning technology was used to prepare the PVDF nanofiber with different applied voltage and flow rate. It is found that the diameter of the PVDF fibers did not change significantly. However, the distribution of diameters was affected by these processing conditions. The FTIR spectroscopy was used to characterize the crystal structures of the electrospun PVDF nanofiber, the data indicated that the initial α-phase of the raw material was converted to β-phase PVDF during electrospinning. It is also suggested that there was an optimized conditions for the most β-phase. The piezoelectric response of the PVDF nano-fibrous membrane is recorded, and the results are consistent with the β-phase formation.

By proper design and fabrication, a flexible tactile sensor could be achieved based on this PVDF nano-fibrous membrane, which would be a good candidate for human-related tactile sensor and other electronic applications.

References

[1] Nalwa H. S. Ferroelectric polymers: chemistry, physics, and applications. New York: Marcel Dekker Inc., 1995
[2] Dargahi J, Parameswaran M, Payandeh S. A micromachined piezoelectric tactile sensor for an endoscopic grasper - theory, fabrication and experiments. J. Microelectromech. Sys., 2000, 9: 329-335
[3] Lang S B, Muensit S. Lesser-known piezoelectric and pyroelectric applications of electroactive polymers. in: Proc. Mater. Res. Soc. Symp., 2006: 01.1- 01.12
[4] Zheng J F, He A, Li J X, Han C C. Polymorphism Control of Poly (vinylidene fluoride) through Electrospinning. Macromol. Rapid Commun, 2007, 28: 2159–2162
[5] Fang J, Niu H T, Lin T, Wang X G. Applications of electrospun nanofibers. Chinese Science Bulletin, 2008, 53: 2265-2286
[6] Choi S W, Kim J R, Ahn Y R, Jo S M. Characterization of electrospun PVDF fiber-based polymer electrolytes. Chem. Mater, 2007, 19: 104-115
[7] Gopal R, Kaur S, Ma Z W, Chan C, Ramakrishna S, Matsuura T. Electrospun nanofibrous filtration membrane. J. Membr. Sci., 2006, 281: 581-586
[8] Yoon S, Prabu A A, Ramasundaram S, Kim K J. PVDF Nanoweb Touch Sensors Prepared Using Electro-Spinning Process for Smart Apparels Applications. Advan. Sci. Tech, 2008, 60: 52-57

21 Modification of Wool Fiber Using Freeze Treatment

Zhengwei Wang, Ruoying Zhu*, Jianfei Zhang

(School of Textiles, Tianjin Polytechnic University, 63 Chenglin Road, Hedong District, Tianjin 300160, China)

Abstract: Wool fibers were modified by freeze treatment in this study. Chemical and mechanical analyses of the treated samples compared with the untreated one are reported. X-ray investigation showed that the crystallinity of wool fiber with freezing treatment was decreased, assumed to be fiber swollen caused by water migration and ice crystal growth during freezing treatment. Different samples were dyed to investigate the influence of freezing on dyeing properties of wool, the results showed that both dyeing uptake and dyeing speed of freezing-treated fabric were increased evidently compared with the untreated one. The mechanical properties test revealed that tensile strength of treated fabric has no marked change while elongation was increased. Freezing treatment also led to increase in moisture regain and solubility in caustic solution of wool fiber. In addition, the effect of pretreatment of wool fabric with freezing treatment and subsequent protease modification or oxidation with hydrogen peroxide was investigated, experimental results suggested that wool fiber after freezing treatment may be more active to chemicals, and freeze treatment may be a potential effective way for wool modification.

Keywords: Modification; Wool; Freezing treatment; X-ray; Dyeing percentage

1 Introduction

Wool is a high-quality protein fiber and is widely used as a high-quality textile material. Since the very surface of wool fiber is highly hydrophobic because it was covered with cuticle cells, chemicals as well as dyestuff is hard to diffuse into the fiber in absence of any treatments that are usually based on partial removal, softening or uniform coating of the cuticle scales, such as pretreatment of oxidizing agent or chlorine compounds or heated in boiled water [1]. For ecological reasons, physical modifications of wool fiber, for examples, UV radiation, plasma treatment as well as steam exploration, are absorbing considerable research interests for its less chemicals involved compared with chemical ways. We have recently applied freezing treatment to get some desire changes of some wool properties. Wool fiber was a solid-liquid system for the nature of its high water regain, which was expected swell and had a loose construction after frozen due to phase transformation of water held inside the small pores in wool. It was assumed that the size and amount of small pores original existed in wool fiber will increase caused by water migration for ice crystal growth during freezing treatment, and expected a higher adsorption, and less diffuse barrier for chemicals and dyestuff. Previous work studied the effect of cooling rate on freezing of wool-water system using DSC at different cooling rates, and swollen wool is assumed to be a dispersion system consisting of domains defined as structurally favored sites, permitting water migration for ice crystal growth [2]. In this paper we study the effect of freezing treatments of wool on some of its physical and mechanical properties, as well as influence of dyeing properties. In addition, after treatment with a protease or an oxidizing agent based on freezing treatment of wool fabric were also studied. The results of experiments showed freezing treatment may be a very effective method in modifying the properties of wool.

2 Experimental

2.1 Materials

Gray wool valetin was used in the experiments. The average diameter of the wool is about 23.78 μm. The wool samples were treated by freezing 24 hours in home refrigerator.

2.2 X-ray diffraction analysis

The sample (powder) was dispersed onto a stub and placed within the chamber of analytical X-ray powder diffractometer. The sample was then scanned from $2\theta=5\text{-}45°$ [3].

2.3 Test of moisture regains

Moisture regain of the samples under standard conditions (RH 65% and 25℃). The moisture regain of the samples was determined by Moisture regain (%) = $(W_2-W_1)/W_1 \times 100$, where W_1 and W_2 represented

---

* Corresponding author's email: ruoyingzhu@hotmail.com

the dry weight and conditioned weight of the samples, respectively.

2.4 Test of mechanical properties

Tensile property of wool fabric was measured on fabric tensile tester (model YG 065), adjust speed rate to stretch breaking in the 30±5s [4].

2.5 Test of solubility in caustic solution

The samples were dissolved in an aqueous solution of 2.5% NaOH at 65℃, residual sample were fully washed and dried to constant weight. The solubility in caustic solution of the samples was determined by solubility in caustic solution (%) = [W_1 × (1−G) −W_2] / [W_1 × (1−G)] × 100, where W_1, W_2 and G represented the sample weight, residual sample weight and sample moisture [5].

2.6 Determination of dyebath exhaustion

The samples were dyed with neutral dyes (F-GS), acid dyes (T-W) and lanasol dyes (6G) respectively. Dyebath exhaustion was measured by sampling the dyebath solution before and after the dyeing process. The absorbance of the dye solution was measured using a UV–vis spectrophotometer and the percentage of dye uptake (F) was calculated using Eqn (1) [6, 7]:

$$F = (1 - A_t/A_0) \times 100\% \qquad (1)$$

where A_0 and A_t represent the absorbance of dye solution before and after the dyeing process respectively.

2.7 Treated by hydrogen peroxide and protease

The samples were treated 60min by hydrogen peroxide under the condition of H_2O_2 (33%, w/v) 10 ml/L, temperature 50℃, liquid ratio 1:30 [8]. The samples were treated 60min by protease savinase under the condition of savinase 2%owf, temperature 60℃, and liquid ratio 1:30 [9].

3 Results and discussion

3.1 X-ray analysis

Results of X-ray diffraction of the samples are shown in Figure 1. The presence of two peaks at 11° and 22° 2θ, corresponding to the α and β –sheet crystalline respectively [3], were observed for the two samples. Compared with the control sample, the intensity of the two peaks decreased after freezing treatment, suggested that the degree of crystallization of treated sample was decreased. This may be attributed to movement of the water immigration and ice crystal growth inner the wool fiber during freezing treatment. When ice crystals grow and fibers swell, most of the hydrogen bonds in the fibers were broken, and some amorphous and crystalline regions were then destroyed.

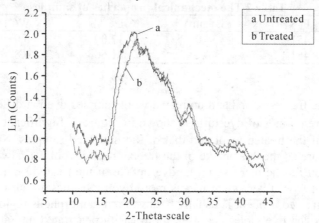

Figure 1 X-ray diffraction intensity curves of the samples

3.2 Moisture regain and Solubility in caustic solution

Moisture regain is one of the important parameter of the amorphous region of the textile fiber, the area accessible for water molecules of textile fiber increased while the amorphous region content of fiber increased [10]. The results of moisture regain test are shown in Table 1. The moisture regain of freeze-treated samples is greater that that of untreated, this is attributed to ability for water adsorption improved due to the surface area of wool fiber increasing when fiber become more looser and porous after freezing treatment. Similar trend is also observed in Figure 2, that is the solubility in caustic solution of treated fabrics increases significantly compared with the untreated one, since the amount of alkaline absorbed by fiber increases as the fiber become more looser, and the destroy effect of alkaline on wool protein molecules become more serious.

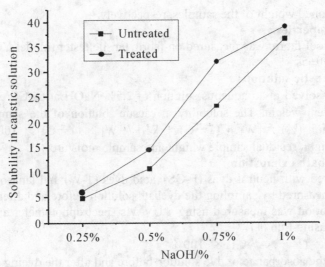

Figure 2 Test of the solubility in caustic solution (60℃, 60min)

Table 1 The moisture regain of samples

Sample	W_2(g)	W_1(g)	R (%)
Untreated	2.48	2.00	18
Treated	2.10	1.60	25

3.3 Mechanical properties

Table 2 shows that the fabric tensile strength remained unchanged and the ultimate elongation increased, which illustrate the abrasion resistance increased after the treatment. Freezing treatment resulted in the fiber swollen, the hydrogen bond and the ionic bond were destroyed, and however, the macromolecular main chain was not affected. Consequently, those lead to the macromolecular main chain slippage more easily and the abrasion resistance increased after the treatment.

Table 2 The mechanical properties of samples

Sample	F(N)	L (mm)	L (%)	W(J)	T(S)
Untreated	451	34.0	17.00	6.0	20.43
Treated	451	43.2	21.60	6.8	25.92

3.4 Dyeing behavior

The dye uptake of the freeze-treated and untreated wool samples dyed at different temperatures of 65℃ and 95℃ with three different kinds of dyestuffs is shown in Figure 3. Significant differences were observed between the dye uptake of the treated wool and that of the untreated samples. No matter what kind of dye used and dyeing temperature at, the dye uptake of the freeze-treated wool was always greater than that of the untreated wool. The reason for the increase in the dye uptake of the treated samples is considered to be the decrease of crystallinity of wool fiber, witch means more dye molecules are absorbed by treated wool fiber compared with that of untreated, since it is usually believed only amorphous region of fiber is accessible for dyestuff molecules. Moreover, dye molecules were more easily penetrated into and diffused in treated fiber because fiber is more porous and looser after treatment, therefore, freeze-treated wool have higher dyeing speed and dyeing percentage even in lower dyeing temperature.

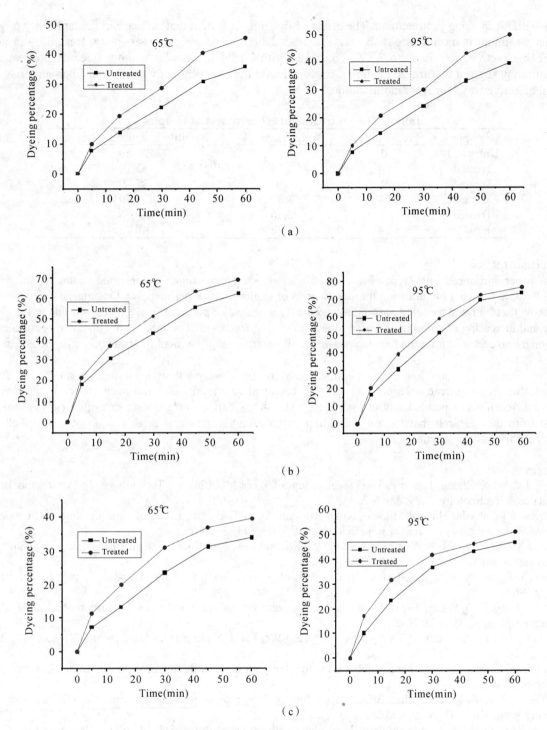

Figure 3 Test of the dyeing percentage at the different temperatures: (a) neutral dyes (F-GS); (b) acid dyes (T-W); (c) lanasol dyes (6G)

3.5 Pretreatment effectiveness analysis

Enzyme modification is an effective way to impart wool fabric some desire properties such as resistance to felting or pilling [8]. Because enzyme is a kind of large molecules, and can only be adsorbed on the fiber surface due to the scale layer of the wool fiber's surface, pretreatment of oxidizing agent or chlorine compounds to destroy the integrity is necessary for the following enzyme degradation [11]. The effect of pretreatment of wool fabric with freezing treatment and subsequent protease modification or oxidation with hydrogen peroxide was investigated and summarized in Table 3. It can be observed from Table 3, weight loss of samples treated with enzyme or hydrogen peroxide with freezing pretreatment is always higher that of samples without freezing pretreatment. The effect of freezing treatment that increase in susceptibility of wool to further treatment is more obvious for enzyme degradation, since weight loss of wool treated by protease with freezing pretreatment is up to 7.33% compared with 0.15% of samples without freezing pretreatment. This result may suggest that freeze treatment could increase the chemical activity of wool fiber, and may be a potential effective ways for wool modification.

Table 3 The pretreatment effectiveness of samples

Sample	Freeze time/h	Weight loss /%	Solubility in caustic solution
Untreated	0	0	23.59
Treated	24	0.15	33.10
Treated	0	1.1	25.09
+ protease	24	7.33	38.79
Treated	0	2.36	30.58
+ H_2O_2	24	4.78	40.23

4 Conclusions

This paper introduced a new type of wool modification technology, namely the freezing treatment, and proved that it is feasible for wool modification through a series of comparative experiments and testing. The experimental results show that both the moisture absorption and the dyeing uptake percentage are increased due to the fiber's swelling and increasing of amorphous region after the freezing treatment. The fabric strength is almost unchanged while elongation deceased after the freezing treatment, illustrating improvement of abrasion resistance of treated fabric.

Wool fiber is a highly hygroscopic, while the wool fiber's crystallinity and the intermolecular forces decreased after freezing treatment, inducing a higher chemical activities, which has been proven in the research that applied freezing as a pretreatment of hydrogen peroxide oxidation and protease enzyme degradation. It is revealed from the research that freezing treatment is an effective physical modification of wool fiber with environment friendly properties.

References

[1] X. F. Lv, W. X. Zheng, L. wang, The Development of Wool Modification Technology [J]. Progress in Textile Science & Technology, 2009, (2): 5-7.
[2] Hidenori Takahashi, Hiroshi Mitomo, Kozo Arai and Shoji Takigami, Effect of Cooling Rate on Freezing of Wool-Water Mixtures [J].Textile Research Journal, 2003, (73):597-605.
[3] W. L. Xu, G. Z. Ke, J. H. Wu, X. G. Wang, Modification of wool fiber using steam explosion [J]. European Polymer Journal, 2006, (42): 2168–2173.
[4] Y. H. Yang, S. T. Yang, J. Z. Liu, Control system of wool drawing machine [J]. Wool Textile Journal, 2007, (2): 58-60.
[5] G. H. Cheng, J. Q. Wang, The difference solubility between cashmere and wool in solution of alkali [J]. Wool Textile Journal, 2002, (2): 20-22.
[6] X. Hao, H. M. Liu, Study of Technology of Dyeing Wool with Neutral Dyes [J]. Liaoning Tussah Silk, 2001, (3): 23-24.
[7] D. Y. Yao, L. H. Jiang, Q. Zhang, Study on the low temperature new dyeing process for wool [J]. Wool Textile Journal, 2005, (1): 25-27.
[8] L.Yao, Y. W. Zhu, J. C. Zhu, Effect of oxidant and proteolytic enzyme treatment on the structure and properties of wool [J]. Wool Textile Journal, 2009, (6): 32-35.
[9] Y. Li, X. L. Ding, W. Li, Finishing of Wool fibers with savinase protease [J]. Wool Textile Journal, 2002, (6): 31-34.
[10] A. H. Guan, J. F. Zhang, Research and Contrast of the Moisture Property between Milk Protein Fiber and Wool Fiber [J]. Cotton Textile Technology, 2007, (12): 728-730.
[11] D. M. Zhou, J. P. He, Review of Wool Modification by Protease Finishing [J]. Textile Dyeing and Finishing Journal, 2006, (8): 12-14.

22 Preparation and Properties Research of Poly(lactide-co-glycolide)/Silk Blend Nanofibrous Membrane

Hongwei Xiao[1], Jie Xiong[1], Ni Li[1], Hongping Zhang[1], Junjun Xie[2]

(1. Key Laboratory of Advanced Textile Materials and Manufacturing Technology, Ministry of Education of China, Zhejiang Sci-Tech University, Hangzhou 310018, China
2. Clinical Medical College, Hangzhou Normal University, Hangzhou 310036, China)

Abstract: Poly(lactide-co-glycolide)/silk blend nanofibrous membrane was fabricated by electrospinning of PLGA/SF blend solutions dissolved in 1,1,1,3,3,3-hexafluoro-2-propanol(HFIP). The morphology, molecule structure, mechanical and hydrophilic properties of PLGA/SF blend membrane were investigated by FE-SEM, FT-IR, Multi-Purpose Tensile Tester and Drop Shape Analyzer. The results showed that the fiber diameter decreased and uniformity changed for the better with the increase of SF. There were miscibility between the molecules and hydrogen bonding interactions between PLGA and SF but the mechanical properties of blend membrane became very poor when the content of SF was more than 40%. Nevertheless hydrophilic properties of blend membrane was improved markedly because of SF.

Keywords: Poly(lactide-co-glycolide); Silk; Electrospinning; Nanofibrous membrane

1 Introduction

Tissue engineering material is one of alive tissue in vitro which is researched and developed by life sciences and engineering science[1]. The key of tissue engineering is the 3D composite construction of cell and biomaterial which can capture nutrition, exchange gas for cell. Non-woven nanofibrous membranes prepared by electrospinning possess the nanostructure of natural extracellular matrix(ECM) and can imitate ECM. This kind of material may be the possible candidates of tissue engineering scaffold for future applications[2].

Regenerated silk fibroin fiber as a natural protein fiber has excellent properties besides the physical characteristics (such as, light weight, high tensile strength, thermal stability), also has good biocompatibility, oxygen and water vapor permeability, biodegradability, weak inflammatory response[3]. Now the researcher has prepared electrospun silk fibroin fiber scaffold used for bone, ligaments, tendons, blood vessels and cartilage tissue repair successfully[4]. Electrospinning protein fiber is soluble in water and the elasticity will be lost after methanol treatment which limits its practical application.

PLGA has good biocompatibility, biodegradability and mechanical properties and its degradation rate can be adjusted but the biocompatibility and biological activity of PLGA have a certain gap compared with the natural macromolecular materials. Anything else it has poor hydrophilicity and cell adsorption capacity is weak, lack of cell recognition sites[5].

In this article we prepared PLGA/SF blend nanofibrous membrane by electrospinning, hoping to combine their good mechanical properties and biological properties and characterized the morphology, structure and mechanical properties of fibrous membrane.

2 Experimental part

2.1 Reagents and Instruments

PLGA(LA:GA=75:25) molecular weight 10,0000, viscosity IV(dl/g)range: 0.15~3.0(Jinan Dai Gang Biological Material Co., Ltd.), silk, Hexafluoroisopropanol (HFIP, Yancheng Dong Yang Biological Products Co., Ltd.), methanol(analytically pure, Hangzhou Gao Jing Fine Chemical Co., Ltd.), high-voltage power supply (U.S. Glassman Corporation), micro-injection pump (KDS-220 type, U.S. KDS Scientific Inc).

2.2 Electrospun SF/PLGA nanofibrous membrane

PLGA and SF were dissolved in HFIP with a mass ratio of 10:0, 8:2, 6:4, 5:5 and 0:10 respectively to prepare spinning solutions with the solution concentration of 10%. Electrospinning parameters were the applied voltage 15kV, the spinning rate 0.01ml/min, the receiving distance between the copper collecting plate and the needle tip 12cm. The blend fibrous membranes from the plate receiving device were treated in methanol water solution with volume ratio of 90:10 and then dried for 48h in the vacuum.

2.3 The structure and performance of PLGA/SF blend fibrous membrane

The morphology of PLGA/SF fibrous membrane was tested by Hitachi S4800 field emission scanning electron microscopy (FE-SEM) after gold plating and the fiber diameter was analyzed by Image-Pro Plus image analysis software. We used Image-ProNicolet5700 Fourier transform infrared spectrometer (FT-IR) to analyze the structure changes of blend fibrous membrane. Mechanical properties of random nanofibrous membrane were determined by KES-G1 Multi-functional tensile tester (Japan Kato-Tech Company) at the elongation rate of 0.2 mm/s. The surface contact angles of both pure PLGA and PLGA/SF blend fibrous membranes were measured by Drop Shape Analyzer (DSA-10, Germany Kruss188 company).

3 Results and discussion
3.1 PLGA fibrous morphology with different ratios of SF

Figure 1 SEM micrographs of PLGA/SF blend fibrous membrane

Figure 1 shows the morphology of PLGA and SF blend fibrous membrane under the same spinning conditions. We can see the morphology of the pure PLGA and SF fibers is different mainly in the fiber diameter from A and E. The average diameter of pure PLGA and SF is 1026nm and 532nm respectively. So the diameter of PLAG fiber is much larger than SF, whereas the average diameter of PLGA and SF blend spinning is smaller than two, especially the average diameter will reduce to 273nm when PLGA/SF blending ratio is 6:4 such as C shown. This may be due to the formation of intermolecular hydrogen bond between PLGA and SF, which made the composite fibrous diameter decreased by means of the total synergistic effect. But when the SF content reach 50% the fiber average diameter is between pure PLGA and SF, and the uniformity of fiber is poor in Figure D as Zhang [6] described PLA and SF blend spinning. This illustrates that there will be a severe phase separation between PLGA and SF when the content of SF is 50%.

3.2 Structure of composite fiber

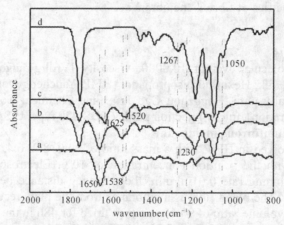

Figure 2 The infrared spectra of PLGA/SF composite nanofiber
a-PLGA/SF(0:10) b-PLGA/SF(5:5) untreated by methanol solution c- PLGA/SF(5:5) treated by methanol treatment d- PLGA/SF(10:0)

In Figure 2 comparing infrared spectra a and b, we found the addition of PLGA did not influence on the random coil structure of the silk fibroin molecular at Silk I (1650cm^{-1}) and Silk II (1538cm^{-1}) in PLGA/SF composite fiber, but there was a absorption peak change in Silk III (moved from 1230cm^{-1} to 1267cm^{-1}), which showed the change of molecular structure from random coil to β sheet. This may be due to the formation of hydrogen bonds between the ester group of PLGA and the amide group of fibroin, then made the molecular chain of fibroin being ordered in a certain extent [7]. Compare b and c there was a marked β transition in Silk I and Silk II of fibroin, while there existed both characteristic peaks of SF and PLGA showing the formation of PLGA and SF compound.

3.3 Mechanical properties of Silk/PLGA composite fibrous membrane

The fracture elongation of PLGA/SF composite fibrous membrane reduced significantly compared with pure PLGA however the fracture stress and Young's modulus increased in Figure 3 and Table 1. Compared with pure PLGA the fracture stress and strain changed little when the ratio PLGA and SF was 8:2. This is because PLGA is a continuous phase, while the small amount SF as a dispersed phase in the composition. So PLGA played a major role in the drawing process. We found the stress and strain of fibrous membrane changed greatly when SF content increased to 40%, especially the strain declined sharply. One reason was the intermolecular hydrogen bonds between SF and PLGA, another there was the crystallization of SF molecule after methanol treatment. These reasons would limit movement of PLGA molecular during stretch to reduce the strain greatly. The stress increased due to the emergence of large-scale crystal structure fracture. We continued to increase the content of SF to 50%, the stress-strain curve of blend membrane was similar to pure SF and it showed the SF played a leading role in the blend membrane. These results indicate that different tissue engineering scaffold material can be prepared by mixing the biocompatible synthetic polymers such as PLA, PCL, PLGA, PGA and natural polymers with a proper ratio [8-10].

Table 1 Mechanical properties of PLGA/SF nanocomposite membrane

PLGA/SF (ratio)	Stress (MPa)	Strain (%)	Young's modulus (GPa)
10:0	23.38	288.35	0.318
8:2	24.77	227.55	0.542
6:4	26.73	42.65	0.730
5:5	28.11	4.70	1.164
0:10	31.27	4.00	1.284

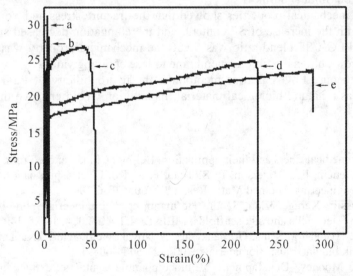

Figure 3 Stress/strain curve of electrospun PLGA/SF nanocomposite membrane
a)PLGA/SF(0:10) b)PLGA/SF(8:2) c)PLGA/SF(6:4) d)PLGA/SF(5:5)
e)PLGA/SF(10:0)

3.4 The hydrophility of PLGA/SF blend membrane

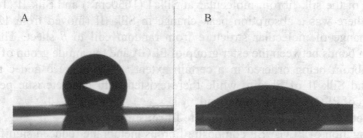

Figure 4 Photographs of water droplets were also taken immediately after contacting (A-pure PLGA nanofibrous membrane, B-PLGA/SF60:40 nanofibrous membrane)

PLGA poor hydrophilicity will limit its application as a cell scaffold material for tissue engineering. To solve the problem Wan and Park [11, 12] used plasma surface treatment method to modify PLGA fibrous membrane surface and cultured mouse 3T3 fibroblasts on the treated PLGA. The surface hydrophilicity of the composition would improve greatly when SF was added to PLGA. Figure 4 is a comparison of surface contact angle between pure PLGA and PLGA/SF blend fibers. We know the contact angle of pure PLGA fibrous membrane is 123°, and it dropped to 30° when SF was added to 40% in PLGA fibrous membrane from Figure 4. The result shows the preparation of PLGA/SF composite fiber can overcome the hydrophobic defects of pure PLGA fiber.

4 Conclusion

This article prepared PLGA/SF blend fibrous membrane with different ratios by electrospinning, and characterized the properties of composite fiber by FE-SEM, FT-IR, Mechanical Stretch and Contact Angle. The results were as follows:

(1) The diameters of PLGA/SF blend fibers prepared by electrospinning were smaller and more uniform than that of pure PLGA and SF. However, the fiber diameter was similar to the pure SF when SF content increased to 50%.

(2) FT-TR analysis showed the presence of PLGA in blend membrane could make Silk III do β transition but there was no effect on Silk I and Silk II.

(3) The results of mechanical properties showed that the fracture stress and Young's modulus of blend membrane increased with the increase of SF content, and the elongation decreased significantly at the same time, especially when PLGA/SF blend ratio was 50:50 the mechanical property was similar to the pure SF and yet the brittle nature would make fibrous membrane to lose applying value.

(4) Compared with pure PLGA fibrous membrane, the hydrophilicity of PLGA/SF blend fibers was better. In addition, SF as a natural biological material, its addition would enhance the biological activity of scaffolds greatly.

References

[1] Xin Yuan Shen. Biomedical fibers and their application. Beijing: China Textile Press, 2009: 4-6.
[2] W. J. Li, C. T. Laurencin, E. J. Caterson, R. S. Tuan, F. K. Ko. Electrospun nanofibrous structure: A novel scaffold for tissue engineering. Biomed Mater Res, 2002, 60: 613-620.
[3] Lorenz Meinel, Vassilis Karageorgiou, Sandra Hofmann, et al. Engineering bone-like tissue in vitro using human bone marrow stem cells and silk scaffolds. MEINEL ET AL, 2004, 12:25-34.
[4] Yongzhong Wanga, Hyeon-Joo Kima, Gordana Vunjak-Novakovic, David L. Kaplan. Stem cell-based tissue engineering with silk biomaterials. Biomaterials, 2006, 27:6064-6082.
[5] B. S. Kim, D. J. Mooney. Development of biocompatible synthetic extracellular matrices for tissue engineering. REVIEWS, 1998, 16:224-230.
[6] Feng Zhang, Yougang Dai, Paul Baoqi Zuo, Lun Bai. Morphology and Thermal properties of SF/PLA nanofibe by electrospun. Polymer Materials Science and Engineering, 2009, 25 (5): 75-78.
[7] Xiaoying Wang, Donghao Sun, Zhengyu Wu. Preparation of SF/PU blend membranes and properties. Soochow University: Engineering Science, 2002 (2):16-19.
[8] Mengyan Li, Mark J. Mondrinos, Xuesi Chen, et al. Co-electrospun poly(lactide-co-glycolide), gelatin, and elastin blends for tissue engineering scaffolds. LI ET AL, 2006, 31:963-973.
[9] Hoi-Yan Cheung, Kin-Tak Laua, Yu-Fung Powc, Yong-Qing Zhaod and David Huie Biodegradation of a

silkworm silk/PLA composite. Biomaterials, 2010, 41:223-228.
[10] Choi J S, Lee S J, Christ G J, et al. The influence of electrospun aligned poly(ε-caprolactone)/collagen nanofiber meshes on the formation of self-aligned skeletal muscle myotubes. Biomaterials, 2008, 29: 2899-2906.
[11] Yuqing Wana, Xue Qua, Jun Lub, Chuanfeng Zhu, et al. Characterization of surface property of poly(lactide-co-glycolide) after oxygen plasma treatment. Biomaterials, 2004, 25:4777-4783.
[12] Honghyun Park, Kuen Yong Lee, et al. Plasma-Treated Poly(lactic-co-glycolic acid) Nanofibers for Tissue Engineering. Macromolecular Research, 2007, 15(3):238-243.

23 Effect of Low Temperature Plasma Treatment on Surface Properties of Polysulfonamide Fiber

Jianzhong Yang*, Lei Ren

(School of Textile and Material, Xi'an Polytechnic Univ., Xi'an 710048, China)

Abstract: The performance of the polysulfonamide fiber (PSA) has been investigated by low temperature glow discharge plasma in this paper. According to observation of scanning electron microscope (SEM), the surface roughness degree of PSA fiber increase after treated by low temperature plasma. The experimental results show that under the treatment of different plasma conditions, the PSA fiber surfaces appear the varying degree physical and chemical etching. It's found that the friction coefficient and hydrophilicity of PSA fiber treated by low temperature plasma improve obviously and X-ray photoelectron spectroscopy (XPS) shows that the reason is the etching and the increase of the ratio O_{1s}/C_{1s} on the surface of PSA fiber. Variations of tensile breaking strength of PSA with different parameters are analyzed.

Keywords: Hydrophilicity; Surface modification; Low temperature plasma; Polysulfonamide

1 Introduction

Polysulfonamide fiber (PSA), providing maximum protection against high temperature, is developed by Shanghai Textile Research Institute and Shanghai Synthetic Fiber Research Institute with great efforts. PSA fiber, with the sole intellectual property in China[1], can be widely used in protective textiles, filtrations, insulation materials, composite materials and so on[2]. The PSA fiber has excellent integrated performance, but the high hydrophobicity limited its application as textile fiber [3]. The material surface after low temperature plasma treatment produces the multiple physical and chemical reaction, or being roughness because of etched effect [4,5], or forming compact cross linking layer, or bringing oxygen polar groups, which can improve the hydrophility, cohesiveness, dyeing, biological compatibility and the electric conductivity [6,7]. Therefore, this article mainly analyzed the morphology and the composition of treated and untreated fiber by Scanning Electron Microscopy (SEM) and X-ray photoelectron spectroscopy (XPS). It also studied friction performance, tensile properties and hydrophility of PSA fiber treated by low temperature plasma.

2 Experimental

2.1 Fiber specification

The PSA fiber linear density was 1.67 dtex, and fiber cut length was 38mm for sample fiber, supplied by Shanghai Tanlon Fiber Co., Ltd.

2.2 Surface modification of fiber

The surface modifications of the PSA fibers were obtained using glow discharge low temperature plasma equipment (HD-1B type). The fibers were placed in the vacuum chamber, which was evacuated to different time, power and vacuum degree of the equipment in air.

2.3 Instruments and methods

(1) The surface morphology of fiber was observed using a scanning electron microscope (Tescan VegⅡXmuinca).

(2) The change in chemical composition and the chemical state on the surface of fiber were measured using an Axis Ultra, Kratos (UK). The voltage and current were 15 kV and 10mA. The limited vacuum degree was 1×10^{-9}Pa.

(3) The friction performance was measured using the Y151 fiber friction coefficient tester. The tensile properties of fiber were measured using YG001N electron single fiber force tester.

(4) Capillarity performance change of PSA fabric was measured using the tester, which was the length of the sample of 20cm.

(5) If the friction coefficient increases 1%, we can gain one point; If the breaking tenacity and elongation at break damage 1%, we can got rid of two points.

---

* Foundation item: Supported by key science research foundation of office of education of Shannxi province (No.08JZ25); Corresponding author's email: yangjianzhong23@yahoo.com.cn

3 Results and discussion

3.1 The test of tensile properties and friction performance of PSA fiber

It shows that from Table 1 the breaking tenacity and elongation at break of PSA fiber decline lightly treated by low temperature glow plasma. The main reason may be that etch effect of low temperature plasma make the loss in fiber weight. But he breaking tenacity and elongation at break of PSA fiber did not show statistically significant differences after treatment by low temperature plasma.

The fiber surface treated by the low temperature plasma was etched and became roughness, which make the friction coefficient of PSA fiber increase. According to the evaluation of 6 indexes to friction coefficient, breaking tenacity damage rate and elongation at break reduces rate and so on, we carried the orthogonal analysis according to the test; finally we found that the better test were No.2[#].

Table 1 Friction performance and tensile properties of fiber and capillarity wicking altitude

No.	Vacuum degree /Pa	Power /W	Time /S	Breaking tenacity /(cN/dtex)	Elongation at break /%	Coefficient of static friction	Coefficient of dynamic friction	Capillarity wicking-altitude/cm
0				3.78	18.5	0.4495	0.3396	1.5
1	30	70	60	3.37	17.1	0.5245	0.3781	13.9
2	30	100	90	3.18	15.5	0.5210	0.3656	14.0
3	30	130	120	3.51	17.2	0.5293	0.4069	13.4
4	40	70	90	3.67	17.3	0.5430	0.3569	13.1
5	40	100	120	3.54	17.1	0.5662	0.4023	12.7
6	40	130	60	3.50	17.4	0.5299	0.3752	12.2
7	50	70	120	3.47	17.2	0.5446	0.3945	12.1
8	50	100	60	3.50	17.4	0.5258	0.3679	12.4
9	50	130	90	3.63	16.9	0.5645	0.4238	12.8

3.2 Capillarity performance change of PSA fabric

Using capillarity wicking altitude to evaluate the influence of surface hydrophility treated by low temperature plasma, we tested capillarity performance change of PSA fabric under the different modification condition. Table 1 also shows that wicking is improved obviously treated by low temperature plasma. It can be seen from Table 1 that the capillarity wicking altitude is 1.5 (untreated PSA fabric), but when the pressure is 30Pa, power is 100W, time is 90S, and the capillarity wicking altitude is up to 14.0cm. The better test plan is No. 2[#]. The improving of moisture absorption and guiding performance was due to that the particle in low temperature plasma bombards PSA surface, which produces etching, cross-linking, oxidation, and introduce hydrophilic groups. These hydrophilic groups improved moisture absorption ability of PSA fiber. At the same time, the pit and crack of fiber could improve moisture absorption ability further because of the low temperature plasma modification.

3.3 The change of surface morphology of PSA fiber

SEM photographs of treated and untreated PSA fibers are shown in Figure 1. For the untreated PSA fibers, the fiber surface is smooth. In contrast, there are pit and the slight crack after treated by the low temperature plasma. The phenomena may be due to the active particles which own highly excited and unstable state in the plasma, which produced etched, crossing linking, groups introduction, roughen and so on, so as to the fiber was modified.

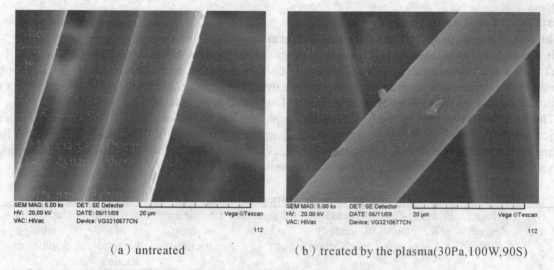

(a) untreated　　　　　　　　　(b) treated by the plasma(30Pa,100W,90S)

Figure 1 Surface morphology of PSA fiber (amplify 5000 times)

3.4 XPS analysis

The low temperature plasma modification not only caused the change of surface morphology, but also it made the chemistry composition change on the surface of fiber. Figure 2 is XPS spectrogram of treated and untreated PSA fiber by the low temperature plasma.

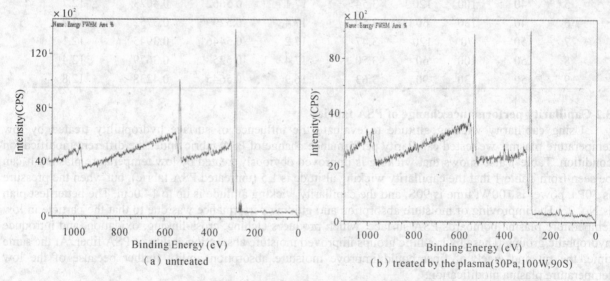

(a) untreated　　　　　　　　　(b) treated by the plasma(30Pa,100W,90S)

Figure 2 XPS spectrogram of PSA fiber

Figure 2 (a) and (b) shows that the binding energy of O_{1s} (untreated sample) is 535.590eV, but the binding energy of O_{1s} (treated sample) moved to 536.174eV; and the binding energy of N_{1s} also moved 402.789eV (untreated) to 403.094eV (treated); and the binding energy of C_{1s} also moved 288.113eV (untreated) to 289.380eV (treated). The move of binding energy of O_{1s} and C_{1s} shows that low temperature plasma modification can cause the change of O and C polar groups. Figure 3 shows the peak separation of C_{1s} of PSA fiber treated by low temperature plasma.

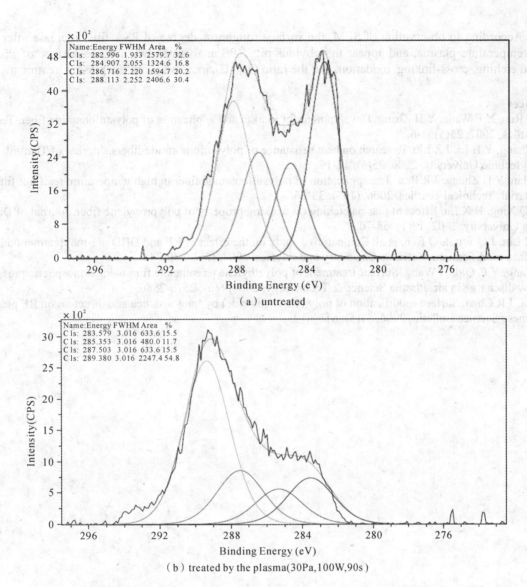

Figure 3 The peak separation spectrogram of C_{1s} of PSA fiber

Table 2 XPS analysis of treated by the plasma (30Pa, 100W, 90S) and untreated PSA fiber

Treatment	Relative intensity,%					Relative peak composition,%			
	C_{1s}	O_{1s}	N_{1s}	S_{2p}	O_{1s}/C_{1s}	C—H	C—OH	C—O	C=O
Untreated	80.38	15.68	3.64	0.31	19.52	32.6	16.8	20.2	30.4
Treated by the plasma	75.35	19.83	4.57	0.25	26.32	15.5	11.7	18.0	54.8

Figure 3 shows the C_{1s} spectra of original and treated by the plasma (30Pa, 100W, 90S) PSA fiber. As shown in the figure, the spectrum has four different peaks (283.0eV, 285.0eV, 287.0eV, 289.0eV) corresponding to the C—H, C—OH, C—O, and C=O groups[6], respectively. Table 2 summarizes the relative chemical composition of the XPS spectra of PSA fiber. The ratio O_{1s}/C_{1s}, which has a good relationship with wetness, increases with the plasma treatment, indicating that PSA fiber surface is modified to produce hydrophilic properties.

4 Conclusions

The research of low temperature plasma modification on PSA fiber shows:

(1) The friction coefficient and hydrophilicity of PSA fiber treated by low temperature plasma improve obviously, and when the power is 100W, time is 90S, pressure is 30Pa, the capillarity wicking altitude is up to 14.0cm. The better test plan is No. 2[#].

(2) According to observation of SEM, the surface roughness degree of PSA fiber increase after treated by low temperature plasma, and appear the obvious pit; XPS analysis shows that the surface of PSA fiber produced etching, cross-linking, oxidation, and the ratio O_{1s} / C_{1s} increases with the plasma treatment.

References

[1] J R Ren, X F Wang, Y H Zhang. Development for market and application of polysulfonamide fiber. Technical Textiles, 2007, 25(5): 1-6.
[2] L Zhang, Y B Li, J Z Liu. Research on heat resistance of polysulfone amide fibers. Journal of Tianjin Polytechnic University, 2006, 25(6): 17-19.
[3] Y Ban, Y H Zhang, J R Ren. The application of polysulfonamide fiber in high temperature resistant filtration material. Technical Textiles, 2006, (195): 33-36.
[4] M Q Xing, B X Lu. Effect of plasma etching on wicking property of polypropylene fiber. Journal of Dong Hua University, 2002, 19(1): 68-70.
[5] C Z Liu, J Q Wu, L Q Ren, et al. Comparative study on the effect of RF and DBD plasma treatment on PTFE surface modification. Materials Chemistry and Physics, 2004, (85): 340-346.
[6] Z Fang, Y C Qiu, H Wang. Surface treatment of polyethylene terepthalate film using atmospheric pressure glow discharge in air. Plasma Science & Technology, 2004, 6(6): 2576-2580.
[7] R Li, J R Chen. Surface modification of poly (vinyl chloride) by long-distance and direct argon RF plasma. Chinese Science Bulletin, 2006, 51(5): 615-619.

24 Rheological Behavior and Spinning Performance of Cellulose/[BMIM]Cl Solutions Prepared by Two-Steps Dissolving Process

Huihui Zhang, Tao Cai, Huili Shao, Xuechao Hu

(State Key Laboratory for Modification of Chemical Fibers and Polymer Materials, College of Material Science and Engineering, Donghua University, Shanghai 201620, China)

Abstract: In this paper, the two-steps dissolving process, i.e. firstly swelling the cellulose pulp to the maximum in aqueous ionic liquids and then dissolving it by vacuum evaporation to remove the excessive water, was used to prepare cellulose spinning dope by using 1-butyl-3-methylimidazolium chloride ([BMIM]Cl) as the solvent. The rheological behavior of the cellulose/[BMIM]Cl solution was investigated and the regenerated cellulose fibers were prepared by a dry-wet spinning method. The results showed that the spinnability is better for the cellulose/[BMIM]Cl solution prepared by two-steps dissolving process according to the calculation of inhomogeneity. The relaxation time spectra showed cellulose chains in solution prepared by two-steps dissolving process were easy to flow than those prepared by direct dissolving process. Moreover, regenerated cellulose fibers with better mechanical properties could be prepared by two-steps dissolving process.

Keywords: Cellulose; Ionic liquid; Fibers; Swelling; Dissolving

1 Introduction

As the most abundant renewable resource in the world, cellulose has attracted more and more attentions due to the consumption and over-exploitation of non-renewable resources. There are various solvents known for cellulose, however, only N-methylmorpholine- N-oxide monohydrate is applied commercially.

In 2002, Swatloski et al [1] firstly reported that cellulose could be dissolved in ionic liquids 1-butyl-3-methyl imidazole chloride ([BMIM]Cl), which opened up a new way for the development of a class of cellulose solvent systems. Zhang et al [2] also found that ionic liquid 1-allyl-3-methylimidazole chloride ([AMIM]Cl) had very strong ability to dissolve cellulose. Moreover, 1-ethyl-3-methylimidazolium acetate ([EMIM]Ac) and 1-(2-hydroxyethyl)-3-methyl imidazolium chloride ([HeMIM]Cl) were also proved to be the solvents of cellulose [3,4].

In all of these studies, cellulose was dissolved in ionic liquids directly. However, it is found that there are a few particles in the cellulose spinning dope prepared by the above direct dissolving process in our previous study by using [BMIM]Cl as the solvent[5]. To avoid these particles in the cellulose/[BMIM]Cl spinning dope, a two-steps dissolving process was developed in our lab. Firstly the cellulose pulp was swelled but not dissolved in the aqueous [BMIM]Cl solution, after cellulose was swelled to the maximum; the excessive water was removed at a certain temperature in vacuum. Finally, the dissolution would gradually happen at inside and outside of cellulose simultaneously. This method was found to produce cellulose dopes with lease particles and improve the spinning performance of the cellulose dopes[6].

In this paper, cellulose was first swelled in the [BMIM]Cl with 2 wt% water, then cellulose/ [BMIM]Cl spinning dope was prepared by the two-steps dissolving process, the rheological behavior of cellulose dope was investigated. Then the fibers were dry-wet spun from the spinning dope, and the mechanical properties of regenerated cellulose fibers were studied.

2 Experimental
2.1 Materials

Cotton pulp (DP=686.5, α-cellulose content 95.5%, Baoding Chemical Fiber Co., China) was used in this paper. [BMIM]Cl was purchased from BASF (Germany).

2.2 The two-steps dissolving process

A certain amount of cellulose was added into the aqueous [BMIM]Cl with 2wt% water, and then the mixture was heated to 100 ℃. After swelled in [BMIM]Cl solution for 15min to reach the maximal swelling ratio, cellulose/[BMIM]Cl dopes were prepared from slurry of cellulose in the aqueous [BMIM]Cl by removing the excessive water at elevated temperature, vacuum and high shearing rates. Finally, 11wt% cellulose/[BMIM]Cl solution can be obtained.

2.3 Rheological measurements

The rheological measurements were made on RS150L Rheometer (Thermo Haake, Germany), and a cone plate (Ti, 35/1°) was used. The dynamic rheological properties were determined at 80, 85 and 90℃, respectively. The master curve for G', and G'' at the reference temperature of 85℃ was obtained via superimposition method. Then the relaxation time spectra could be calculated from the master curve. In addition, the zero-shear viscosity of cellulose/[BMIM]Cl solution was tested in creep recovery mode. All data analysis was performed using Rheowin Pro Data Manager and RheoSoft software.

2.4 Preparation of regenerated cellulose fibers

The cellulose/[BMIM]Cl spinning dope with the temperature of 90℃ was extruded by a gear pump and then through a spinneret with 48 orifices (0.3mm in diameter). After being passed through an air gap of 50 mm in length, the dope was immersed into a coagulation bath and a washing bath successively to remove [BMIM]Cl completely, and then taken-up at the spinning speed of 50, 60 and 70m/min to obtain the regenerated cellulose fibers, respectively.

2.5 Measurements of the mechanical properties of fibers

The mechanical properties of fibers were measured with XQ-1 Tensile Tester (China Textile University, China). The sample length was 20mm and the extension rate was set at 5 mm/min. The statistical results came from more than 20 measurements for each specimen. All measurements were performed at 20℃ and 65% relative humidity.

3 Results and discussion

3.1 Dynamic rheological properties of cellulose/[BMIM]Cl solution

The dynamic rheological properties of 11% cellulose/[BMIM]Cl solution prepared by two-steps dissolving process were determined at 80, 85 and 90℃, respectively, and 85℃ was chosen as the reference temperature. By using time-temperature equivalence principle, the curves obtained at other temperatures can be superimposed on the reference curves. The resulted master curve for G' and G'' at the reference temperature of 85℃ for cellulose/[BMIM]Cl solution is shown in Figure 1. Moreover, the complex viscosity $\eta^*(\omega)$ was obtained as well according to its relation with G' and G'' (equation 1), the curves of complex viscosity of cellulose solution is also illustrated in Figure 1.

$$\eta^*(\omega) = \frac{\sqrt{(G')^2 + (G'')^2}}{\omega} \quad (1)$$

It can be seen that the complex viscosity curve showed typical shear thinning behavior, which is consistent with most polymer melts or solutions. Moreover, at low deformation rate, the cellulose solution behaves most likely as a viscous liquid (loss modulus G'' is larger than storage modulus G'), but at high angular frequency the elastic properties exceed the viscose properties. The two domains of viscoelastic behavior are separated by the so-called crossover point.

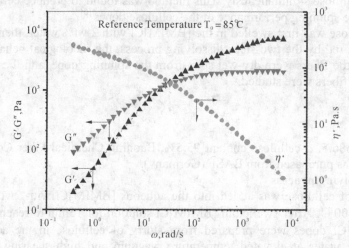

Figure 1 Dynamic moduli master curve and complex viscosity of 11% cellulose/[BMIM]Cl solution prepared by two-steps dissolving process

Michel et al [7] found that the solution state of the viscoelastic spinning dope in the molecular range can be described reasonably with relatively easy accessible rheological parameters like the zero shear viscosity η_0, the storage modulus G' and loss modulus G'', and other parameters derived from that data. The inhomogeneity U_η was used to evaluate the spinnability of cellulose/ NMMO·H$_2$O spinning dope by Michel. He had measured the inhomogeneity of different cellulose in NMMO·H$_2$O solutions, including cotton linter pulps, wood pulps and cellulose mixture. For the cotton linter pulp, it is found that the U_η=2.5-4.2 was suitable for the spinning of Lyocell process[7].

The inhomogeneity had a significant influence on the kinetics of the fiber forming. An increasing inhomogeneity will lead to worse spinning performance and fiber parameters.

According the method reported by Michel[7], the molecular inhomogeneity of the 11wt% cellulose/ [BMIM]Cl solution prepared by two-steps dissolving process was calculated (listed in Table 1). It can be found that U_η=2.92 for the cellulose solution prepared by two-steps dissolving process, which is less than that of cellulose solution prepared by the direct dissolving process (U_η=3.90), although two values are all in the range of U_η reported by Michel [7]. In other words, the spinnability is better for the cellulose/[BMIM]Cl solution prepared by two-steps dissolving process, compared to the direct dissolving process.

Table 1 The rheological parameters of cellulose[BMIM]Cl solution prepared by different dissolving processes

process	η_0 (Pa.s)	$\eta^{\#}$ (Pa.s)	U_η
Direct dissolving process	7797	1628	3.90
Two-steps dissolving process	4645	1185	2.92

3.2 Relaxation time spectra of cellulose/[BMIM]Cl solution

Figure 2 shows the relaxation time spectra ($H=f(\lambda)$) of 11wt% cellulose/[BMIM]Cl solution prepared by two-steps dissolving process. The relaxation time at the frequency maximum was 0.93s for this cellulose solution, which is less than that of cellulose solution prepared by direct dissolving process (2.88s). Therefore, the flow is easier for the cellulose//[BMIM]Cl solution prepared by two-steps dissolving process, which is helpful to the spinning process.

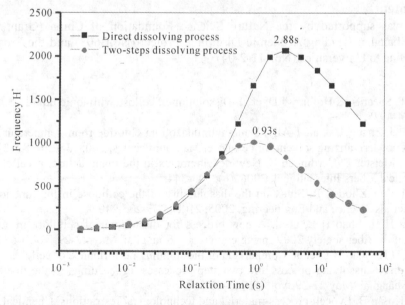

Figure 2 The relaxation time spectra of cellulose/[BMIM]Cl solution prepared by different dissolving processes

3.3 Mechanical properties of cellulose fiber from cellulose/[BMIM]Cl solution prepared by two-steps dissolving process

Table 2 The mechanical properties of regenerated cellulose fiber from the 11wt% cellulose/[BMIM]Cl spinning dope prepared by two-steps dissolving process

Spinning speed (m/s)	Fineness (dtex)	Tensile strength (cN/dtex)	Initial modulus (cN/dtex)	Elongation at break (%)
50	7.51	3.64	51.4	7.2
60	7.24	3.82	53.7	6.9
70	5.64	4.04	55.3	6.7

Table 2 lists the mechanical properties of the regenerated cellulose fiber from the 11wt% cellulose/[BMIM]Cl spinning dope prepared by two-steps dissolving process. It can be found that with the increase of the spinning speed, the tensile strength and initial modulus of cellulose fibers improved, whereas the elongation at break decreased. The reason is that the draw ratio was increased with the increase of the spinning speed, which results the increase of the orientation of cellulose chains, then mechanical properties of cellulose fibers improved.

In general, the tensile strength of Lyocell fiber and viscose rayon was in the range of 3.8~4.2cN/dtex and 2.2~2.6 cN/dtex, respectively [8]. It can be seen from Table 2 that the mechanical properties of regenerated cellulose fibers from the cellulose/[BMIM]Cl spinning dope prepared by two-steps dissolving process are similar to those of Lyocell fibers, and higher than viscose rayon. Therefore, it is an effective way to prepare the cellulose fibers with good mechanical properties by using the two-steps dissolving process.

4 Conclusions

In this paper, two-steps dissolving process was used to prepare cellulose/[BMIM]Cl spinning dope. The results showed that the spinnability is better for the cellulose/[BMIM]Cl solution prepared by two-steps dissolving process according to the calculation of inhomogeneity. The relaxation time spectra showed cellulose chains in solution prepared by two-steps dissolving process were easy to flow than those prepared by direct dissolving process. Moreover, the mechanical properties of cellulose fibers from cellulose/[BMIM]Cl spinning dope prepared by two-steps dissolving process are similar to those of Lyocell fibers, and higher than viscose rayon. Therefore, it is an effective way to prepare the cellulose fibers with good mechanical properties by using the two-steps dissolving process.

Acknowledgement

This work was supported by the Nature Science Foundation of China (Grant No. 50903015 and 50873024), the Shanghai Leading Academic Discipline Project (No.B603) and the Program of Introducing Talents of Discipline to Universities (No.111-2-04).

References

[1] Swatloski R P, Spear S K, Holbrey J D, et al. Dissolution of cellose with ionic liquids. J Am Chem Soc, 2002, 124(18): 4974-4975.
[2] Zhang H, Wu J, Zhang J, et al. 1-Allyl-3-methylimidazolium chloride room temperature ionic liquid: a new and powerful nonderivatizing solvent for cellulose. Macromolecules, 2005, 38(20): 8272-8277.
[3] Hermanutz F, Meister F, Uerdingen E. New developments in the manufacture of cellulose fibers with ionic liquids. Chemical Fibers International, 2006, (6): 343-344.
[4] Luo H M, Li Y Q, Zhou C R. Study on the dissolubility of the cellulose in the functionalized ionic liquid. Polymer Materials Science and Engineering. 2005, 21(2): 233-235,240
[5] Cai T, Zhang H H, Shao H L, et al. A new process for dissolution of cellulose in aqueous ionic liquids. Proceedings of the fiber society 2009 spring conference. Shanghai, May 27-29, Vol. I : 395-397
[6] Cai T, Zhang H H, Shao H L, et al. Comparison of the spinning performance of cellulose/[BMIM]Cl solutions prepared by direct dissolving process and two step processes. Proceedings of the fiber society 2009 spring conference. Shanghai, May 27-29, Vol. I : 261-263
[7] Michels Ch, Kosan B. Lyocell process-material and technological restrictions. Chemical Fibers International, 2000, 50: 556-561.
[8] Liu R G, Hu X C, Zhang T L, et al. A new generation of cellulose fibers Lyocell. Synthetic Fiber in China, 1997(4): 23-28.

25 Fabrication and Application of Carbon Nanotube/Magnetite Composites

Li Zhang[1], Qing-Qing Ni[2], Yoshio Hashimoto[1]

(1. Faculty of Engineering, Shinshu University, Japan;
2. Faculty of Textile Science & Technology, Shinshu University, Japan)

Abstract: A facile synthesis process is proposed to prepare multi-walled carbon nanotubes/magnetite (MWCNTs/Fe_3O_4) hybrids. The MWCNTs/Fe_3O_4 hybrids were characterized with respect to crystal structure, morphology and element composition by X-ray diffraction (XRD), transmission electron microscopy (TEM), X-ray photoelectron spectroscopy (XPS). XRD and TEM results show that the Fe_3O_4 nanoparticles with diameter in the range of 20-60 nm were firmly assembled on the nanotube surface. As for an important application, we prepared the nanocomposite filled with MWCNTs/Fe_3O_4 hybrids, and measured the electromagnetic interfere (EMI) shielding properties.

Keywords: Facile synthesis process; MWCNTs/Fe_3O_4 hybrids; The electromagnetic interfere; Fe_3O_4 nanoparticles

1 Introduction

Carbon nanotubes (CNTs) have been widely applied as a promising material in many areas of nanoscience due to their outstanding physical and electrical properties such as high tensile strength, elastic modulus, excellent thermal and electrical conductivity [1-3]. With the great progress achieved in this field, CNTs have been used to synthesize not only various polymer/CNTs composites [4,5] but also various CNTs/metal oxide hybrids [6,7], where CNTs can serve as high-performance supporting materials. The decoration of CNTs with metal oxide nanoparticles can give them new properties and potentials for various applications [8-10]. Among them, the CNTs coated with Fe_3O_4 nanoparticles have attracted much increasing interest because of the excellent magnetic and biocompatible properties.

The electromagnetic interfere (EMI) shielding of radio frequency radiation has become a serious concern in modern society. Light weight and effective EMI shielding is needed to protect the workspace and environment from radiation coming from computers and telecommunication equipment as well as for protection of sensitive circuits. Electrically conducting polymer composites have received much attention recently compared to conventional metal-based EMI shielding materials [11-13], because of their light weight, resistance to corrosion, flexibility and processing advantages. Therefore, lots of attempts have been made to introduce CNTs into the matrices to prepare nanocomposites for using as EMI materials [14]. However, a major drawback of nanocomposites that contain CNTs or other fillers (nickeled and stainless steel fibers) is a high propensity to reflect the electromagnetic radiation rather than to absorb it [14]. Recently, soft ferrite materials [15,16] or magnetic materials [17,18] are considered as promising absorbers. Therefore, we prepared the PU-MWCNTs/Fe_3O_4 nanocomposite, expecting for both reflection and absorbance of electromagnetic radiation.

This work prepared CNTs/Fe_3O_4 hybrids using a cheap and green in-situ synthesis method, which eliminates both harsh treatment conditions and the need for organic solvent. In first step, we prepare water-soluble CNTs by one-step functionalization route, which was designed by ourselves [19]. In second step, we used these functionalized CNTs to prepare CNTs/Fe_3O_4 hybrids by a hydrothermal process using ferrous sulfate hydrate ($FeSO_4 \cdot 7H_2O$) as a single iron precursor (with electromagnetic stirring). Finally, we prepared the PU-MWCNTs/Fe_3O_4 nanocomposite, which could both reflect and absorb electromagnetic radiation.

2 Experimental details

2.1 Materials

All reagents were used of analytical grade without further purification. Multiwalled carbon nanotubes with a diameter of 20-30 nm, purchased from Wako Pure Chemical Industries Ltd. (Japan), were used as received.

2.2 Assembling Fe_3O_4 nanoparticles on water-soluble MWCNTs

We prepare water-soluble CNTs by one-step functionalization route according the previous research of our group [19]. The assembling process of Fe_3O_4 on MWCNTs was carried out by a hydrothermal. In a typical procedure, water-soluble MWCNTs (30 mg) were dispersed into 100 ml deionized water by sonication. Then $FeSO_4 \cdot 7H_2O$ (86.9 mg) was added to the solution with stirring for 2 hours. After that, $N_2H_4 \cdot H_2O$ (62.5 mg) were added with stirring for 5 min. The solution was transferred into high-pressure microreactor. The microreactor was maintained at 150°C for 8 h with stirring and cooled naturally to room temperature. The black products in the solution were collected after being rinsed with deionized water repeatedly and dried at 60°C for 8 h.

2.3 Preparation of PU–CNT/ Fe₃O₄ composite film

The CNT/Fe₃O₄ composites were dispersed in DMF solution and mixed [ultrasonic mixing (100 W, 45 kHz) and mechanical mixing] for 1 hour. The PU matrix resin solution was then added to the CNT/ Fe₃O₄ solution, and the combined mixing process was continued for another 2 hours. The mixture was cast into a container, and dried at 70°C for 60 hours to evaporate the solvent and produce a uniform film.

2.4 Characterization

The X-ray diffraction pattern (XRD) of the products was recorded on a Rigaku Geigerflex 2028 diffractometer (Rigaku, Japan) with CuKα radiation (1.5418 Å). The XRD data were collected over a 2θ range from 10° to 80°. Transmission electron microscopy (TEM) and high-resolution transmission electron microscopic (HRTEM) images of the samples were observed through JEM-2010 electron microscope with an accelerating voltage of 200 kV. X-ray photoelectron spectroscopy (XPS) analyses were performed on a Kratos Axis Ultra DLD X-ray photoelectron spectrometer with a standard Mg Kα (1256.6 eV) X-ray source operated at 10 mA and 15 kV. All binding energies were referenced to Au (4f7/2) at 84 eV. The EMI SE of PU-CNT/ Fe₃O₄ composites were analyzed using near-field antenna measurement systems.

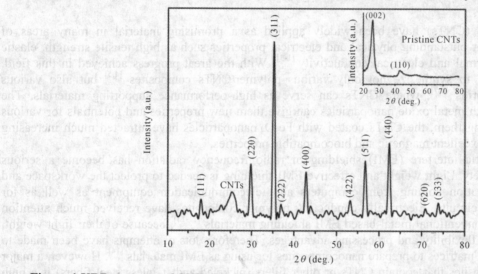

Figure 1 XRD patterns of CNTs/Fe₃O₄ hybrids (the inset shows the XRD pattern of treated CNTs)

3 Results and discussion

After the hydrothermal process of assembling the functionalized MWCNTs with Fe₃O₄, XRD patterns of the products were recorded (see Figure 1). The result shows that the product was a mixture of two phases, cubic Fe₃O₄ and MWCNTs. The diffraction peak at 26° can be indexed to (002) reflection of the MWCNTs (see the inset of Figure 1), similar to the result in the literature [20]. All other diffraction peaks can be well indexed to the magnetic cubic structure of Fe₃O₄ (JCPDS 79-0419) with lattice constants of a=8.396 Å. No other peaks due to the hematite were detected in the XRD patterns, indicating the formation of the pure magnetic products. The strong and sharp diffraction peaks reveal the good crystallization of the product.

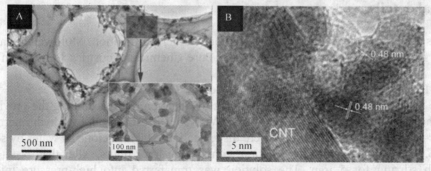

Figure 2 (a) Typical TEM image of CNTs/Fe₃O₄ hybrids; (b) The HRTEM image of individual hybrid structure

The structure and morphology of the obtained products were investigated by TEM. Figure 2a gives the representative TEM image of the products, which indicates that the MWCNTs were assembled with magnetite nanoparticles, and the diameters of nanoparticle are in the range of 20-60 nm. Although the products had been sonicated in ethanol before TEM measurements, a great amount of magnetite nanoparticles are still found on the nanotube surface and this indicates the strong interaction between MWCNTs and magnetite nanoparticles. More detailed structural information of the products was provided by the high-resolution TEM (HRTEM) analysis, and Figure 2b shows the HRTEM image taken from an individual hybrid structure. Clear lattice fringes are observed in HRTEM image, indicting that crystalline particles were formed in the hydrothermal process. Moreover, the measured spacing of the crystallographic planes is about 0.48 nm, which is close to that of the (111) lattice planes of magnetite crystals.

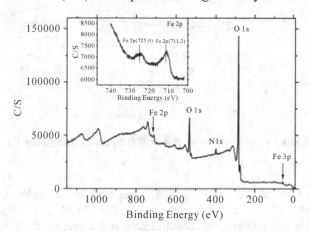

Figure 3 XPS spectrum of CNTs/Fe$_3$O$_4$ hybrids (the inset shows the high-resolution scan of Fe 2p region)

X-ray photoelectron spectroscopy (XPS) was used to explore the surface chemistry of the MWCNTs/Fe$_3$O$_4$ hybrids. The typical survey spectrum of the products (see Figure 3) shows the presence of Fe, O, N and C. The high-resolution Fe 2p spectrum (see the inset of Figure 3) shows that the binding energies of the Fe 2p$_{3/2}$ and Fe 2p$_{1/2}$ are located at 711.2 and 725.0 eV, respectively. The results agree with literature data for magnetite [21,22], thus substantiating the XRD results. Moreover, the Fe/O atomic ratio is about 3/4 based on the ratio of the peak areas and relative sensitivity factor.

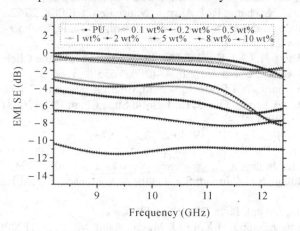

Figure 4 The shielding effectiveness in X-band frequency range for PU nanocomposites at the thicknesses of 1 mm

EMI shielding in the range of 8.2 to 12.4 GHz (the so-called X-band) is very important for military and commercial applications. Doppler, weather radar, TV picture transmission, and telephone microwave relay systems lie in X-band [23]. In this part, we examine the EMI SE of PU nanocomposites for X-band and the results are shown in Figure 4. The results demonstrate that the EMI SE of PU composites increased as the CNTs/Fe$_3$O$_4$ content increased. The composites filled with 10 wt% CNTs/Fe$_3$O$_4$ exhibited SE ~11 dB in the X-band. For the value of 11 dB of EMI SE, this means about 90% of incident signal is blocked. For comparison purposes with PU nanocomposites with 10 wt% MWCNTs/Fe$_3$O$_4$ ($W(CNTs):W(Fe_3O_4) = 1:0.96$), the PU composites with blent fillers of 5.05 wt% CNT and 4.95 wt% Fe_3O_4 were prepared. According to the EMI SE data (see Figure 5): the SE results of ~8 dB in the X-band were obtained with 10 wt% blent filler with CNT and Fe_3O_4. This result is lower than the PU nanocomposite with 10 wt% CNT/Fe_3O_4, which is possibly due to synergistic effect.

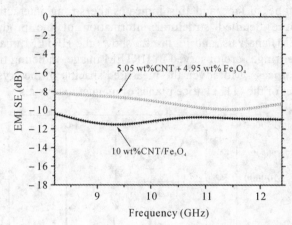

Figure 5 The shielding effectiveness in X-band frequency range for PU nanocomposites

4 Conclusions

The MWCNTs/Fe_3O_4 hybrids were successfully prepared by a facile hydrothermal process with electromagnetic stirring. The CNTs/Fe_3O_4 hybrids, the XRD, TEM and XPS results revealed that the MWCNTs were assembled with magnetite nanoparticles, and the diameters of nanoparticle in the range of 20-60 nm. The PU nanocomposites show excellent EMI shielding effect (EMI SE), and the EMI SE of the nanocomposites increased as the CNT/Fe_3O_4 content increased. The composites filled with 10 wt% CNT/Fe_3O_4 have a good EMI SE of 11dB in X-band.

References

[1] R.H. Baughman, N.A. Zakhidov, W.A. de Heer, Science 297 (2002) 787.
[2] D. Tasis, N. Tagmatarchis, A.Bianco, M. Prato, Chem. Rev. 106 (2006) 1105.
[3] J.N. Coleman, U. Khan, W.J. Blau, Y.K. Gun'ko, Carbon 44 (2006) 1624.
[4] L. Ci, J. Suhr, V. Pushparaj, X. Zhang, P.M. Ajayan, Nano. Lett. 8 (2008) 2762.
[5] Y.K. Choi, K.I. Sugimoto, S.M. Song, Y. Gotoh, Y. Ohkoshi, M. Endo, Carbon 43 (2005) 2199.
[6] Z. Liu, B. Han, Adv Mater 21 (2009) 825.
[7] C. Li, N. Sun, J. Ni, J. Wang, H. Chu, H. Zhou, M. Li, Y. Li, J. Solid State Chem. 181 (2008) 2620.
[8] D.R. Kauffman, A. Star, Angew. Chem. Int. Ed. 47 (2008) 6550.
[9] B. Yu, F. Zhou, G. Liu, Y. Liang, W.T.S. Huck, W. Liu, Chem. Commun. 22 (2006) 2356.
[10] X. Wang, B. Xia, X. Zhu, J. Chen, S. Qiu, J. Li, J. Solid State Chem. 181 (2008) 822.
[11] D.D.L. Chung, Carbon 39 (2001) 279.
[12] J. Joo, A.J. Epstein, Appl. Phys. Lett. 18 (19949 2278.
[13] Y.L. Yang, M.C. Gupta, K.L. Dudley, R.W. Lawrence, Adv. Mater. 17 (2005) 1999.
[14] Y. Yang, M.C. Gupta, K.L. Dudley, R.W. Lawrence, Nano Lett. 5 (2005) 2131.
[15] A. N. Yusoff, M.H. Abdullah, S.H. Ahmad, S.F. Jusoh, A.A. Mansor, S.A.A. Hamid, J. Appl. Phys. 92 (2002) 876.
[16] M. Matsumoto, Y. Miyata, J. Appl. Phys. 79 (1996) 5486.
[17] B.S. Zhang, G. Lu, Y. Feng, J. Xiong, H.X. Lu, J. Magn. Magn. Mater. 299 (2006) 205.
[18] S.S. Kim, S.T. Kim, Y.C. Yoon, K.S. Lee, J. Appl. Phys. 97 (2005) 10F905.
[19] L. Zhang, Q,Q, Ni, Y. Fu, T. Natsuki, Appl. Surf. Sci. 25 (2009) 7095.
[20] B. Jia, L. Gao, J. Sun, Carbon 45 (2007) 1476.
[21] Y. Sahoo, A. Goodarzi, M.T. Swihart, T.Y. Ohulchanskyy, N. Kaur, E.P. Furlani, P.N. Prasas, J. Phys. Chem. B 109 (2005) 3879.
[22] Z. Qian, Z. Zhang, Y. Chen, J. Colloid Interf. Sci. 327 (2008) 354.
[23] Y. Huang, N. Li, Y. Ma, F. Du, F. Li, X. He, X. Lin, H. Gao, Y. Chen, Carbon 45 (2007) 1614.

26 Preparation and Regeneration of Bioplasts for Biomodification of Polyester

Weiling Zhang, Jianfei Zhang*, Zheng Li, Jixian Gong, Qiujin Li, Cheng Chen, Zhaolong Hu

(Key Laboratory of Advanced Textile Composites, Tianjin Polytechnic University, Ministry of Education, Tianjin 300160, China)

Abstract: Bacilli B-F could biomodify polyester fabrics at moderate temperature and thermophilic bacteria S-R could grow well at high temperature. The two strains might be fused together to generate a new strain which could modify polyester under high temperate. The first important step of cell fusion was the preparation and regeneration of bioplasts. Experiments were carried out to study the best preparation and regeneration condition of bioplasts for B-F and S-R. Experimental factors included the concentration of the enzyme, the processing time and temperature. The results showed that the best preparation and regeneration condition for B-F was to digest the cells at 37℃ for 3 h in the presence of lysozyme(12 g/L) and the best condition for S-R was to digest the cells under 60℃ for 1 h in the presence of lysozyme(1.5 g/L). Under these conditions the preparetion ratio of bioplasts was relative high and the bioplasts could regenerate well. The research created good foundation for next step the fusion of the stains.

Keywords: Polyester; Biomodification; Microbe; Cell Fusion; Bioplast

0 Introduction

Cell fusion technic is a recent development biotechnology. It can make two different parent stains fuse to generate new strains with the characterization and function of both parents[1-3]. The basic method was listed below: enzymes capable to digest the cell wall were added into hyperosmotic stable solution to digest cell wall and prepare bioplasts; two kinds of parent bioplasts coherence and fused together in the presence of fusion accelerator PEG[4-5]; the fused cellplasts were cultured in selective regeneration medium to recover the original cell phase and propagate[3, 6]; the fused cells suitable for breeding purpose were chosen out and used for biomodification.

Bacilli B-F and thermophilic bacteria S-R were used as parent stains in the research. B-F were suitable for moderate temperate and were able to biomodify PET fabrics by improving the property of hydrophilicity, anti-static and anti-pilling which had been studied already[7]. But the modification effect was relatively slight because the temperature suitable for B-F was about 37℃, which was far away from the glass temperatue of the fabrics (about 78℃), so it's difficult for the enzyme secreted by B-F to permeate into the fabrics. If the stains could endure high temperature, the modification effect may be improved. So S-R were chosen as the second parent stains. S-R could grow well at 60℃ but can't modify polyester. If the two species could be fused, the fused cell were hoped to own biomodification effect under high temperate.

The first important step of cell fusion was preparation and regeneration of separate bioplasts. Experiments were carried out to study the best preparation and regeneration condition. Experimental factors included the concentration of the enzyme, digestion time and temperature. The research created good foundation for next step the fusion of the stains.

1 Materials and methods
1.1 Materials

Bacilli B-F was separated from waste water; Thermophilic bacteria S-R (00998) was offered by China center of industrial culture collection; Chemical agents (analytically or biochemically pure) were purchased from Kermel Co. and Probe Co. Tianjin.
1.2 Methods
1.2.1 Preparation of solutions:

Medium A: terephthalic acid (TA) 1 g/L, microelements

Medium B: beef extract 3 g/L, peptone 5 g/L, sodium chloride 5 g/L

Medium C: glucose 10 g/L, beef extract 5 g/L, peptone 10 g/L, sodium chloride 5 g/L, agar 20 g/L

---

* Corresponding author's email: zhangjianfei@tjpu.edu.cn; zhangweiling1983@163.com

Medium D: 1 g/L TA, agar 20 g/L, microelements

Medium E: glucose 10 g/L, beef extract 5 g/L, peptone 10 g/L, NaCl 5 g/L, sucrose 0.5 mol/L, $MgCl_2$ 0.02 mol/L, agar 20 g/L

Medium F: 1 g/L TA, sucrose 0.5 mol/L, $MgCl_2$ 0.02 mol/L, agar 20 g/L, microelements

Hyperosmotic solution: sucrose 0.5 mol/L, $MgCl_2$ 0.02 mol/L

Lysozyme solution: 0.5, 1, 1.5, 2, 4, 8, 12 g/L

1.2.2 Preparation and regeneration of bioplasts:

Bacterial strains B-F and S-R were cultured in medium A and B until log phase. Then 4 ml broth was centrifuged. Thallus was collected, washed twice with hyperosmotic solution, centrifuged and suspended in 4 ml hyperosmotic solution[1, 8, 9, 13]. 1 ml of the suspension was diluted and spread on plates of medium C and D for the determination of cell concentration.

2 ml suspension was mixed with 2 ml lysozyme solution and digested at 37℃ and 60℃ separately for different time. Then the mixture was centrifuged. Sediment was collected, washed with hyperosmotic solution and suspended in hyperosmotic solution as prepared bioplasts.

1 ml of bioplasts of B-F or S-R were diluted with physiological saline and spread on plates of medium C and D to test the preparation rate of bioplasts.

1 ml of bioplasts of B-F or S-R were was diluted with hyperosmotic solution and spread on plates of medium E and F to test the regeneration rate of bioplasts[1, 11].

2 Result and analysis

2.1 Shapes of bioplasts in the process of preparation

The preparation of bioplasts was carried out in the presence of hyperosmotic solution. Lysozyme solution was added to digest cell wall[10-12]. Cells without the surrounding of cell wall would naturally contract and form bioplast balls. The two types of bacterial strains in the experiment were both bacilli whose shapes were showed in Figure 1. After the digestion with lysozyme, cell walls were removed, leaving bioplast balls, as showed in Figure 2.

Basically, higher preparation rate of bioplasts was beneficial to the fusion of bacteria. However, lysozyme was toxical to cells[1]. When it was used with high concentration or for long time, it would cause bioplasts to aggregate and affect the regeneration ability of bioplasts. Different bacilli could resist lysozyme in varying degrees. For bacillus B-F, aggregation would occur when the concentration of lysozyme reached 12 g/L and digestion time exceeded 4 h. While for S-R, the same thing would happen with concentration reaching 2 g/L and digestion time exceeding 2 h as showed in Figure 3. Thus, the concentration of lysozyme and digestion time must be controled strictly.

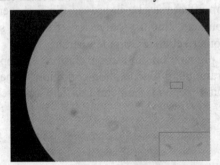
Figure 1 Original Patterns of Bacteria

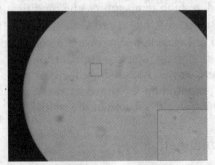
Figure 2 Patterns of Bioplasts

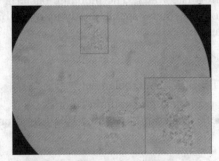
Figure 3 Aggregated Bioplasts

2.2 Preparation and regeneration of bioplasts

Prepared with different concentration lysozyme or digestion time, cell walls were gradually disolved. Claviform bacteria would transform into round bioplasts. Preparation rates of bioplasts of bacillus B-F and thermophile bacteria S-R were showed in Figure 4 and Figure 5, respectively. Bioplasts with high preparation rates were spread on medium E or F to test their regeneration capability, as showed in Figure 6 and Figure 7. Figure 4 and Figure 5 showed that, regeneration rate of bioplasts would increase as the concentration of lysozyme and digestion time was increased. When the concentration of lysozyme reached 8 g/L or 1.5 g/L, the preparation rate of B-F or S-R could reach 70%, respectively. The higher the preparation rate was, the more chances there could be for cell fusion and birth of new bacteria. Whereas, the toxicity of lysozyme would reduce the regeneration rate as the preparation increased. Thus, a balance must be acquired between the preparation and regeneration rate. So the two rates were multiplied to indicate the balance between the two factors, as showed in Figure 8 and Figure 9.

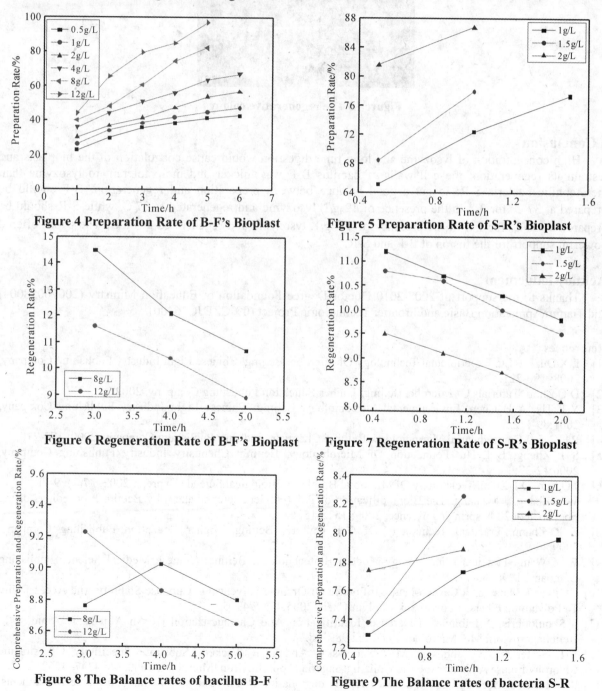

Figure 4 Preparation Rate of B-F's Bioplast
Figure 5 Preparation Rate of S-R's Bioplast
Figure 6 Regeneration Rate of B-F's Bioplast
Figure 7 Regeneration Rate of S-R's Bioplast
Figure 8 The Balance rates of bacillus B-F
Figure 9 The Balance rates of bacteria S-R

As observed in Figure 8 and Figure 9, to get a good balance between preparation and regeneration, B-F need to be prepared at 37℃ for 3h in the presence of 12 g/L lysozyme and S-R need to be prepared at 60℃ for 1h in the presence of 1.5 g/L lysozyme.

2.3 The regeneration of bioplasts

The bioplasts were spread on regeneration medium E and F and they could regenerate new cell walls, renew to complete cell, breed and form even colony finally, as showed in Figure 10.

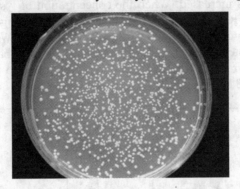

Figure 10 The regenerative colony

3 Conclusion

High concentration of lysozyme and long time digestion would cause coagulation of the bioplasts and restrain its regeneration; the cell walls of bacillus B-F was thicker and more tolerant to lysozyme than thermophilic bacteria S-R; to get a good balance between preparation and regeneration, B-F should be prepared at 37℃ for 3 h in the presence of 12 g/L lysozyme and regenerated at 37℃ while S-R should be prepared at 60℃ for 1 h in the presence of 1.5 g/L lysozyme and regenerated at 60℃. The paper offered powerful support for the fusion of B-F and S-R.

Acknowledgement

Thanks to the support of 2007-2010 Doctor Degree Foundation of Education Ministry (20070058004) and Tianjin Application Basic and Frontier Technologic Project (09JCZDJC26400).

References

[1] L X Du, F P Lu. Experimental Technology Of Microbe. Beijing: Chinese Light Industry Publishing Company, 2008: 190-221.
[2] D Q Zhou. Tutorials Of Microbe. Beijing: Higher Education Publishing Company, 2002:35.
[3] Y Z He, A H Chen. Environmental Mircobiology. Beijing: Chinese Light Industry Publishing Company, 2001:89-112.
[4] P L Cen, J Cai. Industrial Microbiology. Beijing: Chemistry Industry Publishing Company, 2000: 177-210.
[5] S L Zheng, B L Hu. Foundation Of Microbiology. Beijing: Chemistry Industry Publishing Company, 2006:12-37.
[6] J Q Sun, D P Zhou. Technology Of Microbial Bioplasts. Biological Science Express, 2002, 37(7): 9-11.
[7] W L Zhang, J F Zhang. The Biomodification of Poly(ethylene terephthalate) by Bacilli. Proceedings of the fiber society 2009 spring conference, 2009: 1522-1526.
[8] Z R Zhang. Practical Technology Of Cell Culture. Beijing: Higher Education Publishing Company, 2009:89-95.
[9] W Z Wang. Practical Microbiology-Modern Biotechnology. Beijing: Chinese Medical Science Publishing Company, 1996: 56-71.
[10] C Silva, T Matama, A Cavaco-Paulo. Influence Of Organic Solvents On Cutinase Stability And Accessibility To Polyamide Fibers. J Polym Sci A Polym Chem 2005;43: 2749–53.
[11] S Skouloubris, A Labigne, H Dereuse. Identification And Charagerization Of An Alip Hatic Amidase In Heliobagerpylori. Mol Microbiol 1997, 25: 989–98.
[12] J L Seffernick, G. Johnson, M J Sadowsky, And L P Wackett. Substrate Specificity Of Atrazine Chlorohydrolase And Atrazine-Catabolizing Bageria. Appl. Environ. Microbiol. 2000, 66: 4247–4252.
[13] R J Muller, I Kleeberg, And W D Deckwer. Biodegradation Of Polyesters Containing Aromatic Constituents. Biotechnol, 2001: 86-95.

27 Preparation and Properties Characterization of Butyl-methacrylate Copolymer Absorptive Functional Fiber

Zhongjuan Zhang, Changfa Xiao, Naiku Xu

(Tianjin Key Laboratory of Fiber Modification and Function Fiber, Tianjin Polytechnic University, Tianjin 300160, China)

Abstract: The copolymer resin which can absorb organic matter made up of butyl methacrylate (BMA), acrylonitrile (AN), β hydroxyethyl methacrylate (HEMA) was synthesized by suspension polymerization, then the absorption fiber was prepared by gelation-spinning. The effects of consumption of acrylonitrile (AN) on the saturated oil absorbency and the gel fraction were studied. Fourier-transform infrared spectroscopy (FTIR) and Dynamic thermomechanometry (DMA) were used to research the chemical structure and dynamic mechanics performance of butyl-methacrylate copolymer fiber. The results showed that the mass fraction of AN was a main factor which affected saturated oil absorbency, gel fraction and heat-resisting property was better when butyl-methacrylate copolymer fiber had appropriate mass fraction of AN.

Keywords: Methacrylate copolymer; Twin screw spinning machine; Gelation-spinning; Absorption

1 Introduction

Oil pollution of marine environment is becoming a more serious issue with the growth of the off-shore petroleum industry and the necessity of marine oil transportation. As a newly functional polymer material, organic matter absorptive resin has attracted the interest of many researchers[1-2]. Compared with traditional oil-absorptive material, it has a three-dimensional network structure. However, because of its shape limit[3], organic matter absorptive resin have many disadvantages, such as smaller specific-areas, lower absorptive rate and absorptive abilities, compared with fibrous material and this restricted its applications. Although there were a lot of reports about oil-absorptive material, the oil-absorptive fibers were relatively few reported. Feng and her co-workers prepared absorptive fiber by semi-IPN and wet spinning technology[4]. Xu and her co-workers synthesized absorptive fiber to introduce hydroxyethyl methacrylate which was easy to form hydrogen bonding interaction between each other into main chain as structural unit[5]. The purpose of this research was to prepare fibers with oil-absorptive function. We synthesized granular resin by suspension polymerization. The resin could be melted easily by heating when it absorbed some organic matter so as to provide possibility for preparation of fibrous resin. The fiber was prepared by gelation-spinning technique. A detailed study of swelling properties of butyl-methacrylate copolymer fibers was conducted and the structures were characterized through FTIR and DMA.

2 Experimental

2.1 Materials

Butyl methacrylate(BMA); acrylonitrile(AN); hydroxyethyl methacrylate(HEMA); benzoyl peroxide (BPO); N,N-dimethylacetamide(DMAc); Trichloroethylene; Toluene; Kerosene; Poly(vi-nyl alcohol) (PVA)

2.2 Copolymer and fiber preparation

First, 0.5wt% of PVA was dissolved in a flask. A mixture of BMA, AN and HEMA, while mass fraction of HEMA in monomer feed ratio was 15wt% and mass fraction of AN in monomer feed ratio was 0wt%, 10wt%, 20wt%, 30wt%, was added in a beaker. The mixture of BMA, AN, HEMA, 0.5wt% of BPO and 0.5wt% of PVA was stirred to form solution. The solution was reacted under a nitrogen atmosphere for 2.5h when temperature reached 85℃, then the temperature was enhanced to 95℃ for 3.5h. Finally, granular resin was obtained. The granular resin was mixed with DMAc and sealed to deposit for 48～72h at room temperature. In this process, the linear copolymer was dissolved but the three-dimensional network copolymer was only swelling. Through gelation-spinning technique, the butyl-methacrylate copolymer fiber was prepared and dried in oven.

2.3 Test and analysis of BUTYL-METHACRYLATE COPOLYMER fibers

Oil absorbency: A weighed quantity of fiber was immersed in oil until equilibrium was reached, and the residual oil was then removed by dropping for 10 min. The equilibrium was determined by measuring the oil absorbency at each time until it reached a limiting value. The oil absorbency W, was determined by weighing the swollen gel and calculated according to the following equation:

$$W=(W_1-W_0)/W_0$$

where, W_1 is weight of swollen gel and W_0 is weight of dried fiber.

Gel fraction: A weighed quantity of fiber was put in Soxhlet extractor and continuously extracted for 8h, using butanone as solvent. The gel fraction, R, was calculated according to the following formula:

$$R=W_a/W_b \times 100\%$$

where, W_a is the weight of dried fiber after extraction and W_b is the weight of dried fiber before extraction.

FTIR analysis: The spectra were recorded on a Bruker Tensor-37 spectrophotometer with DTGS detector.

DMA analysis: Dynamic mechanics performance of fibers was determined by an instrument NETZSCH DMA242C dynamic thermal mechanical

3 Results and discussion

3.1 Effect of mass fraction of AN in monomer feed ratio on oil absorbency

Figure 1 showed the relationship between saturated absorbency of butyl-methacrylate copolymer fiber and copolymerizing monomer composition. It could be found from Figure 1 that when the mass fraction of AN was 0wt%, adsorption capacity of butyl-methacrylate copolymer fibers on kerosene, toluene and trichloroethylene were the maximum. While the mass fraction of AN was 10wt%, the adsorption decreased. The introduction of –CN increased the force of intermolecular and reduced the gap between molecules. When the mass fraction of AN was 20wt%, the adsorption increased a little. When the mass fraction of AN was 30wt%, the adsorption decreased. Over all, the change can be omitted. Figure 1 demonstrated that absorbency for kerosene was lower, but saturated absorbency for toluene or trichloroethylene was higher. Kerosene with nine to sixteen carbon atoms, toluene with seven carbon atoms, and trichloroethylene has two carbon atoms, but the side chain of polymer has three or five carbon atoms. It is well known that if the length of side chain of polymer is similar to that of organic matter molecule, then absorbency of polymer for this organic matter is higher[6].

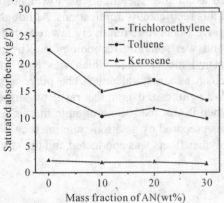

Figure 1 Relation between oil absorbency and the mass fraction of AN

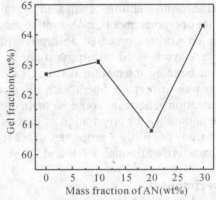

Figure 2 Relation between gel fraction and the mass fraction of AN

3.2 Effect of mass fraction of AN in monomer feed ratio on gel fractions

Figure 2 presented the relation between gel fraction and the mass fraction of AN. Gel fraction was commonly used as a parameter representing the degree of crosslinking of a polymer material[7] When the mass fraction of AN was 0wt%,the gel fraction of butyl-methacrylate copolymer fibers was about 62.8%.When the mass fraction of AN was 10wt%,the gel fraction of butyl-methacrylate copolymer fibers reached 63.2%. Molecules were linked closely together, when it introduced the -CN. When the mass fraction of AN was 20wt%,the gel fraction of butyl-methacrylate copolymer fibers decreased to 60.8%. The number of -CN and steric hindrance suddenly increased. The two ways led to the crosslinked structural decrease. When the mass fraction of AN was 30wt%,the gel fraction of butyl-methacrylate copolymer fibers increased. This was because in the BMA-AN-HEMA copolymer system, BMA reactivity was higher than AN (rAN = 0.29, rBMA = 0.98).While the consumption of AN increased, the cross-termination rate of copolymerization increased. That resulted less average chain length and higher degree of crosslinked.

3.3 FTIR analysis

Figure 3 indicated IR spectra of butyl-methacrylate copolymer fibers. The peak induced by stretching vibration of hydroxyls appeared at 3512.9 cm^{-1} but it was not very obvious when mass fraction of AN was 0wt%,then appeared at 3520.0 cm^{-1} and 3532.1 cm^{-1} when mass fraction of AN was 10wt%,even appeared at 3539.5 cm^{-1} when mass fraction of AN reached 20wt%,finally appeared at 3547.8 cm^{-1} when mass fraction of AN reached 30wt%.It is well known that the peaks at 3550~3450 cm^{-1} and 3400~3200 cm^{-1} are due to hydroxyls forming intermolecular hydrogen bonds while the peaks at 3570~3450 cm^{-1} may be also induced by hdyroxyls forming intramolecular hydrogen bonds[8]. Thus, butyl-methacrylate copolymer fibers contained hydrogen bonds and introduced the physical cross-linked.

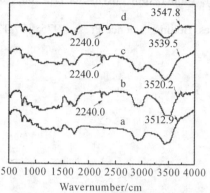 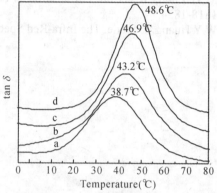

Figure 3 FTIR spectra of butyl-methacrylate copolymer fibers
a AN 0wt%   b AN 10wt%
c AN 20wt% d AN 30wt%

Figure 4 DMA curves of butyl-methacrylate copolymer fibers
a AN 0wt%   b AN 10wt%
c AN20wt%   d AN30wt%

3.4 DMA analysis

Polymer material has viscoelastic and their mechanical properties would change greatly with time, frequency and temperature. Therefore, we studied the mechanical properties of butyl-methacrylate copolymer fibers by dynamic viscoelasticity meter. We could clearly see from Figure 4 that loss peak shifted to higher temperature field as mass fraction of AN was increased. Also, it meaned the glass transition temperature rose with the increased of mass fraction of AN. When we introduced polar -CN group to the macromolecular chains, density of polar groups and the force between molecules increased. That segments moved more difficult and the segment movement needed more energy. Meanwhile, intramolecular hydrogen bonds were enhanced and physical crosslinking between molecules increased.On the other hand,it confirmed that the content of AN in the butyl-methacrylate copolymer was increased.

4 Conclusions

The mass fraction of AN was a main factor affecting absorbency for organic matter and gel fraction. The absorbency for organic matter bagan with decrease, then bagan to increase, and reached minimum value when mass fraction of AN was 30wt%.The highest absorbency of butyl-methacrylate copolymer fibers to various oil was 22.6g(trichloroethylene)/g(fiber), 15.1g(toluene)/g (fiber), and 2.5 g(kerosene)/g(fiber), respectively. Likewise, the gel fraction was changed with increasing mass fraction of AN. Intermolecular and intramolecular hydrogen bond could be found by FTIR. Mass fraction of AN had great effect on dynamic mechanics performance of butyl-methacrylate copolymer fibers. Heat-resisting property of butyl-methacrylate copolymer fibers could be improved by introducing AN into macromolecules.

Acknowledgement

The authors wish to acknowledge the financial support provided by the National Natural Science Foundation of China (Project number: 50673077).

References

[1] J Jang, B S Kim. Studies of crosslinked styrene-alkyl acrylate copolymer for oil absorbency application Ⅱ. Effects of polymerization conditions on oil absorbency. Appl Polym Sic [J], 2000, 77(4): 914-920.
[2] Z X Weng. The synthesis and application of high oil-adsorbing resin. China Synthetic Resin and Plastics [J], 2003, 20(6): 66-69.
[3] J Jang, B Kim. Studies of crosslinked styrene-alkyl acrylate copolymers for oil absorbency application. I.

Synthesis and characterization. Appl Polym Sci [J], 2000, 77(4): 903-913.
[4] Y Feng, C F Xiao. Journal of Applied Polymer Science [J], 2006, 101: 1248-1251.
[5] N K Xu, C F Xiao, Y Feng.Preparation and characterization of absorptive functional fiber copolymerized by n-butyl methacrylate with hydroxyethyl methacrylate. Chemical Journal of Chinese Universities [J], 2008, (8): 1677-1683.
[6] S J Ji, W Q Dai, H S Di, X L Zhu, J M Lu. Study on high oil absorptive resin of copolymer of hydroxyethyl acrylate and (2-ethylhexyl) acrylate. Chem. Eng. Chinese [J]. 2002, 16: 446-449.
[7] M H Zhou, W J Cho. Oil absorbents based on styrene-butadiene rubber.Appl Polym Sci [J], 2003, 89(7): 1818-1824.
[8] W Y Huang, C S Nie. The Infra-Red Spectra of Complex Molecules. Beijing: Science Press, 1975: 107-127.

28 Solubility of Bacterial Cellulose in LiCl/DMAc Solvent System

Chuanjie Zhang, Ping Zhu*, Liu Wang

(Key Laboratory of Green Processing and Functional Textiles of New Textile Materials of Ministry of Education, Wuhan Textile University, Wuhan 430073, China)

Abstract: Activated bacterial cellulose can dissolve in LiCl/DMAc solvent system. The article studied the influences of different activating conditions on the dissolubility of bacterial cellulose (BC), such as the concentration of 1, 2-ethylenediamine, activating time and temperature, as well as the influences of dissolving conditions, like the concentration of LiCl, agitating time and temperature. It discussed the dissolving mechanism of BC in LiCl/DMAc solvent system, and analyzed the infrared spectra of the original and regenerated BC. The results showed that BC could dissolve in LiCl/DMAc solvent system after activated by ethylene diamine. But the dissolubility reduced if the concentration of ethylene diamine exceeded 10%. Agitating temperature and extending of activating time had little effect on the solubility of BC. Contrarily, high agitating temperature would darken the color of BC solution. BC couldn't dissolve completely when concentration of LiCl was lower than 8% and higher than 10%, or at a lower temperature. While the temperature reached 80℃, BC could dissolve in 8% LiCl/DMAc solvent system completely. Though, extending agitating time can improve the solubility of BC, it was better to choose three hours stirring. Therefore the optimum activating conditions were: 10% ethylene diamine, room temperature and 90 minutes activation. The optimum dissolving conditions of BC in LiCl/DMAc solvent system were: 10%LiCl, dissolving at 80℃, agitating for three hours. What's more, in LiCl/DMAc solvent system, bacterial cellulose only swollen when heated and changed into transparent solution when it was cooled to room temperature.

Keywords: Bacterial cellulose; Activation; solubility; LiCl/DMAc solvent system

1 Introduction

Bacterial cellulose (BC) is micromolecule carbohydrate as a kind of matter by means of microorganism fermentation [1]. It exists in pure cellulose, several superfine fiber makes up network structure. Compared to conventional vegetal cellulose, BC possesses many excellent properties, such as high purity, high degree of polymerization, high degree of crystallinity, high hydrophilicity, high Yang's model, high strength, nano grade fiber, biocompatibility and biodegradation [2-4]. Therefore, extensive scientific attention was paid to BC as new-style microbial composite material, it has extensive application in food industry, biomedicine, paper making, acoustic equipment and oil production [5-7].

Because of high degree of polymerization of BC, there are problems to dissolve BC. LiCl / DMAc solvent system is a new solvent for cellulose, the property of this solvent system has been studied at home and abroad [8]. This paper researched the solubility of BC in LiCl / DMAc solvent system, including the influences of different activating conditions and dissolving conditions on the dissolubility of bacterial cellulose (BC), and the dissolving mechanism of BC in LiCl/DMAc solvent system. It's beneficial to further development of BC board its application.

2 Experimental

2.1 Materials

BC was supplied by Hainan Yida Food Co., Ltd with polymerization degree(DP) of around 10,000. LiCl·H_2O(AR), DMAC(AR), ethylenediamine (AR) and methyl alcohol(AR) were purchased from Sinopharm Chemical Reagent Co., Ltd.

2.2 Activation of BC

Immerge 2.0g anhydrous BC into a given concentration of 100ml ethanediamine solution at a temperature for a while, rinse with a lot of distilled water and methyl alcohol respectively, parch to constant weight in vacuum dryer [9].

2.3 Dissolution of BC

Ensure Lithium chloride is absolutely anhydrous at 150℃ in vacuum drying oven. Pour in 100 ml

---

* Corresponding author's tel: +86-27-87770382, email: pzhu99@163.com

DMAc and a certain amount of anhydrous LiCl respectively to a three-necked round flask, heat up it in oil-bath pan. when LiCl is fully dissolved, add 1.5g activative BC, keep heating and stirring at 300r/min for hours, observe after holding it at room temperature for a while. It is necessary to heat and stir again for 1hour after holding for 24 hour but insufficiently dissolved, holding at room temperature until completely dissolved.

2.4 Characterization

FTIR spectra of BC membrane before and after activation were observed with Nicolet 5700 spectrometer.

3 Results and discussion
3.1 Infrared spectra of BC and activated BC

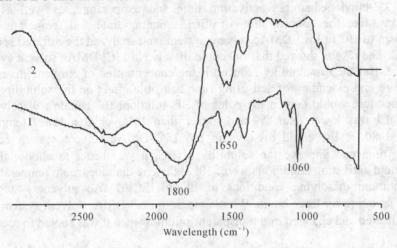

Figure 1 Infrared spectra of BC (1) and activated BC (2)

The Structure of BC is intricate, so it is difficult to dissolve BC directly without activation. Activating is helpful to impair hydrogen bond between molecules and destroy the super molecular structure of BC, which makes it easier to dissolve BC in LiCl/DMAc solvent system. The infrared spectra of BC and activated BC are shown in Figure 1. The absorption peak at 1060cm^{-1} is caused by stretching vibration of carbon-oxygen bond, which is a characteristic peak of cellulose. The absorption peak at 1650 cm^{-1} is attributed to stretching carbonyl and the peak at 1800 cm^{-1} indicates the presence of carbonyl group of cellulose. Figure 1 shows BC and activated BC both have absorption at above three characteristic peaks, but absorption peak at 1060 cm^{-1} of activated BC shifts and the peak area reduces. Therefore, activation to BC makes hydrogen bonds destroyed partly and the degree of polymerization decreased but without derivatization.

3.2 Influence of activating conditions on solubility of BC

When cellulose go swelling in ethanediamine solution, the basal reaction units are chain chips rather than individual molecular chains. Pass through gap between two chips, break hydrogen bonding so that hydroxide radical linked with the third and the sixth carbon atoms are exposed. Hydroxide radical are exposed makes the attainability of solution increases but the degree of array order goes down. Therefore, dissolving capacity turns better.

3.2.1 Concentration of 1, 2-ethylenediamine

Table 1 The influence of concentration of ethylene diamine on solubility of BC

Concentration of ethylene diamine (%)	Agitating for 2h	Placing for 24h	Agitating for 1h again	Placing for 36h again
10	Swelling intensely	Dissolving completely		
20	Swelling intensely	Dissolving partly	Dissolving completely	
30	Swelling intensely	Dissolving partly	Dissolving partly	Dissolving completely
40	Swelling mostly	Dissolving partly	Dissolving partly	Dissolving completely
50	Swelling mostly	Dissolving partly	Dissolving partly	Dissolving completely

Annotation: activating time is 4h, activating temperature is 40℃.

The concentration of 1, 2-ethylenediamine is influential for the degree of activation, and results in different solubility in LiCl/DMAc solvent system. Results are shown in Table1. The solubility of BC in LiCl/DMAc solvent system is improved after activation and BC dissolves completely when the concentration of ethylene diamine is 10%. However, its solubility will decrease when the concentration of ethylene diamine is above 10%, because the polymerization degree of BC is too high and it's difficult for the permeation of high concentration ethylene diamine into BC.

3.2.2 Activation time

Table 2 The influence of activation time on solubility of BC

Activation time (h)	Agitating for 2h	Placing for 24h	Agitating for 1h again	Placing for 36h again
0.5	Swelling slightly	Dissolving partly	Dissolving partly	Dissolving partly
1	Swelling	Dissolving partly	Dissolving partly	Dissolving completely
1.5	Swelling intensely	Dissolving partly	Dissolving completely	
2	Swelling intensely	Dissolving partly	Dissolving completely	
3	Swelling mostly	Dissolving partly	Dissolving completely	
4	Swelling mostly	Dissolving partly	Dissolving completely	

Annotation: concentration of ethylene diamine is 10%, activating temperature is 40℃.

The influence of activation time on solubility of BC is shown in Table 2. It can be seen from Table 2, optimal activation time is 90min. There is no influence to dissolution of BC when continue to prolong activation time. Because it takes some time for ethylene diamine to enter BC and BC can be swollen enough in 90 minutes. Continuing to prolong time can't improve activation of BC.

3.2.3 Activation temperature

Table 3 The influence of activation temperature on solubility of BC

Activation temperature (℃)	Agitating for 2h	Placing for 24h	Agitating for 1h again	Placing for 36h again
25	Swelling intensely	Dissolving completely		
30	Swelling intensely	Dissolving completely		
40	Swelling intensely	Dissolving partly	Dissolving completely	
50	Swelling intensely	Dissolving partly	Dissolving completely	
60	Swelling lightly	Dissolving partly	Dissolving completely	

Annotation: concentration of ethylene diamine is 10%, activating time is 90 min.

Table 3 shows the influence of activating temperature on solubility of BC. When the concentration of ethylene diamine is 10% and activating time is 90min, activating temperature has a small impact on the solubility of BC. Further, the color of BC solution is getting dark gradually while temperature increases. The optimum activating conditions were: concentration of ethylene diamine was 10%, room temperature and activation time was 90 minutes.

3.3 Influence of dissolving conditions on solubility of BC

3.3.1 Concentration of lithium chloride

Table 4 The influence of concentration of lithium chloride on solubility of BC

Concentration of lithium chloride (%)	Agitating for 2h	Placing for 24h	Agitating for 1h again	Placing for 36h again
4	Swelling	Dissolving partly	Dissolving partly	Dissolving partly
6	Swelling intensely	Dissolving mostly	Dissolving partly	Dissolving completely
8	Swelling intensely	Dissolving mostly	Dissolving completely	
10	Swelling intensely	Dissolving mostly	Dissolving completely	
12	Swelling	Dissolving partly	Dissolving partly	Dissolving partly

Annotation: agitating time is 2h, agitating temperature is 100℃.

Lithium chloride plays a important role in dissolving cellulose and the influence of concentration of lithium chloride on solubility of BC is shown in Table 4. It can be seen from Table 4, activated BC can dissolve in LiCl/DMAc solvent system. when the concentration of lithium chloride is relatively low or high, BC dissolves partly in LiCl/DMAc solvent system. There is a best concentration of LiCl for BC's dissolving [10-13] and BC dissolves completely only when the concentration of LiCl is 10%.

3.3.2 Agitating temperature

Table 5 The influence of agitating temperature on solubility of BC

Agitating temperature (℃)	Agitating for 2h	Placing for 24h	Agitating for 1h again	Placing for 36h again
60	Swelling	Dissolving partly	Dissolving partly	Dissolving partly
80	Swelling intensely	Dissolving mostly	Dissolving completely	
100	Swelling intensely	Dissolving mostly	Dissolving completely	
120	Swelling intensely	Dissolving mostly	Dissolving completely	

Annotation: concentration of lithium chlorid is 8%, agitating time is 2h.

The influence of agitating temperature on solubility of BC is shown in Table 5. It can be seen from Table 5, the solubility of BC increases with the increase of dissolving temperature and BC dissolves completely at 80℃. When dissolving temperature exceeds 80℃, the influence of dissolving temperature can be neglected.

3.3.3 Agitating time

Table 6 The influence of agitating time on solubility of BC

Agitating time (h)	Agitating for 2h	Placing for 24h	Agitating for 1h again	Placing for 36h again
1	Swelling	Dissolving partly	Dissolving partly	Dissolving partly
2	Swelling intensely	Dissolving partly	Dissolving completely	
3	Swelling intensely	Dissolving completely		
4	Swelling intensely	Dissolving completely		

Annotation: concentration of lithium chlorid is 8%, agitating temperature is 100℃.

The influence of agitating time on solubility of BC is shown in Table 6. It's beneficial to extend dissolving time for solubility of BC and BC can dissolve completely under conditions of stirring for 3h at 100℃ and placing for 12h at room temperature.

4 Conclusions

Bacterial cellulose can be dissolved in LiCl/DMAc solvent system after activation and the optimum activating conditions were: 10% ethylene diamine, room temperature and 90 minutes activation.

The optimum dissolving conditions of BC in LiCl/DMAc solvent system were: 10%LiCl, dissolving at 80℃, agitating for three hours. What's more, in LiCl/DMAc solvent system, bacterial cellulose only swollen when heated and changed into transparent solution when it was cooled to room temperature.

Compared the infrared spectra of BC and the regenerated BC, it indicated that during the activating reaction and dissolution, there was no derivatization on the BC, only intermolecular hydrogen bonds of the cellulose broke leading to the dissolution of the cellulose.

Acknowledgement

This work was financially supported by National Natural Science Foundation of China (50773032), Doctoral Fund of the Ministry of Education of China (20061065002) and Natural Science Foundation of Hubei Province of China (2009CDA033).

References

[1] B Yu, H L Zhou. Progress in research of bacterial cellulose. Biotechnology bulletin, 2007, (2): 87-90.
[2] C Z Ma, Z R Gu. Biological and physicochemical properties and commercial application of bacterial cellulose produced by acetobacter xylinum. Acta Agriculturae Shanghai, 2001, 17(4): 93-98.
[3] A Bodin, H Backdahl, H Fink, et al. Influence of cultivation condition on mechanical and morphological properties of bacterial cellulose tubes. Biotechnology and Bioengineering, 2007, 97(2): 425-434.
[4] H P Cheng, P M Wang, W T Wu, eta1. Cultivation of acetobacter for bacterial cellulose production in a modified airlift reactor. Biotechnology Applied Biochemistry, 2002, 35:125-132.
[5] F Yashinaga, N Tonouchi, W Kunihiko. Research progress in production of bacterial cellulose by aeration anti agitation culture and its application as a new industrial material. Bioscience biotechnology biochemistry, 1997, 61(2):219-224.
[6] C M Hao, H Luo. Bacterial cellulose—a new biological material. Cellulose science and technology, 2002,

6(2):56.

[7] D P Sun, J D Zhang, et al. Production of bacterial cellulose with acetobacter xylinum fermentation. Journal of Nanjing University of Science and Technology, 2005, 29(5): 601-604.

[8] T Q Wang, J Y Yuan, S Q Yan, et al. Solution characteristics of cellulose in lithum chloride and N,N-dimethylacetamide. Journal Cellulose Science and Technology, 1996, 4(3): 7-13.

[9] Z Q Shao, S Men, et al. The changes of structure and dissolution in DMAc/LiCl of cellulose after pretreated with different methods. Applied Chemical Industry, 2006, 35(8): 587-591.

[10] N G Tsygankova, D D Grinshpan, A O Koren. Mndo modeling of complex formation in N,N-dimethylacetamide-lithium chloridecellulose dissolving system. Cellulose Chemistry Technology, 1996 (30): 357-373

[11] F Albin, B Turbak. Recent developments in cellulose solvent systems. Tappi, 1984, (67): 94.

[12] C L McCormicck, P A Callais. Solution studies of cellulose in Lithium Chlorideand, N-Dimethylacetamide. Macromolecule, 1985, (18): 2394.

[13] M Comick, L Charles. Novel cellulose solutions. US 4278790, 1981.

29　Photocatalytic Properties of TiO$_2$ Supported on Pd-modified Carbon Fibers

Yaofeng Zhu[1], Yaqin Fu[2], Qing-Qing Ni[1]

(1. Graduate School of Science and Technology, Shinshu University, 3-15-1 Tokida, Ueda 386-8576, Japan
2. Key Laboratory of Advanced Textile Materials and Manufacturing Technology, Ministry of Education of China, Zhejiang Sci-Tech University, Hangzhou 310018, China)

Abstract: Pd-modified carbon fibers (CF) are obtained by a facile oxidation-reduction method and then dip-coated in a sol-gel of titanium dioxide (TiO$_2$) to form supported TiO$_2$/Pd-CF photocatalysts. The morphology of the Pd-modified CFs is characterized by field emission scanning electron microscopy. X-ray diffraction is used to investigate the crystal structures of the TiO$_2$ photocatalyst. Acid orange II is used as a model contaminant to evaluate the photocatalytic properties of the composites under UV irradiation. TiO$_2$/Pd-CF exhibits higher catalytic activity than TiO$_2$/CF towards the degradation of acid orange II.

Keywords: Titanium dioxide; Pd particles; Carbon fiber; Photocatalytic; Acid orange II

1 Introduction

Since Fujishima and Honda discovered the photoelectrocatalyzed decomposition of water at n-type semiconductor titanium dioxide (TiO$_2$) electrodes [1], TiO$_2$ has been studied extensively by researchers in different fields. Recently, the problem of separating TiO$_2$ powder catalysts from the working system has been resolved empirically by preparing photocatalytic thin films of immobilized TiO$_2$ [2,3]. However, the immobilized TiO$_2$ thin films exhibit relatively low photocatalytic activity because of their low surface area.

In the present study, the catalytic activity of immobilized photocatalytic materials were improved by taking two measures: 1) determining the best process for fabricating a thin film of TiO$_2$ and modifying its surface, and (2) selecting the appropriate carrier material. The catalytic activity of a material can be improved by a synergistic effect in the composite formed between a carbon-like material and a TiO$_2$ thin film. Several methods have been used to enhance the photocatalytic activity of TiO$_2$ such as preparing as nano-titanium dioxide [4,5], metal doping and coating with metals [6,7], and semiconductor-coupled modification [8]. It has been found that the modification of TiO$_2$ surfaces with a noble metal is one of the most effective ways to improve the photocatalytic efficiency [9].

In the present investigation, polyacrylonitrile carbon fiber (PAN-CF) was used as a catalyst support for photocatalytic TiO$_2$. Pd-modified CF was fabricated through a facile oxidation-reduction method and photocatalytic TiO$_2$ films supported on Pd-modified CF (TiO$_2$/Pd-CF) were prepared by sol-gel dip-coating. The photocatalytic performance of the composite photocatalyst materials is evaluated.

2 Experimental
2.1 Materials
PAN-CF (T300C), with a diameter of about 5-6.5 μm, was provided by Toray Inc. Tetrabutyl orthotitanate (CP grade) was provided by Wuxi Zhanwang Chemical Reagent Co. Ltd., China. Palladium chloride (AR grade) was purchased from Shanghai Jiuling Smelting Co. Ltd., China. All commercial chemicals were used as received without further purification.

2.2 Preparation of TiO$_2$ supported on Pd-modified CF
PAN-CF (0.1 g per fiber was oxidized in concentrated nitric acid (160 ml) by heating under reflux for 4 h in an oil bath at 115°C. After cooling and washing with demineralized water until the pH was >6, the fibers were dried in an oven at 80°C. Initial concentrations of PdCl$_2$ solutions were 0.1, 0.3, 0.5, 0.7 and 0.9 g/L, which were prepared by dissolving PdCl$_2$ in concentrated hydrochloric acid (37%). The oxidized CFs were dipped into PdCl$_2$ solutions (100 ml) for 10 min and then dried in an ambient atmosphere at room temperature. The samples were then immersed in a solution of SnCl$_2$(10 g/L, 100 ml) to reduce the palladium ions. Residual ions on the surface of the substrate were removed by washing with deionized water. The CFs modified with 0.1, 0.3, 0.5, 0.7 and 0.9 g/L PdCl$_2$ solutions are denoted as M1-Pd-CF, M3-Pd-CF, M5-Pd-CF, M7-Pd-CF and M9-Pd-CF, respectively.

A TiO$_2$ sol was prepared using tetrabutyl orthotitanate as the titanium precursor and alcohol as the solvent.The Pd-modified CF substrates were dip-coated in the TiO$_2$ sol for 10 min and pulled out slowly with a uniform pulling rate. The substrate was allowed to dry in an ambient atmosphere at room temperature. Finally, the coated CF substrates were calcined in a nitrogen environment at 600°C for 2 h to give the composite photocatalysts of TiO$_2$ thin films supported on Pd-modified CFs (TiO$_2$/Pd-CF). The composite photocatalysts are denoted as M1TiO$_2$/Pd-CF, M3TiO$_2$/Pd-CF, M5TiO$_2$/Pd-CF, M7TiO$_2$/Pd-CF and M9TiO$_2$/Pd-CF, depending on the concentration of the PdCl$_2$ solution used (0.1, 0.3, 0.5, 0.7 and 0.9 g/L, respectively).

2.3 Characterization

The surface morphology of the Pd-modified CFs was investigated by field emission scanning electron microscopy (FE-SEM, Hitachi S-4800). The crystal structure of the TiO$_2$ was determined using a Bruker AXS (Bruker AXS, D8-Discover) X-ray diffractometer with Cu-Kα radiation. The accelerating voltage and the applied current were 40 kV and 35 mA, respectively.

Evaluation of photocatalytic activity and interface performance. The photocatalytic activity of the prepared samples was evaluated from the photocatalytic degradation of acid orange II in aqueous solution. The sample was placed in a Pyrex tube containing an aqueous solution of acid orange II (100 mg/L, 25 ml) at pH 3. The sample was then irradiated with a 500 W mercury lamp as a UV source. To measure the acid orange II concentration, the absorption at 484 nm, which corresponds to the maximum adsorption of acid orange II, was monitored using a UV-Vis spectrophotometer. The interface performance of the composites was evaluated by using the samples repeatedly.

3 Results and discussion

3.1 Surface morphology of Pd particles deposited on CF

The morphology of the oxidized and Pd-modified CFs is shown in Figure 1. It can be seen that a number of particles are deposited on the surface of the Pd-modified CFs compared with the CFs oxidized with nitric acid. Furthermore, the particles deposited on the CFs uniformly with no aggregation of the Pd particles. This indicates that the facile oxidation-reduction process was used successfully. The uniform distribution of the Pd particles may contribute to stable performance of the composite photocatalysts.

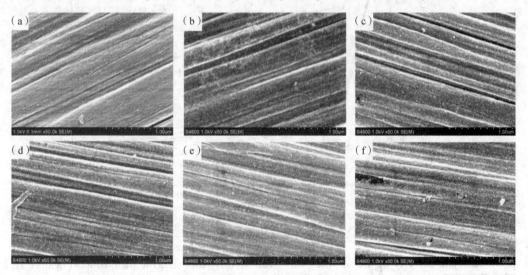

Figure 1 Surface morphology of CF with deposited Pd particles.
(a) CF, (b) M1-Pd-CF, (c) M3-Pd-CF, (d) M5-Pd-CF, (e) M7-Pd-CF, and (f) M9-Pd-CF

3.2 Crystal structure of TiO$_2$

A model TiO$_2$ powder was prepared using the same heat treatment process as the TiO$_2$ films and then used to analyze the crystal structure of the sample. The X-ray diffraction (XRD) pattern of the TiO$_2$ powder calcined at 600°C for 2 h is presented in Figure 2. The XRD pattern shows that only the anatase phase is present, as confirmed from JCPDS NO.21-1272. The average crystal size was determined from parameters in the XRD pattern according to the Scherrer equation (2).

$$D = 0.89\lambda / \beta \cos\theta \qquad (2)$$

where D represents the average crystal size, β is the full width at half-maximum (FWHM), and θ is the diffraction angle. The average crystal size of TiO$_2$ was determined to be 15.2 nm calculated using the characteristic anatase peak at $2\theta=25.3°$.

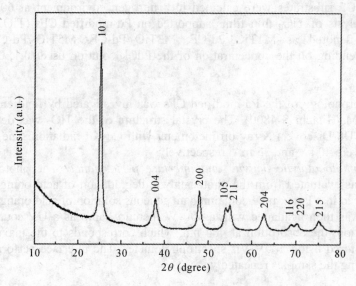

Figure 2 XRD pattern of TiO$_2$ powder

3.3 Photocatalytic activity of the samples

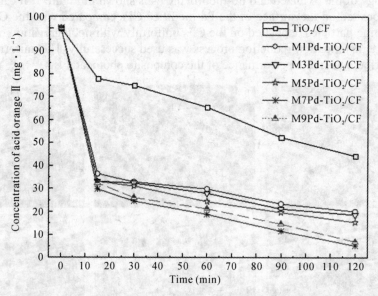

Figure 3 TiO$_2$/CF photocatalytic degradation of acid orange II

The photocatalytic activity of the samples was evaluated by their degradation of acid orange II, a typical azo dye. The TiO$_2$/CF photocatalytic degradation of solutions of acid orange II is shown in Figure 3. The catalytic activity of the composite photocatalysts is influenced significantly by number of Pd particles deposited on the surface of the CFs. After irradiation for 15 min with a 500 W UV light, 22.1% of acid orange II was degraded by TiO$_2$/CF, whereas >63.5% was degraded by the Pd-modified TiO$_2$/CF samples. M7TiO$_2$/Pd-CF showed the optimum catalytic performance of the Pd-modified samples; 70% of acid orange II was degraded by M7TiO$_2$/Pd-CF after irradiation for 15 min. However, as the reaction proceeded, the degradation rate slows because the concentration of the reaction solution decreases continuously, which may be the factor dominating the catalytic reaction rate. As the surface morphology of TiO$_2$/Pd-CF is almost the same as that of TiO$_2$/CF, differences in the structure and properties (such as the physical adsorption properties, catalytic active site and the absorption of UV light) of the samples before and after modification should be minor. It is believed that the deposited Pd particles act as trapping sites to capture photogenerated electrons from the TiO$_2$ conduction band, separating the photogenerated electron-hole pairs. This should improve the photocatalytic activity of the samples, accelerating the degradation of acid orange II. After irradiation for 2h, M7TiO$_2$/Pd-CF still showed the highest photocatalytic activity of the samples, with a degradation ratio of acid orange II that was 70% greater than that of unmodified TiO$_2$/CF. The above analysis indicates that the catalysts possess good stability and photocatalytic activity, exhibiting the potential for TiO$_2$ supported on Pd-modified CF to be used as a composite photocatalyst for industrial organic wastewater treatment.

4 Conclusions

Pd-modified CFs were obtained through a facile oxidation-reduction method and then coated with TiO$_2$ by sol-gel dip-coating. The structure and properties of the resulting TiO$_2$/Pd-CF photocatalysts were studied. The the Pd particles dispersed uniformly on the CFs. The photocatalytic activity of the Pd-modified TiO$_2$/CF photocatalysts for the degradation of acid orange II was significantly improved compared with unmodified TiO$_2$/CF. The M7TiO$_2$/Pd-CF exhibited the optimum catalytic efficiency for acid orange II, which was about 70% higher than that of TiO$_2$/CF, revealing the potential of such a composite photocatalysts for industrial organic wastewater treatment.

Acknowledgement

This work was supported by the Grant-in-Aid Global Center of Excellence (COE) program from the Ministry of Education, Culture, Sports, Science and Technology, Japan and the Natural Science Foundation of Zhejiang Province, China (Grant number: Y406310)

References

[1] Fujishima A, Honda K. Electrochemical photolysis of water at a semiconductor electrode. Nature. 1972; 238: 37-38.
[2] Fernández A, Lassaletta G, Jiménez VM, Justo A, González-Elipe AR, Herrmann JM, Tahiri H, Ait-Ichou Y. Preparation and characterization of TiO$_2$ photocatalysts supported on various rigid supports (glass, quartz and stainless steel) comparative studies of photocatalytic activity in water purification. Applied Catalysis B: Environmental.1995; 7(1-2):49-63
[3] Xua DP, Huang ZH, Kang FY, Inagakic M, Kod TH. Effect of heat treatment on adsorption performance and photocatalytic activity of TiO$_2$-mounted activated carbon cloths. Catalysis Today. 2008; 139(1-2):64-68
[4] Wu ZB, Wang HQ, Liu Y, Gu ZL. Photocatalytic oxidation of nitric oxide with immobilized titanium dioxide films synthesized by hydrothermal method. Journal of Hazardous Materials. 2008; 151(1): 17-25.
[5] Maira AJ, Yeung KL, Soria J, Coronado JM, Belver C, Lee CY, Augugliaro V. Gas-phase photo-oxidation of toluene using nanometer-size TiO$_2$ catalysts. Applied Catalysis B: Environmental. 2001; 29(4): 327-336.
[6] Brezová V, Blažkov A, Karpinský Ľ, Grošková J, Havlínová B, Jorík V, Čeppan M. Phenol decomposition using Mn+/TiO$_2$ photocatalysts supported by the sol-gel technique on glass fibers. Journal of Photochemistry and Photobiology A: Chemistry. 1997; 109(2): 177-183.
[7] Xie BP, Xiong Y, Chen RM, Chen J, Cai PX. Catalytic activities of Pd-TiO$_2$ film towards the oxidation of formic acid. Catalysis Communications. 2005; 6(11): 699-704.
[8] Hou LR, Yuan CZ, Peng Y. Synthesis and photocatalytic property of SnO$_2$/TiO$_2$ nanotubes composites. Journal of Hazardous Materials. 2007; 139(2): 310-315.
[9] Sheng ZY, Wu ZB, Liu Y, Wang HQ. Gas-phase photocatalytic oxidation of NO over palladium modified TiO$_2$ catalysts. Catalysis Communications. 2008; 9(9): 1941-1944.

30 Impregnation of Metal Complex into Epoxy Insulation Materials Using Supercritical Carbon Dioxide and Its Application for Copper Plating

H. Ohnuki[1], S. Sumi[2], T. Hori[1]

(1. Fiber Amenity Engineering Course, Graduate School of Engineering, University of Fukui, 3-9-1 Bunkyo, Fukui 910-8507, Japan

2. Guangzhou Meadville Electronics Co., Ltd. No.1, Xinle Rd. Science City, Guangzhou, Guangdong Province, China)

Abstract: Metal plating of epoxy polymer is widely applied for industrial products for long time, especially in the field of Printed Circuit Boards (PCB's). This technique is one of the most important technologies in terms of electronics devices with high reliability and guaranteed quality. The authors are developing a new concept of PCB's for next decade generations to improve the technique including more fine line circuitry, high density and narrow space conductor lines. An essential subject for this technology is to improve the weak copper adhesive peel strength on epoxy insulation materials. To get the good adhesion property of copper plating, it is now widely applying Pd Colloid Solution method. But to achieve more excellent adhesion for next decade generation PCB's, we are investigating Super Critical Fluid (SCF) method. In this paper, an attempt has been done to impregnate some metal complexes into epoxy resin and decomposed them to produce free metal in the resin by reduction. Using the deposited metal an efficient electro-less Cu plating is achieved. We will discuss on the selection of the metal complexes and impregnation condition of the complexes as well as peel strength of the plating.

Keywords: Printed Circuit Board's (PCB's); Epoxy Materilas; Super Critical Fluid (SCF); Pd complex; Copper peel strength

1 Introduction

PCB's consuming market is rapidly shift to the high density and high reliability circuit technologies[1][2]. Especially in Portable Handy (Cellular) Phone and Laptop Computer markets, they are strongly requesting those latest technology applications. Those technical demands background, there are challenging to satisfy with more smaller size, more light weight and more high multiple function equipments. Then PCB's also applies to change into more fine lines and high dencity circuits boards. Those fine circuitly condition demands for next decade period PCB's Era are strongly requesting to excellent copper conductor peel strength and reliability. On those market requesting background to satisfy with the excellent PCB's quality level, we are investigating on Super Critical Fluid (SCF) method and an attempt has been done to impregnate some metal complexes into epoxy resin and decomposed them to produce free metal in the resin by the reduction of complex.

Currently PCB's main plating technology is the Pd Colloid Solutins process combining with electro-less plating and electric plating for connecting the both side circuits and inner layer circuitries. This plating technology has strong point for cost performance and very popular using, but weak point is not excellent copper peel strength on epoxy direct surface plating.

Using the SCF deposited metal an efficient electro-less Cu plating is achieved [3][4][5].

In this paper we are presenting the SCF application development to epoxy insulation which is the first approach to impregnate metal complex. After some pre-experiments a few Pd complexes were selected and suitable conditions for their impregnations into epoxy polyemr plate were investigated, considering the peel strength of the Cu plated polymer plate.

2 Experimental

The impregnated materials is an Epoxy C stage plate. This epoxy resin for test materials is applied with Bis Fenol A-type resin (Liquid type property) which is widely using in the PCB's industlies. Figure 1 is the Bis Fenol A-type chemical formation.

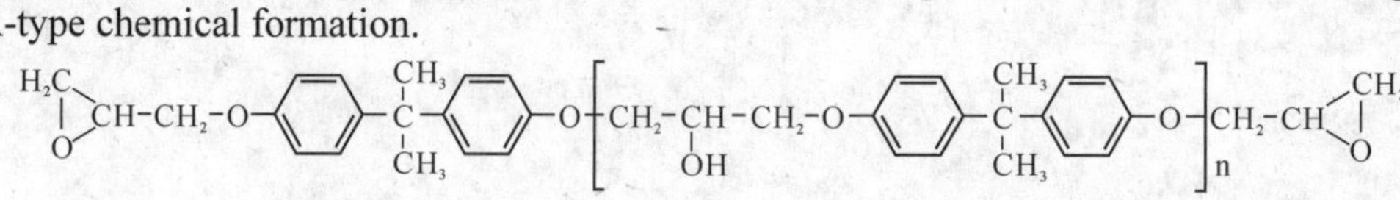

Figure 1 Bis Fenol A-type epoxy resin (Liquid type)

As impregnation medium, Supercritical Carbon Dioxide [$SC(CO_2)$] was used. Temperature, pressure and treatment time were changed for metal complex impregnation.

Table 1 is the $SC(CO_2)$ variable factors combination condittins for epoxy impregnation test.

Table 1 Impregnation conditions

	Level1	Level2	Level3	Level4
Temperature(Deg.)	60	100	140	150
Presssure(Mpa)	10	15	20	25
Time(Min)	30	60	90	120
Leakage time(Min)	1	10	30	

Selected Pd complexes are Pd(II)hexafluoro-acetylacetonate (Complex A), Bis-acetylaceto-nate Pd(II) (Complex B) and Pd(II)acetate (Complex C).

Figure 2 is appearance of the selected complexes before and after $SC(CO_2)$ treatment and their chemical structures. The epoxy plates were treated under various $SC(CO_2)$ conditions to impregnate Pd complex into them. Pd complex impregnated epoxy samples are further treated to decompose the complexes to produce free metal Pd. Then Pd impregnated epoxy plates were applied to electro-less and electric copper sulfate plating processes for metallization. Impregnated Pd complex samples and plated samples are checked by PST (Peel Strength Test), SEM (Scanning Electron Microscope), XPS (X-ray Photoelectron Spectroscopy), EDX (Energy Dispersive X-ray Fluorescence Spectrometer) and TEM (Transmission Electron Microscope) to get the Pd complex existing conditions in epoxy material and to compare the current plating peel strength.

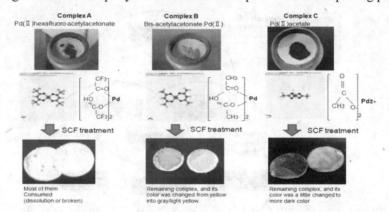

Figure 2 Three typical type of complexes conditions before and after $SC(CO_2)$ treatments

3 Results and discussion

We will discuss in this paper, if the $SC(CO_2)$ is effective or not existing for Pd complexes impregnation into epoxy materials and which kind of complexes is effective property for high amount of complex impregnation into epoxy material to obtain high performance Cu plating.

As the first step, we checked the effect of $SC(CO_2)$ to epoxy materials and to the adsorption of complex on epoxy material.

Firstly, a blank treatment of epoxy material without both $SC(CO_2)$ and complexes on epoxy material was checked.

Secondly, epoxy material was treated with $SC(CO_2)$ and without complexes.

Thirdly, the materials was treated with both $SC(CO_2)$ and some complex.

Figure 3 is one of the test results. Excellent results was found with good copper metal plating performances using $SC(CO_2)$. In the other two processes, no good the copper metal plating on the epoxy resin surfaces was given.

It is clear that SC(CO_2) is one of good technical approach for impregnating the complexes onto epoxy materials, and also SC(CO_2) makes better copper plating adhesive performances compared with blank treatments.

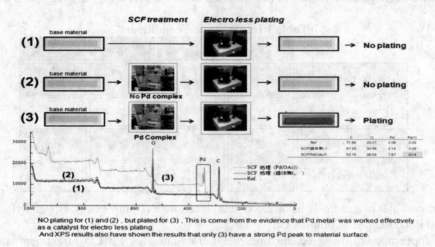

Figure 3 Verification check for impregnation test

As the second step, the complex impregnation using SC(CO_2) was carried out for three kinds of complexes for longer time. We used these complexes singly or mixed parameter. The Pd complex impregnated samples were put into the electro-less plating solution.

Figure 4 shows the appearance of electro-less copper plating after impregnation of one kind of complexes or some mixture complexes.

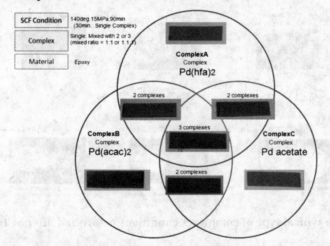

Figure 4 Electro-less plating visible conditions after three complexes combination impregnations

Cu plating was carried out easily. To increase the thickness of the Cu plating layer, the electric plating has been done by treating the electro-less plated plate into copper sulfate plating solution with 2 ASD for 30 minutes. The thickness of the Cu plating was controlled around 15 μm.

Figure 5 shows the results of copper peel strength test after electric copper plating.

In the case of single complex Pd(II)acetate (complex C), the highest peel strength of copper plate was obtained.

In the case of mixed state with two or three complexes, good peel strength was found using a mixture of Pd(II) hexafluoroacetylacetonate and Pd(II)acetate (ComplexA+C).

It is very interesting that better copper peel strength was obtained in the case of mixed state with 2 complexes comparing in the case of single complex implegnation.

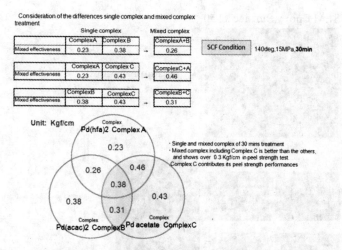

Figure 5 Copper plating average peel strength deviations for three complexes impregnations territory

Finally, we compared Cu plated sample using $SC(CO_2)$ and the current plating method.

As results, the better copper peel strength by Cu plating appling $SC(CO_2)$ method was obtained comparing with the current plating method (Pd Colloid solution plating).

The strong peel strength complex was gotten mainly in the case of Pd(II)acetate (Complex C) or a mixture of Pd(II)hexafluoroacetylacetonate and Pd(II)acetate (Complex A+C).

From XPS analysis, we found the metal state of Pd was located not on the polymer surface, but in a little deeper region of the polymer plate.

Figure 6 shows XPS analysys for mixed complex(A+C) which applied impregnated condition and after Ar gas etching treatments removing skin surface contaminations. Pd intensity strength is little bit increased results. Then Pd complexs are located on near the skin surfaces on Epoxy materials.

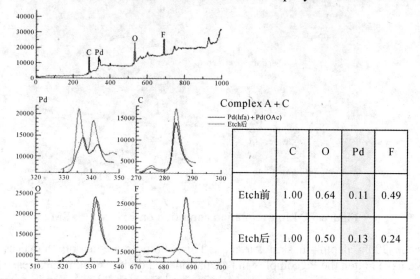

Figure 6 XPS analysis of mixed complex (A+C)

Figure 7 shows EDX mapping connditions result for mixed complex(A+C). Pd is also rocated Epoxy surface and very strongly and naturaly exsisting on its resin surfaces.

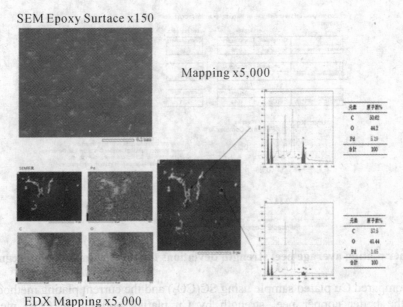

Figure 7 Pd EDX Mapping condition of Complex (A+C)

Finaly we observed TEM annalysis for visible check.
Figure 8 show TEM cross section analysis. The Epoxy surface Pd exsists on both shiny surface and mat sufrace.

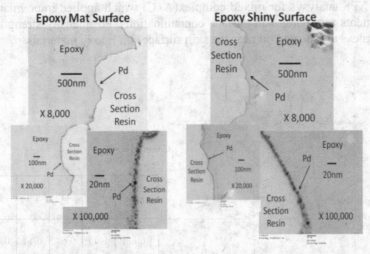

Figure 8 Pd Impregnation condition on Epoxy by TEM

Pd thickness is 20nm depth on skin surfaces. This 20nm impregnation thickness was called ***Skin Impregnation***. We have gotten the Pd complex impregnation area on Epoxy skin surfaces.

In Figure 9, copper plating peel strength was summerized for each Pd complex impregnation and current plating cases. This peel strength level is approximatly 0.4~0.5kgf/cm. These values about 2.0~2.5 times stronger, compared than current sulfate copper direct plating peel strength on bare epoxy surfaces plating.

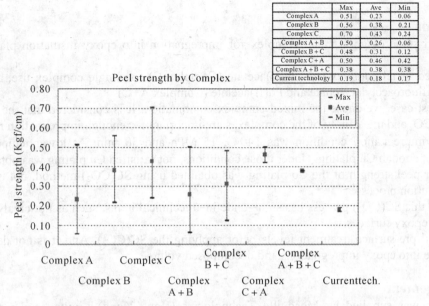

Figure 9 SC(CO$_2$) and current Cu plating peel strength

Figure 10 shows all of the summarized results. Better Cu plating peel strength was gotten by using SC(CO$_2$) under both selected SCF conditions and selected metal complexes.

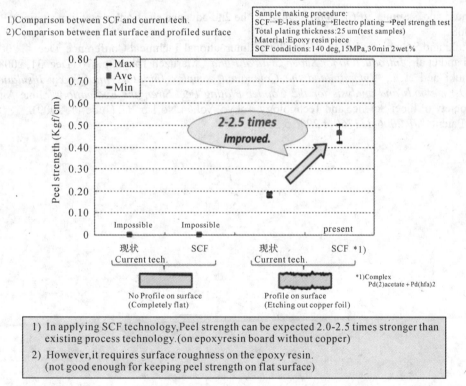

Figure 10 SC(CO$_2$) and current Cu plating peel strength and it surface conditions impact

But there are un-expected results that are impacted Epoxy surface treatment conditions.

It is clear that profile surface (mat) is good adhesion result and non profile surface (shiny) is poor or not peel strength result by SC(CO$_2$) impregnations. Pd complex can be impregnated both surfaces by SC(CO$_2$). But the copper peel strength is affected from Epoxy surface materials roughness. Plated copper peel strength is affected from its anchor phenomenon.

These anchor phenomena consist from Epoxy surface roughness.

4 Conclusions

We found some of good metal complex for impregnation into epoxy insulation plate using $SC(CO_2)$ treatment.

The selected metal complex was Pd(II)acetate (Complex C) for single complex use, and for mixed use case, Pd(II)hexafluoroacetylacetonate and Pd(II)acetate (Complex A+C).

As the best case, we chose the mixed complex combination. It is considered that the dissolution of the complexes in CO_2 and the affinity of the complexes are important for enough impregnation into epoxy resin.

The best impregnation condition was 140℃, 15 MPa and 30 min. At lower temperature and lower pressure are not enough Cu plating. Under those conditions, not satisfied Cu plating was obtained.

The better peel strength of the Cu plating was obtained using $SC(CO_2)$ method, not using conventional Pd Colloid Solution processes.

Currently this $SC(CO_2)$ process has some undefined technical items, such as not crealy fixed the surface condittions on epoxy surfaces.

It is called pre-surface treatment levels befor applying the $SC(CO_2)$. And it is not detected complexes amount volume into epoxy impregnations and decomposed volume.

Acknowledgement

The work was supplied by Meadville Technologies (Hong Kong) Limited and Graduate School of Engineering, University of Fukui under agreement with Mutual Research and Development Agreements (MRDA).

References

[1] H.Ohnuki and S.Sumi, "*JPCA 2008 Seminar*", the 2nd advance electronics pavilion seminar, P71~P88, June 11, 2008.
[2] H.Ohnuki and S.Sumi, "*HKPCA & IPC 2008*" International Technical Conference, Dec. 4, 2008.
[3] T.Hori and et al, "*Introduction for Super Critical Fluid*", Maruzen Tokyo, p.176 (Dec. 31, 2008).
[4] H.Ohnuki and et al, "*Impregnated the Compound complex Initiators into Epoxy Insulation Materials by SC(CO₂) methods and analysis for the Copper Plating Peel Strengths Evaluations*" June Annual meeting of The Society of Fiber Science and Technology, Tokyo, Vol.64,No.1 p.262 (June 11, 2009).
[5] Japan Patent, "*JP 2006-131769 A 2006.5.25*".

31 Synthesis, Characterization and Application of Polyurethane Modifying Polyether Block Polysiloxane

Qiufeng An, Fan Yang, Kefeng Wang, Ge Li

(Key Laboratory of Auxiliary Chemistry and Technology for Chemical Industry, Ministry of Education, Shaanxi University of Science and Technology, Xi'an 710021, China)

Abstract: In this article, a HO-terminated polyether-block-polysiloxane intermediate (PESO) was firstly synthesized by the hydrosilylation of Si-H terminated polydimethylsiloxane with an allyl polyether, then a polyurethane modifying polyether-b-polysiloxane (TESO) was prepared by the reaction of PESO with 2,4-toluene diisocyanate (TDI). Chemical structure and film-forming ability of TESO as well as its application on fabrics were characterized and investigated by IR, ^{1}H-NMR, scanning electronic microscope (SEM) and other instruments. Experimental results indicate TESO had a good compatibility with cationic/anionic resins or auxiliaries, and could form a hydrophilic, relatively smooth film on the cotton fabric. The binding rigidity of the cotton fabric treated by TESO clearly decreased from 146.5[warp (w)] and 318.8[fill (f)] mN to 106(w) and 189(f) mN, meanwhile the winkle recovery angle of the treated fabric increased from 129° to 220.8°. But the wettability of the treated cotton fabric is only 2"68. All indicate TESO offers a good softening property, elasticity and an excellent wettability to the treated fabrics.

Keywords: Polyurethane; Polyether; Polysiloxane; Fabric

In textile finishing, polyether polysiloxane could confer the textiles with suitable softness, excellent wettability and antistatic property as well as soil-release performance since the good hydrophilic and antistatic property when coated on fiber surfaces[1-3]. These performance properties make the functionalized polysiloxanes have great potential in mimic-natural treatment of synthetic fibers and their blends. Compared with the polysiloxanes with pendant polyether side groups, the $(AB)_n$ type, alternate polyether-block-polysiloxane can confer an improved softness, smoothness and superior water wettability to the treated fabrics. However, limited by synthesis methods and skill, only several polyether-block-polysiloxanes are manufactured in feasible way in the past years.

Using active diisocyanate to react with HO-terminated polyether-block-polydimethylsiloxane (PE-b-PDMS) can conveniently change the blocked siloxane polymers into the expected, alternate $(AB)_n$ polyether functional polysiloxanes of high molecular weight. To reach this target, a HO-terminated POE-b-PDMS was first synthesized in the research by hydrosilylation, followed by esterification of the -OH group with the -NCO from 2,4-toluene diisocyanate (TDI). Finally, a novel Polyurethane modifying (POE-b-PDMS)$_n$, namely TESO, was successfully obtained in the research. In view of the fact that micro-morphology of polysiloxanes has an effect on the performance, the film morphology of $(AB)_n$ POE-b-PDMS as well as their performance on the treated fabrics were also investigated in the work.

1 Experimental
1.1 Materials and reagents

Si-H terminated polydimethylsiloxane (α,ω-PHMS, Scheme 1) with a polymeric degree about 30-33 was bought from Qinyang Fine Chemical Co, China. Allyl polyoxyethylene (CG-5) with an average molecular weight (Mn) of 498, was commercial product provided by Nanjing Well Chemical Corp., China. 2,4-Toluene diisocyanate (TDI), isopropanol and ethyl acetate, were all analytical reagents used as received. Alkyl polyoxyethylene ethers AEO$_3$ and AEO$_9$ (the number of ethylene oxide units are respectively 3 and 9), were kindly supplied by BASF Ltd.

A 100% cotton fabric with yarn counts of 474×235 (the warp × fill yarns for 10cm×10cm), obtained from Huarun Dye & Finishing Co., China, was ultrasonicated with deionized water and acetone respectively at 25°C for 20 min to remove the slurry and contaminants adhered on the fiber surface, then dried at 100°C for 5 min. Silicon wafers, kindly supplied by Jinghua Electronic Material Co., Shanghai, China, were washed and dried according to the references [4, 5, 6].

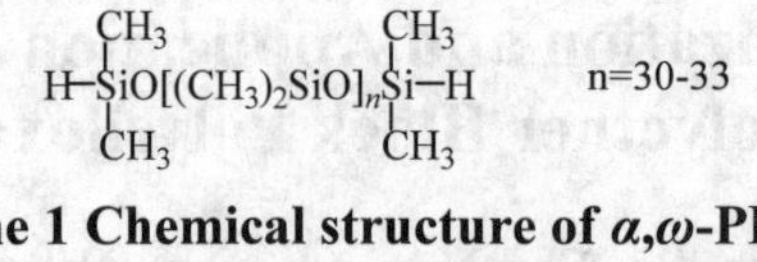

$$H-\underset{\underset{CH_3}{|}}{\overset{\overset{CH_3}{|}}{Si}}O[(CH_3)_2SiO]_n\underset{\underset{CH_3}{|}}{\overset{\overset{CH_3}{|}}{Si}}-H \quad n=30\text{-}33$$

Scheme 1 Chemical structure of α,ω-PHMS

1.2 Synthesis of polyurethane modifying polyether-block-polysiloxane

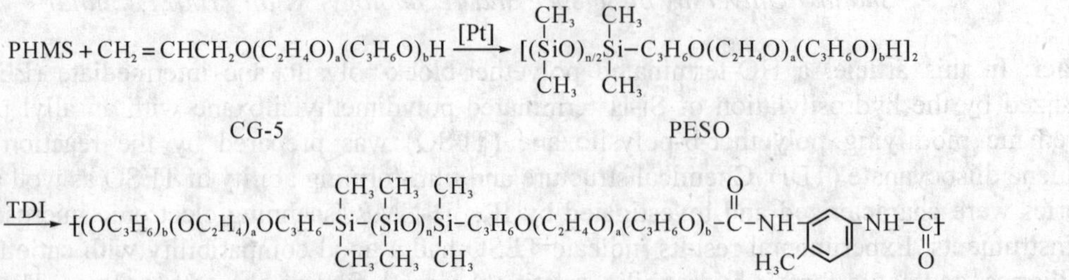

$$PHMS + CH_2=CHCH_2O(C_2H_4O)_a(C_3H_6O)_bH \xrightarrow{[Pt]} [(SiO)_{n/2}\underset{\underset{CH_3}{|}}{\overset{\overset{CH_3}{|}}{Si}}-C_3H_6O(C_2H_4O)_a(C_3H_6O)_bH]_2$$

CG-5   PESO

$$\xrightarrow{TDI} \{(OC_3H_6)_b(OC_2H_4)_aOC_3H_6-Si-(SiO)_nSi-C_3H_6O(C_2H_4O)_a(C_3H_6O)_b-\overset{O}{\overset{\|}{C}}-NH-\underset{H_3C}{\bigcirc}-NH-\overset{O}{\overset{\|}{C}}\}$$

Figure 1 Synthesis of TESO

In a three-necked flask equipped with a mechanical stirrer, a reflux condenser and a thermometer, α,ω-PHMS and CG-5 were added in mole ratio of n(Si-H): n(CG-5) =1:1, gently stirred and bubbled by N_2 for 10 minutes. When the mixture was heated to 60-120℃, a catalytic amount of platinum in alcohol solution was dropped. Then, the mixture was maintained at the temperature to react for 3-4h. At the end of the reaction, low-boiling impurities were removed by vacuum distillation. And a transparent, hydroxyl-terminated polyoxyethylene- block-polydimethylsiloxane (POE-b-PDMS) was obtained and used as an intermediate in the following procedure.

In intermediate POE-b-PDMS system, a stoichiometric TDI was dropped in ratio of n(–NCO):n(-OH)=1:1.2. The mixture was stirred and maintained at 50-60℃ to react for 2h. A small amount of a butanone oxime was then added to terminate the polymer chain propagation. After stripped off the low boiling-point impurities through vacuum, a colorless to light yellow, polyurethane modifying POE-b-PDMS (TESO) in alternate structure was obtained.

1.3 Characterization

Infrared (IR) spectrum was acquired on a Brucker VECTOR-22 spectrometer, KBr liquid film. ^{1}H-NMR spectra were recorded by INOVA-400 spectrometer (Varian), $CDCl_3$ (Sigma-Aldrich Inc.) used as a solvent and tetramethylsilane (TMS) as an internal standard.

1.4 Film morphology and characterization of TESO on cotton substrates

1.4.1 Sample Preparation

TESO film on the silicon wafer was prepared according to the references[5, 7].

1.4.2 Morphology observation

AFM images of TESO coated wafer samples were observed with a Nanoscope IIIA atomic force scanning microscope (Digital Instruments, USA) at 22℃ and relative humidity of 48%. All the scanning was performed in tapping mode. SEM observation was carried out on an S-570 scanning electron microscope (Hitachi, Japan) after the treated fabrics were coated with gold in a vacuum.

1.5 Performance evaluation

30g TESO, 9g of alkyl polyoxyethylence ether AEO-3 and AEO-9 (W_{AEO-3} : W_{AEO-9}=1:1) were homogenized with a high speed mixer to form a turbid fluid. Then, 117g of deionized water was added to stir for 10min until a transparent TESO micro-emulsion with a solid content of 25 wt% was obtained.

25wt% TESO emulsion was diluted with water to form a finishing bath containing about 1% polysiloxane. Fabric samples were prepared and measured referring to the method in the literatue[7,8]

2. Result and discussion

2.1 Characterization of TESO

As an effective route to bond polyether group in polysiloxane molecule, hydrosilylation of polyhydromethylsiloxane with allyl oxyalkylene ether has been widely adopted in previous researches[7, 9]. From Figure 1, IR spectra of POE-b-PDMS and TESO, it is discovered the absorption band at 2150 cm^{-1} attributed to Si-H groups of α,ω-PHMS almost completely disappeared. As a result, a chemical shift peak at δ 0.45 due to Si-C**H**$_2$- appeared in the ^{1}H-NMR spectrum of POE-b-PDMS, indicating polyoxyethylene group were bonded into the intermediate molecule via hydrosilylation. Besides, there were a series of new absorption bands respectively occurring at 3502 (broad, v_{O-H}), 3307 (v_{N-H}, weak) and 1730 ($v_{C=O}$, sharp, weak) and 1602~1413 (due to the substituted phenyl) in the IR spectrum of target TESO. Correspondingly, a group of chemical shift signals respectively at δ 2.3, 6.9 and 7.1 due to -C$\underline{H}_3$ linked in substituted phenyl, b and b' $\underline{H}$ in the phenyl ring as well as the d$\underline{H}$ derived from -CON$\underline{H}$- were also observed from the amplified ^{1}H-NMR spectra of TESO. All indicated that the -CONH- as well as substituted phenyl groups have presented in TESO molecule.

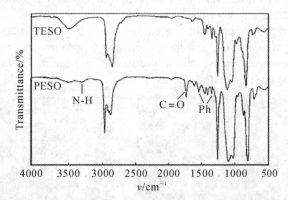

Figure 2 IR spectra of TESO and the intermediate PESO

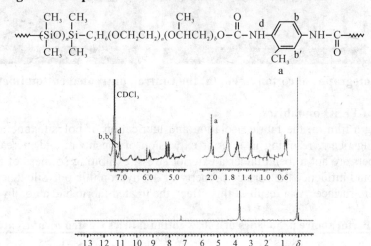

Figure 3 ^{1}H-NMR spectra of TESO and its amplification

2.2 Film morphology and surface properties of TESO on different substrates

Polysiloxanes used as softening agents in textile industry could impart different and diverse finishing styles or hands to fabrics [5, 10], which is believed to be associated with the molecular structures of the polysiloxanes, their orientations and the characteristic film morphology as well as the properties of the polysiloxane films on fiber surfaces [10,11,12].

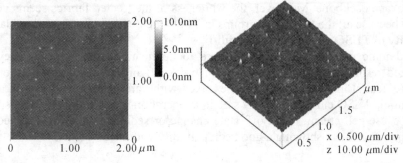

Figure 4 AFM images of TESO film with a root mean square roughness of 0.213 nm in 2×2 μm^2 field

AFM images of TESO film on the silicon wafer were shown in Figure 4. Obviously, influenced by the blocked hydrophilic polyoxyethlene groups, the alternate structural TESO exhibited film morphology very different from those of polysiloxanes bearing pendant polyether side groups (PSPE). In previous studies[7, 9], the PSPEs were usually found to form inhomogenous morphology with many gigantic polyether groups curled into humps or threads on the film surface. This did not occur in morphology of polyether-block-polysiloxanes again, especial when PDMS and polyoxyethylence segments were short in length. Therefore, a large number of small, bright dots which may be derived from the blocked polyoxyethylence were observed from TESO micro-image.

Figure 5 shows the SEM photographs of the cotton fibers treated or untreated by TESO. Clearly, the untreated cotton fiber surface was uneven, and there were some grooves on its surface. However, the treated cotton fiber surface was smooth and most of the grooves disappeared. Obviously, this resulted from the macroscopic smooth polysiloxane film coated on the fiber.

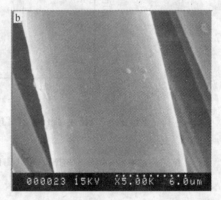

Figure 5 SEM photographs of cotton fibers (a) the control, untreated cotton fiber; (b) treated by TESO

2.3 Performance of TESO on fabrics

When forming a film on the fabric, isolation and lubrication of polysiloxane around the fiber helps to separate the bonding fibers, resulting in interior expansion of the fibers and increased gap between fibers. Complete loose fibers are apt to bend under a exterior stress, exhibiting softness of the fabrics. Furthermore, bumpy structure could influence the surface property of the hydrophilic polysiloxane film and improve water spreading on the film surface. As a result of this effect, the treated fabric had a good wettability of 2.68s.

Table 1 Performance properties of 100% cotton fabrics treated or untreated with TESO

	BR (mN)		WRA (°)			Whiteness	Wettablity(s)
	w	f	w	f	w+f		
TESO	106.0	189.0	100.6	120.2	220.8	87.5	2.68
Control	146.5	318.8	53.0	76.0	129.0	87.0	0.24

BR, bending rigidity; WRA, wrinkle recovery angle; w, warp; f, fill; Control, the untreated cotton fabrics

In addition, other performance properties of TESO were investigated on cotton fabrics again. Like other functional polysiloxanes in performance, TESO also has a good softening-fabric ability to minimize the BR of the treated fabric to some extent because of the flexibility and friction coefficient reducing property of polydimethylsiloxane backbone. Moreover, the whiteness of the treated fabrics seems very near to that of the untreated cotton because all the functional groups in TESO are durable to light and oxidation.

2.4 Compatibility of TESO with resin and additives

Cationic and ionic polysiloxanes are sensitive to application environment. They can react with the positively charged resins or polymers resulting in break off or sedimentation of the emulsions. Nonionic polysiloxanes are neutral in nature. They can be mixed with various resin and aids in application. Shown as Table 2, the nonionic TESO emulsion is very safe in cooperation with cationic/anionic softeners or auxiliaries. TESO does not cause gel, cream, break off, state change or sedimentation when mixed with amino silicone emulsion, shampoo and so on, showing a good compatibility with these agents.

Table 3 Compatibility of TESO with resins and additives

Emulsion	TESO
Cationic	
25% amino silicone emulsion	◎
5% distearyl dimethyl ammonium chloride aqueous dispersions ()	◎
20% fluorinated polyacrylate emulsion	◎
Anionic resin and auxiliaries	
10% polyalkyloxyethylene phosphate aqueous solutions	◎
1.0% sodium dodecyl benzene sulfonate solution	◎
0.5% fluorescent whitening agent VBL aqueous solution	◎

◎ indicates no changes and good compatibility of the system.

3 Conclusion

A polyurethane modifying polyether-b-polysiloxane(TESO) was prepared by the reaction of HO-terminated polyether-block-polysiloxane (PESO) with 2,4-toluene diisocyanate (TDI). Chemical structure and film-forming ability of TESO as well as its performance on fabrics were characterized and investigated by IR, ^{1}H-NMR, AFM and other instruments. The results reveals that TESO film on the silicon is inhomogeneous with some bright spots dispersed in continuous phase and the bright spots are ascribed to the polyether segments. When anchored on the fabrics, TESO caused the fiber surface to be smoother and the original grooves on the fiber surface almost disappeared completely. TESO not only offered a bulky, soft hand to the treated fabrics, but also increased elasticity of the fabrics. In addition, TESO emulsion could be blended with kinds of ionic resins and additives and the stability as well as compatibility of the mixed system are far better than the common used finishing agents.

Acknowledgement

We would like to address our appreciation to National Natural Science Foundation Committee of China (50373025) and Doctoral Foundation, Ministry of Education of China (200807080002) for financial supports of our research.

References

[1] David T. Floyd and Klaus R. Jenni. Silicone, polymers, organo-modified[M]. Polymeric materials encyclopedia, Joseph C. Salamone CRC Press, Inc. New York, 1996, (10): 7677-7688.
[2] Randal M Hill. Silicone surfactant——new developments[J]. *Current Opinion Colloid & interface science,* 2002, 7: 255-261.
[3] Guido Kickelbick, Josef Bauer, Nicola Husing, et al. Spontaneous Vesicle formation of short-chain amphiphilic polysiloxane-b-poly (ethylene oxide) block copolymers[J]. *Langmuir,* 2003, 19: 3198-3201.
[4] Song Xiao-yan, Zhai Jin, Wang Yi-lin, Jiang Lei. Self-assembly of amino-functionalized monolayer on silicon surfaces and preparation of superhydrophobic surfaces based on alkanoic acid dual layers and surface roughing [J]. Journal of Colloid and Interface Science, 2006.298:267-273.
[5] Qiufeng An, Linsheng Li, Liangxian Huang, et al. Film morphology and characterization of functional polysiloxane softeners[J]. AATCC Review, 2006, 6(2): 39-43.
[6] Bai Tao, Cheng Xian-Hua. Preparation and characterization of lanthanum-based thin films in sulfonated self-assembled monolayer of 3-mercaptopropyl trimethoxysilane [J]. Thin Solid Films, 2006, 515: 2262-2267.
[7] Qiufeng An, Qianjin Wang, Yan Wang, Liangxian Huang. Synthesis, Film Morphology, and Performance of Functional Polysiloxane Bearing Polyether and Benzophenone Derivative Side Groups[J]. Fiber and Polymers, 2009, 10(1): 40-45.
[8] An Qiufeng, Li Linsheng, Huang Liangxian, et al. Preparation of Carboxyl Alkyl Silicones by Functional Modification and Studies of their Softness on Fabrics[J]. Journal of Functional Polymers, 2001, 14 (2): 163-168.
[9] Qiufeng An, Gang Yang, Qianjin Wang. et al. Synthesis and morphology of carboxylated polyether-block-polydimethylsiloxane and the supramolecular self- assembled from it[J]. Journal of Applied Polymer Science, 2008, 110(5): 2595 -2600.
[10] J.Y. Li, T.D. Li, Q.S. Zhang. Appl. Chem. Ind. 2001, 30 (6): 17-19.
[11] H. Roy, Y.F. Alexander. Langmuir, 2002, 18: 8924–8928.
[12] Burrell M C, Butts M D, Derr D, et al. Angle-dependent XPS study of functional group orientation for aminosilicone polymers absorbed onto cellulose surface[J]. Applied Surface Science, 2004, 227(1): 1-6.

32 Preparation of Bacterial Cellulose Nanofiber with Silver Nanoparticles by In Situ Method

Zhijiang Cai[1,2], Guang Yang[1]

(1. School of Textiles, Tianjin Polytechnic University, Tianjin 300300, China)
(2. Key Laboratory of Advanced Textile Composites, Ministry of Education of China, Tianjin 300160, China)

Abstract: Bacterial cellulose nanofibers with silver nanoparticles were prepared using silver nitrate and sodium borohydride aqueous solution by in situ method. The prepared nanocomposites were characterized by scanning electron microscopy (SEM), X-ray diffraction (XRD), and thermogravimetric analysis (TGA) test. The morphology of these nanocomposites indicated that the silver nanoparticles have been formed and distributed uniformly on the surface of the BC nanofibers. It was also found that the incorporation of silver nanopartilces onto BC nanofibers affected the crystal structure of BC nanofibers. Meanwhile, the thermal degradation temperature has been increased from approximately 260°C to 340°C for nanocomposites. The average size of silver nanoparticles is around 15 nm, which is bigger than previous report. This may be due to the aggregation of small nanoparticles, especially at high $AgNO_3$ concentration.

Keywords: Bacterial cellulose nanofiber; Silver nanoparticles; Nanocomposite

1 Introduction

Bacterial cellulose (BC) nanofiber, which is produced by strains of the bacterium *Acetobacter xylinum*, differs from plant cellulose with respect to its high purity, high crystallinity, ultra-fine network structure, high water absorption capacity, high mechanical properties and biocompatibility [1]. Due to its unique properties, BC has long been used in a industrial applications and biomedical applications such as artificial skin or wound healing material [2], artificial blood vessels [3], tissue regeneration scaffold [4]. Recently, various attempts have been made to produce BC based composites with high multi-functionality to broaden the biomedical applications of BC. For instance, BC nanocomposites with improved mechanical properties were created by soaking BC on polyacrilamide and gelatin solutions [5]. BC-hydroxyapatite scaffolds for bone regeneration have been developed by immersing the BC gel in simulated body fluid (SBF) or in both calcium and phosphate solutions [6]. Furthermore, BC-polyester and BC-PVA nanocomposites were developed for potential applications as vascular implants [7].

Silver nanoparticles are well-known antimicrobial materials which are highly toxic to microorganisms, and their antimicrobial activates have been studied by many researchers. But silver nanoparicles have only recently studied in the fabrication of novel materials with polymeric matrices for medical and biotechnical applications such as dressing, dental resins and antimicrobial fibers [8~10].

In this paper, we fabricate BC nanofiber with silver nanoparticels by in situ method. In the result, the distribution of silver nanoparticles on the BC nanofibers is expected to be homogeneous throughout the whole volume.

2. Materials and methods

2.1 Materials

$AgNO_3$ and $NaBH_4$ aqueous solution were purchased from Sigma. Other chemicals of the highest purity available were used and were purchased from Sigma-Aldrich, USA.

2.2 Biosynthesis of bacterial cellulose pellicles

Gluconacetobacter xylinum BRC-5 was obtained from Yonsei University and used to produce the BC pellicles. The bacterium was cultured on Hestrin and Schramm (HS) medium, which was composed of 2% (w/v) glucose, 0.5% (w/v) yeast extract, 0.5% (w/v) bacto-peptone, 0.27% (w/v) disodium phosphate, and 0.115% (w/v) citric acid. All the cells pre- cultured in a test tube containing a small cellulose pellicle on the surface of the medium were inoculated into a 500 ml. Erlenmeyer flask containing 100 ml of the HS medium. The flasks were incubated statically at 30°C for 14 days. The cellulose pellicles were dipped into 0.25 M NaOH for 48 h at room temperature in order to eliminate the cells and components of the culture liquid. The pH was then lowered to 7.0 by repeated washing with distilled water. The purified cellulose pellicles were stored in distilled water at 4°C to prevent drying.

2.3 Preparation of bacterial cellulose/silver nanoparticles composite

The resultant bacterial cellulose pellicle was soaked in 0.01 M $AgNO_3$ aqueous solution for about 12 h in a black box to prevent additional photoreduction companied sonication process. Then, the BC/Ag^+ membrane was reduced by 0.01 M aqueous NaBH$_4$ solution with sonication process. Finally, the resultant membrane was washed with a large amount of deionized water to remove the excess chemical, the obtained sample was frozen at −40°C and dried in a vacuum at −50°C. The same experimental process was performed with 0.05 M AgNO$_3$ and NaBH$_4$ solution.

2.4 Characterization of bacterial cellulose/gelatin composite

The morphology of BC/silver nanocomposite and size of silver nanoparticles were observed by scanning electron microscopy (SEM, Hitachi S-4200). X-ray diffraction (XRD) patterns were recorded on an X-ray diffractometer (D/MAX-2500, Rigaku), by using Cu Kα radiation at 40 kV and 30 mA. The diffraction angle was ranged from 5° to 40°. Thermogravimetric analysis (TGA) was carried out with a NETZSCH STA 409 PC/PG system under a dynamic nitrogen atmosphere between 30 and 1000°C.

3 Results and discussion

Statically cultured BC formed a pellicle. This pellicle was composed of a small amount of nanofibrils holding about 99% of water. The ultrafine, highly entangled nanofibrils, which have length in the range of 1 to 9 μm, form a dense reticulated structure, stabilized by extensive hydrogen bonds. BC has high specific surface area endowing the surface of nanofibers with a great deal of hydroxyl groups. These hydroxyl groups make up of active sites for metal ion adsorption. The absorbed silver ions were bound to BC microfibrils probably via electrostatic interactions, because the electron-rich oxygen atoms of hydroxyl groups. After reduction in aqueous NaBH$_4$, silver ions were reduced to form silver nanoparticles in situ. The original white BC was turned to yellow or gray. Finally, it was dried by the freeze-drying method to maintain the original structure of BC and lock up the content of silver nanoparticles.

3.1 SEM observation

The SEM images shown in Figure 1 gives an overview of the Ag nanoparticles formed on the surface and cross-section of BC prepared under different condition. The SEM images clearly show well-dispersed Ag nanoparticles on the surface and cross-section. The nanoparticles appear to have a multi-scale size ranged from several to several tenths of nanometers. In fact, bigger nanometric particles show up as agglomerates with a broad size distribution. Smaller Ag nanoparticles are also observed adsorbed into the BC scaffold. The preparation condition and some properties of BC/silver nanopartilces are listed in Table 1.

Table 1 BC/silver nanocomposites preparation condition and properties

Code	AgNO$_3$ (M)	NaBH$_4$ (M)	Sonication time (min)	Degradation temperature (°C)	Silver content (wt%)
BC	0	0	0	265	0
BCAg1	0.01	0.01	0	343	8.85
BCAg2	0.01	0.01	30	338	31.70
BCAg3	0.05	0.05	30	330	51.78

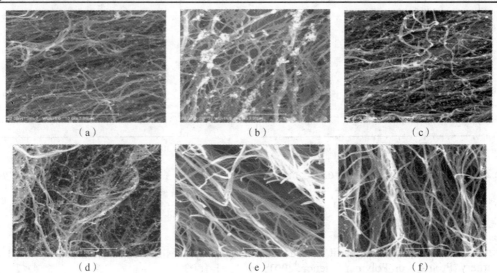

Figure 1 SEM images of BC/silver nanocomposites (a, b, c: surface morphology and d, e, f: cross-section morphology of BCAg1, BCAg2 and BCAg3)

3.2 XRD analysis

The X-ray diffraction (XRD) was used to examine the crystallographic nature of the cubic structure to confirm the formation of silver nanoparticles. Compared with the XRD of blank BC sample (Figure 2a), the diffraction peaks at 14.64°, 16.78° and 22.64° in the freeze-dried BC/silver nanocomposite (Figure 2b) are assigned to the crystallographic plane of (1 $\bar{1}$ 0), (1 1 0) and (2 0 0) reflexions of cellulose I. Another characteristic four major peaks at 38.16°, 44.17°, 64.66° and 77.37° correspond to the (1 1 1), (2 0 0), (2 2 0) and (3 1 1) crystal planes of the face centered cubic (FCC) structure of the metallic silver nanoparticles. It was also found that the incorporation of silver nanopartilces into BC affected the crystal structure of BC. The height of the (1 $\bar{1}$ 0) peak decreased and the ratio of the (1 $\bar{1}$ 0) peak to the (0 2 0) peak was also decreased from 0.50 for BC to 0.38 for BC/silver nanocomposite. Obviously, the presence of silver particles affected the preferential orientation of the (1 $\bar{1}$ 0) plane during the water removal from BC pellicle.

3.3 TGA test

Figure 3 shows the TGA curves obtained for pure BC and BC/silver nanocomposites. For pure BC (Figure 3a), weight loss can be divided into two stages, which are from ambient to 150°C and from 150 to 400°C. The first weight loss stage is due to BC dehydration. Physically absorbed and hydrogen bonded linked water molecules can be lost during this stage. The other stage corresponding to a broad peak (150 – 400°C) can be attributed to BC pyrolysis. Figure 3b, c, d shows the TGA curves of BC/silver nanocomposites. The residual mass obtained for temperatures above 600°C is related to silver nanoparticles. The thermal degradation temperature has been increased from approximately 260°C for pure BC to 340°C for BC/silver nanocomposites. And the weight percent of silver nanopartilces seems to have no significant effect on thermal stability. Comparing with that of BC, we can see inclusion of silver nanoparticles into BC in situ improved thermal stability of BC.

4 Conclusion

The objective of this study was to investigate a novel method using BC as template to prepare silver nanocomposite with silver nanopartilces homogeneously distributed throughout whole bacterial cellulose scaffold. Sonication seems to be an effective way for ions to penetrate into BC and thus the weight percent of silver nanoparticles can be greatly increased. This would be a way to control the weight ratio of silver nanoparticles in the nanocomposite especially for high silver content. However, the average size of silver nanoparticles is around 15 nm, which is bigger than previous report. This may be due to the aggregation of small nanoparticles, especially at high $AgNO_3$ concentration.

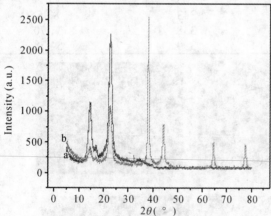

Figure 2 X-ray diffraction patterns of BC and BC/silver nanocomposite
(a: blank BC; b: BCAg-2)

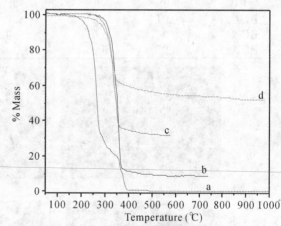

Figure 3 TGA curves for BC and BC/silver nanocomposites
(a: pure BC; b, c, d: BCAg-1, BCAg-2 and BCAg-3)

References

[1] D Klemm, D Schumann, U Udhardt, S Marsch. Bacterial synthesized cellulose-artificial blood vessels for microsurgery. Progress on Polymer Science. 2001, 26: 1561-1603.
[2] W Czaja, A Krystynowicz, S Bielecki, R M Brown. Microbial cellulose-The natural power to heal wounds. Biomaterials, 2006, 2: 145-151.

[3] D Klemm, U Udhardt, S Marsch, D I Schumann. BASYC-Bacterial Synthesized Cellulose: Miniaturized Tubes for Microsurgery. Polym News. 1999, 24: 377-379.
[4] G Helenius, H Backdahl, A Bodin, U Nannmark, P Gatenholm, B Risberg. In vivo Biocompatibility of Bacterial Cellulose. J Biomed Mater Res Part A. 2006, 76: 431-438.
[5] S A Hutchens, R S Benson, B R Evans, H M O'Neill, C J Rawn. Biomimetic Synthesis of Calcium-Deficient Hydroxyapatite in a Natural Hydrogel. Biomaterials. 2006, 27: 4661-4670.
[6] L Hong, Y L Wang, S R Jia, Y Huang, C Gao, Y Z Wan Hydroxyapatite/bacterial cellulose composites synthesized via a biomimetic route. Mater Lett. 2006, 60: 710-1713.
[7] L E Millon, H Mohammadi, W K Wan. Anisotropic polyvinyl alcohol hydrogel for cardiovascular applications. J Biomed Mater Res Part B. 2006, 79: 305-311.
[8] T Maneerung, S Tokura, R Rujiravant. Impregnation of silver nanoparticles into bacterial cellulose for antimicrobial wound dressing. Carbohydrate Polymer. 2008, 72: 43-51.
[9] H S Braud, C Barrios, T Regiani, R F C Marques, M Verelst, J Dexpert-Ghys, Y Messaddeq, S J L Ribeiro. Self-supported silver nanoparticles containing bacterial cellulose membranes. Mater Sci Eng C. 2008, 28: 515-518.
[10] L C de Santa Maria, A L C Santos, P C Olivira, H S Braud, Y Messaddeq, S J L Ribeiro. Synthesis and characterization of silver nanoparticles impregnated into bacterial cellulose. Mater Lett. 2009, 63: 797-799.

33 Study and Development of New Motorcyclists' Racing Suit Protector

Nannan Cao, Jianwei Ma, Shaojuan Chen
(Qingdao University, 266071, Qingdao, China)

Abstract: This paper introduces a new production method of motorcyclists' racing suit protector, which is made of thermal-bond fiber and three-dimensional crimp hollow polyester fiber by hot air compression molding. Study shows that when the weight is 60g, the content of three-dimensional hollow polyester crimp fiber is 20% and the content of vertical setting fiber is 45%, the shoulder protector can reach BS EN1621-1 and BS EN1621-2's requirements. This protector has the characteristics of environmental protection, air permeability and comfort properties. It is of positive significance for the development of back protector, chest protector and kneepad. In addition, it lays the foundation of the development of new motorcyclists' racing suit protector.

Keywords: Thermal-bond fiber; Motorcyclists' racing suit; Protector

1 Introduction

Traditional sports protector is major of rubber products. The rubber products can play a part in protection, but it has bad air permeability, peculiar smell, easy aging and can cause environmental pollution. Nowadays, it is urgently necessary to find a new material to replace rubber in order to product more dexterous, convenient and comfortable protector[1]. The design and exploitation of protector is mainly pursued by several big companies at home and abroad, such as Dow Corning, D3O Lab and so on.

Dow Corning introduces a new impact resistance technology patent—Dow Corning Active Protection System in Textile-Symposium which is held in Atlanta in March 2006. This patent consists of a specially coated three-dimensional textile spacer material. This "intelligent" textile material maintains flexibility and toughness in normal circumstance, but it hardens instantly when it is subject to impact. The material restores to softness after impact disappearing[2].

A special cloth is developed by D3O Lab and the researchers of Innovation Center, University of Hertfordshire. It will quickly harden under impact to reduce impact force. Then immediately soften, it would not restrict the flexibility of wearing. The stronger the impact force is, the faster the reaction occurs. The first product of this intelligent cloth is flexible kneepad for athletes. More products are being developed, such as Skiing dress, sports helmet and the shoes adapted to load changes while running[3].

At present, the design and development of these companies mainly focus on material comfort, protection of intelligent, functional materials, and environmental protection. It means to develop an effective system to restrict strong impact without sacrificing flexibility, breathability and easy care. Traditional hard shell products are neither flexible nor comfortable. Soft foam and gel products which are sold in market lack of air permeability and adequate security[4-5].

Therefore, this paper chooses thermal-bond fiber and three-dimensional crimp hollow polyester fiber as raw material to produce new motorcyclists' racing suit protector by hot air compression molding. Thermal-bond fiber achieves adhesion between the fibers in protector. Three-dimensional crimp hollow polyester fiber enhances cushioning performance of the protector. The development of this new method provides references for the study and exploitation of new motorcyclists' racing suit protector.

2 Protector preparations

The schematic diagram of molding press shows in Figure 1. When the mold temperature reaches a certain temperature, the paved fiber nets put into the mold. Then the press machine and hot-air heating device are turned on, meanwhile the mold closes and hot-air heating device begins to blow hot-air to heat the deep fiber net. After achieving a certain pressure, the press machine stops and maintains the pressure. But hot-air heating device continues to blow hot-air, and deeper thermal-bond fiber can be heated entirely to melt bonding. The protector is shown in Figure 2.

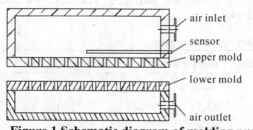

Figure 1 Schematic diagram of molding press

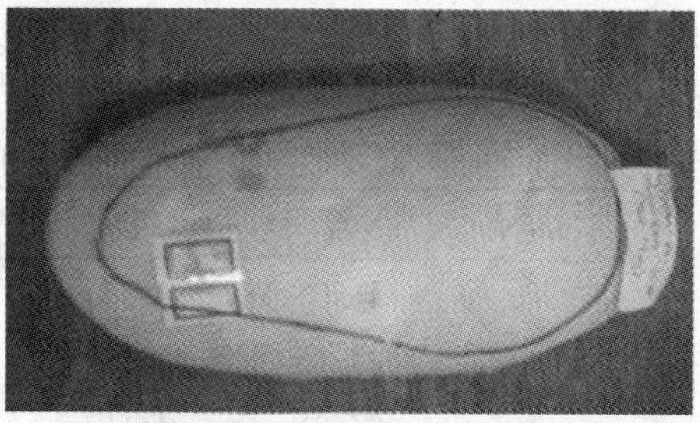

Figure 2 Shoulder protector

Protector performance depends on mixing ration of thermal-bond fiber and three-dimensional crimp hollow polyester fiber. In addition, the different fiber orientation has a significant effect on protector cushioning properties. Vertical setting fiber can enhance the protector cushioning properties. The mass also affects protector performance significantly. In view of these, orthogonal experiment is established to determine the manufacture scheme of protector. The factors are the mass of protector, the content of three-dimensional hollow polyester crimp fiber and the content of vertical setting fiber. Each of the factors is taken in four levels. Experimental design is shown in Table 1.

Table 1 Orthogonal design

	A Mass (g)	B Content of 3D hollow polyester crimp fiber (%)	C Vertical setting fiber (%)
1	50	15	0
2	55	20	15
3	60	25	45
4	65	30	100

3 Impact resistance tests of protectors

Test apparatus is FYFC-50J, which contains dropping apparatus, drop striker and Anvil. The anvil shall be mounted so that during impact testing the whole force between the anvil and the massive base of the apparatus passes through a quartz force transducer in line with its sensitive axis. The force transducer shall have a calibrated range of no less than 200kN and a lower threshold of less than 1kN. The output of the force transducer shall be processed by a charge amplifier and displayed and recorded on suitable instruments. Testing shall be carried out, at three different points an each sample piece, at least 50 mm apart tested. The impact resistance test is in strict accordance with the European standard BS EN1621-1 and BS EN1621-2. The European standard BS EN1621-1 and BS EN1621-2 stipulates clearly that the average peak force of shoulder protector should be below 35kN, and no single value would exceed 50kN[6]. The results are shown in Table 2.

Figure 3 Test apparatus: FYFC-50J

After low velocity mechanical impact, there are depression in the impact surface of the protector (permanent deformation). The protector has been layered internally. Invisible damages observe by scanning electron microscope image (shown in Figure 4). Some bonding points are damaged. The damaged bonding points adhere to fiber surface. There is no fiber breakage basically.

Table 2 Results of impact resistance test

Test No.	A	B	C	Average peak force (kN)	Max force (kN)
1	1(50)	1(15)	4(100)	42.6	51.9
2	2(55)	1(15)	1(0)	45.5	55.1
3	3(60)	1(15)	2(15)	38.5	46.2
4	4(65)	1(15)	3(45)	38.0	46.7
5	1(50)	2(20)	3(45)	36.4	45.1
6	2(55)	2(20)	4(100)	34.6	39.9
7	3(60)	2(20)	1(0)	33.4	42.0
8	4(65)	2(20)	2(15)	25.9	32.5
9	1(50)	3(25)	2(15)	45.8	50.9
10	2(55)	3(25)	3(45)	30.1	36.7
11	3(60)	3(25)	4(100)	29.4	30.4
12	4(65)	3(25)	1(0)	29.4	44.6
13	1(50)	4(30)	1(0)	42.5	49.3
14	2(55)	4(30)	2(15)	34.7	41.4
15	3(60)	4(30)	3(45)	33.7	39.5
16	4(65)	4(30)	4(100)	27.8	29.4

Figure 4 Fiber morphology after mechanical impact

3.1 The influence of protector mass to the resistant impact

With the increasing of the mass, penetrating power is decreasing in turn, which means better cushioning properties. The density of the protector increases. Fibers contact with each other densely, and would not separated under external force [7]. The bonding points increase in unit volume, which are able to absorb more impact kinetic energy. So the protector exhibits better cushioning properties. In the test results, the average penetration of 50g and 55g is respectively 41.8kN and 36.23kN and neither of them doesn't come up to European Standards. In consideration of protective and comfortable properties, the mass of the protector should be more than 60g. The mass of the same specification rubber protector in the market is more than 100g, so this new protector has significant advantages in mass.

3.2 The influence of the content of three-dimensional crimp hollow polyester fiber to the resistant impact

With the content of three-dimensional crimp hollow polyester fiber increasing from fifteen percent to twenty percent, penetrating power decreases gradually because of its good recovery capability of tension deformation and bending deformation. With the content of three-dimensional crimp hollow polyester fiber continuing to increase, penetrating power is rising slowly. With the content of three-dimensional crimp hollow polyester fiber increasing, the content of thermal-bond fiber decreases and bonding chances descend. The protector will inflate after leaving mould, so the density of protector will descend. The increasing of the content of three-dimensional crimp hollow polyester fiber can lead to bad edge bonding. Giving consideration to protection and the feasibility of the production process, the content of three-dimensional crimp hollow polyester fiber should be controlled at around 20%.

3. 3 Influence of the content of vertical setting fibers to the resistant impact

Vertical setting fibers are arranged perpendicular to protector surface, and the structure of fiber net changes from two-dimensional to three-dimensional. A fiber can be run through several fiber layers, so it increases the bonding probability. When molding impact, there are more bounding points to absorb energy, and then cushioning properties and flexibility enhanced significantly. The relationship between penetrating power and the content of vertical setting fibers is linear. With the increasing of the content of vertical setting fibers, penetrating power decreases. It means that the higher content of vertical setting fibers is, the better resistant impact should be.

4 Air permeability test and innocuousness test of protector

4.1 Air permeability test of protector

Fabric air permeability apparatus is Y 561. The test is in accordance with National Standard GB5453-85. Under prescriptive differential pressure, the air flow, which is vertically through the given area of the sample, can be obtained during certain period. Then aerated rate of the protector is figured out. The air permeability of the protector can't be obtained directly because of the irregular shape of the protector. So, different thickness of plates in the same mass are produced to measure air permeability of protector. The relationship between protector thickness and air permeability is shown in Figure 5.

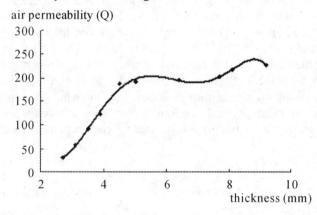

Figure 5 Relationship between protector thickness and air permeability

The air permeability increases with the thickness increasing. As the air transmission also belongs to mass transmission, it depends on the air pressure of material on both sides to force air molecules through the material. So with the thickness of plates increasing, the density decreases. And the channels for gas transmission increase relatively and the air permeability increases. Compared with rubber protector, this kind of protector has better permeability.

4.2 Innocuousness test of protector

Referred to the innocuousness test of European Standard EN340:2003, pH value of this protector is 6.5, which is in accordance with standard.

5 Promotion of the protector

Research on the shoulder protector shows that impact resistance can meet the requirements when weight is 60g, the content of three-dimensional hollow polyester crimp fiber is 20% and the content of vertical setting fiber is 45%. Based on the results, back protector is produced by the same method (shown in Figure 6). Results show that impact resistance of the back protector can reach the second level protection of European Standard BS EN1621-1 and BS EN1621-2 when the mass of the back protector is about 400g.

Table 3 Results of impact resistance test on back protectors

mass/g	Average peak force (kN)	Max force (kN)
230	24.8	26.3
300	19.5	23.5
400	16.2	20.3

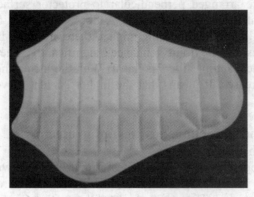

Figure 6 Back protector

6 Conclusions

Results show that the mass of protectors have a significant affection on impact resistance, and penetrating power decreases with the increasing of mass. The shoulder protector which is of weight 60g, the content of three-dimensional hollow polyester crimp fiber 20% and the content of vertical setting fiber 45% can reach BS EN1621-1 and BS EN1621-2's requirements. For different types of protector, the protectors can be produced by changing the mass and fiber content because of the different requirements of impact resistance. The development of shoulder protectors will be helpful to the development of back protector, chest protector and kneepad, and has significant guiding to the development of the whole motorcyclists' racing suit protectors. The protector has good air permeability and cushioning properties, no smell, and recoverable, so it has good feasibility. With the development of motorcyclists' racing, motorcyclists' racing suit protectors will have infinite prospects.

References
[1] Li Hongmei, Song Yujin, Wang Yi. The green design of personal protector equipment. China Personal Protective Equipment [J], 2005, (5): 28-30.
[2] The company of Dow Corning developed a new anti-impact technology. China Textile Leader [J], 2006(5): 43.
[3] England developed a new smart protective material. Shanghai Chemical Industry [J], 2004, 9: 43.
[4] Graham Budden. New impact protection textile provides defense and comfort. China Textile Leader [J], 2006 (9): 62 - 67.
[5] Zheng Haiqi, Yu Xuezhong. Fiber, property and application prospect of spray-bonded nonwoven fabric. Technical Textiles [J], 2000, (2): 38-39.
[6] EN1621-1:1997. Motorcyclists' protective clothing against mechanical impact2Part 1: Requirements and test methods for impact protectors[S].
[7] Chen Shaojuan, Wang jing, Ma jianwei, Cao nannan. Factors influencing impact resistance of motorcyclists' racing suit protector. Journal of textile material [J], 2009, (7): 103-106.

34 Study on the Morphology and Damping Properties of the Organic Hybrids of Chlorinated Polyethylene and Hindered Phenol

Xinbo Ding[1], Tao Liu[1], Jian Han[2]

(1. College of Material & Textile Engin., Zhejiang Sci-Tech University, Hangzhou 310018, China
2. Key Laboratory of Advanced Textile Materials and Manufacturing Technology, Ministry of Education of China, Zhejiang Sci-Tech University, Hangzhou 310018, China)

Abstract: In this article, the miscibility, intermolecular interaction and damping properties of hybrids consisting of 2'-methylene-bis-(4-m-ethyl-6-cyclohexyl-phenol) (ZKF) and chlorinated polyethylene (CPE) have been investigated by DSC, FTIR and DMA. The results of DSC displayed that the ZKF and CPE were miscible in the amorphous state. The analysis of Kwei equation implied that there was a strong intermolecular interaction between ZKF and CPE. A small molecule ZKF may form two hydrogen bonds with two CPE chains at the same time, and the hydrogen bonding network was existed in the hybrids. The energy dissipation due to dissociation of the intermolecular hydrogen-bond network was larger than that of due to general friction between polymer chains, the loss peak increased dramatically and the position of the loss peak shifted high temperature region.

Keywords: Hydrogen bonding interactions; Damping properties; Organic hybrids

1 Introduction

Recently, materials with high damping properties find numerous applications, especially in skyscrapers, automobiles and industries. However, due to the damping efficiency for these materials is not very good, there applications are, to some extent, limited [1-2]. The excellence of organic hybrids consisted of polar polymers and a bi- or multi-functional small molecule are particularly easy flexible in blending and easy processing in compounding [3-5]. For damping materials, it is very important to choose a proper polar polymer and organic filler with bi- or multi-functional groups [6-7].

In order to further understand the mechanism of the hybrid materials, a series of blends of polymer and organic small molecule substance were prepared. The study of this article may be benefit to further understand and explicate such elementary problems as to what extent and on what condition the hybrid effect on the damping behavior of the hybrid materials can really function.

2 Experimental

2.1 Hybrid materials sample preparation

The CPE used in this study, with a chlorination degree of 35 wt%, was a commercial grade (from Weifang Yaxing Chemical Plant). The low-molecular-weight organic filler ZKF (with the melting point (Tm) of 138.5℃), used as dispersing material, are commercially available antioxidant.

The above-mentioned materials were first mixed on a two-roll mill at 65℃. For film samples, the mixtures were then molten at 160℃ by a sulfuration machine at a pressure of 1-2 MPa for 5 min and then at a 15 MPa for 10 min. Finally, the mixtures were quenched into ice water to obtain a film with the thickness of 0.5 mm.

2.2 DMA measurements

Dynamical mechanical measurements are carried out on a dynamical analyzer (Perkin-Elmer DMA 7e) in tension mode at frequency of 1 Hz and a varied temperature from –60 to 100℃ with a heating rate of 5℃/min. The dimensions of sample are about 16mm long, 4mm wide and around 0.5 mm thick.

2.3 DSC testing

DSC measurements were carried out with a DSC-7 Perkin-Elmer calorimeter. Samples about 8mg in weight and sealed in aluminum were heated from 35 to 160℃ at a scanning rate of 10℃/min under the nitrogen atmosphere to investigate the melting temperature and degree of crystallinity of CPE/ZKF and CPE/EBP hybrids. The melting temperature was obtained at the top of the DSC endothermic peak in the heating scan.

2.4 Scanning electron microscopy (SEM)

The surface of the sample was fractured in liquid nitrogen, sputter-coated with gold and observed with a JSM-5600LV (JEOL) scanning electron microscope (SEM) using an accelerating voltage of 10-20 kV.

2.5 Fourier Transform Infrared (FTIR) spectroscopy

Infrared spectra were obtained from an accumulation of 32 scans at a resolution of 2 cm^{-1} with a Nicolet FTIR NEXUS 670 spectrometer. The spectra were carried out at ambient temperature in KBr. The films used in this study were sufficiently thin (about 0.1mm) to be within an absorbency range and were compressed to form disk.

Table 1 The thermal properties of CPE hybrid materials with low molecular compounds

Sample (CPE/ZKF)	T_m (℃)	Melting Enthalpy (J/g)
100/0	0	0
95/5	0	0
85/15	0	0
75/25	0	0
70/30	0	0
65/35	126.64	1.391
60/40	127.848	1.702
0/100	138.5	98.681

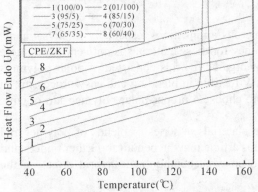

Figure 1 DSC traces of CPE/ZKF hybrids recording during the first heating scans

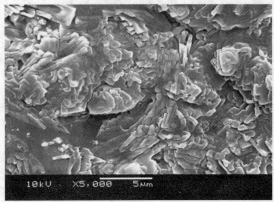

Figure 2 SEM micrographs of CPE/ZKF(60/40) hybrids

3 Results and discussion

Figure 1 shows the DSC traces of CPE/ZKF hybrid materials recording during the first heating scans and the Tm enthalpy of fusion (ΔH_m) thus obtained are summarized in Table 1. As seen from Figure 1, the pure ZKF are crystalline materials and has a melting temperature at 138.5℃. As analysis of the DSC traces of CPE/ZKF samples as shown in Figure 1 and the thermal properties of that listed in Table 1, neither a crystallization peak nor a distinct melting peak is observed in the thermograms of CPE/ZKF hybrids with ZKF contents from 0 to 40 wt%, implying that all of the sample are almost in the amorphous state and that the crystallization of the ZKF component is completely suppressed in the CPE/ZKF hybrids. The disappearance of ZKF crystal also suggests the hybridization of CPE and ZKF during the hot-press process. The representative SEM photographs of the CPE/ZKF two-component hybrids with the mass ratio of 60/40 are shown in Figure 2. The results shows that the CPE/ZKF (60/40) hybrid is almost in the amorphous state and there is only a small amount of crystallites dispersed in the amorphous phase. This phenomenon implies as follows. Firstly, the ZKF could be sufficiently dispersed into CPE matrix. Secondly, the ZKF particles are nearly dissolved in matrix CPE completely from a hot-pressing process and almost amorphous. The third one is that ZKF in CPE/ZKF composites might exist mainly as molecules/clusters. Moreover, Figure 2 also suggests that the crystallization of the ZKF component is suppressed in the CPE/ZKF hybrids, due to the strong intermolecular interactions between ZKF and CPE chains.

We examined the additive effects of ZKF molecules on the DSC curves in the CPE/ZKF hybrids. Figure 3 shows the DSC thermograms of CPE/ZKF hybrids with various compositions recorded during the second heating scan. As seen in Figure 3, a single glass transition of ZKF clearly appears at 46.8℃ during the second heating scan. Moreover, as the ZKF content increases, there is only one glass transition (T_g) for all the CPE/ZKF hybrids. A single Tg implies that blends of CPE and ZKF are miscible in the amorphous phase. T_g's evaluated from the DSC is plotted against the ZKF content in Figure 4. In addition, the maximum Tg deviation is obtained when blend composition at CPE/ZKF=60/40 (weight ratio). Kwei equation describes the effect of Tg hydrogen bonding interaction between polymers as shown in Eq. (1) [8]:

$$T_g = \frac{W_1 T_{g1} + kW_2 T_{g2}}{W_1 + kW_2} + qW_1 W_2 \qquad (1)$$

where W_1 and W_2 are weight fractions of the components, T_{g1} and T_{g2} represent the component's glass transition temperatures. Both k and q are fitting constants. In general, the parameter q may be considered as a measurement of the specific interaction in a polymer blend system. When the intermolecular interaction is stronger than intramolecular interaction in a binary blend, the value of q will be positive; otherwise, q will be negative. When the q value is larger, it represents that the interaction is stronger than the self-interaction of the blend. After the 'best fitting' by the Kwei equation as shown in Figure 4, k=1 and q=90 were obtained. In this study, a large positive q value of 90 indicates that a strong intermolecular interaction exists between ZKF and CPE.

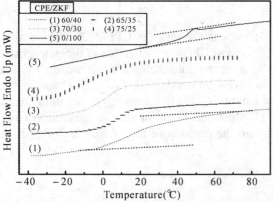

Figure 3 DSC traces of CPE/ZKF hybrids recording during the second heating scans

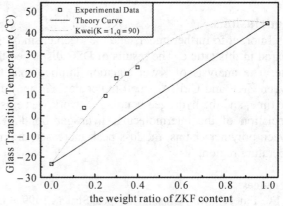

Figure 4 Tg versus composition curse based on CPE/ZKF hybrids

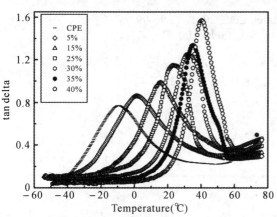

Figure 5 Temperature dependence of tanδ at 1Hz for CPE and CPE/ZKF

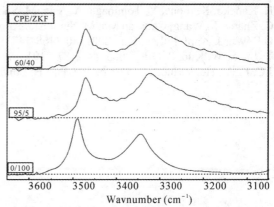

Figure 6 FTIR spectra in the hydroxyl vibration regions of ZKF and CPE/ZKF hybrids

Figure 5 shows the temperature of the loss factor at 1 Hz for CPE/ZKF hybrids. It can be seen from Figure 5, there are only one loss peak for all the samples of the CPE/ZKF hybrids, indicating that the ZKF is compatible with the CPE matrix and the CPE/ZKF hybrid has only one phase, which is consistent with the result observed from the DSC thermograms of CPE/ZKF hybrids with various compositions recorded during the second heating scan as shown in Figure 3. With the increase of ZKF content in the CPE/ZKF hybrids, the position of the loss peak shifts to higher temperature and the height of that increases drastically. This behavior is opposite to that of plasticized poly (vinyl chloride), in which the maximum value of tanδ decreases and its peak shifts to a lower temperature as the plasticizer content increases. Hence, this phenomenon of CPE/ZKF hybrids implies that the presence of an anti-plasticization-like phenomenon and the ZKF molecules can be called an anti-plasticizer.

For further understand the mechanism of the hybrid materials, the FTIR spectra in the hydroxyl vibration regions of ZKF and CPE/ZKF hybrids are displayed in Figure 6. The hydroxyl vibration band of the hybrids show a tendency to shift to a low wave number with the increase of the ZKF content and the width of band is broadened. A small molecule ZKF may form two hydrogen bonds with two CPE chains at the same time, implying that the intermolecular hydrogen bonds exist between CPE and ZKF within hybrids, which is consistent with the result of analysis of Kwei equation. Because of the formation of the intermolecular hydrogen bond, some ZKF molecules might act as a physical bulk side group of the CPE chain in the hybrids. Furthermore, because of the formation of a hydrogen-bond network, ZKF might also act as a physical crosslinking agent in the hybrids. Both the physical bulk side group and crosslinking network can lower the flexibility of the CPE chain [9], resulting in a shift of the position of the loss peak of CPE/ZKF hybrid to higher temperature. The energy dissipation due to dissociation of the intermolecular hydrogen-bond network is larger than that of due to general friction between polymer chains. Therefore, the increase of the height of the loss peak can be attributed to the dissociation of the intermolecular hydrogen bonds between CPE and ZKF within hybrids.

4 Conclusions

In order to further understand the mechanism of the hybrid materials, a series of CPE/ZKF hybrids were prepared in this article. The results of DSC displayed that the ZKF and CPE were miscible in the amorphous phase. The analysis of Kwei equation implied that there was a strong intermolecular interaction existed between ZKF and CPE. A small molecule ZKF may form two hydrogen bonds with two CPE chains at the same time, and the hydrogen bonding network was existed in the hybrids. Due to the energy dissipation due to dissociation of the intermolecular hydrogen-bond network is larger than that of due to general friction between polymer chains, the loss peak increased dramatically and the position of the loss peak shifted high temperature region.

References
[1] I.C. Finegan, R.F. Gibson, Compos. Struct., 1999, 44:89-98.
[2] C. Zhang, J.F. Sheng, C.A. Ma, M. Sumita, Mater. Lett., 2005, 59: 3648-3651.
[3] X. Yan, H.P. Zhang, M. Sumita, Journal of Dong Hua University (Eng. Ed.), 2001, 18: 11-13.
[4] M. Sumita, Funct. Mater., 1995, 15, 12-19.
[5] H. Kaneko, K. Inoue, Y. Tominaga, A. Asia, M. Sumita, Matel. Lett., 2001, 52:96-99.
[6] C. Zhang, P. Wang, Chun-an Ma, M. Sumita, J. Appl. Polym. Sci., 2006, 100:3307-3311.
[7] C.F. Wu, J. Appl. Polym. Sci., 2001, 80:2468-2473.
[8] Y. Su, S. W. Kuo, Polym., 2003, 44:2187-2191.
[9] J.C. Li, Y. He, Y. Inoue, J. Polym. Sci. Part B: Polym. Phys., 2001, 39: 2108-2117.

35 Study on Energy Absorption of Epoxy Resin Composite Targets Enforced with SiC Powder and Twaron Staple Subjected to Bullet Impact

Liping Guan[1,2*], Jianchun Zhang[1,3]

(1. College of Textile, Donghua University, Shanghai 201620, China
2. Ningbo Advanced Textile Technology & Fashion CAD Key Laboratory, Zhejiang Textile & Fashion College, Ningbo 315211, China
3. The General Logistics Department of PLA, Beijing 100088, China)

Abstract: In this paper, the property of energy absorption of epoxy resin composite target enforced with SiC powder and twaron staple was studied. The investigation showed that it have better energy absorption than the epoxy resin composites enforced by only SiC powder or twaron staple with the same volume percentage. The mechanism of energy absorption is the cracking of epoxy matrixes, the tensile destroy of the twaron staple, the pulled-out of twaron staple and SiC powder and the distortion of bullet. The volume percentage of staple and powder interrelated with the energy absorption of the composites.

Keywords: SiC powder; Twaron staple; Epoxy resin; Composite; Energy absorption

1 Introduction

Fiber enforced composites have good length, module, et al. So it was used widely. As an excellent material, the composites enforced by aramid fiber were used in the field of space navigation, ballistics, architecture and others. The properties of the composites were studied [1-3].

Silicon carbide (SiC) is one of the most important non-oxide ceramic materials, which is produced on a large scale in the form of powders, molded shapes and thin film [4]. It has wide industrial applications due to its excellent mechanical properties, high thermal and electrical conductivity, and excellent resistance of chemical oxidation. And it has potential application as a functional ceramic or a high temperature semiconductor [5-7]. SiC powder enforced resin matrix composites used in the ballistic field were investigated widely [8-10].

In this paper, the property of energy absorption of epoxy resin composite targets enforced with Twaron staple and SiC powder subjected to bullet impact were investigated. And the principle of energy absorbing was discussed.

2 Experimental

2.1 Materials

The epoxy resin is EP 3323 come form K.K. Enterprise of China. Twaron staple made from twaron filament (AKZO Ltd., Netherlands, the size is 1680 dtex/1000f) by the way of shearin. The main length of the staple is 10mm. The SiC powder produced by Guanzhoug Jiechuang trading Ltd. of China, the diameter of the powder is form 10 μm to 100 μm, and the content of. β-SiC is 93%.

2.2 Execution of the targets

Twaron staple and SiC powder with a certain volume was put into epoxy resin with decided volume. The blending was churned up to make the three materials uniformity. Then the blending was transferred into a mould. Take it into platen vulcanizing press, and the blending was solidified out into a plate at 145 ℃ temperature. The pressure is 0.5MPa. The solidifying time is 30 minutes. The volume percentages of samples showed in Table 1.

Table 1 The volume percentages of each component

Numbers of sample	Epoxy resin (V%)	SiC powder (V%)	Twaron staple (V%)
1#	70	0	30
2#	70	5	25
3#	70	10	20
4#	70	15	15
5#	70	20	10
6#	70	25	5
7#	70	30	0

* Corresponding author' email: lpguan@mail.dhu.edu.cn

2.3 Ballistic impacting test

The testing equipment of ballistic impact was shown in Figure 1. The gun is droven by high pressure with 14.5mm caliber. The bullets are solid with a taper head (Figure 2). It made of steel, weight is 35.6g. The velocity of the bullet is about 200m/s, which dominated by the pressure of nitrogen and showed on the arithmometer 1, it was tested by photoelectriec sensor. The residual velocities showed on the arithmometer 2 and were induced by a set of aluminum screens. The size of target is 180mm×180mm×2mm. There are 5 tests for each sample. And the data is the average of the 5 times.

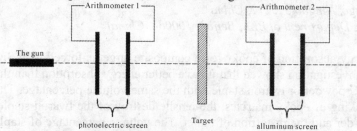

Figure 1 Sketch map of ballistic impact equipment

Figure 2 The sharp of bullet

3 Results and discussion

3.1 Analysis of energy absorbing mechanism

In the experimental, we can see that the targets were cracked when it was projected. The fragments of the targets are bigger if it has bigger volume percentage of twaron staple (v_t). It denoted that the twaron staple promoted the energy transmission. Some twaron staple and SiC powder were pulled out and some staple were snapped. The head of bullets distorted, and some SiC powder have penetrated into the bullet. The distortion of the head of bullet increased following the volume percentage of SiC (v_{SiC}) increased. The SiC grains can distort and penetrate into the bullet because the rigidity of SiC is larger than the steel's.

So the manner of energy absorption of the epoxy resin composites enforced by SiC powder and twaron staple include the cracking of epoxy matrix, the tensile destroy of the twaron staple, the pulled-out of staple and SiC powder and the distortion of bullet.

3.2 The influence of bullet residual velocity from different volume percentage of components

The average residual velocity and penetration energy absorption of the samples with different volume percentage of components were showed in Table 2.

Table 2 The data of residual velocity and energy absorption

Samples	Impactive velocity (m/s)	Residual velocity (m/s)	Energy absorption (J)	Energy absorption percentage (%)
1#	201.2	110.3	50.4	69.9
2#	199.6	86.9	57.5	81.0
3#	197.3	72.7	59.9	86.4
4#	200.4	68.4	63.2	88.4
5#	198.2	66.6	62.0	88.7
6#	196.8	77.5	58.2	84.5
7#	199.7	93.8	55.3	77.9

In Table 2, the energy absorption (E_a) and energy absorption percentage (E_a) were calculated by formula (1) and formula (2) respectively, which based on the law of conservation of energy:

$$E_a = \frac{1}{2} m_b \left(v_i^2 - v_r^2 \right) \quad (1) \qquad E_a\% = 1 - \frac{v_r^2}{v_i^2} \quad (2)$$

In the formulas, m_b is the weight of bullet, v_i is the impactive velocity, and v_r is the residual velocity.

Figure 3 is the scheme of energy absorbed percentage varied with the volume percentage of SiC powder changed. It showed that the epoxy resin composites was enforced by SiC powder and Twaron staple have better energy absorption percentage than it was enforced by only SiC powder or twaron staple if the impactive velocity is similar, and the SiC powder have better efficacy of energy absorption than twaron staple's when they were used in epoxy resin composite. The $E_a\%$ increased following the volume percentage of SiC powder (v_{SiC}) increased from zero to 20%. But the volume percentage of twaron staple (v_t) decreased following the

v_{SiC} increased because the volume percentage of epoxy resin was confirmed (70%) in the samples. So the $E_a\%$ decreased following the v_t decreased when the v_{SiC} exceeded 20%. It implied that the SiC powder and the twaron staple in the composites enforced each other. We can obtain a target with good energy absorption if the v_{SiC} and v_t were controlled advisably.

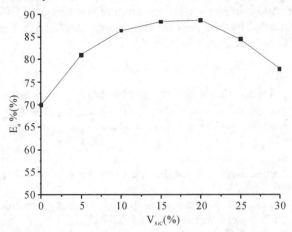

Figure 3 Relationship scheme between volume percentage of SiC powder and absorbed energy percentage

3.3 The deflexed angle of bullet

In the projecting experimental, the flying track of bullet was supposed to be a level line. But the track can be changed when the bullet permeated through the target in fact. So some error will be occurred in the testing and calculating results of bullet residual velocity and energy absorption. So the deflexed angle (θ) of bullet was tested and calculated. The principle was showed in Figure 4. The distance (l) between the first aluminum screen and the last aluminum is confirmed. The vertical space (h) of bullet pores can be tested. θ was calculated by formula (3). And the v_r was corrected by formula (4). The v in formula (4) is the data from the arithmometer 2. The corrected results were showed in Table 3.

$$\theta = ac \tan \frac{h}{l} \quad (3) \qquad v_r = \frac{v}{\cos \theta} \quad (4)$$

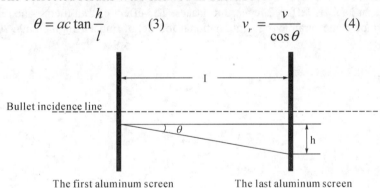

Figure 4 Sketch map of the deflexed angle of permeated bullets

Table 3 The corrected data of residual velocity and energy absorption

Samples	v_i (m/s)	θ	v_r (m/s)	E_a (J)	E_a (%)
1#	201.2	12°32′	113.0	49.4	68.5
2#	199.6	16°57′	90.9	56.3	79.3
3#	197.3	9°08′	73.6	59.3	86.1
4#	200.4	13°45′	70.4	62.7	87.7
5#	198.2	7°26′	67.2	61.9	88.5
6#	196.8	6°15′	78.0	57.9	84.3
7#	199.7	11°41′	95.9	54.6	76.9

In Table 3, we couldn't find any relation between the deflexed angle of bullet and the volume percentage of SiC. The data of energy absorption of the target is lower than that haven't been corrected.

4 Conclusion

The target of epoxy resin composites enforced by SiC powder and twaron staple with different volume percentage of SiC were investigated. It showed that the energy absorption of the composites contribute to the cracking of epoxy matrixes, the tensile destroy of the twaron staple, the pulled-out of staple and SiC powder and the distortion of bullet. And it interrelated to the volume percentage of SiC powder and twaron staple. The epoxy resin composites enforced by SiC powder and twaron staple have better absorbed energy percentage than it enforced by only SiC powder or twaron staple.

Acknowledgement

The authors thank Ningbo Advanced TextileTtechnology &Ffashion CAD Key Laboratory for imbursement and Dr. Xiong Jie of Zhejiang Sci-Tech University for helping with the manufacturing of composites.

References

[1] Yue C Y P, Padmanabhan K. Interfacial studies on surface modified Kevlar fiber/epoxy matrix composites. Composites: Part B. 1999, 30: 205-217.
[2] Tarantili P A, Andreopoules A G. Mechanical properties of epoxies reinforced with chloride-treated aramid fibers. Journal of Applied Polymer Science. 1997, 65: 267-275.
[3] Menon N, Frank D B, Harani L R. Use of titanate coupling agent in Kevlar-phonolic composites. Journal of Applied Polymer Science. 2004, 154: 113-123.
[4] H O Pierson. Handbook of Refractory Carbides and Nitrides, William Andrew, Noyes, 1996: 137.
[5] Andrievskl R A, Glezer A M, Size effects in properties of nanomaterias. Scripta Mater. 2001, 44: 1621-1624.
[6] Jinjun Shi, Yongfa Zhu, Xinrong Zhang. Recent developments in nanomaterial optical sensors. Trends in Analyical Chemistry, 2004, 23: 1-10.
[7] Ning Sheng Xu, Shao Zhi Deng, Jun Chen. Nanomaterials for field electron emission: preparation, characterization and application. Ultramicroscopy. 2003, 95: 19-28.
[8] M Baki Karamis. Tribology at high-velocity impact. Tribology International. 2007, 40: 98-104.
[9] Hideki Tamura, Yasuo Mutou. Quantitative analysis of debris clouds from SiC fiber reinforced silicon nitride bumpers. International Journal of Impact Engineering. 2005, 31: 1192-1207.
[10] Sai Sarva, Sia Nemat Nasser, Jeffrey Mcgee, Jon Isaacs. The effect of thin membrane restraint on the ballistic performance of armor grade ceramic tiles. International Journal of Impact Engineering. 2007, 34: 277-302.

36 Research on Acoustic Performance of Basalt Filament Diaphragm Fabric

Zongfu Guo, Zhili Zhong, Ruibin Yang

(School of Textile, Tianjin Polytechnic University, Tianjin 300160, China)

Abstract: This paper focused on a comparative study on elastic modulus and absorption coefficient between basalt filament diaphragm fabric and glass fiber diaphragm fabric. Several different types of basalt filament diaphragm fabrics were woven with CSW-03 sample loom, using the Werner Mathis AG LTF97885 coating machine to coat the fabrics, using 3380 Instron-based multi-function tester respectively to test and analysis different coating parameters of basalt filament diaphragm fabric's elastic modulus, and to compare with glass fiber diaphragm fabric. The absorption coefficient of the basalt filament diaphragm fabric and the glass fiber diaphragm fabric were measured by VA-Lab4 IMP-AT Materials Testing. Matlab7.0 mathematical processing software was used to analysis the effect relationship between the coating thickness and the average absorption coefficient of basalt diaphragm filament, and to get the optimal relationship points between coating thickness and absorption coefficient. The research showed that the basalt filament was more suitable for basalt filament diaphragm fabrics, and when the coating thickness was 0.16mm, the basalt filament diaphragm's acoustic performance was best. The results provided a scientific basis for the development of the basalt filament diaphragm and must promote the product development of basalt fiber, and increase the market competitiveness.

Keywords: Basalt filament diaphragm fabric; Glass fiber diaphragm fabric; Elastic modulus; Absorption coefficient

0 Introduction

The high-quality loudspeaker diaphragm material should have two characters: firstly, elastic modulus is high, which can assure low distortion of the loudspeaker. Secondly, the loudspeaker diaphragm material should have high internal damping, which can reduce the vibration of the loudspeaker[1]. The common materials are paper diaphragm fabric, plastic diaphragm fabric, and mental diaphragm fabric. The plastic diaphragm fabric, made of polypropylene has achieved great improvements in motivation[2], detailed performance and sound producer. With the development of various new fabrics, the diaphragm fabric based on these fabric also makes improvements. Boron fibers, Dupont.co Kevlar and honeycombed synthetic fiber have been used in the production of speaker louder with the advantage of intensive strength, high efficiency of 98db/w[3], light weight.

Basalt filament is a new type of fiber with the similar character of glass fiber with the advantage of light weight, low noisy and high corrosion resistance[4]. Comparing with glass fiber and aramid fiber, the basalt filament is characterized by low production cost, less pollution. Therefore, the production of loudspeaker diaphragm material made from basalt filament is an important research aspect. This paper discussed the elastic modulus and absorption coefficient of the basalt filament diaphragm fabric. It also focused on a comparative study on elastic modulus and absorption coefficient between basalt filament diaphragm fabric and glass fiber diaphragm fabric. It also proved that basalt filament was available as loudspeaker diaphragm material. It discussed the effect of coating thickness on the absorption coefficient.

1 Materials and instruments for experiment

Basalt filament of 41tex was from TuoXin Ltd. ChengDu., the diameter was 7 μm, and the density was 3g/m^2. The density of the glass fiber is 71tex. Different types of fabrics were woven on CSW-03 Weaving sample loom. Coating was made of acrylic ester and the Werner Mathis AG LTF97885 coating machine are used to coat the fabric. The 3380Instron-based Multi-fictional tester and VA-Lab4 IMP-AT Materials Testing are adopted for testing.

2 Character test of the sample
2.1 Production of the sample

Basalt filament fabrics and glass fiber fabrics were woven with CSW-03 sample loom. They are shown as Figure1.

(a) 41tex Basalt filament plain weave   (b) 71tex Glass-fiber plain weave

Figure 1 Basalt filament plain weave and Glass-fiber plain weave

Acrylic ester is qualified damping material with the advantage of dynamic mechanism and high damping (tanδ>0.3) [5]. In the experiment, acrylic ester was chosen as the material for the production of coating, which can promote the internal damping of the basalt filament and the strength of the basalt filament diaphragm fabric.

According to the procedure, the basalt filament fabric and glass fiber fabric were coated on the Werner Mathis AG LTF97885. According to the experiment, when the curing temperature was 120℃, and the curing time was 30 to 40min, the curing effect was best.

2.2 Test

The 3380Instron multi-functional machine was used to test the elastic modulus of the basalt filament diaphragm fabric and glass fiber diaphragm fabric. This paper figured out the absorption coefficient of the basalt filament diaphragm fabric by using the VA-Lab4 IMP-AT absorption coefficient testing system, which was based on the transfer function method. The transfer function method contributed a lot to the test of the absorption coefficient test, which not only changed the test from single-way to multi-way to save the time, but also promoted the speed and accuracy of the absorption coefficient[6].

The refection power in the transfer function is decided by the H figured out by the two transfers[7]. Incident pressure is $\Pr(X)$, and $\Pr(X) = \Pr \times e^{jkx}$, reflected pressure is $P_f(X)$, $P_f(X) = P_f \times e^{jkx}$, $P_f = r \times P_r$. P_r and P_f is the diaphragm when x=0. P stands for amplitude, and k is the number of complex wave. Thus, the sound pressure on the two speaker louder is $P(X_1) = P_r \times e^{jkx_1} + P_r \times e^{-jkx_1}$, $P(X_2) = P_f \times e^{jkx_2} + P_f \times e^{-jkx_2}$.

The transfer function of the incident wave is $H_r = P_r(X_2)/P_r(X_1) = e^{-jks}$, while the transfer function of the reflected wave is $H_f = P_f(X_2)/P_f(X_1) = e^{jks}$.

S was the distance between the two speaker louder. The general transfer function was $H = P(X_1)/P(X_2) = (e^{jkx_2} + r \times e^{-jkx_2})/(e^{jkx_1} + r \times e^{-jkx_1})$. Thus, the reflecting power is r, and $r = (H - H_r) \times e^{2jkx_1}/(H_f - H)$. From the formula, a conclusion was made that r was figured out by the transfer function of the speaker louder and their position of x_1、x_2. Therefore, the absorption coefficient can be worked out as follows, $\alpha = 1 - |r|^2$.

3 Results and discussions

3.1 Elastic modulus

The comparisons of basalt filament diaphragm fabric and glass fiber diaphragm fabric with similar warp density and weft density were shown in Table 1.

Table 1 Comparison of elastic modulus

Sample	Elastic modulus (GPa)	Elongation at break (%)	Breaking strength (N)
Basalt filament fabric	30.542	12.283	371.042
Basalt filament diaphragm fabric	99.4005	4.159	1247.631
Glass-fiber fabric	21.48	6.362	609.398
Glass-fiber diaphragm fabric	43.62	3.76	758.89

From Table 1, it was shown that there was great change of the elastic modulus for the basalt filament diaphragm fabric and glass fiber diaphragm fabric after the coating. What is the more important is that the change is more obvious for the basalt filament diaphragm fabric. Under the similar conditions, there is higher elastic modulus and extension for the basalt filament diaphragm fabric but less splitting. Further more, based on the similar character of the two fibers, the basalt filament diaphragm fabric can also be used to make diaphragm.

In order to figure out the effect of the coating thickness on the basalt filament diaphragm fabric, this experiment tested the elastic modulus of various thick basalt filament diaphragm fabric. The results were shown as Table 2.

Table2 Comparison of elastic modulus with different coating thickness

	Sample	Coating thickness (mm)	Breaking strength (N)	Elastic modulus (GPa)	Elongation at break (%)
$1^{\#}$	Basalt filament fabric	0	371.043	30.543	12.284
$2^{\#}$	Basalt filament diaphragm fabric	0.05	1247.632	99.4002	4.157
$3^{\#}$	Basalt filament diaphragm fabric	0.095	1253.013	57.6534	3.950
$4^{\#}$	Basalt filament diaphragm fabric	0.29	1161.002	39.734	3.553

From Table 2, a conclusion can be drawn that the strength is more when the thickness become greater. When the coating thickness is 0.05mm, the elastic modulus is at its peak. While the coating thickness surpasses 0.05mm, the elastic modulus is smaller. When the thickness is 0.1mm, the coating thickness is the elastic modulus will be largerest when the coating thickness is 0.05mm. When the coating thickness is bigger than 0.05mm, the bigger the coating thickness is, the smaller the elastic modulus becomes. When the coating thickness is 0.1mm, which is the 1/2 of that of the fabric, the diaphragm can not only show the characteristics of the basalt filament diaphragm fabric, which are the high elastic modulus and high strength performance, but also can develop the enhanced function of acrylic ester to the abrasion resistance and the structral stability of the fabric. So the load of the diaphragm fabric is much higher. When the coating thickness is 0.29mm, which is thicher than that of the diaphragm fabric, the diaphragm fabric is under the condition of tension failure, the performance shown here is the mechanical properties of the Acrylic modal. The functions of the basalt filament diaphragm fabric cannot fully developed, which can lead to the reducing of the breaking tenacity and the falling of the elastic modulus of the diaphragm fabric.

3.2 Influences of coating thickness on absorption coefficient

The absorption coefficient is an important index in measuring sound insulation of the material. There are something in common between the performances of the glass fiber and those of the basalt filament. Using glass fiber fabric as the reference object, we compared the absorption coefficient between the basalt filament diaphragm fabric and the glass fiber diaphragm fabric. Using the VA-Lab4 IMP-AT material absorption coefficient measuring system, we measured the absorption coefficient of the glass fiber diaphragm fabric sample and that of the basalt filament diaphragm fabric, then recorded the results, then compared the absorption coefficient under 2500Hz high frequency sound wave and the average absorption coefficient, see Figure 2 and Figure 3.

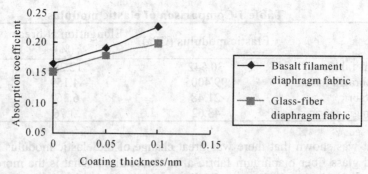

Figure 2 Comparison of high-frequency sound absorption coefficient

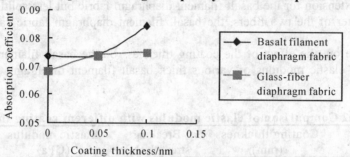

Figure 3 Comparison of the average sound absorption coefficient

As is shown in Figure 3, the changing of average absorption coefficient of the basalt filament diaphragm fabric is similar to that of the glass fiber diaphragm fabric, which is the thicker the coating thickness is, the bigger the absorption coefficient is. This suggests that there are something in common between the basalt filament diaphragm fabric and the glass fiber diaphragm fabric in terms of absorption coefficient. Besides, under the conditions of the same or similar paint, fabric thickness, coating thickness and fabric density, the high frequency sound absorption coefficient and the average absorption coefficient of the basalt filament diaphragm fabric are higher than those of the glass fiber diaphragm fabric. Thus, basalt filament diaphragm fabric is better than glass fiber diaphragm fabric in terms of absorption coefficient. At the same time, the damping coefficient of the basalt filament diaphragm fabric is higher than that of the glass fiber diaphragm fabric. So the basalt filament diaphragm fabric can be used to produce speaker diaphragm, for it has the properties of diaphragm material: high elastic modulus and high damping coefficient.

In order to analysis the effect relationship between coating thickness and the absorption coefficient of basalt filament diaphragm fabric, Matlab7.0 was used to analysis the relationship between the overall absorption coefficient and coating thickness of basalt filament diaphragm fabric, and the relationship was shown in Figure 4.

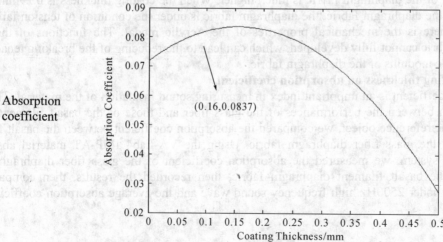

Figure 4 The relationship between coating thickness and absorption coefficient
$y=0.1503x - 0.4792x^2 + 0.0719$

From the analog curve in Figure 4, we could see that as the coating thickness X increases, the average absorption coefficient of basalt filament diaphragm fabric increases. When x=0.16, the average absorption coefficient of basalt filament diaphragm fabric reaches its maximum, which is the summit of the curve. when x>0.16, as X increases, the average absorption coefficient of basalt filament diaphragm fabric decreases. When the coating thickness is 0.16mm, the sound-absorbing performance of the basalt filament diaphragm fabric is the best.

4 Conclusion

From the comparisons of elastic modulus and absorption coefficient between basalt filament diaphragm fabric and glass fiber diaphragm fabric, the conclusion can be drawn that basalt filament diaphragm fabric can be used to produce diaphragm. Under the conditions of the same or similar paint, fabric thickness, coating thickness and fabric density, the sound-absorbing performances of the basalt filament diaphragm fabric are better than that of the glass fiber diaphragm fabric. The influences of coating thickness on the elastic modulus and absorption coefficient of absalt filament diaphragm fabric are significant. When the coating thickness is 0.16mm, the sound-absorbing property of the basalt filament diaphragm fabric is the best.

References

[1] Y W Zhou. Speaker of the diaphragm. Audio-visual Technology J, 1998(7): 37.
[2] Mayr Hans.Theory of low-frequency loudspeaker.Alta Frequenza J, 1974, 43(12): 1023-1032.
[3] B X Cai, M L Jing.Fabric Construction and Design. Beijing: China Textile & Apparel Press, 2004, 8-29.
[4] J C Lei, Y L Lei, H B Wang, etal. Progress of study on the preparation of basalt fiber and composite material of basalt fiber. Materials Review J, 2006(4): 382-385.
[5] W Zeng,S C Li. Latex interpenetrating polymer network damping materials. Polymer Bull J, 2001, (1): 68-72.
[6] Su,Yong; Lin, Wei-zheng. Measurements of material's acoustic velocity and attenuation by signal analysis [J].Journal of Building Materials, 2001, 4(1): 65-69.
[7] Z B Zhang.Expermental Research on Acoustic Material's.Shanghai:Shanghai Jiaotong University D, 2007: 32-37.

37 Preparation and Characterization of Polypropylene Master-batches Containing Electret for Meltblown Nonwoven

Jian Han*, Xiangbing He, Bin Yu, Guoping Xu, Xingbo Ding

(Key Laboratory of Advanced Textile Materials and Manufacturing Technology, Ministry of Education of China, Zhejiang Sci-Tech University, Hangzhou 310018, China)

Abstract: The bicomponent master-batches were prepared by co-extrusion of polypropylene and tourmaline fillers, and were characterized by MIA, PI, SEM, DSC and WAXD. Their melting index and viscosity of blends decreased with tourmaline input and temperature decrease. SEM results showed that inorganic particles in blends had relatively good dispersibility. It could be seen from DSC and WAXD curves that Crystallinity of master-batches with electret raised firstly then decrease with the rising of tourmaline content. Crystal structure of blends relative to its pure polypropylene did not change.

Keywords: Meltblown nonwoven; Polypropylene; Electret; Characterization

1 Introduction

Meltblowing is a one-step process to make microfiber nonwovens directly from thermoplastic polymers with the aid of high-velocity air to attenuate the melt filaments[1]. It has become one of the most important industrial techniques in nonwovens because of its ability to produce fabrics of microfiber structure suitable for air filtration media. Polypropylene (PP) is widely used in this process. Some researches showed that the filtration efficiency of PP melt-blown nonwovens charged by corona-charging could be improved greatly[2-3]. However, their filtration efficiency decreased with the time due to surface density of charge attenuation. In this study, it was researched how to design a meltblown nonwoven with high filtration efficiency, and to prepare the bicomponent PP master-batches and tourmaline filler. The fluidity of polymer was also analyzed.

2 Experimental

2.1 Materials

Tourmaline provided from Beijing Central Iron & Steel Research Institute was used as electrets. The mean diameter of tourmaline was approximately 1 micrometer. Polypropylene chips (isotactic PP for meltblown) was provided from Jiangyin Bairuijia Platics Science and Technology Co.Ltd, China. For easier spinning process, PP/fillers master-batches were prepared by a conventional twin-screw extruder.

2.2 Fluidity test

Melting index(MI) of master-batches were carried out using the Melting Index Apparatus(MIA)RZY-400 in 170℃ and 230℃ with 2.16Kg. Complex viscosity of polymer was tested by plate rheometer(PI) Physica MCR301.

2.3 Morphology observation

The cross-section structure of the bicomponent master-batches were observed using Scanning Electron Microscope (SEM, FETX3 7426). The SEM samples were goldsputtered before observation.

2.4 Thermal analysis

For thermodynamic experiment, dynamic scanning calorimerer(DSC, Perkin-Elmer DSC-7) equipped with a cooler was used under the nitrogen atmosphere. All the samples were heated from 0℃ to 250 at 10℃/min. From this procedure, apparent enthalpies of fusion were calculated from the area of the endothermic peak. The percent crystallinity of polypropylene was evaluated using the following equation:

$$crystallinity(\%) = \frac{\Delta H_f}{\Delta H_f^0 \cdot \omega_f}$$

where ΔH_f is the heat of fusion of PP fibers, ω_f the weight fraction of PP in the blends, and ΔH_f^0 the extrapolated value of the enthalpy corresponding to the heat of fusion of 100% crystalline PP taken as 209 kJ/kg from the literature[4].

---

* Corresponding author's email: hanjian8@zstu.edu.cn

2.5 Wide angle X-ray diffraction (WAXD)

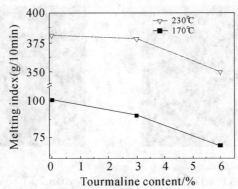

Figure 1 The change of melting index of master-batches with the tourmaline particles content increase

The microstructure and crystallization of samples were analysis by Rigaku D/MAX X-ray diffractmeter, using CuK_α radiation in the 2θ range of 10-30°. The accelerating voltage was 40 kV, and the tube current was 150 mA. All measurements were performed at room temperature.

3 Results and discussion
3.1 Fluidity of polymer

Figure 1 indicated the change curves of melting index of master-batches. The results showed that the input of the tourmaline particles made the melting index of the polymer fall and the temperature rise caused the melting index increase sharply. So the little tourmaline particles inputting and higher process temperature can improve the bicomponent polymer fluidity.

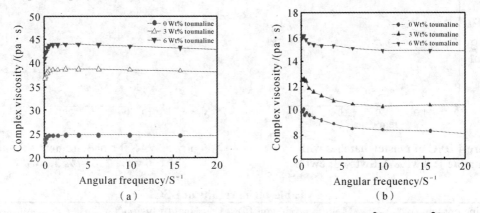

Figure 2 The complex viscosity of the master-batches (a) 180℃ (b) 200℃

The complex viscosity change of blends with various tourmaline content and temperature was shown in Figure 2. The viscosity of master-batchs increased with the decline in the content of tourmaline in 180℃ and 200℃, which was same as the change of melt index. Viscosities of blends rise with the increase in angular frequency in the lower angular velocity which exhibited the phenomenon of shear thickening, as shown in Figure 2(a). The reason may be that the entanglement rate of melt-blown polypropylene macromolecular is faster than that of ease of macromolecular in lower temerperature. The blends showed the shear thinning characterization with temperature increase (Figure 2(b)) which resulted to faster ease rate of macromolecular than that of entanglement.

3.2 Cross sections of fibers observation

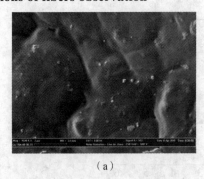 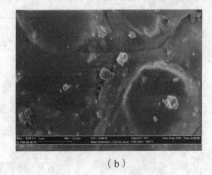

(a) (b)

Figure 3 SEM photographs (×3.00k) of the master-batches with 2(a) 3 Wt% tourmaline and 2(b) 6Wt% tourmaline

SEM is a good way to present a real-space image of the particles filled in the polymer materials. Hence, the cross-section of the fibers was observed on SEM. In Figure 3, SEM micrograph showed the cross-sections of master-batches filled with 3(a) the 3Wt% tourmaline particles and 3(b) the 6Wt% tourmaline particles. In Figure 3(b), some conglomerations of the particles and holes were observed. The SEM photograph of master-batches filled with 3Wt% particles had seldom conglomeration and hole in Figure 3(a). It was demonstrated that the compatibility of the PP with the particles became bad with the particles content increase which will affect the other properties of the nonwoven sometimes.

3.3 Thermal property

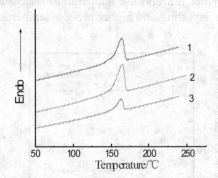 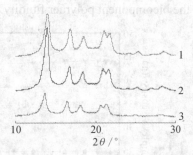

Figure 4 DSC of master-batches with tourmaline 1: 0 wt% 2: 3wt% 3: 6wt%

Figure 5 WAXD curves of neat PP and blends 1: 0 wt% 2: 3wt% 3: 6wt%

Table 1 The results of DSC

Tourmaline content (wt%)	Melting point Tm(℃)	Heat of fusion(J·g⁻¹)	Crystallinity of PP (%)
0	163.98	91.12	43.62
3	164.91	92.24	45.55
6	162.71	77.99	39.75

Differential scanning calorimetry thermograms of blends containing various content tourmalines are depicted in Figure 4. Table 1 summarized the data from DSC of Figure 4. There was a negligible relationship between the melting point of blends and the variation of fillers content. In DSC graphs, the melting peak area of blends including tourmaline was slightly higher firstly and then decreased with the fillers content rise than that of the pure PP polymer. The crystallinity of blends including tourmalines particles showed the same change law with the melting peak area of blends (Table 1). The wt% crystallinity of PP was calculated via the standard heat of crystallization which was taken to be 209 Jg⁻¹. The increase in crystallinity of PP revealed that the lower content filler particles made for the crystallization of PP. Thus, it was supposed that the less content tourmalines in the PP matrix acted not as impurities but as nucleating agents, and the higher ones did as impurities.

3.4 WAXD analysis

Figure 5 WAXD curves were of neat PP and blends. From these curves, it was observed that five distinct crystalline peaks of α-form of PP could be seen on every curve of neat PP and blends. It indicated that inputting of tourmalines did not affect on the crystalline structure of PP. X-ray diffraction intensity of WAXD curves increased firstly, and then decreased with fillers content rise. The results were same as that of DSC in Table 1.

4 Conclusion

Bicomponent master-batches were prepared by a conventional twin-screw extruder with polypropylene chips and tourmalines particles. The less tourmaline particles inputting and the higher process temperature could improve the bicomponent master-batches fluidity. SEM micrographs showed that seldom particles had aggregated. The results of the DSC thermogram and WARD indicated that the crystallinity of polypropylene including particles was slightly increased firstly, and then decreased with increasing of the particles content compared with that of pure polypropylene polymer.

Acknowledgement

This work is supported by Science and Technology Plan Projects of Science Foundation of Zhejiang Provincial (2007C11155), Zhejiang Provincial Key Laboratory (2006E10009) and Training Foundation of Outstanding Young talent of "Textile Subject" (Grant No. 2008QN08).

References

[1] Z.Z. Yang. Particle Filtration with an Electret of Nonwoven Polypropylene Fabric. Textile Research Journal, 2002, 72(12): 1099-1104
[2] R Raghavendra.Nanoparticle Effects on Structure and Properties of Polypropylene Meltblown Webs, Journal of Applied Polymer Science, 2010, 115:1062–1072
[3] P P. Tsai. Different electrostatic methods for making electret filters. Journal of Electrostatics. 2002, 54,333–341
[4] J. Brandrup, E. H. Immergut and E. A. Grulke, Polymer Handbook, John Wiley & Sons Inc., New York, 1999

# 38 Consolidation of Fragile Silk Fabrics with Fibroin Protein and Ethylene Glycol Diglycidyl Ether by Spaying

Xiaofang Huang[1], Zhiwen Hu[1], Zhiqin Peng[1], Jing Zhang[1], Xiaoye Cao[1],
Yang Zhou[2], Feng Zhao[2]

(1. Key Laboratory of Advanced Textile Materials and Manufacturing Technology, Ministry of Education of China, Zhejiang Sci-Tech University, Hangzhou 310018, China
2. China National Silk Museum, Hangzhou 310002, China)

Abstract: Fibroin protein / ethylene glycol diglycidyl ether (EGDE) consolidation treatment of fragile silk fabrics by spaying has been studied in this paper. It is found that the breaking strength, elongation at break and handle of the fragile silk fabrics were improved obviously with the EGDE introducing. TG results show that the decomposition temperature of the modified silk fabrics increased under the influence of EGDE. The structure of silk fabrics was investigated by FTIR and amino acid analysis. The amino acid analysis shows that EGDE exhibited a slightly higher reactivity toward tyrosine, lysine and histidine.

Keywords: Silk fabrics; Silk fibroin solution; EGDE

1 Introduction

Silk culture is an important component of Chinese civilization. The existing ancient silks are indispensable material evidences to research ancient civilization. However, a large number of existing ancient silks need to be consolidated urgently, because they are already very fragile when they are unearthed. At present, the main methods of culture relic consolidation include weave, mount, silk screen, resin, adhesive[1] and Parylene C consolidation[2] and graft copolymerization, and so on. But these methods all have deficiencies. Their applications are subject to certain restrictions.

Fibroin protein, which is homologous and compatible with fragile silk, is used in this study to consolidate silk. EGDE is one of the most important modifying agents applied in industry. Researchers have analyzed the reactivity of epoxide with silk. It is found that the modification of silk with epoxide has influence on the silk fiber and fabric materially, which improves its properties such as physical property, moisture absorption, chemical resistance, and wash and wear property.[3,4]

However, there is no article about the consolidation of fragile silk fabrics with fibroin protein and ethylene glycol diglycidyl ether according to our knowledge. Artificial ageing silk was used in stead of silk relics and EGDE was chosen as consolidation auxiliary reagent in this study. A technique to consolidate fragile silk by fibroin protein / EGDE was also developed in this study.

2 Experimental

2.1 Materials and sample preparation

White silk habutae fabrics (Bombyx mori, plain weave) were supplied by Sichuan Nanchong Liuhe(Group) Corp(Nanchong, China). Bombyx mori silk fiber was supplied by Zhejiang Misai Silk Co., Ltd(Jiaxing, China). EGDE was bought from Nagase ChemteX Corporation. Other chemicals are analytical grade reagents.

Fragile silk fabrics were prepared from white silk habutae fabrics, which were immersed in 50g/L NaOH solution 9 hours under 35℃ and 50% relative humidity (RH).

Bombyx mori silk fiber was degummed by treating two times with 5g/L Na_2CO_3 solution at 98–100℃ for 30min each and air-dried. Deionized water was used throughout the study. 20g of the fibers was dissolved in 1000ml of 50% $CaCl_2$ solution at 96-98℃ for 90min. The solution was dialyzed in a cellulose tube (molecular cutoff of 14,000) against water for 3 days, and aqueous silk fibroin solution with 3.0wt% fibroin was obtained.

The fragile silk fabrics were first spayed with silk fibroin solution. Then, the fragile silk fabrics were spayed with EGDE solution after 10 min. The samples were conditioned in air at 25°C for 2 days.

2.2 Measurement

Tensile properties of silk fabrics were measured with a YG065 strength tester of electronic fabric (LaiZhou Electron Instrument Co., Ltd.,China) using the standard technique at 20°C and 65% RH at a gauge length of 100 mm and strain rate of 100 mm/min. The chromatic aberration of silk fabrics was measured with a SC-80C automatic colorimeter (Beijing kangguang Instrument Co., Ltd., China). The flexural stiffness of silk fabrics were measured with a LLY-01 Electronic Instrument Stiffness (LaiZhou Electron Instrument Co., Ltd., China).

Dried silk fabrics were hydrolyzed by heating in 6M HCl for 24h at 110±1°C. The amino acid composition of silk hydrolyzate was dissolved in a buffered solution (pH2.2) and determined by using a Hitachi L-8800 Type Rapid Amino Acid Analyzer.

Infrared absorption spectra were obtained by the KBr method using a Nicolet 5700 FT-IR Spectrophotometer.

The surface of the treated silk fabrics was examined after gold coating with a JEOL JSM-5610LV scanning electron microscope at 5 kV acceleration voltage.

Thermogravimetric analyses (TGA) were run under nitrogen on a PYRIS 1 instrument programmed under isothermal conditions, raised at 20°C/min to 650°C.

3 Results and discussion
3.1 Mechanical behavior

The tensile strength and elongation at breaking of the silk fabrics are shown in Table I. The strength and elongation of the fragile silk fabrics is 1.54N and 2%, respectively. Strength and elongation of the silk fabrics treated with the fibroin protein / EGDE increased rapidly. Their strength and elongation increased up to 15.8 N and 8%, respectively. The increase in the strength of the silk fabrics treated with fibroin protein / EGDE seems to imply the formation of the cross-links between the adjacent silk fibroin molecules, which was in accord with the estimation from the Amino Acid Analyzer measurement shown later. Compared with fragile silk fabrics, this process is effective in improving handle of silk fabrics. Moreover, chromatic aberration proved that the fibroin protein / EGDE treatment did not affect the color of silk fabrics directly.

Table 1 Mechanical and physical property of silk fabrics

Sample	Strength/N	Elongation/%	Chromatic aberration	Flexural stiffness/(10^{-2}mN.m)
a	1.54	2.0	/	1.08
b	1.8	2.2	1.42	1.07
c	15.8	8.0	0.97	0.71
d	7.5	6	0.91	0.90

a— Fragile silk fabric; b—Spaying 15 g/L silk fibroin solution in fragile silk fabrics;
c— Spaying 15 g/L silk fibroin solution in fragile silk fabrics, and then spaying 50 g/L EGDE solution after 10 min; d— Spaying 50 g/L EGDE solution in fragile silk fabrics.

3.2 Thermal property

The results of TG measurement are shown in Figure 1. The two figures proceed parallel to each other over the temperature range examined, displaying two inflection points; the first at about 100°C, corresponding to the loss of moisture regain, and the second at 280°C. The TGA curves show that the modified silk specimens exhibited a slightly higher thermal stability. The peak at the biggest weight loss of silk fabric treated with fibroin protein / EGDE (c) appeared at 332.99°C, which is 6.58°C higher than that of fragile silk fabrics (a). These results indicate that the crosslinking covalent bonds formed by the reaction between silk and EGDE may be lead to a more thermally stable structure in silk fabrics.

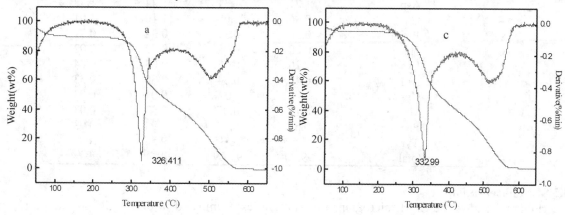

Figure 1 TGA curves of (a) fragile silk fabric and (c) silk fabric treated with fibroin protein / EGDE

3.3 FTIR spectroscopy

Figure 2 shows the FTIR spectra of EGDE and silk fabrics in the range of 700 to 4000 cm^{-1}. Silk fabric treated with fibroin protein / EGDE (c) and silk fabric treated with EGDE (d) appear a new peak at 1105cm^{-1}, which is the characteristic absorption peak of infrared spectroscopy of EGDE presenting the unsymmetrical stretching vibration of -CH_2-O-CH_2- group. This result proves that silk fibroin is cross-linked with EGDE.

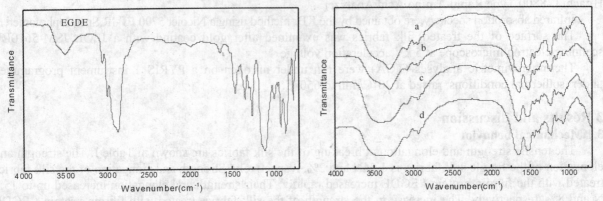

Figure 2 FTIR spectra of EGDE (Left), (a) fragile silk fabric, (b) silk fabric treated with fibroin protein, (c) silk fabric treated with fibroin protein / EGDE and (d) silk fabric treated with EGDE (Right)

3.4 Amino acid analysis

Amino acid content was analyzed to evaluate the reactivity of the functional amino acid residues of silk fibroin. As previously reported[5-7], this analytical technique is useful for studying the formation of epoxide adducts by reaction with tyrosine and basic amino acid residues (lysine, histidine, arginine). Other amino acid with functional groups reactive towards epoxides (aspartic and glutamic acids) should react as well, but the covalent bond formed with epoxide molecule is easily broken during acid hydrolysis. Table II lists the amino acid composition of silk fabrics. Compared with fragile silk fabric (a), the concentration of tyrosine decreased significantly in silk fabric treated with EGDE/ fibroin protein (c). The Tyr content of silk fabric treated with fibroin protein / EGDE (c) is 9.58%, while that of fragile silk fabric (a) is 11.52%. Among the basic amino acid residues, lysine and histidine decreased markedly.

Table 2 Amino acid composition of silk fabrics

Amino acid (mol%)	a	b	c	d
Gly	37.14	36.99	37.36	37.56
Ala	30.54	30.70	30.85	31.01
Val	3.04	2.54	2.63	2.63
Leu	0.08	0	0.51	0
Ile	0.55	0.54	0.61	0.60
Ser	12.08	12.09	12.63	12.03
Thr	0.79	0.98	1.05	0.96
Asp	1.21	1.31	1.44	1.32
Glu	0.86	1.00	1.10	1.01
Pro	0.28	0.31	0.32	0.31
His	0.20	0.22	0.02	0.02
Lys	0.20	0.17	0.03	0.04
Arg	0.47	0.54	0.60	0.53
Cys	0.00	0	0.01	0.03
Met	0.11	0.07	0.03	0.03
Phe	0.63	0.88	0.94	0.93
Tyr	11.52	11.35	9.58	10.65
total	99.70	99.69	99.71	99.66

3.5 SEM observation

Figure 3 shows the SEM micrographs of silk fabrics. Compared to the fragile silk fabric (a) and silk fabric treated with fibroin protein (b), surfaces of the silk fabric treated with fibroin protein / EGDE (c) are more smooth and straight. This indicates that EGDE may work as cross-linking sites between the fibroin protein and silk fiber, which could bond fibers at larger distance together in case of fibroin protein / EGDE treatment; while fibroin protein may just be adsorbed on the surfaces of silk fibers in case of fibroin protein treatment only.

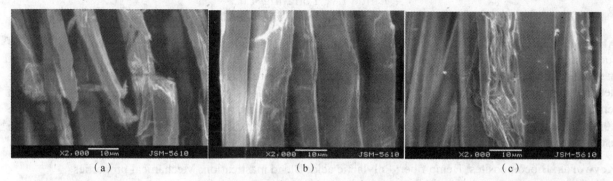

Figure 3 SEM micrographs of (a) fragile silk fabric, (b) silk fabric treated with fibroin protein and (c) silk fabric treated with fibroin protein / EGDE

4 Conclusions

Fibroin protein / EGDE consolidation technique can be applied for consolidating silk relics. The strength and elongation of the silk fabrics treated with EGDE increased by 14.26 N and 6%, respectively. TG results show that the silk fabrics treated with fibroin protein / EGDE have better thermal stability. FTIR results prove that EGDE could react with fibroin protein. Amino acid analysis shows that the EGDE exhibited a slightly higher reactivity toward tyrosine, lysine and histidine.

Silk fibroin consolidation can improve the strength of fragile silk. At the same time, it has no negative impact on other properties, which meets the needs of "the old, such as repair of old" in relic protection. This technique is very helpful in maintaining the damaged silk relics at their original styles. It provides a new way for the protection of cultural relics.

References

[1] H Zh W. Cultural relics protection materials. Xian: Northwestern University Press, 1995: 80
[2] X L Zhang, J J Tang. Strengthening the protection of heritage in the process of composite. Sciences of Conservation and Archaeology, 1999, 11, 23-29
[3] H Shiozaki, Y Tanaka. Reactivity of mono-epoxides toward silk fibroin. Die Makromolekulare Chemie, 1971, 143(1): 25-45
[4] M Tsukada, H Shiozaki. Physical Properties of Silk Fibers Treated with Ethylene Glycol Diglycidyl Ether by the Pad/Batch Method. J Appl Polym Sci, 1993, 50: 1841-1849
[5] H Shiozaki, M Tsukada. Chemical Structure and Dynamic Mechanical Behavior of Silk Fibers Modified with Different Kinds of Epoxides. J Appl Polym Sci, 1994, 52: 1037-1045
[6] M Tsukada, M Nagura. Structural Characteristics of Silk Fibers Treated with Epoxides. J Appl Polym Sci, 1991, 43: 643-649
[7] M Tsukada, H Shiozaki. Effect of epoxide treatment on the fine structural changes of silk fibroin. J Seric Sci Jpn. 1989,58(1): 15-19

39 Study on Forming Technology and Mechanical Properties of Biodegradable Hemp Fiber/ Polylactic Acid Composites

Lihua Lv, Yongling Yu

(Green Applied Fiber Technology Institute, Dalian Polytechnic University, Dalian 116034, China)

Abstract: Fully biodegradable, environment-friendly, hemp fiber/polylactic acid green composites were successfully fabricated using blend mastication and characterized for its tensile, bending and impact mechanical properties. The optimized processing conditions were concluded through a lot of experiments(the detailed experiments were elided in this paper): PLA mass fraction 60%, Mastication time 7min, Mastication temperature 190℃, Heat pressing pressure 15Mpa, Heat pressing temperature 185℃, Heat pressing time 10min. The mechanical properties were good under this condition. The hemp fiber/PLA composites may be used as green composites for certain indoor applications.

Keywords: Biocomposites; Hemp fiber; Polylactic acid; Blend mastication; Mechanical properties

1 Introduction

Due to many environmental problems, the discarding method for glass fiber reinforced plastics (GFRP) and their recycling has seriously been realized in the world. As is widely known, GFRP has excellent thermal and mechanical properties, but these properties make difficult to carry out suitable disposal processing. For example, incineration of discarding GFRP generates a lot of black smoke and bad smells and often gives damage to incinerator by fusion of glass fibers. Reclamation processing generates also a large environmental load, since GFRP is not decomposed easily. For the last decade, therefore, new bio-based composites have been developed[1-2]. Natural fibers play an important role in developing high performing fully biodegradable 'green' composites, which will be a key material to solve the current ecological and environmental problems.

Natural fibers exhibit many advantageous properties as reinforcement for composites. They are a low-density material, yielding relatively lightweight composites with high specific properties. Natural fibers also offer significant cost advantages and benefits associated with processing, as compared to synthetic fibers. Finally, they are a highly renewable resource, which reduces the dependency on foreign and domestic petroleum oil[3]. Recent advances in the use of natural fibers (e.g., flax, cellulose, jute, straw, switch grass, kenaf, coir, bamboo, etc.) in composites have been reviewed by several authors[4-9]. In all previous work, the natural fiber composites were not degradable for the matrix. Based on this background, the main purpose our research is the development of fully green composites with high strength.

Blend mastication means to choose some fiber of certain length as reinforced materials and merge with polylactic acid(PLA) granules used as matrix; and to fabricate the gross material of fibrous composites; and then to form fibrous composite after heat pressing. Small flow, easy processing and easy to control the parameter are the character of this technology[10-11].

The main objective of this work is to successfully manufacture large-scale composite components, from natural resources such as hemp fiber used as reinforced fiber, PLA used as matrix and blend mastication used as the method to form composites at low material and operational cost. Mechanical properties of the composite were also carried out and good. The hemp fiber/PLA composites may be used as green composites for certain indoor applications.

2 Experimental
2.1 Raw Materials

Hemp fiber bundles with a length of 2-3mm were used in this study and the picture of hemp fiber was shown in Figure 1. The density of PLA granules was 1.27g/cm³ and the picture of PLA granules were shown in Figure 2.

Figure 1 Picture of hemp fiber

Figure 2 Picture of PLA granules

2.2 Equipment and Instrument

For main equipment and instrument, please refer to Table 1.

Table 1 Equipment and instrument

Name	Type	Producer
Twin Roller Mastication Machine	SK-160B160×320	Shanghai Machine Factory of Rubber and Plastic
Flat Vulcanizer Machine	QLB-50D/Q	Zhongkai Plastic Machine Factory in Wuxi, Jiangsu
Multifunctional Sampling Machine	RGT-5	Experiment Machine Factory in Chengde, Hebei
Socle Beam Pounding Laboratory Machine	UJ-40	Experiment Machine Factory in Chengde, Hebei
Electronic Multifunctional Experiment Machine controlled by computer	RGT-5	Reger Instrument Company in Shenzhen

2.3 Technology

(1) Technical process: Charging mixture → Blend mastication → Pre-warming → Pressing under constant temperature → Cooling and forming → Removing the mold → Composite plate.

(2) Mastication: PLA granules were put into the gap of the two rollers of the pre-warmed Twin Roller Mastication Machine to cover the roller. After it melded, hemp fibers were also put inside. Under the running of the Twin Roller Mastication Machine they were adequately mixed to form gross materials.

(3) Heat Pressing: Gross material was put on the mold and sent into Flat Vulcanizer Machine to form by heat pressing.

3 Results and discussion

Many factors can affect the mechanical properties(tensile, bending and impact properties) of the composites formed by blend mastication, such as mass fraction of matrix (PLA), mastication temperature, mastication time, heat pressing pressure, heat pressing time and so on.

3.1 Optimization of processing conditions
3.1.1 Effect caused by PLA mass fraction

The proportion of PLA was small, which could not soak the fibers completely, and when it was formed into a plate after heat pressing, some parts among the fibers contained no PLA. So the mechanical properties decrease accordingly. When PLA mass fraction was too high, the proportion of hemp fiber, i.e. reinforced materials reduced, therefore the mechanical properties decreased too. The detailed experiments were elided in this paper. When PLA mass fraction was 60%, the mechanical properties of the composites became comparatively excellent.

3.1.2 Effect caused by mastication time

Mastication time was the lowest in experiments, compared with other factors. Although increasing mastication time was helpful for sticking between PLA and hemp fiber so as to enhance the mechanical properties, long mastication time may destroy the inside structure of PLA to lose the function as matrix under high temperature. Then composites of excellent characters cannot be formed. To save energy and cost, the mastication time 7 minutes in actual producing was adopted.

3.1.3 Effect caused by mastication temperature

PLA's melting temperature is about 175℃, so the rising of mastication temperature is helpful for PLA to melt, which is used for sticking. This makes PLA to fully mixed with hemp fibers to achieve a satisfied gripping effect so that the mechanical properties can be improved. When the mastication temperature came to 190℃, the mechanical properties of composites became comparatively excellent. So, 190℃ was superior temperature.

3.1.4 Effect caused by heat pressing pressure

According to experiments, the rising of heat pressing pressure made the structure of composites tighter, which helped to improve the mechanical properties. When the pressure was kept rising, PLA was pressed out of the gross material, so it could not grip the fiber, which decreased the mechanical properties. 15MPa was appropriate.

The optimized processing conditions were concluded through a lot of experiments (the detailed experiments were elided in this paper): PLA mass fraction 60%, Mastication time 7min, Mastication temperature 190℃, Heat pressing pressure 15Mpa, Heat pressing temperature 185℃, Heat pressing time 10min. The mechanical properties were good under this condition. The original sample was shown in Figure 3.

Figure 3 Photograph of original sample

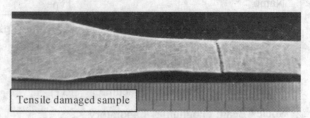

Figure 4 The failure of tensile sample

3.2 Mechanical properties of hemp fiber/polylactic acid composites
3.2.1 Tensile mechanical properties

Tensile mechanical properties were done according to 'Standard test method for tensile properties of plastics' ASTM D 638-2008. The numerical values obtained from three samples under the best processing conditions were shown in Table2. The tensile damaged sample was shown in Figure 4.

Table 2 Tensile mechanical properties

Mechanical property	Area of samples/mm^2	Maximum loading/N	Tension strength/MPa	Modulus of elasticity/GPa	Elongation at break/%
Tensile mechanical property	42.51	2667.14	62.74	7.97	1.23

3.2.2 Bending mechanical properties

Bending mechanical properties were done according to ''Standard test method for bending properties of plastics 'ASTM D790-2003 The numerical values obtained from three samples under the best processing conditions were shown in Table3. The bending damaged sample was shown in Figure 5.

Table 3 Bending mechanical properties

Mechanical property	Area of samples/mm^2	Maximum loading/N	Bending strength/MPa	Modulus of elasticity/GPa	Strength at break/MPa
Bending mechanical property	35.58	166.71	122.92	7.17	28.28

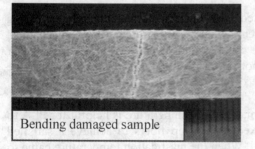

Figure 5 The failure of bending sample

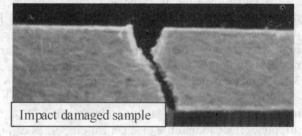

Figure 6 The failure of impact sample

3.2.3 Impact mechanical properties

Impact mechanical properties were done according to 'Plastics-determination of Izod impact strength', ISO 180-2000. The numerical values obtained from three samples under the best processing conditions were shown in Table4. The impact damaged sample was shown in Figure 6.

Table 4 Impact mechanical properties

Mechanical property	Thickness of samples/mm	Width of samples/mm	Gap of samples /mm	Energy absorption / J*m^{-2}
Impact mechanical property	3.84	9.50	2	14.25

To sum up, the bio-composites composed of hemp fiber used as reinforced fiber and PLA used as matrix, was formed by blend mastication. The mechanical properties such as tensile property, bending properties and impact property were good.

4 Conclusion

Biodegradable composites from polylactic acid(PLA) and hemp fiber were fabricated by blend mastication. The optimized processing conditions were concluded: PLA mass fraction 60%, Mastication time 7min, Mastication temperature 190℃, Heat pressing pressure 15Mpa, Heat pressing temperature 185℃, Heat pressing time 10min. Under the best conditions, the tensile property, bending properties and impact property were good.

Acknowledgement

The authors of this paper gratefully acknowledge the supports of Education Department of the Government of Liaoning Province (LT2010011, 2006T027).

References

[1] K Goda, M S Sreekala, A Gomes. Improvement of plant based natural fibers for toughening green composites—Effect of load application during mercerization of ramie fibers. Composites: Part A, 2006, 37: 2213–2220.
[2] S Singh, A K Mohanty. Wood fiber reinforced bacterial bioplastic composites: Fabrication and performance evaluation. Composites science and technology, 2001, 67: 753–1763.
[3] M A Dweib, B Hu, A O Donnell, H W Shenton, R P Wool. All natural composite sandwich beams for structural applications. Composite structures, 2004, (63): 147–157.
[4] A K Bledzki, J Gassan. Composites reinforced with cellulose based fibres. Progress in polymer science,1999, 24: 221–74.
[5] L Y Mwaikambo, M P Ansell. Chemical modification of hemp, sisal, jute and kapok fibers by alkalization. Journal of applied polymer science, 2002, 84(12): 22–34.
[6] S Mishra, S S Tripathy, A K Mohanty, S K Nayak. Noveleco-friendly biocomposites: biofiber reinforced biodegradable polyester amide composites—fabrication and properties evaluation. Journal of reinforced plastics and composites,2002, 21(1): 55–70.
[7] M A Dweib, B Hu,H W Shenton, R P Wool. Bio-based composite roof structure: Manufacturing and processing issues. Composite structures, 2006, 74: 379–388.
[8] O Shinji. Development of high strength biodegradable composites using Manila hemp fiber and starch-based biodegradable resin. Composites: Part A, 2006, (37): 1879–1883.
[9] G Alexandre, M Takanori, O J Jun. Development and effect of alkali treatment on tensile propertiesof curaua fiber green composites. Composites: Part A, 2007, (38): 1811–1820
[10] X M Zhang,X Y Liu. Study and application of mixed material with thermoplasticity of reinforced fiber. Beijing: Chemical industry publishing house, 2007: 1-5.
[11] X Y Liu, H Q Xie. Technology and equipment of mixed material. Wuhan: Publishing house of Wuhan PolytechnicUniversity, 1997: 20-23

40 Rheological Behavior of Spinning Dope of Polyvinyl Alcohol/attapulgite Nanocomposites

Zhiqin Peng[1*], Jinchao Yu[1], Hong Xu[2], Zhiping Mao[2]

(1. College of Materials and Textile, Zhejiang Sci-Tech University, Hangzhou 310018, China
2. College of Materials Science and Engineering, State Key Laboratory for Modification of Chemical Fibers and Polymer Materials, Donghua University, Shanghai 200051, China)

Abstract: The influences of attapugite (AT) nano-rods on the rheological behaviors of spinning dope of polyvinyl alcohol (PVA) have been evaluated by using a RS150L rotational rheometer. The effects of AT nano-rods on the temperature and shear rate dependence of the apparent viscosity (η_a), non-Newtonian index (n), structure viscosity index ($\Delta\eta$), as well as apparent activation energy (ΔE) of the PVA spinning dope are discussed. It has been found that η_a decreases with the addition of AT nano-rods. n and $\Delta\eta$ are more temperature dependent after the addition of AT nano-rods. The temperature dependence of η_a can be well described by Arrhenius Equation. ΔE increases obviously after the addition of AT into PVA spinning dope, showing a higher temperature dependency. Special nano-rod like structure of AT, dispersion and alignment effect of AT nano-rods in PVA, and the interaction between AT and PVA are all discussed as affecting factors.

Keywords: Attapulgite; PVA/AT nanocomposite; Rheological behavior; Spinnability

1 Introduction

Attapulgite(AT), a type of silicate with special nano-rod like microstructure, was used on the study of the crystallization behavior of PVA/AT nanocomposites[1] and the modification of PVA fibers[2] in our previous works. As a part of the whole work, we share our research results on the rheological behaviors of PVA/AT nanocomposite spinning dope in this paper, which actually is a foundation for spinning of PVA/AT nanocomposite fibers. Investigation on rheological behavior of the spinning dope will offer proper evidences to improve or adjust the spinnability. In previous studies, most people focused on the effects of different molecular weights, syndiotactic or atactic structure, variety of solvents and concentrations, gelatinization and hydrogen bonding on the rheological behaviors of PVA [3-9]. Hydrogen bonding has especially significant influence on the rheological behavior of PVA aqueous solutions according to these studies. The influence of attapulgite on the rheological properties of polymer melt has got some attention. According to the study of Liang Shen[10], a grafting-percolated fibrous-silicate network structure is formed in polyamide-6/AT nanocomposites melt. In our previous work [11], the orientation behavior of AT nanoparticles in poly(acrylonitrile)/AT solutions were investigated by rheological analysis. A model was put forward to describe the orientation and percolation behavior at different AT loading. To our knowledge, there are no reports on the rheological behavior of spinning dope of PVA/AT nanocomposites. In this paper, the effect of AT addition on the rheological behavior of PVA/AT spinning dope is studied, which is compared with that of pure PVA spinning dope under the influence of temperature and shear rate.

2 Experimental

Poly (vinyl alcohol) (PVA), supplied by Shanghai Petrochemical Co. Ltd, was 99.8% hydrolyzed with an average number molecular weight of 79,000. The crude AT was 200-mesh and provided by Jiangsu Junda Attapulgite Material Co.Ltd with purity greater than 90%. PVA grains were dissolved in water by stirring at 98℃. AT nano-rods suspensions, which were prepared according to the method described in our previous study [14], were then added into the fully dissolved PVA solution. The blends were mixed uniformly by strong stirring and put into an oven for deaeration at 80℃ for about 24hs. The solution concentrations were controlled within 14-16%. Then the blend solutions were ready for rheological experiment. Rheological experiments were performed on a Rotational Rheometer (Model RS150L) made by Germany HAAKE Co. The cone angle was 0.1 in radius and the diameter was 25 mm for both the cone and plate. The cone /plate gap setting was 0.053 mm. Steady shear tests were performed with the shear rate from 0.01 to 1200s^{-1} and testing

---

* Corresponding author's email: pengqiao6858@126.com

temperatures at 70, 80 and 90℃ respectively. The temperature control was accurate within ±1℃. The samples were PVA spinning dope with concentration at 14~16%. Samples were kept in an oven for about 20 min at same temperature with the testing temperature before each test. The contents of AT in the dopes were 0 and 0.5wt% of pure PVA.

3 Results and discussion

Generally, the rheological behavior of PVA is greatly affected by molecular parameters of PVA, such as the molecular weight, DS and stereoregularity. Since the same PVA grains were used in this study, factors mentioned above could be ignored. The dispersion of AT nano-rods in PVA is very good according to the morphology analyses in our previous works [13,14], so we did not discuss the dispersion effects of AT nano-rods in followings. This work focused mainly on the effect of AT nano-rods on the spinnability of PVA spinning dope and related indexes.

3.1. Temperature and shear rate dependence of the apparent viscosity

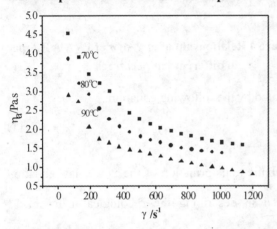 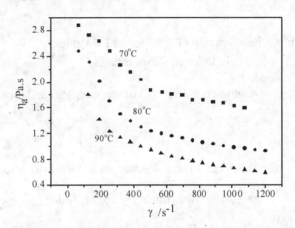

Figure 1 Curves of shear rate (γ) and apparent viscosity(η_a) of the PVA/AT dope

Figure 2 Curves of shear rate (γ) and apparent viscosity(η_a) of the pure PVA dope

The flow curves of pure PVA, PVA/AT dope with different temperatures are shown in Figure 1and Figure 2. It is clear that the apparent viscosities of both kinds are decreased with the increase of shear rate, which shows typical shear-thinning behavior of most polymers with spinnability. The reason for this, on the one hand, is due to the decrease of entanglement concentration of PVA chains with the increase of shear rate. On the other hand, the PVA chains tend to oriented with the increase of shear rate. Thus momentum transporting ability of the fluid decreases, that is, the flow resistance is lower. It also can be seen from these figures that the curves tend to lower with the increase of temperature. This indicates that the apparent viscosity of the fluids decreases with the increase of temperature. As we known, the molecular energy of polymer will increase with the increase of temperature. In addition, the bulk of the fluids expand and thus more space between polymer molecules is free with the increase of temperature. All of these respond to the above results. Compared Figure 2 with Figure 1, one will find that the apparent viscosity of PVA/AT dope is lower than that of pure PVA dope in the same shear rate and temperature. It means the addition of AT into PVA matrix is beneficial to the decrease of apparent viscosity. The alignment of AT nano-rods induced by shear stress may contribute to higher orientation of PVA chains, which results in the decrease of apparent viscosity.

3.2 Non-Newtonian indexes

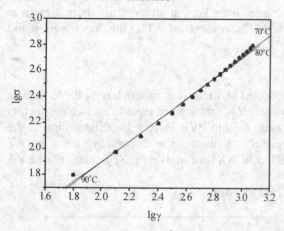

Figure 3 Relationship of lg γ -lgσ of PVA dope in different temperatures

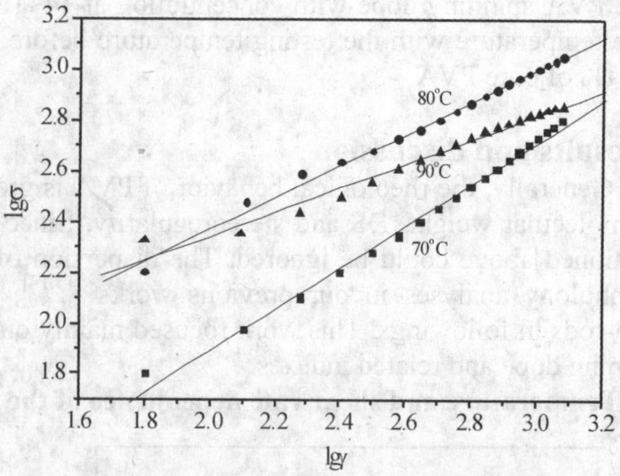

Figure 4 Relationship of lg γ -lgσ of PVA/AT dope in different temperatures

Non-Newtonian indexes (*n*) of the samples can be calculated by the following equation:

$$n = \frac{\lg \sigma}{\lg \gamma} \quad (1)$$

where σ is the shear stress; γ is the shear rate. It is a logarithmic transformation of the Power-law equation ($\sigma = K \gamma^n$; K is the thickness of a fluid) widely used in describing the rheological behavior of non-Newtonian fluid. Results calculated according to the slopes of lgγ -lgσ from Figure 3 and Figure 4 are listed in Table1. It is clear that the non-Newtonian index of PVA dope shows little temperature dependency while that of PVA/AT dope has obviously temperature dependency. With the increase of temperature, the non-Newtonian index is decreased in the case of PVA/AT dope, which is an opposite tendency with normal polymer fluid. Since the movement of polymer chains quickens up with the increase of temperature, which weakens the interaction between polymer chains and lightens the temperature dependency of apparent viscosity on shear rate, the non-Newtonian index is increased in normal cases. So some kind of interaction (eg. hydrogen bonding) may exist between PVA and AT nano-rods which interferes with the movement of polymer chains and thus decreases the non-Newtonian indexes.

Table 1 Non-Newtonian index (*n*) of PVA and PVA/AT dopes

Sample \ Temperature	70℃	80℃	90℃
Pure PVA	0.819	0.801	0.816
PVA/AT	0.838	0.629	0.531

3.3 Structure viscosity indexes

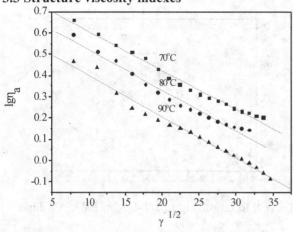

Figure 5 Relationship of $\gamma^{1/2}$-lgη_a of PVA dope in different temperatures

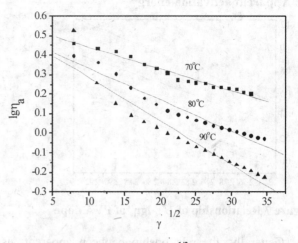

Figure 6 Relationship of $\gamma^{1/2}$-lgη_a of PVA/AT dope in different temperatures

Structure viscosity index ($\Delta\eta$) can be used to characterize the structure of spinning fluids. It is an important parameter to measure the spinnability of a spinning fluid. The calculation of $\Delta\eta$ is according to the following equation:

$$\Delta\eta = (-\frac{d\lg\eta_a}{d\gamma^{1/2}})\times 10^2 \tag{2}$$

According to the slopes of $\gamma^{1/2}$-lgη_a in Figure 5 and Figure 6, values of $\Delta\eta$ of PVA and PVA/AT dope can be determined. The results are listed in Table 2. It is well known that value of $\Delta\eta$ is bigger than zero in the case of shear-thinning fluid under the non-Newtonian area. The lower is the value, the better is the spinnability. In many cases, $\Delta\eta$ decreases with the increase of temperature, which means the spinnability is better in a higher temperature. However, things are quite different here. One can see from Table 3 that the values of $\Delta\eta$ increase slightly in pure PVA dope, while they increase obviously in the case of PVA/AT dope. It indicates that the special structure of PVA chains, which we believe is the existence of hydroxyls, is sensitive to temperature and the sensitivity is strengthened after the addition of AT nano-rods. Therefore, one can not choose an optimal spinning temperature and evaluate the spinnability of PVA dope on the base of $\Delta\eta$ value only, especially in the case of PVA/AT dope. As we can see from the flow curves of γ-η_a in Figure 5 and Figure 6, the curves are not so smooth at 70℃, which means the fluids are not rheologically stable in this temperature. However, the $\Delta\eta$ values are lowest in this temperature as shown in Table 2. If one follows the principle of "the lower, the better", then the optimal temperature for spinning will be 70℃. This is obviously not the case. Taking all of the situations into consideration, we believe that temperature around 80℃ is optimal for PVA fiber spinning in our study. And we may say that the spinnability of PVA/AT dope is better than that of PVA dope in one way or another according to the $\Delta\eta$ values and flow curves in this temperature. $\Delta\eta$ values increase markedly in the case of PVA/AT dope, which changes from 1.12 to 2.24. It indicates that some kinds of crosslink or entanglement between PVA chains are strengthened by the addition of AT nano-rods under the influence of temperature. In conclusion, it is not a good measure to improve the spinnability of PVA spinning dope by increasing temperature only, especially in the case with AT loading.

Table 2 Structure viscosity index ($\Delta\eta$) of PVA and PVA/AT dopes

Sample \ Temperature	70℃	80℃	90℃
Pure PVA	1.83	1.91	1.96
PVA/AT	1.12	1.62	2.24

3.4 Apparent activation energy

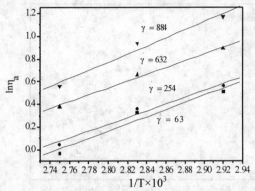

Figure 7 Relationship of 1/T-lgη_a of PVA dope

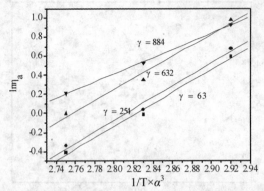

Figure 8 Relationship of 1/T-lgη_a of PVA/AT dope

Generally, the relationship between apparent viscosity and activation energy of a polymer fluid is described by Arrhenius Equation:

$$\eta = A e^{E_\eta / RT} \tag{3}$$

where η is the viscosity of a polymer fluid; A is a constant related to quality of the material; R is gas constant; T is Kelvin temperature; E_η is activation energy of a fluid.

A logarithmic transformation of the above equation is as the following

$$\ln \eta = \ln A + E_\eta / RT \tag{4}$$

Thus, apparent activation energy (ΔE) of a fluid also can be calculated according to the slopes of $1/T$ and $\ln\eta$. Figure 7 and Figure 8 show the relationship between $1/T$ and $\ln\eta_a$ of PVA and PVA/AT dope. Basically, a linear relation is seen between those two parameters. It means that the temperature dependency of these two fluids is in line with Arrhenius Equation. The values of ΔE calculated in different shear rates are presented in Table 3. One can find that ΔE of PVA dope ranges in 25-30kJ/mol, while that of PVA/AT dope ranges in 35-50kJ/mol. It is well known that the activation energy of a fluid is the energy needed to prevent the molecules from movement. The value reflects the sensitivity of viscosity to temperature. So the temperature dependency of PVA/AT dope is higher than that of PVA dope, which is consistent with the results of structure viscosity indexes.

Table 3 Apparent activation energy (ΔE) of PVA and PVA/AT dopes

Sample \ γ / s^{-1}	884	632	254	63
Pure PVA	29.68	24.86	24.78	26.19
PVA/AT	35.25	48.22	50.05	49.55

4 Conclusion

Spinning dope of PVA/AT nanocomposites were prepared by blending. The apparent viscosity decreases with the addition of AT due to the alignment effect of AT nano-rods on the orientation of PVA chains. Spinning dopes of PVA/AT nanocomposite also exhibit shear thinning effect. The Non-Newtonian indexes of PVA/AT nanocomposite spinning dope are decreased dramatically with the increase of temperature. The optimal temperature for spinning is about 80℃ according to the results of structure viscosity indexes and flow curves. PVA/AT spinning dope shows higher temperature sensitivity due to possible strengthening effect of AT nano-rods on crosslink or entanglement of PVA chains under the influence of temperature. The different behaviors between pure PVA spinning dope and PVA/AT spinning dope might due to the interaction within PVA chains or between PVA chains and AT nano-rods. Special nano-rod like structure of AT, well dispersion and alignment effect of AT nano-rods in PVA dope are all related to the results.

References

[1] Peng. Z.Q, Chen. D.J. Study on the nonisothermal crystallization behavior of poly(vinyl alcohol)/attapulgite nanocomposites by DSC analysis. J. Polym. Sci. Part B: Polym. Phys, 2006, 44(3): 534-540.

[2] Zeng. Z.Q, Chen. D.J. J. Polym. Alignment effect of attapulgite on the mechanical properties of poly(vinyl alcohol)/attapulgite nanocomposite fibers. Sci. Part B: Polym. Phys, 2006, 44(14): 1995-2000.

[3] Kim. S. S, Seo. I.S, Yeum. J. H et al. Rheological properties of water solutions of syndiotactic poly(vinyl alcohol) of different molecular weights. J. Appl. Polym. Sci, 2004, 92(3): 1426-1431.

[4] Choi. J. H, Lyoo. W. S, Ko. S. W. Rheological properties of syndiotacticity-rich ultrahigh molecular weight poly(vinyl alcohol) dilute solution. J. Appl. Polym. Sci, 2001, 82(3): 569-576.

[5] Ricciardi. R, Gaillet. C, Ducouret. G et al. Investigation of the relationships between the chain organization and rheological properties of atactic poly(vinyl alcohol) hydrogels. Polymer, 2003, 44(11): 3375-3380.

[6] Lee. E. J, Kim. N. H, Dan. K. S, Kim. B. C. Rheological properties of solutions of general-purpose poly(vinyl alcohol) in dimethyl sulfoxide. J. Polym. Sci. Part B: Polym. Phys, 2004, 42(8): 1451-1456.

[7] Ivanov. A. E, Larsson. H, Galaev. I. Yu, Mattiasson. B. Synthesis of boronate-containing copolymers of N,N-dimethylacrylamide, their interaction with poly(vinyl alcohol) and rheological behaviour of the gels. Polymer, 2004, 45(8): 2495-2505.

[8] Song. S.I, Kim. B. C. Characteristic rheological features of PVA solutions in water-containing solvents with different hydration states. Polymer, 2004, 45(7): 2381-2386.

[9] Briscoe. B, Luckham. P, Zhu. S. The effects of hydrogen bonding upon the viscosity of aqueous poly(vinyl alcohol) solutions. Polymer, 2000, 41(10): 3851-3860.

[10] Shen. L, Lin. Y, Du. Q.G, Zhong.W, Yang.Y.L. Preparation and rheology of polyamide-6/attapulgite nanocomposites and studies on their percolated structure. Polymer, 2005, 46(15): 5758-5766.

[11] Yin. H, Mo. D, Chen. D. J. Orientation behavior of attapulgite nanoparticles in poly(acrylonitrile)/attapulgite solutions by rheological analysis. J. Polym. Sci. Part B: Polym. Phys, 2009, 47(10): 945-954.

41 Finite Element Calculation on the Quasi-static Impact Performance of 3D Integrated Hollow Woven Structure Composites

Wei Tian, Chengyan Zhu*

(College of Materials and Textiles, Key Laboratory of Advanced Textile Materials and Manufacturing Technology, Ministry of Education of China, Zhejiang Sci-Tech University, Hangzhou 310018, China)

Abstract: 3D integrated hollow woven structures are formed together by a number of typical 3D woven structures layers which are connected by the binding yarns in thickness direction. To predict the quasi-static impact process of the 3D integrated hollow structure composites, a FE preprocessing program was builded in this paper. Through comparing the simulated results with the test results, it can be seen that the model is reasonable. And observing from the global effect of the destory, the damaged condition is different for the direction of warp and weft. The damage level of warp direction is more obvious than the one of the weft direction.

Keywords: Composites, Integrated hollow structures, Quasi-static Impact, Finite element calculation

3D integrated hollow woven structures are formed together by a number of typical 3D woven structures layers which are connected by the binding yarns in thickness direction[1]. There are some hollow structures in the inner of the structure and this kind of 3D structure also owns great integrated performance.

M.Styles etc.[2] modeled the flexural behaviour of a composite sandwich structure with an aluminium foam core using the finite element (FE) code LS-DYNA. The deformation and failure behaviour predicted by the FE model compared well with the behaviour observed experimentally.

Parabeam@, which are produced from velvet-weave sandwich-fabric preforms, provide a new type of sandwich structure with a high skin-core debonding resistance and the potential for cost-elective sandwich construction. H.Judawisastra etc.[3,4] developed a finite-element preprocessing program to predict the mechanical performance of the cores of woven sandwich fabric panels. In the model linear static analyses for the basic core properties (flexral shear and flat compression) have been performed. The results for the shear modulus as a function of the pile shape were very good. Model predictions at higher resin contents and for the compression modulus were less accurate. Xincai Tan and Xiaogang Chen etc.[5-7] studied systematically the structural distortion and energy absorption of the hollow structure composites with different structure parameters. They also designed and manufactured this kind of composites successfully.

On the basis of the geometric model[8], this paper will develop a finite-element preprocessing program to predict the quasi-static impact process of the 3D integrated hollow structure composites. Through comparing the process predicted by the FE model with the process observed experimentally, the energy absorption performance of the 3D integrated hollow structure composite can be analyzed and studied, and which will provide a theory direction for the design of the composites.

1 Test of quasi-static impact

The test of quasi-static impact was carried out in the Key Laboratory of Advanced Textile Materials and Manufacturing Technology, Ministry of Education, Zhejiang Sci-Tech University.

1.1 Descriptionfor the impact system

The dimension of the indenting hammer and the photo of the composite panel can be seen in Figure 1.

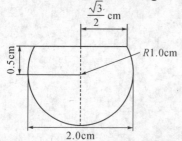

Figure 1 The profile of indenting hammer

Figure 2 Photograph of the composite panel

---

* Sponsoring fund: Program for Changjiang Scholars and Innovative Research Team in University (No.IRT0654), Zhejiang Sci-Tech University Research Project(0801802-Y); Corresponding author: Chengyan Zhu, Professor, ZSTU, tel.: +86 571 86843253, fax: +86 571 86843250, email: cyzhu@zstu.edu.cn

1.2 Test equipment

The tests were conducted on RGM-2000A electronic universal testing machine at a rate of 4 mm/min. After the test, the load-displacement curves will be output from the computer directly. From these curves, the energy-displacement curves and the value of the energy absorption for different situation can be obtained.

2 Finite element model

This paper model the quasi-static impact performance of a 3D integrated hollow structure using the finite element (FE) code LS-DYNA.

2.1 Combined model for indenting hammer and composite panel

According to the dimension of the sample, the structure model can be established, seen as Figure3.

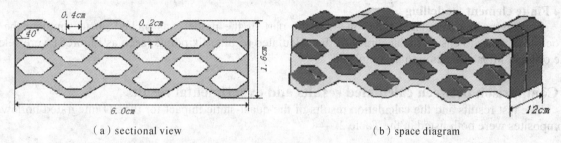

(a) sectional view   (b) space diagram

Figure 3 Model of the 3D integrated hollow woven structure composites

2.2 Characterization of mechanical property

The Material Type 22 (i.e. Composite Damage Model) in Ls-Dyna is chosen to define the material characterization and the model is based on Chang-Chang criterion. Parameters of the material characterization that should be defined are listed in Table 1. Indenting hammer is regarded as rigid body, the moduli is 210Gpa and the density is 7.85g/cm^3.

Table 1 Mechanical properties of one of composites in FE model [9]

Densityρ(g/cm3)	RO	1.58
Moduli(Gpa)	EA	7.4E+11
	EB	7.4E+11
	EC	3.53E+11
Poisson's ratio	PRBA	0.0209
	PRCA	0.0209
	PRCB	0.33
Shear moduli (GPa)	GAB	2.3E+10
	GBC	7.5E+9
	GCA	7.5E+9
Bulk modul (GPa)	KFAIL	0.000
Shear strength (GPa)	SC	3.5E+8
Tensile strength(Mpa)	XT	8.9E+9
	YT	8.9E+9
Compressive strength(Mpa)	YC	3.5E+9
Parameters of shear stress	ALPH	0.000

Comment: A,B and C stand for the warp direction, the weft direction and the vertical direction, respectively.

2.3 Mesh scheme

The mesh scheme for one of the indenting hammer and the 3D integrated hollow composite is shown in Figure4. The element type SOLID 164 (8-node hexahedron element) on Ls-Dyna is used in meshing. There are 2744 elements and 3221 nodes in the indenting hammer model. There are 28772 elements and 38408 nodes in the composite model.

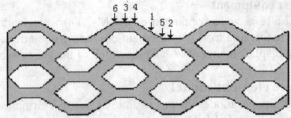

Figure 4 Mesh scheme of finite element model of panel and indenting hammer

Figure 5 Initial location of the impact process

2.4 Finite element modelling

For saving the calculated time, one-half the part of the model was picked up to simulate the impact process. In order to ensure the accuracy of the calculated results, six points (seen as Figure5) were selected as the original impact location to be simulated.

3 Comparison between calculated results and experimental results

The test results and the calculation results of the quasi-static impact for the 3D integrated hollow woven composites were both listed in the Table 2.

Table2 Comparison for the calculated value and measured value

Impact Location		1	2	3	4	5	6
Total absoption energy(J)	Measured Value	184.66	193.85	195.39	199.04	227.76	218.45
	Calculated Value	196.1371	218.3766	203.1255	213.0946	241.1509	242.9178
Difference		6.2%	12.7%	3.9%	7.1%	5.9%	11.2%

As shown in Table 2, there are good agreements in comparison of total energy absorption obtained from experiment and calculation. It illustrates that the calculated method is reasonable to predict the energy absorped-performance of the composites. According to this method, the perforation process, the damage form and the damage mechanism all can be simulated effectively.

4 Damage process of the composites

Figure6 shows the damage process of the composites when it was perforated by the indenting hammer and displays the damage situation at different stage of the impact process.

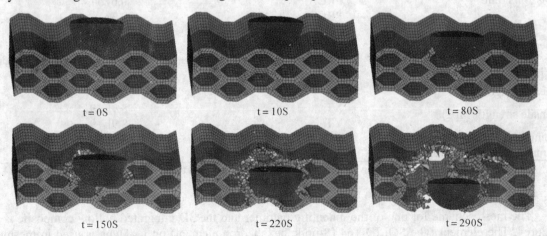

Figure 6 Damage evolution of the composite panel under quasi-static impact (1# Location)

During the impact process, the indenting hammer contacts with the composites at the first time. Then the material elements around the contact point come to fail and the first layer will be destroied. With the impact going on, the indenting hammer will be in touch with the inner part of the panel. The break will appear on the binding part. It can be noticed that during above damage process, the influence between layers is very little, e.g. when the destory is happened in the upper layer, there are few breakages in the next layer. In the end, as the impact further continued, the last layer will be impacted and come to fail. On the other hand, observing from the global effect of the destory, the damaged condition is different for the direction of warp and weft. The damage level of warp direction is more obvious than the one of the weft direction. The results are in accordance with the measured results.

5 Conclusion

This paper established a finite-element preprocessing program to predict the quasi-static impact process of the 3D integrated hollow structure composites on the basis of a geometric model. There are good agreements in comparison of total energy absorption obtained from experiment and calculation. It illustrates that the calculated method is reasonable to predict the energy absorped-performance of the composites. The perforation process and damage situation of composites under quasi-static impact also simulated by the model. The predict results are also in accordance with the measured results.

Acknowledgement

The authors express their sincere gratitude to the support from the Key Laboratory of Advanced Textile Materials and Manufacturing Technology, Zhejiang Sci-Tech University, Hangzhou 310033, China.

References

[1] Ding Xin, Yi Honglei. A geometric model of three dimensional woven structures [J]. Acta Materiae Compositae Sinica, 2003, 20 (5):108-113.
[2] M.Styles, P.Compston, S.Kalyanasundaram. Finite Element Modeling of Core Thickness Effects in Aluminium Foam/Composites Sandwich Structures Under Flexural Loading[J]. Composite Structures 2008, 86: 227-232
[3] H.Judawisastra, J.Ivens, I.Verpoest, K.Swinkels. 3D Woven Composites as Core Materials for Railway Vehicle Applications[J]. Proc. 2nd International Conference on Composite in the Rail Industry, 28 October 1999, Birmingham, UK, 1-15
[4] A.W.van.Vuure, J.Pflug, J.A.Ivens, I.Verpoest. Modelling the Core Properties of Composite Panels Based on Woven Sandwich-fabric Preforms[J]. Composites Science and Technology 2000, 60: 1263-1276
[5] Xincai Tan, Xiaogang Chen. Parameters Affecting Energy Absorption and Deformation in Textile Composite Cellular Structures[J]. Materials and Design 2005, 26: 424-438
[6] Xincai Tan, Xiaogang Chen, Paul P. Conway, Xiu-Tian Yan. Effects of Plies Assembling on Textile Composite Cellular Structures[J]. Materials and Design 2007, 28: 857-870
[7] Xiaogang Chen, Ying Sun, Xiaozhou Gong. Design, Manufacture and Experimental Amalysis of 3D Honeycomb Textile[J]. Composites Part I: Design and Manufacture. Textile Research Journal Vol78 2008, (9): 771-781
[8] Tian Wei, Zhu Chengyan, Wang Shanyuan. A Geometric Model of Three Dimensional Integrated Cellular Woven Structures[J]. Textile Research Journal, 2009, 79(9): 844-852 (2009)
[9] Chen Wei. The study of bulletproof mechanism and ballistic property of fiber-reinforced composite target. Nanjing: Institutes Of Technology Of Nanjing, Scientific Dissertation of Master, 2006

42 Effect of Different Fabric Skin Combinations on Predicted Sweating Skin Temperature of a Thermal Manikin

Faming Wang*, Kalev Kuklane, Chuansi Gao, Ingvar Holmér
(Thermal Environment Laboratory, Division of Ergonomics and Aerosol Technology, Department of Design Sciences, Faculty of Engineering, Lund University, Lund 221 00, Sweden)

Abstract: In this study, a knit cotton fabric skin and a Gore-tex skin were used to simulate two sweating methods. The Gore-tex skin was put on top of the pre-wetted knit cotton skin on a dry heated thermal manikin 'Tore' to simulate senseless sweating, similar to thermal manikins 'Coppelius' and 'Walter'. Another simulation involved the pre-wetted fabric skin covered on top of the Gore-tex skin in order to simulate sensible sweating. This type of sweating simulation can be widely found on many thermal manikins worldwide, e.g. 'Newton'. Two empirical equations to predict the wet skin surface temperature were developed based on the mean manikin surface temperature, mean fabric skin surface temperature and the total heat loss. The prediction equations for the senseless sweating and sensible sweating on the thermal manikin 'Tore' were $T_{sk}=34.05-0.0193HL$ and $T_{sk}=34.63-0.0178HL$, respectively. It was found that the Gore-tex skin limits moisture evaporation and the predicted fabric skin temperature was greater than that in the G+C skin combination. Further study should validate those two empirical equations, however.

Keywords: Thermal manikin; Sweating skin; Skin temperature; Empirical equation

1 Introduction

Clothing evaporative resistance determines how much sweat could be evaporated through clothing ensembles to the environment. As one of the most important physical parameters for a garment, it is widely used as the basic input for human thermal comfort models. Currently, there is only one international standard regarding to how to measure clothing evaporative resistance using a thermal manikin [1]. Based on this standard, clothing evaporative resistance can be calculated by two options: heat loss method and mass loss method. Since current manikin technology could detect the exact amount of heat loss accurately, also, the mass loss rate from the sweating skin can be measured by a high accuracy weighing scale. Therefore, the calculation accuracy on water vapor gradient between wet skin surface and the ambient directly determines the calculation accuracy of clothing evaporative resistance.

The prevailing method to calculate water vapour resistance of the sweating skin assumes the relative humidity on the wet skin surface is 100%. According to Antonio's equation, the measurement accuracy on sweating skin surface temperature directly determines the accuracy of clothing water vapour resistance. However, there is no feedback between the fabric skin surface and the manikin regulation system, i.e., the wet skin surface temperature is not controlled by the regulating system. Therefore, the calculation on clothing evaporative resistance in various laboratories [2-4] is based on the manikin surface temperature rather than the sweating fabric skin surface temperature. Obviously, this is not correct. Wang et al. [5] investigated the calculations of evaporative resistance based on these two surface temperatures and found that an error of up to 35.9% could be introduced by the temperature difference. They [6] developed an empirical equation to predict the sweating skin surface temperature for thermal manikins at 34°C and this equation was successfully validated by adding four functional clothing ensembles on top of the sweating skin. Recently, this empirical equation was further developed and a universal equation was presented and also validated [7]. This universal equation can be used to predict the wet skin surface temperature on most of thermal manikins at a testing temperature range of 25 to 34°C. Obviously, all those measurements were performed with the same fabric skin. As a result, there is still a gap in the effect of different fabric skins on the predicted wet skin temperature.

The main aim of this paper is to investigate the effect of difference fabric skins on the predicted sweating skin surface temperature. Two empirical equations were developed and compared. Finally, the possibility of integrating those two empirical equations to one equation was discussed.

---

* Corresponding author's email: faming.wang@design.lth.se

2 Methods
2.1 Thermal manikin
A dry heated 17-segment thermal manikin Tore was used in this study. This manikin is made up of plastic foam with a metal frame inside to support the body. The total surface area is 1.774 m² and weighs 30 kg. The whole manikin was placed in a controllable climatic chamber.

2.2 Fabric skins
A knit cotton fabric skin and a Gore-tex fabric skin were used. Those two skins are tight-fitting to the manikin. The areal weight of the Gore-tex fabric skin and knit cotton fabric skin are 241 and 228 g/m², respectively. The cotton fabric skin was rinsed in a washing machine (Electrolux W3015H) and centrifuged for 4 s to ensure no water was dripped during the test period. This cotton skin was covered on top of the Gore-tex skin over the thermal manikin to simulate sensible sweating, while it was worn under the Gore-tex fabric skin to simulate senseless sweating, an example thermal manikin for such sweating methods is SAM [8]. The fabric skin covers 12 segments of the thermal manikin, except the head, hands and feet.

2.3 Test conditions
Six temperature sensors (SHT75, Sensirion Inc., Switzerland, accuracy: ±0.3°C) were attached on six sites of the skin outer surface (chest, upper arm, stomach, back, thigh, and calf) by using white thread rings(Resårband Gummilitze Elastic Braid, Sweden) to record the skin surface temperature. For each skin combination, twelve tests were performed at three different ambient temperatures: 34, 25 and 20°C. These three temperature levels could avoid moisture accumulation on the outer skin surface due to dew points are always lesser than those ambient temperatures. As a result, the effect of accumulation on the observed fabric skin temperature can be neglected. The air velocity was maintained at 0.33±0.09 m/s.

3 Results and discussion
The total heat loss and averaged skin surface temperature of the sweating area were calculated for each skin combination. The total heat loss from the thermal manikin ranges from 32.9 to 242.4 W/m². Also, the averaged fabric skin temperature ranges from 29.9 to 33.6°C. Two scatter charts were plotted for those two skin combinations. The linear regression lines were also drawn using Origin v.8.0 (OriginLab Corporation, USA). The results are displayed in Figure 1. The empirical equations for G+C and C+G skin combinations are expressed as follows

$$T_{sk}=34.05-0.0193HL$$
$$T_{sk}=34.63-0.0178HL$$

where, T_{sk} is the mean fabric skin surface temperature, °C; HL is the total heat loss from the thermal manikin.

It can be deduced from Figure 1 that the G+C fabric skin combination has greater influence on the skin surface temperature than the C+G fabric skin combination. This is because the outer Gore-tex fabric skin layer in the combination C+G constraints the moisture transfer to the environment due to its limited pore size of the laminated membrane. Therefore, less evaporation makes the mean skin surface temperature greater than the G+C combination at the same test condition. The findings are similar to the sweating situation of a human. For our human body, sweating glands only release sensible sweat due to exercise, or environmental factors. However, insensible sweat continuously evaporates from the human body under all conditions [9].

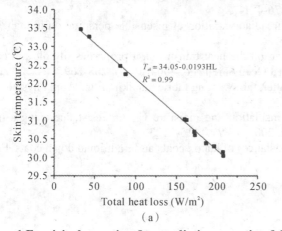
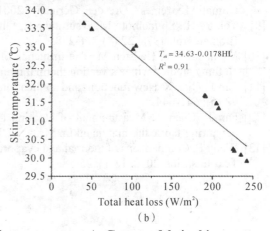

Figure 1 Empirical equation for predicting sweating fabric skin temperature. A: Gore-tex fabric skin + cotton fabric skin combination (G+C); B: Cotton fabric skin+ Gore-tex fabric skin combination (C+G).

Furthermore, we can easily find that the outer fabric surface temperatures in C+G skin combination are always greater than that those in G+C skin combination. The observed amount of sweating ranges from 150 to 348 g/h. On the other hand, for thermal manikins such as Walter and Coppelius, the amount of senseless sweating per hour reported in previous studies [10-13] is much greater than the value on a human body. Therefore, the simulation of senseless sweating on a thermal manikin using a piece of waterproof but permeable fabric is still questionable.

Finally, by comparing those two empirical equations for different skin combinations, we suggest that using different empirical equations for different fabric skin combinations. Otherwise, the predicted fabric skin temperature may not be in an acceptable range ($\pm 0.5^{\circ}$C).

4 Conclusions

In this study, the effect of different skin combinations on the predicted sweating fabric skin temperature was examined. It was found that the Gore-tex skin as an outer layer could limit sweat evaporation to the environment and the predicted fabric skin temperature on the thermal manikin is always greater than that in the G+C skin combination. Also, using current available waterproof but permeable fabric skin to simulate senseless sweating is questionable. Moreover, those equations are not validated on the thermal manikin. The prediction accuracy is still not clear. Therefore, further studies on validation of those equations on the manikin are needed, however.

Acknowledgement

This study was supported by Taiga AB in Varberg, Sweden.

References

[1] ASTM F 2370. Standard Test Method for Measuring the Evaporative Resistance of Clothing Using a Sweating Manikin. American Society for Testing and Materials, USA, 2010.
[2] Wang F., Gao C., Kuklane K., Holmér I. A study on evaporative resistances of two skins designed for thermal manikin Tore under different environmental conditions. J Fiber Bioeng Inform, 2009, 1 (4):301-305.
[3] McCullough E., Kenney W. Thermal insulation and evaporative resistance of football uniforms. Med Sci Sport Exer, 2003, 35 (5): 832-837.
[4] McCullough E., Zuo J. Thermal insulation and evaporative resistance values for sports apparel. Med Sci Sport Exer, 2003, 35 (5):s48.
[5] Wang F., Kuklane K., Gao C., Holmér I. Effect of temperature difference between manikin and sweating skin surfaces on clothing evaporative resistance: how much error is there?. Indus Health, submitted.
[6] Wang F., Kuklane K., Gao C., Holmér I., Havenith G. Development and validation of an empirical equation to predict sweating skin surface temperature for thermal manikins. Textile Bioengineering and Informatics Symposium Proceedings, Li Y., Qiu Y., Luo X., Li J. (eds), Shanghai, China, 2010: 1225-1230.
[7] Wang F., Kuklane K., Gao C., Holmér I. Development and validity of a universal empirical equation to predict skin surface temperature on thermal manikins. J Therm Biol, 2010, 35(4):197-203.
[8] Richards M., Mattle N. A sweating agile thermal manikin (SAM) developed to test complete clothing system under normal and extreme conditions.RTO HFM Symposium on "Blowing Hot and Cold: protecting Against Climatic Extremes", Dresden, Germany, 2001, MP 076-4: 1-7.
[9] Tokura H., Shimomoto M., Tsurutani T., Ohta T. Circadian variation of insensible perspiration in man. Int J Biometeorol, 1978, 22 (4):271-278.
[10] Meinander H., Hellsten M. The influence of sweating on the heat transmission properties of cold protective clothing studied with a sweating thermal manikin. Int J Occu Safe Ergo, 2004, 10(3): 263-269.
[11] Fan J., Qian X. New functions and applications of Walter, the sweating fabric manikin. Eur J Appl Physiol, 2004, 92 (6): 641-644.
[12] Fan J., Chen Y. Measurement of clothing thermal insulation and moisture vapour resistance using a novel perspiring fabric thermal manikin. Meas Sci Technol, 2002, 13(7): 1115-1123.
[13] Wang F. Comparisons of thermal and evaporative resistances of kapok coats and traditional down coats. Fibres Text East Eur, 2010, 18(1): 75-78.

43 Preparation of A Novel Anti-bacterial Wool Fabrics

Junhua Wang[1]*, Zaisheng Cai[2]

(1. Department of Textile and Clothes, Wuyi University, Guangdong 529020, China
2. College of Chemistry, Chemical Engineering and Biotechnology, Donghua University, Shanghai 201620, China)

Abstract: β-cyclodextrin (β-CD) was modified by sulfonated reaction. The sulfonated-β-CD was grafted onto wool fabrics. A novel kind of anti-bacterial fabrics was prepared by filling the anti-bacterial agent (miconazole nitrate) into β-CDs' caves fixed onto the fabrics. The optimum grafting conditions for sulfonated-β-CD onto wool fabrics were: sulfonated-β-CD 60 g/L, Na_2SO_4 5 g/L, pH=2~3, dipping the fabrics into solution at 90~100 ^{0}C for 50~60 min. Compared with the unmodified fabrics, the fabrics modified with sulfonated-β-CD increased the uptake of antibacterial agent, enhancing the antibacterial properties of the modified fabrics. The binding of β-CD onto the wool fabrics improved the resistance of the entrapped antibacterial agents to washing cycles, prolonging the antibacterial effect. The antibacterial abilities of the unmodified fabrics was almost lost when washing 5 times, while the modified fabrics with sulfonated-β-CD kept 60%~70% even after washing 10 times.

Keywords: Miconazole nitrate; β-cyclodextrin; Inclusion complex; Antibacterial wool fabrics

1 Introduction

In recent years, much interest has been concentrated on natural molecular encapsulant, β-CD, which is cyclic glucose oligomers having seven glucose units, linked by 1,4-α-glucosidic bonds, forming a toroidal shape with a hydrophilic outer surface and an internal hydrophobic hollow interior. The cavity can entrap a vast number of lipophilic compounds to form inclusion complexes [1-3]. The physical, chemical and biological properties of molecules, which are encapsulated by β-CD, may be drastically modified, such as masking undesired taste, improving the shelf life and protecting sensitive substrates and so on. Due to these advantages, β-CD has been used extensively in the cosmetics, food and pharmaceutical industry [4,5]. Recently, some researches in textile industry have also taken advantage of β-CD to prepare the fragrant fabrics and have gained the desirable results [6].

Miconazole nitrate, a kind of safe, high-efficiency and broad-spectrum antibiotic external drug, performing the functions such as killing bacteria, diminishing inflammation and relieving tickle, etc. [7], which can be included into β-CDs' cavities [8]. Some literatures have demonstrated that β-CD fixed onto matrix did not lose complexing power to form inclusion with other molecules [9,10]. In this paper, β-CD was modified by sulfonation reaction, and then the sulfonated-β-CD can be "dyeing" onto wool fabrics through electrovalent bond in acid solution. A novel antibacterial wool fabric can be obtained by filling miconazole nitrate into cavities of sulfonated-β-CD fixed onto wool.

2 Experiment
2.1 Materials
Pure wool fabrics (22.8tex×22.8tex) were supplied by Wuxi Wool Factory; β-CD was purchased from Shanghai chemical reagent Co. Ltd. Miconazole nitrate was kindly provided by Zhejiang Shengda Pharmaceutical Co. Ltd. All other reagents were of analytical grade. The water used was double-distilled and deionized.
2.2 Apparatus
380 Fourier Transform infrared spectrophotometer (FTIR) (Thermo Nicolet Corporation, USA); Aspetic operating board 4HC-24 (Sun-great Technology Co.Ltd., Shanghai, China); DZF-6020 vacuum dryer (Precision apparatus factory Shanghai, China).
2.3 Preparation of antibacterial wool fabrics
2.3.1 Modifying β-CD by sulfonation reaction
Putting 90% (mass fraction) H_2SO_4 into 100 ml round flask, in which β-CD was added slowly, then kept stirring 2h in 0~5℃ ice water bath. Reaction solution was poured into water and $CaCO_3$ added, then filtering

---

* Corresponding author's tel.: 13536185727, email: huash@mail.dhu.edu.cn

CaSO₄ precipitation, washing and consolidating filter solution. The solution adding 95% ethanol was kept at 0~5℃ for a whole night. Filtering precipitation and adjusting pH to 10.5 by Na₂CO₃ and filtering again, adjusting pH to 7.0 by HAc, then vacuum filtration, condensation and adding ethanol, a vast number of white deposit was then generated. Filtering, washing and drying deposit to obtain sulfonated-β-CD.

2.3.2 Grafting of wool fabrics with sulfonated-β-CD

The grafting procedure (Figure 1) consisted in soaking the fabric samples for 10 min at 40℃ in the finishing solution (sulfonated-β-CD x g/L, Na₂SO₄ 5 g/L, pH=3, bath ratio 1:10), then raising the bath temperature at rate of 1℃ /min up to T, and keeping T for t min. The fabric was picked up, soap washed firstly, and then washed under running water to remove any unreacted sulfonated-β-CD and dried in vacuum for 8 h at 60℃. The quantity of sulfonated-β-CD bonded to wool fabric was estimated by the weight difference of the sample of fabric before and after the fixing process described. Conveniently, wool fabrics bonded sulfonated-β-CD will be designated hereafter as sulfonated-β-CD-fabric.

Figure 1 Chemical grafting of sulfonation-modified-β-CD onto wool fibers (a); Scheme of host-gust inclusion complex grafted onto the textile surface (b)

2.3.3 Preparation of antibacterial wool fabrics

Sulfonated-β-CD-fabric and unmodified fabric samples were treated with antibacterial agent by dipping the textiles at room temperature for 4 h under stirring in ethanol solution containing 5% (w/v) miconazole nitrate with a liquor ratio 1:10. The samples were then roll-squeezed and washed three times with 30% (v/v) ethanol-water solution to remove the absorbed antibacterial agent from the fabric surface and then rinsed with running tap water for 10 min and dried.

2.3.4 Antibacterial property of the fabrics loaded miconazole nitrate

The determination of antibacterial property of the different fabric samples were carried on according to the Downing's shaker method. Three kinds of fungus—*C.albicans, S. aureus* and *E. Coli* were selected to investigate the antibacterial property of miconazole nitrate impregnated into the fabrics. In addition, in order to investigate the durability of antibacterial performance, the washing fastness of antibacterial fabrics was also tested after washing 0, 5 and 10 times with 2 g/L soap solution at 60℃ for 10 min.

3 Results and discussion

3.1 Characterization of sulfonated-β-CD

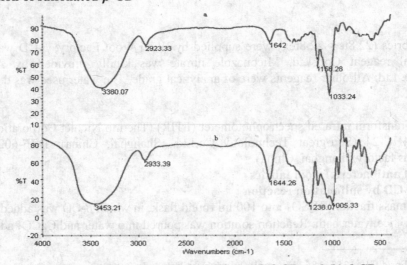

Figure 2 IR spectra of: (a) sulfonated-β-CD and (b) β-CD

Figure 2 shows the IR spectra of sulfonated-β-CD (a) and β-CD (b). As we know, 3400~3700 cm^{-1} belongs to vibrational absorption band of –OH, which is much broad in Figure 2b because of large number of –OH in β-CD. The peak of –OH in Figure 2a is narrower than that of Figure 2b, which means the parts of –OH being substituted by sulfo group. In addition, the characteristic absorption peak of sulfo group located in 1033 cm^{-1} and 1156 cm^{-1} appears in Figure 2a, which indicates β-CD being modified by sulfonation successfully.

3.2 Factors influencing the amount of sulfonated-β-CD grafted onto wool fabrics

The quantity of sulfonated-β-CD on the fabrics was determined by gravimetric measurements. The amount of sulfonated-β-CD grafted onto the fabric was mostly dependent on the sulfonated-β-CD concentration, pH, temperature and time, etc.

The concentration of sulfonation-β-CD in finishing solution influenced the grafted amount on the fabric, as shown in Figure 3a. The rate of weight gain of the fabrics grows up with increasing sulfonation-β-CD concentration. However, no obvious increase tendency appears when concentration is over 60 g/L. The phenomenon is very similar to the mechanism of position absorption of dyeing wool with acid dyes, which changes little when "dyeing position" being occupied completely.

Figure 3b shows the effect of pH on the grafted sulfonation-β-CD onto the fabric. The rate of weight gain increases with lowering pH, which means more sulfonation-β-CD being "fixed" onto the fabrics. The reason is that the stronger the acidity of the solution, the more the positive charge on the wool fibers, which could attract more sulfonated-β-CD with negative charge onto the fabrics. However, the damage to wool is great when acidity is too hard. So, it is desirable to keep pH at 2~3.

The effect of temperature is shown in Figure 3c. The bonded amount of sulfonated-β-CD on the fabrics increases with raising temperature. However, the high temperature may be harmful for the wool fibers. So, 90~100℃ is a suitable temperature span.

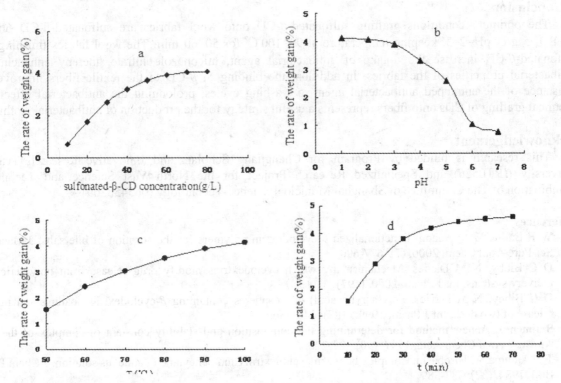

Figure 3 Factors influencing the amount of sulfonated-β-CD grafted onto wool fabrics

The time of keeping temperature was also investigated. The longer the time, the more the amount of bonded sulfonated-β-CD grafted onto the fabric. Figure 3d demonstrates the trend of the rate of weight gain with extending the time. From it, we can see 50-60 min is enough to obtain the desirable result.

Considering the above factors comprehensively, we achieved the optimum conditions grafting sulfonated-β-CD onto wool fabrics: sulfonated-β-CD 60 g/L, Na_2SO_4 5 g/L, pH=2~3, keeping temperature at 90~100℃ for 50~60 min.

3.3 Antibacterial property of the treated wool fabrics

The antibacterial fabrics can be obtained by means of the method 2.3.3. The antibacterial abilities were tested using shaker flask method, and the results are shown in Table 1

Table 1 Antibacterial ability of sulfonated-β-CD-fabrics and unmodified fabrics and their durabilities for washing

Washing times	Anti-bacterial rate (%)					
	Sulfonated-β-CD wool fabrics loaded miconizole nitrate			Un-modified wool fabrics loaded miconizole nitrate		
	C.albicans	S. aureus	E. coli	C.albicans	S. aureus	E. coli
0	93	90	88	62	57	51
5	82	74	69	19	1	0
10	65	61	56	0	0	0

From Table 1, we can see that the wool fabrics loaded miconizole nitrate have the antibacterial ability to C.albicans, S. aureus and E. Coli. The antibacterial activity of sulfonated-β-CD-fabric is better than that of the unmodified samples. This phenomenon was resulted from the different existing forms of antibacterial agents on the fabrics. For the sulfonated-β-CD-fabric, the antibacterial agents were included in the cavities of β-CD, while absorbed physically for the unmodified ones, which made the antibacterial agents easy to wash from the fabric. Based on this, the durability of antibacterial property of the former is much better than the latter. The antibacterial abilities of sulfonated-β-CD-fabric still kept more than 60%~70% even after washing 10 times, while the antibacterial activity was almost lost for the unmodified ones after washing 5 times.

4 Conclusion

The optimum conditions grafting sulfonated-β-CD onto wool fabrics are sulfonated-β-CD 60 g/L, Na_2SO_4 5 g/L, pH=2~3, keeping temperature at 90~100℃ for 50~60 min. The wool fabrics finishing with sulfonated-β-CD increase the uptake of antibacterial agent, miconazole nitrate, thereby enhancing the antibacterial properties of the fabrics. In addition, the binding of β-CD to the textile fibers improves the resistance of the entrapped antibacterial agent to washing cycles, prolonging the antibacterial effect. The chemical grafting of CDs onto fibers represents a useful strategy for the production of antibacterial clothing.

Acknowledgement

This research is funded by Program for Changjiang Scholars and Innovative Research Team in University (IRT0526) and Specialized Research Project for the North-West Science and Technology Combination of The Committee of Shanghai Municipal Science-Technology (065458209)

References

[1] W E Baille, W Q Huang. Functionalized β-cyclodextrin polymers for the sorption of bile salts. J.Macromol Sci-Pure Appl Chem, 2000, (7): 677-690.
[2] D C Bibby, N M Davies. Mechanism by which cyclodextrins modify drug release from polymeric drug delivery systems. Int J Pham, 2000, (197): 5-11.
[3] D C Bibby, N M Davies. Poly(acrylic acid) microspheres containing β-cyclodextrin: loading and in vitro release of two dyes. Int J Pham, 1999, (187): 243-250.
[4] S Jimenez. A new method for determining the composition and stability constant of complex of the form AmBn. J Anal Chem Acta, 1997, (90):223-228.
[5] T Nakajuma, M Hirobasbi. Complex between cyclodextrins and bencylane in aqueous solution. J Chem Pharm Bull, 1984, (32): 383-389.
[6] H J Buschman. New Textile Application of Cyclodextrins. J Incl Phenom Macro Chem, 2001, (40): 169-170.
[7] Z A Pan. The pharmacopeia for the People's Republic of China. Beijing: Chemical Industry Press, 2006: 3984.
[8] J H Wang, Z S Cai. Investigation of inclusion complex of miconazole nitrate with β-cyclodextrin. Carbohydrate Polymers, 2008, 72(2): 255-260.
[9] H J Buschmann, U Denter & D Knittel. Removal of residual surfactant deposits from textile materials with the aid of cyclodextrins. Melliand Textiberichte, 1995(9): 214-216.
[10] P Nostro, L Frantoni. Modification of a cellulosic fabric with β-cyclodextrin for textile finishing applications. J Incl Phenom Macro Chem, 2002, (44): 423-427.

44 Experimental Study of the High Temperature Resistant Filtration Materials Bonding by of Hydroentanglement Technique

Xiangqin Wang*, Lili Wu, Xiangyu Jin, Qinfei Ke

(Key Laboratory of Textile Science & Technology, Ministry of Education, Donghua University, 201620, China)

Abstract: Hydroentanglement is a versatile process for bonding loose arrays of fibers or enhancing fabrics by fine, closely spaced, high velocity water jets. This paper studies the effect of hydroentangling process parameter on the properties of high temperature filtration materials made from PPS and P-84 fibers and woven fabrics by experimental measures. Different hydroentangling process with the pressure from 120 to 200bar and the pass number from 2 to 8 are tested. The experimental results 160bar and 6 passes are critical values for hydroentangling pressure and pass number not only for tensile strength and elongation but also for pore distribution. The results also indicated that the 140bar and 6 passes is the most proper process parameters for filtration efficiency of the fabric used in this study.

Keywords: High temperature resistant; Filter materials; Hydroentanglement

1 Introduction

The advantage of hydroentanglement over needle punching, the conventional form of mechanical bonding, is that with some fibers water acts as a plasticizer drastically reducing the bending stiffness, flexural rigidity and the torsion rigidity of the fibers. This allows the fibers to easily bend, twist, form loops, entangle, etc. Besides, the used of water jets, instead of metal needles means less friction and hence less damage to the fibers. Application areas of hydroentanglement nonwovens cover a wide range of fabric weights from under 10 g/m^2 to more than 600 g/m^2.

The high temperature resistant filtration materials are designed to keep very impurities from entering a control environment or to prevent them from escaping. The design of consideration for the filter materials depends upon the properties of fibers, fiber arrangement, method of bonding the fibers, extent of filtration required [1]. The aim of the work reported here was to investigate the effects of hydroentanglement process parameters on the properties of high temperature resistant filtration materials with a sandwich structure made of widely used high performance fiber PPS, P-84, respectively.

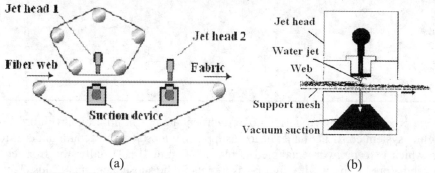

Figure 1 (a) Aquajet T6616 pilot hydroentanglement process scheme, (b) principle of hydroentanglement system

2 Experimental design

Two sets of webs are hydroentangled under numerous sets of processing conditions. It is think the energy transfer distribution is affected by the number of passes, pass speed and double side processing [2]. The processing condition mainly focused on is the pressure and pass numbers and double side processing.

2.1 Materials

The webs we obtained from the procedures have a sandwich structure, the top and bottom layers are carding webs with cross-laid, the middle layer is a woven fabric. The details of the raw materials and the webs are summarized in Table 1 and 2 (In this paper gsm stands for gram per square meter which is the unit for the area density and indicates the weight and thickness of the fabric).

---

* Corresponding author's email: xqwang@mail.dhu.edu.cn

Table 1 Property of raw fibrous materials

Item ID	Material	Linear density (dtex)	Density (g/m^3)	Mean length (g/m^2)	Breaking strength (cN)
1	PPS	3.1	1.34	51	10.93
2	P-84	2.2	1.41	51	7.49

Table 2 Web structures

Item ID	Material	Web density (gsm)	Top layer density (g/m^2)	Middle layer density (g/m^2)	Bottom layer density (g/m^2)
1	PPS	450	150	150	150
2	P-84	450	150	150	150

2.2 Processing conditions

Table 2 Processing conditions

Test No.	Pressure(bar)	First side pass No.	Second side pass No.	Total pass numbers
1	120	1	1	2
2	120	2	2	4
3	120	3	3	6
4	120	4	4	8
5	140	1	1	2
6	140	2	2	4
7	140	3	3	6
8	140	4	4	8
9	160	1	1	2
10	160	2	2	4
11	160	3	3	6
12	180	1	1	2
13	180	2	2	4
14	180	3	3	6
15	200	1	1	2
16	200	2	2	4
17	200	3	3	6

The webs were processed using Aquajet T6616 hydroentanglement pilot plant (Figure 1a). The basic unit of a hydroentangling system consists of a jet head, a supporting mesh (with vacuum under it for water suction), on which a web which is to be hydroentangled is placed (Figure 1b). Usually, the first jet head is used for pre-wetting the web before hitting it with high-pressure jets in the subsequent manifolds. This is necessary for initial consolidation and better entanglement. It is noted that the web used in this work has been pre-treated by needle punching process. Table 2 summarizes details of the processing conditions.

3 Results and discussion
3.1 Tensile strength

Tensile strength is often the determining factor in choosing a fiber for a specific need. Rectangular specimens with a size of 5cm x 25cm were taken from each fabric; five specimens in the machine-direction (MD) and five in the cross-direction (CD). The ISO standard tensile test was used to measure the tensile strength of the each fabric sample.

Figures 2-5 show the variation of tensile strength as a function of hydroentangling pressure for PPS and P-84 filtration materials. The tensile strength of the PPS and P-84 fabrics in MD and CD increases to the maximum at the hydroentangling pressure of 160bar and subsequently levels off within the range of pressure tested here. However, the elongation in of the fabrics in CD decreases to the minimum at the pressure of

160bar, whereas, it increases to the maximum in MD. The tensile strength of PPS filtration materials were treated under the hydroentangling pressure from 120 to 160bar are shown in Figure 4, the tensile value at "0" point means the strength of the woven fabric in the middle layer. Figures 4 and 5 show that there is no more significant effect of passes numbers on the tensile strength and elongation when the fabrics passes more than 6 manifolds.

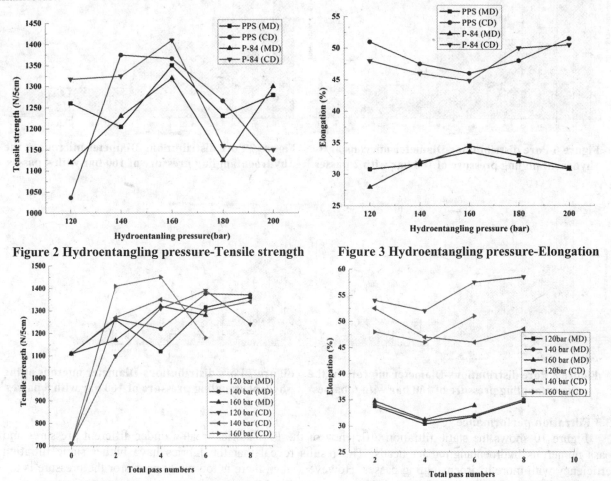

Figure 2 Hydroentangling pressure-Tensile strength

Figure 3 Hydroentangling pressure-Elongation

Figure 4 Total pass numbers-Tensile strength

Figure 5 Total pass numbers-Elongation

In summary, the tensile strength of the fabrics reaches up to the maximum at the hydroentangling pressure of 160bar with 6 passes. The elongation in CD is higher than MD.

3.2 Aperture distribution

The flow porosimetry is based on the principle that a porous material will only allow a fluid to pass when the pressure applied exceeds the capillary attraction of the fluid in largest pore. In this test, the specimen is saturated with a liquid and continuous air flow is used to remove liquid from the pores. At a critical pressure, the first bubble will come through the largest pore in the wetted specimen. As the pressure increases, the pores are emptied of liquid in order from largest to smallest and the flow rate is measured. PSD, number of pores and porosity can be derived once the flow rate and the applied pressure are known [3]. In this study, the CFP-1100A type capillary flow porometer from MPI Inc. is used to measure the pore distribution of the specimens.

The pore distributions of P-84fabrics, which have been treated under the hydroentangling pressure of 160bar with 2 to 8 passes, are shown in Figures 6 to 9.

We noted that only the PPS fabrics treated by 6 and 8 passes have the small pores less than 5 μm, and the PPS fabrics treated by 6 passes has more these small pores than that treated by 8 pores. This suggests that too much hydroentangling treatment by high pressure water will destroy the pore distribution of the fabrics.

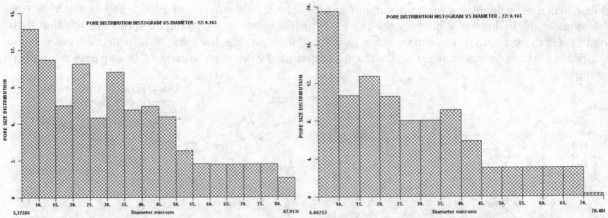

Figure 6 Pore distributions-Diameter microns at the hydroentangling pressure of 160 bar with 2 passes

Figure 7 Pore distributions-Diameter microns at the hydroentangling pressure of 160 bar with 4 passes

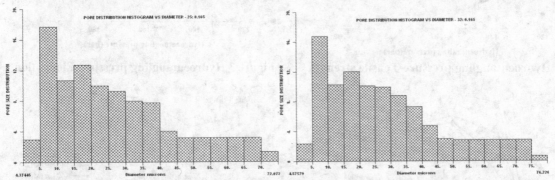

Figure 8 Pore distributions-Diameter microns at the hydroentangling pressure of 160 bar with 6 passes

Figure 7 Pore distributions-Diameter microns at the hydroentangling pressure of 160 bar with 8 passes

3.3 Filtration performance

Figure 10 shows the static filtration efficiency of the PPS fabrics treated under different pressures and pass numerous hydroentangling jet heads. The results reveal that the fabrics have higher static filtration efficiency with more hydroentangling passes. However, when there is too many passes or the pressure is too high, the filtration efficiency goes down. This suggests that 140bar can be a critical pressure for the fabrics used in this study, and 6 passes is the critical hydroentangling pass.

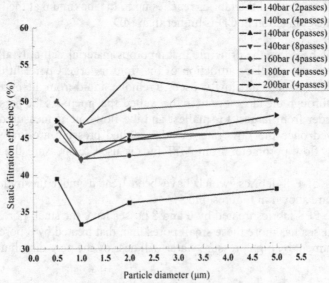

Figure 10 Statir filtration efficiency-Praticle diameter

4 Conclusions

This experimental work presents the effect of hydroentangling process parameter on the properties of high temperature filtration materials made from PPS and P-84 fibers and woven fabrics. The conclusions from this work are as follows: Firstly, the tensile strength of the fabrics reaches up to the maximum at the hydroentangling pressure of 160bar with 6 passes. The elongation in CD is higher than MD. Secondly, the fabric treated by the hydroentangling pressure of 160bar with 6 passes has the best pore distribution with more small pores. Thirdly, the 140bar and 6 passes are suggested to be the critical pressure and hydroentanling pass for getting best filtration efficiency for the fabric used in this study.

References

[1] A M Seyam, D A Shiffler, H Zheng. An Examination of the Hydroentangling Process Variables. International Nonwovens Journal, 2005, 14(1): 25-33.
[2] E Ghassemieh, M Acar, H K Versteeg. Improvement of the efficiency of energy transfer in the hydro-entanglement process. Composites Science and Technology, 2001, 61: 1681–1694.
[3] M. Ziabari, V. Mottaghitalab, and A. K. Hagh, Evaluation of electrospun nanofiber pore structure parameters. The Korean Journal of Chamical Engineering. 2008,25 (4): 923-932.

45 Study on the Intensity of Joints for the Textile Geogrids Manufactured with PET High-performance Yarn

Guoping Xu[1], Xinbo Ding[1], Jian Han[2]

(1. College of Material & Textile Engin., Zhejiang Sci-Tech University, Hangzhou 310018, China
2. Key Laboratory of Advanced Textile Materials and Manufacturing Technology, Ministry of Education of China, Zhejiang Sci-Tech University, Hangzhou 310018, China)

Abstract: In this article, the effects of the content of the PVC coating, the number of the ribs and the joints on the mechanical properties of the textile geogrids manufactured with PET high-performance yarn were studied. It was found that when the PVC coating content was less than 60 wt%, not only the strength of geogrid in warp direction but also that in weft direction increased and the anisotropic mechanical properties was smaller with increasing the PVC coating content. However, when the PVC coating content was more than 60 wt%, the strength of geogrid showed unchanged and the anisotropic mechanical properties were even smaller. Due to the rib structure with the firming yarn in warp direction, the strength of joints was higher than that in weft direction, and the anisotropic become significantly with increasing the number of joints for textile geogrid.

Keywords: Intensity; Joints; Textile geogrids; PET high-performance yarn

1 Introduction

Since the geogrid was developed in England in 1979 as a method of constructing from synthetic materials, its application has increased continuously. Typically, geogrids have grid structures with apertures of 10-100mm between longitudinal and transverse ribs, respectively [1]. There are two types of geogrids that are widely used in the geotechnical and civil engineering fields [2]. One is the sheet type geogrid, which are made from pre-extruded HDPE (high density polyethylene) geomembrane sheets using punching and drawing process. The drawing process not only makes the grid structures, but also enhances the tensile strength of the ribs.

The other is the fabric type geogrid, which are made with a high tenacity polyester filament using a conventional weaving and/or knitting process. Commonly, the fabric type geogrids are then coated with copolymer resins such as PVC (polyvinylchloride), bitumen, PP (polypropylene), acrylic resins, latex and other rubbery materials, which contain a light stabilizer and antioxidant[3,4]. Since the geogrids possess high tensile strengths in nature, they are frequently used as the reinforcement in unpaved/paved road constructions, slopes of waste landfill liner system, embankments and slopes, segmental retaining walls, and so on[5].

In this article, the effects of the content of the PVC coating, the number of the ribs and the joints on the mechanical properties of the textile geogrids manufactured with PET high-performance yarn will be studied. Then, the bearing-force mechanism of the joints will be analyzed and put some suggestions on design.

2. Experimental

2.1 Experiment materials

Geogrids made with a high tenacity polyester filament, which were manufactured by a weaving and knitting process, were used in this study. The transverse ribs (weft) are interlaced and inserted into longitudinal ribs (warp) by a weaving mechanism. The longitudinal ribs were then knitted to create stable junctions. This type of geogrid may be called a weft insertion warp knitting type. After fabrication, the geogrids were coated with PVC resin with additives. Figure 1 shows the schematic diagram of structure of uncoated and coated weft insertion warp knitting type polyester geogrids with PVC. The schematic diagram of structure of the uncoated and coated weft insertion warp knitting type polyester geogrids with PVC are shown as Table 1.

Table 1 The schematic diagram of structure of the uncoated and coated textile geogrids with PVC

Samples	Yarn count (tex)		Number of ribs	weight (g/m2)	coating rate (%)	Grid distance (mm)
	warp	weft				
uncoated	1111.1	1111.1	3 warp	298.7	/	25
coated	1111.1	1111.1	3 weft	376	25.9	25

Figure 1 Schematic diagram of structure of the uncoated and coated textile geogrids

2.2 The instrument and the test method

All samples were beforehand placed in constant temperature and humidity room (T 20℃, Humidity 65%) for 24 hours, and then mechanical properties of the weft insertion warp knitting geogrids were conducted using an electronic universal tensile test device at a tensile speed of 50 mm/min on the samples with an initial length of 100 mm.

3 Results and discussion

The strength of joints is one of the important properties for textile geogrids. Typically, geogrids have grid structures with apertures of 10-100 mm between longitudinal and transverse ribs, respectively. Since the rib structure in warp direction is different from that in weft direction, resulting in that the strength of joints show anisotropic mechanical properties.

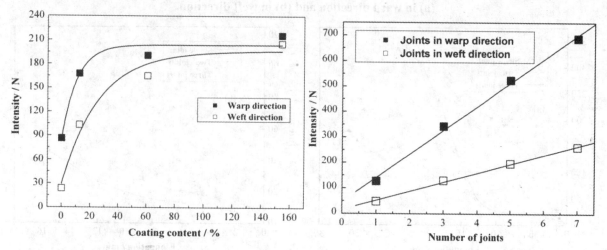

Figure 2 Coating content dependence of the strength of joints for textile geogrids

Figure 3 The number of joints dependence of the strength of the textile geogrids

Figure 2 shows the coating content dependence of the strength of joints for textile geogrid. As can be seen from Figure 2, the strength of the textile geogrid in warp direction is higher than that in the weft direction significantly when without PVC coating. It is mainly attributed to the presence of the firming yarn in the warp direction, increasing the cluster and friction among yarns. When the PVC coating content is less than 60 wt%, not only the strength of geogrid in warp direction but also that in weft direction increase and the anisotropic mechanical properties is smaller with increasing the PVC coating content. However, when the PVC coating content is more than 60 wt%, the strength of geogrid shows unchanged and the anisotropic mechanical properties is even smaller. It is because of that the PVC coating is permeated into yarn and increases the cluster of the yarn. Thus, the mechanical properties of the geogrid are improved. When the PVC coating content is more than the threshold value, the cluster of the yarn cannot be improved further and the mechanical properties of the geogrid show unchanged. Therefore, these results imply that the PVC coating content should not be more than 60 wt%.

The grid structures can interlock the soil, rocks and other filling materials, and the strength of the interlocking effect is related to the strength of joints. Figure 3 shows the number of joints dependence of the strength of the textile geogrids. Due to the rib structure with the firming yarn in warp direction, the strength of joints is higher than that in weft direction, and the anisotropic become significantly with increasing the number of joints for textile geogrid. That is the grid with excellent mechanical properties in the warp direction is better than that in the weft direction. The results can be supported by the broken schematic diagram of the joints for textile geogirds, as shown in Figure 4 (a) in warp direction and (b) in weft direction. As can seen from Figure 4(a), the broken form of the joints in warp direction is that the firming yarn firstly become broken and then the warp rib come into being slipping. However, the broken form in weft direction is that the rib causes slipping as the external force.

**Figure 4 The broken schematic diagram of the joints for textile geogirds:
(a) in warp direction and (b) in weft direction.**

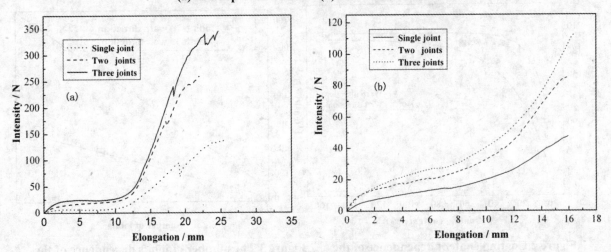

**Figure 5 The number of joints dependence of the ribs for textile geogrids:
(a) in warp direction, and (b) in weft direction**

The mechanical properties of the ribs for textile geogrids are shown in Figure 5. As can be seen from Figure 5(a), the broken form of the ribs in warp direction is not a typical tensile curve. In the second stage of the drawing curse, the mechanical properties shows unchanged, implying that the firming yarn play an important role. Then in the third stage of the drawing curse, which shows the shape of sawtooth, suggests that the firming yarn yield broken firstly and the joints are broken under the external force, then the ribs bear the external force. However, the ribs in the weft direction undertake the external force directly.

4 Conclusions

The effects of the content of the PVC coating, the number of the ribs and the joints on the mechanical properties of the textile geogrids manufactured with PET high-performance yarn were studied. It was found that when the PVC coating content was less than 60 wt%, not only the strength of geogrid in warp direction but also that in weft direction increased and the anisotropic mechanical properties was smaller with increasing the PVC coating content. However, when the PVC coating content was more than 60 wt%, the strength of geogrid showed unchanged and the anisotropic mechanical properties were even smaller. Due to the rib structure with the firming yarn in warp direction, the strength of joints was higher than that in weft direction, and the anisotropic become significantly with increasing the number of joints for textile geogrid.

Ackonwledgement

This work is supported by Science and Technology Plan Projects of Science Foundation of Zhejiang Provincial Key Laboratory (2007C11026) and Training Foundation of Outstanding Young talent of "Textile Subject" (Grant No. 2008QN08).

References

[1] Han Y. J., Seong H. K.,Han K. Y. Assessment of long-term performance of polyester geogrids by accelerated creep test [J]. Polymer Testing 2002, 21:489-495.
[2] GRI Test Methods and Standards, Geosynthetic Research Institute, Philadelphia, USA, (1995).
[3] Ingold T.S. The geotextiles and geoemebranes manual, Elsevier Advanced Technology, 1981:71-246.
[4] Koerner R.M., Desigening with geosynthetics, Prentice Hall Co., 1994: 328-329.
[5] J. A. Finnigan, The creep behavior of high tenacity yarnsand fabrics used in civil engineering applications, in: Proceedings of the International Conference on the Use of Fabrics in Geotechnics, Paris, France, (1977) 645–650.

46 Study on the Morphology, Structure and Mechnical Performances of Electrospun Aligned PAN/Activated Carbon Nanofiber

Hua Xue, Ni Li*, Jie Xiong, Guan Feng Liu

(Key Laboratory of Advanced Textile Materials and Manufacturing Technology, Ministry of Education of China, Zhejiang Sci-Tech University, Hangzhou 310018, China)

Abstract: The morphology, structure and mechnical performances of Polyacrylonitrile(PAN) nanofibers with activated carbon as precursor for carbon nanofibers was investigated. By using rotating drum as the receiving collector, well oriented nanofibers were got at 150rpm. And the ATR-FTIR spectroscopy showed that the chemical composition of PAN nanofiber was unchanged when blending with activated carbon powder. But X-ray revealed that the crystallinity of PAN nanofiber was improved. Moreover, the mechanical properities of PAN nanofibers were decreased with activated carbon powder.

Keywords: Polyacrylonitrile; Activated carbon powder; Electrospinning

1 Introduction

Electrospinning is a novel and simple fabrication process to obtain superfine fibers with diameters ranging from nanometers to sub-microns. In electrospinning, when electrostatic force exceeded the surface tension of the fluid, the charged fluid ejected from the spinneret nozzle. Because of solvent evaporation, the charged jet was solidified and nanofibers were formed and collected on the collector.

Polyacrylonitrile (PAN) is the primary precursor for carbon fibers. This paper is the discussion of our preliminary work. The effect of many kinds of organic or inorganic salts on electrospinning has been investigated[1]. In this work, the single element, activated carbon powder, was selected to add in PAN electrospinning solution for it's cheap cost and purity for the final carbon fiber.

Several researches[2-4] reported that the mechanical strength of electrospun carbon nanofibers was far below the predictions. That was because the precursors were randomly arranged nanofiber mats, and tension could not be applied effectively onto individual nanofiber during the oxidative stabilization in air. In this paper, a rotating drum was used as the receiving device to obtain orientated fibers. Meanwhile the best rotating rate was determined.

2 Experimental section

2.1 Materials

Polyacrylonitrile(PAN) with M_w of 140000 was purchased from Hangzhou Acrylon limited corporation. N, N-Dimethyl Formamide(DMF) was purchased from Hangzhou Gaojing Fine Chemical Industry Co., Ltd. Activated carbon powder was purchased from Hangzhou Huipu Chemical Instruments Co., Ltd.

2.2 Electrospinning

5, 10 and 15 wt% activated carbon powders were blended respectively into 12% PAN/DMF solution to prepare electrospinning solution. To get homogeneous solution the blended solution was oscillated by ultrasonic for 1h. Experimental set-up used for eletrospinning process was show in Figure 1. Spinning voltage was 12kV, the tip-to-collect distance was 12cm, and solution feed rate was 0.01ml/min. All experiments were performed at room temperature.

2.3 Measurements

Hitachi S4800 field emission scanning electron microscopy (FE-SEM) was used to investigate the morphology of electrospinning nanofibers, and Image-Pro ® Plus6.2 Software was used to measure fiber diameter and fiber angle with rotating direction of the drum. The sample number was 50. The structures of electrospinning nanofibers were studied with Nicolet 5700 (USA) infrared spectrometer (FTIR) and ARL-X 'TRA X-ray diffraction testing(XRD). XRD profiles were recorded at the 2θ angles ranging from 5° to 50° at a scanning speed of 5°C / min. The nanofiber membrane was twisted into a 40mm long yarn, then KES-G1 multi-purpose tensile tester(Kato-Tech Company, Japan) was used to test the mechanical properties at 22±1℃ and 65 ± 5% humidity conditions, and each sample tested five times.

---

* Corresponding author's email: lini@zstu.edu.cn

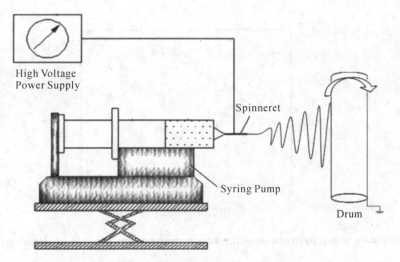

Figure 1 Schematic diagram of experiment set-up

3 Results and discussion
3.1 Effect of rotating rate of the drum on the fiber morphology

SEM images in Figure 2 showed the morphologies of electrospun nanofibers from 12%PAN/DMF solution with different rotating rate of the drum. It was evident that better alignment of the nanofibers with higher rotating rate. The relationship between the fiber angle and the rotating rate was investigated (shown in Figure 3.). As the rotating rate increased, the fiber alignment was significantly increased, but when the rate increased to 150rpm, the orientation of nanofibers on the drum did not change obviously. So for good orientation of nanofibers and energy saving, drum rotating rate was fixed at 150rpm in the following experiments.

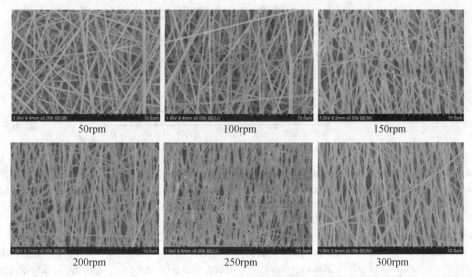

Figure 2 SEM images showing the nanofiber morphologies from 12%PAN/DMF solution with different rotating rate of the drum

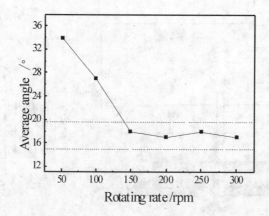

Figure 3 The relationship between rotating rate and the average angle of electrospun nanofibers

3.2 Infrared spectrum

The band at 2236cm^{-1}, 1742 cm^{-1}, and 1231 cm^{-1} is the characteristic band of C≡N, C=O and C-O respectively[1,5]. Figure 4 shows that all the five samples have the same band position. There is no band shift, band disappearance or new band appearance observed, which means the structure of electrospun nanofibers from the solution with or without activated carbon powder is same. That is to say activated carbon has not obvious effect on PAN molecular structure.

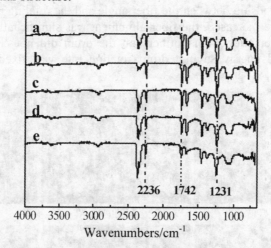

Figure 4 FT-IR spectra of the samples:(a) 12% PAN nanofiber received by plate, (b) 12% PAN nanofiber received by rotating drum at 150rpm, (c) 12% PAN nanofiber with 5% activated carbon at 150rpm, (d) with 10% activated carbon at 150rpm, (e) 15% activated carbon at 150rpm

3.3 X-ray diffraction of nanofibers

The diffraction peaks of PAN nanofiber centered at 2θ angles of 17.1° was attributed to the crystallographic planes of (100)[6]. And around 25°, there was a broad peak. Compared with the curves of b, c, d and e, the diffraction peak of curve "a" was not evident, which implied the crystal structure of electrospun nanofibers received by rotating drum was enhanced because of the traction force of the rotating drum. Compared with the curve b, the diffraction peaks of curves c, d, e were increased, which implied that the addition of activated carbon improved the crystal structure of electrospun PAN nanofiber.

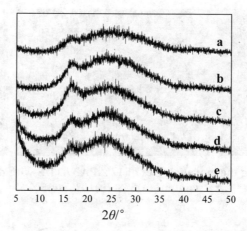

Figure 5 X-ray diffraction patterns of the samples: (a) 12% PAN nanofiber received by plate, (b) 12% PAN nanofiber received by rotating drum at 150rpm, (c) 12% PAN nanofiber with 5% activated carbon at 150rpm, (d) with 10% activated carbon at 150rpm, (e) 15% activated carbon at 150rpm

3.4 Mechanical properties

From the curves a and b in Figure 6, it could be obtained that the mechanical property of the fibers received by rotating drum was better than that of the fibers received by plate. That was because of higher crystallinity(see Figure 5) and much better orientational arrangement for the fibers received by rotating drum. With the increase of activated carbon content, the maximum tensile strength and broken elongation of fiber yarn decreased(shown in Table 1). This was because the carbon content increased, agglomeration of activated carbon powders severely enhanced (shown in Figure 7), making the mechanical properties of fiber yarn deteriorated.

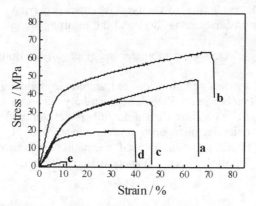

Figure 6. Mechanical properties of the samples: (a) 12% PAN nanofiber received by plate, (b) 12% PAN nanofiber received by rotating drum at 150rpm, (c) 12% PAN nanofiber with 5% activated carbon at 150rpm, (d) with 10% activated carbon at 150rpm, (e) 15% activated carbon at 150rpm.

Table 1 Machanical properties of the five electrospun samples

Sample	Ultimate tensile stress(MPa)	Ultimate tensile strain(%)
a	41.4	60.9
b	71.5	72.6
c	37.8	47.9
d	16.7	45.9
e	3.4	13

Figure 7 SEM images showing the morphologies of 12%PAN/DMF with different activated carbon powder content (a)5%, (b)10%, (c)5%

4 Conclusions

In this study, we identified the best rotating rate of the drum was 150rpm with better fiber orientation. PAN molecular structre was found to be unchanged in FT-IR after the additon of activated carbon powder. But the X-ray diffraction patterns showed that electrospun nanofibers with activated carbon powder have better crystallinity. Meanwhile fiber crystallinity received by the rotating drum was higher than that of nanofibers received by plate. Moreover, with the increase of activated carbon content, agglomeration severely increased. The maximum tensile strength and broken elongation of fiber yarn decreased. Therefore, during the work of the carbonization later, we will select 12%PAN/DMF with 5% activated carbon powder as the precursor, because of its relatively better mechnical property and spinning behavior.

References

[1] XH Qin, SY Wang. Interior structure of polyacrylonitrile(PAN) nanofibers with Licl. Materials Letters, 2008, 62: 1325-1327.

[2] J Liu, P Zhou, L Zhang, Z Ma, et al. Thermo-chemical reactions occurring during the oxidative stabilization of electrospun polyacrylonitrile precursor nanofibers and the resulting structural conversions. Carbon, 2009, 47 (4): 1087-1895.

[3] SC Moon, RJ Farris. Strong electrospun nanometer-diameter polyacrylonitrile carbon fiber yarns. Carbon, 2009, 47(12): 2829-2839.

[4] SY Gu, J Ren, GJ Vancso. Process optimization and empirical modeling for electrospun polyacrylonitrile (PAN) nanofiber precursor of carbon nano fibers. European Polymer Journal, 2005, 41(11): 2559-2569.

[5] ZP Zhou, KM Liu, CL Lai, et al. Graphitic carbon nanofibers developed from bundles of aligned electrospun polyacrylonitrile nanofibers containing phosphoric acid. Polymer, 2010, 51: 2360-2367.

[6] SZ Wu, F Zhang, XX Hou, et al. Stretching-induced orientation for improving the mechanical properties of electrospun polyacrylonitrile nanofiber sheet. Advanced materials research, 2008, 47: 1169-1172.

47 Effect of Fabric Structure on the Sound Insulation Property of Honeycomb Weave Fabric/PVC Composites Material

Tianbing Yang[1], Yaofeng Zhu[2], Bangyong Pang[1], Jin Wang[1], Hao Cen[1], Liyuan Zhang[1], Yaqin Fu[1*]

(1. Key Laboratory of Advanced Textile Materials and Manufacturing Technology, Ministry of Education of China, Zhejiang Sci-Tech University, Hangzhou 310018, China
2. Graduate School of Science and Technology, Shinshu University, 3-15-1 Tokida, Ueda 386-8576, Japan)

Abstract: In order to study the sound insulation property of honeycomb weave fabric/PVC composite material and its mechanical properties, the honeycomb weave fabric/PVC composites were manufactured from cotton yarns fineness with $JC10^S/2$ and PVC resin with high damp. Two-channel acoustic analyzer was used to analyze the sound insulation property. The results shows that different honeycomb weave fabrics of composites have significant influence on the sound insulation property, of which k is 5 (Tissue circulation count) presenting the best noise reduction. The noise reduction of the composite material is superior to that of a single material.

Keywords: Cotton yarn; Honeycomb weave fabric; Polyvinyl chloride; Composite material

0 Introduction

Noise pollution, air pollution and water pollution are considered as the three major kinds of pollution, of which noise hazard attracted more and more attention [1]. In order to make people's lives more comfortable and healthy, sound insulation materials and products have become the main demand of people's lives. Nowadays, equipment and living conditions on the noise requirements demand more efficient and economical ways to make insulation materials. Generally speaking, sound insulation material in industrial applications includes glass fibers, foam materials, inorganic fibers and their composites.

In recent years, textile materials have been widely applied in the field of noise reduction due to its lightweight, flexible and other superior features. However, fabrics are of high permeability usually, which leads to their poor noise reduction as a separate sound-absorbing material or insulation material. Textile materials were introduced to noise reduction in the field of composite materials at home and abroad [2-4]. Hanifi Binici[5]'s research showed that wall intensified by light basalt wall fiber products met the needs of both compressive strength and the noise reduction according to the ASTM and Turkish national standards. This material can be applied to walls of factory to improve insulation performance of walls. Fu Yaqin et al[6-7] have prepared a series of thin, lightweight, flexible glass fiber fabric/PVC composite materials, which have the significant barrier property of low and medium frequency noise by normal pressure pouring process. With the expansion of textile materials and composite materials applications in the field of noise reduction, it is very important to develop and promote new noise reduction textile composite materials. Developments of textile composite materials not only can improve the economic efficiency of textile enterprises, but also can protect people's health from noise and reduce the noise impact on people's work. Different kinds of cotton fabrics have different acoustic properties, of which the tissue of the honeycomb weave fabric and its composite materials also have certain effects on their sound insulation. In order to study the sound reduction of cotton fabric and its composite materials, four different tissues of cotton fabric and PVC resin with high damp were chosen in this study. Researches were carried on in the view of the fabric/PVC composite materials' sound insulation, parameter design and its relation with sound insulation.

1 Experimental materials and methods
1.1 Materials

Polyvinyl chloride resin (EPVC), p-450, is supplied by Tianjin Bohai Chemical co., Ltd. Dioctyl phthalate (DOP), industrial grade, is bought from Hangzhou Jinsheng Plastics Company. Cotton yarn, $JC20^S/2$, is provided by Zhejiang Shuangkeda Textile Co., Ltd.

---

* Corresponding author's email: fuyaqin@yahoo.com.cn

1.2 Sample preparation

ASL2000-20-6 automatic sample loom with JC10^S/2 cotton yarn was used to produce different types of honeycomb weave fabrics (warp density 240/10cm, weft density 150/10cm). Basic materials, including EPVC, DOP, and ESO with mass ratio of 100:130:7, were uniformly mixed. Honeycomb weave fabrics were put into a mold, and the polymer mixing was poured into the fabric samples. Then the mold was laid on an oven in 160℃ for 15 min. Finally, samples were prepared after rapid cooling.

1.3 Performance measurement

1.3.1 Sound insulation

Two-acoustic analyzer BSWA VS302USB of Beijing was used to test the sound insulation and Spectra LAB software was used to analyze the data. VS302USB system, non-directional sound source and BSWA-100 power amplifier and other systems were connected, as shown in Figure 1. Noise reduction of construction and building component of sound insulation were measured according to standards of ISO R140-1 and ISO R140-170. The sound pressure level was selected to 90dB pink noise source. One static speaker volume is 10cm×10cm×10cm. The sample size is 25cm×25cm. Sampling frequency of the sound is 4800. Extraction rate is 1, and fast Fourier transform sample is 4096. The data of measurement used is 1/3 octave.

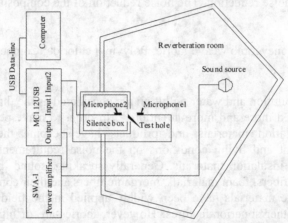

Figure 1 The measurement system of sound insulation

1.3.2 Air permeability

The air permeability of honeycomb weave fabrics was tested by a YG461 type digital style fabric instrument according to GB/T 5435-1997. The test specimen area is 20cm^2, and the pressure difference is 100Pa.

1.3.3 Tensile property

Since there are no test criteria at present, national standard GB/1447-83 (absafil plastic stretch experimental technique) was taken as the basis to test the tensile property of textile composite materials by a RGM200A electronic universal material testing machine. Sample size is 250mm×25mm. Clip distance is 180mm. The speed is 10mm/min.

2 Result and discussion

Basic parameters of the honeycomb weave fabric specimens are shown in Table 1. From Table 1, we can see that the surface density of the honeycomb weave fabrics is basically same. Tissue circulation count has tremendous influence on the thickness and air permeability.

Table 1 Honeycomb weave fabrics related parameter

Tissue circulation count of specimen (k)	Thickness/ (mm)	Surface density/ (g·m^{-2})	Air permeability/ (L·m^{-2}·s^{-1})
5	0.808	315	72.2
6	1.181	312	106.5
8	2.403	307	209.9
9	2.485	318	320.1

2.1 Sound insulation

Figure 2 shows the curves of sound reduction index-Frequency of different tissue of honeycomb weaves fabric. We can see from Figure 2 that sound insulation property in low frequency and high frequency is better than in intermediate frequency. The major factor is that intermediate frequency region is the quality control area, but the fabric surface density is small, which causes the sound reduction index to drop. The air flow resistance of material is one of the major factors of porous material sound insulation property [8]. When the air flow resistance is bigger, the air permeating degree is smaller. So the fabric air permeability is worse and the penetration amount of sound is also smaller, that is, the sound insulation property is better. When k is 5 and 6, the honeycomb weave fabrics have better sound reduction index than k is 8 and 9, compared the four curves in Figure 2. As a result of k is 5 and 6, the yarn arrangement is close, the porosity is small and the air permeability is bad. However, when k is 8 and 9, the yarn arrangement is loose, the air permeability is good and the sound insulation property is bad.

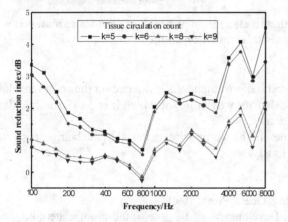 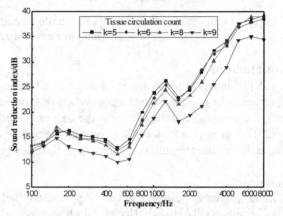

Figure 2 The transmission loss of different tissue circulation count of fabrics **Figure 3 The transmission loss of different tissue circulation count of composite materials**

Figure 3 shows the curves of sound reduction index-Frequency of the composite materials. From Figure 2 and Figure 3 we can see that the sound insulation property of the composite materials surpasses that of the honeycomb weaves fabric obviously. The addition of cotton fiber in the PVC limits the movement of PVC macro-molecule chains. The increase of stress and strain of material is relatively buffered and the modulus of the material enhances distinctly, and its dielectric loss and glass transition temperature also changes correspondently. The sound transmission must overcome a bigger resistance in the material when sound wave drops. The sound energy consumption increases gradually and sound reduction effect is achieved [6]. In addition, Figure 3 shows that sound insulation property of the composite materials increases with the increase of frequency. When k is 5, the sound insulation is best.

2.2 Tensile property

As a sound-proof material, it should have not only a good sound insulation property, but also a good mechanical property. Figure 4 shows tensile property of the composite material which includes the curves presenting the relation between tensile strength-breaking elongation and the tissue circulation count in the warp and weft direction. We can see from Figure 4 that warp tensile strength is higher than weft tensile strength obviously. The primary cause is that the warp density and the weft density are different. The warp density is dense, which means the volume fraction is high in unit volume of cotton fiber on the warp direction. Simultaneously, there are more cotton yarns to withstand tension. Thus the warp direction strength is high, vice versa. From Figure 4(a) we can see that the warp tensile strength and the breaking elongation of the composite material decrease with the increase of the tissue circulation count of honeycomb weave fabrics. This is mainly because that the fabric plays an important role. If the tissue circulation count is small, the yarns arrange is close, the tensile strength is big and the unit volume PVC content is more and the break elongation is big. From Figure 4 (b) we may conclude that the tensile strength trend is invariant and the break elongation increases with the increase of tissue circulation count. Because the weft density is small and the matrix material plays a more important role.

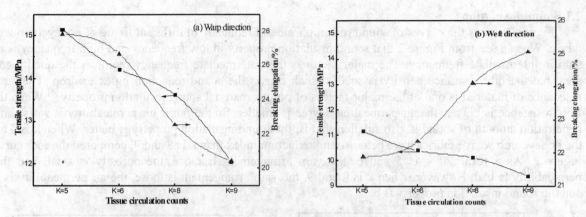

Figure 4 Curves of relation between tensile strength and elongation at break of composite material's sandwich core and fabric tissue circulation count

3 Conclusions

(1) The honeycomb structure of tissue circulation count has tremendous influence on the sound insulation performances of composite materials. When the surface density was same, of which k is 5 (Tissue circulation count), the composite materials show the best noise reduction.

(2)The honeycomb fabrics play enhancement role in the composite materials; the yarn density can enhance the tensile strength of composite materials obviously.

References
[1] Zhou Xinxiang. Noise Control and Application. Beijing. Ocean Press, 1999.
[2] LU Enjie, KURAHASHI Naoya, NI Qingqing, eta1. Development of flexible sound-proof materials. Journal of the Society of Materials Science J, 2006, 55(6): 583-588.
[3] WILSON Adrian. Engineered nonwovens used for automotive acoustic insulation [J]. Technical Textiles International, 2006, 15(8): 11-6.
[4] KINO N, UENO T. Evaluation of acoustical and non-acoustical properties of sound absorbing materials made of polyester fibres of various cross-sectional shapes [J]. Applied Acoustics, 2008, 69(7): 575-582.
[5] Hanifi Binici, Orhan Aksogan, Derya Bakbak et al. Sound insulation of fibre reinforced mud brick walls [J]. Construction and Building Materials, 2009, 23: 1035-1041.
[6] Fu Yaqin, Ni Qingqing, Yao Yuefei, et a1. Sound insulation performance of a glass fibre/PVC composite material [J]. Acta Material Compositae Sinica, 2005, 22(5): 94-99.
[7] Yao Yuefei, Luo Yongbo, Gao Lei, et a1. PVC sound insulation material filled with steel—smelting scoria [J]. Acta Material Compositae Sinica, 2008, 25(2): 74-79.
[8] Zhang Bangjun, Huo Guoqing. Environmental Noise Science[M]. Hangzhou: Zhejiang University Press, 2001, 258-261.

48 Synthesis of Nitrogen/phosphorus/silicon Composite Flame Retardant and Its Charring Properties in Polyester

Lewei Zhang, Huapeng Zhang[*], Jianyong Chen

(Key Laboratory for Advanced Textile Materials and Manufacturing Technology, Ministry of Education, College of Materials, Zhejiang Sci-Tech University, Hangzhou 310018, China)

Abstract: A new charring agent and flame retardant was first designed and synthesized by using 9,10-dihydro-9-oxa-10-phosphaphenanthrene 10-oxide (DOPO) and KH560 with triphenylphosphine as catalyst. A transparent nitrogen/phosphorus/silicon flame retardant resin was then obtained by combination of Benzoguanamine with the product of DOPO and KH560 under 180°C, and the reaction product was investigated by FTIR. By blending this nitrogen/phosphorus/silicon flame retardant resin with polyester, a flame retardant and anti-dripping polyester was obtained, and the flame retarding performance, thermal and charring properties of this polyester were characterized by LOI, TG, SEM and EDS.

Keywords: Flame retardant; Polyester; Charring; Anti-dripping

1 Introduction

Flame retardant is one of the most important additives in functional polymer material, and it is necessary for the flame retardant with no toxic release and no environmental pollution[1]. Among all the flame retardants, flame retardants such as phosphorus–halogen mixtures, ammonium phosphate, and organic phosphorus compounds have been used to impart flame retarding function to ordinary polymers. For quite a long time, the flame retardant with halogen was very popular in flame retarding, but halogens produce problems of smoke, corrosion and carcinogenicity. Therefore it is necessary to develop a new flame retardant to meet these demands[2].

DOPO is a new functional flame retarding intermediates. The flame retardants synthetized with DOPO and its derivatives have the advantages of halogens-free, no smoking, no toxicity, and durable flame retardance, which is an available flame retardant in polyester, polyamide and polyurethane, by the introduction of phosphorus[3].

As the derivative of melamine, benzoguanamine is richer in nitrogen content and is an excellent flame retardant for materials. Compared with melamine, benzoguanamine contains benzene ring, and has higher melting point, which greatly increases its scope of use as a flame retardant.

New intumescent flame retardant mainly contains nitrogen, phosphorus elements in composition, with halogen-free low toxicity, high efficiency in flame retarding and excellent compatibility with other resins, so it is widely used in polymer modification[4,5]. Based on the reaction characteristics of DOPO, a new flame retardant was first synthesized by DOPO with KH560 in this article, and benzoguanamine was then added to this flame retardant, to acquire a nitrogen/phosphorus/silicon composite flame-retardant resin. Then a flame retardant and anti-dripping polyester was obtained by blending with the nitrogen/phosphorus/silicon composite flame-retardant resin.

2 Experimental

2.1 Materials and reagents

DOPO (9,10-dihydro-9-oxa-10-phosphaphenanthrene10-oxide), KH560, triethylamine(TEA), benzoguanamine (BG), triphenylphosphine(TPP), glacial acetic acid(HAc), hydrobromic acid(HBr) were analytical grade, polyester chip with intrinsic viscosity of 7.6 dl/g.

2.2 Preparation of P/Si, N/P/Si Flame Retardant and the blended polyester resin

23.6g DOPO was added into a 250ml three-neck flask, heated to melt and maintained the temperature at 130°C and 180°C, then 21.6g KH560 and a certain amount of TPP as catalyst was added. After certain time a yellow transparent viscous liquid of P/Si flame retardant was obtained. The reaction scheme is illustrated in Figure 1.

---

* Corresponding author's email: zhp@zstu.edu.cn

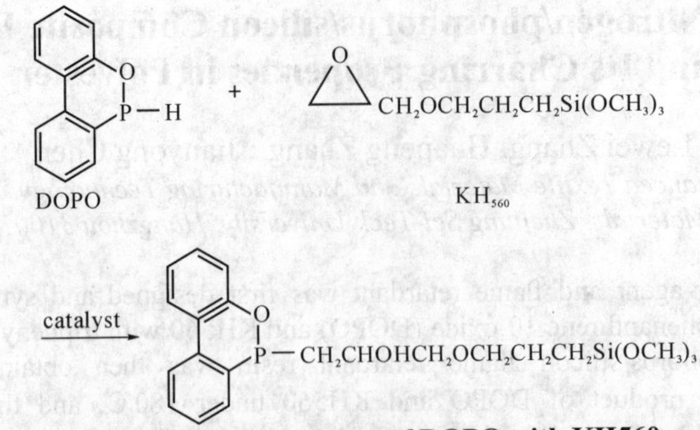

Figure 1 Reaction scheme of DOPO with KH560

BG was added to the phosphorus/silicon flame retardant with the mass ratio of 1:1 at temperature 180℃ for 30min under agitation, and heated to 220℃, then 5ml triethylamine was added dropwise[6], during which a vacuum pump pumped for 15min to remove air bubbles. After cooling, the resultant product was a yellow transparent resin of N/P/Si flame retardant.

The composite N/P/Si flame retardant was ground and then mixed with polyester chips in different mass ratio with a blender, then the blended polyester was injection molded into a sample sized 100mm* 20mm* 2mm (length*width*thickness) at temperature of 270℃.

2.3 Measurements

The reaction between DOPO and KH560 was characterized by Infrared Spectroscopy FTIR with a Nicolet5700 in the range 4000–400cm^{-1}. The thermal properties of the nitrogen/phosphorus/silicon composite flame retardant resin and the blended polyester were tested by thermal gravimetric analysis (Pyris Diamond TGA) with N_2 flow rate 20ml/min, and heating rate 20℃/min. The surface and surface elements of the burned flame retardant polyester was investigated with JSM-5610LV scanning electron microscope equipped with OXFORD instrument corporation INCA alpha ray spectrometer (EDS).

3 Results and discussions

3.1 Synthesis of P/Si flame retardant

The P-H bond is connected with P-O, a strong electron withdrawing group, so the P-H group can react with some active group such as epoxide group. Based on the reaction scheme between DOPO and KH560, the mole ratio of DOPO to KH560 was set as 1:1, and the reaction temperature, catalyst dose and reaction time were experimentally determined. Epoxy group mole content was analyzed by titration method as ASTM D1652, where 0.1M HBr in HAc solution was used. The reaction degree of DOPO and KH560 was expressed by residual epoxy group content, and the initial epoxy group content of KH560 was assumed as 100%. Figure 2 and Figure 3 give the reaction degree of DOPO and KH560 under different reaction conditions.

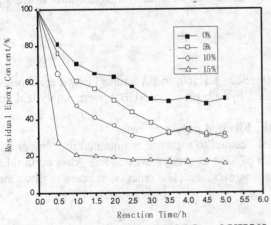

Figure 2 The reaction degree of DOPO and KH560 at 180℃ 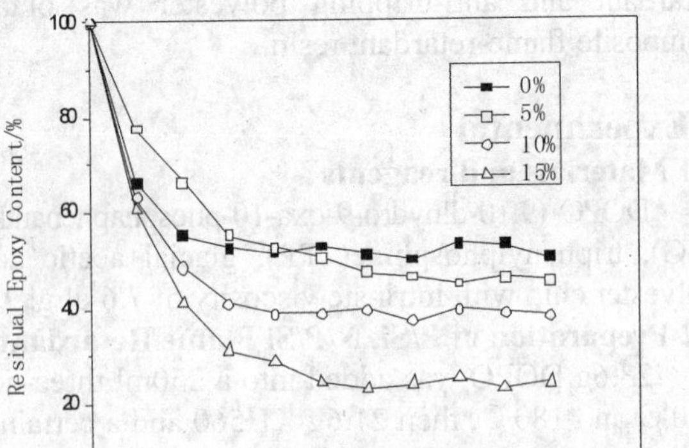   **Figure 3 The reaction degree of DOPO and KH560 at 130℃**

Mass ratio of catalyst from 1 to 4 is 0%, 5%, 10%, 15%

From Figure 2 and Figure 3, it was shown that the reaction degree was very low (43%) without catalyst, and reached 83.7% with 15% catalyst. As far as the reaction temperature was concerned, the reaction degree was higher at 180℃ than at 130℃. The optimal conditions for preparation of the flame retardant were (DOPO):(kh560) mol ratio 1:1, reaction temperature 180℃, reaction time 90min and mass ratio of catalyst 15%.

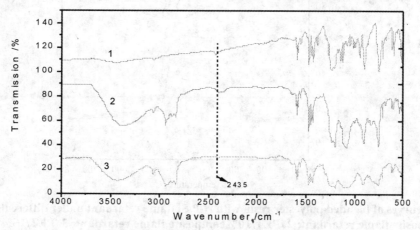

Figure 4 FTIR of P / Si Flame Retardant
1. FTIR of DOPO; 2. Reaction of DOPO and KH560 after 30 min; 3. Reaction of DOPO and KH560 after 90 min

FTIR spectroscopy is used to investigate the reaction product as giving in Figure 4. The P-H absorption band at 2435 cm^{-1} showed a slightly change in Figure 4(2), and disappeared in Figure 4(3). OH absorption band at 3400 cm^{-1} appeared in Figure 4(2) and Figure 4(3), which indicated the reaction of P-H with epoxy group.

3.2 Characterization of nitrogen/phosphorus/silicon composite flame retardant

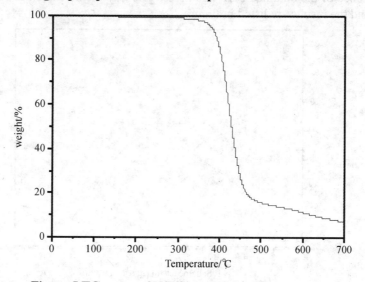

Figure 5 TG curve of N/P/Si composite flame retardant

The thermal gravimetric analysis result of the nitrogen/phosphorus/silicon composite flame-retardant resin is given in Figure 5, from which it was seen the weight loss temperature of 1% and 5% was 290℃ and 380℃, the resin showed rapid weight loss with peak temperature at 410℃, showing better thermal stability. After the temperature raised up to 500℃, the resin was decomposed completely and the residual weight was 11%.

3.3 Performance of blended flame retardant polyester resin

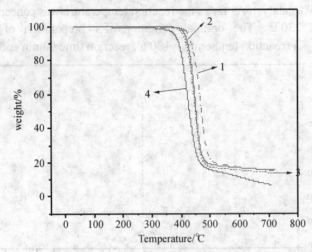

Figure 6 TG curves of blended polyester resin with N/P/Si flame retardant under different blending ratio 1. PET/5%composite flame retardant; 2. PET/10%composite flame retardant; 3. PET/15%composite flame retardant; 4. composite flame retardant

The thermal behavior was investigated by thermo gravimetric analysis (TGA) and the result was given in Figure 6. From Figure 6, PET/15% composite flame retardant blend began to decompose at about 380℃, after the temperature raised up to 480℃, the resin was decomposed completely and the residue was 17%, and meanwhile, PET/5% composite flame retardant blend began to decompose at about 420℃; the temperature raised up to 480℃, the resin was decomposed completely.

The flame retardance of the obtained PET was investigated by the measurement of their limiting oxygen index (LOI) values, and the results are listed in Figure 7.

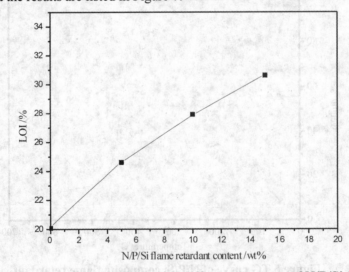

Figure 7 LOI values of the blended PET with different contents of N/P/Si flame retardant

The LOI value of PET was raised from 20.1 to 30.6 with 15% weight of N/P/Si flame retardant, demonstrating a significant improvement with the addition of the N/P/Si flame retardant. The anti-dripping effect of the blended PET is given in Table 1.

Table 1 Anti-dripping effect of the N/P/Si composite flame retardant

	PET	PET/5% composite flame retardant	PET/10% composite flame retardant	PET/15% composite flame retardant
droplets/min	19	12	6	0

The blended PET sample with 15% weight of N/P/Si flame retardant was burned and then the residue surface was observed by SEM, which is illustrated in Figure 9.

Figure 8 Charring layer of pure polyester

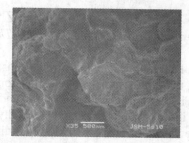
Figure 9 Intumescent layer structure of flame retardant polyeste

Charring layer on the surface demonstrated a large number of bubble-like protrusions as in Figure 9, indicating a very good flame retardant expansion foaming function. The Si-O-C group in the resin can hydrolyze and then crosslink into three-dimensional network structure, preventing polyester melting down, thus obtaining a better charring. Figure8 is the SEM of pure polyester dripped down.

4 Conclusions

A new charring agent and flame retardant was designed and synthesized by using 9,10-dihydro-9-oxa-10-phosphaphenanthrene10-oxide (DOPO) and KH560 in combination with benzoguanamine to obtain a new N/P/Si composite flame retardant. Polyester blended with 15% weight content of this N/P/Si composite flame retardant showed good flame retardance and anti-dripping charring effects.

References
[1] Tanaka Y. In: May CA, editor. Epoxy resins chemistry and technology. New York: Dekker; 1988, 719-82.
[2] Wang C S, Shieh J Y. Phosphorus-containing epoxy resin for an electronic application. Journal of Applied Polymer Science, 1999, 73:353-361.
[3] Ying Ling Liu, Yie Chan Chin, Chuan Shao Wu. Preparation of silicon-phosphorus-containing epoxy resins from the fusion process to bring a synergistic effect on improving the resins thermal stability and flame retardancy. Journal of Applied Polymer Science, 2003, 87:404-411.
[4] Levchik S V, Well E D. Overview of the recent literature on flame retardancy and smoke suppression in PVC. Polymers for Advanced Technologies, 2005, 16(10):707-716
[5] Camino G, Casis L, Martinasso G. Intumescent fire-retardant systems. Polymer Degradation Stability, 1989, 23(4):359-376.
[6] Z D Du, Higher Education Reference Textbook Chemistry of Organic Silicone. China Higher Education Press, 1990:267

49 Computer Evaluation System of Woven Fabric Smoothness Based on 2D Wavelet Transform

Yifan Zhang, Ameersing Luximon*

(Institute of Textiles and Clothing, The Hong Kong Polytechnic University, Hong Kong, China)

Abstract: Fabric appearance is always considered to be one of the most important aspects of fabric quality. In this paper, two-dimensional wavelet transform is used to analyze woven fabric smoothness characteristics like wrinkle. Firstly, fabric image is pretreated and decomposed by wavelet transform; from which the high frequency information is extracted. Secondly, four kinds of wrinkle feature parameters are applied to calculate the feature values of fabric wrinkles with different AATCC templates. Finally, smoothness grade of different types of fabrics were evaluated according to this result by using minimum distance classification. For describing the assessment result quantitatively, the correlation coefficient is calculated between objective assessment and subjective assessment to validate the feasibility of this method. A high correlation of 0.9117 was obtained indicating the usefulness of the wavelet transform for wrinkle evaluation.

Keywords: Woven fabric smoothness, Grade assessment, Wavelet analysis, Feature values extraction

1 Introduction

Fabric appearance is always considered to be one of the most important aspects of fabric quality. How to evaluate the grade of woven fabric wrinkles is an important issue for woven fabrics. Domestic fabric smoothness assessment is primarily tested under the American standard of AATCC 124-2001 [1]. Wrinkle properties of samples are commonly being assessed with standard cards in the standard lighting conditions through visual comparison. The GB/T 13796-1992 [2] is based on the American AATCC standard and still uses the AATCC sample cards; the working environment and assessment methods are similar to AATCC. However, this method is subjective and liable to experimental error.

In recent years, as a means of dealing with image, digital image processing technology has successfully made its debut in textile industry [3-5]. These techniques can be applied for image transformation and feature extraction subsequently to analyze fabric smoothness. Amirbayat and Alagha [6] used triangulation theory and non-contact laser scanning method for detecting fabric exterior grade, identifying the high point through the three points of the geometric objects imaging of the firing point out, projection and imaging point. They calculated the geometric features, such as average length, surface area, volume and the degree of distortion, etc. These indicators have a good correlation with the smoothness grade [6]. Combining band-pass filter and laser scanning technology in addition, proved to be a good evaluation on the assessment tool for analyzing wrinkles and seams smoothness. 2D band-pass filter can avoid fabric surface texture and noise impact; thereby effectively extracting the wrinkle contour shape [7, 8]. 3D laser scanning through progressive scans can be more accurate to acquire 3D fabric contours, high accuracy can be achieved (± 0.01mm), but the process of scanning is slower and the equipment is expensive. In this paper, the image processing technique using wavelet transform is implemented to analyze wrinkles. The woven fabric smoothness after laundering is evaluated by digital image processing.

2 Computer evaluation system of objective assessment

The system of objective assessment includes two parts: hardware and software. The hardware part is the fabric image acquisition device; the software employed Matlab and Visual C++ 6.0 programming environment. Figure 1 shows the diagram of system structure.

---

* Corresponding author's email: tcshyam@inet.polyu.edu.hk

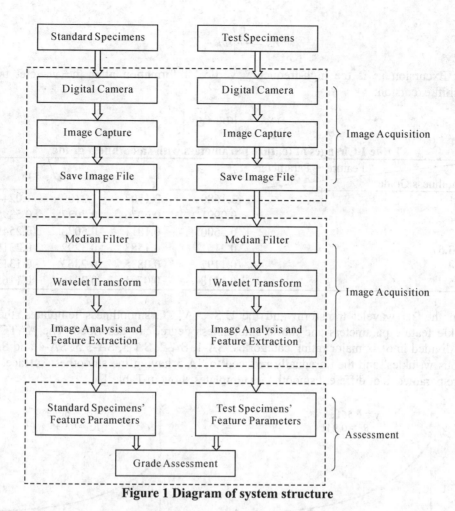

Figure 1 Diagram of system structure

3 2D WT

One of the main advantages of wavelet analysis is that it provides the capacity of local analysis by refining and redefining it. Compared with the traditional signal analysis technology, wavelet analysis are cable of doing the signal compression and de-noising with no apparent loss [9]. The applications of wavelet transform in image processing are mostly done by 2D discrete wavelet transform [10-11].

4 Results and discussion

After wavelet transform, feature parameters that reflect fabric smoothness degree are defined by four kinds of wrinkle feature parameters: horizontal aberrancy, vertical aberrancy, horizontal shift excursion and vertical shift excursion. These four kinds of feature parameters are derived from statistical approach. The aberrancy is obtained by calculating the unbiased variance and shift excursion by calculating the one-order central moment of high-frequency information. The brief descriptions of the four kinds of feature parameters are as follows [12]. Horizontal Aberrancy ha is the high-frequency vertical information after the wavelet transform, then calculate the aberrancy:

$$ha = \frac{1}{N}\sum_{i=1}^{N}(dh_i - \overline{dh_i})^2 \qquad (1)$$

Vertical Aberrancy va is the high-frequency vertical information after the wavelet transform, then calculate the aberrancy:

$$va = \frac{1}{N}\sum_{i=1}^{N}(dv_i - \overline{dv_i}) \qquad (2)$$

Horizontal Shift Excursion hs is the high-frequency horizontal information after the wavelet transform, then calculate the shift excursion:

$$hs = \frac{1}{N}\left|dh_i - \overline{dh_i}\right|$$

(3)

Vertical Shift Excursion *vs* is the high-frequency vertical information after the wavelet transform, then calculate the shift excursion:

$$vs = \frac{1}{N}\left|dv_i - \overline{dv_i}\right|$$

(4)

Table 1 Changes of feature parameters with smoothness grade

Smoothness Grade	ha	va	hs	vs
SA-1	3.1680	2.1507	1.2109	1.0244
SA-2	0.9991	0.6076	0.7037	0.5395
SA-3	0.5600	0.1811	0.5078	0.2548
SA-3.5	0.2161	0.1583	0.3292	0.2311
SA-4	0.1192	0.0835	0.2363	0.1313
SA-5	0.1118	0.0721	0.2215	0.1165

Based on the 2D wavelet transform and the U.S. AATCC smoothness templates, the relationships between wrinkle feature parameters and fabric smoothness degree are described. The AATCC smoothness templates are divided into six major rating categories: SA-1, SA-2, SA-3, SA-3.5, SA-4 and SA-5. SA-1 has the most serious wrinkles, and then gradually reduced to SA-5 having the smoothest surface. Four kinds of wrinkle feature parameters of different smoothness templates show in Table 1.

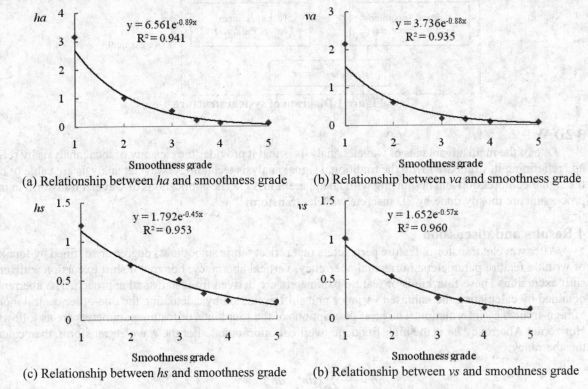

(a) Relationship between *ha* and smoothness grade (b) Relationship between *va* and smoothness grade

(c) Relationship between *hs* and smoothness grade (b) Relationship between *vs* and smoothness grade

Figure 2 Correlation between feature parameter and smoothness grade

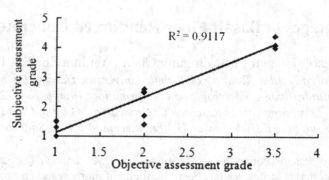

Figure 3 Correlation between objective assessment and subjective assessment

From Table 1, it can be seen that the SA-1 has the most serious wrinkle and the largest feature parameters; the more serious the wrinkle is, the larger parameter it has and vice verse. Correlation coefficient (R^2) is calculated between wrinkle feature parameters and smoothness grades. The results are shown in Figure 2. We selected 13 species with different levels of smoothness of woven fabrics as test subjects. After extracting four kinds of wrinkle feature parameters from each fabric, the wrinkle grades is calculated and classified by the minimum distance classification. Thereafter three experts' subjective assessment results have been taken. The results are shown in Figure 3.

5 Conclusions

Woven fabric image can be filtered through Median filter, and decomposed by wavelet transform; meanwhile, high frequency information is extracted. Four kinds of wrinkle feature parameters could successfully applied to calculate the fabric smoothness feature values with different smoothness templates. Minimum distance classification can be used to evaluate the smoothness grade of 16 different types of knitted fabrics. The correlation coefficient between the subjective assessment and objective assessment was found to be 91.17%. These research results indicate that this method can be successfully applied for wrinkle evaluation of textiles.

Acknowledgement

This work was supported by the Hong Kong Polytechnic University PhD studentship (RU2J).

References

[1] AATCC Test Method 124-2001. Appearance of fabrics after repeated home laundering.
[2] GB/T 13769-1992. Evaluation method of durable press fabric after home washing and drying.
[3] Yu W, Xu B. A sub-pixel stereo matching algorithm and its applications in fabric imaging. Machine Vision and Applications, 2009, 20 (4): 261-270
[4] Abidi N, Hequet E, Turner C, Sari-Sarraf H. Objective evaluation of durable press treatments and fabric smoothness ratings. Journal of Applied Polymer Science, 2005, 96 (2): 392-399.
[5] Turner C, Sari-Sarraf H, Hequet E, Abidi N, Lee S. Preliminary validation of a fabric smoothness assessment system. Journal of Electronic Imaging, 2004, 13 (3): 418-427
[6] Amirbayat J, Alagha, M J. Objective assessment of wrinkle recovery by means of laser Triangulation. Journal of Textile Institute, 1996, 87(1): 349-355.
[7] Fan J, Liu F. Objective evaluation of garment seams using 3D laser scanning technology. Textile Research Journal, 2000, 70(11): 1025-1030.
[8] Fan J, Lu D, Macalpine J M, Hui C L. Objective evaluation of pucker in three-dimensional garment seams. Textile Research Journal, 1999, 69(7): 467-472.
[9] He D J, Geng N, Zhang Y K. Digital Image Processing: Xi'an Electronic and Science University Press, Xi'an, 2005: 165-166.
[10] Cui J T. Wavelet Analysis Introduction: Xi'an Electronic and Science University Press, Xi'an, 1997: 24-28.
[11] Zhang J W, Jiang D Z, Zheng C X. Application of wavelet technology in biomedical imaging. Fascicule of Foreign Medicine in Biomedical Engineering, 1997, 20(4): 204-209.
[12] Yang X B. Study on fabric crease based on 2D wavelet analysis. Proceedings of Suzhou University, 2004, 24(2): 20.

50　Advantages of Basalt Fiber Reinforced Concrete Structures

Dangfeng Zhao[1], Huawu Liu[1], Zhigang Chen[2], Yinhua Zhang[3], Dangqi Zhao[4]

(1. School of Textiles, Tianjin Polytechnic University, Tianjin 300160, China
2. Tianjin Institute of Electronics and Information, Tianjin 300132, China
3. Yueyang Textile Research Center, Hunan 414200, China
4. China Water Resources and Hydropower 11th Engineering Bureau, Sanmenxia 472000, China)

Abstract: A severe weakness of concrete is interior and surface cracks. Preventing the occurrence and propagation of cracks in concrete structures has been the focus of many research projects. Fiber reinforcement is a commonly used method to control crack propagation and enhance the tensile strength of concrete structures. Continuous basalt fiber exhibits high performance in terms of compatibility with cement, tensile strength, stiffness, resistant to erosion, durable to physical wear and aging. The key properties of basalt fiber and other reinforce fibers of concrete, were evaluated to state the advantages of basalt fiber as a new concrete reinforce material. Basalt fiber is better than any other concrete reinforcing fibers in terms of its chemical stability, environment friendly nature, stiffness, strength and compatibility with cement. With the forbid of fiberglass and price soaring of petrol chemical and steel fibers, the application of basalt fiber will increase significantly and benefit civil construction industries.

Keywords: Basalt fiber; Steel and polypropylene fibers; Fiber reinforced concrete.

1 Introduction

Surface and internal cracks lead to the reduction of physical properties of engineering materials. The tensile strength may increase more than 10 times, when the surface cracks of glass material are removed [1,2]. Preventing the occurrence and propagation of cracks in concrete structures has been the focus of many research projects. In order to control crack propagation and enhance the tensile strength of concrete structures, fibers have been applied as reinforcement since 1910. The commonly used reinforcing fibers are glass fiber, carbon fiber, steel fiber, synthetic fibers (polypropylene, nylon, polyethylene, PVA, etc.). Carbon and aramid fibers are too expensive to make them acceptable in construction industries. In addition, carbon and glass fibers both age quickly in the alkali concrete environment. Thirdly, glass and asbestos fibers have been known as white pollutions and barely are applied nowadays [3]. The main fibers used in civil construction are largely polypropylene and steel fibers, though they are not active to cement [4]. Both polypropylene and steel fibers are physically anchored inside the concrete and the interfaces between fibers and concrete are incompatible. Yang Xudong et al. revealed that the strength of polypropylene geotextile decreased to 60% within one year, when the samples were exposed to sunlight [5,6]. We may conclude that polypropylene fibers are not suggested to be used outdoor. Ordinary steel fiber is also a non-active material to cement, easy to be drawn out from the concrete. Owing to their incompatibility, the volume fractions of the currently reinforcing fibers, including steel and polypropylene, are no more than 2%.

The ideal reinforcing fiber should meet the following criteria: high index of activity to ensure an excellent interface adhesion between fibers and concrete, environment friendly for the safety of end users, alkali resistant to fit the concrete interior environment, high stiffness and high strength to bear more load, anti-aging and resistance to physical wear for a long lasting concrete structure. Basalt fiber has many advantages, such as high-strength, high stiffness, thermal insulation, erosion resistant, low moisture sorption, cheap, no harm to the environment so on [7,8]. In addition, basalt fiber and concrete have similar composition and density, thus best dispersion inside of concrete, compared with any other reinforced fibers. As the stated advantages stated above, basalt fibers may be a promising reinforcing fiber for concrete structures.

2 Basalt fiber

The research and development of basalt filament may be traced back to 1922, the preparation of basalt glass-ceramic fiber patented in United States [9]. The furious competition of US and former Soviet Union Defenses in mineral fiber researches resulted in a rapid growth of basalt fiber development in 1960s. It was military confidential of US before the declassification in 1995. Continuous basalt fiber was commercialized in 1990s in Russia and Ukraine [10-12]. Continuous basalt fibers was introduced to China around 2004-2005, through the National 863 Research Plan.

2.1 Characterization of basalt fiber

Basalt rocks contain compositions, such as SiO_2, Al_2O_3, CaO, MgO, Fe_2O_3, FeO, TiO_2, K_2O, Na_2O, and a small amount of impurities. The main compositions are SiO_2, Al_2O_3, CaO and MgO respectively [13]. SiO_2 and Al_2O_3 are the skeleton and the active compositions, accounting for more than half of the total materials. SiO_2, Al_2O_3, TiO_2 MnO and Cr_2O_3 are favorite to chemical stability. The increase of RO contents may lead to lower working temperature, whereas the increase of Fe_2O_3 content significantly rises the working temperature. The basalt fiber is brown or bronze with smooth cylindrical appearance (Figure1), which is because the melting basalt shrinks into the smallest circular circumference due to surface tension.

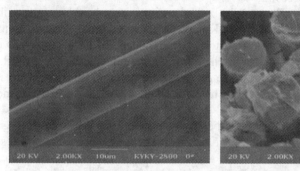

Figure 1 SEM images of basalt fiber and its cross section

2.2 The activity index of basalt fiber

Activity index measures affinity and the content of SiO_2 and Al_2O_3 of basalt fiber, which react with the alkali components of cement on the interface. The higher activity index, the better the composite interface. The composition of a group of basalt rock is shown in Table 1.ho w

Table 1 The major components of a basalt rock

>1% rock composition	SiO_2	Al_2O_3	MgO	CaO
Content wt%	49.16	14.04	7.31	6.68
>1% rock composition	Na_2O	K_2O	Fe_2O_3+FeO	
Content wt%	3.32	2.17	7.81	

Sample rocks were grinded into marble with size below 5mm. Then melted at 1500℃ and drawn into filament. The basalt filament was crushed into particle on ball mill, the size of the particle is below 5 microns, shown as Figure 2. The powder was made into concrete with cement, the activity index on the 7th and 28th days were 38% and 53%, respectively.

Figure 2 SEM images of basalt particle

Standards GB 1344-1999 state that the mass ratio of volcanic ash is between 20% and 50% in the materials mixed by Portland cement and volcanic ash, which means that the volume ratio of basalt fiber in the cement may be up to 50%. The specific addition quantity was determined by the ratio of active components, namely silica and aluminum oxide. Thus basalt fiber reinforced cement composites are far superior to other non-active chemical fiber and steel fiber.

2.3 Environmental friendly basalt fiber

There is no boron and other alkali metal oxides discharged during the basalt rock melting process. No harming substances are discharged from furnace during processing. The waste glass after melting was called clinker. The glass industry usually adds a significant proportion of clinker into the raw material in order to reduce the energy consumption, which means the waste materials are reusable immediately. In addition, the waste material can be grinded into powder, called volcanic ash fertilizer. Summarizing stated above, there is no pollution and toxic wastes during processing.

People may worry about that basalt fiber would be harm to people's health as glass fiber and asbestos fibers when basalt fiber was in its infancy. In 1994, Kogan FM and Nikitina [14] put a number of mice six months in the environment containing 25mg/m^3 asbestos fibers powder and basalt fiber powder respectively. When the concentration of asbestos fibers powder reach 1.7g/kg in the mice bodies, one third of the mice died. When the concentration reach 2.7g/kg, all of the mice died. While the concentration of basalt fibers powder reached 10g/kg, all the mice were still alive. Other researchers also found basalt fiber did not cause any damage to animal or air quality [15, 16].

Basalt is consolidated magma before forming fiber filaments and the gases causing greenhouse effect were emitted millions years ago. The discarded basalt fibers could be weathered into natural soil. Furthermore, basalt is 100% inert, have no toxic reaction with air and water. Thus basalt fiber is a unique environment friendly reinforcing fiber for concrete, without pollution and carcinogenesis [8].

2.4 The important mechanical properties of basalt fiber

Basalt fiber is a high stiffness material, with average elastic modulus 93GPa, tested at China National Test Center of Building Materials (Table 1) and the variation is from 89-110GPa (Table 2).The elastic modulus of polypropylene is between 0.5 and 1.3GPa [17], which is two orders of magnitude lower than basalt fiber. The tensile strength of basalt fiber is between 3500 and 4840MPa, three orders of magnitude higher than that of ordinary concrete (1-4MPa) and one order of magnitude higher than that of polypropylene (894.9MPa) [17]. Concrete reinforced by basalt fiber may reduce crack propagation and enhance the tensile strength of concrete structures. In addition, basalt fiber is a kind of wear-resistant material, with Mohs hardness 7.5 (Table1). Hence, basalt fiber may significantly improve the durability of concrete. The mechanical performances of basalt fiber are as good as other high performance fibers and key mechanical properties of high performance fibers are shown in Table 2.

Table 2 The comparison of mechanical properties of basalt fiber and other high performance fibers[17]

	S275 steel	E-glass fiber	High strength carbon fibers	Aramid fiber	Basalt fiber	Ultra-high molecular weight polyethylene
Tensile strength /MPa	275-430	3100-3800	3300 - 6370	2700 -3000	3000 - 4840	3000 - 4840
Elastic modulus /GPa	205	73-78	230 - 300	124- 130	89-110	100
Density kg/m^3	7900	2560	1800	1380	2750	970

2.5 The alkali resistance performance

The alkali resistance of basalt, polypropylene and glass fibers were shown in Figure 3. These fibers were immersed into 2 mol / L sodium hydroxide solution at 80℃ for 2, 4, 6 and 8 hours, respectively. Observed changes of fiber surfaces in different time are shown in Figure4. It may be concluded that the alkali resistance of basalt fiber is superior to glass fiber, but worse than polypropylene. However, anti-aging performance of polypropylene is worse than that of basalt fiber and the tensile strength of polypropylene decreased up to 60% within one year when exposed to outdoors.

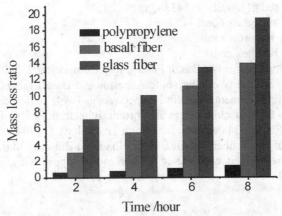

Figure 3 Mass loss ratios in different periods

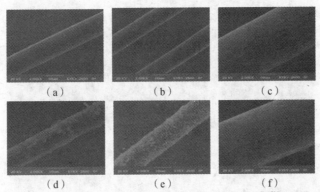

Figure 4 SEM images of untreated and treated filaments

Note: The untreated fiber was basalt fiber, glass fiber and polypropylene from a to c, respectively. The fiber treated by 2 mol / L NaOH for 4 hours was basalt fiber, glass fiber and polypropylene from d to f, respectively.

3. Conclusion

Basalt fiber is recently commercialized in China and its advantages as a concrete reinforcing material have not been recognized by the civil engineers. The compositions, activity index, environmental performance, mechanical properties and alkali resistance were summarized in this study. From the comparisons of key properties, it may conclude that basalt fiber is better than any other currently used concrete reinforcing fibers in terms of its chemical stability, environment friendly nature, stiffness, strength and compatibility with cement. With the forbid of fiberglass and price soaring of steel and petrol chemical fibers, the application of basalt fiber will eventually increase and benefit civil construction industries.

References

[1] X Z Zhao, H R Yin. Glass technology. Beijing: Chemical Industry Press, 2006:123
[2] Loewenstein, K.L. Continuous glass fiber manufacturing process. Beijing: China Standard Press, 2008:25-26
[3] Y X Yong, Q R Yue. Some problems of Basalt fiber and its application. Industrial Construction J, 2007, 37(6):1-4
[4] Y Gong, Z J Xu, Fiber reinforced concrete and fiber mortar application guide. Beijing: China Building Industry Press, 2005:25-28
[5] X D Yang, X Ding, et al. Natural aging behavior of polypropylene geotextile. Donghua University J, 2007, 33(1):57-61
[6] X Yang, D Xin. Prediction of outdoor weathering performance of polypropylene filaments by accelerated weathering tests. Geotextiles and Geomembrance J, 2006,24(2006):103-109
[7] Tamás Deák. Chemical Composition and Mechanical Properties of Basalt and Glass Fibers: A Comparison, Textile Res J, 2009. 79(7): 645-651
[8] Z L Zhong, H W Liu. Study on modification of basalt filaments weaving ability. Industrial Textiles J, 2008,(2):33-36.

[9] Paul Dhe, Filament composed of basalt, US1438428, 1922
[10] Information on http:/www.sudaglass.com
[11] Information on http:/ www.basaltfm.com
[12] Information on http:/www.basaltex.com
[13] E G Xie, Z Y Li. Application Prospect of Basalt Fiber. Fiber Composites J, 2003, (9):17-20.
[14] Kogan F M, Nikitina O V. Solubility of Chrysotile Asbestos and Basalt Fibers in Relation to their Fibrogenic and Carcinogenic Action. Environmental Health Perspectives J. 1994, 102(5): 205-206.
[15] Mcconnell E E., Kamstrup O, Musselman R, et al. Chronic Inhalation Study of Size-Separated Rock and Slag Wool Insulation Fibers in Fischer 344/N Rats. Inhalation Toxicology J. 1994, 6(6): 571-614.
[16] D Q Ye. The comparison of basalt fiber and glass fiber. Shanghai Building Materials J, 2006 (6):8-9
[17] Y Gong, Z J Xu. Fiber reinforced concrete and fiber mortar application guide. Beijing: China Building Industry Press, 2005:28-29.

51 Fabric Drape Prediction and Simulation

Hua Zhou, Yanfang Shao, Quan Wen

(College of Materials and Textiles, Zhejiang Sci-Tech University, Hangzhou, Zhejiang 310018, China)

Abstract: In order to achieve fabric drape shape prediction and simulation, in this paper, we get the mechanical and basic structural parameter properties of fabrics by a large number of experiments, and receive a series of forecast equations about drape shape through linear regression analysis. The regression equations are verified by the computer programming. The results showed that, fabric drape shape are closely related to the mechanical and basic structure parameter properties, especially the shear rigidity in the direction of 45°, the interweave resistance, the thickness and the weft density, their differences will cause tremendous changes in appearance of fabric drape, and influence the sense of overhanging beauty.

Keywords: Fabric; Drape shape; Parameter properties; Mechanical properties; Prediction

0 Introduction

Fabric drape shape is formed by dead weight which hung down smooth and uniform surface curvature form, leads to physical and psychological reflection, is a key factor of affecting the appearance of the fabric [1]. So far, the research for mechanical and structural parameters having the influence on the degree of fabric drape has reached a deep level, but influence on the fabric drape shape is rarely [2,3]. Research on the drape shape prediction related to the development of new fabrics, fashion design choice of materials, production, computer simulation, even to stand-alone or network-based three-dimensional fitting system, and clothing sales and other aspects. Therefore, this article will combine the mechanical and structural parameters of fabric with the drape shape, identify the key factors that affect the drape shape, provide the valuable reference of fabric drape prediction and computer simulation, CAD design clothing.

1 Experimental program

1.1 Experimental materials preparation

Different factors like composition and structure of the fiber fabric, yarn fineness and structure, the structure of fabric will affect the performance of fabric drape [4]. This paper we use 35 kinds of polyester fabric for the basic study, including plain, twill, satin basic ternary organization. The experiments which get the parameters of fabric thickness, square weight, bending rigidity and others are in the temperature of $20 \pm 2\,°C$, relative humidity of $65 \pm 2\%$ for 24 hours, in accordance with "The People's Republic of China textile industry standard FZ65001-1999" test values.

1.2 Laboratory Instruments

Digital fabric thickness gauge (YG (B) 141D), digital electronic display balance (BS124S), fabric dynamic drape style instrument (YG (L) 811-DN), fabric bending rigidity tester (siroFAST- 2), fabric shear rigidity tester (siroFAST-3), fabric strength tester (YG065).

1.3 Select the experimental index

From the perspective of simulation of drape shape [5], the average peak height, the average peak width, the average valley height, wave number, angle of irregularity, the peak of height irregularity, valley height irregularity, peak width of irregularity[6], enough to simulate fabric drape shape, so the eight indicators were studied as the dependent variable Y.

Fabric mechanical index and structure parameter as follows:

X1 shear stiffness (N/m),   X2 weft bending rigidity (uN·m),
X3 warp bending rigidity (uN·m),   X4 weft cutting resistance (N),
X5 warp interweave resistance (N),   X6 square weight (g/m^2),
X7 thickness (mm),   X8 weft density (number/10cm),
X9 warp density (number/10cm),   X10 weft fineness (tex),
X11 warp fineness (tex),   X12 textile weave,

Calculation of textile weave K as follows [7]:

K is determined by the textile weave and used to characterize the degree of interweave. Plain weave fabric is obtained 1, k value of other organizations is expressed as a fraction relative to the plain, and the algorithm is:

$$K = \frac{S}{D} \tag{1}$$

In formula: S—fabrics unit organizational loop along the direction of warp, warp and weft staggered times; D—plain fabric loops along the corresponding organization by the direction of warp and weft yarn interlacing frequency.

Due to space limitations in this list is only part of the data, as shown in Table 1 and Table 2:

Table 1 The data of fabric physical properties

Number	Organization structure	Fineness /tex		Density/nuber /10cm		Thickness /mm	Square Weight /g/m²	Interweave Resistance/N		Bending Rigidity/uN·m		Shear Stiffness/N/m
		warp	weft	warp	weft			warp	weft	warp	weft	
1	2/2 right twill	40.45	49.79	348	252	0.66	262.38	0.47	0.64	14.8	5.55	28.06
2	4/1 right twill	6.17	9.17	904	400	0.16	108.85	0.1	0.16	7.42	1.33	28.93
3	1/3 right twill	37.42	26.08	588	512	0.57	325.34	0.56	0.49	7.71	4.44	27.31
4	plain	36.97	44.09	278	262	0.56	216.5	0.45	0.53	7.29	6.36	28.36
5	plain	13.49	13.43	450	402	0.23	99.8	0.61	0.52	5.63	6.98	31.86
6	plain	10.04	9.53	440	348	0.2	77.54	0.13	0.12	1.01	0.94	25.42
7	plain	41.49	33.37	240	180	0.57	205.14	0.69	0.68	7.49	4.34	25.16
8	2/2 left twill	43.93	41.67	356	296	0.73	284.7	0.74	0.74	12.93	10.74	30.88
9	plain	24.74	25.63	322	318	0.27	161.08	0.56	0.65	4.65	2.85	30.86
10	5-end warp satin step 2	4.27	12.01	1064	354	0.24	84.98	0.02	0.06	2.54	1.19	24.41

Table 2 The data of fabric drape shape indicator

Number	Drape Coefficient	Wave Number	Average Peak Angle	Peak Angle Irregularity	Average Peak Height	Peak Height Irregularity	Average Valley Height	Valley Height Irregularity	Average Peak Width	Peak Width Irregularity
1	0.818	5	72.000	4.192	33.070	5.501	60.313	3.349	17.418	10.441
2	0.763	5	72.000	5.028	35.678	6.305	62.594	3.930	16.823	12.456
3	0.820	6	55.385	5.652	29.771	6.106	60.631	2.915	13.764	9.745
4	0.809	5	72.000	3.869	36.182	5.823	60.738	2.340	16.619	8.968
5	0.679	5	72.000	7.766	22.486	12.633	74.593	4.698	19.661	13.364
6	0.965	6	60.000	5.041	20.794	9.158	60.937	1.286	14.147	14.310
7	0.741	6	65.455	4.792	31.900	5.807	65.437	3.874	16.357	10.900
8	0.768	5	72.000	3.816	31.670	6.341	62.740	3.073	19.542	6.579
9	0.777	5	55.385	6.168	26.073	10.116	65.297	3.106	14.220	10.581
10	0.835	6	60.000	5.347	30.783	7.327	60.194	3.006	14.375	14.684

2 Data analysis

The following will take the average peak height to mechanical and structural parameters for example, do the multiple regression analysis, the remaining indicators drape shape the same as the analysis process to solve the average peak height.

2.1 The average peak height and the Correlation of the parameters

In order to analyze main factors which affect the average peak height, first of all the data are analyzed by SPSS software, and analysis of results from Table 3 can be seen the relationship between the average peak height and shear rigidity, weft bending rigidity, warp bending rigidity, weft interweave resistance, warp interweave resistance, square weight are closer, correlation coefficients were 0.596, 0.634, 0.578, 0.576, 0.531, 0.527. It can be considered indicators of the selected six all having a linear relationship with the average peak height.

Table 3 Variable correlation coefficient

	Y	X1	X2	X3	X4	X5	X6	X7	X8	X9	X10	X11
X1	0.596											
X2	0.634	0.677										
X3	0.578	0.620	0.890									
X4	0.576	0.868	0.676	0.485								
X5	0.531	0.903	0.703	0.419	0.960							
X6	0.527	0.355	0.732	0.756	0.411	0.399						
X7	0.441	0.025	0.625	0.687	0.155	0.123	0.852					
X8	-0.328	-0.09	-0.11	-0.09	-0.23	-0.20	-0.22	-0.31				
X9	0.034	0.150	0.046	0.111	-0.01	0.048	-0.05	-0.21	0.443			
X10	0.365	0.076	0.355	0.384	0.192	0.142	0.627	0.687	-0.678	-0.441		
X11	0.215	0.065	0.403	0.292	0.149	0.109	0.649	0.653	-0.599	-0.593	0.896	
X12	-0.137	0.329	0.069	-0.07	0.291	0.311	-0.25	-0.30	-0.042	-0.178	-0.36	-0.3

2.2 The correlation between the parameters of target

Then observing the impact of the correlation coefficient between variables, some variables have the high degree of correlation. Variables may exist multicollinearity, should remove the relevant variables, in order to ensure the mutual independence among the parameters. Based on this, stepwise regression analysis is used here for variable selection by the SPSS software. By the stepwise selection of variables, we can reach a conclusion: six variables exist multicollinearity, we chose to join the variables for the regression model: X2 weft bending rigidity, X4 weft interweave resistance.

2.3 Model building

On the basis of these variables, the average peak height as the dependent variable, X2 weft bending rigidity, X4 weft interweave resistance as independent variables, do multiple linear regression analysis again to establish linear regression model between them. The regression results in Table 4.

Table 4 Coefficients (The average peak height)

Model		Unstandardized		Standardized Coefficients	t	Sig.	Collinearity Statistics	
		B	Std.Error	Beta			Tolerance	VIF
2	(Constant)	89.022	1.470		60.572	0.000		
	Weft bending rigidity	0.507	0.177	0.455	2.868	0.007	0.693	1.439
	Weft cutting resistance	3.271	1.600	0.324	2.044	0.050	0.695	1.439

The results from Table 4 can get the multiple linear regression equation:
$$Y=89.022+0.507X_2+3.271X_4 \tag{2}$$

In equation the two indicators of tolerance values greater than 0.1, the VIF values are much less than 10, indicating the existence of the two indicators are not linear, can be independently express the average peak height.

2.3.1 Model checking

To test whether the regression equation makes sense, Table 5 is the analysis of variance F statistics. As the model sig <0.01, we believe that the regression equation is statistically significant. Moreover, from Table 5 the partial regression coefficient t, two variables of sig values were less than 0.05, we can consider two independent variables are passed the test of significance, the coefficient of this equation is statistically significant [9].

Table 5 The analysis of variance

Model	Sum of Squares	df	Mean Square	F	sig
Regression	1131.860	2	406.325	13.586	0.000
Residual	1585.253	30	29.908		
Total	2717.113	32			

2.3.2 Other indicators of multiple linear regression equations

Similarly available drape shape of the other indicators of the regression equations:

$$\bar{h} = 89.022 + 0.507X_3 + 3.271X_4 \tag{3}$$

$$\hat{h} = 87.174 + 4.327X_4 + 0.025X_6; \tag{4}$$

$$\bar{d} = 2.437 - 0.068X_8 + 0.042X_1; \tag{5}$$

$$\hat{d} = 16.896 + 0.011X_9 - 6.157X_7 - 0.025X_8; \tag{6}$$

$$\bar{l} = 58.818 + 0.101X_1 + 6.629X_{12}; \tag{7}$$

$$\hat{l} = 8.607 + 0.017X_1 - 0.08X_{11} - 0.014X_8 + 0.001X_9 + 0.043X_3; \tag{8}$$

$$\hat{\theta} = 25.709 + 0.048X_1 - 9.261X_7 + 2.131X_5 + 0.008X_9 - 0.029X_8 - 0.056X_{11}; \tag{9}$$

$$N = 11.09 + 0.001X_8 + 0.328X_5 - 0.209X_1 - 0.016X_3 - 0.17X_4; \tag{10}$$

In equation: $\bar{h}$ —Average Peak Height; $\hat{h}$ —Peak Height Irregularity; $\bar{d}$ —Average Peak Width; $\hat{d}$ —Peak Width Irregularity; $\bar{l}$ —Average Valley Height; $\hat{l}$ —Valley Height Irregularity; $\hat{\theta}$ —Peak Angle Irregularity; N—Weave Number.

2.3.3 Model analysis

From the established model results, we can get the following analysis:

(1) From the regression analysis, between the fabric drape shape and mechanical and structural indicators having a close relationship, we can intentionally change the parameters such as thickness, fineness, density and so on, and change the fabric mechanical index to predict the fabric the drape shape, meet different requirements.

(2) Weft density (X8) has shown a negative correlation to the average peak width, peak width irregularity, valley height irregularity, peak angle irregularity, indicating that the increase of the weft density, the average peak height and the average peak width decreases, overhanging area reduced, vertical fabric to show a good sense; while peak width irregularity, peak angle irregularity and valley height irregularity decreased evenness increase, the fabric to show a good symmetry, and enhance the aesthetic appearance.

(3) Shear stiffness (X1) has shown a positive correlation to peak angle irregularity, the average valley height, peak width irregularity, valley height irregularity, showing the larger of 45°the shear stiffness, the more difficult fabrics is bent, symmetry is relatively worse. Therefore, the drape shape is easier to get good when the shear stiffness is small.

(4) Thickness (X7) has shown a negative correlation to peak angle irregularity, peak width irregularity, peak height irregularity. When increasing the thickness, the fabric becomes difficult bend, resulting in uneven rates decrease, the fabric to show a good aesthetic appearance.

(5) Textile weave (X12) has a great impact on and a high positive correlation to the average valley height, meaning that the closer to the plain weave fabric, the average valley height is greater, and the fabric drape shape is worse.

3 Model validation

Equation of regression analysis's ultimate goal is for forecasting and computer simulation that by entering mechanical and structural parameters, access the drape shape indices, and then restore the fabric drape shape. Therefore, based on the regression equation, application VC++ programming to realize this function, and thus test the practicality of the regression equation. Realization process is as follows:

(1) Obtain drape shape index and reproduce the weave expand image by formula (2)-(10).

(2) Use cubic spline algorithm to restore the drape edges and amend the area according to the drape coefficient.

(3) Calculate contour fringes under the two-dimensional shape [8].

(4) Establish mathematical model of the drape shape of the Bezier surface [9-10], to three-dimensional reconstruction of the drape images. Figure 1 is a simulation map and original image contrast.

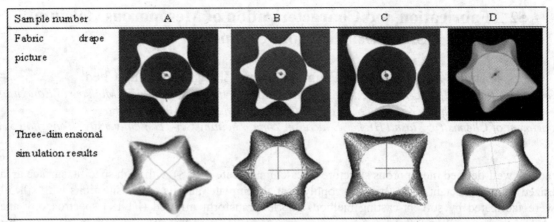

Figure 1 Comparison between simulation and original picture

4 Conclusion

Drape shape and basic structural and mechanical properties are closely related, especially the 45°shear stiffness, interweave resistance, thickness, weft density will cause different changes in the fabric drape, impact draping aesthetic. By validate the prediction equations we can get the drape shape of computer simulation. Article only discuses the polyester fabric, but also for other types of fabric through extensive laboratory analysis and research we can establish a more accurate mathematical prediction equation, and constantly improve the drape shape simulation program, to make the fabric drape prediction system more comprehensive.

References

[1] Chen Lixi, Zeng Xiuru, Zhao Wenxian. Research on the drape evaluation methods and indicators. Journal of Xi'an Polytechnic University, 1991, 17(1): 25-28.
[2] JinLian,Hu, Yuk-Fung,Chan. Effect of Fabric Mechanical Properties on Drape. Text Res.J, l998, 68(1): 57-64.
[3] P.Fisher. Simulate the drape behavior of fabric. Textile Research Journal, 1999, 69(5): 331-334.
[4] Xu Jun, Jiang Xiaowen, Yao Mu. The relationship discussion between fabric structure and aesthetics of fabric drape. Shan Xi: Journal of Xi'an Polytechnic University, 2001, 15(2): 126-128.
[5] Stylios, G.K., T.R. Wan. The Concept of Virtual Measurement 3D Fabric Drape. International Journal of Clothing Science and Technology, 1999, 11(1): 10-18.
[6] Chen Ming, Zhou Hua, Yang Lanjun, Li Bin. Three-dimensional fabric drape shape test indicators and three-dimensional reconstruction. Journal of Textile Research, 2008, 29(9): 51-55.
[7] Wang Ling. Shear properties of woven fabrics. Journal of DongHua University, 2005.
[8] L.Can. A study of fabric deformation using nonlinear finite elements. Textile Research Journal, 1995, 65(11): 660-668.
[9] Shen Yi, Liu Xuanmu, Wang Shoubing. Measurement and reconstruction of 3D draping shape of woven fabric. Journal of Textile Research, 2006, 27(6):12-16.
[10] Wu Xiuchun, Guo Dongming, Wang Xiaoming. Study of algorithm of reconstructing 3D figures from 2D gray image and its realization. Journal of DaLian Sci-Tech University, 2002, 42(6): 701-7.

52 Fabrication and Characterization of Mesoporous Calcium Silicate/silk Fibroin Composite Films

Hailin Zhu[1,2], Xinxing Feng[1], Huapeng Zhang[1], Jianyong Chen[1]*

(1. Key Laboratory of Advanced Textile Materials and Manufacturing Technology, Ministry of Education of China, Zhejiang Sci-Tech University, Hangzhou 310018, China
2. Department of Chemistry, Xiasha Higher Education Zone, Zhejiang Sci-Tech University, Hangzhou 310018, China)

Abstract: A well-defined mesoporous structure of calcium silicate (MCS) with high specific surface area was synthesized using surfactant P123 (triblock copolymer) as template, and its composite films with silk fibroin (SF) were fabricated by solvent-casting method. Fourier transform infrared (FT-IR) spectroscopy analysis showed that random coil and β-sheet structure co-existed in the SF films. The bioactivity of the composite film was evaluated by soaking in a simulated body fluid (SBF), and formation of a hydroxycarbonate apatite (HCA) layer was determined by XRD and FESEM. The results showed that the MCS/SF composite film was bioactive as it induced the formation of HCA on the surface of the composite film after soaking in SBF for 7 days. Consequently, the incorporation of MCS into the SF film can enhance the bioactivity of the film, which suggests that the MCS/SF composite film may be a potential biomaterial for bone tissue engineering.

Keywords: Silk fibroin; Mesoporous calcium silicate (MCS); Composite film; Bioactivity

1 Introduction

Tissue engineering is an interdisciplinary field that is drawing more and more attention in materials science, cell biology, and biotechnology to develop effective strategies for the repair or replacement of damaged or diseased tissues[1]. Naturally derived polymers are widely used in many tissue engineering applications, such as in dental repairs, ligament reconstruction, orthopedic fixation devices, as well as many others[2,3]. Among these naturally derived polymers, silk fibroin has recently attracted much interest because of its outstanding properties including good biocompatibility, water vapor permeability, biodegradability, and minimal inflammatory reactions[4,5]. However, poor bioactivity of pure SF makes them not suitable for bone tissue engineering. For solving this problem, an important strategy is to combine SF with inorganic bioactive materials so that the resulting hybrid materials possess improved bioactive and biological properties.

Calcium silicate (CS) is a typical bioactive inorganic material. When it is implanted in human body, a biologically active hydroxycarbonate apatite (HCA) layer is formed on the surface. Subsequently, the bioactive materials spontaneously bond to and integrate with living bone[6]. Especially, the synthesis of highly ordered mesoporous calcium silicate (MCS) with high specific surface area and pore volume, which has a greatly enhanced bone-forming bioactivity as compared with conventional CS, improves the properties of the CS significantly[7]. Considering the bioactivity of MCS, it is assumed that the incorporation of MCS into SF matrix may improve the bioactivity of composite.

In this study, novel MCS/SF composite films were prepared by solvent-casting method. The morphology, structure, and *in vitro* bioactivity evaluation of the composite films were investigated.

2 Experimental

2.1 Preparation and characterization of MCS

Well-ordered mesoporous calcium silicate (MCS) (molar ratios TEOS: $Ca(NO_3)_2$: $4H_2O$: P123 = 1:1:0.013) was prepared by sol–gel method[8]. Briefly, 3.0 g of P123 was dissolved in 120 ml of 2mol/L HCl and 30 ml of distilled water solution while stirring at 35°C in water bath until the solution became clear. 8.50 g of TEOS and 9.64 g of $Ca(NO_3)_2 \cdot 4H_2O$ were then added into the solution. The mixture was stirred at 35°C for 24 h, and the resulting precipitate was dried at 100°C for 20 h in air without any filtering and washing. After that, the MCS was calcined at 600°C for 3 h to remove the surfactant template and obtain the final products. Calcium silicate (CS) without well-ordered mesoporous structure was also synthesized by an almost identical process but without using surfactant (P123) as control.

---

* Corresponding author's fax: 86-571-86843169; tel: 86-571-86843622; email: cjy@zstu.edu.cn

2.2 Preparation of MCS/SF composite films

The regenerated SF solution was prepared according the literature [9]. A certain amount of MCS particles was added into the SF solution followed by ultrasonification for 30 min in order to disperse the MCS particles uniformly. Finally the products were cast on polystyrene Petri dish surfaces and dried for 48 h at 30°C and 50% RH. The thickness of obtained films was about 40–60 μm. In this work, three kinds of samples were prepared with the weight ratio of 0:100, 10:90, and 20:80 (MCS:SF, w/w), respectively.

2.3 Characterization

The XRD data were obtained in an ARL-X'TRA diffractometer using Cu Kα radiation, in 2θ range of 10–70° with a scanning speed of 2°/min. The room-temperature measurements were performed with the samples spread on a conventional glass sample holder. The microstructures of the composite films were observed with a field emission scanning electron microscope (FESEM) (S-4800, Hitachi, Japan). The specimens were gold sputtered. The FT-IR spectra of the samples were measured with a FT-IR (Nicolet 5700, America) spectrophotometer. Each spectrum of the sample was acquired in transmittance mode by accumulation of 64 scans with a resolution of 2 cm^{-1} and a data collecting range of 4000–400cm^{-1}.

2.4 Films soaking in SBF

SBF solution was prepared according to the procedure described by Kokubo[10]. The films were soaked in the SBF solution at 37°C for 7 days and 30 ml SBF was used for each sample. After soaking, the films were removed, washed in deionized water, and finally freeze-dried. XRD and FESEM were used to monitor the structure and composition of formed HCA on the surface of the films.

3 Results and discussions

3.1 Characterization of the MCS powders

MCS was synthesized by using nonionic block copolymers as structure-directing agents through an EISA process. TEM analysis shows that MCS powders possess highly ordered one-dimensional channel structure with a pore size of about 6 nm (Figure 1). N$_2$ sorption isotherms of MCS and CS are shown in Figure 2. The BET surface area, pore volume, and pore size were calculated to be 398 m^2/g, 0.49 cm^3/g, and 5.8 nm, respectively. For comparison, the calcined CS synthesized without surfactant P123 had a much lower BET surface area and pore volume of 38 m^2/g and 0.04 cm^3/g, respectively. The inset in Figure 2 gives the pore size distribution curve calculated from the adsorption branch by the BJH model. The pore size distribution of MCS is narrow, indicating the prepared MCS has a uniform mesoporous structure and the pores are homogeneous. The nitrogen sorption result is consistent with the TEM image of the MCS.

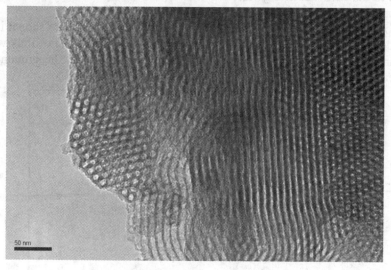

Figure 1 TEM images of MCS powders

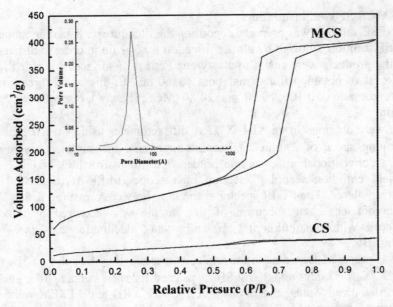

Figure 2 N$_2$ adsorption–desorption isotherms of the MCS and CS and pore size distribution of MCS (inset)

3.2 Structural analysis and surface morphologies of the composite films

IR spectroscopy is a frequently used method to monitor the secondary structure of SF. The absorption bands observed for SF at 1655 ± 5 cm^{-1} (amide I), 1540 ± 5 cm^{-1} (amide II) and 1235 ± 5 cm^{-1} (amide III) are assigned to random coil structure. The absorption bands for SF treated with methanol have frequencies of 1625 ± 5 cm^{-1} (amide I), 1525 ± 5 cm^{-1} (amide II) and 1265 ± 5 cm^{-1} (amide III), the characteristic of β-sheet structure[11, 12]. Figure 3 shows the FT-IR spectra of the pure SF film, the SF/MCS and the pure MCS powder in the spectral range of 2000–400 cm^{-1}. Both the pure SF film and the composite films showed the characteristic absorption bands at 1630 cm^{-1} (amide I), 1530 cm^{-1} (amide II), 1265 cm^{-1} (shoulder peak, amide III) and 1235 cm^{-1} (amide III), indicating that random coil and β-sheet structure co-existed in the composite films. On the other hand, MCS and CS showed the absorption bands centered at 1075 cm^{-1}, 1033 cm^{-1}, and 945 cm^{-1}, attributed to the Si-O stretching vibration mode. The absorption band at 460 cm^{-1} was assigned to Si-O-Si bending vibration mode[13, 14].

The surface morphologies of the composite films examined using SEM are shown in Figure 4. Pure SF film exhibited the dense and uniform morphology. In the case of the composite films with 10 and 20 wt% filler, the MCS particles were dispersed homogenously in the films, and the amount of MCS particles increased with increasing MCS content.

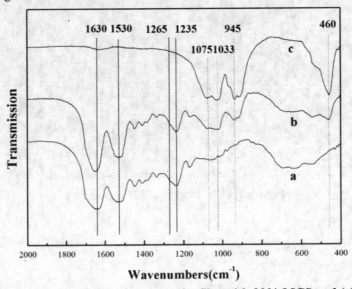

Figure 3 FT-IR of (a) pure SF film, (b) composite film with 20% MCS and (c) MCS powder

(a) (b) (c)

Figure 4 SEM images of the surfaces of the MCS/SF composite films: (a) pure SF film, (b) composite film with 10 wt% MCS, and (c) composite film with 20 wt% MCS

3.4 Evaluation of the composite films in vitro bioactivities

The bioactivity of biomaterial to form bone-like apatite could reflect their potential for bonding with bone[15, 16]. Figure 5 shows the thin film-XRD patterns of the composite films after soaking in SBF for 7 days. After immersion for 7 day, the diffraction peaks ($2\theta = 25.9°$ and $31.8°$) of crystalline apatite on the surface of the composite films could be obviously detected. The relative intensity of apatite peaks increased with increasing MCS content in the composites. The peaks were very broad resulting from the amorphous apatite granules. However, it was noted that the intensity of apatite peaks was very weak for the pure SF films. It was suggested that the incorporation of MCS into SF significantly enhanced its apatite-formation ability in SBF. SEM images of the composite films after soaking in SBF for 7 days are shown in Figure 6. A layer of apatite particles with ball-like shape forms on the surface of the composite films. A higher magnifivation examination showed that the sizes of the crystals were 200-400 nm in length.

4 Conclusion

The silk fibroin/mesoporous calcium silicate composite films were prepared by solvent-casting method. FT-IR analysis showed that random coil and β-sheet structure co-existed in the SF films. The MCS particles can disperse homogenously in the SF films These composite films were bioactive, confirmed by the formation of the HCA layer on the surface of the composites after immersing in SBF for 7 days. These results reveal that SF mixed with MCS is applicable as a biomaterial with good bioactivity.

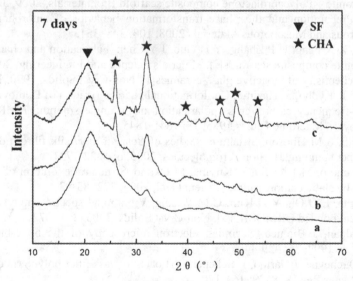

Figure 5 Thin film-XRD of the composite films after soaking in SBF for 7 days: (a) pure SF film, (b) composite film with 10 wt% MCS, and (c) composite film with 20 wt% MCS

Figure 6 SEM images of the composite films after immersion in SBF for 7 days: (a) pure SF film, (b) composite film with 10 wt% MCS, (c) composite film with 20 wt% MCS

Acknowledgements

We gratefully thank the National Natural Science Foundation of China (Grant No. 50903073, 50973096), and Zhejiang Natural Science Foundation (Grant Y407295) for support of this program.

References

[1] M J Mondrinos, R Dembzynski, K C Byrapogu Venkata, D M Wootton, P I Lelkes, J Zhou. Porogen-based solid free form fabrication of polycaprolactone–calcium phosphate scaffolds for tissue engineering[J]. Biomaterials 2006, 27: 4399–4408.
[2] L Ma, C Y Gao, Z W Mao, J Zhou, J C Shen, Enhanced biological stability of collagen porous scaffolds by using amino acids as novel cross-linking bridges[J]. Biomaterials, 2004, 25: 2997–3004.
[3] L Y Jiang, Y B Li, X J Wang, L Zhang, J Q Wen, M Gong, Preparation and properties of nano hydroxy apatite /chitosan/carboxymethyl cellulose composite scaffold[J]. *Carbohydrate Polymers* 2008, 74: 680–684.
[4] Y Z Wang, D J Blasioli, H J Kim, H S Kim, D L Kaplan, "Cartilage tissue engineering with silks caffolds and human articular chondrocytes," [J].Biomaterials, 2006, 27: 4434- 4442.
[5] Q Lv, S J Zhang, K Hu, Q L Feng, C B Cao, F Z Cui, Cytocompatibility and blood compatibility of multifunctional fibroin/collagen/heparin scaffolds[J]. Biomaterials 2007, 28 (14): 2306–2313.
[6] K Ohura, T Nakamura, T Yamamuro, T Kokubo, Y Ebisawa, Y Kotoura, Oka. Bone-bonding ability of P_2O_5-Free CaO- SiO_2 glasses [J]. Biomed Mater Res 1991, 25: 357-365.
[7] J Wei, F P Chen, J W Shin, H Hong, C L Dai, J C Su, C S Liu. Preparation and characterization of bioactive mesoporous wollastonite – Polycaprolactone composite scaffold Biomaterials, 2009, 30: 1080–1088.
[8] W Xia, J Chang. Preparation and the phase transformation behavior of amorphous mesoporous calcium silicate[J]. Microporous and Mesoporous Materials 2008, 108: 345–351.
[9] H L Zhu, J Y Shen, X X Feng, H P Zhang, Y H Guo, J Y Chen. Fabrication and characterization of bioactive silk fibroin/wollastonite composite scaffolds[J]. Materials Science and Engineering, 2010, 30: 132–140.
[10] T Kokubo, Surface chemistry of bioactive glass-ceramics[J]. Non-Cryst Solids, 1990, 120: 138-157.
[11] A. Tetsuo, K. Akio, T. Ryoho, S. Hazime, Conformational characterization of Bombyx mori silk fibroin in the solid state by high-frequency carbon cross polarization-magic angle spinning NMR, x-ray diffraction, and infrared spectroscopy[J]. Macromolecules,1985, 18: 1841-1845.
[12] S W Ha, A E Tonelli, S M Hudson, Structural studies of *Bombyx mori* silk fibroin during regeneration from solutions and wet fiber spinning[J]. Biomacromolecules, 2005, 6: 1722-1731.
[13] K Shimoda, H Miyamoto, M Kikuchi, K Kusaba, M Okuno, Structural evolution of $CaSiO_3$ and $CaMgSi_2O_6$ metasilicate glasses by static compression[J]. Chem Geol, 2005, 222: 83-93.
[14] H Y Jung, R K Gupta, E O Oh, Y H Kim, C M Whang, Vibrational spectroscopic studies of sol-gel derived physical and chemical bonded ORMOSILs[J]. Non-Cryst Solids, 2005, 351: 372-379.
[15] J E. Davies, N Baldan, J Biomed. Scanning electron microscopy of the bone-bioactive implant surface [J].Biomed.Mater Res, 1997, 36: 429-440.
[16] M Marcolongo, P Ducheyne, J Garino, E Schepers, "Bioactive glass fiber polymeric composites bond to bone tissue"[J]. Biomed Mater Res,1998, 39: 161-170.

53 Comparison of Fiber Configurations between Low Torque, Compact and Ring Spun Yarns

Ying Guo[1,2], Xiaoming Tao[2*], Bingang Xu[2], Jie Feng[2], Tao Hua[2], Shanyuan Wang[1]

(1. College of Textiles, Donghua University, Shanghai, China
2. Institute of Textiles & Clothing, The Hong Kong Polytechnic University, Hong Kong, China)

Abstract: Yarn structure plays a key role in determining yarn physical properties and characteristics of resultant fabrics. The yarn structure changes with the yarn production technology, process parameters and fiber parameters. Ring spinning system is one of the most important spinning technologies in the textile industry and the properties and structure of ring spun yarn have been considered as bench mark for decades. This paper is aimed to investigate the structures of spun yarns produced by two types of spinning system using fiber 3D trajectories technology, namely low torque yarn spinning and compact spinning system, and make a comparison between them and the one produced by ring spinning system. In the low torque yarn spinning system, a false twisting device was incorporated into a ring frame for producing a low torque and soft handle singles yarns. In this paper, the internal structure of the above three yarns was analyzed in terms of three-dimensional configuration of fibers, 2D FFT analysis on fiber paths and other migration parameters based on tracer fiber technique. Experimental results reveal that low torque yarn has irregular structure and unique migration behavior comparing with compact and ring spun yarns. Meanwhile, results demonstrated the negative orientation angle of fibers exists in low torque yarns.

Keywords: Fiber migration; Low torque yarns; Tracer fiber technique; 3D configuration

1 Introduction

Ring spinning is the most widely used method for the production of short staple fibers in the textile industry. In recent years, novel spinning technology has developed. The low torque spinning technology is one kind of novel spinning technology, which was developed by our group [1-3] to remove or reduce the residual torque in singles yarns and thus improve the skewness of denim fabrics and spirality of plain knitted fabrics. Previous studies have indicated that low torque yarn has lower wet snarling, less hairiness with comparable yarn strength at the low twist level [2].

Yarn structure is a key factor in determining yarn physical properties and characteristics of resultant fabrics. The yarn structure changes with the spinning technology. In the low torque spinning technology, the incorporation of the false twisting device into a ring frame changes yarn structure, thus yarn physical properties. In this paper, a comparative study is carried out to analyze typical features of fiber configurations in the low torque, compact and ring spun yarns in terms of 3D fiber configurations, FFT analysis on 2D fiber paths, fiber radial positions and fiber spatial orientation angle.

2 Experimental

2.1 Preparations

In this experiment, Tencel fiber was adopted due to their regular cross-section and optical isotropy for easier observation and operation. Tencel fiber has a linear density of 1.5 Denier with the fiber length of 38mm and fiber diameter of 12.08 μm. Also the roving of Tencel fiber (0.97Ne) with 0.32% black-dyed tracer fiber inside was used for production of various yarns for comparison. Three types of 20Ne yarns, ring spun yarn, compact yarn and low torque yarn, were spun with the twist multipliers of 2.5 and 3.5, respectively, for evaluating yarn structure. Here a modified Zinser 351 ring spinning system with the false twisting device installed was employed for producing low torque yarn.

2.2 Methods

In order to acquire clearer images of the tracer fiber, the mixture of Turpentine oil and Bromonaphthalene at the ratio of 1:1 was chosen as observation liquid since this chemical solution has the refractive index close to that of uncolored Tencel fiber. In the experiment, Tencel yarns with tracer fibers inside were immersed into observation liquid and examined under a CCD camera (SAMSUNG SCC-131BP), which is attached to a pairs of zoom lens with the magnification of 0.5×, 0.7~4.5, for clear observation. A

---

* Corresponding author's email: tctaoxm@inet.polyu.edu.hk

continuous measurement system[4] was used in the investigation. From the polished metal mirror set at the inclination angle of 45° relative to horizontal line, yarn can be observed from two perpendicular planes and then analyzed three-dimensionally.

3 Results and discussion
3.1 Typical tracer fiber images

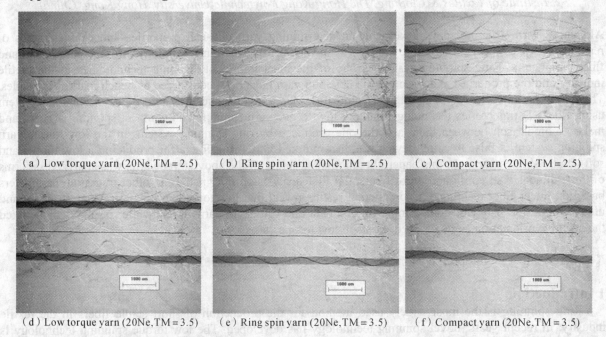

(a) Low torque yarn (20Ne,TM = 2.5)　　(b) Ring spin yarn (20Ne,TM = 2.5)　　(c) Compact yarn (20Ne,TM = 2.5)

(d) Low torque yarn (20Ne,TM = 3.5)　　(e) Ring spin yarn (20Ne,TM = 3.5)　　(f) Compact yarn (20Ne,TM = 3.5)

Figure 1 Typical tracer fiber images in different yarn samples

Figure 1 show typical tracer fiber images obtained by the measurement system[4]. Two yarn images in each picture reveal the two perpendicular plane of yarns along the yarn axis. The ring spun yarns and compact yarns have regular fiber paths while the fiber trajectory in the low torque yarn is more complex. Meanwhile, the fiber in low torque yarn seems to have more turns at same length along yarn axis than the ring yarn and compact yarn with same twist level, indicating more compact structure of low torque yarns.

3.2 3D configurations of fibers

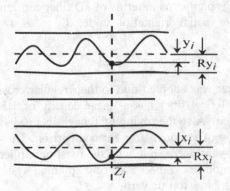

Figure 2 A sketch to determine 3D fiber coordinates

From the tracer fiber images, spatial coordinates of each point on the tracer fibers can be determined[4], as shown in Figure 2, in which x_i/Rx_i and y_i/Ry_i were adopted to define the radial position of the point i ($i=1,2,...n$) corresponded to the instantaneous yarn diameter observed from the two planes. Thus a composite 3D configuration of tracer fiber can be plotted. Figure 3 shows typical 3D configurations of fibers in different yarn samples. The fibers in ring spun and compact yarn nearly follow a perfect coaxial helical path under both two twist levels. By contrast, the fibers in the low torque yarn reveal different migratory pattern in which the fibers do not follow a concentric helix and it migrates from the outside to the centre of the yarn frequently with certain random features. Meanwhile, some small fiber migrations are involved in the big fiber migrations. The unique feature on structure of low torque yarn can increase the frictional force between fibers, and thus minimize the chance of fiber slippage during yarn rupture process, which may explain why low torque yarn can maintain comparable high strength at the lower twist level.

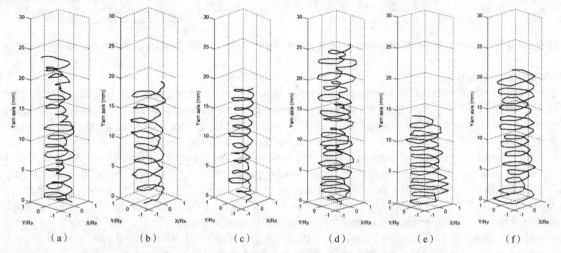

Figure 3 Typical 3D configurations of fibers in different yarn samples
(a) Low torque yarn (20Ne,TM=2.5); (b) Ring spin yarn (20Ne,TM=2.5); (c) Compact yarn (20Ne,TM=2.5);
(d) Low torque yarn (20Ne,TM=3.5); (e) Ring spin yarn (20Ne,TM=3.5); (f) Compact yarn (20Ne,TM=3.5);

3.3 2D FFT analysis

The Fast Fourier Transform (FFT) is capable of breaking down the paths of single fibers in a spun yarn into a wide range of periodicities. Thus it is used to convert the image into a complex function in the frequency domain, and then analyze the frequency of the 2D tracer fiber paths so that we can decide if we use one or several specific functions to describe the 2D traces of fibers in yarns. The process is accomplished by Matlab^{TM} software. Figure 4 shows the results of 2D FFT analysis in terms of frequency-amplitude curves of the fiber specimens corresponding to the 3D configurations in Figure 3.

For the ring spun yarn, the period of its 2D fiber path is relatively even and stable at the twist multiplier of 2.5 and 3.5 (see Figure 4(b) and 4(e)). The peak on the FFT spectrum means the fiber path can be simulated by using a curve with certain frequency and magnitude. There is also a dominating peak on the spectrum of compact yarn sample. The results obtained under 2.5 twist level reveals the fiber in compact yarn has bigger migration frequency than that in ring spun yarn. However, the difference is not so significant under 3.5 twist level. The fluctuation of amplitude is also related to the position of the fiber in yarns. For the low torque yarn, there are several weak peaks under both 2.5 and 3.5 twist multipliers, which reveal that fiber path is a combination of many functions which have various wavelengths and magnitudes. It also confirms that fibers in ring spun yarn and compact yarn follow periodical curves while fibers in low torque yarn undergo dramatic and irregular path.

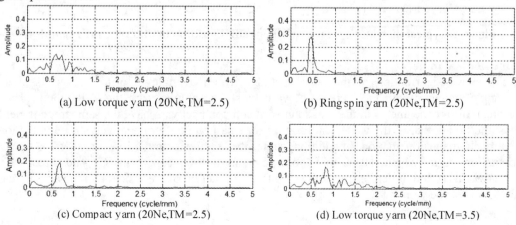

(a) Low torque yarn (20Ne,TM=2.5)   (b) Ring spin yarn (20Ne,TM=2.5)
(c) Compact yarn (20Ne,TM=2.5)   (d) Low torque yarn (20Ne,TM=3.5)

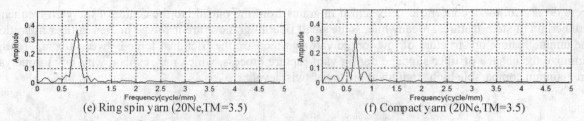

(e) Ring spin yarn (20Ne,TM=3.5)   (f) Compact yarn (20Ne,TM=3.5)

Figure 4 Frequency-amplitude curves of typical fiber paths on 2D planes in different yarn samples

3.4 Fiber radial position

Fiber radial position is an important parameter to describe the internal structure of yarn, which represents overall tendency of a fiber to be near the surface or near the centre of yarn. Figure 5 shows typical distribution curves of fiber radial positions along the yarn axis, in which the fiber specimens are corresponded to those in Figure 3. Under both 2.5 and 3.5 twist multiplier, radial position of the fiber in ring yarn samples only varies within a very small range around the mean position, far away from the yarn centre. Generally, the distribution of fiber radial position in the compact yarn has a similar path with smaller amplitudes than that in ring spun yarns. Comparatively, the low torque ring yarn has the lower mean fiber position among the three types of yarn with same twist level. Furthermore, the fiber alters its position from the centre of yarn to the surface in low torque ring yarn in the highest rate and more frequently when compared with the other two types of yarn under both 2.5 and 3.5 twist multiplier. It can enhance cohesion and frictional force between fibers, thus increase yarn strength and minimize the chance of fiber slippage during yarn rupture.

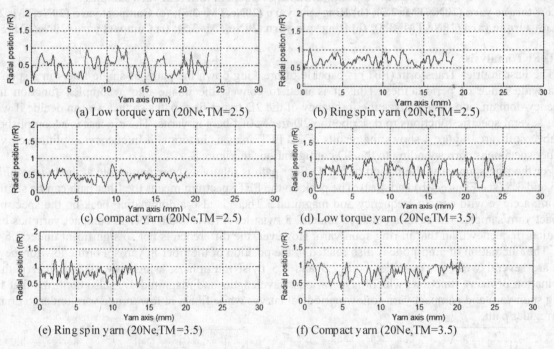

(a) Low torque yarn (20Ne,TM=2.5)   (b) Ring spin yarn (20Ne,TM=2.5)

(c) Compact yarn (20Ne,TM=2.5)   (d) Low torque yarn (20Ne,TM=3.5)

(e) Ring spin yarn (20Ne,TM=3.5)   (f) Compact yarn (20Ne,TM=3.5)

Figure 5 Typical distribution curve of fiber radial position in different yarn samples

3.4 Fiber inclination angle

Figure 6 shows distribution of the fiber spatial orientation angle in different yarn samples. Fiber orientation angle increases with the increase of yarn twist for samples with the same count. The average orientation angle of fiber segments in the ring spun yarn and compact yarn have the same trend under both 2.5 and 3.5 twist multiplier, that is, the values varies in a very small range. Comparatively, those in the low torque yarn have a much greater fluctuation. In addition, some fiber segments in the low torque yarns have negative spatial orientation angle, which means that those segment follows a path in opposite direction with the original yarn twist direction. It demonstrates the significant reduction of the residual torque and increased yarn strength for a low torque yarn.

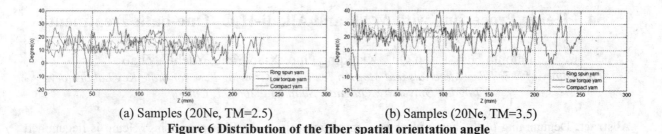

(a) Samples (20Ne, TM=2.5)　　　　　　(b) Samples (20Ne, TM=3.5)

Figure 6 Distribution of the fiber spatial orientation angle

4 Conclusion

This article reports a comparative study on fiber migration behavior between low torque yarn, compact yarn and ring spin yarn with various spinning parameters. The analyzed results indicated that the fiber in the low torque yarn follows a deformed helix which is not a concentric structure like in ring spun and compact yarns. 2D FFT analysis reveals the dramatic and irregular structures on low torque yarn. The fiber in low torque yarn migrates from the surface to the centre of the yarn frequently with some random features and small fiber migrations are included in the big fiber migrations. Under the same twist level, the fiber in low torque yarn has the lower mean radial position compared with compact and ring spun yarn. Results also reveals some fiber segments in low torque yarn have an opposite orientation angle with yarn twist direction which explains the reduction of yarn residual torque experimentally.

Acknowledgement

The authors wish to thank the Donghua University and the Hong Kong Polytechnic University for supports.

References

[1] X M Tao, B G Xu. Manufacturing Method and Apparatus for Torque-Free Singles Ring Spun Yarns. US Patent, 2005 No: 6,860,095.
[2] Yang K, Tao X M, Xu B G, and Lam J. Structure and properties of low twist short-staple singles ring spun yarns. Textile Research Journal, 2007, 77(9): 675-685.
[3] Xu B G, Tao X M. Techniques for torque modification of singles ring spun yarns. Textile Research Journal, 2008, 78(10): 869-879.
[4] Y Guo, X M Tao et al. A continuous Measurement System for Yarn Structures by Optic Method. submitted to Measurement Science and Technology

54 Hemp Processing with MAE and Alkali-H_2O_2 One-bath Treatment

Guojun Han, Lijun Qu*, Xiaoqing Guo, Xiao Yang, Yuehua Zhao

(Laboratory of New Fiber Materials and Modern Textile, the Growing Base for State Key Laboratory, Qingdao University, Qingdao, Shandong 266071, China)

Abstract: Degumming of the hemp plays very important role in the fiber processing. Hemp is degummed and bleached in an alkali-H_2O_2 one bath with microwave-assisted extraction (MAE). MAE is a new separation technique that combines microwave and traditional solvent extraction. MAE has many advantages, such as shorter time, higher extraction rate and lower cost etc. In this paper, barium peroxide (BaO_2) as an oxidant, bleaching agent and catalyst is introduced and the mechanism of alkali-H_2O_2 one bath with MAE of degumming and bleaching of hemp is analyzed. Microwave pretreatment time, MAE time, the NaOH and BaO_2 concentration are discussed in the experiments. The results show that the alkali-H_2O_2 one bath with MAE has uniform and effective removal of various kinds of components such as hemi-cellulose, pectin, lignin and so on. It has significant effect on fiber handle and fiber whiteness features. The key technological parameters are also obtained. The alkali-H_2O_2 one bath with MAE is a prospective method for hemp degumming.

Keywords: Hemp; MAE; Alkali-H_2O_2 one bath; BaO_2; Residual gum content

1 Introduction

The development of fibrous plants for textiles and other applications has been continuously motivated by the worldwide demand for non-food crops [1]. Bast fibers like hemp, flax, banana and jute tend to be potential candidates as renewable cellulosic fiber sources. However, degumming of bast fibers for textiles processing and applications is a very complicated process, it is closely related to the environment issues, ecological concerns, production costs, resource balance, and future demand for clothing and food [1].

Compared with flax and ramie, the content of non-cellulose in hemp is as high as 40%. It makes the hemp degumming more difficult than flax and ramie. However, the degumming effects influence the process of spinning, weaving and dyeing directly. The outstanding and comfortable properties of hemp fiber such as absorbent quality, moisture permeability, antibacteria property etc. are all influenced by the quality of degumming, so the studies on degumming method with high efficiency is necessary [2].

Usually, hemp fibers are processed with bacterial, chemical and enzymatic methods. Bacterial processing depends on the weather, water quality, and other factors and often results in inconsistent fiber quality [6]. Enzymatic processing method was developed for flax in the 1980s, but may not be a feasible method for processing hemp fibers. Chemical processing of these fibers is effective for removing noncellulosic substances, but it wastes too much water and pollutes the environment. Recently, explosion method and ultrasonic treatment come into the view of researchers. These methods can effectively separate the fiber bundles bonded by the gum, but can not remove the pectin from the bast thoroughly. And explosion and ultrasonic methods are still in exploratory stage [2, 9].

In this paper, a new method was used to obtain hemp fibers. Microwave-assisted extraction (MAE), also called microwave extraction, is a new separation technique that combines microwave and traditional solvent extraction. MAE of biologically active compounds was first presented by Ganzler as a novel and effective sample preparation technique in 1986 [5]. Early reports on the application of MAE were focused on extraction of effect components from herbs and natural plants. Studies showed that MAE has many advantages, such as shorter time, less solvent, higher extraction rate, better products with lower cost. The apparatus of MAE is simple and MAE method can be used to process more materials with less limit of the polarity of extractant. The microwave technique of materials can produce a volume heating effect [10].

In this paper, in order to achieve the quality requirements of spinning, MAE combined with alkali-H_2O_2 one bath chemical process was used to obtain hemp fiber. Barium peroxide (BaO_2) as an oxidant, bleaching agent and catalyst is introduced and the mechanism of alkali-H_2O_2 one bath with MAE of degumming and bleaching of hemp is analyzed. The properties of the degummed hemp fibers are tested. The alkali-H_2O_2 one bath with MAE can remove the gum more effectively with shorter time [3,4].

---

* Corresponding author's email: lijunqu@126.com

2 Experiments

2.1 Materials and apparatus

The hemp material used for experiment was obtained from Yunnan province. The middle part of hemp was chosen and cut into 15cm length. The reagents, such as NaOH, H_2SO_4, H_2O_2, BaO_2, stabilizer of H_2O_2, $MgSO_4$ and so on, are involved in experiments.

In the experiments, APEX microwave chemical worktable, TG328A electronic analytic balance, Y802 constant temperature oven, etc. are used.

2.2 Degumming process

The hemp material was fully pretreated with microwave-assisted extraction, then degummed by alkali-H_2O_2 one bath with MAE, then washed and dried for the measurements, i.e. Hemp → Microwave pretreatment → alkali-H_2O_2 one bath with MAE → Washing → Drying → Degummed hemp fibers

2.3 Gum removal content after being pretreatment

The gum removal content W (%) by the pretreatment is calculated by Equation 1:

$$W = \frac{G_0 - G_1}{G_1} \times 100\% \qquad (1)$$

where G_0 and G_1 are the weight of the hemp before and after pretreatment, respectively.

2.4 Residual gum content

Referring to the GB5889-86 "ramie chemistry composition quantitative analysis method"[7]. Residual gum content M (%) is calculated by Equation 2:

$$M = \frac{G_a - G_b}{G_a} \times 100\% \qquad (2)$$

where G_a and G_b are the weight of the sample before and after alkali cooking, respectively.

3 Results and discussions

3.1 Effects of the microwave pretreatment

Pretreatment plays very important role in the hemp degumming. Microwave heating not only accelerates the decomposition of the gum, but also could reduce the damage to the hemp. The solution concentration, temperature and pretreatment time are the key technological parameters that could affect the pretreatment effect. Microwave heating could be used for the degradation of the non-cellulose materials under acid condition.

In this experiment, technological parameters during the pretreatment are listed as follows:

Microwave pretreatment: H_2SO_4 (98%) 1ml/L, temperature 40-50°C, bath ratio 1:20, time: 5, 10, 15, 20, 25, 30min.

Alkali-H_2O_2 one bath: NaOH 10 g/L, H_2O_2 10 g/L, stabilizer of H_2O_2 2.7g/L, $MgSO_4 \cdot 7H_2O$ 0.1g/L, bath ratio 1:20, temperature 99°C, cooking time 130 min. The results are shown in Figure 1.

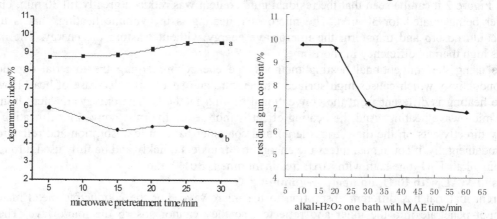

Figure 1 The relation curve between the microwave pretreatment and degumming index

Figure 2 The relation curve between the MAE and residual gum content

In Figure 1, a is the gum removal content after microwave pretreatment, b is the residual gum content of the degummed hemp fibers after pretreated by microwave and degummed by alkali-H_2O_2 one bath. The gum removal content increases slowly and the residual gum content of the degummed hemp fibers declines sharply within 15 minutes. After 15 minutes, the residual gum content of the degummed hemp fibers declines slowly to a plateau. Due to the polar materials decomposed with the high thermal effect, the gum may decompose slowly with the evaporation of water [8]. Considering all these factors, the optimum microwave pretreatment time is 15-20 min.

3.2 Comparing alkali-H_2O_2 one bath with alkali-H_2O_2 one bath with MAE treatments

The hemp after being pretreated by microwave was then degummed by alkali-H_2O_2 one bath process and alkali-H_2O_2 one bath with MAE method respectively. Table 1 shows the residual gum content degummed by these two methods.

Table 1 Contrast alkali-H_2O_2 one bath and alkali-H_2O_2 one bath with MAE treatments

Degumming method	Bath ratio/g: ml	Degumming time/min	Residual gum content/%
Alkali-H_2O_2 one bath with MAE	1:20	40	6.79
Alkali-H_2O_2 one bath	1:20	130	8.18

Table 1 shows that the residual gum content after being treated by alkali-H_2O_2 one bath with MAE is lower than treated only by alkali-H_2O_2 one bath process, and the treating time of the former method is much shorter than the latter. As the frequency of microwave is 300 MHz-300 GHz, it can freely penetrate the inside of fiber during MAE. As a result of absorbing microwave energy, the polar molecule of radiated substance can swerve rapidly and arrange oriented, thus results in tearing and mutual friction causing the material temperature rise rapidly. Once the rising temperature makes the inner pressure overweigh the capacity of cells could be received, then the cell was broken, and the gum was removed from the fiber [11-12].

Besides, the electromagnetic field caused by microwave accelerates the gum decomposing from inner side of cell to extraction solvent and therefore greatly improves degumming efficiency, and meanwhile contributes to saving degumming time and quality guaranteed [12-13]. Therefore, alkali-H_2O_2 one bath with MAE has practical application value for hemp degumming.

3.3 Effects of treating time on residual gum content by alkali-H_2O_2 one bath with MAE

In alkali-H_2O_2 one bath process, alkali and H_2O_2 work simultaneously to remove most proportions of the non-cellulosic components efficiently. The effective parameters are microwave radiation time, temperature, catalyst concentration, bath ratio etc. Microwave radiation time is especially investigated in this paper.

The technological parameters of microwave pretreatment and alkali-H_2O_2 one bath with MAE degumming are listed as follows:

Microwave pretreatment: H_2SO_4 (98%) 1ml/L, temperature 40-50 ℃, bath ratio 1:20, time 15 min.

Degumming by alkali-H_2O_2 one bath with MAE: NaOH 10 g/L, H_2O_2 10 g/L, stabilizer of H_2O_2 2.7 g/L, $MgSO_4 \cdot 7H_2O$ 0.1 g/L, bath ratio 1:20, temperature 99 ℃, time: 10, 15, 20, 30, 40, 50, 60 min. The results are shown in Figure 2.

From Figure 2 it can be seen that the residual gum content was reduced greatly till 30 min, but reduced slowly after being treated for 30 min. The microwave heating is the volume heating, that is to say, the materials could absorb and transform the microwave energy without wasting the energy. The microwave heating has high thermal efficiency in short time.

While using the conventional heating methods, the energy immerging the material depends on the thermal conductivity which causes high surface temperature and successive decrease of heat. However, the microwave heating is different from the conventional heating methods. The energy transfers from inner to outside in microwave heating. And the heating effect is more even than the conventional heating methods. The energy directly acts on the dielectric, the air and container have little absorption and reflection for the energy. Fundamentally, it could guarantees the energy that transfers quickly and be fully used. From Figure 2, the optimum alkali-H_2O_2 one bath with MAE treatment time is 30-40 min.

3.4 Effects of MAE with BaO_2 on hemp degumming

Different material has different absorption to the microwave. For the conductive metal materials, the waves cannot penetrate into the inner and reflected, so they cannot absorb the microwave. The material composed of polar molecule could absorb the energy sufficiently. Water molecules have strong polarity, is the best dielectric for the microwave absorption. So the materials that contain water must be able to absorb microwave [13]. At the same time, the dielectric materials have different absorption of the microwave. Namely the material with large dielectric constant absorbs the microwave easier. One new dielectric material-BaO_2 is introduced and used in this paper. Figure 3 shows the relationship between the BaO_2 concentration and residual gum content.

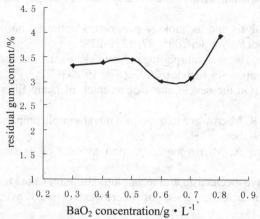

Figure 3 The relation curve between the BaO_2 and residual gum content

The results indicate that BaO_2 could improve the degumming efficiency obviously. BaO_2 with high dielectric constant was used to speed up its internal thermal motion of molecules and accelerate the dissolution of gum. Microwave heating is a dielectric heating. Under the microwave electromagnetic, the heating molecules become dipoles due to polarization, these dipoles would occur disturbances followed the changing of microwave electromagnetic. The disturbance will be disturbed and embarrassed by the interaction with the thermal motion between the molecules, and then causing the friction effects. At last, part of the energy transferred into molecular motion kinetic energy and showed with the form of temperature [14]. The temperature of the fiber is improved by microwave radiating and this can accelerate the dissolution of the gum. So the degumming effects are improved efficiently.

4 Conclusions

In the present investigation, a successful attempt has been made to hemp degumming, using a combination of "alkali-H_2O_2 one bath and MAE", several conclusions are achieved as follows,

(1) Microwave pretreatment has evident influence on the hemp degumming, it is a potential method for hemp degumming under acid condition. When the pretreatment time is 15-20 min, the effect of the degumming is ideal;

(2) Compared with the alkali-H_2O_2 one bath process, the alkali-H_2O_2 one bath with MAE can remove the gum more effectively in short time.

(3) BaO_2 as an oxidant, bleaching agent and catalyst with large dielectric constant used in the experiment could improve the efficiency of degumming.

(4) To achieve the degumming purpose, alkali-H_2O_2 one bath with MAE is a prospective method for hemp degumming with shorter time, higher extraction rate and lower cost. The key technological parameters are obtained as the following: microwave pretreatment time: 15-20 min, MAE time: 30-35 min, the auxiliary agents BaO_2: 0.6-0.7 g/L.

Acknowledgement

This work was supported by National Basic Research Program of China (2009CB626606).

References

[1] Postle R., Wang H.M. The quality and mechanical properties of hemp after chemical processing. Pro. Conf Bast Fibrous Plant on the Turn of Sec Third Millennium 2001: 18-21.
[2] Qu L.J., Wang X.Y., You R.H., Wang S.Y. Study on alkali-H2O2 one bath process of degumming and bleaching of hemp. 83rd TIWC 2004, 5: 721-725.
[3] Wang H.M., Wang X. Hemp processing with microwave and ultrasonic treatments. 83rd TIWC 2004, 5: 779-782.
[4] Yang Y.X., Jiang Y.K., Zhang S.C. Research on apocynum venetum degumming with microwave and ultrasonic. Wool Textile Journal 2006, 9: 27-30.
[5] Thostenson E.T., Chou T.W. Microwave processing: fundamentals and applications. Composites: Part A 1999, 30: 1055-1071.

[6] China textile standard compilation, basic standards and methods of the standard volume (one), Text Ind Res Inst Stand 424-463.
[7] Qu L.J. The mechanism analysis and technology parameters optimization of alkali-H2O2 one-bath cooking and bleaching of hemp. J Appl Polym Sci 2005, 97: 2279-2285.
[8] Nikki Sgriccia M.C. Hawley. Thermal, morphological, and electrical characterization of microwave processed natural fiber composites. Compos Sci Technol, 2007, (67): 1986-1991.
[9] Liu J., Cheng H.Z., Li Z.H. On the development of research of hemp fiber degumming. J Plant Fiber Prod 2002, 24(4): 39-42.
[10] Ganzler K., Salgo A., Valko K. Microwave extraction-a novel sample preparation method for chromatography. J Chromatogr 1986, 371: 299-306.
[11] Hao J.Y., Huang R.H., Deng X. Microwave extraction of seed of passionflower. J East China Univ Sci Technol 2001, 27: 117-220.
[12] Zhao K.Y., Xu F.Y. Microwave mechanism and technology. Beijing, China text publ company 2005.
[13] Xie M.Y., Chen Y. The research progress of microwave-assisted extracting technology. J Food Sci Biotechnolo 2006, 25: 105-114.
[14] Wang S.L. Application of microwave heating technology-drying, sterilization. Beijing, China Mach Press 2004: 1-5.

55 Design of Multicolored Warp Jacquard Fabric Based on Space Color Mixing

Qizheng Li, Jiu Zhou, Gan Shen, Chenyan Zhu
(College of Materials and Textiles of Zhejiang Sci-Tech University, Hangzhou 201003, China)

Abstract: Multicolored warp jacquard fabric is one of the typical traditional jacquard fabrics presenting color pattern on the surface of fabric. With the development of electronic jacquard technology, numerous color effects can be produced on jacquard fabric by limited dyed yarns. The design principles and methods are different from the traditional ways. Moreover, this paper probes into analyzing principles of color presentation and structure design of fabric. The 6-warp 3-weft type is selected in this paper, as the most representative type of multicolored warp jacquard fabric. Then, a new design method is introduced, which enables the multicolored warp jacquard fabric to express 302 mixed colors based on space color mixing.

Keywords: Multicolored warp; Jacquard fabric; Weave design; Space color mixing

0 Introduction

Multicolored warp fabric is classified into woven fabric, which express colorful patterns via the intertexture of colored warp and weft threads. Moreover, this type of fabric intertexture of low density and heavy bold yarn can be applied to wall hanging and upholstery fabrics and so on. Multicolored warp products are firmly thick and water-washable with remarkable decorative functions which have been extensively applied in many realms [1]. Generally, multicolored warp jacquard fabric have 4 to 6 colorful yarns in warp which are mainly polyester thread, while two bold wefts (white and black) are made up of cotton yarn, and one thin weft is the same as warps. Essential design methods of multicolored warp jacquard is: the necessary warp is applied in the surface warp while the others is used in back lining. With coordination with weft (deep and shallow) produces the effect of darkness or lightness. The thin weft is mainly responsible for the balance of the weave structure. Its structure has been integrated with double weft and multi-layer changes. In fact, it is regarded as a changeable structure of double-layer weaves [2~4]. Traditional design of multicolored warp jacquard in general is color separation and then the weave structure can be generated and designed according to colors, the fabric shall only express the inherent colors of weft and warp yarn. [4] Based on principles of space color mixing[5~7], an innovative design method for multicolored warp jacquard is proposed in the paper, which only use yarns of six colors, respectively including red, yellow, green, blue, black and white. Through the interwoven of threads, 302 fabric textures can be mixed up and supported by different color effects.

1 Color Principles of multicolored warp jacquard design
1.1 Determination of the basic colors

As all factors are taken into practical considerations, the group number of the warp and weft should be as few as possible. According to the color theory of Hering, a German physiologist, there are three pairs of visual elements in the retina of the human eye, as: white-black, red-green, and yellow-blue. It is the creation and destruction of those three pairs that generates nerve impulses of a person's color feeling. Thus, red, yellow, blue, green, black and white are inherent basic colors in human vision perception. With the mixed six colors, theoretically, all colors can be created from the basic colors[8]. The study of color psychology also indicates that four psychologically primary colors: red, yellow, green, and blue are most likely to produce the visual beauty in the nature of people. In addition, it is also found that the yellow mixed by red and green, the green mixed by yellow and cyan in fabric are not sufficiently fresh brilliant. Thus, colors of red, yellow, and blue, green, black and white are selected in this paper as six basic colors for mixing, which are further applied to fabric design of color combination for weft and warp.

1.2 Principles of color mixing

According to the principles of color mixing, there are mainly two types of color mixing principles, namely, additive mixing and subtractive mixing, both of which cannot be applied to the principle of fabric color performance. The colors of fabric are mixed together for the interwoven of warp and weft threads, where the principle is the mixing of color spaces. When the distance between colorful threads or dots is set in a certain visual degree, if their projection size in the retina is smaller than the diameter of the perception of chromaticity cells, the human eye could not be able to distinguish them, thus giving rise to color mixing. The farther the distance is, the more obvious the outcome will be. The principle shall be in compliance with the following laws [8]:

(1) By mixing any two non-complementary colors, intermediate colors can be produced where the brightness is only the average of the brightness of the primary colors. In addition, by mixing complementary colors in different proportions, the corresponding gray color can be obtained.

(2) By mixing the colorful color and colorless color, the intermediate color can be generated. Mixing it with white color, light colors of high brightness can be obtained. Mixing it with black color, dark colors of low brightness can be obtained.

2 Design method of multicolored warp jacquard
2.1 Define the basic weave unit

Basic weave unit is the interwoven weave equipped with the minimum repeat of the single group of warp and weft. The larger the repeat is, the more weaves can be generated from the basic unit. Yet, it also gives rise to promoted the design difficulty and error rate. In addition, if the repeat is too small, it may not be able to resemble the desired color effects of the weave. The traditional weave unit of multicolored warp jacquard mostly constitutes two threads, which obtains fairly scarce color effects and monotonous hand-feel. Under the circumstance of digital technologies, greater weave repeat may be adopted to enrich the fabric effects such as 4 threads, 6 threads, and 8 threads. Weave repeat of 4 threads is selected in this paper with the consideration of practical factors. Take the 6-warp and 3-weft type as an example, combination of the basic weaves of the repeat of 4 threads could generate effective weaves up to 5400. In order to simplify and regulate the weave design, the weave design is divided into "Hue", "Brightness", and "Saturation".

2.2 Hue design

As human vision has the ability to distinguish "Hue" in a limited degree which is not high, 8-hue is eligible to meet the fundamental requirements of identifying colors of the fabric by the human eyes. The "Red", "Yellow", "Green", and "Blue" series are necessarily selected for the corresponding colorful warp, which is then applied to surface warp. The "Orange", "Green-Yellow", "Cyan" and "Purple" series are mixed in compliance with the color space mixing principle. Moreover, the primary colorful warp is selected as surface warp, e.g. orange =Red + Yellow, Cyan=Green+ Blue, as shown in Figure 1. Gray series can be selected only by black and white warp as surface warp.

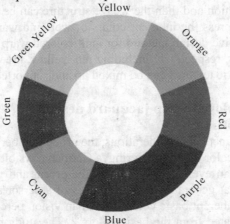

Figure 1 Multicolored warp jacquard hue circle

2.3 Brightness design

The brightness is primarily and mostly determined by the area ratio of black yarn on the surface layer. For dark colors, the black weft is selected as the surface weft; for bright colors, the white weft is selected as the surface weft. In case of additional changes of saturation, changes of brightness shall be as clarified as possible. As shown in Figure 2, the vertical screw thread warps represents one group of colorful warps. With the different arrangement of warp and weft that are black and white, 6 obvious brightness changes of surface layer can be designed.

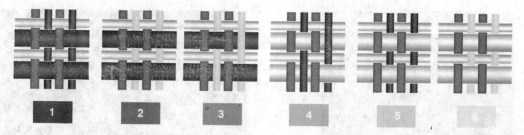

Figure 2 Gradual changes of brightness of surface weave of multicolored warp jacquard fabric

The grayscale variations of the multicolored warp jacquard fabrics can be classified into at least 14 grades. As shown in Figure 3, white wefts are applied as the surface weft in the former 7 gray weaves, and the others, black wefts are applied as the surface weft.

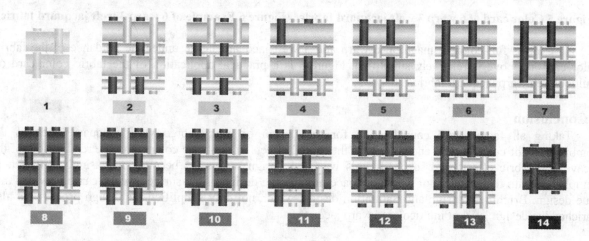

Figure 3 Grayscale variations of surface weave of multicolored warp jacquard fabric

2.4 Saturation design

"Saturation" of colors (purity) indicates the proportion of gray in color [8], where the saturation of pure gray is "0". Thus, with the adjustment of the area ratio of colorful warp in the surface layer, the changes of saturation of fabric can be generated. Moreover, as the area ratio of colorful warp on the surface of the fabric is reduced, the color of this fabric gradually fades into gray. As shown in Figure 4, the vertical screw thread warps represent the colorful warp while there are 6 kinds of saturation changes marked with "X", "B", "D", "E", "F", "G" in the figure.

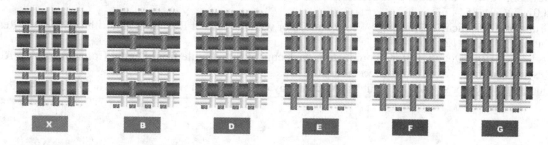

Figure 4 Gradual changes of saturation of surface weave of multicolored warp jacquard fabric

3 Application and examples

Based on the principles as mentioned above, a fabric color card is made which includes 302 weaves, most constituted by Red, Orange, Yellow, Green-Yellow, Green, Cyan, Blue, Purple series and 14 grayscales from dark to light (in the center). Each color series provide the brightness changes vertically and the saturation changes horizontally.

Figure 5 Color card of 6 warp 3 weft jacquard fabric   **Figure 6** Example of 6 warp 3 weft jacquard fabric

To express full-colored image on 6 warp and 3 weft jacquard fabric under the guidance of the fabric color card as above is relatively simplified. Figure 6 is a practical application of the fabric color card of multicolored warp jacquard fabric.

4 Conclusion

Taking all factors into consideration for the design of multicolored warp jacquard, such as the combinations of colorful weft and warp, and the mixed effects of the basic colors, and the relationship of the weave and fabric effects and so on. In this way, the design process is become increasingly difficult. In compliance with the principle of the color space mixing, the design of the surface weave can be divided into Hue design, Brightness design and Saturation design, which not only simplified the design process but also enriched the design level of multicolored warp jacquard.

References
[1] Yuefei Weng. Technics and designs of Jacquard Fabric [M]. Beijing: Publishing of Textile industry, 2002, 254-267.
[2] Jiu Zhou, Qizheng Li. Standardized Design on Structure of Digital Multi- warp Jacquard Fabric[J]. Journal of Textile Research, 2005, (5): 52-54.
[3] Qizheng Li, Jiu Zhou. Research on Color Model of Digital Multi-warp Jacquard Fabric Design[J]. Silk, 2005, (5): 14-16.
[4] Qizheng Li. Zhoujiu, Shengan. Principle and Method of Digital Six-Warp Jacquard Fabric Design[J]. Journal of Zhejiang Sci-Tech University, 2006, (2): 32-35.
[5] Dawson, R.M., Color and Weave effects with some small weave repeat sizes[J], Textile Research Journal, 2002, 72(10): 854-863.
[6] Grundler, D., Rolich, T., Matching Weave and Color with the help of Evolution Algorithm[J], Textile Research Journal, 2003, 73(12): 1033-1040.
[7] Osaki, K., High Quality Color Reproduction on Jacquard Silk Textile from Digital Color Images[J], AUTEX Research Journal, Vol. 3, No. 4, December 2003.
[8] Guoxing He. Color science [M]. Publishing company of Donghua University, 2004: 16-19, 33, 138-142.

56 Design and Development of a Detection System for Recognising Emotions towards Creation of Interactive Fashion

XIA W.J., NG. M.C.F.

(Institute of Textiles and Clothing, The Hong Kong Polytechnic University, Hung Hom, Kowloon, Hong Kong, China. Email: 07900557R@polyu.edu.hk)

Abstract: Today, interactivity has become an incisive point that fuses art and design with technology. Interactivity is ever becoming in the 21st Century, when increased emphasis is being put on people-oriented interaction. In this connection, there is huge potential in deploying interactivity in art and design as well as in fashion. i.e., Interactive Fashion (IF). Immutable fashion norm is being overtaken by a surge of searching for new, dynamic clothing models. Interactive Fashion emerged as the times required. This paper is to report on a study of Interactive Fashion. Emphasis is on the design of a detection system applied into Interactive Fashion for recognising emotions via comprehensive measurement of physical data. The aim is to approach creation of Interactive Fashion that is able to sense the change in human emotions and display them via visual mode(s).

Keywords: Interactive fashion, Detection System, Recognising emotions, Physical data

1 Introduction
1.1 The background of Interactive Fashion

In a dynamic and ever-changing world today, any immutable fashion norm is being overtaken by a surge of searching for new, dynamic clothing models. At present, the prevailing concept of clothing is one that clothing is perceived often as static and mortal. A piece of clothing item is 'active' only when it is activated by the movement of the wearer while it is being worn. Nevertheless, this kind of actions and/or changes is unilateral, initiated but by the wearer. Scare past attempts have been made to expand and invent fashion that go beyond this unilateral 'action' to becoming 'interactive' (Interaction is broadly defined as presence of interdependent actions at the mutual initiation between two or among more subjects). It is envisaged that the successful creation of Interactive Fashion (IF) will expand the invent visual dimensions of fashion which in turn, attributive to the ultimate re-definition of art (fashion), humanities and technology by which our culture and lifestyle are re-shaped, and will further enhance research in this area.

1.2 Concept source: Interactivity in Art and Design

Prior to experimentation for creation of Interactive Fashion, a comprehensive review of the development and applications of interactivity in various disciplines of art and design is essential. Literature review reported that interactivity was first expressed in art as early as the 60s [1], and it started to develop into various design disciplines such as installation, architecture, product, as well as fashion and textiles since 1980s [2].

1.3 Interactivity in Art – Interactive Art

With the development of computer science and digital technology, multimedia technologies have been incorporated into artwork which gave rise to new art forms, i.e., New Media Art [3]. Being the main characteristic of new media art, interactivity detached and developed into a medium of new art form – Interactive Art [4]. It is a form of installation-based art that involves spectator in some ways. Artworks frequently feature computer and sensors to respond to motion, heat, meteorological changes or other types of input to which their makers programme them to respond [5]. Interactivity being used as a communicative tool between artist and spectator as an art form originated in the late 60s. In 1968, Jasia Reidhardt arranged The Landmark Computer Art Exhibition Cybernetic Serendipity at the Institute of Contemporary Art (ICA) in London [6] when it was the first exhibition ever to attempt to demonstrate all aspects of computer-aided creative activity: art, music, poetry, dance, sculpture, and animation. Since then, Interactive Art has gradually become an important expressive and communicative means of new art.

1.4 Interactivity in Design – Interaction Design

Since electronic products and user experience needs were becoming complicated, requirements of interactivity have been evidenced increasingly in the 80s [7]. It is a discipline which defines the behaviour of products and systems with which a user interacts. The practice typically centres on complex technology systems such as software, mobile devices, and other electronic devices. Interactivity has become a design form to create dialogues between human and artifact [8]. The term interaction design was first proposed by

Bill Moggridge and Bill Verplank in the late 1980s. To Verplank, interaction design is an adaptation of the computer science term user interface design to the industrial design profession. To Moggridge, it was an improvement over soft-face, which he had coined in 1984 to refer to the application of industrial design to products containing software [9].

The 21st Century saw a continuous advancement of human minds. New concepts are increasingly people-oriented. Emphasis has been put on human intra-action and interaction with the environment. It is under this setting that Interactive Fashion has become a new form of design expression that attracts considerable attention and popularity in the fashion arena [10].

1.5 Identification of Interactive Fashion

Conventional fashion lacks feasibility, motility and vitality, whose fashion language expression is limited through a single wearing mode. Modern society today demands individuals to change their roles and emotions against the needs of different social situations. To this end, a dynamic and changeable fashion language is required. On the other hand, there has been an imminent need to monitor and regulate our psychology and emotion in a society of ever-increasing pressure. Being a medium of frequent and intimate contacts with human, fashion is enriched with deeper meanings and mission through activating its fashion language. Thus, the concept of Interactive Fashion is presented based on these needs. Yet, what is Interactive Fashion?

Interactive Fashion is a new fashion language when non-interactive fashion is regarded as the traditional. It is the fashion language that could be expressed on its own initiative. It possesses activeness, self-initiation, motility and reasoning by which direct and visible interaction or communication among fashion, wearers and others is created.

Expression of activeness: Interactive Fashion is enlivened with liveliness. It is no longer a layer of still cover that wraps round the body or a kind of fashion language that cannot speak much for itself. With the activeness, fashion has transformed itself from its introvert being to an expressive extrovert one. Interactive Fashion responses to human emotions such as: happiness, sadness, anger etc. and/or human intention through the major elements of fashion language like colour, pattern and style for simultaneous interactions; Besides colour, pattern and style, Interactive Fashion can also uses other forms of expressions such as sound and music to express its fashion language. Interactive Fashion makes delivery of fashion language more direct and expression more lively.

Expression of reasoning: to a certain extent, Interactive Fashion is like fashion with brain of reasoning ability. From active perception to corresponding responses is a course of reasoning. It reacts with corresponding responses according to the various inputs it received after some kind of consideration in its mind. Thus, the responses it offers is a result of reasoning. The reasoning Interactive Fashion processes literally makes fashion language a language that communicates.

Expression of self-initiation: Interactive Fashion changes fashion from totally passive to simultaneously active. Because Interactive Fashion has considerable reasoning power, it can determine its own form and content of responses through its 'mind', and forms interaction and functions communication ultimately.

Expression of motility: Interactive Fashion is capable of instant reception and reaction. It means that Interactive Fashion could change to appropriate colour, pattern or style at once when it is aware of human needs by their informs. And, in turn, Interactive Fashion can offer appropriate responds such as sound, temperature and also the traditional elements changes immediately when it detects human emotional changes.

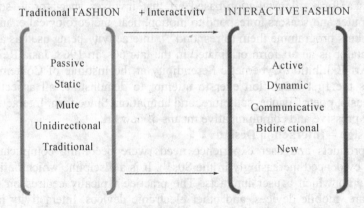

Figure 1 Comparison between traditional fashion and Interactive Fashion

2 Design of Detection System for Interactive Fashion Recognising Emotions

2.1 Design concept

Emotional reaction - Human emotion can trigger human biological changes. More obvious changes include breath rate, heartbeat, sphygmus, skin temperature, sudation, etc. [11].

Interactive relation - The key point in Interactive Fashion is the relationship between human and IF in the course of interaction process.

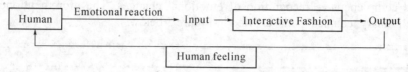

Figure 2 The Interactive relation

Emotion is not only an important research area in psychology; it is also a social concern. Normally the majority of people under emotional fluctuations cannot conduct immediate self-regulation by themselves, they need external influence to catalyse emotional change [12]. The concept behind the design of Interactive Fashion which can recognise emotions is that not only will it detect human emotion of the wearer, but also will it regulate with corresponding emotional responses to the wearer or others interactively. Such interaction enlivens clothing from one that is without live to offers emotional comfort to the wearer. In addition, it can also act as communicative tool of emotion. Often, slight emotional fluctuation of a person cannot be detected, which can cause misunderstanding and regret among people as a result of unnecessary speculation. It is the good intent of Interactive Fashion to make transparent hidden or unnoticed emotional changes through visual manifestations on it. And, the detection system is to recognize human emotional reactions via comprehensive measurement of physical data. This is particularly helpful for those who are too shy to 'talk' their emotion out or who needs emotion care and health administering.

2.2 Design methodology

Above all is to understand the classification of collecting emotional reaction information in detail. It could be broadly summarized as the following types.

2.2.1 Electrical reaction of skin

when emotion changes, the expansion and contraction of blood vessels in the skin and the secretion of sweat glands will trigger change in the electrical resistance of skin. It is these changes that the skin electrical reactor is used to measure emotional reactions of nerves system in plants. As for the electrical resistance of skin, focus is put on the analysis of the tendency of the electrical resistance of skin and the changing states of emotion instead of their corresponding definite value[13].

2.2.2 Circulation system indicator (e.g., electrocardio, blood vassal volume, blood pressure, etc.)

Electrocardio (pulse) is a good indicator of emotional changes. When one is in a state of being content and happy, his heart beat will be normal. When he is in a state of anxiety, fear or anger, his heart beat will increase. The change in the capacity of blood vassals is a result of the expansion and contraction of smooth muscles on the inner lining of the arterial wall controlled by the autonomic nerve system. In other words, it is a result of partial vascular expansion and contraction. Anxious hard-working, angry, fear, abnormal excitement, etc. will all cause vascular contraction; when a person feels embarrassed or shame, his blood vessels in skin will relax. Blood pressure is related to volume of blood vessels; they both reflect the condition of circulation system and share a similar biological mechanism [14].

2.2.3 Respiration

Change in emotion will result in change in speed and depth of respiration. For examples, breath rate increases when a person is in agony, it stops when he is in great fear, convulsed when he is in exultation or grief[15].

2.2.4 Voice

A voice is vocalised by small vibration when air hits the vocal duct while passing through the vocal organ. When a person is nervous, the normal vibration of his vocal organ will be somewhat restrained. So the vibration cannot be manipulated as it is intended to. Yet, this measuring method has gradually been eliminated by a more advanced computing detecting model. It uses Guassian Mixture Model (GMM), Principal Component Analysis (PCA), and Artificial Neural Network (ANN) to analyse and recognise vocal emotions. The analysis focuses primarily on the vocal speed, mean value, range and magnitude of

fundamental tones, mean values and variance range between mean variances and resonance peaks of fundamental tones to detect emotions[16].
2.2.5 Brainweave
2.2.6 Other physical indicators (e.g., body temperature, oxygen in blood, etc.)

Based on these understandings and taking into account the specific requirement for Interactive Fashion and feasibility of available technologies, the authors will concentrate on selected major physical indicators that consist of heart beat, blood pressure, respiration, skin electricity resistance and body temperature for an enhanced accurate comprehensive recognition of emotional changes. The aforementioned tasks are to be realised by the preliminary technical paths charted in Figure 3 below.

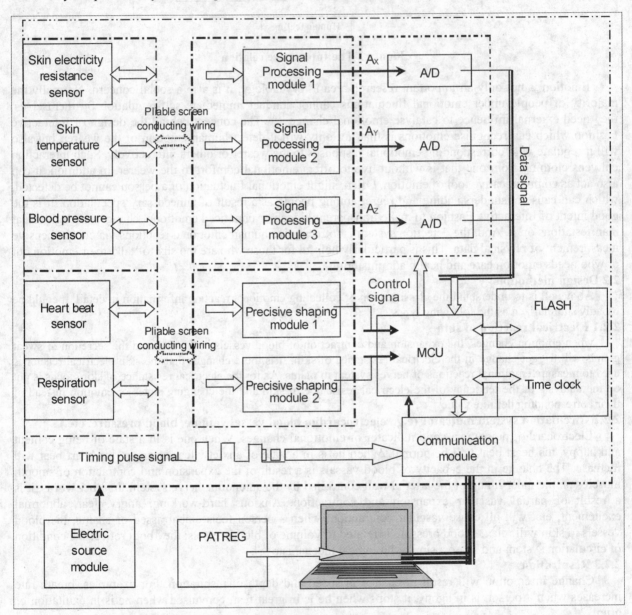

Figure 3 Detection system for recognizing emotions via comprehensive measurement of physical data

Sensory modules (sensors) are crucial components in this experiment. Existing sensory modules will be identified or modified and where necessary, customer-made for application on Interactive Fashion. Either piezoelectric belt or piezoresistant semi-conductor one will be carried on chest or belly for detecting respiration; piezoresistant semi-conductor pulsate pressure device will be carried on wrist for detecting blood pressure; non-invasive oscillation blood pressure unit will be carried on arm for detecting blood pressure; thermoelectrical infrared sensor will be carried on wrist for measuring temperature; since detection of skin electrical resistance may involve applying electrical shock and electrode, its safety requires further study at this stage.

Signal processing unit is primarily used to filter the weak signals sent from various sensors. After magnifying and shaping, such signals will be collected by an A/D or a frequency register. The A/D will adopt multi-channel collection between the ranges of 1-50 kHz with the degree of accuracy of 16. The surveyed data are stored in a non-volatile FLASH memory of not less than 16 Mbit. The time-signal is measured via a calendar clock and will be stored in the same memory. The data in the memory can be sent via a RS232 port to the host computer for further processing. Battery is used to power the system, thus lower power consumption for the whole system will need to be resolved. Artificial Neural Network (ANN) will be merged with fuzzy pattern recognition for data processing and comprehensive evaluation. Currently, the data will be processed in the host computer; DSP can be deployed for processing once the expected target is met.

Once the trial system has been established, data will be collected from voluntary human subjects who form the crucial part of the experiment. The emotional states of individual subjects will be recorded in details. The size of human subjects will not be less than 10, and each of not less than 10 emotional condition samples to be recorded.

Since prior attempts in designing and developing sensors for fashion remain scare, the current proposal serves as a pilot study of the feasibility of detecting emotions via physical bodily changes. Special consideration will be put on the wearability, convenience and comfort of the devices. The effectiveness of the data so obtained is dependent on reliability of the device developed as well as the successful procurement of available specimens in the market.

3 Conclusion

Designing and developing a detection system for recognising emotions via comprehensive measurement of physical data towards creation of Interactive Fashion which will enable and expedite the realisation of interactivity in fashion design as well as to improve our understanding of the inter-relationships between interactivity and humanity in Interactive Fashion. It is envisaged that the successful creation of Interactive Fashion will not only expand the aesthetical and technological dimensions of fashion attributive to the subsequent redefinitions of fashion as object d'art as well as of utility, humanities and technology, but also will it reshape our lifestyle and cultural context in which we live.

Acknowledgment

The authors gratefully acknowledge financial support from the Postgraduate Research Fund from The Hong Kong Polytechnic University.

References

[1] Chris, C. 2003, The Art of Interactive Design, San Francisco: No Starch Press.
[2] Liu, Yuping and Shrum, J.J. 2002, What is Interactivity and is it Always Such a Good Thing? Implications of Definition, Person, and Situation for the Influence of Interactivity on Advertising Effectiveness, Journal of Advertising, 31 (4): 53-64.
[3] Rada, R. 1996, Interactive Media. New York, Berlin, Heidelberg, London, Paris, Tokyo, Hong Kong, Barcelona, Budapest: Springer-Verlag press.
[4] Paul, C. 2003, Digital Art (World of Art series). London: Thames & Hudson.
[5] Reichardt, J. 1968, Cybernetic Serendipity: the computer and the arts. London: Studio International.
[6] Dan, Saffer. 2006, Designing for Interaction, New Riders.
[7] B., Moggridge. 2007, Designing Interactions, MIT Press.
[8] Xia, W.J.C. and NG, M.C.F., 2009, A Study of Interactivity in Art and Design, Research Journal of Textile and Apparel. In press.
[9] James, W. 1884, What is Emotion. Mind 9: 188–205. http://www.psych.gov.cn/article/article_view.asp?id=1806.
[10] Cacioppo JT, Berntson GG, Larsen JT, et al. 2000, The psychophysiology of emotion. Handbook of Emotions, In Lewis M, Jeannette M, & Haviland J. New York: Guilford Press, 2000
[11] Collet C, et al. 1997, Autonomic nervous system response patterns specificity to basic emotion. Journal of Autonomic nervous system, 1997.

57 Design Creations of Black-and-white Simulative Effect Digital Jacquard Fabric

Jiu Zhou[1,2], Frankie NG[2], Yejin Jiang[1]

(1. Key Lab of Advanced Textile Materials and Manufacturing Technology of Ministry of Education, Zhejiang Sci-Tech University
2. Institute of Textiles and Clothing, The Hong Kong Polytechnic University, Hong Kong, China)

Abstract: Research on structural design of Black-and-White simulative effect jacquard fabric is the key to obtaining nice black-and-white shading effect of fabric. In this paper, by merging the characteristics of traditional manual design method and the application of digital design technology, three design methods of black-and-white simulative effect jacquard fabric under layered-combination design mode were introduced with design illustrations. Differs from the effect of traditional black-and-white simulation jacquard fabric, the design creations of black-and-white simulative fabric with different digital fabric structures are capable of producing black-and-white simulative fabric accurately. It is envisaged that such technical advancement will have a tremendous potential in commercial applications.

Keywords: Fabric structure; Black-and-white simulation; Digital, jacquard fabric; Black-and-white shading

1 Introduction

Black-and-white simulative fabric refers to fabric whose visual effects such as pattern and color are capable of imitating that of given grey images. In this study, with the application of layered-combination design mode, three design methods on black-and-white simulative fabric under layered-combination design mode were proposed. The main design problems during structural design processes were identified and solved effectively. It is of great importance for the innovation of black-and-white simulative fabric by using digital technologies.

The structural design of black-and-white simulative fabrics was approached in a manual manner a hundred years ago in both ancient China and Europe [1-2]. In ancient Europe, black-and-white simulative fabric was produced with single-layer fabric structure in cotton or linen, often of rather coarse fabric effect. However, in ancient China, black-and-white simulative fabric was a normally silk product featuring elaborate pattern and color effects. It was constructed with two-weft backed structure and threads of pure silk, and was given the name *Xiangjin* - fabric with portrait and landscape simulated motifs[3-4]. In order to meet the technical requirement of balanced interlacement of fabric construction, the *Bangdao* device, a kind of warp lifting device whose working principle is the same as that of a heddle frame in a dobby machine and *Bangdao* weave were employed when producing *Xiangjin*.

At present, the digital jacquard machine is no longer installed with the bounding format of having "one harness corresponding to two or more ends". The *Bangdao* device thus cannot be employed in digital production processing. For this reason, the design method of traditional Chinese black-and-white simulative jacquard fabric has lost its popularity over time. The design methods for black-and-white simulative jacquard fabric need to be innovated. The innovative design concept suggested in this paper is combining digital image processing method with digital structure design method[5-6]. Three methods were introduced to creating black-and-white simulative fabrics under layered-combination design mode.

2 Design principles, methods and creations

The design innovation of black-and-white simulative jacquard fabric via the application of digital design and production technology has three directions. The first one is to design black-and-white fabric by tailor-made gamut weaves based on a colorless design mode and single-layer structure. The fabric structure is designed according to the varied brightness of grayscales of the digital gray image. The second method is to produce the uncovered fabric structure with the support of proper accessorial threads based on the colorful design mode and compound fabric structure. The last method is to design black-and-white fabric directly from a successful colorful simulative design.

2.1 Design principles, methods and creations with digital single-layer structure
2.1.1 Design principles
The colorless design mode of layered-combination design mode can be applied directly to develop

black-and-white simulative fabric with single-layer structure. The pattern design of black-and-white simulative fabric can be achieved by using the CAD system. And any 8-bit digital gray image that shows a maximum of 256 grades of gray can fulfill the design requirement of black-and-white simulative fabric.

The weave design should be approached with an image of less than a range 256 grays to form shaded gamut weaves based on detailed technical specification of desired fabric[5,7]. In general, there are three methods of adding interlacing points that are utilized to design shaded gamut weaves upon selected primary weaves: vertical (warp-wise), horizontal (weft-wise) and diagonal transitions[4-6]. However, due to the restriction of balanced interlacement and mutual covering effect after production, the design of shaded gamut weave cannot be approached mechanically, as it needs to be varied frequently along with the change of technical parameter of the desired fabric and the nature of the applied image.

2.1.2 Design methods

During fabric production, an advanced weaving loom is used together with an electronic jacquard machine. Black-and-white simulative fabric designed with single layer structure is interlaced by one series of warp and one series of weft threads so that the design of the fabric structure is correspondingly regular. The major design input during this process is the image and weave design in the jacquard textile CAD system. The series of tasks carried out are specified in Figure 1.

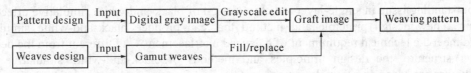

Figure 1 Design processes for black-and-white simulative fabric with single-layer structure

2.1.3 Design creations

Based on the design method introduced above, the design creation of simulated effect fabric was approached efficiently by using only one series of warp thread and one series of weft thread with tailor-made digital gamut weaves. Its simulated effects adequately achieved with the reproduction of a given image.

The key technical parameters of fabric are detailed in the technical specification table (Table 1). Due to the restriction of technical condition in fabric making, the parameters of the black-and-white simulative fabric based on the production devices cannot be modified during the design process. According to the key technical parameters shown in Table 1, the final fabrics were designed by using 12-thread gamut weaves with lower fabric density. Polyester threads were employed in both warp and weft direction to facilitate ease of production. Design creations are shown in Figure 2.

Table 1 Technical specification of black-and-white simulative fabric with single-layer fabric structure

	Parameters	
	Warp	Weft
Materials	1/166.7 dtex polyester (white)	1/166.7 dtex polyester (dark/dark/light/light)
Density	45 threads/cm	60 threads/cm
Composition	Polyester 100%	
Weave structure	12-thread gamut weaves, 41 gray grades	
Design repeat	1248 needles × 1680 fillings	
Pattern repeat	22.01 cm (width) × 28.07 cm (length)	
Weight	238 g/m²	

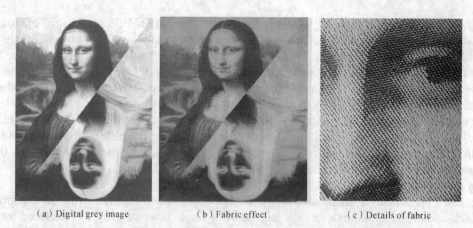

(a) Digital grey image　　(b) Fabric effect　　(c) Details of fabric

Figure 2 Design creation of black-and-white simulative fabric with single-layer structure

2.2 Design principles, methods and creations with compound structure

Due to outstanding issue in single-layer structure design of black-and-white simulative effect fabric, it is imperative for a new design principle and method that can produce black-and-white simulative fabric accurately and meet the technical requirement of balanced interlacement[8-9]. Different from the design method of single-layer structure, the design principles and methods of black-and-white simulative fabric with compound structure are based on a new design concept devised from layered-combination design mode.

2.2.1 Design Principles

The basic design principle is to add a supplementary thread between the two adjacent threads of single-layer structure. By doing so, the mutual covering among coloring/floating threads arranged in juxtaposition can be avoided. Details are as follows:

(1) The fabric is constructed by two series of threads arranged in juxtaposition and one series of thread in another direction. One group of juxtaposed threads is used as coloring/figuring threads, whereas the other group of juxtaposed threads is applied to forming a compound fabric structure as joint threads. The latter serves as a supplementary thread and functions to prevent the coloring thread from slipping and covering one another. The joint threads are not visible from the face side.

(2) Coloring threads are capable of expressing figured effect with smooth black-and-white shading through the deployment of shaded gamut weaves similar to that of single-layer fabric structure. Due to the existence of supplementary threads, the coloring threads will not move and be covered by each other.

(3) For color consideration, the coloring threads should employ black or dark color while the supplementary threads should be of the same color as that of the threads interwoven in the other direction (white or light colors). In addition, in order to reduce the color influence on figuring effect, the supplementary threads are preferred to be finer than the coloring threads.

2.2.2 Design methods

The structure design of black-and-white simulative fabric with compound structure involves three parts: primary weave design, gamut weaves design of figuring threads and weave design of supplementary threads. The primary weave design is the same as that of full-color compound structure design. It involves the selection of primary weave and the design of full-color technical points upon primary weaves. Taking 16-threads satin as an example (see Figure 3), the primary weave is designed with 16-threads 5 step weft-face sateen, whose valid full-color technical points is 16-threads 13 step warp-face satin.

Figure 3 Primary weave and its full-color points

Based on the primary weave and its full-color technical points[10], the design of gamut shaded weaves named as basic weaves for figuring threads can be approached. The approach to the joint weave design for supplementary thread is based on full-color technical points. Full-color points can be applied as joint weave directly. For combination method of fabric structure, the combination proportion is 1:1 across weft between coloring thread and supplementary thread. Due to the application of supplementary threads and corresponding full-color structure, the adjacent figuring threads arranged in juxtaposition cannot be covered by each other and the entire weave variations are colored on the face of fabric. It enables jacquard fabric to express the black-and-white shading of grayscales accurately, and thus the true-to-original simulated effect of black-and-white fabric can be achieved.

2.2.3 Design creation

The key technical parameters of fabric are shown in Table 2. Similar to the design with single-layer fabric structure, based on valid technical parameters, there is no restriction for digital image used for simulative design of black-and-white fabric. Table 2 reveals higher thread density in the final fabric than that designed with single-layer fabric structure. Two silk threads were employed in weft direction. One weft served as figuring thread that had a contrasting color to the warp thread while the other served as supplementary thread that had a similar color to those of the warp thread. In order to reduce the color influence by the supplementary threads, the figuring thread is 2.5 times thicker than the supplementary thread. Moreover, in order to achieve higher thread density, larger repeat gamut weaves, i.e., 24-thread weaves were applied which doubled those used for the design with single-layer structure. Design creations are shown in Figure 4.

Table 2 Technical specification of black-and-white simulative fabric with compound fabric structure

	Parameters	
	Warp	Weft
Materials	22.2/24.4dtex×2 silk (white)	22.2/24.4dtex×5 silk (black) 22.2/24.4dtex×2 silk (white)
Density	115 threads/cm	(45+45) threads/cm
Composition	Pure silk 100%	
Weave structure	24-thread gamut weaves, 45 gray grades	
Design repeat	12000 needles×3744 fillings	
Pattern repeat	104.3 cm (width)×41.6 cm (length)	
Weight	139.9 g/m²	

(a) Digital grey image   (b) Fabric effect   (c) Details of fabric

Figure 4 Design creation of black-and-white simulative fabric with compound fabric structure

2.3 Simulation design from colorful effect to black-and-white effect

In addition to the two design methods of black-and-white simulative fabric introduced above, following the design method of colorful simulative fabric, it is possible to design black-and-white simulative fabric based on designed compound fabric structure of colorful simulative fabric directly. As to the design of the digital image, decolorizing a colorful digital image can produce a gray digital image. Similarly, when weft colors of colorful simulative fabric are transferred into grayscales with the same levels of brightness and under the same fabric structure, black-and-white simulative fabric would be produced from a colorful simulative fabric without any additional structural design procedure. Following this idea, the appropriate design illustration was approached to produce the black-and-white simulative fabric directly from the structure of colorful simulative fabric. Table 3 shows the technical specification of fabrics with both true-color effect and black-and-white effect.

Table 3 Technical specification of fabric from colorful to black-and-white

	Parameters	
	Warp	Weft
Materials	22.2/24.4 dtex×2 silk	22.2/24.4×2 dtex silk (CMYK-grays)
Density	115 threads/cm	160 threads/cm
Composition	Pure Silk 100%	
Weave structure	24-thread gamut weaves, 85 gray grades	
Design repeat	4000 needles×4416 fillings	
Pattern repeat	34.8 cm (width)×27.6 cm (length)	
Weight	141.5 g/m^2	

The fabric was made by pure silk threads and has higher thread density in both warp and weft directions. The design processes are the same as that of colorful simulative fabric, accomplished by color separation and combination of CMYK primary colors; however, in the course of fabric production, four gray colors were employed to replace the CMYK primary colors respectively with the same levels of brightness, and applied to fabric production. That is to say, four gray colors instead of CMYK primary colors produced the fabric notwithstanding the fact that the fabric structure was designed in CMYK true-color mode. Figure 5 shows the fabric effect of both true-color and black-and-white simulative fabric.

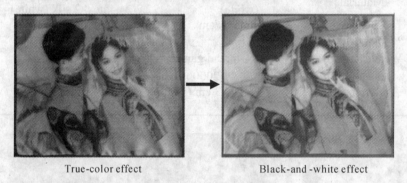

True-color effect　　　　　　　　Black-and-white effect

Figure 5 Design creations of simulative effects from true-color to black-and-white

3 Conclusions

In this paper, based on the application of layered-combination design mode, three available innovative design methods on true-to-original effect jacquard fabric of black-and-white simulative effect were introduced with illustrations. The first method is based on a colorless design mode and single-layer structure. The tailor-made gamut weaves is necessary for design creation in order to obtain fine simulative effect and meet the balanced interlacement of fabric construction. The second and third methods are based on full-color compound structure. The second method is to produce the uncovered fabric structure with the support of proper accessorial threads in compound fabric structure. The third method is to design black-and-white fabric directly from a successful colorful simulative design by using four grey threads. To conclusion, the first method is suitable for the design creation with lower density due to mutual covering effect in single-layer structure. And the second and third methods are most suitable for the design creation with higher density. The results achieved from this study hold promising prospects for wide creative and commercial applications of black-and-white simulative effect fabric.

References

[1] Bell T.F., Jacquard Weaving and Designing, London: Longmans Green and Co., 1895: 117.

[2] Grosicki, Z.J., Watson's Advanced Textile Design, London: Newnes-Butterworths, 1977: 83-102.
[3] Weinsdorfer, H., 50 years of weaving technology, International textile bulletin, 2004, (3): 54-56.
[4] Li, J. l., Modern Silk Woven Photograph and Its Structure Design Features, Silk Monthly, 2004, (03), 11-13.
[5] Zhou, J., Ng, Frankie, Practice and analysis on structural design for colorless digital jacquard fabric, Journal of Textile Research, 2007, 28 (10): 34-37.
[6] Zhou, J., Ng, Frankie, Innovative Principle and Method for Digital Jacquard Fabric Designing, Journal of Dong Hua University (Eng.Ed), 2007, 24(3): 341-346.
[7] Zhou, J., Ng, Frankie, Innovative Principle and Method of Colorless Digital Jacquard Fabric Design, Journal of Textile Research, 2006, 27 (4): 1-4.
[8] Zhou, J., Ng, Frankie, Principle and method of weave design for woven fabrics using digital technology, Journal of Textile Research, 2008, 28 (4): 53-56.
[9] Ng, Frankie, Zhou, J., Innovative Layered-Combination Mode for Digital Jacquard Fabric Design, Textile Research Journal, 2009, 79(8): 737-748.
[10] Zhou, J., Ng, Frankie, All-coloring compound construction of digital jacquard fabric, Journal of Textile Research, 2007, 28 (6): 55-58, 65.

58 Thickness Variations of Length-distributed Slivers and Draft Conditions in Roller Drafting

Jong S. Kim[1], Jung Ho Lim[2], You Huh[3]*

(1. Laboratory of Intelligent Process and Control, College of Engineering, Kyunghee University, Yongin, 449-701, R. O. Korea
2. Department of Textile Engineering, Graduate School, Kyunghee University, Yongin, 449-701, R. O. Korea
3. Department of Mechanical Engineering, College of Engineering, Kyunghee University, Yongin, 449-701, R. O. Korea)

Abstract: Roller drafting is an important operation in staple yarn production. The drafting conditions affect the product qualities strongly, depending on the fiber length of the constituent fibers. This study reports on the thickness variations of the output slivers according to the draft conditions such as roller adjustment and draft ratio for the slivers with a uniform fiber length or with a fiber length distribution. Experimental results revealed that the output linear density was influenced by the fiber length distribution and process conditions. The output sliver was more regular in low draft ratio level for the length distributed fibers than the uniform length fibers. In the level higher than 30 of the draft ratio, the length distribution effect disappeared. There was a critical condition under which the irregularity began increasing abruptly. The bundle with a uniform fiber length had the critical condition at lower draft ratio and longer roller adjustment than the length distributed. For the dynamic model describing the bundle flow, the bear diagram was taken into consideration. And the output linear density was simulated, while a random variation for the input linear density was assumed. The simulation results coincided well with the experiments, which confirmed the feasibility of the model equipped with fiber length distribution.

Keywords: Fiber length distribution; Roller adjustment; Draft ratio; Irregularity; Bundle flow; Dynamic model

1 Introduction

Roller drafting is an important operation for staple processing, employing a simple principle of the velocity difference between the input and output rollers. This operation is used for all kinds of yarn formation technologies. This operation, however, is very sensitive to the processed fiber feature; fiber length, fiber elasticity, fiber surface feature, crimp, etc., which are also influenced by the pre-processes to mix or arrange fibers along their axes or to individualize fibers. The roller drafting is adopted in the final yarn formation process as a continuous operation device, or in sliver drawing as a unit process. But the roller drafting is a disturbing source in the evenness of the bundle thickness, because the operation added some specific thickness irregularities to the sliver or bundle. Especially, the fiber length distribution is the most important factor to decide the sliver irregularity, which is influenced with the roller setting.

There have been many researchers on the irregularity of slivers or yarns. A few of important researches can be mentioned here such as those by Balasubramanian et al [1], Dutta and Grosberg [2], and Mandl and Noebauer [3]. But theoretical descriptions of the roller drafting dynamics are not so much systemized as to be used in engineering. Recently Huh and Kim [4-8] have reported on the dynamic state of the bundle in process and out of the process. Since the relationship between the fiber length distribution and the roller setting is a vital issue for the engineers in the industrial field, this research is focused on systemizing the effect of the fiber length distribution on the sliver irregularity and on estimating the bundle irregularity according to the roller setting on the basis of a dynamic model describing the bundle flow in a drafting zone. Employing beard diagrams for the fiber length distributions, we established an experimental plan for different levels of roller adjustments and draft ratios. The experimental results were then compared with those from simulations based on a theoretical model. An adequate model could be assured to provide information on the dynamic characteristics of the roller drafting operation, which could only be explained by experience.

2 Experiment
2.1 Experimental rig
In order to identify the effects of fiber length distribution on the irregularity of output sliver thickness, an

---

* Corresponding author's email: huhyou@khu.ac.kr

experimental system was constructed and established. The experimental rig consisted of 3 over 3 roller arrangement, while each roller was driven by individual servomotor that was controlled by a computer (Figure 1).

Figure 1 Experimental rig used in this research

2.2 Measurement of the output sliver thickness

The output sliver was guided through a trumpet. To measure the sliver thickness a leaf spring was employed. The sliver in the trumpet neck was pressed slightly by the leaf spring. A laser beam measured the position change at a point of the leaf spring on a real-time mode, which was caused by the output sliver thickness change (Figure 2).

Figure 2 The on-line measuring device of the sliver thickness

2.3 Raw material and experimental conditions

Fibers at a point between the two clamping lines move with the speed of either the front roller or the back roller. Floating fibers move with an intermediate speed. Therefore the fiber length distribution in relation with the drafting zone length plays a decisive role in the movement of the individual fibers, which leads to the thickness variation of the output sliver.

In this research we considered the fiber length distribution in drafting in terms of two kinds of beard diagram; a uniform fiber length and a distributed fiber length, which can be described in forms of a power form as

$$\gamma(x) = \left[\frac{l_{max} - x}{l_{max}}\right]^i \cdot u(l_{max} - x) \quad \text{for} \quad i = 1 \text{ or } 2 \tag{1}$$

where l_{max} is the maximal fiber length, and $u(x)$ denotes the unit step function.

The raw material with a uniform fiber length used for experiments was rayon and the staples with a fiber length distribution were cotton. Figure 3 and Figure 4 show the beard diagrams obtained from the specimens; a uniform length staple bundle and a cotton staple bundle, respectively.

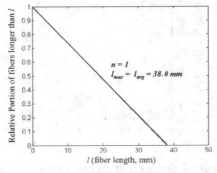

Figure 3 Beard diagram of the square-cut rayon staples

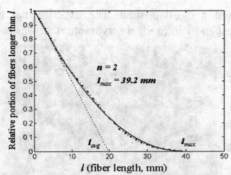

Figure 4 Beard diagram of the cotton used for the experiment

The specifications of the fibers and experimental conditions are summarized in Table 1.

Table 1 Material specifications and experimental conditions

Variables		Value
Material	Rayon	Fiber length: 38 mm (uniform) Fineness: 1.2 den Sliver thickness: 3.78 kTex
	Cotton	Fiber length: avg. 19.6 mm max. 39.2 mm (distributed) Fineness: 4.65 micronaire Sliver thickness: 3.42 kTex
Roller adjustment	main	40, 45, 50 (mm)
	break	45 mm
Draft ratio	main	3.85, 7.69, 11.54, 5.38, 19.23, 23.08, 26.92, 30.77
	break	1.3
	total	5, 10, 15, 20, 25, 30, 35
Process speed		200 mm/s
Number of doubling		9

3 Experimental results

The sliver thickness irregularity is expressed in terms of coefficient of variation in percentage. Followings show the effects of the drafting conditions on the irregularity of the output slivers attained from experiments.

3.1 Effect of the draft ratio

Figure 5 shows the CV(%) of the drafted sliver measured according to the draft ratio in roller adjustment of 40 mm.

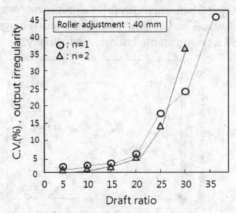

Figure 5 Irregularity of the drafted sliver vs. draft ratio for different fiber length distributions (roller adjustment: 40 mm)

As the draft ratio increases, the drafted slivers have worse irregularities. However, the draft ratio has an apparently different influence on the output thickness variation. Especially the higher draft ratio over 20 shows a dramatic increase in the irregularity, while in the low draft ratio level, the irregularity is very small.

The slivers of length-distributed fibers have better regularity in the thickness in the low level of draft ratio, which appears oppositely in high level of draft ratio.

If the roller distance is adjusted to 45 mm, then, the difference in affecting the irregularity by the draft ratio according to the fiber length distribution becomes smaller as given in Figure 6

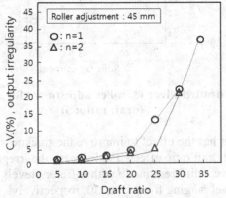

Figure 6 Irregularity of the drafted sliver vs. draft ratio for different fiber length distributions (roller adjustment: 45 mm)

If the roller adjustment is conducted with relatively long roller interval of 50 mm, the fiber length distribution shows almost the same behavior of the drafted sliver for the draft ratio, which is demonstrated in Figure 7.

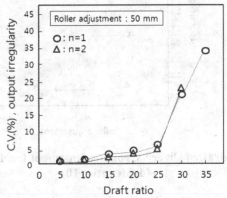

Figure 7 Irregularity of the drafted sliver vs. draft ratio for different fiber length distributions (roller adjustment: 50 mm)

There seems to be a level of draft ratio in which a dramatic change in the thickness irregularity starts to occur. At the roller adjustment of 45 mm, for example, the change in draft ratio from 25 to 30 for the cotton sliver (with a fiber length distribution) causes an abrupt irregularity increase in the output sliver, while the rayon sliver (with a uniform fiber length) experiences a big change in the thickness irregularity at the roller adjustment of 50 mm.

3.2 Effect of the roller adjustment

The sliver irregularity is known to be very sensitive to the roller adjustment, because the fibers not clamped by the rollers at the back or at the front of the draft device float in the bundle flow field, which is closely related with the fiber length distribution.

Figure 8 shows the experimental result, how the roller adjustment has the influence on the irregularity of the drafted slivers

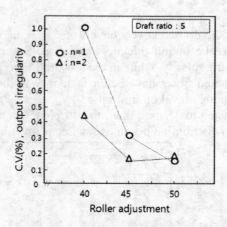

Figure 8 Irregularity of the drafted sliver vs. roller adjustment for different fiber length distributions (draft ratio: 5)

The longer roller adjustment has the effect to improve the thickness regularity in a low level of the draft ratio. Considering that the higher draft ratio affected to increase the irregularity, this irregularity behavior due to the roller adjustment could have a different mode with different levels of draft ratio. Figures 9 to 13 are the results obtained for draft ratio level ranging from 10 to 30, respectively.

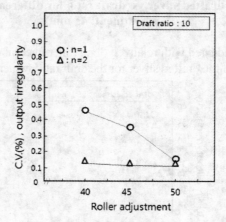

Figure 9 Irregularity of the drafted sliver vs. roller adjustment for different fiber length distributions (draft ratio: 10)

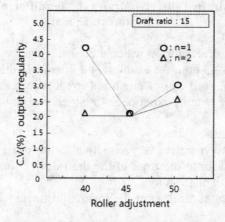

Figure 10 Irregularity of the drafted sliver vs. roller adjustment for different fiber length distributions (draft ratio: 15)

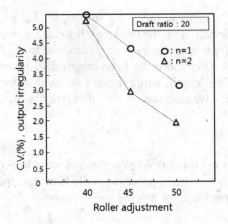

Figure 11 Irregularity of the drafted sliver vs. roller adjustment for different fiber length distributions (draft ratio: 20)

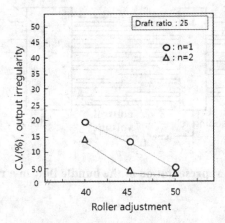

Figure 12 Irregularity of the drafted sliver vs. roller adjustment for different fiber length distributions (draft ratio: 25)

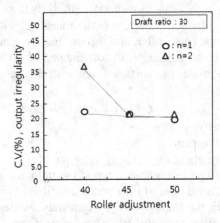

Figure 13 Irregularity of the drafted sliver vs. roller adjustment for different fiber length distributions (draft ratio: 30)

Details of individual factors acting on the irregularity look different. The roller adjustment in a very low level of draft ratio shows higher irregularity and also influences the sliver irregularity more sensitively for the uniform fiber length than for the fiber length distribution, which is not, however, so much significant as in the high level draft ratio, because the irregularity is negligibly small. For the draft ratio of 25 the Rayon sliver, the uniform fiber length, for example, shows a steady change in irregularity, while cotton sliver, the length-distributed fibers, shows a rapid change between 40 mm and 45 mm and then a steady value of the irregularity. For the draft ratio of 30 or more the roller adjustment shows too large irregularity which seems not to provide any meaningful information on the thickness fluctuation due to the unsuitable draft conditions.

It can be confirmed that the sliver irregularity is very sensitive to the form of the fiber length distribution. In the high level of the draft ratio 30 or more the fiber length distribution seems to cause the output sliver more irregular than the uniform fiber length, while in the low level of the draft ratio, on the other hand, the length-distributed fibers are more advantageous to the sliver regularity than the square-cut uniform fiber length.

4 Theoretical considerations

In order to explain the results obtained from experiments we adopted a theoretical dynamic model that was suggested in our previous study [4]. The in-process bundle flow constrained by two distant roller pairs can be described schematically as Figure 14.

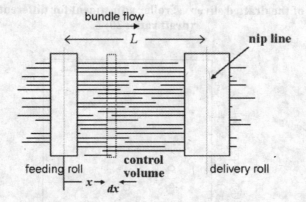

Figure 14 Schematic representation of the bundle flow in a roller drafting process

The model consists of
the continuity equation;

$$(l_b)_t = -(l_b \cdot v)_x \tag{2}$$

the equation of motion;

$$(l_b \cdot v)_t = -(l_b \cdot v^2)_x - (l_b \cdot Var[v_i])_x + (f)_x \tag{3}$$

where l_b denotes the linear density of fiber bundle, v is the mean velocity of fibers, t, x, and f denote time, distance in the flow direction from the back roller, and the surface force acting on fibers, respectively. $Var[v_i]$ stand for the velocity variance of fibers,. Subscripts denote the partial differentiation.

The constitutive equation that relates the surface force with the deformation can be supposed to have a linear form as

$$f \propto l_b \cdot (v)_x = \mu \cdot l_b \cdot (v)_x \ , \tag{4}$$

where μ stands for the bundle viscosity.

4.1 Velocity variance and beard diagram

Equation (3) is including a term about the velocity variance of the individual fibers. If a solution is to be attained from the model, the model can be complete, when the velocity variance is specified or described in the unknown variables. In general the number of the constituent fibers of a staple yarn is at least more than 50, the velocity variance distribution in the bundle flow system may be described in terms of the fiber length distribution of the bundle and the fiber velocity. We assumingly formulated the velocity variance with the velocity v and a velocity variance shape function $g(x)$ in such a proportional way as

$$Var[v_i] = a_0 \cdot v \cdot g(x) \ , \tag{5}$$

a_0 represents a parameter related to the length distribution.

Since the velocity variance is associated with the fiber length distribution of the bundle clamped at a point, this relationship in Equation (5) should satisfy the following conditions; the fibers clamped at rollers move with the linear velocity of the roller surface with the zero velocity variance. Therefore the velocity variance shape can be described in consideration of the fiber length distribution as follows;

Given the fiber beard diagram $\gamma(x)$, the velocity variance at position x, is expressed in terms of the portion of the floating fibers as

$$g(x) = 1 - \gamma(x) - \gamma(L-x) \qquad (6)$$

where L is the draft zone length.

The beard diagrams in Figure 3 and Figure 4, which could be described in a power form as given in Eq. (1), were substituted into Eq. (6) to make the model complete. The boundary conditions at the both ends of the draft zone were determined as

$$v(t,0) = v_0, \quad l_b(t,0) = l_{b0}, \quad v(t,L) = D_R \cdot v_0 \qquad (7)$$

4.2 Simulation

On the basis of Equations (2)~(7), while taking two representative examples for the beard diagrams given in Figure (2) and Figure (3), simulations were conducted.

For simulation of the output linear density out of the process, while the roller drafting zone length and the draft ratio were adjusted in various levels, we took a test signal that represented the input sliver thickness: a random signal with a 1% magnitude variation around the average, as given in Figure 15. The simulation conditions were taken as given in Table 2.

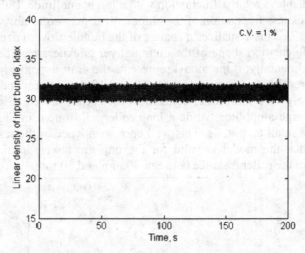

Figure 15 Test signal representing a random change in the input sliver thickness

Table 2 Simulation conditions

Parameter	Value	Parameter	Value
v_L, mm/s	200	l_{max}, mm	39.2, 38.0
D_R	5, 10, 15, 20, 25, 30	n	1, 2
		μ, mm^2/s	2.5 x 10^7
L, mm	40, 45, 50	a_0, mm/s	2.0 x 10^5

5 Simulation results
5.1 Effect of the roller adjustment
5.1.1 The case of a uniform fiber length

Figure 16 illustrates a simulation result about the response of the output sliver thickness drawn with the roller adjustment of 40 mm to the random perturbation in the input sliver thickness, provided the sliver consists of staples of a uniform fiber length.

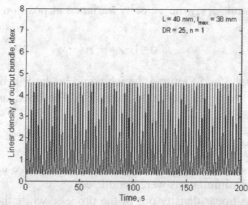

Figure 16 Linear density of the output sliver with the uniform fiber length distribution ($L = 40$ mm, $l_{max} = 38$ mm $D_R = 25$) simulated

The output thickness displays regular fluctuation with a large amplitude. If the roller adjustment changes, however, to 50 mm, the thickness fluctuation of the output sliver gets noticeably attenuated, even though the draw ratio remains (Figure 17). This visualizes a change of the bundle flow dynamics due to the drafting zone length. The change in the fluctuation shape of the output sliver thickness resulting from a theoretical model needs to be identified experimentally, if the model represents the draft dynamics to such an extent as the real process. The experimental results are given in Figures 18 and 19 that correspond to the simulation results in Figures 16 and 17 respectively. In a short roller adjustment the output sliver in Figure 16 reveals a fluctuating thickness behavior with a large amplitude, while a long roller adjustment in Figure 17 yields the thickness fluctuation with a relatively small amplitude. This well correspondence of the results from theoretical model and experiment indicates that the model is valid for the uniform sliver drafting and a drastic change in dynamics could take place in the roller distance between 40 mm and 50 mm.

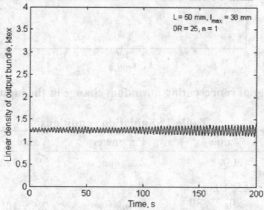

Figure 17 Linear density of the output sliver with the uniform fiber length distribution ($L = 50$ mm, $l_{max} = 38$ mm $D_R = 25$) simulated

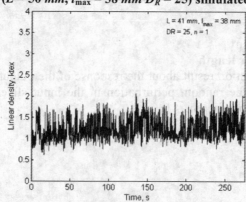

Figure 18 Linear density of the output sliver with the uniform fiber length distribution ($L = 40$ mm, $l_{max} = 38$ mm $D_R = 25$) measured

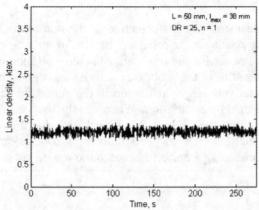

Figure 19 Linear density of the output sliver with the uniform fiber length distribution ($L = 50$ mm, $l_{max} = 38$ mm $D_R = 25$) measured

5.1.2 The case of a fiber length distribution

If the sliver consists of staples with a fiber length distribution, the effect of the grasped fibers can be different from the sliver with a uniform fiber length. Based on the model including the beard diagram in a 2nd-power form, the output thickness was simulated, while the roller adjustment was changed. Figure 20 shows a result, when the roller distance was adjusted to 40 mm for the maximum fiber length of 39.2 mm: a very close roller adjustment by comparison with the maximum fiber length. In the same manner as the sliver with a uniform fiber length, the output sliver has a large and regular change in the thickness. The fluctuation amplitude of the output thickness, however, is decreased in a noticeable way (Figure 21). When the roller span is adjusted to a longer distance of 50 mm, the fluctuation in the output thickness becomes by far weaker.

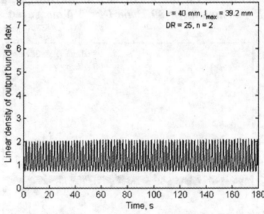

Figure 20 Linear density of the output sliver with the fiber length distribution ($L = 40$ mm, $l_{max} = 39.2$ mm $D_R = 25$) simulated

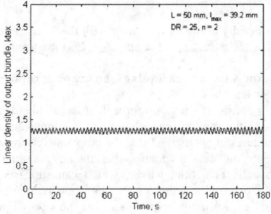

Figure 21 Linear density of the output sliver with the fiber length distribution ($L = 50$ mm, $l_{max} = 39.2$ mm $D_R = 25$) simulated

The effect of the roller adjustment on the fluctuation in the output thickness was also investigated by experiments. Figure 22 shows the result for the roller adjustment of 40 mm, corresponding to the simulation result in Figure 20. Figure 23 is a result for the roller adjustment of 50 mm, comparable with the simulation result in Figure 21. The sliver output from the process reveals a very similar behavior to the simulation; a 40 mm roller adjustment yields a relatively large fluctuation in the output thickness, and a 50 mm adjustment, however, results in a very small thickness variation. When the fluctuation amplitudes are compared for the two different fiber length distributions, the sliver with a fiber length distribution has less fluctuation than the sliver having a uniform fiber length, which suggests drawing a conclusion that slivers with fiber length distribution is advantageous for producing a uniform sliver to the square-cut length slivers.

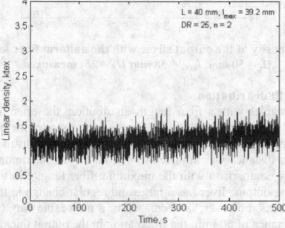

Figure 22 Linear density of the output sliver with the fiber length distribution
($L = 40$ mm, $l_{max} = 39.2$ mm $D_R = 25$) measured

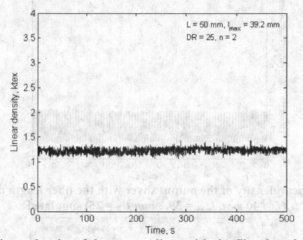

Figure 23 Linear density of the output sliver with the fiber length distribution
($L = 50$ mm, $l_{max} = 39.2$ mm $D_R = 25$) measured

5.2 Response of the output thickness in various levels of the drawing ratio
5.2.1 The case of a uniform fiber length

Given a roller adjustment for the sliver with a uniform fiber length, the thickness responses of the output sliver for two draft ratio levels to the random perturbation in the input sliver thickness are illustrated in Figures 24 and 25 respectively. Interesting is the fact that the output sliver drafted with a low draft ratio of 10 (Figure 24) has a less amplitude of thickness fluctuation than the sliver drafted with a high draft ratio of 30 (Figure 25). A low draft ratio delivers a thick sliver, which can supposedly lead to a large thickness fluctuation. However, the simulation results demonstrated that the irregularity magnitude increases, as the draft ratio increases, which could be confirmed experimentally as shown in Figures 5, 6, and 7.

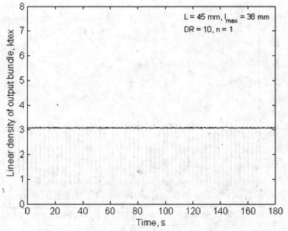

Figure 24 Linear density signals of the roller drafted slivers with the uniform fiber length distribution ($L = 45\ mm$, $l_{max} = 38\ mm$ $D_R = 10$) simulated

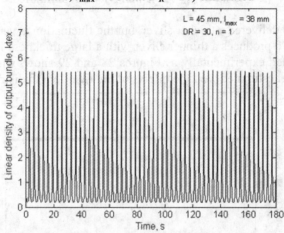

Figure 25 Linear density signals of the roller drafted slivers with the uniform fiber length distribution ($L = 45\ mm$, $l_{max} = 38\ mm$ $D_R = 30$) simulated

5.2.2 The case of a fiber length distribution

When the input slivers with a fiber length distribution and a random thickness change are drafted with different draft ratios, the output sliver thickness can reveal the effects of the draft ratio. Figures 26 and 27 give the simulation results.

In the same manner as in the sliver with a uniform fiber length, the sliver with a fiber length distribution reveals a thickness fluctuation different from what can be expected.

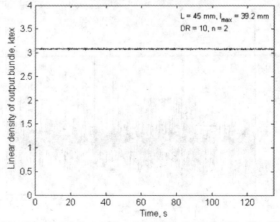

Figure 26 Linear density signals of the roller drafted slivers with the fiber length distribution ($L = 45\ mm$, $l_{max} = 39.2\ mm$ $D_R = 10$) simulated

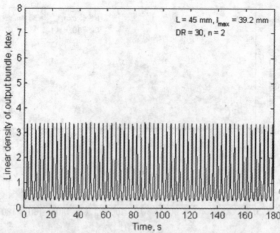

Figure 27 Linear density signals of the roller drafted slivers with the fiber length distribution ($L = 45$ mm, $l_{max} = 39.2$ mm $D_R = 30$) simulated

A lower draft ratio of 10 delivered a thicker sliver, but the fluctuation is very small as given in Figure 26, while a higher draft ratio of 30 produces a thinner sliver with a large thickness fluctuation (Figure 27). These results could also be confirmed experimentally as Figures 28 and 29 show, which implies that a simulation based on the theoretical model estimated well the effects of the fiber length distribution.

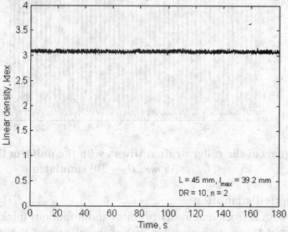

Figure 28 Linear density signals of the roller drafted slivers with the fiber length distribution ($L = 45$ mm, $l_{max} = 39.2$ mm $D_R = 10$) measured

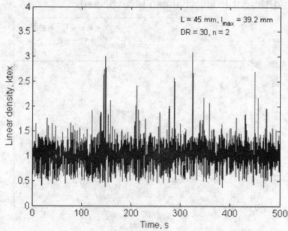

Figure 29 Linear density signals of the roller drafted slivers with the fiber length distribution ($L = 45$ mm, $l_{max} = 39.2$ mm $D_R = 30$) measured

The dynamic characteristics seems to change dramatically in a small range of draft ratio between 25 and 30, leading to a large thickness fluctuation in the output sliver, if a fiber length distribution is given, as interpreted from Figures 5, 6, and 7. The same phenomenon was also observed for the sliver with a uniform fiber length, but the range of the draft ratio leading to a dramatic increase in the thickness fluctuation seems to be shifted to the lower side than 25.

6 Conclusions

This research targeted to show the effects of the fiber length distribution on the irregularity of roller drafted slivers in consideration of the roller adjustment and draft ratio as process variables.

We confirmed that the form of the fiber length distribution exerted a strong influence on the sliver thickness fluctuation. Especially, the fiber length distribution led the output sliver to more regular thickness than the uniform fiber length in the low level of the draft ratio.

The fiber length distribution exerted different effects on the output thickness irregularity through the roller adjustment; for a given draft ratio, the uniform length fibers result in a drastic change in thickness fluctuation at a longer roller adjustment than the length-distribution fibers.

For a given roller adjustment, the draft ratio can also cause a different fluctuation behavior in the output thickness. An abrupt fluctuation increase in the output sliver thickness occurs, for example, between the draft ratios of 25 and 30 for the length- distributed fibers at the roller adjustment of 45 mm. For a uniform fiber length it takes place in the lower-shifted draft ratio.

In this research a mathematical model describing the dynamic behavior of the bundle flow inside of a roller draft zone was applied, while the beard diagram was taken into consideration. For a random change in the input sliver thickness, the output sliver thickness simulated was compared with experimental results for various levels of roller adjustments and draft ratios. The simulation results agreed well with those from experiments, which indicated that the model impregnated with the beard diagram described the roller drafting operation well.

References

[1] H. Balasubramanian, P. Grosberg, Y. Turkes, Studies in Modern Yarn Production. Textile Institute, Manchester, p 169-180 (1968).
[2] B. Dutta and P. Grosberg, "The Dynamic Response of Drafting Tension to Sinusoidal Variations in Draft Ratio under Conditions of Sliver Elasticity in Short-staple Drafting", *J. Textile Inst.*, **64**, 534-542 (1973).
[3] G. Mandl and H. Noebauer, The Influence of Cotton-Spinning Machinery on the Random Irregularity of Sliver and Yarns - Part I, II, III. J.T.I., 68, 387-393, 394-399, 400-406 (1977).
[4] You Huh and Jong S. Kim, "Modeling the Dynamic Behavior of the Fiber Bundle in a Roll-Drafting Process". Textile Research Journal, 74 (10), 872-878 (2004).
[5] You Huh and Jong S. Kim, "Steady-Flow Characteristics of Bundle Fluid in Drawing", Transaction of the KSME, Vol B, 30(7), 612-621 (2006).
[6] You Huh and Jong S. Kim, "Effect of Material parameters and Process conditions on the Roll-Drafting Dynamics", *Fibers and Polymers*, 7(4), 424~431 (2006).
[7] Jong S. Kim and You Huh, "Stability of Bundle Flow Dynamics", J. Korean Fiber Soc., 45(2), 89-96 (2008).

59 The Study on Comprehensive Comfort Property of Garment with Knitted Underwear Fabrics during Exercise

Shan Cong*

(Shanghai University of Engineering Science, Shanghai, China)

Abstract: Comfort is one of the most important attributes demanded by consumers during exercise. It reflects the psychological feeling of a wearer, featured by three latent independent sensory factors: thermal-wet comfort and tactile comfort. The aim of this paper is to presents a detailed discussion of knitted underwear fabrics in garment to achieve maximum comprehensive comfort during exercise. The 6 T-shirts made of different kinds of fibers was investigated in wear trials in a standard environmental chamber. The experimental results show that the comfort of the 6 kinds of T-shirt is varied due to the fiber types, which mainly influence the heat and moisture transfer during exercise. The relationship between the feelings of thermal, moisture and tightness are also discussed in this paper. Based on the experimental results, a predictable model of comprehensive comfort during exercise is developed using mainly sensory factors, which is proved valid in comparison wear trials.

Keywords: Comprehensive comfort; Thermal-wet comfort; Tightness sensory; Exercise; Knitted underwear fabrics

0 Introduction

Comfort is one of the most important attributes demanded by modern garment consumers. It reflects the psychological feeling of a wearer. Comfort researchers recognize that garment comfort has two main aspects that combine to create a subjective perception of satisfactory performance: thermo physiological and sensorial comfort. The first relates to the way garment buffers and dissipates metabolic heat and moisture [1, 2], whereas the latter relates to the interaction of garment with the senses of the wearer, particularly with the tactile response of the skin, which includes moisture sensation on the skin [3,4]. A. S. Wong and Li studied the contributions of each factor to the comprehensive comfort by carrying out a series of subjective wear trials, found that thermal-wet comfort, tactile comfort and pressure comfort were the three main sensations perceived by subjects during exercise, thermal-wet comfort being strongest [5].

The garment comfort can be divided into three groups as follows: thermal, psychological and tactile comfort in some research [6, 7]. Comfort is a state of multiple interactions among psychological sensations that are determined by physical, psychological and environmental factors in their opinion. However the relationship between the sensory factors and comprehensive comfort model before, during and after exercise are not studied. Hence, the comprehensive comfort of garment during exercise is the main issue discussed in this paper.

Some hygroscopic fabrics are popular in modern underwear textile market, which are chosen in this experiment include natural fabrics such as cotton, as well as regenerated cellulosic fibers. Knitted cotton as traditional underwear fabric possess high performance properties in the garment market; Regenerated cellulosic fibers offer many advantages over synthetic ones, comprising a significant share of the man-made textiles market, such as modal, soybean and bamboo fiber. This subject studied the comfort of the 6 kinds of T-shirt is made by the newly developed fabric, which mainly influence the heat and moisture transfer during exercise. The psychological sensory process is obtained by carrying out a series of subjective wear trail in a standard environmental chamber.

1 Experimental
1.1 Test subjects

20 female college students were selected as subjects to take part in a psychological sensory running trial conducted in an environmentally controlled laboratory. The wear trial was divided into two parts: objective measurements and subjective perceptions. The subjects between 18 and 22 were invited to have a pre-trial before formal trials, to obtain training and an understanding of the questions and the procedures involved. All

---

* Corresponding author's email: congsun@sina.com. This project is supported by Shanghai science fund for nature, Fund NO. 10ZR1412800

subjects wore garment of the same size. Physical characteristics of the subjects are medium height and weight.

1.2 Test fabric and garment

The 6 T-shirts made of different kinds of fibers, which are various fibers content, the same blend composition and fabric weave. All fabric construction is weft flat knit. Wear trials were designed to investigate the thermal and humidity comfort of these fabrics during sweating in wear conditions, and to establish optimal combinations of multiple fabrics to achieve maximum thermal and humidity comfort during wearing.

The fabrics were put together to produce matching long-sleeved round neck T-shirts in sizes to fit the subjects. During the trials, all subjects wore the same underpants. The T-shirts were of the same color to ensure that the subjects could not visually distinguish between the garments. The basic properties of the knitted underwear fabrics are shown in Table 1. Before the garments were used in the test, they were conditioned in a controlled room with temperature at $20\pm2\,^\circ C$ and humidity of $65\pm3\%$ for 24 hours.

Table 1 Performance experimental data of fabric

Sample code	Fiber type and composition	Fabric weight /g·m^2	Loop length /mm	Fabric thickness /mm	Clo (clo)	Moisture regain /%
1#	Cotton 95% Lycra 5%	172.15	2.80	0.81	0.362	4.12
2#	Model 96% Lycra 5%	189.12	2.90	0.74	0.321	7.82
3#	Soybean95% Lycra5%	176.91	2.20	0.58	0.261	5.78
4#	Bamboo95% Lycra 5%	173.82	3.00	0.71	0.346	7.65
5#	Chatoyant fiber Lycra 5%	163.13	2.55	0.76	0.278	6.21
6#	Model 33% cotton 60% Lycra 7%	187.14	3.10	0.69	0.341	7.86

1.3 Experimental procedure

Before the wear trial each subject rested in the environment for 15 minutes to reach equilibrium. Then the subject was asked to fill in a questionnaire by rating the three sensations (thermal, moisture, tight) and comprehensive comfort a five-main-point scale rated from -2 to +2. During each trial, each subject changed into the test garments with the test T-shirts on the top, bottom, with the uniform underwear on the bottom. The test conditions were controlled at temperature $21\pm2\,^\circ C$ and humidity $65\pm3\%$ with an air velocity varying around 0.50 m/s.

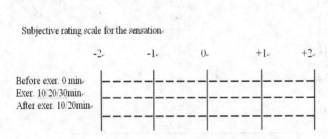

Figure 1 Five points scales for measuring the sensations of thermal, moisture and tightness

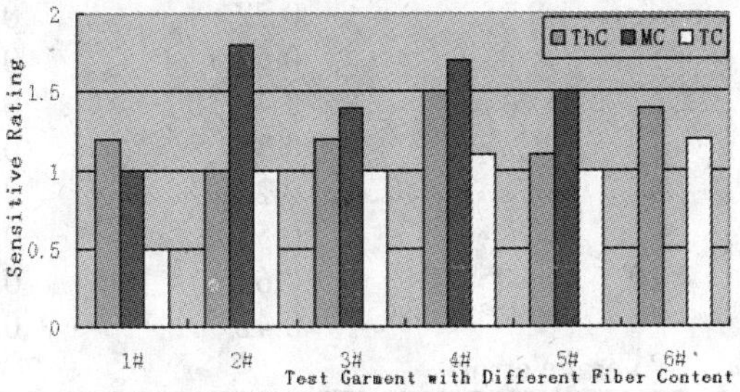

Figure 2 Subjective rating for the different test garment

After end of the equilibrium the subject was asked to walk on a treadmill (DF-1160) for 30 minute of walk. Each subject was asked to fill in a questionnaire at the time =10, 20 and 30 by rating the three sensations and comprehensive comfort. At the end of the walk and after 10 and 20 minutes of walk the subject was also asked to fill the questionnaire by rating the three sensations and comprehensive comfort. The trial aimed to investigate the influence of garment to the subjects for a normal walk time. Therefore, the three states were recorded at equilibrium, exercise and rest time. Figure 1 shows four samples of the scale. Figure 1 shows sample of Subjective rating scale for the sensation. In the scale cold is equal to -2, moderate is equal to zero, and too hot is equal to +2. The description of moisture is from too dry to too damp, and tightness is from too tight to too loose.

2 Results and discussion

Figure 2 is the subjective rating for the different test garment to the thermal comfort (ThC), moisture comfort (MC) and tightness comfort (TC) during the whole exercise process. Subjects perceived a more excellent performance to the thermal-wet comfort with soybean, bamboo and model than cotton, because these new fiber *mainly influence the heat and moisture transfer during exercise. The sample 1#* exhibited the lowest tightness sensation. Because perspiration from the human body may build up with unfit garment, the too tight and the evaporation of water can cause discomfort. Cotton exhibited the highest moisture after exercise: wool the lowest. The others are between the two, whereas bamboo/cotton, model and bamboo are closer to the cotton.

2.1 The correlative analysis between sensory factors

Graphs were used in comparison the changes of each factor at three states. SPSS (Statistical Package for the Social Sciences) were carried out to analysis the correlation, Pearson coefficient, two-tailed test and the significant level at 5% in Table 2.

Table 2 Correlations between mean comfort scores

		Thermal comfort	Moisture comfort	Tightness comfort
Thermal comfort	Pearson Correlation	1	.662(*)	–0.121
	Sig. (2-tailed)	.	.001	.002
	N	120	120	120
Moisture comfort	Pearson orrelation	.662(*)	1	-.362(**)
	Sig. (2-tailed)	.001	.	.001
	N	120	120	120
Tightness comfort	Pearson Correlation	–0.121	-.362(**)	1
	Sig. (2-tailed)	.002	.001	.
	N	120	120	120

* Correlation is significant at the 0.05 level (2-tailed).

Table 3 Result and error analysis of predicted comprehensive comfort model

subject no.	Actual value	Prediction value	Relative error/%
1	0.47	0.46	2.27
2	0.52	0.55	2.64
3	0.62	0.53	2.30
4	0.52	0.50	1.92
5	0.52	0.51	1.92
6	0.42	0.41	2.38
7	0.32	0.31	3.12
8	0.24	0.25	4.16
9	0.53	0.54	1.88
10	0.63	0.63	0
	0.42	0.41	2.38
Average error			2.49

By using the correlative analysis, mean tightness sensory perception has negative correction with moisture comfort (r=-.362 at the 0.05 level) at the whole state. And the tightness has slight negative correlation with the thermal sensory perception (r=–0.121at the 0.05 level). Table 2 shows the correlation between mean thermal comfort and mean moisture comfort correlation coefficient r is equal to 0.662 and the probability is equal to 0.001. therefore the moisture comfort has positive relativity with thermal comfort. The results indicate tight sense would result in cold and dry sense. The mean damp and thermal sensor is positive relative during exercise. But the three sensory factors conduct differently at the different stages. Figure 2 illustrate the behavior of the three factors sensory perception against time.

2.2 The behavior analysis mainly sensory factors against time periods

In order to compare the behavior of different sensory factors during different time period, the three sensations factors during different time period, the factor scores are plotted against time periods. Figure 3 shows the changes of thermal, moisture and tightness at 3 different states of the time, which are equilibrium, movement and after movement time at time = 0, 10, 20, 30, 40 and 50 minute.

The mean factor score of thermal comfort has increased significantly with time in the exercise period as

Figure 3 showed. But the scores of moisture comfort and tightness comfort change little, though with an increasing trend. Moisture factor scores increase and reach a peak at the 30 minutes, then begin to decline towards the end of 50 minutes of the trial. This suggests that after half an hour, moisture comfort became the main perception. The reason of the changes that exercise makes human sweated slightly has significant influence on moisture sensory comfort.

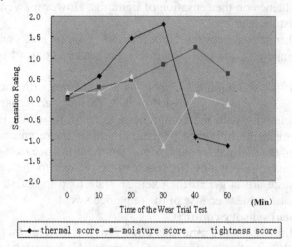

Figure 3 Comparison the values of 3 sensory comfort at different time states

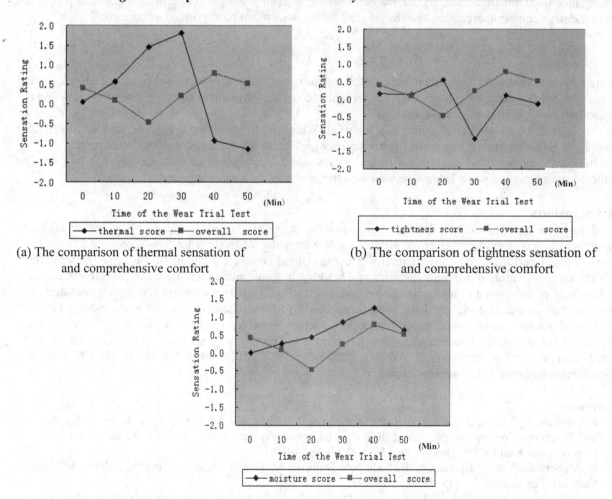

(a) The comparison of thermal sensation of and comprehensive comfort

(b) The comparison of tightness sensation of and comprehensive comfort

(c) The comparison of moisture sensation of and comprehensive comfort

Figure 4 Comparisons analysis the relationship of mainly sensory factor with comprehensive comfort

Thermal and moisture comfort values increased obviously just finishing movement and decreased 10 minutes rest after movement. Yet the values of 10 minutes after movement are high than the values at the beginning of the trial. However the values of tightness are approximate under three different states.

The tightness comfort change at three conditions show that movement that makes human sweated slightly has no significant influence on the sensation of tightness. However sweat, even slight, has significant influence on thermal and moisture comfort.

In order to analyze correlation between three sensory factors and comprehensive comfort during different time period, the compare factor scores with comprehensive comfort separately are plotted against time periods in Figure 4.

The analyses shows moisture and tightness sensory have significant influence on comprehensive comfort at 21±2℃ and humidity, 65±3%, which is consistent with the former researches [8,9]. But the thermal sensory has more negative relativity with the comprehensive comfort, which is different with the result of Wong's [5]. Just after movement the influence of thermal and moisture enhanced and tactile decreased or human body is not sensitive to the thermal sensory. Moreover the influence of tightness and moisture comfort is more significant after the exercise minutes with each value recover to the comprehensive comfort value. However the values of moisture are larger in the exercise state than in the static values and the values of roughness are less than the static values because of the remained sweat.

2.3 Model to simulate garment sensory comfort

From the results and discussions we find the three sensory factors, thermal, moisture and tightness comfort factors, determine the comfort performance of garment. A model to simulate the over comfort performance from individual sensory factors is proposed; with an assumption of a linear relationship between comprehensive comport perception and the sensory factors which can be expressed as equation 1:

$$P_{CS} = \sum V_F W_F \tag{1}$$

$$P_{CS} = -0.498 F_T + 0.275 F_M + 0.112 F_{TI} \tag{2}$$

Where P_{CS} s the simulated comprehensive comfort performance, V_F is the individual sensory factor score and W_F is the weight of variance explained by the performance of the garment. Weighting coefficient is set on the basic of the findings from regressive analysis.

The P_{CS} can be denoted as equation 2 in our trial. Where F_T = Simulated comprehensive comfort factor score; F_M = Simulated moisture comfort factor score, F_{TI} = Simulated tightness comfort factor score.

In order to examine the validity of this model, equation 2 is applied to 10 other new set of trial data. The predicted sensory scores were Table 3 was showed the predicted sensory scores and error analysis.

3 Conclusions

This paper presents a detailed discussion of knitted underwear fabrics in garment to achieve maximum comprehensive comfort during exercise Knitted underwear garment made from new reproduced fibers have more excellent performance to the thermal-wet comfort than cotton. By conducting a series of wear trials with 20 females wearing 6 different garments and walking in a controlled climatic chamber, subjective rating on 3 sensory perceptions and the comprehensive comfort of the garments are obtained during and after their 20 minutes of the exercise trial. A linear predictable model to comprehensive comfort was developed using 3 factors and relative contribution of corresponding factors as weights. By comparing the comfort score predicted from the models with the actual comfort score rated by the subjects. Good predictable ability of the model is found, average relative error (%) at 2.49%, indicating that the comprehensive comfort performance is predictable from individual sensory factor.

References

[1] K L Hatch, S. S Woo, R L. Barker. Radhakrishnaiah, P., Markee, N. L., Maibach, H. I.; In Vivo Cutaneous and Perceived Comfort Response to Fabric Part I:Thermo-physiological Comfort Determinations for Three Experimental Knit Fabrics, Textile Res. J. 1990, 60, pp. 405-412.

[2] D. M, Scheurell,. S. M Spivak., N. R. S Hollies; Dynamic Surface Wetness of Fabric in Relation to Clothing Comfort, Textile Res. J. 1985, 55, pp. 394-399.

[3] R. F Goldman.; Evaluating the Effect of Clothing on the Wearer, 'Bioengineering, Thermal Physiology and Comfort', Cena, K. and Clark, J. A., Eds., Elsevier Scientific Publishing Co., 1981, Chap. 3, Amsterdam.

[4] A. M Plante, B. V Holcombe, Stephens, L. G.; Fibre Hygroscopicity and Perception of Dampness: Part 1.

Subjective Trials, Textile Res. J. 1995, 65, pp. 293-298.
[5] A. S Wong, W., Li, Y.; "Psychological Requirements of Professional Athletes for Active Sportswear," in "Proc. 5th Asian Textile Conference," 1999, Vol. 2, pp. 843-846.
[6] A S Wong, Y. Li and K W Yeung, Statistical simulation of Psychological Perception of Garment Sensory Comfort, J. T. I. 2002, 93(1):108-119.
[7] Y. Li, Garment Comfort and Development of Products [M], First version, Beijing: China Textile Press, 2002, 129.
[8] Nanliang Chen. Research on new testing method and instrument of garment pressure, Journal of Donghua University, 1999, 16(4):104-107.
[9] W Yu, J Fan,, A Soft Mannequin for Predicting Girdle Pressure on Human Body, Sen'I Gakkaishi, 2004, 60(2):57-64.

60 Preparation and Application of a Novel Fluoroalkylpolysiloxane Fabric Finish with Waterproofing and Washing Resistant Properties

Lifen Hao, Qiufeng An, Wei Xu, Qianjin Wang

(Key Laboratory of Auxiliary Chemistry and Technology for Chemical Industry, Ministry of Education, Shaanxi University of Science and Technology, Xi'an 710021, China)

Abstract: A novel fluoroalkylpolysiloxane fabric finish(FAS-PFMS), bearing pendant long chain perfluoroalkyl, stearyl acrylate, and epoxy, was synthesized by hydrosilylation of polytrifluoropropylhydromethylsiloxane (PFHMS), perfluorooctylethene (PFOE), stearyl acrylate (SA) and allylglyeidyl ether (AGE). The chemical structure of FAS-PFMS was characterized by infrared spectrum (IR) and proton nuclear magnetic resonance (^{1}H-NMR). The film morphology of FAS-PFMS on cotton fiber surfaces was observed by field emitting scanning electron microscope (FESEM). Results show that fiber treated by FAS-PFMS shows a correspondingly smooth macro-morphology under FESEM observation. Grade of washing resistance of the treated samples could reach to 90 and CA of water utmost reach to 132° by a mass concentration of 0.5% FSA-PFMS ethyl acetate solution. Softness of the treated cotton fabric is relatively increased and whiteness is comparable to the blank sample.

Keywords: Fluoropolysilioxane copolymer; Waterproofing property; Washing resistant property; Fabric finishing

0 Introduction

Long fluoroalkyl functionalized polysiloxanes possessing merits of organosiloxane and organofluoro chemicals, such as excellent environmental stability, water and oil repellency, low coefficient of friction, excellent thermal stability and chemical resistance and low interfacial free energy [1-2], etc., can present the treated fabrics outstanding water and oil repellency, antisoiling properties and elevated softness, good permeability as well and have recently been attracted much attention. Involved researches are mainly focused on long fluoroalkyl polysilanes [3-5], fluorinated acrylate copolymer modified by organosilicon [6] and fluorinated polysiloxanes. Water and oil repellency of long fluoroalkyl polysilanes is excellent, however, due to its low molecular weight, compact hydrophobic film is difficult to form on fiber substrates and its high cost disadvantages broad application. Some fluorinated polysilanes can overcome the flaws of small molecular self-assembly method but its washing resistant property is poor owing to single fluoroalkyl group in backbone or side chain. Poly[methyl (3,3,3-trifluoropropyl)siloxane] (PMTFPS) is a favorable performance polymer and widely used as various functional coatings such as fabrics, paper, leather, wood, building and metal, etc. Nevertheless there do not contain reactive groups in its molecules and it can not react with substrates in chemical bonds or static effects, i.e., the fabrics pretreated by those reagents do not possess wash durability.

To improve the above-mentioned drawbacks of the PMTFPS and obtain anticipative hydrophobic property and wash durability, we successfully prepared a novel fluoro-silicon copolymer (FAS-PFMS) through hydrosilylation of PFHMS, perfluorooctylethene, stearic acrylate and allylglyeidyl ether. Then hydrophobic property and wash durability of the treated fabrics by FAS-PFMS were chiefly reported in this manuscript.

1 Materials and methods

1.1 Materials

Polymethyltrifluoropropylhydrosiloxane (PFHMS) was prepared in our laboratory with a Si-H content of 0.11% (expressed by the moles of Si-H groups contained in 1 grams silicone). Allylglyeidyl ether (AGE), perfluorooctylethene (PFOE), stearyl acrylate (SA), all industrial grades, was purchased from Jintan Huadong Inc.and Haerbin Xuejia Inc. respectively. Acetic acid and acetic ether, all A.R, were from Xi'an Chemicals, all in China.

1.2 Synthesis and characterization of FAS-PFMS

In a three-necked flask equipped with a mechanical stirrer, a reflux condenser and a thermometer, PFHMS, SA, PFOE and AGE were added according to the stoichiometric ratio, gently stirred, and then bubbled by N_2 for 10 minutes. When mixture was heated to 95-105°C, a catalytic amount of H_2PtCl_6 was dropped. Then, the mixture was kept at this temperature for an additional 3-4 h. At the end of the reaction, low-boiling residues were removed by vacuum distillation. Finally, a clear, viscous fluid, stearyl acrylate/perfluorooctylethene/epoxy modified polysiloxane (FSA-PFMS), was obtained.

Samples were cross-examined using different instruments combined with chemical analysis method. An infrared spectrum was recorded on a Bruker VECTOR-22 FT-IR Spectrometer. ^{1}H NMR spectrum were recorded at 26°C on INOVA-400 with $CDCl_3$ as solvent and tetramethylsilane (TMS, δ=0 ppm) as internal standard.

1.3 Film morphology of FAS-PFMS on cotton fabrics and its application evaluation

1.3.1 Samples treatment and application technology

100% cotton fabrics with yarn counts of 474×235 (the number of warp and filling yarns, 10cm×10 cm) was used for FESEM investigation and application experiments. Prior to the treatments, all cotton samples were ultrasonicated with deionized water and acetone respectively at 25°C for 20 min, then dried at 100°C for 5 min. This process was to remove the slurry and contaminants on the fiber/fabric substrates.

FSA-PFMS was dissolved in redistilled ethyl acetate to form a FSA-PFMS solution with a mass concentration of about 0.2%, 0.3%, 0.4%, 0.5% and 1.0%, respectively. Degrease cotton pretreated according to reference[7] method were immersed in the above FSA-PFMS solutions for several seconds, and then dried at 100°C for 5 min and cured at 170°C for 2 min. Finally, the treated samples were kept in a desiccator to balance at room temperature for 24 h.

1.3.2 Morphology observation

FESEM observation was carried out on a SIRION 200 field emitting scanning electron microscope (FEI Co, Hollan) after the treated fabric sample of cotton fabrics were coated with gold in a vacuum.

1.3.3 Application performance evaluation

Bending rigidity (BR) was measured with a Kawabata Evaluation System (KES) instrument. Whiteness was determined by an YQ-Z-48B fluorescent whiteness tester. Wettability was evaluated by contact angle of H_2O (WCA) on the FSA-PFMS treated cotton fabrics measured by JC2000 angle contact measurement made in Powereact, Shanghai, China and values were the average of five determinations. Washing resistance was analyzed according to the normal AATCC135-1987 method before which fabric was washed under 50°C and then was ironed to dryness.

2 Results and discussion

2.1 IR and ^{1}H NMR analysis

Figure 1 and 2 showed the typical spectrum of IR and ^{1}H-NMR of FSA-PFMS, respectively. According to Figure 1, there are typical adsorption bands appeared separately at 2963-2857 cm^{-1}, 1737 cm^{-1}, 1128 cm^{-1}, 1211 cm^{-1}, 1129-1024 cm^{-1}, which indicated functional groups such as enormous methyls and methylenes, -C=O and O=C-O, C-F groups in FSA-PFMS molecule. Adsorption band at 2163 cm^{-1} disappeared basically denoted that the Si-H addition reaction was accomplished during the modifying process. A broad and enhanced band at 1129-1024 cm^{-1} should be resulting from adsorption band overlaps between Si-O-Si and C-O-C. Owing to very low content of the material AGE, the adsorption bands of epoxy group could not be observed.

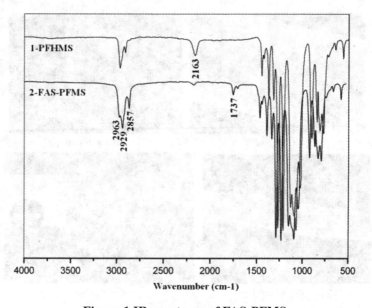

Figure 1 IR spectrum of FAS-PFMS

Figure 3 is ^{1}HNMR spectra of FSA-PFMS. From ^{1}H NMR spectrum and peaks assignments analysis, it was clear that there were functional groups such as -Si-CH$_3$, -Si-CH$_2$-CH$_2$-CF$_3$, -Si-CH$_2$CH$_2$CH$_2$OCH$_2$CHCH$_2$O, -Si-CH$_2$CH$_2$COOCH$_2$(CH$_2$)$_{16}$CH$_3$, -Si-CH$_2$CH$_2$CF$_2$(CF$_2$)$_6$CF$_3$, and residual -SiOCH$_3$ in FSA-PFMS molecule. In conclusion, combined the results of IR and ^{1}H NMR analysis, a novel polysiloxane was prepared containing pendant stearyl acrylate, perfluorooctylethene, epoxy, and trifluoropropyl side groups.

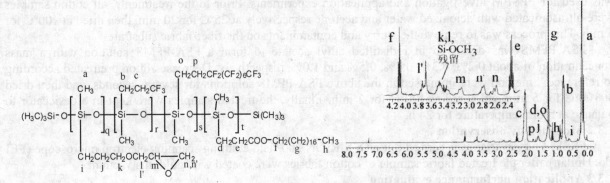

Figure 2 ^{1}H-NMR spectrum of FAS-PFMS

2.2 FESEM analysis of FAS-PFMS on cotton fabrics

The direct way to study the morphology of functional groups modified polysiloxane on fabrics is to observe the treated fabric surface instrumentally. However, SEM cannot reveal the clear microstructure of the siloxane film on the fiber surface. Hence, a new powerful instrument FESEM was utilized in our experiment. The results are shown in Figure 3. Known to all of us, polysiloxanes are low surface energy materials which makes them easily spread and form a film of molecular dimension on the hydrophilic fiber surface. As such a film sheathes the fibers; it must lead to a modification in morphology of the fibers. Therefore, observation of the appearance of the treated fiber and comparing it with that of the control, the untreated fiber, will help us to get the information about the siloxane softener on fiber/fabric substrate. From Figure 3, it was clearly observed that the control, the untreated fiber surface was uneven, there were many deep grooves or slender concaves on its surface, and the edges on some locals of the fiber were slightly sharp. Whereas, the treated fiber surface was relative smooth, most of the concaves disappeared and some local edges of the treated fiber surface became blunt. It is obvious a polysiloxane resin film was deposited on the treated fiber surface which would influence properties of the treated fibers. So we can see that surface of the treated fibers is changed from hydrophile to hydrophobe with water contact angle of 132°.

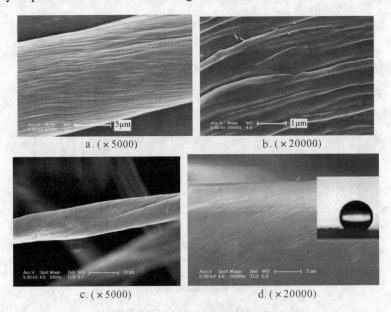

Figure 3 FESEM photographs of cotton fibers (a) (b), the untreated; (c) (d), treated by PSA-PFMS and θ=132°

2.3 Application properties of FAS-PFMS

Table 1 shows the performance properties of the cotton fabric treated or untreated with FSA-PFMS. According to the table, binding rigidity of the fabric treated by FSA-PFMS is gradually decreased with increase of the FSA-PFMS dosages. While the contact angle also increased with the FSA-PFMS dosages but to 132° it was no longer increased when the dosage of FSA-PFMS was greater than 0.5 g, moreover, grade of washing resistance could reach to 90. Along with augment of the FAS-PFMS dosage, whitenesses of cotton fabrics treated by different concentration polysiloxanes were all slightly diminished and had no effects on fabrics application. All of the above facts indicated that FSA-PFMS not only offered a bulky, soft hand to the treated fabrics, but also increased waterproofing and washing resistant properties of the fabrics. Besides, FSA-PFMS provided the treated fabrics with a desirable whiteness.

Table 1 Effects of FAS-PFMS dosages on fabrics performances

FAS-PFMS dosages/g*	The treated cotton fabrics				
	CA/°	Grade of washing resistance	Whiteness/%	BR/mN	
				f	w
Blank fabric	0	—	83.67	454	290
0.2	118.7	70	81.38	438	266
0.3	128.6	80	81.74	426	264
0.4	130.5	80	81.18	414	250
0.5	132.0	90	81.47	388	241
1.0	132.1	90	81.72	366	239

Remarks: g * denotes the mass of FAS-PFMS (g) in 100g acetic ether solution; BR: bending rigidity; CA: contact angle; w: warp, f: fill.

From the viewpoint of our previous researches [7-11] concerning microcosmic film morphology of functional polysiloxane softeners, in virtue of different arranging manners of functional polysiloxanes, the fabrics treated by those would show corresponding styles and performance properties. In FSA-PFMS molecule, there exist perfluoroalkyl, long chain stearyl acrylate and reactive epoxy groups as well, thus the treated fabrics will display good waterproofing, washing resistant properties.

3 Conclusions

A novel fluoropolysiloxane fabric finish (FAS-PFMS), bearing pendant long chain perfluoroalkyl, stearyl acrylate, and epoxy, was synthesized by hydrosilylation of polytrifluoropropylhydromethylsiloxane (PFHMS), perfluorooctylethene(PFOE), stearyl acrylate(SA) and allylglyeidyl ether(AGE). FESEM observation discovered that the treated fibers showed correspondingly smooth macro morphology. Grade of washing resistance of the treated samples could reach to 90 and CA of water utmost reach to 132° by a mass concentration of 0.5% FSA-PFMS ethyl acetate solution. Softness of the treated cotton fabric is relatively increased and whiteness is comparable to the blank sample.

Acknowledgement

We would like to address our appreciation to the national natural science foundation committee of China (50373025), the science and technology bureau of Zhejiang province, China (2008C11113) and natural science foundation of Shaanxi University of science and technology (ZX09-11) for financial supports of our research.

References

[1] B Ameduri, B Boutevin, G Kostov. Fluoroelastomers: synthesis, properties and applications. Prog Polym Sci 2001, 26(1): 105.
[2] Y Y Li, J Z Fang, G Zheng. Study on preparation and properties of pseudo-perfluoroalkyl polysiloxanes. New chemical materials, 2006, 34(6): 49-53.
[3] Y Hu, K Kazutomo, T Kaoru. High oil-repellent poly (alkypyrrole) films coated with fluorinated alkylsilane by a facile way. Clolloids and surfaces A: Physicochemical and Engineering Aspects, 2007, 292(1): 27-31.
[4] S A Kulinich, M Farzaneh. Hydrophobic properties of surfaces coated with fluoroalkylsiloxane and alkylsiloxane monolayers. Surface science, 2004, 573(3): 379-390.
[5] Y Norio, S Tomohiko, M Kensuke, et al. Synthesis of novel highly heat-resistant fluorinated silane coupling

agents. Journal of fluorine chemistry, 2006, 127(8):1058-1065.

[6] Z H Luo, T Y He. Synthesis and characterization of poly (dimethylsiloxane)-block-poly (2, 2, 3, 3, 4, 4, 4-heptafluorobutyl methacrylate) diblock copolymers with low surface energy prepared by atom transfer radical polymerization. Reactive and Functional Polymers, 2008, 68(5): 931-942.

[7] Q F An, G W Cheng, L S Li, et al. Film morphology and orientation of poly(dimethylsiloxane) on cotton fiber and its cellulose model substrates. Acta Polymerica Sinica, 2007, 6: 524- 530.

[8] Q F An, Q J Wang, L S Li, L X Huang. Study of Amino Functional Polysiloxane Film on Regenerated Cellulose Substrates by Atomic Force Microscopy and X-ray Photoelectron Microscopy. Textile Research Journal, 2009, 79(1): 89-93.

[9] Q F An, Q J Wang, Y Wang, et al. Synthesis, film morphology, and performance of functional polysiloxane bearing polyether and benzophenone derivative side groups. Fibers and Polymers, 2009, 10(2): 40-45.

[10] Q F An, G Yang, Q J Wang, et al. Synthesis and morphology of carboxylated polyether-block-polydimethylsiloxane and the supermolecule self-assembled from it. Journal of Applied Polymer Science, 2008, 110(5): 2595-2600.

[11] Q F An, L S Li, L X Huang, et al. Film morphology and characterization of functional polysiloxane softeners. AATCC Review, 2006, 6(2): 39-43.

61 Research on Pattern Grading of Apparel Shoulder Slope

Canyi Huang[1], Lina Cui[2]

(1. Deparment of Business and Information, Quanzhou Normal University, Quanzhou, Fujian 362000, China
2. Deparment of Arts and Design, Quanzhou Normal University, Quanzhou, Fujian 362000, China)

Abstract: It's necessary to master the changing law of human body shape before we make pattern grading of apparel. In this research we withdraw human measurements through the advanced non-touched 3D body scanner, then utilize SPSS software to analyze the data gathered, and then obtain the law of shoulder slope changing with body size.Futher more, the stepping data of shoulder slope in pattern grading is concluded.

Keywords: Apprael pattern; Shoulder slope; Pattern grading; Stepping data

Apprael pattern grading is the production of the development of apparel industry. As living standard improves, people's requirement towards garment fitting is more and more increasing. At the same time, according to the requirements of garment batch production, the same style of the garment has to adapt to different body shape of people for wearing. This requires the same style of clothing must be produced in accordance with many specifications and different size series. Patterns uesd for garment industry are the main technical material and also the foundation of garment cutting in mass production, and in order to ensure the sample size is accurate, at the same time the work is efficient, when making the pattern design, we always first develop one or two standard patterns, then have this pattern zoom in and out, so all patterns with different sizes we need are made. This process is called pattern grading.The technology of pattern grading is an important element in apparel manufacturing, it not only affects garment enterprise's productivity, but also determines the apparel fitting and appearance [1].

1 The shortcomings of pattern grading in garment industry at present

The graphical form of apparel pattern grading is not simply graphics displacement, but through system processing according to the stepping data of different apparel detailed specifications, and after that we can get pattern grading graphic which meets body fitting requirement. The determination of apparel detailed specifications has to take "National Apparel Size Standard of the People's Republic of China" as the basis, "The National standard" is completed on the basis of scientific investigation to human body, and concluded by using mathematical statistics and analysis methods. "The National standard" shows us the stepping data of some major parts of human body. Table 1 shows "GB/T1335.1-1997" National Men's apparel size standard, the values and steppig numbers of control sizes of 5.4A body shape.The values are the basic and important reference data when making pattern grading [2].

We can see easily from Table 1 what we can obtain from the "National Apparel Size Standard" are the data of different steps of human basic control parts and stepping data, But the sizes of apparel pattern depend on dimensions of different body detail parts commonly.So when we confront the problem of pattern grading to the detail parts, such as shoulder slope, collar neck and armhole etc, we have no data for reference. We can only depend on our experience to set and adjust the stepping data and pattern. Therefore this will lead to lacking of regulatory and larger randomness on pattern grading. This research mainly focuses on the pattern grading on shoulder slope.

Table 1 GB/T1335.1-1997 National Men's apparel size 5.4A (unit: cm)

Parts	Data							Stepping data
height	155	160	165	170	175	180	185	5
cervical height	133.0	137.0	141.0	145.0	149.0	153.0	157.0	4
cervical height (sitting)	60.5	62.5	64.5	66.5	68.5	70.5	72.5	2
arm length	51.0	52.5	54	55.5	57	58.5	60	1.5
waist height	93.5	96.5	99.5	102.5	105.5	108.5	111.5	3
chest circumference	72	76	80	84	88	92	96	100
neck circumference	32.8	33.8	34.8	35.8	36.8	37.8	38.8	39.8
shoulder across	38.8	40.0	41.2	42.4	43.6	44.8	46	47.2
waist circumference	58	62	66	70	74	78	82	86
hip circumference	77.2	80.4	83.6	86.8	90.0	93.2	96.4	99.6

2. Analysis on pattern grading of shoulder slope

Shoulder slope of human body has the effect of supporting the costumes, enhancing the beauty of our body and dress, and at the same time, apparel shoulder links directly to the harmony and performance of the overall style, so pattern grading of shoulder slope has a crucial impact on apparel overall effect.

In this research, we compare a dozen of apparel manufacturing enterprises' references document of apparel grading and finally finds, different manufacturing enterprises have different methods and data on pattern grading of shoulder slope.The method most enterprises use is keep the angle of shoulder slope the same in all sizes.That's when grading on two controlling endpiont of apparel pattern,shoulder piont and side neck piont,we keep the two points have the same stepping data on Y-axis direction.we can see from Figure 1. And some other enterprises use the method of the two points using different stepping data on Y-axis direction.

As we know, apparel pattern is restricted by the shape of human body, that only follow the body shape changing laws will the apparel pattern be practical. The grading of the two controlling endpiont of apparel pattern, shoulder piont and side neck piont, should according to the law which body shoulder slope changing with the body size. Therefore, this study carries on relevant research such as body measurement experiment related to this question in order to find out this changing law.

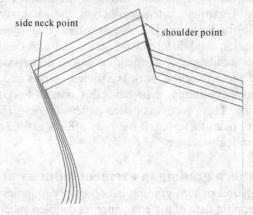

Figure 1 Commonly used method of pattern grading on shoulder slope

3 Experiment on anthropometric measurement of 3D

3.1 Experiment equipment

This experiment utilizes non-touched 3D human body laser scanner made by German TecMath Corporation to carry on human body data acquisition, this scanner can scan the 2.1m high region in 8~10 seconds, the resolution may reach 5mm, measuring accuracy for ±2mm.

3.2 Experiment scope

The survey scope is 400 aged from 19 ~26 years old male university students.

3.3 Body measurement Project

This research withdraws data of 6 measurements (two human bodies foundation measurement: height and chest circumference; 4 shoulder related detail sizes), the detail items can be seen in Table2.

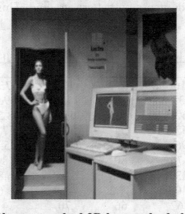

Figure 2 TechMath non-touched 3D human body laser scanner

Table 2 Body measurement project

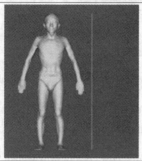
height

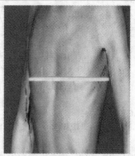
chest circumference

left and right shoulder widths

left and right shoulder slope angles

3.4 Paired-sample T test

As the human body is not completely symmetrical individual, we extract the measuring data corresponding to the left or right of human body parts alone. Then through annalysis by SPSS software, we can find out whether there is obvious difference between the symmetrical body parts. Table 3 shows the results of Paired-sample T test. From the table we can see the significance level of T test (two-tailed) of two pairs of items (left shoulder width, right shoulder width, left shoulder slope angle and right shoulder slope angle) are 0.01 and 0.00, both smaller than 0.05, so we consider there is significant difference on the mean of the two pairs. In accordance with the common processing method on the symmetry parts when there is obvious difference in apparel manufacturing, this research select the parts whose values are greater as the basis and easy for rectification. But for shoulder slope angle, we choose the part whose value is smaller. So in this research, we left one part for the symmetry parts, they are small shoulder width and shoulder slope angle.

Table 3 Paired-sample T test

	Paired difference					T	df	Sig.(two-tailed)
	mean	Std.Deviation	Std.Error mean	95%Confidence Interval of the difference				
				lower	Upper			
Pair 1 left shoulder width-right shoulder width	0.210	1.1619	0.0822	0.048	0.372	2.553	399	0.11
Pair 2 left shoulder slope angle-right shoulder slope angle	-1.959	3.3592	0.2375	-2.427	-1.490	-8.246	399	0.000

4. Analysis on the law of Shoulder slope changing with the body

As we can see from Figure 3, the right angled triangle reflects the distribution relations of small shoulder width L, shoulder slope H and shoulder slope angle α. And we can conclude from the trigonometric functions:

$$H = L \times SIN\alpha \quad (1)$$

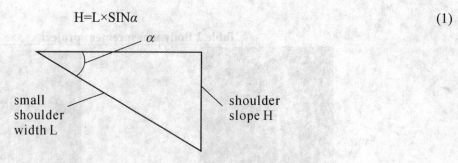

Figure 3 Right angled triangle abridged general view

Then we carry on statistical analysis by SPSS software for data regression, and conclude one independent variable regression equation between shoulder slope and height, shoulder slope and chest circumference, dulity regression equation between shoulder slope and height, chest circumference.

$$\begin{cases} \text{Shoulder slope} = -0.019 \times \text{body height} + 8.26 & (2) \\ \text{Shoulder slope} = 0.019 \times \text{chest circumference} + 3.25 & (3) \\ \text{Shoulder slope} = -0.02 \times \text{body height} + 0.02 \times \text{chest circumference} + 6.52 & (4) \end{cases}$$

Table 4 Regression analysis

Model	Unstandardized coefficients		Standardized coefficients	T	Sig.
	B	Std.Error	Beta		
constant	8.264	2.423		3.411	0.001
height	-1.98E-02	0.14	-0.104	-1.389	0.166

Dependetn variable: Shoulder slope

Model	Unstandardized coefficients		Standardized coefficients	T	Sig.
	B	Std.Error	Beta		
constant	3.250	1.305		2.491	0.014
chest circumference	1.879E-02	0.15	0.095	1.267	0.207

Dependetn variable: Shoulder slope

Model	Unstandardized coefficients		Standardized coefficients	T	Sig.
	B	Std.Error	Beta		
constant	6.523	2.604		2.620	0.010
height	2.27E-02	0.14	-0.119	-1.584	0.115
chest circumference	2.204E-02	0.15	0.111	1.477	0.141

Dependetn variable: Shoulder slope

5. The grading method of apparel pattern shoulder slope and feasibility validation
5.1 The grading method of apparel pattern shoulder slope

By the analysis in this research,we know human's shoulder slope will changes with the variation of human body 's height and chest circumference.So keeping the shoulder slope invariant in pattern grading in most apparel manufacturing enterprises at present is unscientific. The experimental data indicates the shoulder slope of bigger size apparel will become smaller and smaller, and smaller size apparel will become bigger and bigger if using the unscientific method.And if the more are the specifications the more obvious difference will appear. It not only mismatches the body characteristics but also influences apparel's style and comfortableness.

We can conclude the calculation formula of stepping data of shoulder slope by the above analysis:

$$\begin{cases} \text{Stepping data of shoulder slope} = -0.019 \times \text{stepping data of height} & (5) \\ \text{Stepping data of shoulder slope} = 0.019 \times \text{stepping data of chest circumference} & (6) \\ \text{Stepping data of shoulder slope} = -0.02 \times \text{stepping data of height} + 0.02 \times \text{stepping data of chest circumference} & (7) \end{cases}$$

If apparel size designation mode is synchronization on height data and chest circumference data in apparel manufacturing enterprises, and according to National Apparel Size Standard 5.4 system, human height's stepping data is 5cm, chest circumference is 4cm, Then the stepping data of shoulder slope in apparel pattern grading can be: the stepping data of shoulder slope=$-0.02 \times 5 + 0.02 \times 4 = 0.02$;

If apparel size designation mode is one height data matching different chest circumference data in apparel manufacturing enterprises, Then the stepping data of shoulder slope in apparel pattern grading can be: the stepping data of shoulder slope=$0.019 \times 4 \approx 0.08$;

If apparel size designation mode is one chest circumference data matching different height data in apparel manufacturing enterprises, Then the stepping data of shoulder slope in apparel pattern grading can be: the stepping data of shoulder slope=$-0.019 \times 5 \approx -0.1$.

5.2 Feasibility validation

This research extract several group of data whose height differ for about 5cm or chest circumference differ for 4cm successively from the 400 samples, applying the calculation functions concluded by this research to individual sample, and finally we find there is ignoring difference between the individual stepping data of shoulder slope and the data by the conclusion of this research.

6 Conclusion

The grading of apparel pattern is a key technology and an important section in garment industrial production process. The grading of apparel pattern should be on the basis of the understanding the shape and specifications of human body, and mastering the changing law of body shape. In this research we withdraw human measurements through the advanced non-touched 3D body scanner, then utilize SPSS software to analyze the data gathered, and then obtain the law of shoulder slope changing with body size. Futher more, the stepping data of shoulder slope in pattern grading is concluded. And this conclusion may be good reference for people who undertakes the grading of apparel pattern.

References

[1] Pan B. Industrial apparel pattern design. Beijing: China Textile Press, 2000, 18-35
[2] The National Standard GB/T1335.1-1997 apparel size—men, State Bureau of Technical Supervision, 1997-11-13, 7-8.
[3] Lu W D, SPSS for Windows statistical analysis. Beijing: Publishing House of Eectronics Industry, 2002, 141.

62 Study on the Elastic Recovery of PLA/Cotton Fabric

Peng Liu, Wei Tian, Yanqing Li, Zhaohang Feng, Chengyan Zhu*

(College of Materials and Textiles, Key Laboratory of Advanced Textile Materials and Manufacturing Technology, Ministry of Education of China, Zhejiang Sci-Tech University, Hangzhou 310018, China)

Abstract: A tester named YG541 is adopted to test the elastic recovery which is characterized by elastic recovery angle. Elements such as fiber content, weaving coefficient, compactness of the fabric whose influence on the elastic recovery of the fabric were studied here by regressive analysis. The results show that parabolic connection could be found between the elastic recovery of the fabric and every element mentioned above.

Keywords: The content of fiber; Compactness of fabric; Weaving coefficient; Regressive analysis

1 Introduction

Crease recovery refers to the capability that a fabric resists crinkle, and it is also one of the most significant performances of the fabric. Normally, elastic recovery angle is used for describing the recovery capability which refers to the angle between the two sides which were created when the fabric with certain shape and size was puckered under prescriptive conditions took shape after a period of time that external force was unload [1-3].

In this paper, elements such as fiber content, weaving coefficient, compactness of the fabric whose influence on the elastic recovery of the fabric have been analyzed respectively, it helps to get to know more about elastic recovery which also makes great sense for improvement.

2 Experiments
2.1 Fabric specifications

Fabrics are required to divide into three groups (A, B, C) for different purposes. Tencel staple was adopted as warps of all fabrics with its fineness of 9.72tex×2. Wefts are made up of three different kinds of materials which are designed to make certain arrangements in order to meet the demands of the experiments. The three kinds of materials used as wefts are given below: a: PLA/Cotton (60/40) blended yarn, 9.72tex×2; b: cooldry filament, 16.67tex; c: Cotton/ Metal fiber covering yarn,18.22tex. More details about the specification of all the fabrics are shown in Table 1.

Table 1 Specification of all the fabrics

Fabric number	Weaving	Weft a	Weft b	Weft c	Weaving Density (picks/10cm)	Weaving Compactness (%)
A1	5/2 satin	0	4	1	360×316	72.5
A2	5/2 satin	1	3	1	360×316	72.5
A3	5/2 satin	2	2	1	360×316	72.5
A4	5/2 satin	3	1	1	360×316	72.5
A5	5/2 satin	4	0	1	360×316	72.5
A6	5/2 satin	1	1	0	360×316	72.5
A7	5/2 satin	1	0	0	360×316	72.5
B1	Plain	2	2	1	360×316	72.5
B2	2/2 twill	2	2	1	360×316	72.5
B3	5/2 satin	2	2	1	360×316	72.5
B4	8/3 satin	2	2	1	360×316	72.5
B5	Mesh	2	2	1	360×316	72.5
C1	2/2 twill	2	2	1	360×221	70.75
C2	2/2 twill	2	2	1	360×253	73
C3	2/2 twill	2	2	1	360×285	75.25
C4	2/2 twill	2	2	1	360×316	77.5
C5	2/2 twill	2	2	1	360×348	79.75
C6	2/2 twill	2	2	1	360×380	82

---

* Sponsoring fund: National Program (2008C11071-2), Program for Changjiang Scholars and Innovative Research Team in University (No.IRT0654); Corresponding author's tel.: +86 571 86843253; fax: +86 571 86843250, email: cyzhu@zstu.edu.cn

2.2 Experimental apparatus and method

Equipment: YG541 tester, includes a set of elastic blocks, protractor, glass lens, heavy punch etc.

Method: 5 pieces of fabrics are cropped for measurement each type, implementation of the experiment strictly follows the norm GB/T3819—1997[4].

3 Experimental results and analysis

Based on the method for testing elastic recovery, every fabric was tested by 5 times for its average. Experimental results are shown in Table 2.

Table 2 Elastic recovery angle of all the fabrics

Fabric number	Rapid elastic angle (°)			Slow elastic angle (°)		
	Warp	Weft	Warp & Weft	Warp	Weft	Warp & Weft
A1	98.33	60.10	158.43	118.07	65.07	183.13
A2	99.03	70.60	169.63	119.63	76.03	195.67
A3	113.07	77.20	190.27	132.93	81.60	214.53
A4	101.23	65.60	166.83	120.13	69.83	189.97
A5	98.47	52.50	150.97	115.17	60.60	175.77
A6	99.37	128.93	228.30	116.63	134.97	251.60
A7	97.68	109.67	207.35	108.83	118.33	227.17
B1	72.57	53.40	125.97	89.10	65.87	154.97
B2	98.27	71.60	169.87	115.47	77.27	192.73
B3	113.07	77.20	190.27	132.93	81.60	214.53
B4	106.67	60.54	167.21	120.42	75.87	196.29
B5	75.36	63.84	139.20	90.25	70.12	160.37
C1	92.58	60.52	153.10	108.34	67.49	175.83
C2	94.10	78.20	172.30	110.63	85.60	196.23
C3	91.30	89.17	180.47	109.30	97.21	206.51
C4	98.27	71.60	169.87	115.47	77.27	192.73
C5	86.43	65.83	152.27	104.23	70.87	175.10
C6	80.50	67.17	147.67	99.10	70.37	169.47

3.2 The influence the content of PLA on elastic recovery

According to the data in Table 2, the connection between the content of PLA involved in fabrics of series A and elastic recovery was studied here by regressive analysis, the curve is shown in Figure 1.

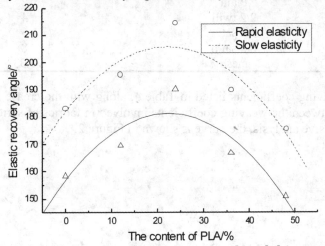

Figure 1 Relationship between elastic recovery angle and the content of PLA

Relationship between elastic recovery angle and the content of PLA along with related coefficients are given below in Table 3:

Table 3 Relationship between elastic recovery angle and the content of PLA along with related coefficient

Elastic recovery	Equation	related coefficient
Rapid elasticity	Y = 156.74143 + 2.19043x − 0.04871x^2	R^2=0.82
Slow elasticity	Y = 182.05514 + 2.13698x − 0.04807x^2	R^2=0.82

According to Figure 1, whatever rapid elasticity or slow elasticity it is, the change of elastic recovery angle releases a parabolic connection with the content of PLA. The larger fabric crease recovery angle it is, the better performance of crease recovery means.

The crease recovery capability increases as the growth of the content of PLA when it is lower than 24%. As known, cotton fiber performs a better Elastic recovery than PLA. The crease recovery is mainly influenced by the content of PLA in such condition. Once the content of PLA exceeds 24% but lower than 48%, the crease recovery presents a trend of decrease which is considered that cotton fiber plays a more important role in influencing the crease recovery. So, an appropriate selection for Proportion of PLA and cotton will leave a great influence on the crease recovery of the fabric.

In addition, as shown in Table 2, the Elastic recovery angle in the warp direction is bigger than that of the weft direction which is mainly caused by the addition of the mental fiber.

3.2 The influence weaving coefficient on elastic recovery

The weaving coefficient can be calculated by formula (1) and (2).

$$F = \frac{n_T \times n_W}{t} \quad (1)$$

$$t = \frac{t_T + t_W}{2} \quad (2)$$

F—weaving coefficient;

n_T, n_W —the numbers of warps and wefts in an integrated fabric weave

t —total weaving points of warps and wefts interwoven in an integrated fabric weave;

t_T —weaving points of warps and wefts interwoven in the warp direction;

t_W —weaving points of warps and wefts interwoven in the weft direction;

Based on the formula for calculation of weaving coefficient, weaving coefficients of fabrics are shown in Table 4

Table 4 Weaving coefficients of the fabrics of series B

Fabric number	weave	weaving coefficient
B1	Plain	2
B2	2/2 twill	4
B3	5/2 satin	5
B4	8/3 satin	8
B5	mesh	5.33

According to the weaving coefficients listed in Table 4, along with the crease recovery angle shown in Table 2, the connection between the weaving coefficients involved in fabrics of series B and elastic recovery was studied here by regressive analysis, the curve is shown in Figure 2.

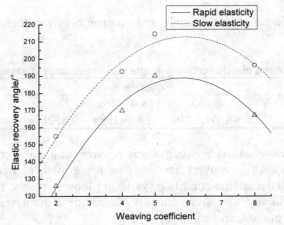

Figure 2 Relationship between elastic recovery angle and weaving coefficient

Relationship between elastic recovery angle and weaving coefficient together with related coefficient are shown in Table 5.

Table 5 Equation between elastic recovery angle and weaving coefficient along with related coefficient

Elastic recovery	Equation	related coefficient
Rapid elasticity	$Y = 39.94747 + 51.3302x - 4.41967x^2$	$R^2=0.98$
Slow elasticity	$Y = 78.6076 + 45.13107x - 3.79267x^2$	$R^2=0.97$

According to the Figure 2, the equation between elastic recovery angle and weaving coefficient turns to be a parabola with a high correlation. Yarns in the fabric interweave more frequently when the weaving coefficients are designed small to certain extent which consequently leads to an increase of weaving points. As the external force removed, it will not be an easy task for the yarns in the fabric to return to the original state. As a result, elastic recovery increases as the growth of weaving coefficient within certain scope. However, if weaving coefficient is big enough, both wefts and warps in the fabric are able to move more freely, thus yarns in the fabric can hardly turn back to the original state while in a large space caused by too much long floating length. Therefore, improper weaving coefficient is not conducive for the yarns in the fabric to return to the original state. It can be concluded that appropriate selection of weaving coefficient helps to improve the elastic recovery of the fabric.

Among the three basic weave plain, twill and satin, plain has the largest numbers of weaving points with the worst performance of elastic recovery while satin has the least numbers of weaving points with the best performance of elastic recovery.

Due to the special formational principle of mesh, the elastic recovery of the mesh is different from others and not discussed here.

3.3 The influence compactness on elastic recovery

According to the data in Table 2, the connection between the compactness involved in fabrics of series C and elastic recovery was studied here by regressive analysis. The curve is shown in Figure 3.

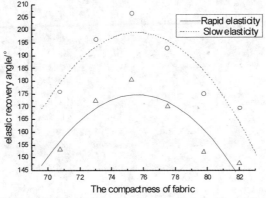

Figure 3 Relationship between elastic recovery angle and compactness

Relationship between elastic recovery angle and compactness together with related coefficients are Shown in Table 6:

Table 6 Relationship between elastic recovery angle and compactness along with related coefficient

Elastic recovery	Equation	related coefficient
Rapid elasticity	$Y = -4300.3271 + 118.41445x - 0.78335x^2$	$R^2=0.82$
Slow elasticity	$Y = -4670.68336 + 128.89288x - 0.85287x^2$	$R^2=0.81$

According to Figure 3, the equation between elastic recovery angle and compactness of the fabric also turns to be a parabola. The elastic recovery increases as the growth of compactness within the scope between 70% and 75.25%. Under such a condition, yarns are loosely placed in the fabric, too much space between yarns only adds difficulty for movements, as compactness increases, spaces between yarns gets smaller, thus elastic recovery of the fabric comes up to a rise.

As the compactness exceeds 75.25%, the elastic recovery of the fabric turns to decrease, this is mainly because of the increased weft density which leads to a rise of tangential resistance, once the external force acting on the fabric removed, yarns can hardly move, and this contributes to the decrease of elastic recovery. So, excessive or deficient compactness of the fabric both hamper the promotion of elastic recovery.

4 Conclusions

Obviously, parabolic connection can be found between the elastic recovery of the fabric and every element mentioned above. Plain performs relatively inferior than the other two weaves in the aspect of elastic recovery and satin is considered to be the best among the three. Appropriate selection of weaving coefficient helps to improve the elastic recovery of fabric.

Acknowledgement

The authors express their sincere gratitude to the support from the Key Laboratory of Advanced Textile Materials and Manufacturing Technology, Zhejiang Sci-Tech University, Hangzhou 310033, China.

References

[1] Xiaoting Zhang, Weidong Gao. The relationship between elastic recovery angle and mechanical property[J]. Journal of Textile Research.2008, 29(6): 29-31.
[2] Jun Jin, Shufeng Wang. Regressive Analysis On Cockle-resist Capability of Fabric[J]. Shan Xi Textile. 2006, 4(72): 8-10.
[3] Lili, WuJun He, Jianyong Yu. Crease resistance of soybean protein fiber / cotton blended fabric[J]. Journal of Textile Research.2005, 26(5): 91-93.
[4] Norm: GB/T3819-1997, Testing Method of Elastic Recovery Angle[S].

63 Functional Properties of Hemp Union Fabrics for Home Textiles

Lin Lou, Xiaohang Zhu[*], Hongyan Xu, Jianfang Wang, Xinghai Pei, Jianliang Li

(Zhejiang Sci-Tech University, No.5 2nd Street, Hangzhou 310018, China)

Abstract: By adjusting the yarn component, blend ratio and weave structure, fabrics with hemp fibers achieved good light stability, flexibility, pilling resistance, air permeability, antibacterial function, and flame retardant function. Those union fabrics were suitable for home textiles, such as curtains, sofa cloths, table cloths, etc. The existence of modacrylic fiber improved the light stability of the fabric. The fabric with sateen weave and with hemp/cotton blended weft yarn had a better flexibility. The fabrics with no less than 40% of hemp fibers in the weft yarns had a better pilling resistance, especially when its structure was plain weave or twill weave. Hemp fiber contributed to the air permeability of the fabric. Sateen fabric had a better air permeability than plain or twill fabric. The fabric with hemp fibers had good antibacterial function. When its weft yarn had no less than 80% of modacrylic fibers, the fabric met the requirements of flame retardancy. The flame retardant ability of the fabric was better when burnt in the direction perpendicular to the flame retardant yarns than when burnt along those yarns. The sateen fabrics provided a better flame retardant stability than plain or twill fabrics.

Keywords: Hemp fiber; Home textile; Functional application; Aantibacterial; Flame retardant

1 Introduction

Home textile products with bast type style are increasingly favored by customers these years. Hemp fiber is a natural green material with high economic and environmental value. Industrial hemp fiber has multiple functions and advancced properties but with almost no narcotics, fitting the contemporary trend of functionalizing, high-ranking and naturalizing. Researches on hemp fiber home textile products are seldom reported.

Light stability, flexibility, pilling resistance, air permeability, antibacterial function and flame retardant function are important to home textile products such as curtains, sofa cloths, table cloths, etc. It is necessary to study on and to develop functional fabrics according to the characteristics of hemp fiber and the end uses of different products.

2 Experimental

2.1 Materials

A series of weft yarns were fabricated with a linear density of 58.3±3 tex. They were woven with interlaced polyester warp yarns into home textile fabrics, with warp density of 60 ends/cm and weft density of 15 picks/cm. Other detailed parameters are listed in Table 1.

2.2 Mechanical test and outdoor exposure experiment

The tensile properties of the fabrics both before and after outdoor exposure were tested according to GB/T 3923.1-1997 'Textiles-Tensile properties of fabrics-Part 1: Determination of breaking force and elongation at breaking force-Strip method' (YG065), with gauge length of 200mm±1mm. The outdoor exposure experiment was carried out during Jul. 20th ~ Aug. 28th for 40 days and nights in the summer of 2009 in Hangzhou.

2.2 Bending length test

The bending lengths of the fabrics were tested and flexural rigidities were calculated according to GB/T 18318-2001 'Textiles-Determination of bending length of fabrics' (LLY-01B).

2.3 Pilling resistance test

The pilling resistance was tested according to GB/T 4802.2-1997 'Textiles-Assessing the rate of pilling of fabrics-Martindale method'. Four samples were tested for each fabric group, and each sample was rubbed for 1000 times.

2.4 Air permeability test

The air flow was tested and the air permeability rate was calculated for each fabric group according to GB/T 5453-1997 'Textiles-Determination of the permeability of fabrics to air' (YG(B)461D), with testing area of 20cm^2 and pressure drop of 100Pa.

2.5 Antibacterial activity test

The antibacterial activity was tested according to GB/T 20944.3-2008 'Textiles-Evaluation for antibacterial activity-Part 3: Shake flask method' with the bacteria of Staphylococcus aureus (ATCC 6538).

Table 1 Fabric parameters

Fabric Group	Weave Structure	Fabric Weight /(g/m^2)	Composition of Weft Yarn	Linear Density of Warp Yarn /(tex)
A	Sateen $\frac{8}{3}$	160±5	Modacrylic	11.1
B	Sateen $\frac{8}{3}$	160±5	H10/M90	11.1
C	Sateen $\frac{8}{3}$	160±5	H20/M80	11.1
C-T	'Z' Twill $\frac{1}{3}$	160±5	H20/M80	11.1
C-P	Plain Weave	160±5	H20/M80	11.1
D	Sateen $\frac{8}{3}$	160±5	H30/M70	11.1
E	Sateen $\frac{8}{3}$	192±5	H30/C70	16.7
F	Sateen $\frac{8}{3}$	192±5	H40/C60	16.7
G	Sateen $\frac{8}{3}$	192±5	H55/C45	16.7
H	Sateen $\frac{8}{3}$	192±5	Hemp	16.7
H-T	'Z' Twill $\frac{1}{3}$	192±5	Hemp	16.7
H-P	Plain Weave	192±5	Hemp	16.7
I	Sateen $\frac{8}{3}$	192±5	Cotton	16.7
J	Sateen $\frac{8}{3}$	192±5	Polyester	16.7

Note: In the Composition of Weft Yarn, H stands for hemp fiber, M stands for modacrylic fiber, and C stands for cotton fiber (e.g. H20/M80 means 20% of dry weight of the blended yarn are hemp fibers, and 80% are modacrylic fibers).

2.6 Flame retardancy test

The vertical flammabilities of Fabric Groups of A, B, C, C-T, C-P, D, and H were tested according to the requirements of GB/T 17591-2006 'Flame retardant fabrics' to decorative fabric Grade B$_2$. The limiting oxygen indexes of Fabric Groups of A, B, C, D, and H were tested according to GB/T 5454-1997 'Textiles-Burning behavior-Oxygen index method'.

3 Results and discussion
3.1 Mechanical property and light stability

In Figure 1, containing certain amount of modacrylic fibers, Fabrics A~D had more stable weft tensile strengths, with small strength losses less than 9.5% after sunlight exposure. The nitrile groups (-CN) in acrylonitrile monomers of modacrylic molecules absorbed the energy from ultraviolet, which tried to oxidatively disintegrate the fibers, and transformed it into vibrational energy, maintaining the integrity and stability of the chemical structure[1]. Fabrics E and I kept better strengths after sunlight exposure. The weft tensile strength of Fabric J dropped 23.5% after sunlight exposure due to the bad light stability of polyester molecules. The main chains of polyester molecules have many carbonyls (C=O) that tend to absorb infrared and ultraviolet energy and cause chemical bond disintegration.

Figure 2 shows Fabrics A~D had relatively small warp tensile strength losses, between 24.1% to 37.7%, while others had much greater ones after 40d exposure to sunlight. The existence of modacrylic fibers in weft yarns also contributed to the stability of warp tensile strength.

Each direction of every fabric group had tensile strengths larger than 350N both before and after sunlight exposure, qualified for seat cloth of First Grade and drapery and cover cloth of Superior Grade[2]. Fabrics A, B, C, D, E, and I had better light stabilities and were more suitable for curtain, sofa cover, table cloth by the window, etc.

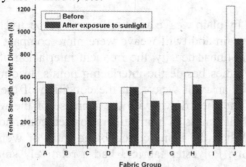

Figure 1 Tensile strengths of weft direction before and after sunlight exposure

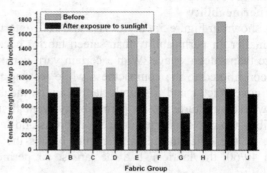

Figure 2 Tensile strengths of warp direction before and after sunlight exposure

3.2 Flexibility

As shown in Figure 3, the flexural rigidities of warp direction lay between 137~176 mN·cm, similar among different fabric groups. As to the flexural rigidities of weft direction, it reached the highest point when in plain weave structure of Fabric H-P. As the yarn float grew, the fabric compactness decreased, the flexural rigidities of weft direction in twill weave H-T and in sateen weave H dropped significantly. Hemp fiber had a higher rigidity than cotton or polyester fiber. The more hemp fibers the fabric had, the more rigid it could be in the direction. Fabric H had a much greater flexural rigidity than Fabric G because there was much more gum in pure hemp yarns in Fabric H than in hemp/cotton blended yarns in Fabric G.

To increase the flexibility of the fabrics, sateen weaves and hemp/cotton blended yarns could be applied. Fabrics with better flexibility were suitable for drapery and covering use, such as curtain and table cloth.

3.3 Pilling resistance

The degrees of pilling for different fabrics are shown in Figure 4. Hemp fiber, with a greater strength and a higher rigidity, is neither easy to break nor tending to curl or protrude from the fabric surface. And the remainder gum in the technical fiber enhances the cross-sectional bondings and prevents the single fibers from loosening and protruding. Thus Fabrics H-P, H-T, and H had apparently better degrees than other fabrics. Fabric H-P, with plain weave structure, and fabric H-T, with twill weave structure, were compactly interlaced with shorter floats. As a result, the hairiness was difficult to occur by rubbing the fabric surface. As the ratio of hemp fiber fell, the pilling resistance of the fabric decreased. Fabrics G and F shared a pilling degree of 2.5, when hemp fibers took 55% and 40% of the weight of weft yarns respectively. When the ratio of hemp fiber reduced to 30%, Fabric E shares the same degree of 1 with Fabric I, whose weft yarns were pure cotton. Polyester fabric J had a higher pilling resistance than fabrics I, E, F, and G because polyester fibers were stronger than cotton fibers and were not easy to form hairy fabric surface in a certain extent. On the other hand, cotton fibers broke under friction, form hairiness, and entangle with stronger polyester fibers to create pilings that were hard to rub off.

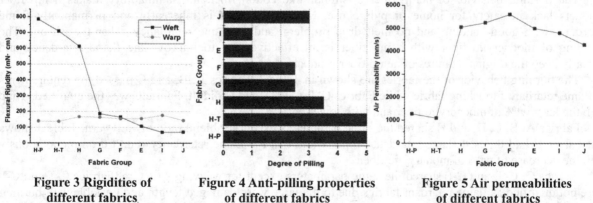

Figure 3 Rigidities of different fabrics

Figure 4 Anti-pilling properties of different fabrics

Figure 5 Air permeabilities of different fabrics

The union fabrics of pure hemp yarn and interlaced polyester yarn had good pilling resistance, especially when with compact weaving structures and short floats. The pilling resistance was also satisfying when the fabrics had no less than 40% of hemp fibers in the weft yarns. Those fabrics were suitable for sofa cloths that are rubbed often.

3.4 Air permeability

As shown in Figure 5, Fabrics H-P and H-T, interlaced by plain weave and twill weave respectively, had much smaller air permeability than sateen fabrics because plain and twill weave were more compact in 3D structure than sateen weave. With a certain yarn density and linear density, the frequent interlacing of plain weave contributed to the compactness while leaving more holes beside the interlacing points, leading to a smaller cover factor and a slightly larger air permeability than twill weave. In the condition of the same sateen weave structure, the air permeability went first up and then down with the increase of the wt% of hemp fiber. On one hand the surface roughness and inner hollowness of hemp fibers formed continuous air tunnels, on the other the remainder gum blocks the air. When the hemp/cotton blend ratio of weft yarn reached 40/60, the fabric F had the largest air permeability. Polyester fabric J had a relatively small air permeability.

Sateen weave was a better structure than plain weave or twill weave to improve the air permeability of the fabric. The existence of hemp fiber enhanced the air permeability. Fabric F with 40/60 blend ratio of hemp/cotton weft yarn had the best air permeability among the union fabrics.

3.5 Antibacterial activity

In Table 2, the bacteria reduction of Fabric H, hemp and polyester union fabric, was 12.73% higher than that of Fabric I, cotton and polyester union fabric, showing a good antibacterial ability with no need of any functional finishing. The porousness of hemp fiber makes more air staying inside the fiber. The continuousness of the pores let the air flow and makes the anaerobes hard to live. When the fabric gets wet, the moisture and the liquid could also be easily transferred by the porous tunnels in the hemp fibers, making it hard to maintain the wet environment which bacteria prefer. Besides, in the hemp fiber there are small quantities of phenols, which can impede the metabolism and physiologic activity of the bacteria and eventually kill the bacteria[3].

Table 2 Antibacterial Activity of Different Fabrics

Microorganism	Staphylococcus Aureus (ATCC 6538)	
Fabric Group	H	I
Number of Bacteria from Control Specimen at "0" Contact Time (CFU/ml)	2.8×10^4	2.8×10^4
Number of Bacteria from Control Specimen after Shaking 18h (CFU/ml)	1.1×10^6	1.1×10^6
Number of Bacteria from Test Sample after Shaking 18h (CFU/ml)	1.2×10^5	2.6×10^5
Reduction (%)	89.09	76.36

With natural antibacterial property, hemp union fabrics were suitable for healthy table cloths, doilies, sofa cloths, etc.

3.6 Flame retardancy

The burning behavior of hemp fiber is similar like cotton fiber. Its flammability makes hemp fiber products lack in safety for home or public use. Modacrylic fiber is inherently and permanently flame retardant, with good handing and no melt drop problem, and its flame retardancy doesn't wash out. The blending of modacrylic fiber with hemp fiber is an effective way to improve the flame retardancy and partially keep the original features of hemp fiber product.

The burning behavior of the fabrics are shown in Table 3. Fabrics A, B, and C passed the requirements of flame retardant furnishing fabric while Fabrics C-T, C-P, D, and H didn't. Fabric C was the qualified group with the least wt% of modacrylic fiber and with lower cost.

Fabrics A, B, C, D, and H shared the same parameters except for blend ratio of weft yarn. Table 3 shows that Fabric B had the best flame retardant property and that the flame retardancy didn't drop monotonously with the decrease of wt% of modacrylic fiber.

Fabrics C, C-T, and C-P shared the same parameters except for weaving structure. Fabrics C-T and C-P were not qualified as flame retardant fabrics due to the large warp damaged length and unstable weft burning behavior respectively. The increase of interlacing frequency might reduce the flame retardancy of the fabrics.

The weft damaged lengths were larger than the warp ones for Fabrics A, B, C, and C-P. The nonflammable weft yarns acted like barriers when the fire burnt towards warp direction, especially when the combustible polyester warp yarns melted and shrank along warp direction, which increased the local weft density and enhanced the fire retardancy by increasing the wt% of nonflammable fibers in the area. On the other hand, when the fire burnt towards the weft direction, the shrinkage of warp yarns caused gaps between weft yarns and decreased the weft density. For Fabric C-T, with an average float length of 2, the $\frac{1}{3}$ twill weave formed favorable warp passages for the flame or glow. And the shrinkage length in warp direction of the twill fabric was much larger than that of plain and sateen fabrics, which might be another reason for the large warp damaged length. To develop fabrics with better flame retardancy, certain fabric structures such as $\frac{1}{3}$ twill weave in this example should been avoided.

Table 3 Burning behavior (vertical method)

Test Item		Requirement (B_2 Furnishing Fabric)	Average Results for Different Fabric Groups							
			A	B	C	C-T	C-P	D	H	I
Afterflame Time /(s)	Warp	≤15	0	0	0	0	0	/	×	×
	Weft		0	0	1.9	1.2	/	/	×	×
Afterglow Time /(s)	Warp	≤15	0.3	0	9.2	6.0	6.2	/	×	×
	Weft		1.1	1.4	5.5	1.0	/	/	×	×
Damaged Length /(mm)	Warp	≤200	153	142	179	210	156	/	×	×
	Weft		165	153	195	148	/	/	×	×
If the Cotton Can Be Burnt	Warp	Not Required	No	No	No	No	No	/	×	×
	Weft		No	No	No	No	/	/	×	×
	Conclusion		Pass	Pass	Pass	Fall	/	/	Fall	Fall

Note: 5 samples were tested for each item of each group. "×" means all 5 samples were burnt through. "/" means some of the 5 samples were burnt through. The cotton not being burnt means that there was no melt drop from the sample or the melt drop didn't start flame or glow in the cotton.

Fabrics A, B, and C were suitable for flame retardant home textiles such as curtains, sofa cloths, table cloths, doilies, and beddings, etc. Fabric C was the qualified flame retardant group with the lowest cost. Fabrics with sateen weave had more stable flame retardant properties than plain or twill ones.

4 Conclusions

Home textile hemp union fabrics with better light stability, flexibility, pilling resistance, air permeability, antibacterial function, and flame retardant function could be achieved by adjusting the yarn component, blend ratio, and weave structure.

a) Fabrics of all groups were qualified in tensile strengths both before and after sunlight exposure. Modacrylic fibers in weft yarns contributed to the light stability. Fabrics A, B, C, D, E, and I were more suitable for lightproof products.

b) Sateen weave with long floats and hemp fiber yarn blended with cotton should be adopted to improve the flexibility of the fabric. Fabric E with 30/70 blend ratio of hemp/cotton weft yarn had the best flexibility among hemp union fabrics, but not as good as cotton union fabric I and polyester fabric J.

c) Hemp/polyester union fabric had good pilling resistance, especially when interlaced with plain or twill weave structure. Fabrics H-P, H-T, H, G, and F, with no less than 40% of hemp fiber in the weft yarn, were good at pilling resistance.

d) Hemp fiber and sateen weave structure could effectively improve the air permeability of the fabric. Fabric F with 40/60 blend ratio of hemp/cotton weft yarn had the best air permeability among the union fabrics, followed by fabrics G, H, and E.

e) Hemp/polyester union fabric H had a better antibacterial ability.

f) Hemp/modacrylic/polyester union fabrics A, B, and C met the requirements of flame retardant fabrics. The flame retardancy of plain weave fabric or twill weave fabric, especially the latter one, was not as stable as that of sateen weave fabric.

References

[1] W D Yu. Textile Materials. Beijing: China Textile & Apparel Press, 2006: 155
[2] GB/T 19817-2005, Textiles-Fabrics for furnishings.
[3] J C Zhang, et al. Structure and Properties of China Hemp Fibre. Beijing: Chemical Industry Press, 2009: 126

64 Effects of Annealing Atmosphere on the Structures and Photocatalytic Properties of Gd-Doped TiO$_2$/CF Photocatalysts

Bangyong Pang[1], Yaofeng Zhu[2], Tianbing Yang[1], Yaqin Fu[1*], Hao Chen[1], Liyuan Zhang[1], Jin Wang[1]

(1. Key Laboratory of Advanced Textile Materials and Manufacturing Technology, Ministry of Education, Zhejiang Sci-Tech University, Hangzhou 310018, China

2. Graduate School of Science and Technology, Shinshu University, 3-15-1 Tokida, Ueda 386-8576, Japan)

Abstract: In the present investigations, sol-gel method was utilized to prepare Gd-doped TiO$_2$/CF precursor and then the Gd-doped TiO$_2$ precursor was loaded on the PAN-carbon fiber (PAN-CF) substrates via the dip-coating process. Superheated steam, nitrogen and air atmosphere were chosen for the annealing treatment of the material, respectively. Field emission scanning electron microscopy (FE-SEM) and X-ray diffraction (XRD) were used to characterize the surface morphology of the composites and crystal phase of Gd-doped TiO$_2$. The photocatalytic properties of the Gd-doped TiO$_2$/CF annealed under different atmosphere were investigated by degrading acid orange II under UV irradiation. The results indicate that the annealing atmosphere has significantly effects on the structures and photocatalytic activity of the thin films. The Gd-TiO$_2$/CF annealed under superheated steam atmosphere exhibits the highest photocatalytic activity. The XRD analysis suggests that all of the samples were in the anatase form after calcinations at 550°C.

Keywords: Gd-doping; Photocatalysis; Annealing ambient; Acid orange II; Sol-Gel

0 Introduction

Heterogeneous photocatalysis utilizing titanium dioxide (TiO$_2$) is becoming more and more focused in the environmental treatment and purification purposes recently[1]. Sol-gel method has been taken as the first choice to prepare TiO$_2$ precursor due to its simple process, good dispersibility, and fineness in product. As is known that the post-heat treatment plays a decisive role in the structure and performance of TiO$_2$. At present, sol-gel process was employed in the following different annealed atmosphere of preparation for TiO$_2$: Firstly, vacuum make it easy for oxygen atoms to escape from crystal lattice of TiO$_2$, and thus vacancy is formed, namely a "to e-center". It can depress the photocatalytic properties of TiO$_2$ obviously by reducing the concentration of photo-hole in photocatalytic reaction[2]. Secondly, annealing in air[3-4], in which the drawback of first method can be retrieved, but the TiO$_2$ films can be oxidized easily in the annealed oxygen atmosphere. Thirdly, heat-treatment in nitrogen atmosphere[5-6] may not only makes the surface of TiO$_2$ films crack, but also produces a new compound of O-Ti-N because a small amount of oxygen atoms are substituted by nitrogen atoms in the lattice of TiO$_2$.

In the present investigations, the photoactive TiO$_2$ was homogeneously mounted on PAN-based carbon fiber by a dip-coating process, then annealed in air, nitrogen and superheated steam ambient, respectively. The effects of annealing atmosphere on the structures and photocatalytic properties of Gd-Doped TiO$_2$/CF photocatalysts were discussed.

1 Experimental

1.1 Materials

PAN-carbon fiber(T300C), with the diameter of about 6-6.5 μm, were commercial products of Toray Inc.; Tetrabuty-titanate (CP) and Nitric acid(65%-68%) were purchased from Hangzhou Gaojing chemical industry Co., Ltd., China; absolute alcohol, Acetyl acetone and acetic acid were all analytical reagent and used as received. Gadolinium Nitrate was purchased from Shanghai jinchun chemical industry Co. Ltd.

1.2 Surface treatment of the Carbon fiber

PAN-carbon fiber (0.1g/ beam, 40 beams) was oxidized in refluxing concentrated nitric acid (160 ml) for 4h in 115°C oil bath, then washed with demineralized water until pH>6 and parched in the 80°C oven.

---

* Corresponding author's email: fuyaqin@yahoo.com.cn

1.3 Preparation of the Photocatalyst

The Gd-Doped TiO_2 sol was prepared by using tetrabutyl orthotitanate as Titanium source and alcohol as the solvent. The preparation process were as follows: 17.08 ml $Ti(O_4H_9)_4$ was dissolved in the mixture solution of 54.85ml EtOH and 2.86ml HAC. Then 0.5ml AC and 0.052g $Gd(No_3)_3 \cdot 6H_2O$ dissolved in 10 ml Nitric acid(65%-68%) were added. After magnetic stirring for 1h, the obtained solution was hydrolyzed by adding 0.9 ml water and 10 ml EtOH dropwise under stirring for another 1h. The Gd-Doped TiO_2 sol was obtained.

The surface modified PAN-carbon fiber (PAN-CF) substrates were dip-coated with sol-gel mixtures, and then dried at room temperature. Finally, the coated carbon fiber substrates were calcined in superheated steam, air and nitrogen ambient respectively with the ramp rate of 3°C /min to 550°C for 2h. Then the Gd-Doped TiO_2/CF photocatalyst with different annealing ambient were obtained. In addition, the Gd-Doped TiO_2 sol was treated by the some process, then the Gd-Doped TiO_2 powder with different annealing ambient were obtained.

1.4 Characterization

The surface morphology of Gd-Doped TiO_2/CF photocatalysts were observed by field emission scanning electron microscopy(S-4800). The crystallinity of the Gd-Doped TiO_2 was determined by X-ray diffraction (D8-Discover Cu-Ka λ=1.540562Å).

1.5 Photocatalytic experiments

The photocatalytic activity of the prepared Gd-Doped TiO_2/CF was evaluated by the photocatalytic degradation of acid orange II under UV irradiation of 500 watt. The acid orange II after degradation were measured by UV-spectrophotometer at the wavelength of 484nm, which were corresponded to the maximum adsorption of acid orange II.

2 Results and discussion

2.1 Surface morphology of Gd-Doped TiO_2/CF photocatalyst

(a) annealed in N_2 ambient (b) annealed in air (c) annealed in steam ambient

Figure 1 FE-SEM photograph of Gd-TiO_2/CF annealed in different atmosphere

Figure 1 shows the FE-SEM photographs of the Gd-Doped TiO_2/CF with different annealing ambient. As shown in Figure 1(a), serious crack occurred for the Gd-Doped TiO_2/CF annealed in N_2 ambient. It is believed that the crack will deteriorate the photocatalytic activity of the photocatalyst. Figure 1(b) shows the FE-SEM photographs of Gd-Doped TiO_2/CF annealed in air. It can be found that there are some small fragments on the smooth face. From Figure 1 (c) one can see that the Gd-Doped TiO_2/CF annealed in steam ambient possess smooth, continuous and evenly surface, resulting from steady annealed temperature provided by the superheated steam ambient.

2.2 Photocatalytic properties

The photocatalytic activity of Gd-Doped TiO_2/CF annealed in different atmosphere were evaluated by degradation of acid orange II (80mg/L), which is a typical azo dye and used as model compound to test the photo-degradation capability of photocatalysts, and the results were shown in Figure 2. From the figure we can deduce that the reaction is a typical photo-degradation when compare to the blank sample. It could be seen that the annealing ambient has significant effects on the photocatalytic activity of the prepared Gd-Doped TiO_2/CF. From Figure 2 we can find the composite annealed in superheated steam ambient possess the highest photocatalytic activity and that annealed in N_2 ambient shows the lowest ability to degrade the acid orange II. Authors believed that it is because the TiO_2/CF annealed in different atmosphere possesses different type and size of crystal, surface morphology and the defects. In addition, comparing curve (b), (c), (d) with (a) in Figure 4 we can find that Gd-Doped TiO$_2$/CF exhibit lower photocatalytic activity than that of pure TiO$_2$/CF. It is believed that too much Gd doped on the TiO$_2$/CF precursor can become the center of electron and hole recombination and, consequently, result in the decrease of the photocatalytic activity of the composite.

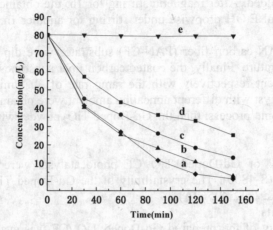

(a) pure TiO$_2$/CF annealed in steam ambient (b) Gd-TiO$_2$/CF annealed in steam ambient (c) Gd-TiO$_2$/CF annealed in air (d) Gd-TiO$_2$/CF annealed in N$_2$ ambient (e) blank

Figure 2 Photocatalytic degradation of acid orange II solution

2.3 Crystal phase analysis of Gd-doped TiO$_2$

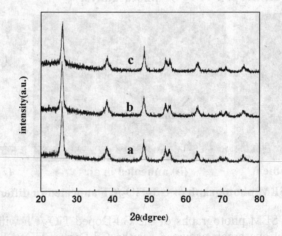

(a) annealed in steam ambient (b) annealed in air (c) annealed in N$_2$ ambient

Figure 4 XRD patterns of Gd-TiO$_2$ annealed in different Atmosphere

Gd-Doped TiO$_2$ powder was used for the crystal phase analysis instead of Gd-Doped TiO$_2$ film because the signal of Gd-Doped TiO$_2$ film was too weak. The XRD patterns of the sol-gel Gd-Doped TiO$_2$ powders calcinated in different ambient were presented in Figure 3. We find that all of the samples were in the anatase form after calcinations at 550°C. The Characteristic spectrum of Gd doesn't appear in Figure 3, which maybe due to the too low content of Gd and X-ray diffraction can not detect it.

3 Conclusions

Gd-doped TiO$_2$/CF precursor was prepared by sol-gel method, and the Gd-doped TiO$_2$ precursor was loaded on the PAN-carbon fiber (PAN-CF) substrates via the dip-coating process. The materials were annealing treated under superheated steam, nitrogen and air atmosphere, respectively. The photocatalytic properties of the Gd-doped TiO$_2$/CF annealed under different atmosphere were investigated by degrading acid orange II under UV irradiation. The results indicated that the annealing atmosphere has significantly influenced on the structures and photocatalytic activity of the thin films. The Gd-TiO$_2$/CF annealed under steam atmosphere exhibited the highest photocatalytic activity and Gd-Doped TiO$_2$/CF possessed lower photocatalytic activity than that of pure TiO$_2$/CF. The XRD analysis suggested that all of the samples were in the anatase form after calcinations at 550°C.

References
[1] XI Beidou, LIU Chunxin, ZHOU Yue xi, *et al*. Photocatalytic oxidation of sodium pentachlorophenolate (pcp-Na) using TiO$_2$ powder prepared by tetrabutyl titanate hydrolsis[J]. China Environmental Science, 2000, 20(5): 449-452
[2] FU Xiaorong, ZHANG Xiaogang, SONG Shigeng, *et al*. Influence of Annealing Atmosphere on the Structures and Photocatalytic Properties of Ag-Doped TiO$_2$ Thin Film[J]. Acta chimica sinica, 1998, 56: 521-526.
[3] Yu Jiaguo, Zhao Xiujian, Zhao Qingnan. Photocatalytic activity of nanometer TiO$_2$ thin films prepared by the sol-gel method[J]. Materials Chemistry and Physics, 2001, 69: 25-29.
[4] Ryuhei Yoshida, Yoshika zu Suzuki, Susumu Yoshikawa. Syntheses of TiO$_2$(B) nanowires and TiO$_2$ anatase nanowires by hydrothermal and post-heat treatments[J]. Journal of Solid State Chemistry, 2005, 178: 2179-2185.
[5] JU Yongfeng, ZU Xiaotao, XIANG Xia. Optical properties of TiO$_2$ film after annealing in nitrogen[J]. High power laser and particle beams, 2008, 20(1): 143-146
[6] P.N. Gunawidjaja, M.A. Holland, G. Mountjoy *et al*. The effects of different heat treatment and atmospheres on the NMR signal and structure of TiO$_2$-ZrO$_2$-SiO$_2$ sol-gel materials. Solid State Nuclear Magnetic Resonance, 2003, 23: 88-106.

65 Study on Preparation Techniques and Function of Thermochromic Microcapsule

Zanmin Wu*, Wenzhao Feng, Xiaozhu Sun

(Institute of Textile, Tianjin Polytechnic University, key Laboratory of Advanced Textile Composites, Tianjin Polytechnic University, Ministry of Education, Tianjin, China)

Abstract: Preparation technics of thermochromic finishing auxiliaries have very important effect on function of thermochromic. Organic thermal materials were wrapped by microencapsulation, it can enhance the stability and wear of color changing. The function is relational with the ratio of core and shell, the rate of stirring super ultrasonic, the emulsifier, time and the temperature of reaction. Using melamine resin as the wall materials, the crystal violet lactone, stearic acid and tetradecanol as the core materials, and the emulsifying agent, xanthan gum aqueous as the protective colloid, have prepared the thermochromic microcapsule. Its color changes reversibly from purple to light yellow. The temperature of color changing of the microcapsule is 46℃, time of color changing is 25s, and the temperature and the time of color fading are 40℃ and 49s.

Keywords: Thermochromic material; Microcapsule; Embedding efficiency; Reversible sensitivity of color changing

1 Introduction

Presently, the reversible thermochomic materials have become a hotspot of researching, and have high practical valued. Thermochromic materials are the intelligent materials, whose color can change with temperature. Organic reversible thermal materials have characteristics of color changing at low temperature and high sensitivity to change color, which make it the material with bright prospects of applications[1-3]. However, organic phenol is selected to be the traditional core material in general, so the chemical durability of color materials is poor[4,5]. In this paper, stearic acid is used to replace organic phenols to reducing toxicity and we microencapsulate the thermochromic materials to separate it from outside environment, improve chemical stability and enhance the reversibility and environmental adaptability.

2 Experiment
2.1 Reagents

Crystal violet lactone (Changshu Dyestuff Factory), Emulsifier (industrial), xanthan, melsmine, formaldehyde (analytically pure, Tianjin Development Zone Letai Chemical Co., Ltd.), stearic acid, tetradecanol (Tianjin FuChen chemical reagent Factory)

2.2 Analysis method
2.2.1 Calculation of encapsulation rate

(1) Encapsulation Rate=[(A–B) /A]×100%　(1)

A:total mass of core material; B: mass of unencapsulated core material

(2) Reversible thermal sensitivity:　　$K = t_f/|t_f - t_c|$　　......... (2)

Where, K is reversible thermal sensitivity, t_c is heating time of thermochromic materials, t_f is color-fading time of natural cooling. In general, $t_f > t_c$. When $t_f \gg t_c$, K is close to 1 which indicates the chromotropic procedure of thermochromic materials is irreversible. The nearer t_f approaches to t_c, the smaller the absolute value of $t_f - t_c$ is, the greater K value is, the better the chromotropic procedure of thermochromic materials is.

2.2.2 Chromotropic temperature and time of microcapsule test

The heating samples at speed 2℃/min, test the chromotropic temperature and discolor time. The samples changes from purple to slight yellow under the chromotropic temperature.

2.2.3 Color-fading temperature and time

Keep the temperature at 23±2℃ under ventilation and nature cooling conditions. Test the color-fading temperature and　time when the color of samples changes from slight yellow to purple.

3 Results and discussion
3.1 The effect preparation of microcapsule on thermochomic efficiency
3.1.1 The effect shell-core rate on thermochomic efficiency

The reversible thermochomic performance of microcapsule is caused by electron transport mainly, when the core material is heated and the solvents melt. And the shell material only plays a protective role for core material.Therefore, the thermochomic performance of core material is the primary factor effecting on thermochomic performance of microcapsule.

The different property of finishing agents can be produced by the core materials which have different ratio of crystal videt lactone-stearic acid-tetradecy in three substances, microcapsules are produced following the ratio in Table 1.The effect of core material in microcapsules on thermochomic efficiency is showed in Figure 1

Table 1 The ratio of different thermochromic core materials

the ratio of core material	1	2	3	4	5
crystal violet lactone–stearic acid–tetradecanol	1:1:10	1:2:30	1:3:30	1:4:40	1:5:50

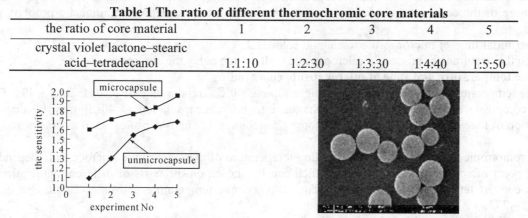

Figure 1 The effect core materials on thermochomic efficiency

Figure 2 SEM photo of thermochromic microcapsule

Figure 1 compares organic composite with its corresponding microcapsules, which indicates the sensitivity of reversible color changing between the two is different. The chromogenic properties of microcapsules are mainly due to core material chromogenic properties, which are changed after microencapsulation, but the influence is less than that of core material.

3.1.2 Appearance and performance of microcapsule

The micro-morphology and core-shell ' structure is showed in Figure 2.From Figure 2, it can be seen that the shape of thermochromic microcapsule approximate spherical shape, the surface is smooth, adhesion between each other is less. The particle size are similar. It has a uniform dispersion, and the diameter is approximately 3~10 μm. It has good size homogeneity and dispersion stability.

3.1.3 Grain size analysis of microcapsule

Using biomicroscopy and image analysis software HORIBA LA–300 to measure and analysis the size and distribution of microcapsule particle(the ratio of core and shell is 2:1, the concentration of emulsifier is 2%, the time is 30min, the rate of stirring is 720r/min, the temperature is 70℃), the result is showed as followed in Figure 3.

Figure 3 shows that, the particle size of microcapsule is about 1-10μm, and the distribution range is narrow.

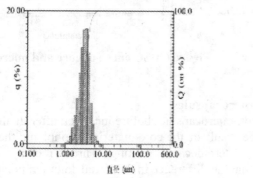 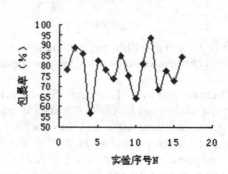

Figure 3 The particle size of microcapsule prepared

Figure 4 Embedding efficiency of microcapsule

3.1.4 Encapsulation rate of microcapsule

The encapsulation rate of microcapsule synthesis using different process in Figure 4.

As can be seen from Figure 4, the encapsulation maximum rate of 12th sample (4% of emulsifier, emulsifying time is 40min, core-shell ratio is 1:2, speed of stirring is 240r/min, the reaction temperature is 70℃) is 93.22%, followed by samples No.2, 3 and No.8, and their parcel rates are 88.65%, 85.77% and 84.61%, while sample 4 has the lowest inclusion rate, only 56.84%. Inclusion rate is one aspect of micro-capsules' evaluation. Chromogenic properties and grain size are major influencing factors of the microcapsule performance to the thermochromic microcapsule.

3.2 Chromogenic properties of thermochromic microcapsule

3.2.1 The temperature and time of microcapsule color changing

The initial temperature of microencapsulation to change color is 46℃, which is higher than the temperature of the core material by 5℃, this may be due to a certain degree of adiabatic property of shell material.

The initial time of microcapsule to change color is 25s, while the time of core material is 15s, this may be due to the heat transfer process of microcapsule wall, which delay the chromogenic time.

3.2.2 The temperature and time of microcapsule color fading

The temperature of color fading of coating sample is 40℃, and the time of color fading is 49s. The time of the composite material which did not change into microencapsulation is 40s. It may be due to heat conduction of the wall of microcapsule, and some shortcoming in micro-capsules, such as cracks and uneven coating.

Chromotropic temperature and color-fading temperature of thermochromic microcapsule depend on the melting point of solvent in core material, which can be chosen on the basis of different application range, such as convert tetradecyl to hexadecanol. The chromotropic temperature and color-fading temperature is about 45-47℃.

3.2.3 Sensitivity of reversible color change microcapsules

According to the formula (2), the sensitivity of reversible color in kinds of microcapsules prepared by different thermal processes is calculated, shown in Figure 5.

As shown in Figure 5, the shortest fading time is connected with the integrity of coated microcapsule wall and the appropriate thickness. Such as the 3# sample (2% of emulsifier, emulsifying time is 30min, core-shell ratio of 2:1, stirring speed of 720r/min, the reaction temperature is 70℃) and 8# sample (3% of emulsifying agent, emulsifying time is 40min, core-shell ratio of 2:1, stirring speed of 480r/min, the reaction temperature is 50℃) thermal discoloration of the microcapsule, reversible color sensitivity is 2.04 and 1.96, which have the best color sensitivity and reversibility, While 6#, and 9# samples is poor and have poor reversibility.

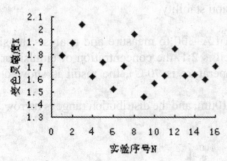
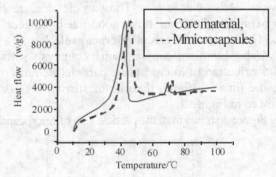

Figure 5 The reversible color changing sensitivity of microcapsules made using vary techniques

Figure 6 DSC curve of core and microcapsules

3.2.4 Thermal stability of thermal discoloration microcapsules

By the analysis of Figure 6, thermal discoloration microcapsules before inclusion after inclusion have an endothermic peak near 40℃ and 70℃, which is the result of the core material alcohol and the stearic acid melting. Temperature change is due to the solvent tetradecanol melting. Through comparing the first endothermic peaks of the two curves, it can be seen that the difference in the initial color temperature between unencapsulated sample and encapsulated one is about 5℃, the major reason is that color component is covered with the melamine resin and the color temperature is higher. However, it has little effect on the discoloration performance of the components. And the shell material will separates the core materials and the external environment, which plays a role of protecting color changing material.

3.3 Chemical stability of thermal discoloration microcapsules
3.3.1 The microcapsule's resistance to alkali and acid

Table 2 The acid or alkali resistant properties of core and microcapsule

acid-base indicator	core material			microcapsule		
	solubility	reversible thermal sensitivity (K)		solubility	reversible thermal sensitivity (K)	
		before immersion	immerse 2h		before immersion	immerse 2h
Acetate (10%)	partial dissolution	1.5	lose chromogenic performance	partial dissolution	1.96	1.89
Sulfate (2%)	partial dissolution	1.5	lose chromogenic performance	partial dissolution	1.96	1.98
Hydrochloride (5%)	partial dissolution	1.5	lose chromogenic performance	partial dissolution	1.96	2.00
Sodium hydroxide (2%)	partial dissolution	1.5	lose chromogenic performance	partial dissolution	1.96	1.93
Detergent (2%)	partial dissolution	1.5	lose chromogenic performance	partial dissolution	1.96	1.94
Soap flake (5%)	partial dissolution	1.5	lose chromogenic performance	partial dissolution	1.96	1.90

In the practical applications of micro-capsules, thermal discoloration microcapsules often encounter with the solvent or additives which are alkali or acid, so the resistance of product to alkali and acid is very important. Sensitivity of reversible color changing microcapsules in alkali and acid is shown in Table 2.

It can be seen from Table 2 that after immersion in acid or alkali solution, the core group lose the color changing characteristics, but the core material's reversible sensitivity of color changing vary little. A small number of microcapsule flaws may lead to changes in color sensitivity.

3.3.2 The microcapsule's resistance to organic solvent

The reversible color sensitivity of the finishing agent in organic solvent before and after microencapsulation under naturally air-dried condition is shown in Table 3.

Table 3 The organic solvent resistant properties of core and microcapsule

Organic solvent	core material			microcapsule		
	solubility	reversible thermal sensitivity (K)		solubility	reversible thermal sensitivity (K)	
		before immersion	immerse 2h		before immersion	Immerse 2h
alcohol	dissolution	1.5	Lose chromogenic performance	insolubilization	1.96	1.94
acetone	dissolution	1.5	Lose chromogenic performance	insolubilization	1.96	1.93
chloroform	dissolution	1.5	Lose chromogenic performance	insolubilization	1.96	1.96
toluene	dissolution	1.5	Lose chromogenic performance	insolubilization	1.96	1.90
cyclohexane	dissolution	1.5	Lose chromogenic performance	insolubilization	1.96	1.92
ethyl acetate	dissolution	1.5	Lose chromogenic performance	insolubilization	1.96	1.98

As can be seen from Table 3, the organic solvents could dissolve the thermal composite material and make it lose color changing performance. In contrast, after microencapsulation, the protective effect of the wall inhibit the dissolution action of organic solvents. At general temperature, the shell of the microcapsule in organic solvent is insoluble, so only a small number of microcapsules coated heterogeneously in organic solvents under a long time will cause defects to reversible sensitivity of color changing. Therefore, the microencapsulation of reversible thermal discoloration core material can greatly enhance the performance of its organic solvent-resistant.

3.3.3 The microcapsule's resistance to humidity

Environmental humidity's impact on the thermal discoloration microencapsulation materials is shown in Table 4.

Table 4 The humidity-fast properties of core and microcapsule

Environmental condition	core material		microcapsule	
	chromotropic time(s)	color-fading time(s)	chromotropic time(s)	color-fading time(s)
Sunny	15	40	23	48
Rainy	14	42	23	49
Water	unchromotropic	unchromotropic	21	51

As can be seen from Table 4, at different humidity, the fading time of the core material before and after microencapsulation almost don't change, but just add the core material to the water, color-change performance is lost, while it have little effect on the micro-capsules.

4 Conclusions

(1) The paper has prepared the thermochromic microcapsule: emulsifier is in 2%, emulsifying time is 30min, core-shell ratio is 2:1, stirring speed is 720r/min, the reaction temperature is 70℃, or emulsifier is in 3%, emulsifying time is 40min, core-shell ratio is 2:1, stirring speed is 480 r/min, the reaction temperature is 50℃, the inclusion performance of microcapsule is the best. The average particle size is 3 ~ 8 μm, the encapsulation rate is greater than 85.77%.

(2) The temperature when color changing is 46℃, discoloration time is 25s, color fading temperature is 40℃. There is little hysteresis and color fading time is about 49s, reversible color sensitivity is 2.04.

(3) The developed thermal color microcapsule core material increases the thermal stability, has a good package and the protective effect and enhances its chemical stability and environmental adaptability such as organic solvent-resistant and acid resistant.

References

[1] Rossetti G P, Susz B P. Addition compounds of p–substituted derivatives of acetophenone with ZnCl2, Helv.Chim.Acta, 1944, 47(1):289-299.
[2] Scott B, Willett R D. A copper(II) bromide dimer system exhibiting piezochromic and thermochromic properties: the crystal structure and electronic spectroscopy of the two room–temperature phases of bis(tetrapropylammonium) hexabromodicuprate(II),Journal of the American Chemical Society. 1991, 113(14): 5253-5258
[3] Pylkki R J, Willett R D, Dodgen H W. NMR studies of thermochromic transitions in copper(II) and nickel(II) complexes with N, N–diethylethylenediamine[J]. Inorganic Chemistry,1984, 23(5):594-597.
[4] Shulkin A, Ster H. D. Polymer microcapsules by interfacial polyaddition between styrene–maleic anhydride copolymers and amines, Journal of Membrane Science, 2002, 209:421-432.
[5] Makino K, Fujita Y, Takao K. Preparation and properties of thermosensitive hydrogel microcapsules, Colloids and Surfaces B: Biointerfaces. 2001, 21(4):259-263.

66 The Influence of Fabric Structures on the Property of Anti-electromagnetic Radiation of Fabrics with Embedded Silver-plated Fibers

Hongxia Zhang, Zhilei Chen, Lijia Shi, Yanqing Li

(Key Laboratory of Advanced Textile Materials and Manufacturing Technology, Ministry of Education of China, Zhejiang Sci-Tech University, Hangzhou 310018, China)

Abstract: The influences of weave cycles and weave structures on the property of anti-electromagnetic radiation of the fabrics with embedded silver-plated fibers were analyzed in this paper. The results showed that, under the same conditions of other basic parameter, there's a trend of drop of the property with the increasing of weaves cycles. In the fabrics of three-wefts-backed weaves, double weaves and double-connecting and filling weaves, the property of the fabrics which silver-plated fibers of weft yarns are in each layer of the weaves averagely is worse than the property of the fabrics which they are all in the surface layer, middle layer and inner layer of the weaves separately.

Keywords: Silver-plated fibers; Anti-electromagnetic-radiation; Weaves cycles

As is well known, with the development of technology, electronic products such as mobile telephone, television, microwave oven and so on are changing with each passing day. Although they have brought great help of high speed information transmission and convenient life, they have also brought electromagnetic radiation which is regarded as another great harmful pollution source after the pollution of water, air and noise. It impacts not only on communications, but also on human health and even on the basic human right for survival. So it has great realistic meanings to study and develop the fabrics with the property of anti-electromagnetic radiation [1].

1 Property test of the fabrics
1.1 Fabrics design

Under the same conditions of materials, processing and so on, a series of polyester fabrics were designed, in order to research the influences of weave cycles and weave structures on the property of anti-electromagnetic radiation of the fabrics with embedded silver-plated fibers. In the fabrics, fabric density: 800count/10cm×660count/10cm. A warps and A wefts were low-elastic polyester fibers with 8.33tex, B warps and B wefts were nylon silver-plated filaments with 9.34tex. The other basic parameter is in Table 1.

Table 1 Fabrics basic parameter

Fabric number	Weave	A warp: B warp	A weft: B weft	The contents of silver-plated fibers (%)
1A	2/2 twill weave	2:1	1:1	43.6
1B	3/3 twill weave	2:1	1:1	43.6
1C	4/4 twill weave	2:1	1:1	43.6
1D	5/5 twill weave	2:1	1:1	43.6
1E	6/6 twill weave	2:1	1:1	43.6
1F	5/2 satin weave	2:1	1:1	43.6
1G	8/5 satin weave	2:1	1:1	43.6
1H	10/3 satin weave	2:1	1:1	43.6
1I	12/5 satin weave	2:1	1:1	43.6
1J	16/7 satin weave	2:1	1:1	43.6
2A	Weft backed weave, silver in the surface layer	2:1	2:1	35.9
2B	Weft backed weave, silver in the inner layer	2:1	2:1	35.9
2C	Weft backed weave, silver in each layer averagely	2:1	2:1	35.9
3A	Three-wefts-backed weave, silver in the surface layer	2:1	2:1	35.9
3B	Three-wefts-backed weave, silver in the middle layer	2:1	2:1	35.9

(continued)

Fabric number	Weave	A warp: B warp	A weft: B weft	The contents of silver-plated fibers (%)
3C	Three-wefts-backed weave, silver in the inner layer	2:1	2:1	35.9
3D	Three-wefts-backed, silver in each layer averagely	2:1	2:1	35.9
4A	Double weave, silver in the surface layer	2:1	2:1	35.9
4B	Double weave, silver in the inner layer	2:1	2:1	35.9
4C	Double weave, silver in each layer averagely	2:1	2:1	35.9
5A	Double-connecting and filling weaves, silver in the surface layer	2:1	2:1	35.9
5B	Double-connecting and filling weaves, silver in the middle layer	2:1	2:1	35.9
5C	Double-connecting and filling weaves, silver in the inner layer	2:1	2:1	35.9
5D	Double-connecting and filling weaves, silver in each layer averagely	2:1	2:1	35.9

1.2 Test method and the instrument

Depending on <Measurement of shielding effectiveness of materials> and ATSMD4935-99 of American standard, fabrics property of anti-electromagnetic radiation was tested by flange coax (plane wave). The instrument: Network analyzer, FY800 Anti-electromagnetic radiation tester. Calculation methods for shielding effectiveness: shielding effectiveness $SE_{dB}=P_1-P_2$, SE_{dB}—The representation of logarithm for shielding effectiveness (dB); P_1—the real-time spectrum analyzer reading when shielding materials are not in the test jigs (dBm); P_2—the real-time spectrum analyzer reading when shielding materials are in the test jigs (dBm) [2-3].

1.3 Test results

The results of the test are in Table 2.

Table 2 Test values of SE_{dB}

Frequency / Fabric number	30MHz	307.2 MHz	822MHz	1020MHz	1812MHz	3000MHz
1A	50	52	55	62	58	56
1B	49	49	52	50	54	48
1C	47	49	48	58	45	53
1D	41	48	50	53	51	50
1E	46	48	47	53	50	48
1F	50	52	55	57	55	56
1G	45	49	49	51	53	53
1H	45	47	52	50	51	50
1I	42	48	44	50	52	51
1J	33	47	45	51	52	52
2A	40	51	39	52	48	51
2B	43	50	43	47	52	49
2C	45	48	42	53	48	48
3A	53	58	49	50	57	52
3B	52	51	51	54	55	58
3C	52	55	52	56	58	51
3D	49	50	42	45	51	48
4A	46	51	60	51	52	49
4B	53	58	55	48	59	51

(continued)

Frequency / Fabric number	30MHz	307.2 MHz	822MHz	1020MHz	1812MHz	3000MHz
4C	49	48	50	47	47	48
5A	51	53	54	53	54	53
5B	50	51	50	51	56	55
5C	46	53	57	53	58	53
5D	33	49	47	51	48	45

2 Analysis and discussion

2.1 The influence of weave cycles on the property of anti- electromagnetic radiation

Fabrics of 1 series are designed in order to study the relationship between weave cycles and the property of anti-electromagnetic radiation. The contents of silver-plated fibers are all 43.6%, and they have the same embedded rule. The weaves are twill and satin, and t

he cycles are increasing.

In Figure 1 and Figure 2, the property of the fabrics has a degressive current with the increasing of weave cycles. In 1 series, the amount of interwoven points between warp and weft yarns is decreasing with the increasing of weave cycles, so the number of interwoven points among silver-plated fibers is also decreasing. Meanwhile, the silver-plated fibers can't form smaller meshes and can't relate closely in space either. As a result, the property is degraded. Fabrics with 2/2twill weave and 5/2satin weave have the better property relatively. From Figure 2 we know, after the cycle reaches 8, the property is decreasing slowly [4-5].

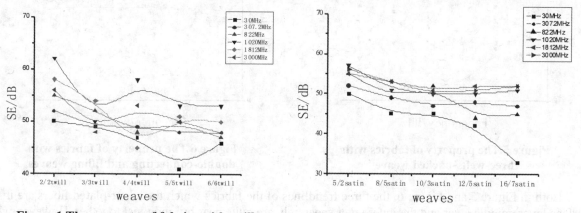

Figure 1 The property of fabrics with twill weave   Figure 2 The property of fabrics with satin weave

2.2 The influence of silver-plated fibers in weft backed weave and double weave on the property of anti- electromagnetic radiation

In 2 and 4 series, the contents of silver-plated fibers are all 35.9%. Here the distributions of silver-plated fibers which are in the surface layer, in the inner layer and in each layer averagely are all in the weft yarns.

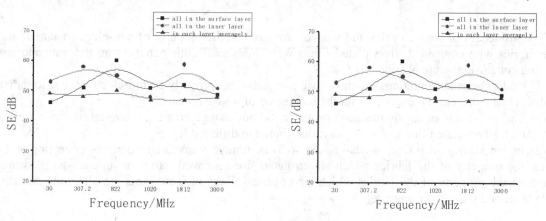

Figure 3 The property of fabrics with weft backed weave   Figure 4 The property of fabrics with double weave

In Figure 3, the influence of the different distributions of silver-plated fibers along the weft in the fabrics of weft backed weave on the property of anti-electromagnetic radiation is not obvious. This is because silver-plated fibers and polyester fibers in the wefts are overlapped together, and also interlace with the silver-plated fibers in the warps closely, and there is no obvious difference among the meshes.

In Figure 4, the impact of the different distributions of silver-plated fibers along the weft in the fabrics of double weave on the property is great, and the data of the property among the three fabrics fluctuates a lot. Because compared with weft backed weave, the fabric structure of double weave is looser, so the property of the fabrics in 4 series is obviously different. In addition, the property of the fabric which silver-plated fibers are in each layer averagely is the worst. Because the meshes become bigger while the silver-plated fibers interact and overlay mutually, thus lead to the decline of the property.

2.3 The influence of silver-plated fibers in three-wefts-backed weaves and double-connecting and filling weaves on the property of anti- electromagnetic radiation

The contents of silver-plated fibers are all 35.9% in fabrics of three-wefts-backed weaves of 3 series and double-connecting and filling weaves of 5 series. Here the distributions of silver-plated fibers which are in the surface layer, in the middle layer, in the inner layer and 1/3 in each layer are all along the weft.

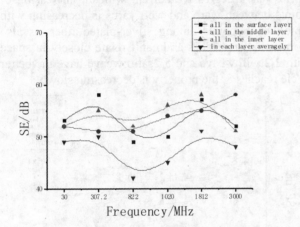
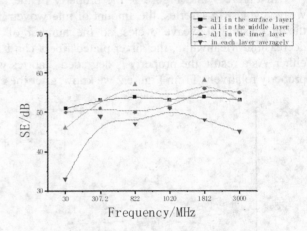

Figure 5 The property of fabrics with three-wefts-backed weave **Figure 6 The property of fabrics with double-connecting and filling weaves**

Both in Figure 5 and Figure 6, the three trendlines of the fabrics which the silver-plated fibers are in the surface layer, middle layer and the inner layer separately are quite near by, and twist together. However, when the silver-plated fibers are in each layer averagely, compared with the others, the average descending range is 7dB, and result in a relatively large deviation between the trendline and the other three trendlines. In fabrics of 3D and 5D, silver-plated fibers are overlapped together, and result in the augment of the meshes and the decrease of the contact between the silver-plated fibers in weft yarns and warp yarns, and finally lead to the decline of the property.

3 Conclusions

The influences of weave cycles and weave structures on the property of anti-electromagnetic radiation of the fabrics with embedded silver-plated fibers were analyzed in this paper. From the tests and analyses, following conclusions were obtained.

(1) Under the same conditions of other basic parameter, the property of anti-electromagnetic-radiation of the fabrics had a decreasing tendency with the increasing of weave repeats.

(2) The anti-electromagnetic radiation property did not change greatly among fabrics with weft backed weave which silver-plated fibers of weft yarns distributed in different layers.

(3) In the fabrics with three-wefts-backed weaves, double weaves and double-connecting and filling weaves, the property of the fabrics which silver-plated fibers of weft yarns are in each layer averagely is worse than the property of the fabrics which silver-plated fibers of weft yarns are all in the surface layer, middle layer and inner layer separately.

References

[1] Zhibing Yao, Hao Jiang, Ting Wu, Jin Xu, Xinyu Shi. The harm and prevention of electromagnetic radiant. Chinese Journal of Social Medicine Res J, 2007, 24(3): 177-179.
[2] Mingjun Cheng, Xiongying Wu, Ning Zhang, Shanyuan Wang. Investigation on test methods for electromagnetic shielding effectiveness of fabrics. Dye Printing Res J, 2003, (9): 31-35.
[3] IEEE Power Engineering Society. IEEE standard method for measuring the effectiveness of electromagnetic shielding enclosures. IEEE Res J. 2006.
[4] Guohua Liu, Wenzu Wang. Development of electromagnetic shielding fabric. Industrial Textiles Res J. 2003, (6): 16-18.
[5] Heyi Chen. The design of new anti-electromagnetic-radiation fabrics. Tianjin: Tianjin Polytechnic University Res D. 2006, 17-24.

67 Study on the Patterns of Small Black-and-white Grid in Costumes

Jian Zhao, Cai Qian Zhang

(Textile and Apparel Institute, Shaoxing University, Shaoxing Zhejiang 312000, China)

Abstract: Small black-and-white grid pattern for apparel design is to use black-and-white lines between the two veined changes, reflecting the unique design and style. This pattern has been warmly welcomed and formed a unique market characteristic and style. The real reason of the classic and eternal effect in fashion design can be found in exploring the discipline of this style and effect. This paper reveals many ways to use the peculiarity and characteristics of its style, researching its basic techniques and design features, and propose conclusions and recommendations on that basis.

Keywords: Blank and white; Patterns; Design

The usage of small black-and-white grid pattern in fashion design majors in subtle aspects of permutations and combinations of black-and-white colors, warp and weft arrangement. It can form a small black-and-white grid pattern in the fabric surface, with the different proportion of black-and-white pattern, white or symmetrical arrangement; or staggered, different combinations of pattern can be formed in different styles. Black-and-white pattern on the fabric surface forms a continuous effect as if being broken, a unique pattern effect, which is, soft to touch. A worsted fabric imitation effect can be created in style using this material[1-2], as well as an effect of prominent black-and-white pattern fabric surface.

1 Introduction

Pattern of small black-and-white grid design of clothing usually brings people a unique aesthetic experience, and people in different locations and ages have different feelings about this. Just like the research and design of other decoration, we need to discuss the reasons and development of it in order to a achieve suitable road of design and marketing fitting in the public aesthetic, we will get valuable information from different design styles after our methods for designers and markets. The black-and-white dot decorative design closures need little consideration for standards, but they always appears different dressing styles with broad applicable scope and applicable people, not only for common people, but also for people pursuiting for strangeness. The target consuming group can be common people or special people, to some extent, this design pattern can even stand for some political attitudes.

1.1 The characteristics

The classical point of the black-and-white dot decorative design is in its conciseness and gender-neutralization. Firstly, it has no clear resuming group deviation, and i.e. it is applicable for folk people, which is one of important reasons for the popularity of it. On this basis, its contrast between black and white brings a strong visual attack, with clean and regular properties. Many people like it because of its conciseness and changeable properties. It can stands for the traditional aesthetic experience to appear normative appearance. At the same time, it is popular among rock stars. Different aesthetic points join in the black-and-white lines to create a unique style.

1.2 Patterns in ancient China

Figure 1 Ancient Chinese clothing

First, we can deduce from the ancient China's similar diamond tread pattern, which is one of the black-and-white patterns with small grid, and it was often presented in official because of its generous decency. "As an abstract geometric pattern, Link Check has a large number of applications in the ancient Chinese silk jacquard pattern, during the previous Qin to the Qing dynasty. With the continuing transformation of the ancient loom, the continuous progress of technology and change in aesthetic needs, Link Check has performed different decorating styles at different times, playing different decorative role in pattern at different times[3]."

1.3 Patterns in foreign countries

It is more prevalent in the west using the patterns of small black-and-white grid than the east, frequently visible in the western building, furniture and even the design of ceramics. Because the geometry has been well developed in the West, and the entire design is based on geometry and even more abstract. By comparison, China's Patterns are more symbolic and complex.

The most famous black-and-white abroad is the Tartan Patterns (Tartan), with small number of grid points. In Scotland, it was the symbol pattern about the clan, identity, status, reputation at first. It appeared as early as 1800 years ago in Scotland, and eventually evolved into its national dress, with thousands of models. Nowadays the style has changed greatly; modern designs have their own colors and patterns because of their different regions, places, organizations and purposes, while in ancient times it was only used to distinguish the clan. So the companies have to register the copyright to make sure the copyright will not be duplicated or stolen.

Another black-and-white pattern for a small grid in the Arab Middle East region, it was made to withstand wind and sun dress scarves and so on, black-and-white grid is very popular in the Levant, while the Jordanians like a red and white cheek style. This small grid pattern for black-and-white is not only popular in folks, but also becoming a political signal of the Middle East, to some extent.

1.4 It is widely used in clothing, bags, perfume bottles and other apparel industry

It is widely used in clothing such as shirts, coats, pants, skirts, etc. Both for men and women, various of styles.

Figure 2 Bags black-and-white lattice   **Figure 3 Grid of black-and-white corsage ornament**

Black-and-white grid pattern can be applied to making small accessories. Such as bags, corsage matching with the clothes, form a unified style, and reinforce the effectiveness and infections of clothing. It is particularly unusual in hat with that Patterns, this is the reason why manufacturers particularly like to do with the hat. There are also used in perfume bottles, umbrellas, etc. including in the garment industry within a large number of apparel products.

2 Style design features

Black-and-white grid pattern for a small clothing design, including both relaxed and casual, can be superimposed like one plus one. So the style makes it particularly charm. Although with a single color, it can fly its own colors in the pattern area because it can be diversified in different areas and its unique pattern.

Table 1 Grid pattern of small black-and-white fabric style list

I	Casual style	mashups + modern innovation → free or leisure; (see Figure 4)
II	Formal style	brand + celebrity model → formal or rigorous (see Figure 5)

Figure 4 IT girls mix and match   **Figure 5 Arafat classic scarf law**

Clothing with a regular black-and-white pattern and the internal seriousness of black-and-white costumes reflects the strict and formal, style which can break the whole rigid and solemn because of its mixed black and white. So, there is a profound reason why this black-and-white pattern has gained special popularity among street youth and successful people. For example, in the 1980s, a group of young Americans with hippie spirit wore black turtleneck sweaters, worn Levi's, high shoes, eight black leather zipper jacket, added collar with a small section of black and white, fully embodies the youth and unruly, rebellious street culture and fashion features.

But from another perspective, some successful people including the celebrities who are extremely favorite to the small grid pattern for black-and-white dress, even as their identifier. The greatest impact on Small grid on the black-and-white pattern should be the former Palestinian leader Yasser Arafat (to Figure 5), but Arafat was not the only celebrity who shows special inferences to black-and-white grid pattern for small soft spot in the history of clothing, besides, designers Chanel and the famous British fashion designer Vivian ·Westwood prefer to using black-and-white pattern for a small grid style to reflect styles of their clothing.

3 Main points of design and technology

Table 2 Grid pattern of small black-and-white fabric style list

Name	Youth	Arabs	Celebrity	Political	Brand	Music
bonding point	hippie	exotic	fashion trend	political signs	1. Chanel 2. Burberry 3. pringle	rock musicians dress

The reasons why eternal popularity in patterns of small black-and-white costumes are as following:

3.1 Meet the aesthetic demands of the times and mass trends

Small black-and-white grid pattern should combine the traditional features, with the aesthetic characteristics of people, especially it can meet the youth's visual aesthetic needs; because its perception is a part of the most sensitive and intuitive aspect. And the small black-and-white grid pattern just satisfies this characteristic. It is a good example for the ever popularity in small grid pattern though a long evolution process.

The aesthetic is always going with times, involving in political factors, environmental factors, and economic factors, the needs and enduring of black-and-white grid pattern for a small clothing show the mass popularity in this era, meeting the public demands and popularity.

It has a deep historical background and deep cultural connotation which lead to the particular popularity of patterns of small black-and-white grid in the Western world. As is known to all, Geometric is the basis of abstract pattern, what is important, an aesthetic genre - Pythagoras, and its founder is a mathematician, so Westerners like this aesthetic form has basis on both of original and profound cultural foundation and the masses. Long-term academic origin and aesthetic geometric makes the black and white with small grid pattern accepted by western people for a long time, and being passed by as a good heritage

3.2 Meet the technical process and cost requirements

It greatly reduces the production costs because of its simple color. In the mean time the change of grid pattern seems to be variable. However, changes in lattice pattern are relatively much smaller, compared with other cumbersome terms; therefore, its design costs lower, Technical process makes production more efficient.

4 Conclusions and recommendations

(1) Fashion degree. As can be seen from the above table, black-and-white pattern for a small grid is closely related to "youth, music, celebrities" and so on. These points are combined with one of fashion elements, so black and white for a small grid patterns has the characteristics of fashion, to some extent. This is one of the very important reasons for a small black-and-white grid pattern's being a classic. This is to be considered first during the development and design of small black-and-white grid pattern for clothing, particularly during the conception process of the product's development and design. Market apparel products can be easily approved with the combination of these essential elements.

(2) Consumer's psychology. Apart from the classical style, considering multi-level consumer's psychology is necessary, because different customers have different "fans" of "celebrities, rock musicians, and politicians," they will bring a lot of consumer groups. This is also the point in developing the design, apparel products would be welcomed only when the time lines correspond with consumer's psychology, or it will fail.

(3) Design innovation. It has great influence on the clothing design such as patterns, fabrics, styles and other technical parameters. It will cause troubles if the fabric in the design is inappropriate. It will lead to low profits and poor sales if the style is out of time.

References
[1] Gu Ping, editor. Fabric Structure and Design Shanghai: East China University Press, 2004:1-210
[2] ZHU Su-kang, Gao Weidong. Woven Science Beijing: China Textile Press, 2008: 1-240.
[3] Ying ancient Chinese silk jacquard fabric evolution of a pattern in the Link Silk A, 2006(8): 51-53

68 Thermodynamics Study of Monomer Adsorption Process for Fabrication of Conductive Textiles

Yaping Zhao[1], Zhaoyi Zhou[2], Xiaolan Fu[1], Zaisheng Cai[2*]

(1. College of Chemistry, Chemical Engineering & Biotechnology, Donghua University, Shanghai 201620, China
2. Shanghai Institute of Fibre Inspection, Shanghai 201600, China)

Abstract: Conductive textiles can be fabricated by a two-stage method involving adsorption of organic monomers by fibers and successive oxidation polymerization. Thermodynamics studies of monomer adsorption process with and without the existence of anionic surfactant or nonionic surfactant were investigated in this article. From thermodynamic analysis results, it can be concluded that the standard affinity ($-\Delta\mu^0$), enthalpy (ΔH^0) and entropy change (ΔS^0) of aniline monomers towards polyethylene terephthalate (PET) fibers are 7.48 kcal/mol and 0.029 kcal/mol·K, respectively. It can be deduced that the anionic surfactant or nonionic surfactant plays a role of increasing the solubility of aniline monomers in aqueous solution for the decrease of standard affinity ($-\Delta\mu^0$) and that the addition of anionic surfactant leads to the increase of standard enthalpy (ΔH^0), with the monomers strongly integrated into the polymer chains of substrates. The results of the thermodynamic study indicated that the adsorption processes are endothermic and spontaneous because of the positive value of ΔH^0 and negative value of ΔG^0.

Keywords: Thermodynamics; Adsorption; Aniline; Surfactant; Conductive Textiles

1 Introduction

Conducting polymer-based conductive textiles have been extensively studied for the application of industrial materials like filters, deelectrifying and electromagnetic interference shielding materials as well as special purpose clothing, which is dust and germ free. Polyaniline (PANI) is one of the most promising conducting polymers for its environmental stability in conducting form, unique redox properties, and high conductivity[1-3]. Conducting polymer-based conductive textiles can be fabricated by in-situ polymerization consisting of adsorption of monomer on the surface of the textiles and consequently oxidations polymerization process[4,5]. The in-situ adsorption process of monomer on the surface of the textiles is very important for preparing uniform coating layer of conducting polymers on the textiles[6]. However, the mechanism of this adsorption step has been concerned by a minority of research groups and previous study in our laboratory have focused this aspect. In addition, the important role for the homogeneous polymerization of aniline that the surfactant played has been reported by several research groups recently[7,8]. The influence of the surfactant on monomer adsorption towards textiles also deserves attention. In this article, we study several physicochemical parameters such as standard affinity ($-\Delta\mu^0$), enthalpy change (ΔH^0), entropy change (ΔS^0) to interpret the thermodynamic behaviors of the aniline in the adsorption system with and without the existence of surfactants.

2 Experimental
2.1 Materials

PET fabric (102×76 yarns/inch, 87±2 g/m^2) was used after scouring in a 20g/l aqueous sodium hydroxide solution at 95℃ for 90min. The liquor ratio was 50:1. All chemicals used in this study including aniline which was distilled at a reduced pressure before use, ammonium persulfate (APS), polyethylene glycol (PEG, M_w=200), sodium dodecylsulfonate (SDS) and hydrochloric acid were purchased from Sinopharm Chemical Reagent Co. (Shanghai, China).

2.2 Adsorption experiments

Equilibration experiments were carried out by agitating a series of bottles containing aniline solutions of different initial concentration (0.1-0.5 mol/L) with PET fabrics at temperature (313-368K) until equilibrium adsorption was obtained. Aniline concentration in solution was then measured at 231nm using a Hitachi UV-3100 spectrophotometer.

---

* Corresponding author's email: zshcai@dhu.edu.cn

2.3 Determination of thermodynamic parameters

The partition coefficient (K) of the monomers between the fibers ($[D]_f$) and the adsorption solution ($[D]_s$) was obtained from the adsorption isotherm and the standard affinity ($-\Delta\mu^0$) was calculated by Eq. (1). The enthalpy change (ΔH^0) in adsorption process was obtained from the relationship between $\Delta\mu^0/T$ and $1/T$ using Eq. (2) and the entropy change (ΔS^0) was calculated by Eq. (3). The Gibbs free energy change, ΔG^0 value was calculated from Eq. (4).

$$-\Delta\mu^0 = -\left(\mu_f^0 - \mu_s^0\right) = RT \ln \frac{[D]_f}{[D]_s} = RT \ln K \qquad (1)$$

$$\Delta H^0 = \frac{\delta(\Delta\mu^0/T)}{\delta(1/T)}, \quad \frac{\Delta H^0}{T} = \frac{-\Delta\mu^0}{T} + C \qquad (2)$$

$$\Delta\mu^0 = \Delta H^0 - T\Delta S^0 \qquad (3)$$

$$\Delta G^0 = \Delta H^0 - T\Delta S^0 = \Delta\mu^0 \qquad (4)$$

Where $-\Delta\mu^0$ is standard affinity (cal/mol); μ_f^0, standard chemical potentials of monomers in the fiber; μ_s^0, standard chemical potentials of monomers in the adsorption solution; $[D]_f$, monomer concentration in the fiber (g/kg fiber); $[D]_s$, monomer concentration in the adsorption solution (g/l); K, partition coefficient; R, gas contant (1.9872 cal/mol·K); T, absolute temperature (K); ΔH^0, heat of adsorption (cal/mol); C, integral constant; ΔS^0, change in entropy (cal/mol·K) and ΔG^0, change in Gibbs free energy(cal/mol).

3 Results and discussion
3.1 Partition coefficient and standard affinity

The standard affinity ($-\Delta\mu^0$) of monomers in adsorption bath towards textile substrates is the most basic thermodynamic parameter, which can measure the tendency of monomer moving from adsorption solution to fabrics under its standard state[9]. With the aim of calculating the standard affinity of monomers towards polyester fabrics, the isotherm data of aniline adsorption were gained at several different temperatures. Figure 1 showed that the adsorption isotherms of the monomer solution with and without surfactants such as polyethylene glycol (PEG) or sodium dodecyl sulfate (SDS) all obeyed Henry's adsorption model, which commonly represents the dyeing mechanism between disperse dye and PET[10,11]. The linear partition isotherms obtained for adsorption of aniline monomers on PET fibers can be considered to support the view that the monomers interacted with the fibers by means of a solid solution mechanism, according to which the monomer molecule dissolved in the fiber as it would be in an organic solvent.

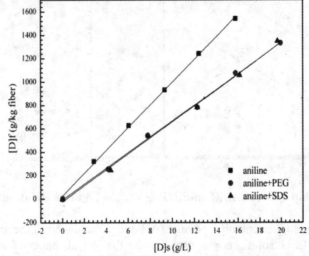

Figure 1 Adsorption isotherms of monomers on PET fabrics at 368K

Table 1 Partition coefficients (K) and standard affinities ($-\Delta\mu^0$) of the monomers between PET fabrics and adsorption bath

Temperature (K)	Adsorption Bath					
	Aniline		Aniline+PEG		Aniline+SDS	
	K	$-\Delta\mu^0$(cal/mol)	K	$-\Delta\mu^0$(cal/mol)	K	$-\Delta\mu^0$(cal/mol)
348	50.94	2718	41.73	2580	37.83	2512
353	59.60	2867	48.00	2716	42.52	2631
358	67.06	2992	53.94	2837	51.47	2804
363	79.01	3152	60.34	2958	60.10	2955
368	92.26	3309	67.18	3077	67.99	3086

The partition coefficients (K) of the monomers between the adsorption bath and the PET fabrics and the values of standard affinity ($-\Delta\mu^0$) calculated by Eq. (1) were summarized in Table 1. From Table 1, it was found that the partition coefficient (K) and standard affinity ($-\Delta\mu^0$) increased in all monomer solutions with the temperature increasing, which indicated this adsorption step was an endothermic process, resulting higher temperature gives a positive effect on the thermodynamic adsorption. The results presented in Table 1 also showed that with the addition of a surfactant, the values of partition coefficient and standard affinity were lower, comparing with that without any surfactant when adsorption occurred. This implied that the aniline monomer was more soluble in water-surfactant mixture than in pure water (free of surfactants), that is, the existence of anionic surfactant or nonionic surfactant can enhance the solubility of aniline monomers in aqueous solution.

3.2 Enthalpy and entropy change

The enthalpy change (ΔH^0) can be utilized to measure the adsorption strength of aniline monomers on the substrates. Figure 2 and Figure 3 showed the linear relationship between $\Delta\mu^0/T$ and $1/T$ and that between $\Delta\mu^0$ and T, respectively. Both the values of enthalpy change (ΔH^0) and entropy change (ΔS^0) were all calculated from the slopes of the straight lines in Figure 2 and Figure 3 according to Eq. (2) and Eq. (3) were listed in Table 2. The positive values of the standard enthalpy change (ΔH^0) and standard entropy change (ΔS^0), which are 7.48 kcal/mol and 0.029 kcal/mol·K respectively, for polyester fabrics in monomer solutions, indicated that the interaction of aniline molecule with fabric was endothermic in nature[12,13], which was distinct from the adsorption behavior of a disperse dye on polyester fibers.

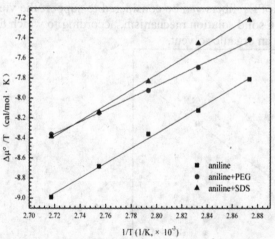
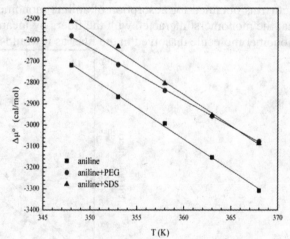

Figure 2 Relationship between $\Delta\mu^0/T$ and $1/T$ **Figure 3 Relationship between $\Delta\mu^0$ and T**

Table 2 suggested the higher enthalpy change (ΔH^0) were achieved in the aniline-SDS bath but the lower obtained in the aniline-PEG solution, compared with that in absence of a surfactant. Commonly, the positively larger value of the enthalpy change represents that the adsorbate molecules were more strongly integrated into the adsorbent. The greater the value of the enthalpy change, the more stable the adsorption would be. In the aniline-SDS adsorption solution, SDS could form micelles in the water phase and aniline monomers with positive charges would be attracted to these micelles with the phenyl group of aniline was exposed to the bulk water to some extent, as Figure 4 showed. One of the adsorption theories suggested that aromatic compounds were adsorbed through sorbate-sorbent π-π interaction[14,15]. Hence, the addition of SDS may lead to the benzene ring approaching the surface, enhancing the adsorption capacity of aniline on fibers, which had a dominant role in the process of monomers adsorbed towards fibers. In the aniline-PEG adsorption solution, as a nonionic surfactant, PEG was added into the adsorption system mainly to increase the solubility of aniline molecules in water medium, resulting the migration of aniline monomers towards fibers restricted.

Table 2 The values of enthalpy change (ΔH^0) and entropy change (ΔS^0) of aniline monomers on PET fibers with and without the surfactant

Adsorption Bath	ΔH^0(kcal/mol)	ΔS^0(kcal/mol·K)	$T\Delta S^0$(kcal/mol) (T=298K)
Aniline	7.48	0.029	8.64
Aniline+PEG	6.01	0.025	7.45
Aniline+SDS	7.73	0.029	8.64

The positive values of the entropy change (ΔS^0) in Table 2 indicated there was an increase of randomness at the solid/solution interface during the adsorption of aniline monomers onto fibers. As a kind of hydrophobic substance, aniline has the tendency to make water molecules form a more regular structure just like cluster conformation in the adsorption bath[16].when aniline molecules moved from adsorption bath to fiber substrates, some water molecules were released, causing the increase of randomness in the sorbate-sorbent system. Secondly, the formation of the highly stable PEG micelles when aniline monomers were adsorbed from water phase by fibers might be an important factor in a lower value for ΔS^0 in the monomer solutions with the presence of PEG.

$$CH_3(CH_2)_{11}OSO_3^- Na^+ + H_3O^+ + Cl^- \Leftrightarrow CH_3(CH_2)_{11}OSO_3^- + H_3O^+ + Na^+Cl^-$$

$$C_6H_5NH_2 + H_3O^+ + Cl^- \Leftrightarrow C_6H_5NH_3^+ + H_2O + Cl^-$$

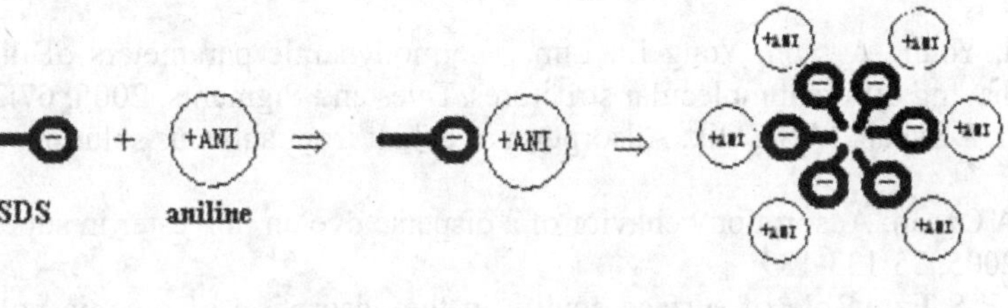

Figure 4 Scheme of aniline dispersion in the SDS micellar system

3.3 Gibbs free energy change

The Gibbs free energy change (ΔG^0) was calculated to evaluate the thermodynamic feasibility of the process and to confirm the nature of the adsorption process. Based on the following Eq. (4), ΔG^0 values can be obtained at constant temperature. Two factors of energy factor ΔH^0 and entropy factor $T\Delta S^0$ contributed to the value of free energy ΔG^0. As Table 2 displayed, the entropy factor $T\Delta S^0$ was positive and the energy factor ΔH^0 was also positive. In this case, the feasibility of the reaction depends on the factor which predominates. As listed in Table 2, when the temperature was at 298K, the numerical value of $T\Delta S^0$ was more than ΔH^0 and the reaction was feasible because the value of ΔG^0 would negative, which indicating that the sorption was spontaneous at all temperatures studied[17,18].

4 Conclusions

The adsorption isotherms of the monomer solution with and without surfactants such as polyethylene glycol (PEG) or sodium dodecyl sulfate (SDS) all obeyed Henry's adsorption model. Thermodynamic adsorption characteristics of polyester textiles in the monomer solutions were investigated and compared with those in monomer-surfactant solutions. For all the fiber-solution adsorption systems, the partition coefficient (K) increased as the temperature increased. The standard affinities($-\Delta\mu^0$) in the monomer solutions with the presence of polyethylene terephthalate (PET) or sodium dodecyl sulfate (SDS) were a little lower than those in the monomer solutions without any surfactant. In the case of enthalpy change (ΔH^0) and entropy change (ΔS^0), polyester fibers showed relatively higher positive values in adsorption bath with or without SDS than those in adsorption bath with PEG. The results showed that the adsorption process was endothermic and spontaneous because of the positive value of ΔH^0 and negative value of ΔG^0.

Acknowledgement

This work was supported by the Key Project of Chinese Ministry of Education (No.109066) and by the Nano-project of Science and Technology Commission of Shanghai Municipality (0852nm00300). It was also funded by Innovation Foundation for PhD Candidate of Donghua University (DHU) (BC200927).

References

[1] X G Jin. Diffusion-deposition of polyaniline onto textiles with high electric conductivity and improved adhesion. J Industrial Textiles, 1996, 26(1): 36-44.
[2] J Wu, D Zhou, M G Looney, P J Waters, G G Wallace and C O Too. A molecular template approach to integration of polyaniline into textiles. Synth Met, 2009, 159(12): 1135-1140.
[3] K Firoz Babu, R Senthilkumar, M Noel and M Anbu Kulandainathan. Polypyrrole microstructure deposited by chemical and electrochemical methods on cotton fabrics. Synth Met, 2009, 159(13): 1353-1358.
[4] K W Oh, K H Hong, S H. Kim. Electrically conductive textiles by in situ polymerization of aniline. J Appl Polym Sci, 1999, 74, 2094-2101.
[5] H H Huang, W J Liu. Polyaniline/poly(ethylene terephthalate) conducting composite fabric with improved fastness to washing. J Appl Polym Sci, 2006, 102: 5775-5780.
[6] M X Wan. Conducting polymers with micro or nanometer structure. Tsinghua University Press, Beijing and Springer-Verlag GmbH Berlin Heidelberg, 2008:78.
[7] N Kuramoto, E M Geniès. Micellar chemical polymerization of aniline. Syn Met, 1995,68: 191- 194.
[8] R Guo, Z C Li and T Q Liu. Electrochemical polymerization of aniline in SDS admicelles. Colloid Polym Sci, 2005, 283: 1063-1069.
[9] T K Kim, Y A Son. Affinity of disperse dyes on polyethylene terephthalate)in non-aqueous media.Part 2:effect of substituents. Dyes & Pigments, 2005, 66: 19-25.
[10] J Alan. The theory of coloration of textiles, 2nd ed. West Yorkshire: Society of Dyers and Colourists, 1989: 263e4.
[11] Tae-Kyung Kim, Young-A. Son, Yong-Jin Lim. Thermodynamic parameters of disperse dyeing on several polyester fibers having different molecular structures. Dyes and Pigments, 2005, 67:229-234.
[12] M Kara, H Yuzer, E Sabah, M S Çelik. Adsorption of cobalt from aqueous solutions onto sepiolite. Water Res, 2003, 37: 224.
[13] A S Özcan and A Özcan. Adsorption behavior of a disperse dye on polyester in supercritical carbon dioxide. J Supercri Fluid, 2005, 35:133-139.
[14] R W Coughlin, F S Ezra.Role of surface acidity in the adsorption of organic pollutants on the surface of carbon. Environ Sci Technol, 1968, 2: 291-297.
[15] Y H Han, X Quan, S Chen, H M Zhao, C Y Cui, Y Z Zhao. Electrochemically enhanced adsorption of aniline on activated carbon fibers. Sep Purif Technol, 2006, 50: 365-372.
[16] T Hori, H Zollinger. Role of water in the dyeing process, Textile Chemist & Colorist, 1986, 18(10): 19-25.
[17] R Coskun, C Soykan. Lead(II) Adsorption from aqueous solution by poly(ethylene terephthalate)-g-acrylamide fibers. J Polym Res, 2006, 13: 1-8.
[18] F A Pavan, I S Lima, E V Benvenutti, Y Gushikem, C Airoldi. Hybrid aniline/silica xerogel cation adsorption and thermodynamics of interaction. J Colloid Inter Sci, 2004, 27: 386-391.

69 An Analytical Model for Ballistic Impact on Textile Based Body Armour

Fuyou Zhu[1], Xiaogang Chen[1], Garry Wells[2]

(1. Textiles and Paper, School of Materials, University of Manchester, Manchester, M13 9PL, UK
2. Physical Science Department, Dstl, Porton Down, Salisbury, Wiltshire, SP4 0JQ, UK)

Abstract: Polymer fibres with high tenacity and modulus such as Kevlar were widely used in personal protection applications. Normally these fibres were manufactured into different fabric structures before being lay-up layer by layer to form personal protection panels. Any guidance generated by analytical study is highly valuable for the later empirical and numerical studies. An analytical model of ballistic impact of multi-layer woven fabrics has been developed based on analytical models of ballistic impact of single yarns available in the literature and Newton's third law of motion. It incorporates the effect of various intrinsic parameters such as yarn linear density and weaving density. It also incorporates the effect of various extrinsic parameters such as projectile diameter and mass. The model also incorporates the phenomenon of strain gradient and its effects on the tensile strain at the edge of the projectile and the angle between the impact line and the yarn. In addition, possible shear failures were also incorporated by using shear strength together with maximum tensile strain as failure criterion. It has been found out that the results from the analytical modelling were very close to the experimental results. Further studies of the analytical modelling results reviewed strain distribution history of the principle yarns in each layer. It has been concluded that during the ballistic impact of multi-layer woven fabrics shear failure occurs before tensile failure for the front layers which causes the cease of the high strength polymer fibres to reach its full energy absorption potential.

Keywords: Body armour; Analytical model; Ballistic; Strain distribution; Textiles

1 Introduction

Textile based body armour panels are normally formed by an assembly of multiple layers of woven fabrics. Woven fabrics made of polymer fibres with high tenacity and modulus were widely used as part of or all the layers in textile based body armour. Thus studies of the ballistic performance of multiple layers of woven fabrics are vital to the success of textile based body armour. Empirical studies are the most straightforward method but are costly and time-consuming and often not efficient due to the large amount of testing parameters needed to be considered. Numerical studies are cost effective but also time-consuming and require high computing power and may produce misleading conclusions due to the lack of clear understanding of physical and mechanical events happened during the impact process. Analytical method becomes handy if close-form mathematic equations can be set up to describe the physical and mechanical phenomenon happened during the impact. Any guidance generated by analytical study is highly valuable for the later empirical and numerical study. Encouraged by this, the focus of this paper was placed on studying the possibility of setting up an analytical model of ballistic impact of multi-layer woven fabrics.

2 Literature review

Analytical modelling of single textile yarns has been well studied by Stone et al [1] and Smith et al [2-5]. If assuming linear elastic stress-strain relation and no effect of poison's ratio, and no creep and stress relaxation, the reaction of a textile filament subjected to transverse impact can be described using equation (1 – 6) where c is the longitudinal stress wave velocity, W is the velocity of inward flow of material towards the impact point, U_{lab} is the transverse wave velocity respect to the laboratory, u is the transverse wave velocity relative to the strained yarn, V is the impacting velocity of the projectile, ε is the strain behind the elastic stress wave generated at the instance when the projectile contacted with the yarn and θ is the angle between the impact line and the yarn. There are only a few studies of analytical modelling of ballistic impact of multi-layer fabrics [6-9] all of which are based on the analytical modelling of single textile yarns. Various assumptions and various general continuum mechanics equations and laws, such as Newton's third law of motion have been used. It was observed that improvements are needed for the existing analytical modelling of ballistic impact of multi-layer woven fabrics in the following two aspects: (1) failure criteria; (2) strain distribution expression.

$$c = \sqrt{\frac{E}{\rho}} \tag{1}$$

$$W = c\varepsilon \tag{2}$$

$$U_{lab} = c\left(\sqrt{\varepsilon(1+\varepsilon)} - \varepsilon\right) \tag{3}$$

$$u = c\sqrt{\frac{\varepsilon}{1+\varepsilon}} \tag{4}$$

$$V = C\sqrt{2\varepsilon\sqrt{\varepsilon(1+\varepsilon)} - \varepsilon^2} \tag{5}$$

$$\cos\theta = \frac{1}{1+\varepsilon}\frac{V}{u} \tag{6}$$

3 Assumptions and formulation of the analytical model

3.1 Overview

An analytical model of ballistic impact of multi-layer woven fabrics based on the analytical model of ballistic impact of single textile yarn by Stone et al [1] and Smith et al [2-5] and Newton's third law of motion has been developed. As shown in Figure 1, principle yarns are in contact with the projectile, i.e. only principle yarns provide the resistance force causing the deceleration of the projectile. Although not being contact with the projectile directly, the secondary yarns which were denoted with green colour in Figure 1 cause the transmission and deflection of longitudinal stress waves travelling along the principle yarns. Figure 2 shows the reaction of the principle yarns incorporated the strain gradient caused by the transmission and deflection of the longitudinal stress waves. In this study, the issue of strain gradient was accounted using a strain express developed by Porwal and Pheonix [10] where the strain at the edge of the projectile where the maximum strain in the principle yarns was related to projectile velocity and the distance travelled by the transverse wave in terms of Lagrangian (material) coordinates. And like in other works [7, 9-11], the decreasing of the longitudinal stress wave in a woven fabric was accounted by using a constant called Roylance coefficient [12], α. The velocity of the longitudinal stress wave in a woven fabric is equal to the velocity of the longitudinal stress wave in the textile yarn divided by α.

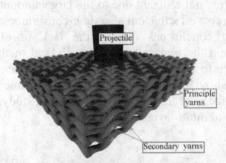
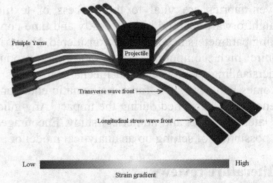

Figure 1 Simulation of ballistic impact of multi-layer woven fabric by cylindrical projectile

Figure 2 Analytical modelling of the reaction of the principle yarns in one layer of fabric in multi-layer textile based body armour

3.2 Governing equations and failure criteria

The distance travelled by the projectile during the impact process was divided into a number of constant distance intervals, h. The total number of distance intervals the projectile travelled until the fracture of the last layer was referred as m_n. Similarly the number of distant intervals travelled by the projectile when ith layer has been perforated was referred as m_i. There are four parameters for each layer in each distance interval, i.e. the time at the end of each distant interval, t_{ij}, the strain of the principle yarns near the edge of the projectile, ε_{ij}, the resistance force on the projectile because of the tensile strain and the ratio between the distance travelled by the transverse stress wave, F_{ij}, and the radius of the projectile at the end of each distance interval ψ_{ij} where $i=1, 2, 3,...,n$ representing the layer number, and $j=i, i+1, i+2,...,m_i$ representing the distant interval number and $j \geq i$.

The velocity of the projectile at the beginning and at the end of each distance interval were denoted as

V_j and $\widetilde{V}_j$. It was assumed that the velocity of the transverse stress wave in terms of Lagrangian (material) coordinates, u_i, in each layer was constant. According to equation (4) and (5), a database of u_i values versus the projectile impact velocity can be set up for each kind of woven fabric. The database was then used to decide the u_i values in a specific impact process in each layers by choosing the closest match in the database according to $V=V_j$ when j=i. Then ψ_{ij} can be calculated as $u_i \cdot t_{ij}/r_p$ and t_{ij} can be expressed as $\sum_{k=i}^{k=j} \Delta t_{ik}$. According the work of Porwal etc, ε_{ij} can be identified by solving equation (7). Combining equation (4-6) results in equation (8) for F_{ij}. The total reactive force on the projectile due to the tensile stress of the principle yarns in all the active fabric layers, F_j, can be calculated as $\sum_{i=1}^{i=n} \varphi_i F_{ij}$ where φ_i is equal to 1 when the fabric layer is in contact with the projectile and 0 when the projectile hasn't reach the fabric layer yet or the fabric has been perforated. Assuming uniform deceleration of the projectile in each distance interval, equation (9) and (10) can be setup. By solving equations (7-10) for each layer in each distance interval, the reaction of the textile based armour can be identified.

$$\varepsilon_{ij} = \left(\frac{V_j}{c\sqrt{2}}\right)^{4/3} \psi_{ij}^{1/3} \left(\frac{\sqrt{\psi_{ij}/\varepsilon_{ij}}(\psi_{ij}-1)}{\ln\left(1+\sqrt{\psi_{ij}/\varepsilon_{ij}}(\psi_{ij}-1)\right)}\right)^{2/3} \quad (7)$$

$$F_{ij} = \frac{2 D_p}{d} \cdot E \cdot S \cdot \varepsilon_{ij} \cdot \sqrt{\frac{2\sqrt{\varepsilon_{ij}(1+\varepsilon_{ij})} - \varepsilon_{ij}}{1+\varepsilon_{ij}}} \quad (8)$$

$$h = V_j \cdot \Delta t_{ij} - \frac{1}{2} \cdot \frac{F_{ij}}{M_p} \cdot \Delta t_{ij}^2 \quad (9)$$

$$\widetilde{V}_j = V_j - \sum_{i=1}^{n} \varphi_{ij} \frac{F_{ij}}{M_p} \cdot \Delta t_{ij} \quad (10)$$

The value of the indicator of the activation, φ_i is dependent on value of two values, i.e. ε_{ij} and σ_{ij}^{Back}. σ_{ij}^{Back} is defined as $\sum_{i=j+1}^{i=n}\left(\frac{\varphi_i F_{ij}}{\pi(D_p/2)^2}\right)$. Once $\varepsilon_{ij} > \varepsilon_{max}$ or $\sigma_{ij}^{Back} > \sigma_{shear}$, φ_i becomes zero where ε_{max} is the maximum strain, and σ_{shear} is the transverse shear strength.

4 Results and validation

For validating purpose, the experimental results of the work of Cork[13] and Gu[7] were used. The dynamic mechanical properties of Kevlar yarns and Twaron yarns in the work of Wang and Xia[14] and Gu[7] were used. The static shear strength of Kevlar fibre described in the work of Yang[15] was used. And the shear strength at high strain rate was calculated as the shear strength at static value multiplied by a parameter called dynamic enhancement factor. Dynamic enhancement factor were calculated according to the experiments results in the work of Wang and Xia[14] and Gu[7].

Table 1 Input parameter for analytical models

Parameter	Kevlar	Twaron
Young's Modulus	125GPa	69GPa
Fibre volumetric density	1440 kg m^{-3}	1440 kg m^{-3}
Distance between yarns	0.00147 m	0.00147 m
Yarn fracture strain	3.85%	5.22%
Distance between layer	0.00005 m	0.00005 m
Diameter of the yarn	0.000384m	0.00054 m
Shear strength	0.18GPa	0.16GPa
Mass of the projectile	0.00106 kg	0.00795kg
Diameter of projectile	0.0066 m	0.0066 m

4.1 Energy loss

Analytical model based on parameters listed in Table 1 was set-up to study the ballistic impact reaction of multi-layer woven fabrics. Penetration tests with a velocity of 500 m/s were performed. Figure 3 shows the kinetic energy loss results of the projectile from both experimental tests and analytical modelling. It can be seen that the kinetic energies loss predicted were very close to the experimental values. The systemic reduction of energy between the predicted and experimental values reflects the energy loss due to friction which was ignored in the analytical model.

4.2 Tensile strain and shear stress at failure

Figure 4 shows the distribution of the tensile strains at failure and the shear stress at failure along the through-thickness direction. It has been observed that the first few layers failed before the activation of the last few layers. Figure 4 shows that only the last few layers of fabric were able to reach tensile fracture strain of the fibre materials and most front layers failed due to shear. This suggested that the shear stresses in the front layers play important roles in the penetration of the front layers. This indicates that in multiple layer textile armour, single layer fabric with good energy absorption due to high tensile strength should be put in the back. Cuniff [16] even stated that the energy absorption of first few layers are very limited, and should be replace by other fabric with low strength for the purpose of saving cost.

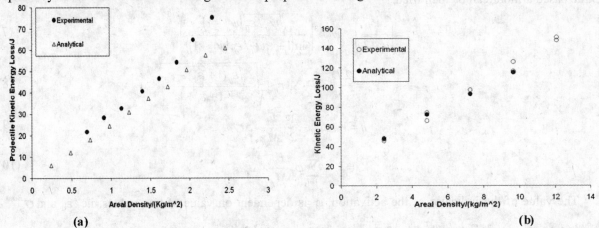

(a)   (b)

Figure 3 Comparison between kinetic energy losses of projectile measured and predicted using an analytical model

5 Conclusions

The aim of this investigation is to develop an analytical model as an approach to study the reaction of ballistic impact of multi-layer woven fabrics. It has found that the analytical model used in this study reviewed a simplified version of the reaction of the multi-layer woven fabrics and achieved reasonable accuracy. With the aid of the analytical model, it can be concluded that (1) Materials high stress wave velocity should be used as front layers; (2) Materials with high tenacity and fracture strain should be used as back layers.

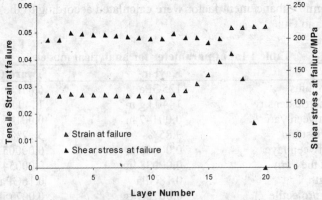

Figure 4 Tensile strains and shear stress at the moment of failure in the principle yarns of different layers in a 20-layer of Tawron plain woven fabrics subjected to 7.95g cylindrical projectile at the speed of 367 m/s

References

[1] Stone, W.K., H.F. Schiefer, and G. Fox, *Stress-Strain Relationships in Yarns Subjected to Rapid Impact Loading: Part I: Equipment, Testing Procedure, and Typical Results1,2*. Textile Research Journal, 1955. 25(6): p. 520-528.

[2] Smith, J.C., F.L. McCrackin, and H.F. Schiefer, *Stress-Strain Relationships in Yarns Subjected to Rapid Impact Loading: Part IV: Transverse Impact Tests*. Textile Research Journal, 1956. 26(11): p. 821-828.

[3] Smith, J.C., F.L. McCrackin, and H.F. Schiefer, *Stress-Strain Relationships in Yarns Subjected to Rapid Impact Loading: Part V: Wave Propagation in Long Textile Yarns Impacted Transversely*. Textile Research Journal, 1958. 28(4): p. 288-302.

[4] Smith, J.C., J.M. Blandford, and K.M. Towne, *Stress-Strain Relationships in Yarns Subjected to Rapid Impact Loading: Part VIII: Shock Waves, Limiting Breaking Velocities, and Critical Velocities*. Textile Research Journal, 1962. 32(1): p. 67-76.

[5] Smith, J.C., J.M. Blandford, and H.F. Schiefer, *Stress-Strain Relationships. in Yarns Subjected to Rapid Impact Loading: Part VI: Velocities of Strain Waves Resulting from Impact*. Textile Research Journal, 1960. 30(10): p. 752-760.

[6] Parga-Landa, B. and F. Hernadez-Olivares, *An analytical model to predict impact behaviour of soft armours*. International Journal of Impact Engineering, 1995. 16(3): p. 455 - 466.

[7] Gu, B., *Analytical modeling for the ballistic perforation of planar plain-woven fabric target by projectile*. Composites Part B: Engineering, 2003. 34(4): p. 361-371.

[8] Porwal, P.K. and S. Leigh Phoenix, *Modeling System Effects in Ballistic Impact into Multi-layered Fibrous Materials for Soft Body Armor*. Chemistry and Materials Science, 2005. 135(1-4): p. 217-249.

[9] Mamivand, M. and G.H. Liaghat, *A model for ballistic impact on multi-layer fabric targets*. International Journal of Impact Engineering, 2010. 37(7): p. 806 -812.

[10] Leigh Phoenix, S. and P.K. Porwal, *A new membrane model for the ballistic impact response and V50 performance of multi-ply fibrous systems*. International Journal of Solids and Structures, 2003. 40(24): p. 6723-6765.

[11] Roylance, D., A. Wilde, and G. Tocci, *Ballistic Impact of Textile Structures*. Textile Research Journal, 1973. 43(1): p. 34-41.

[12] Stempie, Z., *Influence of a Woven Fabric Structure on the Propagation Velocity of a Tension Wave*. FIBRES & TEXTILES in Eastern Europe, 2007. 15(5-6): p. 64-65.

[13] Cork, C.R. and P.W. Foster, *The ballistic performance of narrow fabrics*. International Journal of Impact Engineering, 2007. 34(3): p. 495-508.

[14] Wang, Y. and Y. Xia, *The effects of strain rate on the mechanical behaviour of kevlar fibre bundles: an experimental and theoretical study*. Composites Part A: Applied Science and Manufacturing, 1998. 29(11): p. 1411 - 1415.

[15] Yang, H.M., *1.07 - Aramid Fibers*. Comprehensive Composite Materials. Vol. 1. 2000. 199 - 229.

[16] Cunniff, P.M., *Analysis of the system effects in woven fabrics under ballistic impact*. Textile Research Journal, 1992. 62(9): p. 495-509.

70 Study on Automatic Classification of Size Designation in Clothing MC Based on Improved LBG Algorithm*

Fengyuan Zou, Li Dong, Lifeng Pan, Xiaojun Ding, Minzhi Chen

(Fashion College, Zhejiang Sci-Tech University;
Key Laboratory of Advanced Textile Materials and Manufacturing Technology, Ministry of Education of China, Zhejiang Sci-Tech University, Hangzhou 310018, China)

Abstract: Automatic classification of size designation is a difficult part in clothing MC. It needs the accumulation of years' practical experiences. With the application of the improved LBG algorithm which can simulate the technique and experience of classification into automatic classification of size designation in MC, it realizes the automatic mapping from net body-measurements of MC customers to series of size designation and improves the production efficiency of mass customization.

Keywords: Clothing MC; Classification of size designation; Vector quantization; LBG algorithm

MC (Mass Customization) mode is a completely customer-oriented and highly automatic industrialized method of production. It is the outcome of the development of applying modern science and technology and new consumption patterns into the garment industry and is one of the areas which have the broadest prospect in this field[1]. The core of MC is to identify and seek out the garment specification and pattern which are matched with individuals best according to the data information of the customers' bodies, and thus to produce the clothing that perfectly fits customers' body figures[2]. However, since the generation of series of size designation in MC is generally by manual work with low accuracy rate and slow efficiency, the research on how to obtain the automatic classification through the application of technology becomes one of the keys to achieve a high degree of automation of clothing MC.

Through the abstract and mathematical description of the issue of classification by vector quantization technology, this paper constructs the mathematical model of classification of size designation in MC. Based on analyzing the flow-chart of classical LBG algorithm, it puts forward the improved algorithm based on LBG and thus accomplishes the process of automatic classification of size designation in clothing MC.

1 Mathematical models

1.1 The principle of classification of size designation

One size designation includes the data of "m" parts of human body and the total number of sizes designations stored in the size designation data-base of MC clothing enterprises is "K". Now the number of the given people who need clothing MC is "n". This crowd is divided into "k" groups. Then corresponding to these "k" groups, "k" size designations are selected from the size designation data-base containing "K" number of size designations. The selected "K" size designations represent the size designation of each group of clothing MC. The division of the crowd and the choice of size designation should make each group have the best adaptability to the size designation of its own group, and make the classification of size designation most optimal from the overall point of view[3].

Size designation in the size designation data-base and net body-measurements awaiting classification are regarded as m-dimensional vector, then the template vector of size designation data-base and sample vector can be respectively expressed as: $R = (r_1 \ r_2 \ ... \ r_m)^T$ and $S = (s_1 \ s_2 \ ... \ s_m)^T$. Under the demand of body-fitness, dimensionality weighting calculation is applied to each vector according to the weight of each control position: $S' = (w_1 s_1, w_2 s_2, ... w_m s_m)^T$ and $R' = (w_1 r_1, w_2 r_2, ... w_m r_m)^T$ [4]. Then it is classified according to the nearest neighbor rule, $\{S' | S' \in \Phi_i \ \text{where} \ d[S', R_i'] = \min d[S', R_j'] \ j = 1, 2, ...m\}$, when S belongs to Φ_i and $d = w_1^2(r_1 - s_1)^2 + w_2^2(r_2 - s_2)^2 + ... w_m^2(r_m - s_m)^2$ here [5].

Thus classification of size designation in MC will be abstractly expressed as: there are "n" number of

---

* Foundation items: Hangzhou Economy Committee, Technology (No.140, 2007), "Research and Application of Key Technique of Clothing MC"; Fresh Talent Initiative of Zhejiang Provincial Science & Technology Department (No.14530132661046), "Promotion of Automatic Classification of Size Designation in Clothing MC".

m-dimensional sample muster $\{S_i \mid S_i \in \Phi \ \ i=1,2,...n\}$ and "K" number of template muster $\{R_1, R_2,...R_K\}$. The sample muster is to be divided into "k" groups $\Phi_1, \Phi_2,...\Phi_k$, ($k < K$), and pick "k" template vector from template muster to be corresponding to each group. The aberration d_i in each group when S belongs to Φ_i should be minimized and make the total aberration $d = \sum d_i$ minimum.

1.2 Vector Quantization

The classification of size designation in MC is just aiming to find ways to complete the issue of vector mapping, i.e. mapping a group of m-dimensional vector to another group of m-dimensional vector, which is known as VQ (Vector Quantization, VQ for short). Its mathematical equation is $R_j = Q(S_i)$ [6].

The principles of encoding part of sending-end of VQ is equal to the abstract mathematical description of classification of garment size designation need to be solved. After VQ encoding, one input vector obtains the serial number representing one code word. Then in decoding-end, it is restored to the corresponding code word according to the serial number and output. Certain distortion must exist between the input and output. As far as a group of input vector is concerned, the smaller the average distortion between input and output is, the better the effect of the VQ will be. Hence it needs to find a most optimal VQ quantizer with minimum average distortion, i.e. the most high quality code book[7].

2 Improved LBG algorithm and automatic size-classification method
2.1 Improved LBG algorithm

LBG algorithm is a classic VQ method, also known as K-means algorithm. It can acquire fairly satisfactory code book by several times of iterations, assortments, and seeking center of mass operations based on given training samples. The design of code book in LBG algorithm is based on the following two principles:

1) Nearest neighbor rule. The optimal classification of training vector set can be obtained through mapping each training vector to the code word neareast to its spatial Euclidean distance. It is supposed that a space vector muster is divided into "k" groups and each group is labeled as Φ_i. When the vector S belongs to Φ_i $\{S \mid S \in \Phi_i \ \ where \ \ d[S, R_i] = \min d[S, R_j]\}$ $j = 1,2,...k$

2) Centroid assortment condition. After each classification, the sort centre will be redefined. The centre is figured by the centroid of all training vectors in this sort. If there are n_i number of vectors in the i group, the center of mass of this sort is: $R_i = \dfrac{1}{n_i}\sum_{j=1}^{n_i} S_j$

Generally there are two ways to choose the initial code book: random method and split method. The former one is to randomly select "k" training vectors as the initial code word. The latter is in the hope to expand the distance among the code words as far as possible in order to get better effect of initial classification[8].

The theory of LBG algorithm is strict and easy to implement. It has become the basis of many other improved algorithm. After the initial code book is selected, the initialization condition is set, that is, supposing the number of initial iteration is $m = 0$ or 1; the total number of iteration is L; the average distortion i.e. total aberration of each classification is $D_{total}^m \to \infty$; the threshold of distortion control is $0 < \delta < 1$, in which δ is relative distortion generated from the need of procedure control, and it reflects the extent of reduction between total aberration of two adjacent classifications. If it is less than δ, the procedure ends. Generally we set that $\delta = 0.001$. The chart of improved algorithm based on LBG is shown in Figure 1.

The improved algorithm based on LBG adds one process, that is, mapping the code book to the template vector after the initialized code book is confirmed. Besides, when the new code book is produced after each iteration, the process of mapping the new code book to the template vector is also added. Thus it makes LBG algorithm more conforming to the practical significance of classification of size-designation in MC.

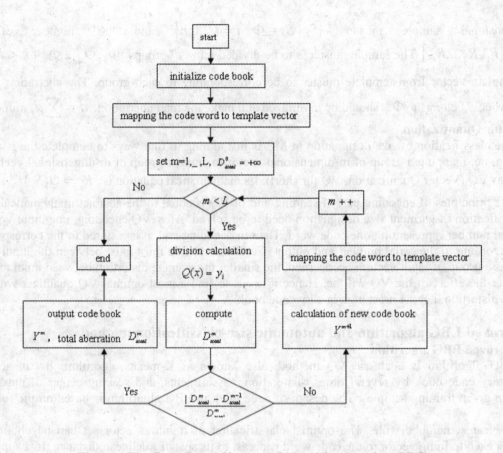

Figure 1 The flow-chart of improved algorithm based on LBG

2.2 Automatic classification method

1) Data pre-processing. As the classic method, first, the dimensionality weighting treatment is applied to the input data of net body measurements (total number of "n") and data of control positions in the garment size designation data-base (the total number of "K").

2) By adopting random method or split method, the code book including "k" initial codes is obtained from the sample muster. Then following nearest neighbor rule, the code book is mapped to unrepeated "k" template vectors. Later, the circular iterative process of improved LBG algorithm works until the last iteration stops and final result is output.

3) It can be known from the content of the algorithm that the final output result of improved algorithm is the final size designation of classification of size designation in MC that is pursued. The use of nearest neighbor rule can map all "n" samples to "k" size designations. Thus the classification of size designation based on improved algorithm completes.

The output code book based on improved algorithm is a subset of the template of the size designation data-base, i.e. the result we seek.

3 Application

3.1 To determine the part of input data

The part of input data includes size designation data-base of MC and the input of samples under test. In this study, the size designation data-base of MC includes 98 size designations containing 16 control positions data, such as 160/84AF2B2. Here 160/84 represents the series of size designation and Y, A, B, C is the national standard body-figures. F represents the Front-side shape and B represents the Back-side figure. 1, 2, 3 represent respectively three protuberance positions from top to bottom. Thus we get the identifications of body-figure subdivisions in the size designation data-base of MC. According to the requirement of body-fitness, 16 control positions which can reflect the characteristics of each body-figure subdivisions include: body height, back neck point height, front length, back length, arm length, waist height, chest height, the front crotch length, the back crotch length, the shoulder width, neck circumference, the front bust, the back bust, waist, front hip and back hip. The samples under test of classification of size-designation in MC are collected by measuring the net body-measurements of 129 female youth aged 18-24 from Jiangsu and Zhejiang Provinces with TC2 3D body measuring apparatus from United States[9].

3.2 Result of experiment and analysis

The output result of automatic classification of size designation in clothing MC based on improved LBG algorithm is a subset of MC size designation data-base, and the number of size designations of test samples is fixed as 11. Then all of the 129 test samples can be mapped to the 11 final size designations by using the nearest neighbor rule. The output part includes the names of 11 selected size designations, the data of control positions of each size designation, as well as all the sample serial numbers suitable for each size designation. The results of experiment are as follows:

size 1: name 170/92AF1B3; suit to the following samples: 62, 95, 96, 101, 102, 121, 123, 125, 129

......

size 11: name 160/84AF2B3; suit to the following samples: 4, 11, 12, 27, 35, 44, 56, 60, 114

Experts' subjective judgment method is adopted to evaluate the result of automatic classification of size designation in MC. The members of expert group are required to have both professional skills and wealthy experience as well. This experiment requires experts with more than 10 years working experiences in fashion industry or in technology research of fashion in universities. Totally, 15 experts are chosen. Their occupations are shown in Table 1.

Table 1 Proportion of experts of subjective judgement

Occupation	Number (unit: person)	Proportion (unit: %)
Quality control	2	13.3%
Production planning	4	26.7%
Technology management	5	33.3%
Associate-professor	3	20%
Professor	1	6.6%
Total	15	100%

Each one of the above-mentioned experts makes judgment separately on the output results of the 129 tested samples using the automatic classification of size designation in clothing MC based on improved LBG algorithm. He/She compares the data of the 16 control positions of every tested sample with the data of the 16 control positions of the size designation it belongs to. If he/she thinks the data of the 16 control positions of the tested sample coincide with the corresponding data of the control positions of the size designation which it belongs to, then the expert puts a √ in the questionnaire. After making judgment on every control position, if the expert thinks the tested sample coincides with the size designation it belongs to, he/she puts a √ in the overall evaluation. Each expert's questionnaire is totalized, thus we get the judgment result indicated in Table 2.

Table 2 Accuracy rate of the result of automatic classification of size designation based on improved LBG algorithm

expert	correct number (unit: person)	wrong number (unit: person)	accuracy rate
expert #1	121	8	93.8%
expert #2	119	10	92.2%
expert #3	123	6	95.3%
expert #4	121	8	93.8%
expert #5	118	11	91.5%
expert #6	119	10	92.2%
expert #7	120	9	93.0%
expert #8	124	5	96.1%
expert #9	121	8	93.8%
expert #10	121	8	93.8%
expert #11	122	7	94.6%
expert #12	119	10	92.2%
expert #13	122	7	94.6%
expert #14	122	7	94.6%
expert #15	120	9	93.0%

It can be seen from Table 2 that the highest accuracy rate is 96.1% with 124 correct numbers and 5 wrong numbers and the lowest accuracy rate is 91.5% with 118 correct numbers and 11 wrong numbers. It can be concluded that the accuracy rate of the result of this experiment using automatic classification of size designation in clothing MC is not below 91.5%. Generally speaking, with the increase of total number of the size designations to be classified, the accuracy rate will be increased gradually as well.

4 Conclusion

Using the improved LBG algorithm model to study and simulate the technology and experience of manual classification of size designation in MC can achieve the automatically computer-generated series of size designation in MC with the purpose of reducing the workload and improving the efficiency of classification of size-designation. Through the experiment on testing samples of classification of size designation in MC, it shows that the improved LBG algorithm has high efficiency and preferable precision.

The subsequent research can continue to extend the capacity of test samples and refine the size designation data-base. Besides, in the dimensionality weighting calculation, the weight distribution only take the influence of each control position on body-fitness degree of body-figure characteristics into consideration, which is only applicable to the operation of well-fitting clothing. So it needs to consider the factor of different types of clothing to improve the mathematical model of nearest neighbor rule and perfect the automatic classification system of garment size designation continously.

References

[1] Li M. Industrial production of customized clothing based on the technology of measuring body features. Donghua University Journal, 2000, 26(4):107-109
[2] Gu X., Yang Z. and Zhang X. Key technologies in clothing mass customization. Textile Journal, 2003, 24(3): 85-87
[3] Nei Y. Research on the optimize allocation of clothing size specifications of adult male in lower reaches of Yangtze River. Donghua University, 2007, Shanghai, 1 July.
[4] Xie J., Liu C. Fuzzy mathematics method and its application. Wuhan: Huazhong University of Science and Technology Publishing House, 2004:136
[5] Wang Y. Artificial intelligence theory and methods. Xi'an: Xi'an Jiaotong University Publishing House, 1998:55
[6] Sun S., Lu Z. VQ Technology and its application. Beijing: Science Publishing House, 2002:90
[7] Linde Y., Buzo A. and Gray R. An algorithm for vector quantizer design. IEEE Trans. Commun, 1980, Com-28:84-95
[8] Li H., Liu H. New algorithm of studying initial code book of vector quantization. Journal of Beijing Post and Telecommunications University, 2006, 29(4):33-35
[9] Dong L., Zou F. and Zhang, Y. Subdivision of body-figures of young female in Jiangsu and Zhejiang based on three-dimensional measuring and its comparative study. Journal of Zhejiang Sci-tech University, 2007, 24(6): 636-639

71 Thermodynamic Behavior on the Binding of the Polymers and Acid Dyes in Inkjet Ink for Textiles

Juyoung Park, Yuichi Hirata, Kunihiro Hamada

(Faculty of Textile Science and Technology, Shinshu University,
s08t113@shinshu-u.ac.jp, 3-15-1 Tokida, Ueda, Nagano 386-8567, Japan)

Abstract: To formulate inkjet ink for textiles, not only dyes or pigments but also additives are needed and the types and amounts of additives are the most important factors. In the present study, the interaction between the three acid dyes (C. I. Acid Red 88, 13, and 27) and additives such as water-soluble polymer, poly(vinylpyrrolidone) (PVP), were investigated. To elucidate the effects of the additives, the binding constants of the dyes with the polymers, K_{bind}, the visible absorption spectra of aqueous dye solutions in the absence and presence of PVP were measured at 15, 25, 35 and 45°C. To discuss the thermodynamic behavior on the binding of the polymers and the dyes, the thermodynamic parameters ΔH_{bind} and ΔS_{bind}, were estimated. The binding constants, K_{bind}, of the acid dyes with PVP were dependent on the dye structure and the molecular weight of PVP. The ΔH_{bind} values for all the dyes were negative, showing that the binding processes are exothermic. The absolute values of ΔH_{bind} decreased in the order of R-1 > R-2 > R-3, which is comparable to the order of the number of sulfonate groups. The increase of PVP molecular weight made the thermodynamic process to be less enthalpic and more entropic.

Keywords: Thermodynamic behavior; Polymer; Acid dyes; Inkjet ink; Visible absorption spectra

1 Introduction

The composition of inkjet ink is very complicated due to the complex nature and very challenging requirements. The ink must have physicochemical properties which are specific to the various printing devices [1-4]. To formulate inkjet ink, not only dyes or pigments but also additives are needed and the types and amounts of additives are the most important factors.

Commonly, the additives in inkjet ink are used to adjust physical properties of ink such as stability, viscosity, surface tension, droplet form and so on [4-6]. Also various types of additives such as surfactants, inorganic salts and polymers have been used in the inkjet ink for textiles in order to solve the other problems which occur in printing processes. For example, to improve the binding force of the ink and the fiber, polymers are used as additives [7-9]. A water-soluble polymer affects the improvement of the water fastness and light fastness since polymers can stabilize and immobilize the dye both physically and chemically [10-12], but causes negative effect on print quality [11]. Besides rapid ink absorption polymers offer high print density [8] and addition of polymer in inkjet ink is progressed the durability of ink and prints [4,13].

Various interactions such as dye-dye, dye-additive and additive-additive interactions are observed in the inkjet ink. The intermolecular interactions include electrostatic binding, π-π interaction, hydrogen bonding, van der Waals force, and so on. The interdependence of dye structure and the type of additives has been shown to play a major role [14]. The cooperative binding of a dye with a polymer chain is a perinent example [11,15]. Furthermore, an azo dye containing sulfonate groups showed good stability in contact with polymer [16] and the selection of the good solubility dye very impotents in the inkjet ink for ink stability [2].

In the present study, the thermodynamic behavior of highly soluble acid dyes (C. I. Acid Red 88, 13, and 27) containing sulfonate groups in the absence and presence of the additive were investigated by means of visible absorption measurements. To elucidate the effects of the polymer additive, the binding constants of the dyes with poly(vinylpyrrolidone) (PVP) were estimated. Furthermore, thermodynamic parameters were determined from the temperature dependence of the binding constants in the absence and presence of PVP. The effects of molecular weight on the interaction between the dyes and PVP were also discussed. From the results, the effects of the polymer additive on the thermodynamic behavior are discussed.

2 Experimental
2.1 Materials

Three acid dyes containing the different number of sulfonate groups, C. I. Acid Red 88 (R-1), C. I. Acid Red 13 (R-2), and C. I. Acid Red 27 (R-3) were used after purification. As additives, three types of poly(vinylpyrrolidone) with different molecular weight (PVP-1; MW~10,000, PVP-4; MW 40,000, PVP-63;

MW~630,000) were used (Figure 1). Three types of poly(vinylpyrrolidone) were purchased from Tokyo Chemical Industry Co., Ltd., and used without further purification (Figure 1).

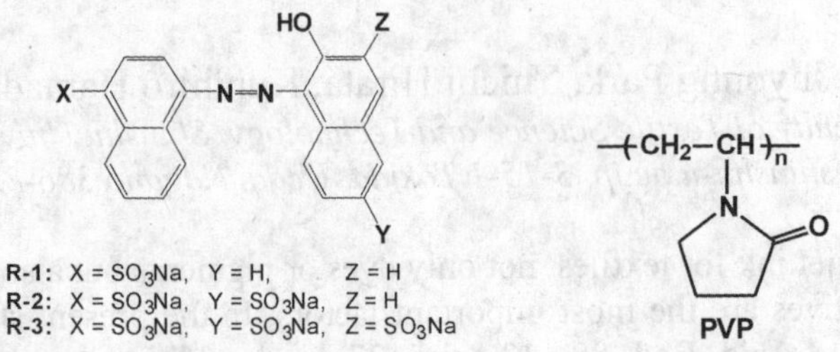

R-1: X = SO$_3$Na, Y = H, Z = H
R-2: X = SO$_3$Na, Y = SO$_3$Na, Z = H
R-3: X = SO$_3$Na, Y = SO$_3$Na, Z = SO$_3$Na

Figure 1 Chemical structure of polymer and acid dyes used

2.2 Visible absorption spectrum measurements

The visible absorption spectra of aqueous solutions having various polymer concentrations (the polymer concentration is expressed in the basis of a monomer unit) and a constant dye concentration (R-1; 2.97×10^{-5} mol dm^{-3}, R-2; 2.81×10^{-5} mol dm^{-3}, R-3; 2.83×10^{-5} mol dm^{-3}) were recorded using a JASCO UV-530 spectrophotometer at 15, 25, 35 and 45°C.

3 Results and discussions

The effect of successive addition of PVP on the absorption spectra of aqueous R-1 solution is represented in Figure 2. Isosbestic points were defined in the polymer concentration region examined, where the polymer concentration is expressed based on monomer units. This fact makes it possible to assume a single equilibrium in the R-1/PVP systems. The similar spectral changes were also observed for the other dyes. To analyze the spectral change with PVP concentration, the extinction coefficients, ε, at the wavelength where the largest difference was observed are useful. These extinction coefficients decreased with increasing polymer concentration for all the dyes, as shown in Figure 3. The binding constants, K_{bind}, are calculated from the change of the extinction coefficients. Furthermore, these extinction coefficients changed on raising the temperature from 15 to 45°C as shown in Figure 4.

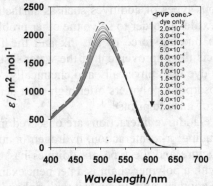

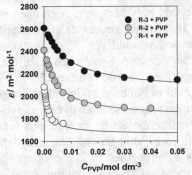

Figure 2 Visible absorption spectra of aqueous R-1 solutions in the absence and presence of PVP at 25°C

Figure 3 Dependence of ε on PVP concentration at 25°C R-1, 492 nm; R-2, 494 nm; R-3, 509 nm

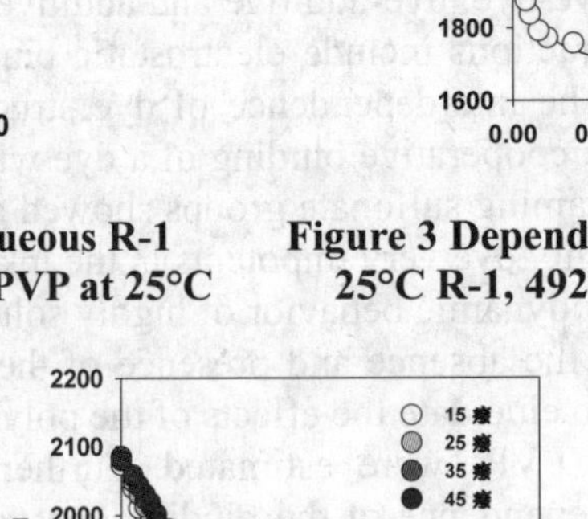

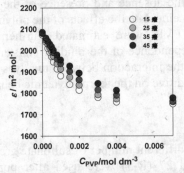

Figure 4 Temperature dependence of ε on PVP-63 concentration at 15, 25, 35 and 45°C. R-1, 492 nm

We assumed that the following single equilibrium exists in the dye/PVP systems because the isosbestic points were observed.

$$D + P \rightleftarrows D \cdot P \tag{1}$$

where D, P and D·P express the dye, PVP and dye-PVP complex, respectively. When the concentrations of the total dye, the total PVP, the bound dye and the free dye are defined as C_0, C_P, C_b, and C_f, respectively, the binding constant of the dye with the PVP (K_{bind}) should be expressed by eq. (2):

$$K_{bind} = \frac{C_b}{(C_P - C_b) \cdot C_f} \tag{2}$$

since $C_0 = C_f + C_b$, eq. (2) can be rewritten by eq. (3):

$$C_b = \frac{1}{2}\left\{ A - (A^2 - 4C_0 C_P)^{1/2} \right\} \tag{3}$$

where $A = C_0 + C_P + 1/K_{bind}$. On the other hand, if ε_f and ε_b are the extinction coefficients of the free and bound dye, respectively, then the observed extinction coefficient, ε, can be expressed as follows:

$$\varepsilon = \frac{C_f}{C_0} \cdot \varepsilon_f + \frac{C_b}{C_0} \cdot \varepsilon_b \tag{4}$$

By substituting eq. (3) and $C_f = C_0 - C_b$ into eq. (4), we can obtain.

$$\varepsilon = \varepsilon_f + \frac{\varepsilon_b - \varepsilon_f}{2C_0}\left\{ A - (A^2 - 4C_0 \cdot C_P)^{1/2} \right\} \tag{5}$$

On the basis of eq. (5), the binding constant, K_{bind}, and the extinction coefficient of the bound dye, ε_b, were calculated by using the nonlinear least-squares method (SigmaPlot, SPSS) [17]. The binding constants, K_{bind}, thus calculated are given in Table 1. The binding constants of the acid dyes with PVP were dependent on the dye structure. The K_{bind} values decreased with an increase in the number of sulfonate groups in the acid dyes at all temperatures. It is worthwhile to mention that the introduction of sulfonate groups into the dye molecule increases the solubility in water, i.e. the solubility increases in the order of R-1 < R-2 < R-3, which corresponds to the decreasing order of K_{bind} (R-1 > R-2 > R-3). This suggests that the affinity to PVP decreases with increasing water solubility.

For all the dyes, the binding constants decreased with an increase in PVP molecular weight. This result should be explained as follows. The molecular weight is strongly concerned with the higher structure of polymers. On the other hand, the anionic groups (sulfonate groups) of the dyes are believed to interact with the positively charged pyrrolidone rings (pyrrolidone carries partially positive charge on the nitrogen atom). This electrostatic interaction is influenced by the higher structures. If the polymer has a compact formation, the cationic loci are hardly exposed to water, and the electrostatic interaction is hindered. As the higher molecular weight makes the conformation to be compact, the binding constants are diminished by increasing molecular weight of PVP.

For further discussion, the thermodynamic parameters were determined from the temperature dependence of K_{bind}, which is related to the enthalpy change, ΔH_{bind} and the entropy change, ΔS_{bind} as shown in eq. (6):

$$\ln K_{bind} = -\frac{\Delta H_{bind}}{R} \cdot \frac{1}{T} + \frac{\Delta S_{bind}}{R} \tag{6}$$

where T is the absolute temperature and R is the gas constant. ΔH_{bind} and ΔS_{bind} were determined from the plot of $\ln K_{bind}$ against $1/T$ (van't Hoff plot) (Figure 5) and the values are shown in Table 1. The slope and intercept of the van't Hoff plot make it possible to calculate the enthalpy and entropy change, respectively, during the binding process.

The ΔH_{bind} values for all the dyes were negative, showing that the binding processes are exothermic. The absolute values of ΔH_{bind} decreased in the order of R-1 > R-2 > R-3, which is comparable to the order of the number of sulfonate groups. This suggests that the enthalpy change is concerned with the solubility of the dyes. The ΔS_{bind} values did not show a tendency with increasing number of sulfonate groups. This might be due to the complicity in dehydration of hydrated water around the dyes and PVP. On the other hand, the increase of PVP molecular weight made the thermodynamic process to be less enthalpic and more entropic. This result supports that the higher molecular weight makes the conformation to be compact leading to the hindrance of the electrostatic interaction. The compensation relationship between ΔH_{bind} and ΔS_{bind} differd among the three dyes (Figure 6). This also supports the above results.

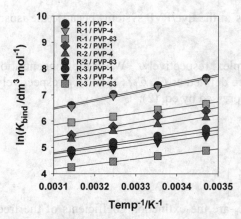

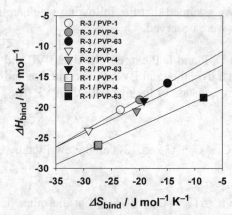

Figure 5 Van't Hoff plots (ln Kbind vs. 1/T) for the three dyes with PVP

Figure 6 Compensation relationship between ΔHbind and ΔSbind

Table 1 The binding constants and thermodynamic parameters for the binding of the dyes with PVP

	K_{bind} (dm^3 mol^{-1})				ΔH_{bind} (kJ mol^{-1})	ΔS_{bind} (J mol^{-1} K^{-1})
	15 ℃	25 ℃	35 ℃	45 ℃		
R-1						
PVP-1	2100 ± 200	1570 ± 100	1120 ± 70	730 ± 60	-26 ± 2	-28 ± 8
PVP-4	1950 ± 170	1480 ± 100	1060 ± 40	690 ± 30	-26 ± 3	-27 ± 8
PVP-63	780 ± 90	630 ± 60	480 ± 40	380 ± 30	-18.4 ± 0.7	-8 ± 2
R-2						
PVP-1	590 ± 30	450 ± 20	327 ± 14	238 ± 12	-24 ± 2	-29 ± 8
PVP-4	464 ± 15	357 ± 9	277 ± 7	204 ± 9	-20.7 ± 1.0	-21 ± 3
PVP-63	266 ± 19	208 ± 14	165 ± 11	125 ± 7	-18.9 ± 0.9	-19 ± 3
R-3						
PVP-1	298 ± 15	230 ± 10	179 ± 7	132 ± 4	-20.4 ± 1.0	-24 ± 3
PVP-4	227 ± 17	180 ± 12	139 ± 8	109 ± 6	-18.8 ± 0.5	-20.0 ± 1.8
PVP-63	132 ± 13	110 ± 10	87 ± 7	70 ± 5	-16.0 ± 0.5	-15.0 ± 1.5

4 Conclusions

Thermodynamic behavior on the binding of the polymers and acid dyes reflects some of their properties in aqueous solutions. The higher the solubility of the dyes in water (increasing in the number of sulfonate groups in the dyes), the lower is the affinity to PVP and the absolute values of ΔH_{bind}. For all the dyes, the binding constants decreased with an increase in PVP molecular weight, suggesting that the polymer conformation affects the binding behavior. The increase of PVP molecular weight made the thermodynamic process to be less enthalpic and more entropic. From the above results, it is concluded that the interactions between the dyes and the additives were dependent on various factors; the dye structures, the molecular weight of polymer, temperature, and so on.

Acknowledgement

This work was supported by Grant-in-Aid for Global COE Program by the Ministry of Education, Culture, Sports, Science, and Technology.

References

[1] Magdassi S. In: Magdassi S, editor. The chemistry of inkjet inks, Chapter 2. Ink requirements and formulations guidelines: World Scientific Ltd, 2008: 19-20
[2] Moon S J. Trends in development of digital textile printing ink, Dyeing and Finishing 2006, 1: 48-54.
[3] Tincher W C, Hu Q, Li X, Tian T, Zeng J. Coloration systems for ink jet pringting of textiles. NIP14: International Conference on Digital Printing Technologies. Society for Imaging Science and Technology. Toronto, Canada, 1998, 243-6.
[4] Schmid C. In: Magdassi S, editor. The chemistry of inkjet inks, Chapter 7. Formulation and properties of waterborne inkjet inks: World Scientific Ltd., 2008: 123-4.

[5] Tyler D J. Textile digital printing technologies, Textile Progress 2005, 37(4): 24-31.
[6] Bae J S, Son Y A. Digital textile printing. Journal of the Research Institute of Industrial Technology 2006,21(1): 33-44.
[7] Wang J, Chen T, Glass O, Sargeant SJ. Light fastness of large format ink jet media. NIP15: International Conference on Digital Printing Technologies. Orlando, USA, 1999, 183-186.
[8] Pinto J, Nicholas M. SIMS studies of ink jet media. NIP13: International Conference on Digital Printing Technologies. Seattle, USA, 1997, 420-425.
[9] Yuan S, Sargeant S, Rundus J, Jones N, Nguyen K. The development of receiving coatings for inkjet imaging applications. NIP13: International Conference on Digital Printing Technologies. Seattle, USA, 1997,413-417.
[10] Alfekri D, Staley G, Chin B, Hardin B, Siswanto C. Ink jet Printed Textiles, US Patent 6,001,137, 1999.
[11] Khoultchaev K. and Graczyk T. Influence of polymer-polymer interactions on properties of ink jet coatings, Journal of *Imaging* Science and *Technology* 2001, 45(1): 16-23.
[12] Clifton A A, Nugent N. Light fastness performance in digital imaging: the role of dye-dye and dye-polymer interactions. NIP16: International Conference on Digital Printing Technologies. Vancouver, Canada, 2000, 762-6.
[13] Hornby J C, Kung K H, Locke J S, Wheeler J W. Inkjet ink set, US Patent 0032098 A1, 2008.
[14] Fryberg M, Hofmann R. Influence of dye structures on permanence, NIP16: International Conference on Digital Printing Technologies. Vancouver, Canada, 2000, 95-8.
[15] Pak S, Ando Y, Koshikawa J, iijima T. Cooperative interaction of carboxymethyl cellulose with C.I. basic red 18 in aqueous solution. Journal of Macromolecular Science, Part B: Physics 1984, 23(1): 85-91.
[16] Naisby A, Suhadolnik J, Debellis A, Pennant D, Renz W. Role of media polymer chemistry on dye/polymer interactions and light stability of ink jet graphics. NIP18: International Conference on Digital Printing Technologies. SanDiego, USA, 2002. p. 749-52.
[17] [Hamada K, Miyawaki E. Interaction between water-soluble polymers and azo dyes containing fluorine atoms. Part 4. Comparison with azo dyes having different chain length of alkyl groups. Dyes and Pigments 1998, 38: 147-56.

72 Applications of UV Curing on Textiles

Shiqi Li, Henry Boyter Jr.
(Institute of Textile Technology, North Carolina State University, Raleigh, NC 27695, USA)

Abstract: Conventional curing is the most common thermal process for textile finishing. During the finishing process, a fabric is usually padded finishers, dried, then, cured by heated to 130-200°C for 1-15 min based on finishers applied, within curing resins or binders can react or crosslink on the fabric (hardening) to perform a special function with certain fastness. Although it is convenient to use a thermal curing in textile finishing processes, the curing is huge energy consumption. For energy saving, a low temperature and fast process for textile finishing should be considered. The low temperature curing processes can use ultraviolet light, electron beam, or laser to polymerize the binder material (monomers and oligomers) to fix finishers on textiles. Among them, the UV curing process needs the simplest equipment, and lowest cost. Examples of application of UV curing on textiles are discussed below.

Keywords: UV; Conventional curing; Textile finishing; Curing resins

1 Pigment dyeing[1]

Textile coloration is one of the most important steps in textile production. Dyes have been used to color textiles for a century. However, there are still many limitations in dye applications. Different textile material should use different dyes with different application methods, conditions, chemicals and equipment; some dyeing procedures are very complex. The properties of dye colored textiles also vary significantly. To overcome the dye weaknesses, textile coloration by pigments is a choice. The advantages[2] of the pigment method are simple procedures, probably omitting washing; applicability to all fabrics; extensive color range, excellent lightfastness; coloration combining with finishing, reducing cost, and much less or no wastewater effluent. The difference between dye dyeing processes and pigmentation is that pigment colored textiles require a curing procedure. Since pigments do not have affinity to textiles, the fixation of pigment on textiles relies on binders, and only with curing process the binders can bind pigments on textiles. Conventional curing is a thermal process. UV curing is an alternative to radiant process. UV-curable formulations contain monomer, oligomer, and a photo-sensitive component ("photoinitiator") that triggers a nearly instantaneous curing reaction upon exposure to ultraviolet light. Thus, UV curing produces a completely dry and finished surface in a second or two, compared with minutes for conventional curing[2].

1.1 Material and methods

The textile materials used in this research included three kinds of woven fabrics: a 100% cotton plain fabric at 153 g/m², a 40/60 polyester/cotton plain fabric at 161 g/m², and a 100% polyester twill fabric at 215 g/m². The size of fabric sample employed was 4" by 6". A padder system was employed for applying chemicals to the fabrics. The selection of these fabric materials and the chemical application system was based on their common use in the textile industry. The chemical applied fabrics were cured on a conveyer belt through a UV area, or held static under a UV light. The linear speed of the conveyer belt was adjustable from 4 ft/min to 23 ft/min. Two UV light sources were used. One was a xenon pulse UV light with linear power 120 W/inch. Its UV light exposure area was 5" long by 2"wide. The other system was an electrode medium pressure mercury arc continuous UV light with linear power 300 W/inch. Its exposure area was 5" long by 6" wide. The difference between the two systems was the xenon pulsed system could deliver high UV power in short time, while its average power output was lower than the continuous UV system. The pigments used as received were four water base dispersions (W-B Blue, W-B Red, W-B Yellow, and W-B Black). Pigments were fixed on the textile substrate through UV-curable resins. The resins were compositions of oligomers, a monomer, and a photoinitiator. The ingredients of oligomers and monomers are listed in Table 1, and the photoinitiator is Ethyl-2,4,6-trimethylbenzoyl-diphenylphosphinate. The selection of the chemicals was according to their chemical structures, functionality, and availability. The temperature on fabrics during UV curing was measured by attaching a paper thermometer on the back of the curing fabric.

Table 1 Oligomers and monomers used in UV curing experiments

Type	Abbreviation	Chemical property	Functionality[1]
Oligomer	PU-PA	Polyurethane / polyacrylate copolymer dispersion (water-based)	x
	UPU	Unsaturated polyurethane (water-based)	x
Monomer	TPGDA	Tripropylengylkol diacrylate	2

1 The number of carbon-to-carbon double bonds in each oligomer or monomer molecule

After coloration, the textiles were evaluated on a spectrophotometer for K/S shade depth values and total color difference (ΔE). The spectrophotometer settings for ΔE measurement were CIELAB, D6500, and 10° (color scale, illuminant, and view angle). AATCC Test Method 8, Method 61-2A, and Method 16-E were used to examine colorfastness to crocking, washing, and light, respectively. ASTM D-1388-96 (cantilever test) was employed to test fabric stiffness. The fabric stiffness was indicated by the flexural rigidity value, with larger values indicating stiffer fabric. After treatment, the increased flexural rigidity values of a fabric from untreated were used for evaluating the treatment properties.

1.2 Water-based UV resin for fabric coloration
1.2.1 Effects of UV light system and curing speed on colorfastness to crocking and fabric stiffness

Water-based UV resin formulas are given in Table 2. Figure 1 displays the colorfastness to dry crocking and fabric stiffness of UV resin colored polyester fabric cured with the two UV light systems and different curing speeds. The mercury UV light yielded higher efficiency than the xenon light in curing the fabric. With a higher fabric speed, the sample curing time decreased, and both dry crockfastness and fabric stiffness decreased significantly.

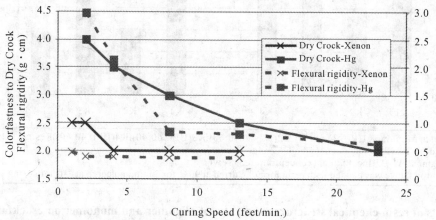

Resin formula: K4, Resin add-on: 20 ± 2%
Pigment: W-B Red, Pigment concentration: 3% owf
Polyester fabric was padded with resin formula, dried at 90℃ for 3 min., and cured at different speed passing a xenon light or a mercury light.

Figure 1 Effects of light system and curing speed on colored polyester crock fastness and increase of flexural rigidity

Table 2 Water-based UV resin formula

Formula	Oligomer + Monomer			Component			
	Oligomer[1]	Monomer	Ratio (w/w)	Oligomer + Monomer	Photoinitiator	Pigment	Water
K1	PU-PA	TPGDA	60/40	20%	0.6%	10%	69.4%
K2	PU-PA	TPGDA	40/60	20%	0.6%	10%	69.4%
K3	UPU	TPGDA	60/40	20%	0.6%	10%	69.4%
K4	UPU	TPGDA	40/60	20%	0.6%	10%	69.4%

The polyester sample cured with xenon light showed lower colorfastness to crocking and lower stiffness values compared to the samples cured with mercury light. As the power output of xenon light was only 40% of that of the mercury light, the intensity of the used xenon light might not be strong enough to cure the UV resin with pigment on the fabric.

1.2.2 Effects of Resin chemical structure and oligomer/monomer ratio on colorfastness to crocking and fabric stiffness

Fabric pigment coloration requires high colorfastness to crocking in conjunction with low stiffness of the treated fabric. Both the crockfastness and fabric stiffness depend on resin chemical structure, ratio of oligomer and monomer, as well as resin distribution on fabrics. Colorfastness to crocking and stiffness values of cotton fabric colored with different chemical types and different ratios of oligomer and monomer of UV-curable resin are given in Figure 2. The resin with oligomer that consisted of polyurethane and polyacrylate and a ratio of oligomer and monomer of 40:60 (K2) yielded the poorest colorfastness to crocking on the cotton fabric. Higher oligomer ingredient levels in the resin (K1, K3) could improve the colorfastness to dry crocking; however, the fabric stiffness also increases. The resin containing only polyurethane as oligomer (K3, K4) had a better colorfastness to crocking and lower stiffness levels than other formulas used on cotton. The resin formula that contained polyurethane as oligomer, and a ratio of oligomer and monomer at 40:60 (K4) displayed the lowest stiffness level and colorfastness to crocking at level 4. Overall, among the 4 formulas, the K4 formula was the most suitable for UV curing pigment coloration on cotton.

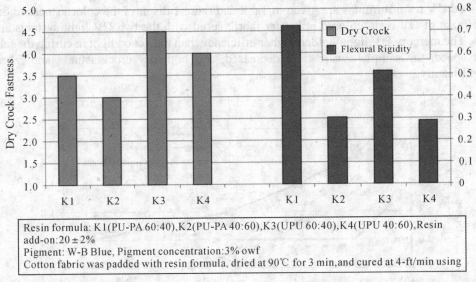

Resin formula: K1(PU-PA 60:40), K2(PU-PA 40:60), K3(UPU 60:40), K4(UPU 40:60), Resin add-on: 20 ± 2%
Pigment: W-B Blue, Pigment concentration: 3% owf
Cotton fabric was padded with resin formula, dried at 90℃ for 3 min, and cured at 4-ft/min using

Figure 2 Effects of resin chemical structures and ratio of oligomer and monomer on crockfastness and increase of flexural rigidity on colored cotton

1.2.3 Colorfastness to washing of UV cured fabrics.

Washfastness and lightfastness are other important properties of colored textiles. Table 3 demonstrates the two fastness properties of UV cured fabrics with different resin formulas and pigments. Except for the blue pigment colored polyester fabric, all of the UV cured fabrics exhibited washfastness to staining levels of 3 or higher. However, there were variations in color change levels from 1.5 to 4.5. In comparing the resin formulas, the blue fabrics finished with the K1 resin formula showed the best washfastness, while those finished with the K3 resin formula demonstrated the worst levels. Among the three UV cured colored fabrics with the K4 resin formula, the cotton demonstrated the highest colorfastness to washing, the polyester/cotton was similar, and the polyester was unacceptable. Different pigments displayed various properties. The black had the highest colorfastness to washing on cotton and polyester/cotton, and the yellow had the lowest colorfastness on all three fabrics.

Table 3 Colorfastness to wash and to light of UV Cured Fabrics

Fabric	Color	Resin[1] Formula	K/S	ΔE^2	Color Change	Wash Fastness Multifiber						Light fastness
						Acetate	Cotton	Nylon	PET	Acrylic	Wool	
Cotton	Blue	K1	8.79	0.95	4.5	5	5	5	5	5	5	5
		K3	10.24	15.82	1.5	5	5	4	5	5	5	NA
		K4	9.47	11.35	2.0	4.5	4	4	5	5	5	NA
	Black	K4	11.14	0.78	4.5	5	5	5	5	5	5	>5
	Red	K4	6.62	3.52	3.5	4.5	5	3.5	5	5	5	>5
	Yellow	K4	11.80	7.44	3.0	5	5	5	5	5	5	4-5
Polyester /Cotton	Blue	K1	10.98	0.88	4.5	5	5	5	5	5	5	>5
		K3	15.25	15.81	2.0	5	5	4	5	5	5	NA
		K4	11.76	11.15	2.0	5	4.5	4	5	5	5	NA
	Black	K4	16.19	0.75	4.5	5	5	5	5	5	5	>5
	Red	K4	9.29	6.93	2.5	4	4.5	3	5	5	5	5
	Yellow	K4	15.59	14.65	1.5	5	5	5	5	5	5	4
PET	Blue	K1	16.00	5.77	3.0	2.5	5	4.5	3	3.5	4	5
		K3	11.88	25.15	1.5	4.5	5	4	4.5	4.5	5	NA
		K4	7.04	20.77	1.0	4.5	5	4.5	4.5	4.5	4.5	NA
	Black	K4	7.43	12.62	1.5	4.5	5	5	4.5	4	4.5	>5
	Red	K4	10.00	15.42	1.5	3.5	5	3	5	5	5	5
	Yellow	K4	9.16	22.32	1.5	4	5	4.5	4	4	4.5	4

1. Resin formulas were the same as indicated in Figure 2 footnote. Curing condition: at 4 ft/min speed passing a mercury UV light
2. Color difference after washfastness testing

1.2.4 Colorfastness to light of UV cured fabrics

The colorfastness to light of UV cured fabrics is also presented in Table 3. The yellow pigment colored fabrics had a fastness level of 4, while the remaining colors had levels of 5 and higher. Pigment coloration on different textile materials fixed by UV resins did not seriously affect the fabric colorfastness to light.

2 Encapsulated aroma finish[3]

The advantages of UV curing include not only limited in energy savings, pollution reduction, and high productivity. A significant property of the UV curing is a quick process at low temperature, in which many heat sensitive materials can be easily finished with full functionality on textiles. Encapsulated aroma finish by UV curing is a good example for durable aroma finishing on textiles. The encapsulation technology is the best way to control release fragrances. It makes a lot of reservoir for perfumes. The surface area of encapsulated aromas is much less than that without encapsulating. So, the evaporation of aromas can be controlled. However, there is no affinity between encapsulated aromas and fabrics. A binder must be used to fix the encapsulated particles on fabrics. Fabrics with binder should pass a curing process to let the binder affect, and the normal curing process requires a finishing fabric going through a high temperature zone during a certain time. Since the high temperature process causes aromas evaporating much fast, less amount of aromas left on fabric greatly reduce the durability of the aroma finishing (less than 30 washes).[4] As well as too much perfume releases during curing would also cause the air pollution in manufactory sites. Therefore, UV curing is reasonable approach to fix encapsulated aromas on fabrics.

2.1 Materials and finishing processes

A bleached and mercerized 100% cotton plain woven fabric (153 g/m^2), a microencapsulated fragrance (lemon fragrance capsule, 45% solids in water), and an UV resin were used throughout the study. The resin formula included unsaturated polyurethane (40% solid/water emulsion) as oligomer, a tripropylene glycol diacrylate (TPGDA) as monomer, and one of four initiators listed in Table 4.

Table 4 Information of initiators used in the encapsulated aroma finish

Abbreviation	Chemical name	CAS number
Initiator A	2-Benzyl-2-(dimethylamino)-1-[4-(4-morpholinyl) phenyl]-1-butanone	119313-12-1
Initiator B	Phenyl bis(2,4,6-trimethyl benzoyl) phosphine oxide	162881-26-7
Initiator C	bis(2,6-dimethoxy-benzoyl) (2,4,4-trimethyl pentyl) phosphineoxide and 1-Hydroxy-cyclohexyl-phenyl-ketone	145052-34-2 and 947-19-3
Initiator D	Diphenyl (2,4,6-trimethylbenzoyl)-phosphine oxide	75980-60-8

The equipment for fabric finishing was the same as the pigment dyeing above, and added an iron UV light for curing. In the finishing process, a cotton fabric was padded through an aquatic finish water bath containing oligomer 6%, monomer 6%, initiator 0.72%, and aroma capsule 5% (in solid). There were four samples finished according to this formula, with different initiators as listed in Table 4, and two control samples: one finished with the same formula without an initiator and the other was padded with the capsule only. The wet pick-up was about 86%. The fabrics were dried at 100°C for 1.5 min, and cut into 5 in × 6 in pieces as specimens for various UV curing. During curing, fabric specimens were put on the belt of the conveyer, and passed through an UV light- illuminated area. Unless otherwise indicated, the belt moving speed was 49 in/min for mercury light and xenon light curing and 91 in/min for iron light curing. The distance between UV lamps and samples was 16 mm for the mercury light, 10 mm for the iron light, and 10 mm for the xenon light. After curing, the cured specimens were repeatedly laundered in a home washer for up to 50 cycles. The washing conditions were: permanent press cycle, medium water level, 10 min warm water wash, cold rinse, and spin. The detergent used in each cycle was 60 g of 1993 AATCC Standard Reference Detergent WOB. The samples were then dried for 30 min in a home dryer set on the permanent press cycle.

2.2 Evaluation methods

Finished samples were tested after 1, 2, 5, 10, 20, 25, 30, 35, 40, 45 and 50 wash cycles, for the presence of fragrance. To detect odor, a judge would use a fingernail to scratch an 'x' on the specimen and then smell the swatch. The response required was a 'yes' or 'no' about the presence of fragrance. Stiffness of fabrics was tested according to ASTMD-1388 (ASTM, 2000). Whiteness of finished samples was evaluated by a BYK Gardner Model 880 spectrophotometer (BYK-Gardner, Inc., Silver Spring, MD). Whiteness index was tested based on ASTM E-313 (ASTM, 2000).

2.3 Durability of fragrant finish

In the finishing process, an UV resin serves as a binder to fix aroma capsules on the fabric. Although the same oligomer and monomer were used in each finish formula, with various initiators and curing conditions, the finished samples demonstrated differences in aroma durability.

2.4 Effect of initiators

Figure 3 indicates the wash durability of encapsulated aroma finished cotton cured under mercury light. For the finished sample with aroma capsule only (the first bar in Figure 3), its wash durability was low (about 5 wash cycles), which indicated that the capsule did not have a good affinity to cotton fabric. In order to obtain high-finish durability, a fixing agent should be used. When a sample was finished with aroma capsules and UV resin without an initiator, it could bear 20 wash cycles after curing. Its laundering durability was improved compared with the 'capsule only' sample. However, with an initiator listed in Table 4, the finished sample could withstand 50 wash cycles. The initiator is critical in UV curing because it is the only component that can generate free radicals under UV light. These free radicals trigger oligomer and monomer cross-linking instantly to form a polymer, which can strongly hold the aroma capsules, and greatly increase the wash durability.

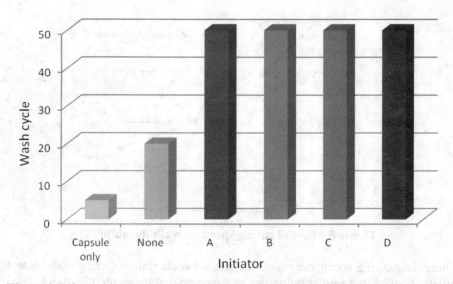

Figure 3 Effect of initiator on durability curing by mercury light at 1.24m/min.

2.5 Effect of UV sources

UV light is the energy source for fabric UV curing. Different UV lights can have different effects on UV resin reactions on fabrics, and change the sample durability. Figure 4 presents the effect of UV light on wash durability. For the fabrics finished with UV resin including an initiator, and cured under either the mercury light or the iron light, these samples could withstand 50 wash cycles. Curing under xenon light, the fabric had different durability when various initiators were used. The fabric containing Initiator A and B could still bear 50 wash cycles, while if Initiator C or D was used, the durability was only 40 and 30 cycles, respectively. One reason that the xenon lamp was not as efficient as the other lamps could be that the linear power of the xenon lamp was 120 W/in, only 40% of the others. Thus, more sensitive initiators may be needed for xenon lamp curing to promote the UV resin fully to polymerization.

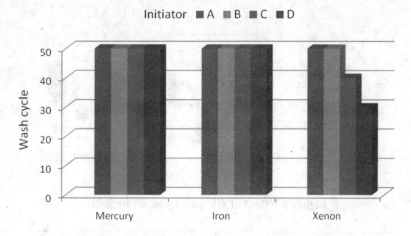

Figure 4 Effect of UV light on wash durability

2.6 Effect of curing speeds

Curing speed can decide how long a fabric needs to be exposed under UV light. The slower the speed, the longer the exposure and the more energy delivered to the sample. As shown in Figure 5, a sample containing Initiator A cured under mercury light, gave high durability if cured at 49 in/min.

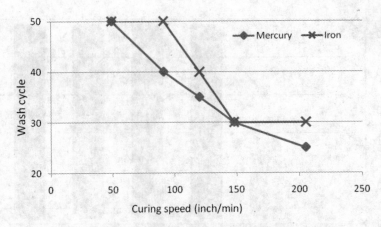

Figure 5 Effect of curing speed on wash durability

However, increasing curing speed, decreased sample wash durability. Curing under iron light, a sample finished with Initiator C exhibited similar properties as the one cured under mercury light. The difference was that iron light curing speeds could be up to 91 in/min and achieve the durability of 50 wash cycles. The iron light is more efficient for curing in this case.

2.7 Whiteness of finished fabrics

Figure 6 displays the whiteness of finished fabrics with or without UV curing. The whiteness index of the original fabric was higher than 80. Padded with the finishing chemicals, the fabric whiteness reduced to between 42 and 67. After UV curing, the whiteness further decreased. A strong yellow effect was measured in samples containing Initiator A. Their whiteness indexes were −16, −39 and −42 cured under xenon, iron, and mercury lights, respectively. Whiteness of samples containing other initiators was also reduced with UV curing, but their whiteness indexes were larger than 20. The highest whiteness among cured samples was that containing Initiator D cured under xenon light. Its whiteness index was 52.

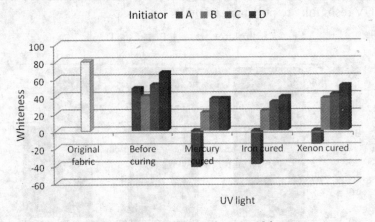

Figure 6 Effect of UV curing on whiteness

3 Conclusion

UV curing is a useful method for textile pigment coloration and encapsulated aroma finishing. Water-based UV resins can be applied on fabrics with traditional wet process equipment, such as a padder. The curing process is simple and easy to control. With suitable formulations and processes, the UV cured pigment colored fabrics demonstrated a deep shade with good washfastness, good lightfastness, and medium to good crockfastness. The encapsulated aroma finishing cotton by UV curing shows excellent wash durability. These examples indicate the possibility of applying UV curing on textile wet processes. Comparing the conventional curing with UV curing processes, the former requires putting a fabric in high temperature zone (130-200°C) for minutes, while the later can do curing at room temperature in seconds. Therefore, Using UV curing can significantly reduce energy consumption in fabric finishing. Besides, UV curing can be used for applying heat sensitive materials on textiles for various functional finishing.

References

[1] Li Shiqi, Boyter, Henry, Stewart, Neil. (2004). Ultraviolet (UV) curing processes for textile coloration. AATCC Review, 4 (8), 44-49.
[2] Lee, J., Dyeing with Pigments, Textile Wet Processing, August 1999, pp4-5.
[3] Li, Shiqi, Boyter, Henry, Qian, Lei. (2005). UV curing for encapsulated aroma finish on cotton. The Journal of the Textile Institute, 96 (6), 407-411.
[4] Li Shiqi, Lewis, Joy E., Stewart, Neil M., Qian, Lei, Boyter, Henry (2008). Effect of finishing methods on washing durability of microencapsulated aroma finishing. The Journal of the Textile Institute, 99 (2), 177-183.

73 A Study on the Kinetics and Thermodynamics of UV-absorber Taken Up to Polyester

Weiguo Chen[1,2], Xiaofang Wang[2], Son Fang[2], Qingqing Hu[2], Yining Cao[2]

(1. College of Chemistry and Chemical Engineering, Donghua University, Shanghai 200051, China
2. Key Laboratory of Advanced Textile Materials and Manufacturing Technology, Ministry of Education of China, Zhejiang Sci-Tech University, Hangzhou 310018, China. Wgchen62@126.com)

Abstract: UV absorbers were used in dyeing of polyester to improve the UV protection and the light fastness of colored fabrics. The kinetics and thermodynamics of UV absorbers taken up to polyester were studied in this paper. The results showed UV-2 with the smallest molecular had the highest uptake and greatest affinity onto polyester, shortest dyeing time, fastest diffusion into the fibres in the selected UV-absorbers.

Keywords: UV absorber; Polyester; Dyeing; Kinetics; Thermodynamics

1 Introduction

UV absorbers can strongly absorb the UV radiation, so that the degradation of polymer materials and dyes can be decreased with the application of UV absorber [1,2]. UV absorbers can be taken up to polyester together with disperse dyes in the high pressure dyeing and then the light fastness of the polyester fabrics is improved [3]. The kinetics and thermodynamics of disperse dyes in the dyeing of polyester have been studied [4,5]. In this paper, the kinetics and thermodynamics of UV absorbers taken up to polyester were studied.

2 Experimental

A plain weaved polyester fabric (220g/m^2) was used. The chemical structures of three benzotriazole type UV absorber used were listed as below:

UV-1   UV-2   UV-3

The UV absorbers were purified with Sohxlet extraction of absolute ethyl alcohol at 90°C for 10 hours and then dispersed stably with dispersing agent. The dispersed UV absorbers were applied to the high pressure dyeing of polyester. The uptake of the UV absorbers onto polyester calculated through the measurement of absorbency of the extraction of UV absorbers from dyed polyester with solvent DMF at 110°C. SHIMADZU UV-2550 UV and visible light spectrophotometer was used in the measurement of absorbency. The kinetics and thermodynamics parameters of UV-absorbers taken up to polyester, such as time of half dyeing $t_{1/2}$, coefficient of diffusion D, partition ratio K and affinity $-\Delta\mu^o$, were calculated according to that of disperse dyeing [4,5].

3 Results and discussion

3.1 The molecular structure of UV-absorbers selected

The software ChemOffice was used to calculate the molecular volume and resulted as Table 1. There are differences in the molecular volume and molecular weight because of the different chemical structures of the selected UV-absorbers. This should be one of the most important factors to determine the uptakes onto polyester.

Table 1 The molecular volume of UV-absorbers

UV-absorbers	UV-1	UV-2	UV-3
Molecular volume /A^3	2238.37	1487.54	3350.50
Molecular weight	447.6	315.8	645.9

3.2 The kinetics of UV-absorbers taken up to polyester

The uptake curves of selected UV-absorbers at 130°C were shown in Figure 1. UV-2 has the smallest molecular volume and molecular weight shown as in Table 1, it has highest uptake onto polyester (curve 1). Figure 2 showed the relationship between t/C_t and the dyeing time. According to the calculation on that of disperse dyes, the kinetics parameters of UV-absorber taken up to polyester were calculated as showed in Table 2.

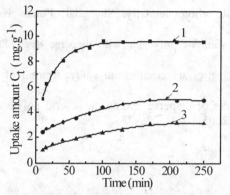

Figure 1 The uptake curve of UV-absorbers at 130°C (1-UV-2;2-UV-1;3-UV-3)

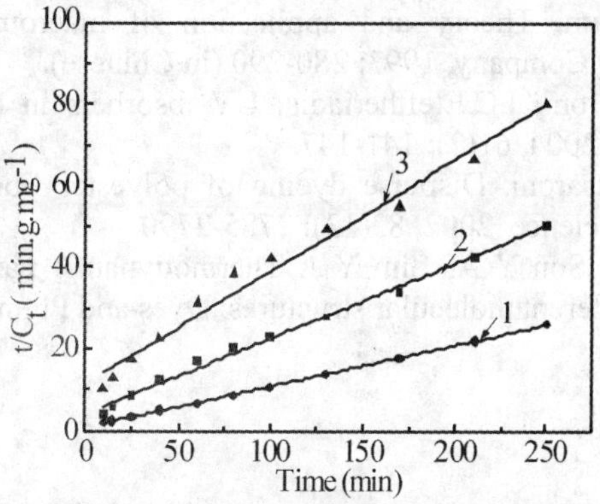

Figure 2 Relationship between t/C_t and the dyeing time (1-UV-2;2-UV-1;3-UV-3)

It is obvious that UV-2 had the greatest C_∞, the shortest dyeing time and the fastest diffusion into the polyester fibres.

Table 2 The kinetics parameters of UV-absorbers taken up to polyester

UV-absorbers 1.5%o.w.f	C_∞ (mg/g)	$t_{1/2}$ (min)	D ($10^{-12} \cdot m^2$/min)
UV-1	5.411	21.847	1.147
UV-2	9.987	8.038	3.119
UV-3	3.644	41.936	0.598

3.3 The thermodynamics of UV-absorbers taken up to polyester

The absorption isotherm of UV-absorbers taken up to polyester was drawn as Figure 3.

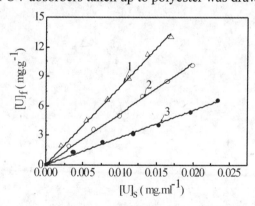

Figure 3 The absorption isotherm of UV-absorbers (1- UV-2; 2-UV-1; 3-UV-3)

According to the calculation on that of disperse dyes, the thermodynamics parameters of UV-absorbers taken up to polyester were calculated as showed in Table 3. UV-2 had the greatest affinity onto polyester fibres.

Table 3 The thermodynamics parameters of UV-absorbers taken up to polyester

UVA	UV-1	UV-2	UV-3
K	518.8	791.0	277.2
$-\Delta\mu^\circ$(kJ/mol)	20.95	22.36	18.85

4 Conclusions

In the selected UV-absorbers, UV-2 with the smallest molecular had the highest uptake and greatest affinity onto polyester, shortest dyeing time, fastest diffusion into the fibres.

References

[1] Wu Mao-ying. Photo-aging and Photostabilization of Polymers and Light Stabilizers (II). Chinese Polymer Bulletin, 2006, 6: 89-96 (in Chinese).

[2] LI Shan-jun. Theory and application of macromolecule photochemistry. Shanghai: Fudan University Publishing Company, 1993: 280-290 (in Chinese).

[3] E.G.Tsatsaroni, I.C.Eleftheriadis. UV-absorbers in the dyeing of polyester with disperse dyes. Dyes and Pigments, 2004, 61(2): 141-147.

[4] E. G. Tsatsaroni. Disperse dyeing of polyester fibers: kinetic and equilibrium study. Journal of Applied Polymer Science. 2002, 83(13): 2785-2790.

[5] Kim T K, Son Y A, Lim Y J. Thermodynamic parameters of disperse dyeing on several polyester fibers having different molecular structures. Dyes and Pigments. 2005, 67(3): 229-234.

74 Improving Dyeing Behavior of PLA Fiber with Plasticizing and Solubilizing System

Xiaonan Dang[1], Jinhuan Zheng[1,2*], Jianjian Fu[1]

(1. Key Laboratory of Advanced Textile Materials and Manufacturing Technology, Ministry of Education of China, Zhejiang Sci-Tech University, Hangzhou 310018, China
2. Engineering Research Center for Eco-Dyeing & Finishing of Textiles, Ministry of Education, Hangzhou 310018, China)

Abstract: Two cationic Gemini surfactants (D821, D1021) and plasticizer(n-amyl acetate) were compounded scientifically to prepare for two kinds of more effective accelerating agents, which were used in improving the dyeing property of Poly(lactic acid) fiber when dyed with disperse dyes. Thanks to the two kinds of accelerating agents added in the dye bath, both the dye-uptake and the color depth were improved significantly compared with that obtained in the absence of the accelerating agents. The experimental results show that there is synergistic action between the cationic Gemini surfactant and n-amyl acetate. FTIR, microscope analysis and dyeing mechanism are also investigated to explain the mechanism.

Keywords: PLA fiber; Disperse dye; cationic Gemini surfactant; *n*-amyl acetate

1 Introduction

Poly(lactic acid) (PLA) is derived from 100% annually renewable sources such as corn and sugar beet[1-2], and it is a kind of synthetic fibers with good performances, which has bring new development space for textile and new challenges for textile dyeing and finishing. PLA fiber is most commonly dyed with disperse dyes due to its hydrophobic structure, but the relatively low dye up-take and poor color fastness cause the waste of disperse dyes, increase the burden on sewage disposal, and then limit its application extension in the textile industry. At present, many methods have been studied to solve the problem of low dye up-take of PLA fiber, such as surface modification, application of ultrasonic dyeing and plasticizer added in the dye bath [3-5]. However, there are still some problems on the above-mentioned methods. The aim of this paper is using two cationic Gemini surfactants and n-amyl acetate to remix rationally to research and produce for two kinds of more effective accelerating agents to improve both the dye up-take and color depth of PLA fiber dyed comprehensively.

2 Experimental

2.1 Materials

Poly(lactic acid) yarn(14.76 dtex) were obtained from HuaYuan Eco-technology CO., LTD, Shandong province, PLA knitted fabrics were knitted by the yarn mentioned above. Disperse dyes were generously supplied by Zhejiang LongSheng Group CO., LTD. D821 and D1021 were supplied by Rugao WanLi Chemical Industry CO., LTD.

2.2 Dyeing method

The fabrics were dyed with disperse dyes of 1%(o.w.f) in a laboratory-scale infrared dyeing machine, using a liquor with fiber ratio(LR) of 50:1 and pH value of 4.5-5.0. The dyeing was started at 20℃, temperature raised to 100℃ at the heat-up rate of 2℃/min, then thermal insulation for 40min, temperature dropped to 60℃ at the rate of 2℃/min at last.

The soaping[6] was carried out in warm water(60℃) for 15 minutes using 1.5 g/L sodium carbonate and 2.0 g/L non-ionic surfactant 209.

2.3 Dye up-take

Absorbance of the dye liquor, before and after dyeing was measured at the maximum absorbance wavelength of the dye, using a PE Lambda 900 UV/Visible spectrophotometer, and the percentage of dye uptake was calculated [7-8]. In each case, the exhausted dye bath and the soaping and washing solutions were combined and a sample of this mixture was diluted at a ratio of 1:1 with acetone.

2.4 Color depth (K/S value)

The color depth of the dyed fabrics was measured using a Data-color Spectrophotometer.

---

* Corresponding author's tel: 0571-86843616, email: hzzjh1968@163.com

2.5 Color fastness
The color fastness to washing was determined according to GB/T 3921.3-1997.
2.6 Infrared spectroscopy
PLA fiber and n-amyl acetate were tested for FTIR spectra, which were recorded from 4000 to 700 cm-1 on a Nicolet Nexus 5700 FTIR spectrometer.
2.7 Fiber diameter analysis
Tested the diameters of PLA fibers with or without n-amyl acetate treated using optical microscope, which were simulated and undergone the dyeing process, and the values were the average of 50 times repeated.

3 Results and discussion
3.1 Effect of cationic Gemini surfactants on PLA fiber dyed with disperse dyes
Effect of two kinds of cationic Gemini surfactants D821 and D1021(0.8g/L) on the dye up-take and color depth of PLA fabric dyed with disperse dyes was shown in Figure 1.

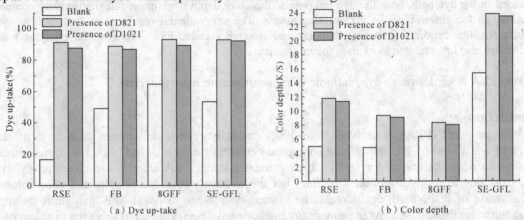

Figure 1 Influence of D821 and D1021 on PLA fiber dyeing properties of various disperse dyes

Both the dye up-take and color depth are increased significantly when the two kinds of cationic Gemini surfactants added in the dye bath. There are two aspects of the principle about the cationic Gemini surfactants, they are increase the solubility of disperse dye and strong adsorption on PLA fiber. Although the dye up-rate and color depth are increased dramatically, the color uniformity and fastness are unsatisfactory, which is due to the disperse dyes adsorbed to the fiber too fast to transfer the dyestuff to the internal of the PLA fiber.

3.2 Effect of plasticizer (n-amyl acetate) on PLA fiber
PLA fiber is a kind of polyester compounds. According to the principle of the similar solubility to those compounds with the similar structure, organic ester compounds can be applied to control the micro-structure of PLA during the dyeing process theoretically, and n-amyl acetate is a kind of the appropriate plasticizer for PLA fiber dyed with disperse dyes.

3.2.1 FTIR spectrum analysis
Infrared spectroscopy can be used to study the molecular structure and chemical bond. The FTIR spectrums of PLA fiber and n-amyl acetate were tested to research the similarity between the two materials.

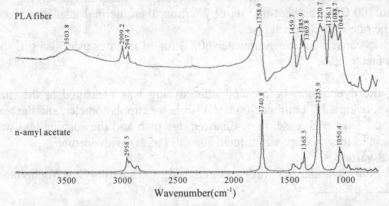

Figure 2 FTIR spectrums of PLA fiber and n-amyl acetate

FTIR spectrums of PLA fiber and n-amyl acetate are similar (Figure 2), indicating good compatibility between the two materials. And both of them have ester group, peaks at 1220.7 cm^{-1}, 1136.1 cm^{-1} and 1088.7 cm^{-1}, 1046.4 cm^{-1} with 1235.9 cm^{-1}, respectively. The compatibility means that as a kind of small molecule, n-amyl acetate could enter into the internal of PLA fiber and swell the fiber, which is attributed to the dyeing.

Furthermore, solubility parameters for PLA and n-amyl acetate were estimated using the group contribution method[9], and the values are 22.7 and 27.7, respectively. The closer solubility parameters also mean better compatibility.

3.2.2 Fiber diameter analysis

The PLA fiber would swell and fiber diameter would increase after n-amyl acetate entered into the interior of the fiber as a kind of plasticizer.

Table 1 Diameter changes and Expansion Ratio of PLA fiber

	None	Presence of n-amyl acetate
Diameter (μm)	11.72	12.86
Expansion Ratio (%)	—	20.38

Both the diameter and sectional area were increased after the treatment with n-amyl acetate, which show the fiber is swelled by the effect of n-amyl acetate.

3.3 Composition of the Plasticizing and Solubilizing system

It can be seen from the above experiments, D821 and D1021 could improve the solubitity of disperse dyes in the dye bath, which is helpful to speed up the rate of dye molecular adsorption on the fiber; and the n-amyl acetate could swell the hydrophobic synthetic fiber, which is useful to speed up the rate of dye molecular diffusion into the fiber. Therefore, consider to composite a plasticizing and solubilizing system that has the role of two.

3.3.1 Compound proportion and emulsifier selected

The compound proportions were selected in the vicinity of the optimum concentration according to the experimental results of the previous study, and the optimum complex ratios of the two accelerating agents are: accelerating agent 1: D821 0.5g/L, n-amyl acetate 6ml/L; accelerating agent 2: D1021 0.3g/L, n-amyl acetate 6ml/L.

Emulsification technology is needed in the dyeing process due to n-amyl acetate is insoluble in water. And the two kinds of surfactant D821 and D1021 were found could play the role of emulsifier during the choice of the emulsifying agent.

3.3.2 Preparation of the two accelerants

Accelerating agent 1: D821 0.5g/L, n-amyl acetate 6ml/L, appropriate water was added according to the proportion, emulsified 10 min at the speed of 10000r/min using a emulsifying machine to ensure a stable system with milk white appearance.

Accelerating agent 2: D1021 0.3g/L, n-amyl acetate 6ml/L, appropriate water was added according to the proportion, emulsified 10 min at the speed of 10000r/min.

3.4 Applicability of the accelerating agents to various disperse dyes

Six kinds of disperse dyes were used to test the dye up-takes, color depth and color fastness to washing to ensure whether the accelerating agents having a general dramatic promoting effect among PLA fiber dyed with different disperse dyes.

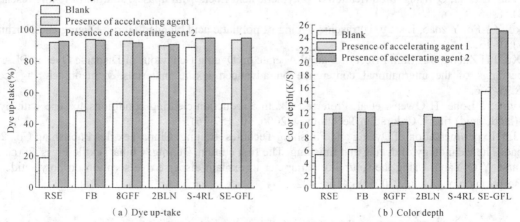

Figure 3 Dyeing properties of PLA fiber with two kinds of accelerating agents

Table 2 Color fastness of various disperse dyed PLA fabrics

Dye	Blank(grade)			accelerating agent 1 (grade)			accelerating agent 2 (grade)		
	Change	Staining		Change	Staining		Change	Staining	
		Cotton	PLA		Cotton	PLA		Cotton	PLA
Disperse blue RSE	4-5	4-5	4-5	4	4-5	4-5	4	4-5	4-5
Disperse red FB	4-5	4-5	4-5	4	4-5	4-5	4	4	4-5
Disperse yellow 8GFF	4-5	4-5	4-5	4-5	4	4	4	4	4-5
Disperse red SE-GFL	4-5	4-5	4-5	4	4	4-5	4-5	4	4-5
Disperse blue 2BLN	4	4-5	4-5	4	3-4	3-4	4	4	3-4
Disperse orange S4RL	4-5	4-5	4-5	4	4-5	4-5	4	4-5	4-5

The dye up-takes (Figure 3) of six kinds of disperse dyes on PLA fibers when the two accelerating agents added in the dye bath achieve more than 90%. Color depths are almost increased to two or three times when compared with the samples of blank. What's more, the dyed fabrics are uniform coloration and soft handle. In terms of the levels of fastness to washing in Table 2, the fastness slightly decrease, but it is consistent with the requirements of production and wear behavior still.

4 Conclusions

1) Two kinds of cationic Gemini surfactants D821 and D1021 could make more disperse dyestuff adsorbed on the surface of PLA fiber; FTIR and fiber diameter analysis show that n-amyl acetate could enter the fiber interior and increase the free volume, which is beneficial to speed up the diffusion of dyestuff into PLA fiber.

2) Through the study of plasticizing and solubilizing system for PLA fiber dyed with disperse dyes, two kinds of more efficient accelerating agents are composite and prepared as follow: accelerant 1-D821 0.5g/L emulsified n-amyl acetate 6ml/L, 10 minutes at the speed of rotation 10000r/min; accelerant 2-D1021 0.3g/L emulsified n-amyl acetate 6ml/L, emulsification method as accelerant 1.

3) There is a good synergistic action between D821(D1021) and n-amyl acetate, which is played different roles in the different periods of PLA fibers dyed with disperse dyes. That is more dyestuffs are absorbed on the PLA fiber by adding the two kinds of cationic Gemini surfactants, then through the faction of n-amyl acetate to open the pore of the PLA fiber to ensure more dyestuff diffuse to the interior of the fiber. It is the coordination of the two compounds to improve both the dye up-take and color fastness significantly of PLA fibers dyed with disperse dyes.

Acknowledgement

This work was supported by the Science and Technology Department of Zhejiang Province (2005C11028-03).

References

[1] B Gupta, N Revagade, J Hilborn. Poly(lactic acid)fiber: An overview. Prog. Polym. Sci, 2007, (32): 455-482.
[2] K David, N Digvijay, Y Q Yang. Effect of disperse dye structure on dye sorption onto PLA fiber. Journal of Colloid and Interface Science, 2007, 106-111.
[3] X R Fan, L L Su, Q Wang. Modification of polylactic acid fibers with lipases. Journal of textile research, 2009, 30(3):58-66.
[4] L J xu, Z J Fu, Y zhao, L s Yu, Ultrasonic dyeing of polylatic acid blended fabric. Dyeing and finishing, 2009, 35(9):10-13.
[5] J H Xu, J H Zheng, S H Sun. Effect of n-amyl Acetate on Dyeing Behaviour of Disperse Dyes in PLA fiber. Proceedings of the international conference on advanced textile materials & manufacturing technology, 2008:99-106.
[6] O Avinc, J Bone, H Owens, et al. Preferred alkaline reduction-clearing conditions for use with dyed Ingeo poly(lactic acid) fibers. Coloration Technology, 2006, 30: 157-161.
[7] X R Jin. Experimentation of Dyeing and Finishing Technics. Beijing: China Textiles Publishing Co., 2001.
[8] T Zhao. The technology of dyeing and finishing. The first edition. Beijing: China Textile Press, 2005: 221.
[9] D Karst, Y Q Yang. Using the solubility parameter to explain disperse dye sorption on polylactide. Disperse dye sorption on polylactide,2005,96:416~422.

75 Surface Resistivity of PET/COT Fabrics Treated with 3,4-ethylenedioxythiophene via Vapor Phase Polymerization

Qinguo Fan[1,2], Okan Ala[1] Jianzhong Shao[2], Jinqiang Liu[3]

(1. Department of Materials & Textiles, University of Massachusetts, Dartmouth, USA
2 Key Laboratory of Advanced Textile Materials and Manufacturing Technology, Ministry of Education of China, Zhejiang Sci-Tech University, Hangzhou 310018, China
3. Engineering Research Center for Eco-Dyeing & Finishing of Textiles, Ministry of Education, Zhejiang Sci-Tech University, Hangzhou 310018, China)

Abstract: Flexible, permeable and electrically conducting textiles can be obtained via vapor phase polymerization (VPP) of 3,4-ethylenedioxythiophene (EDOT). By VPP, the polymerization reaction proceeds in the gas phase via direct oxidation using an ionic dopant. In this VPP process, PEDOT was formed on PET/COT blend fabrics. The resultant surface resistivity on the PEDOT coated fabric materials achieved under optimized conditions is about 50-75 ohm/square, very close to that of ITO, while the textile materials remain soft, flexible, and permeable. This process did not use rigid inorganic conducting materials like ITO or a second polymeric component polystyrene sulfonic acid (PSS).

Keywords: E-Textiles; Conductive textile; PEDOT; Vapor phase polymerization

1 Introduction

With the advent of conductive polymers in 1977[1], the interest in this field has significantly increased due to the lightweight and semiconducting nature of conductive polymers enabling them to be used in applications such as microelectronics, rechargeable batteries, photovoltaic panels, light emitting diodes, electrochromic devices, electromechanical actuators, membranes, antistatic packaging, corrosion protections and biomedical applications[2]. Conductive polymers are usually formed on glass or other rigid substrates in many applications because of their poor mechanical properties. Many conductive polymers are rigid materials and can degrade easily. One of the approaches to address the rigidity issue is to use plastic substrates. But some applications require better mechanical properties, more flexibility, and better permeability as well as certain transparency. Therefore, the idea of using textile materials with their inherent softness, flexibility, permeability, and transparency to make a novel type of electrically conductive materials has emerged.

Among conductive polymers, poly 3, 4-ethylenedioxythiophene (PEDOT) is significantly important due to its combined properties of small band gap, high conductivity and high stability, although it should be noted that these individual properties are not limited to PEDOT[3]. The electrical conductivity of PEDOT layers on different surfaces may not be lower than that of some other electrically conductive polymers such as polyaniline (PANI) and polypyrrole (PPy). Having a small band gap structure plus its excellent processability, one of the most significant advantages of PEDOT, eases the electron movement between energy levels enabling the material to be utilized in electronic applications where high electron transfer capability is required.

PEDOT can be applied on to textiles by utilizing different methods. In most cases, the aqueous dispersion of PEDOT is prepared with the help of poly (styrenesulfonate) (PSS) as a doping agent and solubilizing component. The PEDOT:PSS dispersion can be printed[4] on textile materials utilizing different methods such as (1) screen printing in which the polymer paste is passed through a permeable screen, (2) gravure printing in which the ink pattern is formed on the fabric by engraved cylinders, and (3) inkjet printing in which the ink droplets are jetted onto the textile substrate with great precision. Though inkjet printing PEDOT:PSS on textiles seems to be a preferable way to form conductive textiles, the surface resistivity results are not low enough. The surface resistivities below 100 ohms per square of PEDOT treated textiles are considered low for many electronic applications. The aqueous solutions of PEDOT exhibit short shelf life, bad film forming capability and difficulty in synthesis. The dispersion of PEDOT is favorable over its aqueous counterpart. However, the dispersions of PEDOT (PEDOT:PSS) exhibit higher resistivity and are influenced by water or other common solvents. It is possible to form PEDOT on textile materials by utilizing electrospinning method[5]. Electrospun nanosized PEDOT fibers have the disadvantage of having very low mechanical properties compared to traditional textile materials.

Vapor phase polymerization[6] (VPP) is carried out in the gas phase and is based on direct oxidation by a catalyst solution. The initial experiments were done by Mohammadi et al under the name of Chemical Vapor Deposition (CVD) process for polypyrrole polymerization in which $FeCl_3$ and H_2O_2 were used as oxidants. Then, it was adapted as a well-defined surface patterning method using $CuCl_2$ as the oxidaizing agent in polymerization of polypyrrole. Later, Ueno et al reported that by blending polyvinyl chloride (PVC) and $FeCl_3$, a conducting composite can be produced after exposing the mixture to pyrrole vapor. In 1998, Fu et al replaced the conventional oxidant agent, $FeCl_3$, with ferric tosylate for VPP of pyrrole in polyurethane foam. The first VPP of EDOT using $FeCl_3$ as oxidizing agent was reported[7] by a group of scientists at Hanyang University and Korea Basic Science Institute, in which surface resistance of 500 ohm/sq was reported for PEDOT films with a thickness of 300 nm.

2 Experimental

We used PET/COT (50/50%) blend ready for printing fabric as the substrate. The fabric is a plain weave with 40 x 40 Ne yarn counts, 82 ends and 54 picks per inch, and 94 g/m^2 at a thickness of 0.27 mm. The PET/COT fabric was cut carefully with the dimensions of 3 cm x 2 cm. To remove possible stains and to prepare the substrate, the samples were washed in ethanol/deionized water solution (volume ratio of 30/70%) twice at the room temperature for 5 minutes and then dried at 50ºC in an oven. Subsequent to drying process, the samples were conditioned at room temperature for 2 days before they were treated with the VPP of EDOT which is available from.

Ferric tosylate solution (40%) was used as an oxidant/dopant in the VPP of EDOT (Baytron M V2). A VPP chamber shown below which is based on the design by Winter-Jensen[8] was used for the VPP process.

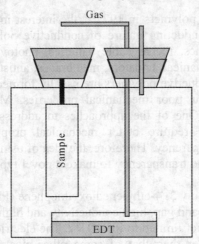

The steps for the VPP process are given below:
1. A VPP chamber was placed onto a heater.
2. The temperature inside the chamber was set to 50ºC and the monomer in a small glass beaker was transferred into the chamber.
3. By using nitrogen gas, 50 ml/sec, the monomer vapor distribution inside the chamber was improved.
4. The weight of the textile sample was measured.
5. By using a dropper, the ferric tosylate solution was applied onto the textile substrate.
6. The weight of the ferric tosylate solution was noted.
7. The textile sample was then hung inside the chamber with the help of a hook.
8. After an hour, the PEDOT-coated textile sample was taken out from the chamber and awaited in the hood for half an hour.
9. To avoid excessive amount of ions, ethanol was used to wash the sample (tea beg method, gently washing in a solution, for 10-15 seconds).
10. The textile sample was then transferred into the oven set at 50ºC.
11. After 15 minutes, the textile material was taken out from the oven and washed again in ethanol at the room temperature for 10 sec.
12. After second washing, the textile material was put into the oven and dried at 50ºC for 15 min.
13. The PEDOT coated textile material was then conditioned in the room temperature for a day. Then resistivity measurements were made at room temperature and relative humidity of 40%.

The surface resistivity is defined as the resistance of a material between two opposite sides over a unit square of its surface and expressed in ohm per square. The thickness of the layer is not considered during measurements[9]. The surface resistivity is calculated using the following equation.

$$R = O \times W / L$$

where R is the surface resistivity in ohm/square; O measured resistance in ohm; M width of the material; L distance between electrodes. It is worth noting that ohm/square is not a valid unit by dimensional analysis. It is used for the sake of avoiding mixing up with the resistance unit. Surface resistivity tests were conducted according to AATCC Test Method 76. To test the electrical resistivity of the materials, silver coated copper electrodes were formed on the bottom PMMA glass plate. The distance between the electrodes was adjusted to 1 cm as shown below.

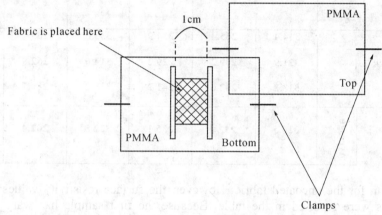

3 Results and discussion

Three preliminary samples were prepared. The details for each sample are given in Table 1.

Table 1 Sample preparation details

	Sample 1	Sample 2	Sample 3
Fabric Weight (g)	0.070	0.071	0.067
Fe (III) Tosy. Weight (g)	0.110	0.105	0.110
EDOT Weight (g)	0.50	0.50	0.50
Polymerization Time (h)	1	1	1
Polymerization Temperature (C°)	50	50	50

Electrically conducting textile materials formed via VVP of EDOT were sandwiched between the two PMMA plates using clamps. Thus, a firm contact between the electrodes and the textile material was provided. The resistance between the electrodes was measured on both sides of the fabric, face and back sides as regulated by the test method. By using the equation mentioned above, the lowest surface resistivity values were noted in Table 2. (Considering the dimensions of the fabric, the resistance values along the fabric length were multiplied by 3 and the resistance values along the width multiplied by 2).

Table 2 Surface resistivity values of three samples

		Sample 1		Sample 2		Sample 3	
		Length	Width	Length	Width	Length	Width
Face		307.8	264.8	224.4	316.4	260.4	285
		295.8	237.8	232.8	292.6	244.5	291
		298.8	266.8	237.3	304.6	252.6	287.6
	Average	300.8	256.4	*231.5*	304.5	*252.5*	287.8
Back		314.4	265.4	234.9	335.6	269.7	307.8
		321.9	249.2	229.2	314.6	262.8	310.2
		313.2	242.6	241.2	318.2	254.7	303.4
	Average	316.5	*252.4*	235.1	322.8	262.4	307.1

Same test was run for the uncoated fabric. However, the surface resistivity values were undetectably high. Thus, no values were reported in the table. Because the first sample has warp yarns in its width direction, it gave lower resistivity values along the width. The other two samples gave lower surface resistivity values in their length direction, again due to warp yarn arrangement. The reason for low surface resistivity obtained along the warp yarn direction is that the higher yarn density providing better intercontact in the structure leading to lower surface resistivity values.

After a few different trials, we found out that if argon gas is used in the VPP process, the surface resistivity of the sample thus prepared can be in the range of 50-75 ohm/square which is comparable with that achievable with indium tin oxide, a commonly used conductive inorganic material in many electronic applications.

4 Conclusions

We have formed a flexible, permeable, electrically conductive textile product. The method includes wetting a textile sample with a predetermined chemical containing oxidant, forming an oxidant enriched textile sample, placing the oxidant enriched textile sample in a vapor phase polymerization chamber having a predetermined inside temperature, providing a predetermined monomer inside the chamber, providing monomer vapor flow inside the chamber, resulting in the oxidant enriched textile sample being contacted by the monomer vapor flow and allowing contact between the oxidant enriched textile sample and the monomer vapor flow for a predetermined time, whereby an electrically conductive textile sample is formed. The optimized surface resistivity values are in the range of 50-75 ohm/square. By a carefully designed approach, our approach can be extended to form flexible, permeable, electrically conductive patterns on textile materials which enable the construction of many textile-based technical and biomedical devices.

Acknowledgement

Authors are grateful for the financial support provided by Zhejiang Provincial Top Academic Discipline of Applied Chemistry and Eco-Dyeing & Finishing Engineering, Zhejiang Sci-Tech University.

References

[1] C.K Chiang, C.R. Fincher, Jr. Y. W. Park, A.J. Heeger, H.Shirakawa, E.J Louis, S.C. Gau, and A.G. MacDiarmid, Electrical Conductivity in Doped Polyacetylene, Physical Review Letters, Vol. 39, No. 17, p. 1098, 1977
[2] O. Ala and Q. Fan, Applications of Conducting Polymers in Electronic Textiles, Research Journal of Textile & Apparel, Vol. 13, No. 4, p.51, 2009

[3] T.A. Skotheim, J.R. Reynolds, Handbook of Conducting Polymers: Theory, Synthesis, Properties and Characterization, CRC Press, 2006.
[4] S. Kirchmeyer, D. Gaiser, Extremely Flat and Flexible, Kunststoffe International, 2007
[5] T.H. Grafe, K.M. Graham, Nanofibers Webs From Electrospinning, Nonwovens in Filtration, Fifth International Conference Stuttgart, Germany, 2003.
[6] G. G. Wallace, G. M. Spinks, L.A.P. Kane-Maguire, P. R. Teasdale, Conductive Electroactive Polymers: Intelligent Polymer Systems, CRC Press, 2008.
[7] J. Kim, E. Kim, Y. Won, H. Lee, K. Suh, The preparation and characteristics of conductive poly(3,4-ethylenedioxythiophene) thin film by vapor-phase polymerization, Synthetic Metals, vol: 139, p. 485-489, 2003.
[8] B. Winter-Jensen, K. West, Vapor-Phase Polymerization of 3,4-Ethylenedioxythiophene:A Route to Highly Conducting Polymer Surface Layers, Macromolecules, vol: 37, 2004
[9] C. A. Harper, E. M. Petrie. Plastics Materials and Processes: A Concise Encyclopedia, Wiley, 2003.

76 Synthesis and Application of MDI Water-borne Polyurethane Fixing Agent

Baozhou Li, Shuling Cui, Rui Li

(Hebei University of Science and Technology, Shijiazhuang 050018, China)

Abstract: The water-based polyurethane fixing agent was synthesized mainly from the reaction of diphenylmethane diisocyanate (MDI), polyethylene glycol (PEG) and dimethylol propionic acid (DMPA). The influence of time, temperature and DMPA dosage on the pre-polymerization was studied. Application of the fixing agent was also discussed. The results of tests showed that the optimum synthesis process is 100℃, 140 ~ 160 min, and DMPA 7%. The fixing agent could significantly improve the wet crocking fastness of colored fabrics dyed with direct dyes, reactive dyes and acid dyes.

Keywords: Water-borne polyurethane; Fixing agent; Synthesis; Application; MDI

1 Introduction

There are two reasons for why the crocking fastness of dyed fabric is low. First, loose color dyes form large amorphous aggregates and gather on fiber surface. The dyes in this state are difficult to be washed off from fiber in general washing conditions. Second, the dyes form dye crystals on the fiber surface. X-ray diffraction shows that there are considerable amount of dye crystals on fiber surface. When rubbed by the crocking fastness tester, these dye crystals could transfer from the dyed fabric to the white test cloth because of the large friction encountered [1].

In 1942, Schlack first synthesized the water-borne polyurethane by adding emulsifier [2]. From then on, water-based polyurethane has been gradually used in textile field because it is non-toxic, non-polluting, energy saving, safe and reliable [3].

According to the literature, the water-based polyurethane which connects with amine compounds can combine with the anionic part of direct dyes, acid dyes and reactive dyes, so it has very good fixing effect [4-6]. But the affinity between water-based polyurethane and fiber is not good[7], so it needs to be modified by using epichlorohydrin, and thus make the fixing agent has a reactive groups so that it can form a stable membrane on the fiber surface and the washing color fastness can get increased[8~11].

In this study, the influence of pre-polymerization time, temperature and dimethylol propionic acid (DMPA) dosage on the reaction was studied. Application effects of the fixing agent in color fabrics dyed with reactive dyes, direct dyes, and acid dyes are discussed.

2 Experimental

2.1 Main chemicals and instruments

Diphenylmethane diisocyanate (MDI), Shanghai Reagent plant; polyethylene glycol (PEG, M_n =1000), Jiangsu Haian Petroleum Chemical Reagent Company; dimethylol propionic acid (DMPA), Shanghai Reagent plants; epichlorohydrin, Tianjin Damao Chemical Reagent Co.; N, N-dimethylformamide amide, acetone, diethylenetriamine, butylamine, toluene, and isopropanol were all chemically pure reagentl bought from chemical market.

Electronic stirrer, Changzhou Guohua Electric Co., Ltd.; temperature control device, Xi'an Galaxy Instrument; constant temperature oven, Chongqing Yinhe Instrument Manufacturing Co., Ltd.; soaping machine, SW-12AⅡ, Wenzhou Daiei Textile Standard Instrument; crocking fastness tester Y571L (A), Wenzhou Daiei Textile Standard Instrument Factory.

2.2 Synthesis of waterborne polyurethane

First, PEG dehydrated with oven for 10 minutes. DMF and DMPA were quantitatively added to a three-neck flask connected with electric stirrer and thermometer. Next, heat the system gradually to 100℃. Then, certain amount of MDI was added to the three-neck flask, and the bulk polymerization goes on for 2.5h. When the content of NCO is close to the theoretical mass, add stoichiometric diethylenetriamine and take the reaction for 30 min.

2.3 Synthesis of waterborne polyurethane fixing agent

Epichlorohydrin was added to above reaction system at 100℃, and take the reaction for 2 h. The brown-yellow emulsion was obtained. That is the product waterborne polyurethane fixing agents.

2.4 Performances test

The content of isocyanate group (-NCO) was tested by the method of Di-n-Butyl Amine (HG/T 2409-1992). Crocking fastness of textiles was determined according to GB 3920-1997. Soaping fastness of textiles was measured according to GB 3921-1997.

3 Results and discussion
3.1 Determination of reaction time

The reaction time has a great effect on the polyurethane pre-polymer. If the time is too short, the degree of polymerization is not enough. The emulsion looks like limewater and is so instable that it easily separates from each other. On the other hand, if the time is too long, the polymerization system easily generates gel. In this study, the pre-polymerizing time was determined by following the content of –NCO in reaction process with the method of Di-n-Butyl Amine. The result is given in Figure 1.

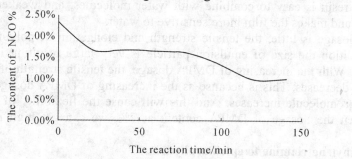

Figure 1 Influence of reaction time on the content of –NCO group

Figure 1 shows that the content of -NCO is no longer decrease further when the reaction time is more than 140 min, thus the reaction time for 140 ~ 160 min is more appropriate.

3.2 Determination of reaction temperature

The reaction temperature has also a great influence on the performance of the polyurethane pre-polymer prepolymer. The temperature was determined by testing the viscosity of the liquid system and the film-forming state of the prepolymer. The result is given in Table 1.

Table 1 Effect of reaction temperature on the viscosity of emulsion

Temperature/℃	Viscosity/mPa·s	Film-forming state	Appearance
80	20	-	yellow emulsion
90	25	-	yellow emulsion
100	35	+	Yellow-brown emulsion
110	45	+	Yellow-brown emulsion

("+" good film-forming state, "-"bad film-forming state)

From Table 1, it can be seen that as the reaction temperature rising, the viscosity of the polymer emulsion increases, but with the reaction temperature rising, the film-forming property of the product is better and that is helpful to color fixing. So 100℃ is chosen as the reaction temperature.

3.3 Determination of DMPA dosage

In the reaction, DMPA as hydrophilic chain extender. The concentration of DMPA has an impact on the water absorption and stability of polyurethane emulsion. The result is given in Table 2.

Table 2 Effect of DMPA dosage on the appearance and stability of the emulsion

Dosage /%	Moisture absorbency/ %	Stability/ month	Emulsion appearance	Crocking fastness before finishing		Crocking fastness after finishing	
				dry	wet	dry	wet
6	6.12	<4	Brown translucen	4	2~3	5	4
7	7.23	>6	Brown translucen	4	2~3	5	4
8	10.56	>6	Brown translucen	4	2~3	5	3~4
9	13.35	>6	Brown translucen	4	2~3	5	3
10	15.32	>6	Brown translucen	4	2~3	5	3

From Table 2, when the content of DMPA lies in the range of 7 ~ 10%, stability and appearance of the emulsion is good. As the increasing of the content of DMPA, moisture absorbency of the film increases gradually, as means that the water resistance of the product is declined. The reason is that carboxyl anionic ion in polyurethane resin is easy to combine with water molecules, and weakens the interaction of the polyurethane chains and makes the film more sensitive to water.

When DMPA dosage is little, the tensile strength and elongation of polyurethane is relatively low because at this condition the size of emulsion particle is very large resulting in uneven film and poor mechanical strength. With the increasing of DMPA dosage, the tensile strength of the film goes up but the breaking elongation decreases. This is because as the increasing of DMPA dosage, the proportion of hard segments in the giant molecule increases. And this will cause the flexibility of film decrease. From the above analysis, when the dosage of DMPA content at 7%, we can get high performance water-based polyurethane.

3.4 The ratio of diethylenetriamine to epichlorohydrin

In the reactions, diethylenetriamine is a chain extender and endcapper. The epichlorohydrine have two main functions: first, it can provide reactive end group in the emulsion giant molecule so as to make subsequent crosslinking with cellulose fibers possible; second, epichlorohydrin can react with amino or imino groups, and produce tertiary amine or quaternary ammonium salt. With the cationic center the polyurethane resin can combined with cellulose fiber by ionic bond. However, the ratio of diethylenetriamine to epichlorohydrin has an impact on the stability and finishing effect of the polyurethane fixing agent. The result is given in Table 3.

Table 3 Effect of ratio of diethylenetriamine to epichlorohydrin on the finishing effect and stability

Use proportion	Emulsion appearance	Organic solvent consumption /g	Stability /month	Dry crocking fastness	Wet crocking fastness
1:1	Yellow-brown translucent	20	>4	5	3~4
1:2	Yellow-brown translucent	30	>6	5	4
1:3	Yellow-brown translucent	60	>6	5	4

From Table 3 it can be known that when the proportion of diethylenetriamine and epichlorohydrin is 1:2, the finishing effect and stability is better.

3.5 The concentration of diethylenetriamine and epichlorohydrin

In the reaction process, the dosage of diethylenetriamine, which was used as a chain extender, determines the size of the molecular weight of polyurethane, and the molecular weight affects the emulsification quality of polyurethane. The result is given in Table 4.

Table 4 Effect of concentration of diethylenetriamine and epichlorohydrin on the finishing effect

Diethylenetriamine dosage /%	Epichlorohydrine dosage /%	Emulsion appearance	Emulsion water-soluble	Dry crocking fastness	Wet crocking fastness
0.7	1.25	Yellow-brown translucent	-	5	4
1	1.8	Yellow-brown translucent	-	5	4
1.5	2.7	Yellow-brown translucent	+	5	4
2	3.6	Yellow-brown opaque	+	5	3~4
2.5	4.5	Yellow-brown opaque	++	5	3
3	5.4	Yellow-brown opaque	++	5	3

(diethylenetriamine: epichlorohydrin = 1:2; "+" good film-forming state, "-" bad film-forming state)

Table 4 shows that the concentration of diethylenetriamine and epichlorohydrin has influence on appearance, water solubility and finishing effect of the fixing agent. When epichlorohydrin reacts with diethylenetriamine, it generates hydroxyl groups to the fixing agent molecule, and these hydroxyl groups can increase water-solubility of fixing agent. From Table 3, we can know that the water-soluble and the finishing effect of the fixing agent is a pair of contradiction, so a combination of factors we choose the dosage of diethylenetriamine for 1.5%, and epichlorohydrin for 2.7%.

3.6 Application process
3.6.1 Dosage of fixing agent

When testing the dosage of fixing agent, other process of the fixing agent is: padding pick-up 100%, baking temperature 140℃, baking time 90s. The result is given in Table 5.

Table 5 Effect of dosage of fixing agent on the finishing effect

Dosage g/L	Dry crocking fastness	Wet crocking fastness	Washing fastness	Soaping fastness
30	3~4	3~4	4	3~4
40	4~5	3~4	4	3~4
50	4~5	3~4	4	4
60	4~5	4	4	4
70	4~5	4	4	4

Table 5 shows when the dosage of fixing agent reached 60 g/L, the finishing effect achieve the best, so the dosage of fixing agent is 60 g/L.

3.6.2 Baking time

When testing the baking time, other process of the fixing agent is: fixing agent 60g/L, padding pick-up 100%, baking temperature 140℃. The result is given in Table 6.

Table 6 Effect of baking time on the finishing effect

Baking time/s	Dry crocking fastness	Wet crocking fastness	Washing fastness	Soaping fastness
30	4~5	3~4	4	3~4
60	4~5	3~4	4	3~4
90	4~5	4	4	4
120	4~5	4	4	4
150	4~5	4	4	4

Table 6 shows that as baking time increasing, crocking fastness increases. When the baking time reaches 90 s, the crocking fastness achieves the best, so the best baking time is 90 s.

3.6.3 Baking temperature

When testing the baking temperature, other process of the fixing agent is: fixing agent 60g/L, padding pick-up 100%, baking time 90s. The result is given in Table 7.

Table 7 Effect of baking temperature on the finishing effect

Baking temperature /℃	Dry crocking fastness	Wet crocking fastness	Washing fastness	Soaping fastness
100	4~5	3~4	4	3~4
110	4~5	3~4	4	3~4
120	4~5	3~4	4	4
130	4~5	4	4	4
140	4~5	4	4	4
150	4~5	4	4	3~4

Table 7 shows that when the baking temperature reaches 130~140℃, the dry crocking fastness and wet crocking fastness of fabric achieves the best. If the baking temperature is too high, it may destroy the polyurethane film fixing on the fiber surface. So the best baking temperature is 130℃.

3.6.4 Mangle expression

When testing the mangle expression, other process of the fixing agent is: fixing agent 60g/L, baking temperature 130℃, baking time 90s. The result is given in Table 8.

Table 8 Effect of mangle expression on the finishing effect

Mangle expression	Dry crocking fastness	Wet crocking fastness	Washing fastness	Soaping fastness
80%	4~5	3~4	4	3~4
90%	4~5	3~4	4	3~4
100%	4~5	4	4	4
110%	4~5	4	4	4
120%	4~5	4	4	4
130%	4~5	4	4	4

From Table 8 we can see that as the mangle expression increase, the crocking fastness also increases. However, when the mangle expression reaches to saturated, the crocking fastness doesn't increase any longer. So the best padding pick-up is 100%.

4 Conclusions

(1) The optimum synthesis process of waterborne polyurethane is: reaction time of pre-polymerization 2.5 h, dosage of hydrophilic chain extender DMPA 7%, reaction temperature 100℃. Divinyl triammonium functions as the chain extender with dosage of 1.5%. Epichlorohydrin is chosen as a modifier with dosage of 2.7%.

(2) When used as the fixing agent on color fabric dyed with reactive dyes, the optimum process of the fixing agent is: fixing agent 60g/L, padding pick-up 100%, baking temperature 130℃, baking time 90s.

References

[1] Mofu Huang, Yiming Wang. The application of crocking fastness promoter CY-3101. Printing and Dying, 2006, 1:23-25
[2] Xiangdong Wang, Huiti Yang. The review and progress of polyurethane industry at home and abroad. China Plastic, 1999, 13(8): 21-22
[3] Lijun Wang, Zhongyin, XinFan Zhang. The evolution of waterborne polyurethane technology. Leather Science and Engineering, 2005, 15(2): 1-2
[4] AjayaK N, Douglas AW, Samy AM, etal. Nanostructured polyurethane /poss hybrid aqueous dispersions prepared by homogeneous solution polymerization. Macromolecules, 2006, 39: 7038-7041
[5] Sanchel M, Bennevault V, Deffieux A. Functionalizable polyurethane networks based on hydroxy-modified poly-(chloroalkylvinyl ethers). Polymer Bulletin, 1994, 33: 59-66
[6] Gooddard R J, Cooper S L. Polyurethane cationomerswith pedanttrialkylammonium groups: Effects of ion content, alkyl group and neutralizing anion. Journal of Polymer Science, Part B, 1994, 32(8): 1557-1571

[7] Zunqiao Huang, Meina zhu, ling Ru. Synthesis and performance of waterborne polyurethane fixing agent. Anhui Chemical Industry, 2007, 33(5): 38-40
[8] Yanxia Xu, Jianqiu zhang, Yueying Wu. Synthesis and performance study of Waterborne Polyurethane fixing agent. Applied Chemical Industry, 2003, 32(4): 9-11
[9] Chunyong Lin, Bitai lu. Fixing properties of waterborne polyurethane fixing agent for ramie dyed fabric. Textile Technology Development, 2008, (1): 71-73
[10] Libo Liu. Synthesis and application of fixing agent NF. Dyeing Technology, 2008, 30(5): 37-39
[11] Yongchun Dong. Chemistry and Applicalion of Textile Auxiliaries. Beijing: Textile Press of China, 2007:165-175.

77 A Novel Approach for Evaluating Cotton Fabric Strength Damage Caused by Bleaches during Laundering

Yongqiang Li[1,2], Jinqiang Liu[1,2], Chunjie Qian[1,2], Liming Deng[1,2]

(1. Key Laboratory of Advanced Textile Materials and Manufacturing Technology, Ministry of Education of China, Zhejiang Sci-Tech University, Hangzhou 310018, China
2. Engineering Research Center for Eco-Dyeing & Finishing of Textiles, Ministry of Education, Zhejiang Sci-Tech University, Hangzhou 310018, China)

Abstract: This paper focuses on the application of a novel approach for evaluating fabric strength damage caused by bleaches during laundering. The tensile strength, copper number as well as the dyeing properties of the fabrics were investigated to characterize the property changes after laundering. Moreover, the relationships between tensile strength loss and color depth of cotton fabrics were further studied. The results showed that dyeing can reflect the dyeing properties changes of cotton fiber after bleaching laundering. The chemical damage of cotton fabrics after bleaching laundering can be evaluated effectively with dyeing method.

Keywords: Cotton; Copper number; Fabric strength damage; K/S value

1 Introduction

Cotton fabrics are easily stained in the daily use, and need regular washing care. Bleach is a popular laundry aid and helps detergents remove soils and stains. Through oxidization, laundry bleaches convert soil into more soluble, colorless or dispersible particles that can be removed by detergent and carried away in the wash water. At the same time, cotton fabrics are bleached during laundering.

In the washing process, washing cotton fabric with bleach may lead to chemical damage and loss of tensile strength. In some instances, chemical damage on the fabric can be reflected in the changes of physical properties and chemical properties. Chemical damage of the fiber may not be apparent until the fibre has been treated with alkali, therefore the determination of tensile strength of bleached cotton fabric alone can not detect the actual extent of the degradation of cellulose. Hence, the effective evaluation of family washing bleach on fabric care is particularly important. The current methods are to test tensile strength loss of fabrics and depolymerization of cellulose after repeated washing[1], but these methods are usually complicated. The purpose of this paper is to find a rapid and more sensitive method to detect the chemical damage of cotton fabrics during home laundering and establish a foundation on the basis of conventional evaluation criteria.

2 Experiments
2.1 Materials and main equipments
2.1.1 Materials

A professional manufacturer of cotton based fabric (123.4g/m^2) selected was supplied by Jingqiu Textile Co. Ltd, Shaoxing, China.

Arel Color detergent (U.S.A,P&G); Arel Bio detergent(U.S.A,P&G); $CaCl_2$; $CuSO_4·5H_2O$; $NaHCO_3$; Na_2CO_3; $NH_4Fe(SO_4)_2·12H_2O$, H_2SO_3; H_2SO_4; $KMnO_4$; Na_2SO_4(AC, Hangzhou Huipu Chemicals Co. Ltd). Reactive dyes such as C.I. Reactive Red 84, C.I. Reactive Red 177, C.I.R.R.194, C.I.R.B.13, Red RB and Blue BB were supplied by Zhejiang Runtu Co., Ltd, Shangyu, China, a professional manufacturer of dye.

2.1.2 Main equipments

Wash-fastness machine (Taiwan, Rapid Co., Ltd); Electrolux washing machine EWS-850 (Electrolux China Co., Ltd); Data color matching machine SF600 PLUS(U.S.A, Dic Co., Ltd); DHG-9070A drying-machine (Shanghai), Dyeing machine.

2.2 Laundering method

Fabrics: cotton fabric. Cycles: to take enough fabrics every 5 cycles to total of 25cycles. Washer: EWS-850. Water: 8 L of water with hardness of 130 ppm Ca. Wash/Rinse temperature: 90℃ (wash) / room temperature (rinse). Detergent: a) 74.0g of Ariel color, b) 74.0 g of Ariel Bio.

2.3 Dyeing method

Dyeing formula for reactive dye: Reactive Dye: 1% (o.w.f); Agent: Na_2SO_4 20g/L; Na_2CO_3 10g/L; Bath ration: 1: 50.

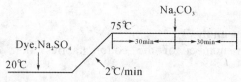

Figure 1 Dyeing process

2.4 Tensile strength tests

According to the method of GB/T3923.1-1997, the tensile strength of fabrics was measured by Instron-2365 fabric tensile strength tester (Instron Corporation, America). The samples were balanced at 25℃ and a relative humidity of 60% for 24 h before testing. The sample was in dimension of 50mm×250mm after drawing out the yarns of 5mm at both sides. Tensile strength of cotton fabrics were only tested in the warp direction and the results were given as the arithmetic means of five different samples. Tensile strength loss ratio = $(a_0-a_1)/a_0 *100\%$ …(1), where a_0 and a_1 represent the tensile strength of control fabric and fabrics after home laundering, respectively.

2.5 Aldehyde group quantitative analysis

Chemical damage from over oxdiation or hydrolysis of cellulose produces aldehyde groups. Aldehyed groups in oxidized cellulose reduce copper ions to copper which is deposited on the fibers as a reddish brown solid. The test can be used as a qualitative spot test or can be done quantitatively. When done quantitatively, the result is called the "copper number." [2]

Copper number was determined according to the method of $Cu^{\#}$ test[3]. The copper number was analyzed in duplicate for each sample giving results with errors of less than ±5%.

2.6 K/S value test

With the use of a color yield parameter K/S value, the dyeing quality was evaluated. The determination of K/S values of the dyed samples was carried out with the SF 600X Datacolor spectrophotometer. In this work, all samples were replicated six times in same condition and the average values were plotted. K/S value loss ratio = $(b_0-b_1)/b_0 *100\%$ …(2), where b_0 and b_1 represent the K/S value of control fabric and fabrics after home laundering, respectively.

3 Results and discussion

3.1 Tensile strength of fabrics after home laundering

The tensile strength of cotton fabrics vs. the laundering cycles is shown in Figure 2. Two types of detergents were used during home laundering. Arel Bio detergents have bleaching effect, but Arel Color detergents have no bleaching effect. Referring to Figure 2, the tensile strength of cotton fabrics gradually decreased with increasing the laundering cycles. The tensile strength of cotton fabrics is highly polynomially correlated to laundering cycles, with R_2 of 0.9612 and 0.9918 for Ariel Color washing cotton and Ariel Bio washing cotton, respectively. As shown in Figure 2, for Ariel Bio washing cotton, the tensile strength loss was much higher than that of Ariel Color washing cotton after correspondent cycles of home laundering. This may be because strength damage for Ariel Color washing cotton fabric was mainly contributed by mechanical impacts during home laundering, which include impact, abrasion, bending, shearing, stretching, and so on. However, strength damage for Ariel Bio washing cotton fabric was caused by bleaching effect except by mechanical impacts. The results showed that chemical damage of cotton fabrics was caused by bleaching during home laundering.

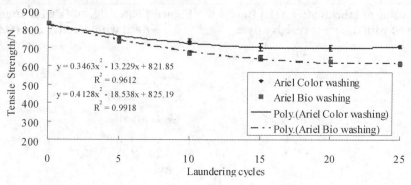

Figure 2 Tensile strength of fabrics after home laundering

3.2 Aldehyde group quantitative analysis

In this study, the copper number of the cotton fabrics after different laundering cycles was experimentally measured.

Table 1 Copper number of fabrics after home laundering

Laundering cycles	Copper number of Ariel Color washing samples	Copper number of Ariel Bio washing samples
0	0.20	0.20
10	0.21	0.27
25	0.19	0.35

It can be seen from Table 1 that the 10 laundering cycles cotton and 25 laundering cycles cotton washed with Ariel Bio detergent had much higher copper numbers compared to the original cotton, meanwhile, the copper number increased with increasing of laundering cycles. For the Ariel Bio detergent washing cotton, the copper number had little significant change after different laundering cycles. This is because the C-2,3 or C-6 hydroxyl groups of the glucose repeating unit of cellulose may be oxidized with increasing laundering cycles[4], during Ariel Bio detergent laundering. More laundering cycles, the hydroxyl groups oxidized more, therefore, the copper number increased. While Ariel Color detergents have no bleaching effect, the copper number had no significant change after Ariel Color laundering.

3.3 Dyeing properties

Figure 3 shows K/S value of fabrics after 10 and 25 laundering cycles dyed with different reactive dyes. As shown in Figure 3, after Ariel Bio laundering, all color depth of cotton fabrics decreased compared to the control cotton fabrics, which dyed with different reactive dyes. There is a distinct downward trend in color depth of cotton fabrics, especially when cotton fabric dyed with C.I.R.R.84.

The K/S value of fabrics after laundering vs. the laundering cycles is shown in Figure 4. From Figure 4, it is observed that the color depth of Ariel Bio washing cotton fabric gradually decreased with increasing the laundering cycles, while the color depth of Ariel Color washing cotton fabric did not change obviously with increasing the laundering cycles. The reason is that the C-2, 3 or C-6 hydroxyl groups of the glucose repeating unit of cellulose may be oxidized, which would reduce reactive dye dye-firing percentage on cotton fabrics[5].

Figure 5 shows K/S value loss ratio and tensile strength loss ratio of fabrics after different laundering cycles. According to Correl Funticon Model, the correlation coefficient is 0.979 between K/S value loss ratio and tensile strength loss ratio of fabrics after laundering. The result shows that there is a good correlation between them.

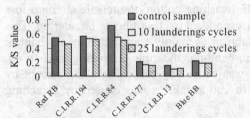

Figure 3 K/S value of fabrics after Ariel Bio laundering dyed with different reactive dyes

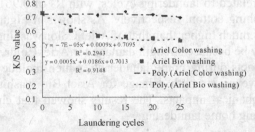

Figure 4 K/S value of fabrics after home laundering dyed with reactive dye (C.I. R. R.84)

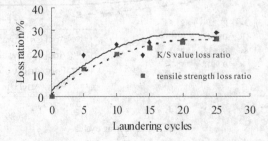

Figure 5 K/S value loss ratio and tensile strength loss ratio of fabrics after different laundering cycles

4 Conclusions

There is a good correlation between the tensile strength of cotton fabrics and laundering cycles, and chemical damage of cotton fabrics was caused by bleaching during home laundering.

Dyeing properties of cotton fabric changed after bleaching laundering. The K/S value of cotton fabrics gradually decreased with increasing bleaching laundering cycles after dyeing. The chemical damage of cotton fabrics after bleaching laundering can be evaluated effectively when the fabrics were dyed with specified reactive dyes. To test color depth is a possible way to evaluate the chemical damage of cotton fabric after bleaching laundering.

Acknowledgement

We want to express our gratitude to Dr. Jiping Wang for his help and valuable suggestions.

References

[1] Xiaofen Ding, et al. A bleachers handbook. Beijing: China Textile Press, 2005:60.
[2] Ro¨hrling J, et al. A novel method for the determination of carbonyl groups in cellulosics by fluorescence labeling.2. Validation and applications. Biomacromolecules 2002, 3(5): 969–975.
[3] Xianxiang Jin. Experiment of Dyeing and Finishing Engineering. Beijing: China Textile Press, 2001:51-54.
[4] Jianguo Zhang et al. Oxidation and sulfonation of cellulosics. Cellulose, 2008, 15: 489–496.
[5] Jinxin He. Color chemistry. Beijing: China Textile Press, 2001:155-157.

78 Dyeing of Polyurethane Fiber and Polyurethane/Nylon Blend with Temporarily Solubilised Disperse Dyes

Hongfei Qian[1*], Xinyuan Song[2]

(1. College of Textiles and Apparel, Shaoxing University, Shaoxing 312000, China
2. College of Chemistry and Chemical Engineering, Donghua University, Shanghai 200051, China)

Abstract: In this study, a temporarily solubilised disperse dye was prepared and applied to polyurethane fibers and polyurethane/nylon blend to investigate a new dyeing process to improve color uniformity and color fastness. The effect of pH on the dyeing of this dye for both the polyurethane fiber and its blend was investigated at different dyeing temperature. The optimum dyeing condition was found respectively. In addition, the dyeing rate curve was measured at 75°C and pH 10 for various components in polyurethane/nylon blend. It was observed that the transfer of dye from polyurethane component to nylon occurred because of difference in substantivity, and the dyeing equilibrium was nearly achieved by increasing the dyeing time to 100 min. Even distribution of dye between polyurethane and nylon as well as satisfactory color fastnesses were obtained with this dye.

Keywords Temporarily solubilised disperse dyes; Polyurethane fiber; Polyurethane/nylon blend; Exhaustion; Distribution

1 Introduction

The application of a temporarily solubilised disperse dye containing β-sulphatoethylsulphonyl groups to polyester/silk blend was previously reported.[1,2] Further studies on the synthesis, hydrolysis behaviour and applied properties of this type of dye have been carried out.[3-6]

Polyurethane(PU) fiber, owing to its excellent flexibility, is widely used in textile products. It is usually blended with other fibers such as polyester, nylon, silk and wool. The low T_g of PU provides easy dye access. PU fiber can be dyed with disperse and 1:2 metal complex dyes at 90°C.[7] In previous study, It was found that most types of disperse dyes were preferentially exhausted on PU fibers rather than PET or nylon components as dyeing of PU/polyester and PU/polyamide blend with dispersed dyes.[8] Heavy staining of disperse dyes on PU fibers decreases the wet fastness of blend owing to the low T_g of polyurethane, which means that adsorbed dyes can easily migrate out from the inner to surface.[9]

PU/nylon blend can be dyed with acid dyes. In this process, there are deviations in hue between PU and nylon[7]. In the present study, a temporarily solubilised disperse dye containing β-sulphatoethylsulphonyl groups was prepared and used to dye PU fibers and its blend with nylon to assess its applicability. In theory, when the β-elimination reaction occurs in alkaline solution or at high temperature, dye 1 can convert into a non-ionic substance, the vinylsulphone (VS) form 2, as shown in Scheme 1, that has substantivity for both PU and nylon as well as can covalently bond on the nylon component with the amine group in polyamide, so this type of dye have the potential capacity to dye simultaneously the two components in PU/nylon blend.

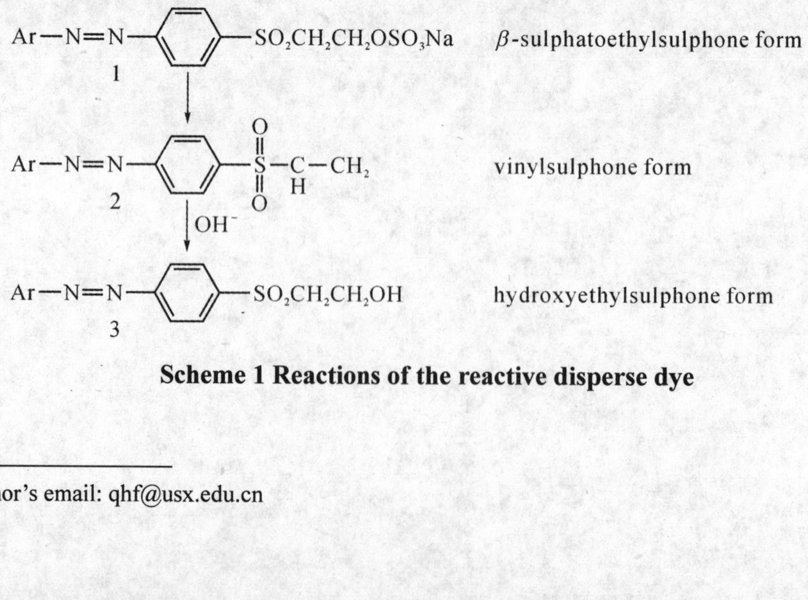

Ar—N=N—⟨phenyl⟩—SO$_2$CH$_2$CH$_2$OSO$_3$Na β-sulphatoethylsulphone form
1

Ar—N=N—⟨phenyl⟩—S(=O)(=O)—C(H)—CH$_2$ vinylsulphone form
2

Ar—N=N—⟨phenyl⟩—SO$_2$CH$_2$CH$_2$OH   hydroxyethylsulphone form
3

Scheme 1 Reactions of the reactive disperse dye

---

* Corresponding author's email: qhf@usx.edu.cn

2 Experimental
2.1 Materials

Polyurethane fiber (44.4 dtex) was provided by Invista DuPont, China. To obtain uniform dyeing, fibers were cut into short lengths of 3-4 cm and scoured in a bath containing 1 g/l non-ionic detergent (Sandozin NIE, Clariant) at 85°C for 30 min, then rinsed in distilled water and dried in the open air. Scoured and bleached 100% nylon 6 knitted fabrics (87.5 g/m^2) and nylon 6 blend with 10% polyurethane fiber (120 g/m^2) were obtained from Hongda Group, China.

Aminophenyl-4-(β-sulphatoethylsulphone) (98% purity) and N-ethyl-N-(2-cyanoethyl)-aniline (98% purity) was kindly supplied by Zhejiang Longsheng Group Co. Ltd., China. All chemicals used in the study [N,N-dimethylformamide (DMF), sodium acetate, acetic acid, disodium phosphate, sodium phosphate, sodium carbonate and sodium bicarbonate] were of reagent grade.

2.2 Synthesis of dye

A reactive disperse dye, as shown in Chart 1, was synthesised and purified using the methods reported by Lee and Kim [3]. The Elemental analysis confirmed the dye to have the composition $C_{19}H_{21}N_4NaO_6S_2$. The ^{1}H NMR spectra were recorded on a Bruker Advance DMX 500 spectrophotometer and the chemical shift values (ppm) of the dye found to be as follows (J values given in Hz): δ_H([$^2H_6$]DMSO)/ppm1.17 (3H, t, J 6.94, CH$_3$), 2.84 (2H, t, J 6.72, NCCH$_2$), 3.56 (2H, q, J 7.01, NCH$_2$), 3.69 (2H, t, J 6.60, α-CH$_2$), 3.77 (2H, t, J 6.75, NCH$_2$), 3.99 (2H, t, J 6.68, β-CH2), 6.95 (2H, d, J 9.25, H-3/5), 7.86 (2H, d, J 9.13, H-2/6), 7.95 (2H, d, J 8.60, H-2'/6'), 8.03 (2H, d, J 8.58, H-3'/5').

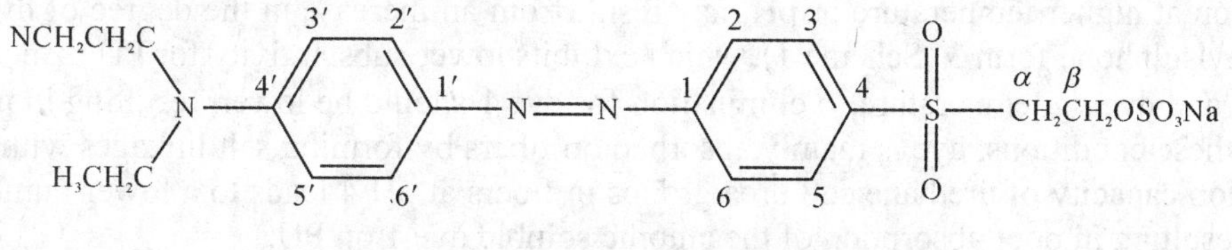

Chart 1 Structure of the synthesised dye

2.3 Dyeing operation for effect of pH at different temperature

A dye bath containing 2% (owf) reactive disperse dye was prepared. Dyeing was carried out in a dyeing machine (Rapid, China) at a liquor ratio of 20:1. The pH was adjusted to a required value using various 0.1mol/l buffer system. PU fibers or mixed samples containing PU fiber/nylon fabric (1:9 mass ratio) were immersed in the liquor. The bath was heated to 60, 75 or 90°C respectively and maintained for 90 min. The dyed samples were removed and then rinsed with water and air-dried.

2.4 Dyeing operation for dyeing rate curve

A series of dye baths containing 2% owf reactive disperse dye at a liquor ratio of 20:1 was prepared and placed in the dyeing machine. When the temperature reached 75°C, the PU/nylon blend samples were added. At intervals of 10 min, a dyed sample was removed, rinsed with water and air-dried.

2.5 Measurement of dye exhaustion

Dyed PU fibers were repeatedly extracted with DMF until they became colourless. The absorbance of the extract was then measured spectrophotometrically at λ_{max} on a Shimadzu UV-2401PC UV/vis spectrophotometer. The dye concentration in the initial bath was determined by measuring the absorbance of the solution before dyeing diluted 20-fold with DMF. The percentage exhaustion (%E) was calculated according to:

$$\%E = (C_2/C_1) \times 100 \tag{1}$$

where C_1 is the dye concentration in the initial bath and C_2 is the concentration of extracted dye.

Dye exhaustion on PU fiber in dyeing of blend was determined using the method described above. Dye exhaustion on nylon components in blend was calculated according to:

$$\%E_n = E_{total} - E_u \tag{2}$$

$$E_{total} = \left[1 - (C_3/C_1)\right] \times 100 \tag{3}$$

where E_{total} is total exhaustion in the dyeing of blend, E_n and E_u are the exhaustion on nylon and polyurethane fibers in blend, respectively. C_3 is the residual dye concentration in the bath, which is determined using the same method as for measurement of the initial dye concentration in the bath (C_1).

2.6 Measurement of fixation on nylon

The dyed nylon component of the blend was treated in the solution of 2g/l non-ionic detergent (Sandozin NIE, Clariant) and 1g/l sodium carbonate for 15min at 95°C using a 50:1 liquor ratio to remove the bonded dyes by salt linkage, followed by repeated refluxing in 50% aqueous DMF (liquor ratio 20:1) to extract the unfixed dye until the extract was clear.[10] The concentrations of the clearing solution and extract were then measured spectrophotometrically at λ_{max}. Dye fixation (F), was calculated using Eqn 4:

$$F\% = [(C_1 - C_3 - C_4)/C_1] \times 100 \qquad (4)$$

where C_4 represents the total concentration in the clearing solution and extract.

2.7 Color fastness testing

The dyed samples with 1:1 standard depth were obtained and then tested according to standard ISO methods, ISO 105-C06, method C1S for colourfastness to washing and ISO 105-X12 for fastness to rubbing.

3 Results and discussion
3.1 Effect of pH value and dyeing temperature

The results for effect of pH at different temperature are presented in Figs. 1 and 2 for dyeing of PU fibers and blend, respectively.

The data in Figure 1 show that the degree of exhaustion is clearly dependent on both temperature and pH. The optimum dye exhaustion was achieved at pH 8 and 75°C. This can be attributed to a high degree of conversion of dye 1 into the non-ionic reactive VS form 2 as a result of β-elimination, which in turn maximises the dye–fiber affinity owing to the inherent hydrophobic properties of PU. On the other hand, lower exhaustion at higher temperature or pH may result from an increase in the degree of dye hydrolysis to the hydroxyethylsulphone form 3 (Scheme 1), which exhibits lower substantivity for PU than the VS form.

At low pH and dyeing temperature, β-elimination for dye 1 should be lower, resulting in poor exhaustion on PU. Under these conditions, dye is mainly absorbed on fibers by forming salt linkages with the fibers. The weak protonation capacity of urethane and urea groups in fibers at pH 4 leads to a lower number of dye sites on the fibers, resulting in poor absorption of the anionic soluble dye 1 on PU.

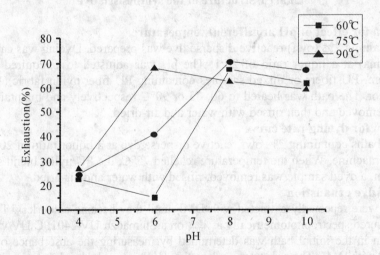

Figure 1 Exhaustion for dyeing of PU fibre at different pH and temperature

Figure 2 shows the effect of pH on exhaustion on both PU and nylon, as well as fixation on nylon in dyeing of blend at different temperature.

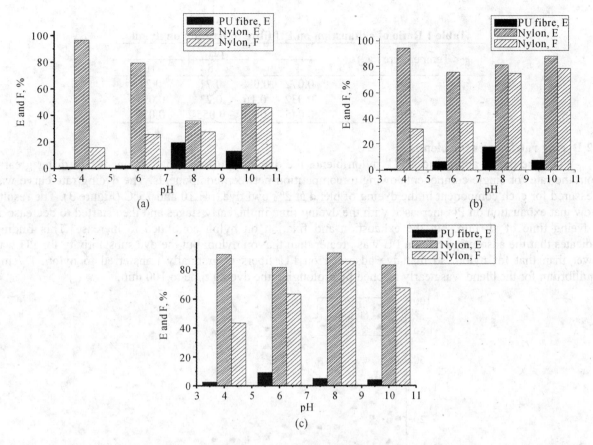

Figure 2 Exhaustion (E) for PU fibre and nylon and fixation on nylon (F) for dyeing of blends at different pH at: (a) 60; (b) 75; and (c) 90°C

The pattern of variations in exhaustion on PU with pH at different temperature is very similar to that for dyeing of PU fiber alone. However, maximum exhaustion on PU was achieved at pH 8 and 60°C for the blend instead of at 75°C for the fibers alone owing to competition with nylon for the dye. At pH 4, soluble dye 1 preferentially adsorbed on nylon, resulting in higher exhaustion on nylon and very low exhaustion on the PU component. This can be attributed to good protonation of amine groups in the nylon polymer, resulting in a considerable number of dye-binding sites but low fixation on nylon. At pH 6, both exhaustion on PU and fixation on nylon increased with temperature. This can be explained by an increase in the degree of conversion of dye 1 into VS form 2 and higher reactivity towards amine groups in nylon with increasing temperature. At pH 8, as the temperature was increased from 60 to 90°C, the dye was preferentially exhausted on nylon, resulting in an increase in exhaustion and fixation on nylon and a decrease in exhaustion on PU. This is attributed to the increasing capacity of amine groups in nylon to react with the VS form. At pH 10, the exhaustion on PU gradually decreased with increasing temperature, with optimum exhaustion and fixation on nylon achieved at 75°C instead of 90°C. To explain this finding, the hydrolysis and reactivity of the dye must be simultaneously considered when the dyeing temperature is increased to 90°C. Hydrolysed dye has lower reactivity towards the fibers, and thus degree of fixation is lower. An increase in dyeing temperature results in an increase in activity and in the degree of hydrolysis for the reactive dye, so variation of exhaustion and fixation on either a component or another with temperature depend on which effect is prominent. The results indicated that when the temperature was raised from 75°C to 90°C at pH 10, the effect on hydrolysis degree of dye was greater than that on activity so that the exhaustion and fixation on nylon decreased.

For dyeing of blend, uniform distribution of the dye is important to achieve solid dyeing on both types of fibers. To determine the optimum dyeing conditions, the ratio of exhaustion on PU relative to fixation on nylon was calculated (Table 1). As the weight ratio of PU relative to nylon in the blend was 1:9, pH 10 and 75°C may be the optimum conditions, under which a distribution ratio of nearly 0.1 was achieved, with fixation on nylon relatively higher.

Table 1 Ratio of exhaustion on PU fibre to fixation on nylon

Temperature (°C)	pH			
	4	6	8	10
60	0.072	0.088	0.71	0.27
75	0.032	0.16	0.23	0.091
90	0.045	0.14	0.055	0.059

3.2 Dyeing rate curves for blend

When the dye is applied to a PU /nylon blend, the dyeing rate for each component will differ greatly from the rate for single components owing to competition between components. The dyeing rate curve was measured for each component in the dyeing of blend at 2% owf dye, pH 10 and 75°C (Figure 3). The results show that exhaustion on PU increased with the dyeing time in the early stages and then started to decrease at a dyeing time of 10 min, but the exhaustion and fixation on nylon continued to increase. This finding indicates that the adsorption rate on PU was greater than that on nylon, but the dye substantivity for PU was lower than that for nylon, so that dye adsorbed on PU fibers is gradually transferred to nylon. Dyeing equilibrium for the blend was nearly reached by prolonging the dyeing time to 100 min.

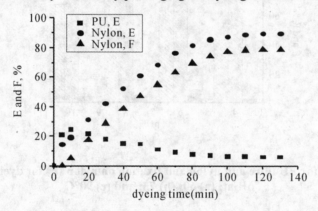

Figure 3 Dyeing rate isotherm for various components in dyeing of blend at pH 10 and 75°C

3.3 Color fastness properties

Table 2 lists washing and wet and dry rubbing fastness results for dyed samples at a standard depth of 1/1. The fastness properties for PU fiber were moderated. The staining rate on nylon was relatively low. Owing to the low PU content in blend, the color fastness results for nylon blend were satisfactory.

Table 2 Color fastness results

Fibre or fabric	Change	Staining		Rubbing	
		Nylon	Cotton	Wet	Dry
PU fibre (1/1 standard depth)	3–4	3	3–4	3–4	4–5
Nylon blend containing 10% PU fibre (1/1 standard depth)	4	3–4	4	4–5	4–5

4 Conclusions

This study demonstrated that temporarily soluble disperse dye containing β-sulphatoethylsulphonyl groups can be used to dye PU fiber and its nylon blend. Satisfactory dyeing exhaustion, fixation and color fastness can be obtained by good control of dyeing conditions. Optimum dyeing conditions were pH 8 and 75°C for single PU fiber, and pH 10 and 75°C for PU/nylon blend, under which the uniform distribution on two components can be achieved.

During dyeing of nylon blend at pH 10 and 75°C, dye adsorbed on PU fibers could be gradually transferred to the nylon fibers and the dyeing equilibrium was nearly reached as the dyeing was maintained for 100 min.

References

[1] M Dohmyou, Y Shimizu and M Kimura. Dyeing of Silk and Synthetic Fibres with Sulphatoethylsulphonyl Reactive Disperse Dyes. J.S.D.C., 1990, 106(12): 395–397.

[2] S M Burkinshaw and G W Collins. The Dyeing of Conventional and Microfibre Nylon 6.6 with Reactive Disperse Dyes. Dyes Pigm., 1994, 25(1): 31–48.

[3] W J Lee and J P Kim. The Dyeing of Conventional and Microfibre Nylon 6.6 with Reactive Disperse Dyes. J.S.D.C., 1999, 115(9): 270–273.

[4] W J Lee and, J Pkim. Dispersant-free Dyeing of Polyester with Temporarily Solubilised Disperse Dyes. J.S.D.C., 1999, 115(9): 370–374.

[5] W J Lee, W H Choi and P J Kim. Dyeing of Wool with Temporarily Solubilised Disperse Dyes. Color. Technol., 2001, 117(4): 212–216.

[6] J Koh, J D Kim and J P Kim. Synthesis and Application of a Temporarily Solubilised Alkali-clearable Azo Disperse Dye and Analysis of Its Conversion and Hydrolysis Behaviour. Dyes Pigm., 2003, 56(1): 17–26.

[7] S D Bhattacharya and B H Patel. Processing of polyurethane Fibre and Its Blends. Man-Made Text. India, 2003, 46(7):248–254.

[8] H F Qian and X Y Song. The Structure of Azo Disperse Dyes and Its Distribution on polyurethane Fibre Blend with Polyester, or Polyamide Fibre. Dyes Pigm., 2007, 74(3): 672–676.

[9] H F Qian and X Y Song. Structure–property relationships for azo disperse dyes on polyurethane fibre. Color. Technol., 2009, 125 (3): 146-150.

[10] A Amousa and Y A Youssef. Dyeing of Nylon 6 and Silk Fabrics with a Model Disulphide Bis(ethylsulphone) Reactive Disperse Dye. Color. Technol., 2003, 119 (4): 225-229.

79 Dyeing Kinetics of Carrier Cindye Dnk in Dyeing Process of Aramid Fiber with Disperse Dyes

Lan Wang[1], Ping Liang[2], Junxiong Lin[3*], Duan Ni[4]

(1. Key Laboratory of Advanced Textile Materials and Manufacturing Technology, Ministry of Education of China, Zhejiang Sci-Tech University, Hangzhou 310018, China
2. Zhejiang Huatai Silk Co., Ltd, Hangzhou 310018, China
3. Engineering Research Center for Dyeing and Finishing of Textiles, Ministry of Education, Zhejiang Sci-Tech University, Hangzhou 310018, China
4. Keyi College of Zhejiang Sci-Tech University, Hangzhou 311121, China)

Abstract: The dyeing kinetics of Cindye Dnk in dyeing process of aramid fiber dyed with C.I. Disperse Yellow 21 was studied. The results indicated that the dyeing rate constant and diffusivity of disperse dye on aramid fiber were increased with the Cindye Dnk as the dyeing carrier, and the half-dyeing time was shorten as well.

Keywords: Aramid fiber; Carrier; Disperse dyes; Dyeing kinetics

1 Introduction

As the glass transition temperature (Tg) of aramid fiber is 275℃, its dyeing with conventional method at 130℃ is more difficult. Dyeing with carrier is a universal method for polyester and aramid fabrics at both home and abroad now. In order to comply with the goal of green dyeing and finishing, choosing an environment-friendly carrier with disperse dyeing on aramid fabric was a convenient, effective, energy-saving and safe dyeing process. The effect of carrier Cindye Dnk on disperse-dyeing process of aramid fiber mainly shows the influence of carrier on dyeing thermodynamics and kinetics. There are only a few researches on the dyeing thermodynamics and kinetics in the disperse-dyeing process of aramid fiber. On respect of dyeing thermodynamics on aramid fiber, for example, Nobuhiko Kuroki considered that disperse dyeing on aramid fiber is generally consistent with Nernst-type adsorption and the diffusing of disperse dye into aramid fiber is consistent with free volume model[1]. However, the research on kinetics in the dyeing process of aramid fiber has not been reported yet.

In the present work, the aramid yarn dyeing with C.I. Disperse Yellow 211 was taken as the research object, and the dyeing kinetics of carrier Cindye in the dyeing process of aramid fiber was studied.

2 Experiments
2.1 Materials

Aramid yarn extracted from aramid fabric was used, acetone (AR), ethanol (AR), DMF (AR), HAc-NaAc buffer solution, carrier Cindye Dnk, $NaNO_3$ (AR) were applied. C.I. Disperse Yellow 211 manufactured from Longsheng Co. was used with further purification as well.

C.I. Disperse Yellow 211

2.2 Methods
2.2.1 Drawing of standard working curve

1 g of C.I. Disperse Yellow 211 weighed precisely was purified by re-crystallization method, and dye solution with DMF was prepared. The dye solution was taken with different volumes according to multiple relationships, then they were transferred into 8 measuring flask of 25ml and fixed the volume with DMF.

---

* Corresponding author's tel.: 86-571-86843605; fax: 86-571-86843250; email: linjunxiong@zstu.edu.cn (J.X. Lin)

Finally, they were shaken evenly and measured a series of corresponding absorbency at wavelength of 380-780nm with UV spectrophotometer. The spectral absorption curve was drawn basing on the wavelength as X-coordinate and absorbency as Y-coordinate. The maximum absorption wavelength of dye λmax was ascertained. The known dye solutions with mass concentration were measured respectively with the mass concentration of dye as X-coordinate and absorbency as Y-coordinate. The standard working curve was drawn [2].

2.2.2 Dyeing rate study

The dyeing solutions of pure C.I. Disperse Yellow 211 of 2% (o.w.f.) were prepared. Specifically, a certain quality of pure disperse dye was first precisely weighed and dissolved with acetone thoroughly; 1.5 g/L solution was prepared. 1ml solution was accurately transferred to the dye cup with suction pipe as well. Then HAc and NaAc (comparison sample needs carrier Cindye Dnk and $NaNO_3$) were added and 100 ml working solution was prepared. Placing the dye cup in the SHA-C digital thermostat water-bath pot, the acetone was evaporated cleanly at about 90℃. Meanwhile, an amount of water was added to keep the working solution at 100ml.

0.1g aramid yarns were precisely weighed. The yarns were put into the working solution in turns after the high-temperature dyeing machine got to the required temperature. And then the dye cups were placed into the dyeing machine quickly in turns. After keeping the temperature at different times, the dye cups were gotten out separately. Thus we placed them in cold water until the yarns were taken out, washed, soaped, washed and naturally dried. After that, the yarns were extracted and stripped in oil-bath pot at 110℃ with DMF. The stripping solution was transferred to the scale-line of 25ml measuring flask and the absorbency at the maximum absorption wavelength was measured. The concentration of disperse dye can be examined according to the standard working curve of pure disperse dye in DMF. Therefore, the disperse dye amount C_t onto the aramid fiber per unit mass can be calculated. The diagram between C_t~t was finally obtained, namely, the dyeing rate curve.

2.2.3 Calculation of adsorption amount of the dye on the fiber

At the end of dyeing, the samples were soup-boiled, washed and dried. The dried samples were weighed and then dyes were extracted using N,N-dimethylformamide at 110℃ until the samples became colorless. The dye concentration in the extracts was measured using spectrophotometer meter and the extent of dye adsorption was calculated [3].

2.2.4 Calculation of dyeing rate constant and half-dyeing time

The half-dyeing time $t_{1/2}$ is the time when the dye is the half amount of equilibrium absorption on the fiber. It is a common rate indicator of dyeing to the equilibrium [4].

There is still no correct measuring method of the calculation about dyeing rate constant so far. There are a lot of empirical formulas which coincide with the dyeing rate curve. However, in the usual data processing, dyeing rate constant (k) is often reckoned from the slope and intercept of the diagram of C_t~t and the half-dyeing time $t_{1/2}$ is calculated from formula $t_{1/2}=1/(k*C_\infty)$.

2.2.5 Calculation of dyeing diffusion coefficient

According to Eq (1) [5,6], the diffusion coefficient was calculated from the plot that shows the relationship between M_t/M_∞ and t in the initial stage of dyeing.

$$\frac{M_t}{M_\infty}=1-4\sum_{n=1}^{\infty}\frac{e^{-V_n^2 Pt/a^2}}{V_n^2}=$$

$$1-4\left\{\frac{1}{5.785}e^{-5.785\,Dt/a^2}+\frac{1}{30.47}e^{-30.47/Dt/a^2}+\frac{1}{74.89}e^{-74.89\,Dt/a^2}\right.$$

$$\left.+\frac{1}{139}e^{-139\,Dt/a^2}+\frac{1}{222.9}e^{-222.9\,Dt/a^2}+\cdots\cdots\right\} \tag{1}$$

M refers to the dyeing exhaustion on the fiber; D is the diffusion coefficient; t is dyeing time; V_n means the number n measured fiber radius; a, refers to the radius of fiber.

3 Results and discussions

3.1 Standard working curve

The λ_{max} of C.I. Disperse Yellow 211 is 436 nm, and the standard working curve is shown as Figure 1. The standard curve equation of C.I. Disperse Yellow 211 fitted by the software Origin is y=-0.00753+74.70313x. In the equation, x is mass concentration of dye, y is absorbency. The value of R^2 is 0.99985. It means the straight line is consistent with the Lambert-Beer law, and the value of absorbency A and dye concentration are a positive correlation.

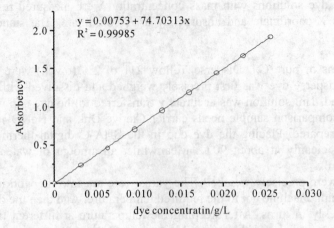

Figure 1 Standard working curve of C.I. disperse yellow 211

3.2 Dyeing kinetics parameters

Dyeing kinetics mainly studies the rate and process of dye onto the fiber [7-10]. Dyeing kinetics parameters include dyeing rate constant, half-dyeing time and diffusion coefficient.

3.2.1 Dyeing rate constant and half-dyeing time

Dyeing rate constant and half-dyeing time reflect the speed of the disperse dye on the fiber to achieve a balance. In this paper, C.I. Disperse Yellow 211 with purification was used, and dyeing kinetics experiment to aramid fiber was carried under the condition of pH value of 5.5, temperature of 130℃, for 0-120 min. Thus, dyeing rate curve of C.I. Disperse Yellow 211 was drawn. The result was shown as Figure 2.

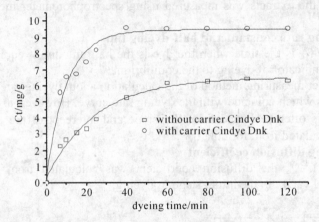

Figure 2 Dyeing rate curve of C.I. disperse dellow 211 on aramid fiber

From Figure 2, in weak acid condition, no matter with carrier Cindye Dnk or not, the dyeing rate of C.I. Disperse Yellow 211 on aramid fiber increases with the increasing in dyeing time until achieving respective balance. And the dyeing rate with Cindye Dnk is significantly higher than that of without it.

Vickerstaff hyperbolic equation can be used for most of the constant temperature dyeing rate curves to describe the adsorption, assuming that dyeing rate of disperse dye on aramid fiber can be described by this hyperbolic equation[11]:

$$\frac{1}{C_\infty - C_t} - \frac{1}{C_t} = kt \qquad (2)$$

or

$$\frac{t}{C_t} = \frac{1}{C_\infty}t + \frac{1}{kC_\infty^2} \tag{3}$$

This equation describes the linear extent of function $t/C_t=f(t)$, if $\text{tg}\alpha$ is the slope of the linear equation and b is the intercept, the value of C_∞, k and $t_{1/2}$ can be calculated by equation (4)~(5).

$$C_\infty = \frac{1}{\text{tg}\alpha} \tag{4}$$

$$k = \frac{1}{bC_\infty^2} \tag{5}$$

When the dye uptake reaches half of the balance, $t = t_{1/2}$, $C_t = C_\infty/2$, we can deduce from the equation (5)-(6):

$$t_{1/2} = \frac{1}{kC_\infty} \tag{6}$$

C_t, dye uptake on fiber at time of t (mg/g); C_∞, dye concentration on fiber when at balance (mg/g); K,-dyeing rate constant (min·g/mg); $t_{1/2}$, half-dyeing time (min).

The $t/C_t \sim t$ curve is shown in Figure 3.

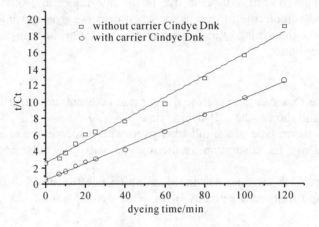

Figure 3 Curve of t/Ct~ t

Dyeing parameters can be calculated from Figure 3 and equation (4)-(6). The results are shown in Table 1.

Table 1 Dyeing rate constant and half-dyeing time of C.I. disperse yellow 211 on aramid fiber

Dyeing methods	Equilibrium adsorption/C_∞/mg/g	Dyeing rate constant	Half-dyeing time/min
Without carrier	6.410	0.006770	19.53
With carrier Cindye Dnk	9.556	0.01798	5.490

From Table 1, we know that the dyeing rate constant and equilibrium of disperse dye on aramid fiber were increased with the carrier Cindye Dnk, and the half-dyeing time was shorten as well. It means that with carrier Cindye Dnk (130℃), the dyeing rate and dye uptake at balance increase obviously. Dyeing rate constant and equilibrium adsorption were increased by 165.58% and 49.08% respectively, and half-dyeing time is shorten by 71.89%. In addition to fiber radius, the adsorption rate of disperse also has relations with diffusion coefficient of disperse dye on the aramid fiber.

3.2.2 Diffusion coefficient

As kinetics parameter of dye onto fiber, diffusion coefficient can be used to evaluate diffusion behavior of dye onto the fiber and the effect of dyeing conditions on diffusivity of dye-fiber. Basing on Table 1, which is deduced from formula (1), $Dt/a^2 = 6.292 \times 10^{-2}$, when C_t/C_∞ is 0.5, $t=t_{1/2}$, the average radius of aramid fiber a measured by optical microscope is 15.977 μm, According to Table 1, $t_{1/2}$ without carrier at 130℃ equals to

19.53 min and $t_{1/2}$ with carrier Cindye Dnk equals to 5.49 min. The results are shown in Table 2.

Table 2 Diffusion coefficient of C.I. disperse yellow 211 on aramid fiber

Dyeing methods	Diffusion coefficient/*10^{-13}
Without carrier	8.218
With carrier Cindye Dnk	29.23

We know from Table 2 that the diffusion coefficient of aramid fiber dyeing with disperse dye and carrier are higher than that of without carrier. It indicates that carrier Cindye Dnk enhances the spread power of disperse dye. And it makes the dye penetrate and spread into the aramid fiber, thus improving the dye uptake. Once the disperse dye shifts into the aramid fiber, the surface of the fiber will be vacated by some dyeing block. It makes more disperse dyes from the dye bath onto the aramid fiber and more dyestuff will be accommodated inside the fiber, thus further increasing the dyeing rate. Because diffusion coefficient mainly depends on the dye dispersion and the viscous block layer thickness of the fiber surface, the dispersing action of carrier Cindye Dnk makes the dye aggregate, breaking down into uniform dispersions, thus increasing the dispersing and penetrating speed, and diffusion coefficient of the dye. The greater of the diffusion coefficient is, the faster of the adsorption rate is. In terms of fiber, treatment with carrier will swell the fiber, which makes the resistance of dye diffusing into the fiber smaller, and adsorption speed faster. In the terms of disperse dye, treatment with carrier can reduce the particle size and aggregation of disperse dye. It can also increase the content of single-molecule disperse dye in the dye liquor and obviously reduce the limit of dyeing speed by disperse dye molecule in dye liquor when disperse dyes get into the aramid fiber[13]. So carrier Cindye Dnk can make the dyeing rate of disperse dye on aramid fiber faster, the dyeing rate constant bigger and half-dyeing time shorter.

4 Conclusions

(1) The carrier Cindye Dnk can increase the dyeing rate constant and diffusion coefficient of disperse dye onto the aramid fiber, and shorten the half-dyeing time.

(2) The adsorption isotherm type of aramid fiber dyed with disperse dye at low concentration was not affected by the Cindye Dnk. Its adsorption isotherm type was generally consistent with Nernst-type adsorption.

(3) The carrier Cindye Dnk can increase the distribution coefficient of disperse dye between aramid fiber and dye bath, dyeing affinity and enthalpy of disperse dye to aramid fiber, while reduce the dyeing entropy.

References
[1] Nobuhiko Kuroki. Chemistry of Dyeing Theory (Book II) [M].Beijing: Textile Industry Press, 1981: 196-216.
[2] JIN Xianrang. Experiment on Dyeing and Finishing [M]. Beijing: Textile Industry Press, 2002: 66-91.
[3] LI Yong, ZHU Quan. The study on dyeing properties of aramid microfiber with disperse dye [J].Printing and Dyeing, 2005(7): 5-7.
[4] SONG Xinyuan, ZHAO Tao, SHEN Liru. Dyeing properties of polyester microfiber and theoretical studies [J].Journal of China Textile University, 1997, 23(3): 1-7.
[5] JIN Xianrang. Experiment on Dyeing and Finishing [M]. Beijing: China Textile Press, 2000: 86-87.
[6] Translated by SHI Xiaoli. The dyeing kinetics and thermodynamic study of disperse dye on aggregation fiber [J].Silk Textile Technology Overseas, 2002, (6): 13-15.
[7] M.R.Fox, W.J.Marshall, N.D. steward; Padding, Drying, Steaming and Baking stages in the Application of Dyes to Cellulosic Fibers and their Blens with Polyester Fibres.J.S.D.C.,1967, 83: 493.
[8] F.J.Wortmann. Pathways for dye diffusion in wool fibers. Text. Res.J., 1997, 67(10): 720-724.
[9] J Chgarra, P Puente et al.Kinetic aspects of dye addition in continuous integration dyeing.JSDC,1989,105(10): 349-355.
[10] G.Alberghina.Diffusion kinetics of direct dyes into cotton fibre. Colourage, 1987, 34(14): 19-23.
[11] TANG Rencheng, XU Sufang. Dyeing kinetics and thermodynamic of PTT fiber [J].Printing and Dyeing, 2006, (16): 1-5.

80 Energy Savings via Fast Drying in Textile Industry Part I: Theoretical Discussion and Technical Strategies

Jiping Wang[1], Jinqiang Liu[2]

(1. The Procter & Gamble Company, Cincinnati, Ohio, USA
2. Zhejiang Sci-Tech University, Hangzhou, China)

Abstract: The total energy consumption of fabric drying for producing 70 billion kg per year of textiles in the global textile industry is about 313 billion kWh or 39 million metric tons of coal. Even with 10% energy savings from the fabric drying at the textile industry will create direct savings of 31 billion kWh, or 4 million metric tons of coal or US$ 520 million per year. Intention in this paper is utilizing chemistry approaches based on interactions between textile fibers/fabrics and processing liquid, i.e. water to improve fabric drying effectiveness, leading to energy savings. Scientific bases and technical strategies are discussed for energy savings via fast drying in textile industry.

Keywords: Fabric drying; Water-fiber interaction; Fiber swelling; Centrifuging; Energy saving

1 Introduction

According to Textile Pipeline published by PCI Fiber, the global textile fiber consumption reaches 70 billion kg in 2010 [1]. In order to produce enough textiles, apparel, carpet, etc. to meet global consumer needs, amount of energy consumed in the global textile and apparel industry is gigantic and hard to imagine. Based on an estimate from Textile World Asia [2], the total energy consumption for producing 70 billion kg of textiles is 1253 billion kWh or 154 million metric tons of coal. Using a 100% cotton shirt as an example, about half of the total energy consumed in the textile industry is for wet processing such as dyeing and finishing which including fabric drying process. Fabric drying is a necessity for the textile wet processing. About 50% of energy consumption in the textile wet processing is for fabric drying. Therefore, the total energy consumption of fabric drying for producing 70 billion kg of textiles in the global textile industry is about 313 billion kWh or 39 million metric tons of coal.

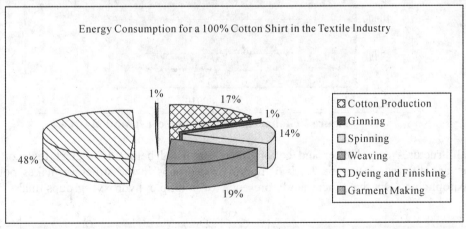

Figure 1 Energy consumption for a 100% cotton shirt in the textile industry

With global sustainability touching every corner of consumer life, business opportunities and environmental impact for saving energy in fabric drying at the textile mills becomes much bigger and more attractive to companies involved in whole textile life cycles from fiber manufacturers, textile mills, apparel companies, carpet producers, textile chemical suppliers and even fabric care consumer product companies. Even with 10% energy savings from the fabric drying at the textile industry it will create direct savings of 31 billion kWh, or 4 million metric tons of coal or US$ 520 million per year assuming 1 metric tons of coal to cost US$130. In addition to direct energy and economic savings, the reduction of the carbon footprint will have huge positive impacts on our communities and global environment.

Many factors impact fabric drying effectiveness in the textile industry, including dryer configurations, energy supply sources, air flow/heat distributions, fiber/fabric types and constructions, exhausting and circulating systems, and heat recovery settings, and many more. Intention in this paper is utilizing chemistry approaches based on interactions between textile fibers/fabrics and processing liquid, i.e. water to improve fabric drying effectiveness, leading to energy savings. Discussed in this paper are scientific bases and technical strategies for energy savings via fast drying in textile industry. The technical feasibility study with right evaluation methods and chemistry will follow.

2 Textile fiber-water interaction (bound water vs. liquid water)

There are many different textile fibers used for making textiles and apparel in the textile and apparel industry, including natural fibers such as cotton, silk, wool, linen, etc. and man-made fibers such as rayon, polyester, nylon, polyacrylic, polyolefin, etc. Based on global textile mill consumptions, polyester fiber market share in all textile fiber applications reached 45% in 2009. It will increase to 47% in 2012. On the other hand, cotton's share will continue to drop from 36% in 2007 to 34% in 2012. Nylon had a share of 5.2% in 2009, Rayon had a share of 3.6% and wool had a share of 1.7%. All of them are expected to reduce their shares in the near future. It is obvious, polyester and cotton are two major textile fibers now and in the foreseeable future. They are textile fibers for the focused discussion in this paper.

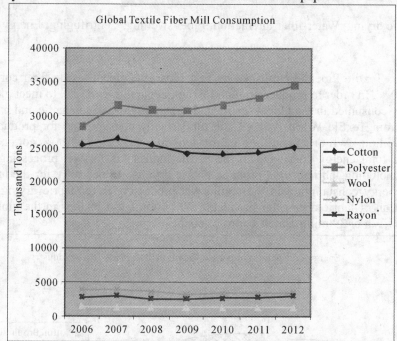

Figure 2 Global textile fiber mill consumption

Chemical structures of polyester and cotton fibers are listed below. It is clear that cotton fiber and polyester fiber are different in nature. Lack of hydrophilic groups in the structure makes polyester fibers extremely hydrophobic. On the other hand, presence of multiple hydroxyl groups makes cotton very hydrophilic.

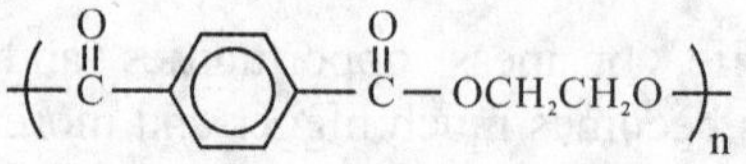

Figure 3 Polyester 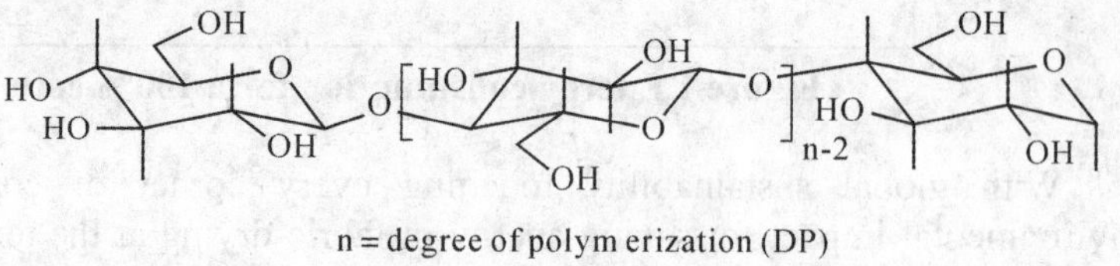　　　　　　　　　　　　　　**Figure 4 Cotton**

n = degree of polymerization (DP)

Cotton, a cellulose fiber is hydrophilic and swells in the presence of water. Water penetrates to inside of cotton fibers, leading to 40-50% of cross-sectional area increase. Since axial swelling increase is very limited for cotton fibers, cotton's total volume normally increase 40-50% with a full saturation with water. Under normal conditions, cellulose-water interactions are considered to occur either in intercrystalline regions (amorphous area) or on the surface of the crystallites. Since water can not penetrate into cotton's crystalline area and the crystalline degree of cotton is about 70%, cellulose-water interaction only happens in about 30% of cellulose molecules inside of cotton fibers. For mercerized cotton, the water accessibility could increase from 30% to 50% [3].

It is well known that at low moisture uptakes, the water associated with the cellulose exhibits properties that differ from those of liquid water and it has been called by 'bound water' or 'hydrate water'. During drying, it requires more energy to separate the bound water from cellulose than the liquid water. From chemistry stand point, the bound water is water molecule that forms direct or the first layer hydrogen bond with a cellulose molecule. Since each cellulose repeat unit has 5 oxygen atoms which may form direct hydrogen bond with 5 water molecules, the ratio of the bound water can be theoretically calculated as below:

$$\% \text{ Bound Water} = \frac{MW_{H2O} \times 5}{MW_{Cell\ unit}} \times \text{H2O Accessible Area} \times 100 = \frac{18 \times 5}{161} \times 0.3 \times 100 = \mathbf{16.8}$$

From a review of the literature, which included determinations by NMR and calorimetry, between 0.10 and 0.20 g/g of the water present in the cellulose fiber appeared to be the bound water [4].

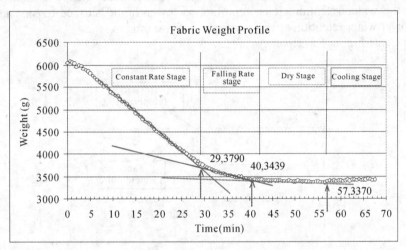

Figure 5 Fabric weight profile

According to P&G internal work, about 0.15 g of water per g of fiber has significantly slower drying rate than other water present in cotton fabrics, indicating about 15% of water is the bound water based on weight of the cotton fabric.

Therefore, theoretical calculation, chemical measurements and actual drying measurement are surprisingly consistent. Amount of the bound water inside of cotton fibers is about 15% of the fiber weight. Since water content of cotton fabrics from regular spinning or padding is about 80-100%, there is about 65% to 85% of the liquid water left on fabrics before drying stage at the textile mills.

Polyester fibers by nature are very hydrophobic with very limited fiber swelling in presence of water. Therefore, the bound water in polyester fibers is very limited. The water content of polyester fabrics is dominated by the liquid water

How to reduce amount of the bound water inside of cotton fibers could be one of the key approaches to increase drying speed for cotton fabrics. Based on the chart above, the drying time can be shorten to <u>35 minutes from 55 minutes, a 36% of drying time saving</u> if we can convert all bound water to the liquid water assuming a same water content. Disruption of the first layer of H-bonding with cellulose fibers could be a right strategy to reduce or to release amount of bound water inside of cotton fibers. Limit water accessibility to hydroxyl groups in cellulose fibers by cross-linking, hydrophobic blocking, etc. could be another effective strategy to reduce amount of bound and liquid water inside of cotton fibers in order to speed up drying process.

3 Centrifuging of wet fabrics

Amount of water carried into dryer by wet fabrics is the most important factor to determine drying time and energy spending in textile mills. According to literature review [5], the water retention of fabrics after centrifuging follows the equation below:

$$\text{Water Retention} = A \frac{\gamma \cos\theta}{gh(\mu\rho)^{1/2}}$$

γ: liquid surface tension
θ: contact angle
μ: fiber mass/unit length
ρ: fiber density
g: centrifugal acceleration
h: thickness of total fabrics

In the equation, A is a constant for a given fabrics. The centrifugal acceleration (g) is mainly determined by the spinning rate and machine design. The fiber linear density and the fiber density are mainly determined by fiber types. The thickness of total fabric mass parallel to the centrifugal field (h) is mainly controlled fabric constructions and processing condition. The liquid surface tension and the contact angle are two major factors which could be controlled by adding other agents chemically or by altering interactions between fibers and liquid. It is well known that the liquid surface tension has some impacts on the contact angle of the liquid on the fabric surface so the liquid surface tension and the contact angle are not independent in terms of their influence on the water retention. It is important to lower liquid surface tension and to increase the contact angle for lower water retention after centrifuging of wet fabrics.

Strategies for High Centrifuging Effectiveness are:

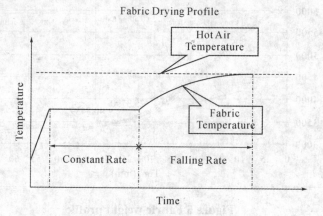

Figure 6 Fabric drying profile

- Reduce water surface tension by surfactants with superior lower surface energy such as silicone-containing and fluoro-containing surfactants.
- Reduce fabric surface energy for larger contact angle. The best technology will be those surfactant mixtures which have hydrophobic component for lowering fabric surface energy by depositing on fabrics and have hydrophilic component for staying in the liquid to lower down water surface tension.
- Since the second strategy may lead to poor wicking, it will be great if the hydrophobic component deposited on fabrics can be evaporated during drying.

4 Drying process – mass and heat transfer

Drying process is a simultaneous heat and mass transfer operation in which the energy and evaporation of a liquid from a solid can be described in the following empirical graph [6]. In the first period, wet fabrics are heated to a temperature which can support equilibrium between heat supply from hot air and heat transfer from water evaporation. At this temperature, the liquid water on the fiber surface and inside of fibers evaporates at a constant rate until the drying reaches the bound water. Since the bound water needs more energy to separate them from fibers, temperature on fabrics begins to increases and drying rate begins to decrease. When all water removes from fabrics, fabric temperature will equal to the hot air temperature.

Technically speaking, the process of drying with the hot air consists of putting energy into the fabric as heat transfer and removing the water as it vaporizes as mass transfer. According to a recent paper titled 'Mathematical Modeling of Combined Diffusion of Heat and Mass Transfer Through Fabrics'[7], the rate of heat transfer can be expressed in the equation below:

$$R_{HT} = h\,A_s\,(T_A - T_F)$$

R_{HT}: rate of heat transfer
h: heat transfer coefficient
A_s: fiber surface area the hot air can touch
T_A: temperature of the hot air
T_F: Temperature of the fabric surface

The rate of mass transfer can be expressed in the equation below:

$$R_{MT} = k\,A_s\,(C_S - C_A)$$

R_{MT}: rate of mass transfer
k: mass transfer coefficient
A_s: fiber surface area the hot air can touch
C_S: water vapor concentration of fabric surface
C_A: water vapor concentration of the hot air

The heat transfer coefficient (h) and the mass transfer coefficient (k) are mainly determined by velocity of the air flow during drying, fiber and fabric characters, and dryer design. For a given dryer system and a given fabric bundle, h and k are constants. T_A, T_F, C_S and C_A are mainly controlled by drying conditions. In order to increase the heat transfer and mass transfer, what chemistry can impact is the fiber surface area the hot air can touch, which are presented in both equations.

Strategy for Increasing Fiber Surface Area the Hot Air Can Touch is to increase fabric fluffiness by static, lubrication etc. to enlarge fiber surface area the hot air can touch to speed up rate of heat transfer and mass transfer during drying. This strategy may be particularly effective at the last drying period with falling drying rate.

5 Conclusion

Based on the theoretical discussions on the fiber-water interaction (bound water vs. liquid water), the centrifuging effectiveness, and drying process in mass and heat transfers, it is feasible to speed up fabric drying and to save drying energy without changing drying machines at textile mills. By controlling amount of bound water, by increasing centrifuging effectiveness via lowering liquid surface tension and raising liquid-fabric contact angle, and by increasing fiber surface area the hot air can touch, a right chemistry approach should be able to increase fabric drying speed, leading to energy savings. Even with 10% energy savings from the fabric drying at the textile industry which is a conservative estimate will create direct savings of 31 billion kWh of energy, or 4 million metric tons of coal or US$ 520 million per year.

References

[1] Textile Pipeline Special, PCI Fibers, June 2009.
[2] Rupp, J., Ecology And Economy in Textile Finishing, Textile World Asia, March, 2009.
[3] Bertoniere, N. R. and Zeronian, S. H., Chemical characterization of cellulose, in The Structures of Cellulose, Atalla, R. H., Ed., ACS Symposium Series, No. 340, American Chemical Society, Washington, DC, 1987, p255.
[4] Zeronian, S. H., Intercrystalline swelling of cellulose, in Cellulose Chemistry and its Applications, Nevell, T. P. and Zeronian S. H., Eds., Ellis Horwood Ltd., Chichester, England and Halsted Press, New York, 1985, chap. 5, p 138-158.
[5] J. R. Kuppers, Text. Res. J., 1961, 31, 490.
[6] H. Hardisty, J. Oil Col. Chem. Assoc., 60, 479 (1977).
[7] B. Etemoglu, Y. Ulcay, M. Can, and A. Avci, Fibers and Polymers, 2009, 10, 252.

81 Utilization of Enzyme in the Degumming Process of Jute Fibers

Weiming Wang[1], Zaisheng Cai[2,3], Jianyong Yu[4]

(1. College of Textile & Apparel Engineering, Shaoxing University, Shaoxing 312000, China
2. College of Chemistry, Chemical Engineering & Biotechnology, Donghua University, Shanghai 201620, China
3. Key Laboratory of Science & Technology of Eco-Textile, Ministry of Education, Donghua University, Shanghai 201620, China
4. Modern Textile Institute, Donghua University, Shanghai 200051, China)

Abstract: The jute fibers were treated with pectinase, cellulase, xylanase and their mixtures with the aim to study their effectiveness as degumming agents. The xylanse and pectinase had a good synergistic effect when their mixture was applied in the degumming process of jute fiber. And the optimum conditions for degumming were established as follows: xylanase concentration 0.8%, pectinase concentration 0.8%, treatment time 180min, pH value 8.0 and temperature 55^{0}C. Furthermore, the SEM analysis showed that the removal of impurities on the jute fiber surface could be remarkably improved when enzyme was applied during the traditional chemical degumming.

Keywords: Jute fibers; Degumming; Bio-degumming; Enzymes

1 Introduction

Jute is a vegetable fiber derived from the bark of an annual plant grown mainly in the subtropical areas of Asia, which occupies the second place in terms of world production levels of cellulosic fibers[1-2]. Jute fiber is basically used for traditional purposes such as making sackings, hessian, carpet backing and ropes due to its coarseness and stiffness[3]. Recently, the demand of natural biodegradable and eco-friendly fibers is rising worldwide day by day. So scientists all over the world started working on this golden fiber and developed some diversified products like furnishing and apparel textiles. But there is a series of wet chemical processes namely retting, degumming and bleaching are needed to improve its spinnability[4-7]. The qualities of the fiber and yarn mostly depend on the degumming effect. So degumming is one of the most important pretreatments of jute fiber for apparel textile.

Generally, there are three methods for degumming namely mechanical, chemical and biological methods[8]. Enzymes play a main role in degumming of bast fibrous plants and bast fibers. But the appropriate selection is the major problem in enzyme utilization[9]. In this work, the role of cellulase, pectinase, xylanase and their mixtures for degumming of jute fibers were investigated. Additionally, the optimal bio-degumming conditions and the combined chemical-biochemical processes were discussed.

2 Materials and methods

2.1 Materials

The jute fiber was obtained from Jiangsu Redbud Textile Technology Co., Ltd, China. Cellulase, pectinase and xylanase were kindly provided by Novozymes. Sodium hydroxide, sulphuric acid, hydrogen peroxide and sodium sulfite were purchased from Shanghai Ruiteliang Chemical Industry Co., Ltd, China. Sodium silicate and penetrate agent JFC were purchased from Shanghai Nuotai Chemical Co., Ltd, China.

2.2 Methods

2.2.1 Bio-degumming

The jute fibers were treated with cellulase, pectinase, xylanase and their mixtures in 250ml flasks. After vibrating at 55^{0}C for 90min, the fibers were washed and dried in an oven.

2.2.2 Chemical degumming

The jute fibers were treated in the prepared solution with sodium hydroxide 16g/L, sodium silicate 3g/L, sodium sulfite 4g/L, penetrate agent JFC 2g/L, and kept at 90^{0}C for 120min with fiber-to-liquor ratio 1:20. The treated fibers were neutralized with 2g/L sulphuric acid, and then thoroughly washed.

2.2.3 Hydrogen peroxide bleaching

The degummed jute fibers were bleached using hydrogen peroxide 8g/L, sodium silicate 3g/L, penetrate agent JFC 1g/L, pH value was 10.5, and kept at 90^{0}C for 90min with fiber-to-liquor ratio 1:20. The bleached jute was neutralized with 2g/L sulphuric acid, and then thoroughly washed.

2.3 Testing
2.3.1 Gum removal
The gum content of jute fiber was tested according to GB/ 5889-86, and the gum removal was calculated using Eq.1:

$$\text{Gum Removal}(\%) = \frac{M_0 - M_1}{M_0} \times 100\% \qquad (1)$$

where M_0 is the gum content of the raw jute fiber, M_1 is the residual gum content of the degummed jute fiber.

2.3.2 Breaking tenacity
The raw jute fibers and degummed jute fibers were kept in a constant 20 ±2 °C temperature and 60 ± 3% relative humidity room for 24h. The fineness was tested according to GB/T 12411.3-90 and the breaking strength was examined according to GB/T 12411.2-90. Then the breaking tenacity can be calculated by Eq.2:

$$BT = \frac{BS}{F} \qquad (2)$$

where BT is breaking tenacity, cN/dtex; BS is breaking strength, cN; F is fineness, Nm.

2.3.3 SEM Analysis
The morphological structures of the raw jute fibers and degummed jute fibers were observed with a DXS-10ACKT scanning electronic microscope.

3 Results and discussion
3.1 Selection of enzyme
Substrate is one of the most important factors in the case of enzymatic hydrolysis[5]. Since the complex compositions of gum in the jute fiber, the selection of appropriate enzyme is necessary. The effect of different enzyme i.e. cellulase, pectinase, xylanase and their mixtures on the degumming efficiency are shown in Table 1.

For the gum removal, the results in Table 1 showed that the effect of single enzyme followed the descending order: cellulase > pectinase > xylanase. Table 1 also indicated that there were good synergistic effects between xylanase and pectinase on the gum removal of jute fibers. The gum removal got an optimal value when the mixture of xylanase and pectinase was applied, and there was no additional effect on the breaking tenacity in comparison with other mixtures of enzymes. Tenkanen et al[10] have shown that the lignin linking with xylan could be removed by xylanase, which confirmed the strict substrate selection of enzyme. So the mixture of xylanase and pectinase was the appropriate selection for the degumming agent of jute fibers.

Table 1 Effect of enzyme on the refining efficiency

No.	Enzyme	Gum removal (%)	BT (cN/dtex)
1	cellulase	23.63	2.65
2	xylanase	16.71	2.82
3	pectinase	20.75	2.56
4	Cellulase + xylanase	23.34	2.75
5	Cellulase + pectinase	26.02	3.35
6	Xylanase + pectinase	29.28	3.45
7	Cellulase + xylanase + pectinase	27.24	3.32

3.2 Effect of enzyme concentration on the degumming efficiency
The concentration is an important factor in the enzyme degumming. The effect of enzyme concentration on the gum removal and BT were tested, and the results are shown in Table 2.

Table 2 Effect of enzyme concentration on the degumming efficiency

No.	Xylanase (%)	Pectinase (%)	Gum removal (%)	BT (cN/dtex)
1	0.6	0.4	22.53	3.29
2	0.6	0.6	24.34	3.21
3	0.6	0.8	25.20	3.18
4	0.8	0.4	25.24	3.18
5	0.8	0.6	28.39	3.09
6	0.8	0.8	29.12	3.04
7	1.0	0.4	25.87	2.94
8	1.0	0.6	28.75	2.89
9	1.0	0.8	29.91	2.90

The results in Table 2 showed that the gum removal value increased with the increase of pectinase concentration when the concentration of xylanase was kept at a constant value, and the same trend of effect of xylanase concentration was observed. At the same time, Table 1 indicated that there was an inverse effect of enzyme concentration on the breaking tenacity. It could be explained that the removal of lignin and hemicellulose could reduce the breaking tenacity, which had been observed by Mukhejee et al[11]. Taking into account gum removal and breaking tenacity, the optimum enzyme concentration was xylanase concentration 0.8% and pectinase concentration 0.8%.

3.3 Effect of treatment time on the bio-degumming efficiency

The results in Figure 1 indicated that the gum removal increased when the treatment time increased in the range of 0~180min. When the treatment time further increased, there is no positive effect on the gum removal. It can be also clearly seen form Figure 1 that the breaking tenacity decreased with the treatment time decreased, but there was a little change of breaking tenacity when the treatment time further increased after 120min. It may be explained that the reaction between the enzymes and the substrates had been completed in 180 min, and there would be little benifit gained by further prolonging the time. So our conclusion was that the optimal treatment time was 180min.

3.4 Effect of pH value on the bio-degumming efficiency

Figure 2 showed the effect of pH value on the bio-degumming efficiency. The results in Figure 2 showed that both the gum removal and the breaking tenacity were remarkably affected by the pH value. The considerable value of the gum removal was obtained when the pH value was at 8. Although the breaking tenacity was seriously affected, but still was more than 3.0cN/dtex. The above discussions confirmed that the pH value was a major important factor on the enzyme activity. So the optimum pH value was 8.0.

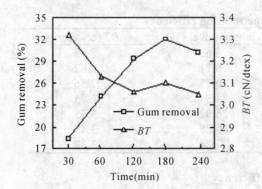
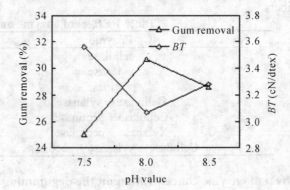

Figure 1 Effect of treatment time on the bio-degumming efficiency

Figure 2 Effect of pH value on the bio-degumming efficiency

3.5 Effect of temperature on the bio-degumming efficiency

Temperature is another important effecting factor for enzymatical treatment. The effect of temperature on the degumming efficiency was presented in Figure 3.

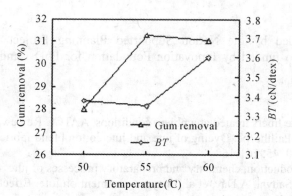

Figure 3 Effect of treatment temperature on bio-degumming efficiency

It can be seen from Figure 3 that the gum removal rapidly increased when the temperature increased from 50°C to 55°C, then there was a little decrease when the temperature further increased. However, there was an inverse effect of temperature on the breaking tenacity. Therefore, the appropriate utilization temperature for the mixture of xylanase and pectinase was 55°C.

3.6 SEM analysis

The main purpose of degumming is to remove non-cellulose materials from the fiber. So the morphological structures of the raw and degummed jute fibers were observed by SEM. The results are shown in Figure 4.

It can be seen from Figure 4 that there were many visible grooves along the vertical section for all samples. The surface of the raw jute fiber was rough and covered by a number of impurities like hemicellulose, pectin, waxy substances and so on. In comparison with the raw jute fiber, the degummed samples showed a smooth and clean surface. Although most of the gum had been removed by chemical degumming and bleaching treatments, the surface was still covered by a layer of compact gum. The morphology of single jute fiber could be clearly observed when the jute fiber was treated by the combined chemical and biochemical method, which indicated that the combined chemical and biochemical method was an effective method in the removal of non-cellulosic materials.

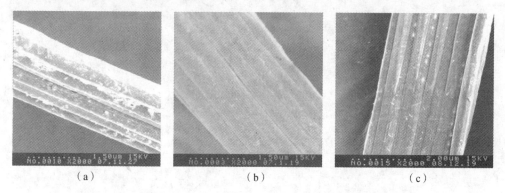

**Figure 4 SEM micrographs of (*a*) raw jute; (*b*) chemical degumming+bleaching;
(*c*) chemical deumming+bio-degumming+bleaching**

4 Conclusion

Since the complex compositions of the non-cellulosic materials of jute fibers, the pectinase, cellulase, xylanase and their mixtures were applied to treat jute fiber with the aim to study their effectiveness as degumming agent. It was found that the mixture of xylanase and pectinase was an effective degumming agent for jute fiber. After the desired experiments, the optimum conditions for degumming were established as follows: xylanase concentration 0.8%, pectinase concentration 0.8%, treatment time 180min, pH value 8.0 and temperature 55°C. Additionally, the morphological structures of the raw jute fiber and the degummed jute fiber were analysised by SEM, the results showed that the combined chemical and biochemical method was an effective method in the removal of impurities of the jute fiber.

Acknowledgement

This research is funded by the Nation Supported Planning Project of Science and Technology (2007BAE41B00). It is also funded by Innovation Foundation for PhD Candidate of Donghua University (DHU) (105-06-0019057).

References

[1] J J Moses. A study of jute fibers treated at ambient conditions. AATCC Review, 2001, 5: 34-37
[2] Y Cai, S K David, M T Pailthorpe. Dyeing of jute and jute/cotton blend fabrics with 2:1 pre-metallised dyes. Dyes and Pigments, 2000, 45: 161-168
[3] D P Chattopadhyay. Introduction, chemistry and preparatory processes of jute. Colourage, 1998, 5: 23-35
[4] S N Chattopadyay, S K Sanyal, A Day, et al. Enzyme treatment on jute: Effect of resin treatment. Colourage, 2001, 5: 20-24
[5] S N Chattopadyay, S K Sanyal, A Day, et al. Enzyme treatment on jute: Effect of preatment. Colourage, 2001, 7: 20-24
[6] G Basu, S S De, A K Samanta. Effect of bio-friendly conditioning agents on jute fiber spinning. Industrial Crops and Products, 2009, 29: 281-288
[7] W M Wang, Z S Cai, J Y Yu. Study on the chemical modification process of jute fiber. Journal of Engineered Fibers and Fabrics, 2008, 3(2): 1-11
[8] Z T Liu, Y N Yang, L L Zhang, et al. Study on the performances of ramie fiber modified with ethylenediamine. Carbohydrate Polymers, 2007, 71: 18-25
[9] R Kozlowski, J Batog, W Konczewicz, et al. Enzyme in bast fibrous plant processing. Biotechnology Letters, 2006, 28(10): 761-765
[10] L S Zheng, Y M Du, J Y Zhang. Degumming of ramie fibers by alkalophilic bacteria and their polysaccharide-degrading enzymes[J]. Bioresource Technology, 2001, 78: 89-94
[11] A Mukherjee, P K Ganguly, D Sur. Structure mechanics of jute: The effect of hemicellulose or lignin removal[J]. Journal of the Textile Research, 1993, 84(3): 348-353

82 Kinetics of Ultrasonic Dyeing of PTT Fiber with Emodin

Xueni Hou, Xiangrong Wang

(National Engineering Laboratory for Modern Silk, Soochow University, Suzhou 215123, China)

Abstract: Comparative dyeing kinetics of poly (trimethylene terephthalate) fiber with emodin natural dyes using conventional and ultrasonic methods were studied in this paper. The plot of time/dye-uptake of both methods showed that the dye-uptake values of ultrasound dyed samples were obviously higher than those prepared by the conventional heating method. The values of dyeing rate constant, diffusion coefficient and half–time of dyeing and activation energy were calculated. The results showed that ultrasonic could evidently enhance the rate constant of dye progress, shorten half-dyeing time, increase the diffusion coefficients and decrease the activation energy, which positively indicated a favorable effect of ultrasonic power on the dyeing process.

Keywords: Dyeing kinetics; Ultrasound dyeing; Emodin; PTT fiber

1 Introduction

Historically natural dyes were produced from plants and animals. Nowadays, many kinds of natural dyes have been registered as food additives from plant, animal, and mineral substance. The major source of natural dyes is plants [1]. Recently a revival interest in the use of natural dyes in textile coloration has been growing. This is a result of the stringent environmental standards imposed by many countries in response to the toxic and allergic reactions associated with synthetic dyes. Natural dyes are friendly to the environment, relatively low toxic and can exhibit better biodegradability and various natural coloring sources [2, 3]. In the past decade, investigations about possible uses of natural dyes in textile dyeing processes have been performed by various research groups. However, most research on natural dyes was focused on silk, wool and cotton fibers [4-6]. Along with the development of synthetic fiber in multiple, top-grade and functional level, it has a broad prospect to dye synthetic fiber with natural dyes. In this article, the emodin was chosen to dye PTT fiber. Emodin (1, 3, 8-trihydroxy-6-methylanthraquinone) is a major constituent of rhubarb [7]. Emodin is a natural yellow dyestuff extracted from the root and rhizome of polygonum cuspidatum and Rheum palmatum [8]. Figure 1 shows the structure of emodin, which is used extensively in the medical field and in cosmetics, as well as in textile dyeing.

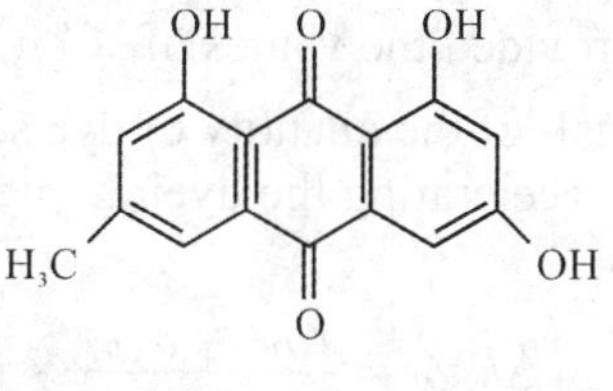

Figure 1 Structure of emodin

The use of ultrasound energy for dyeing textiles as a cleaner production is very well known in the literature [9-13]. Ultrasound produces chemical effects through several different physical mechanisms and the most important nonlinear acoustic process for sonochemistry is cavitation [14]. In textile process, ultrasound can offer a potential benefits for shortening process time, reducing pollution load and getting improvement in product quality [15]. Due to the poor thermal stability of the natural dye, ultrasonic dyeing also can avoid heat decomposition of natural dyes at high temperature.

In this paper, we present a study on the kinetics of PTT fiber dyeing with emodin natural dyes using both conventional and ultrasonic conditions. The dye uptake, dyeing rate, time of half dyeing, diffusion coefficient and activation energy were measured.

2 Experimental
2.1 Materials

Emodin dye used in this work was purchased from Weiwei institute of Phytochemistry, Jiangsu, China. Scoured PTT fiber (83.3dtex/36f) was supplied by Fineyarn Corporation, Jiangsu, China. Before using, the fabric was treated with a solution containing 1g/L of scouring agent, at 95℃ for 60min, and then the fabrics were thoroughly rinsed with water and air dried at room temperature.

2.2 Methods

2.2.1 Dyeing rate of PTT fiber using conventional and ultrasonic method

PTT fabric samples (0.5g each) were dyed with the natural dye emodin at liquor ratio 1:100. Dyeing was carried out at 3% concentrations of dye, pH values was 5, at the temperature of 65℃, 70℃ and 75℃, and duration time (5–90 min) as detailed in the paper. The same conditions of dyeing were carried out using ultrasonic dyeing (US) method with power at 600W. Then the dyed samples were rinsed with cold water and air dried.

2.2.2 Adsorption amount of the dye on fiber

At the end of dyeing, the samples were completely washed with cold water and dried in a oven. The dried samples were weighed and then dyes were extracted using N, N-dimethylformamide at 95℃ until the samples became colorless. The dye concentration in the extracts was measured using spectrophotometer and the extent of dye adsorption was calculated.

2.2.3 Diffusion coefficient

The dye diffusion coefficient was measured by dye-uptake, according Eq. (1),

$$\frac{dC_t}{d_t} = k(C_\infty - C_t)^2 \qquad (1)$$

Where C_t and C_∞ were the concentration of dyes on the fiber at a time equal to t and equilibrium, respectively. K was the rate constant of dye sorption. Taking the deformation of Eq. (1) would lead to Eq. (2).

$$\frac{t}{C_t} = \frac{1}{C_\infty} \cdot t + \frac{1}{C_\infty^2 \cdot k} \qquad (2)$$

Where C_t can be measured from the experiment, C_∞ can be calculated from Eq. (2).

This experiment condition was limited dye bath and unsteady diffusion. The diffusion coefficients (D) was obtained from the relation of M_t/M_∞ and Dt/r^2 (The relation table can be found from the monograph). As a result of $M_t/M_\infty = C_t/C_\infty$, the value of Dt/r^2 can be found in the corresponding database table [16]. The radius of PTT fiber which directly measured by microscope was 2.916 μm.

2.2.4 Dyeing rate constant and half-dyeing time

Eq.2 described the linear dependence of the function $t/C_t = f(t)$, the plots of t/C_t exhibited good linearity, whose slope and intercept provided the values of C_∞, k and half dyeing time ($t_{1/2}$), which was the time required for the fiber to take up half of the quantity of dye sorption at equilibrium.

Ultrasonic efficiency (△k%) in accelerating the dyeing rate was examined by introducing the following equation:

$$\Delta k\% = \frac{k_{US} - k_{CH}}{k_{CH}} \times 100 \qquad (3)$$

where k_{US} and k_{CH} were the rate constant of dyeing with ultrasonic and conventional heating, respectively.

2.2.5 Activation energy of the diffusion

Eq. (3) was known as the Arrhenius Equation. Using Eq. (4), the activation energy of the diffusion was calculated from the relationship between $\ln D_T$ and $1/T$

$$\ln D_T = \ln D_0 - E/RT \qquad (4)$$

D_T, diffusion coefficient at a temperature T; D_0, constant; E, activation energy; R, gas constant (8.3145J · mol^{-1} · K^{-1}); and T, absolute temperature (K).

3 Results and discussions

3.1 Dyeing rate

The dyeing rate was obtained at 65℃, 70℃ and 75℃. As shown in Figure 2, the concentration of dyes on the fiber obtained increased as the time increased in both ultrasonic (US) and conventional (CH) dyeing methods with much higher dye uptake at all points in the ultrasonic case. This could be explained by the cavitation effect, power ultrasound can enhance a wide variety of chemical and physical processes, mainly due to the cavitation effect in liquid mediums (that is the growth and explosive collapse of microscopic bubbles, which causes a large increase in the pressure and temperature) that enhances the rate of dyeing process.

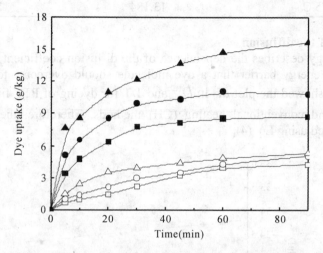

Figure 2 Ultrasonic and conventional dyeing rates of PTT fiber
Dyeing condition: 600W, LR 100:1, owf=3%, pH=5, ultrasonic: at 65℃(■), 70℃(●) and 75℃(▲);
conventional: at 65℃(□), 70℃(○) and 75℃(△)

3.2 Dyeing rate constant and times of half dyeing

The data in Figure 2 could be analysed by using Eq. (1) or Eq. (2). The values of dyeing rate constants k, the diffusion coefficient and half dyeing $t_{1/2}$ were given in Table 1. It can be seen that, the dyeing rate constant of PTT fibers dyed with emodin dye at 65℃, 70℃ and 75℃ clearly increased with ultrasonic dyeing comparison with conventional heating, the values of $t_{1/2}$ of dyeing were obviously less for those samples dyed with ultrasonic than those conventionally dyed. Also, as shown in Table 1, the value of ultrasonic efficiency positively indicated a favorable effect of ultrasonic power on the dyeing process, the final dye uptake of the ultrasonic was much faster than that of conventional dyeing.

Table 1 Dyeing rate constant, efficiency of ultrasonic, time of half dyeing and final dye-uptake by PTT fiber using Emodin dye

Parameter		C_∞ (g/kg)	$t_{1/2}$ (min)	$k \times 10^{-3}$ ($kg/g \cdot min$)	$\triangle k$ (%)
65℃	Conventional	4.2955	26.54	8.7705	27.01
	Ultrasonic	9.6899	9.265	11.139	
70℃	Conventional	4.9456	19.57	10.332	22.91
	Ultrasonic	11.737	6.709	12.699	
75℃	Conventional	6.1125	14.88	10.997	15.37
	Ultrasonic	14.880	5.297	12.687	

3.3 Diffusion coefficient

Ultrasonic also can enhance the diffusion coefficients (Table 2); the dye diffusion coefficient is primarily determined by the dispersion rate of emodin dyes and the thickness of fabric surface boundary layer. Generally, the increase in diffusion coefficients can be explained by fiber swelling, which enhanced dye diffusion. Moreover, the ultrasonic power provided other additional factor of de-aggregation of dye molecules, such as dispersing, degassing and spreading, which led to further enhancement of dye diffusion and better dyeability than conventional dyeing.

Table 2 Diffusion coefficient by PTT fiber using Emodin dye

$D \times 10^{16} m^2/s$	Ultrasonic	Conventional
65℃	8.895	2.876
70℃	11.663	3.699
75℃	13.184	5.231

3.4 Activation energy of the diffusion

The activation energy describes the dependence of the diffusion coefficient on the dyeing temperature and also represents the energy barrier that a dye molecule should overcome to diffuse into the polymer molecules [17]. Figure 3 showed the plot of $\ln D_T$ and 1/T for dyeing of PTT fiber with emodin dye using both ultrasonic (US) and conventional heating (CH) methods. Therefore, the activation energy of the diffusion can be calculated using Eq. (4).

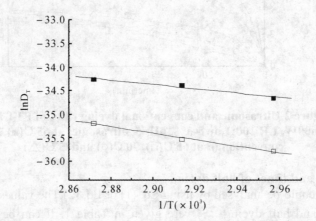

Figure 3 Plots of LnDT versus 1/T of dyed PTT fiber under conventional and ultrasonic condition
Ultrasonic (■), conventional (□)

The activation energy of conventional and ultrasonic were 58.466 J·mol^{-1} and 38.546 J·mol^{-1}, respectively. It can be seen that ultrasound can decrease the activation energy of dye progress with emodin, which indicated that the dyeing was easier under ultrasonic condition. The activation energy with emodin dye on PTT fibre decreased 34.07%. These was mainly attributed to that the ultrasound increased the dispersion of dye and decreased its particle size, which made it easier for dyes entering into the fibre. On the other hand, ultrasound can decrease the crystallinity of fibres and enlarge the amorphous region, which reduced the dyeing power boundary layer and in favor of the diffusion of dye. In a word, the promotion of ultrasound in dyeing was the combination of its effects on fibers and the dye bath.

4 Conclusions

Comparative dyeing kinetics of PTT fiber with emodin natural dyes using conventional heating and ultrasonic conditions were studied in this paper. The time/dye-uptake isotherms of both methods showed that the dye-uptake values of ultrasound dyed samples were obviously higher than those prepared by the conventional heating method.

Ultrasound can evidently enhance the rate constant of dye progress, shorten half-dyeing time, increase the diffusion coefficients. The value of ultrasonic efficiency (△k%) positively indicated a favorable effect of ultrasonic power on the dyeing process.

Because the promotion of ultrasound in dyeing was the combination of its effects on fibers and the dye bath, ultrasound can decrease the activation energy of dye progress with emodin. The activation energy with emodin dye on PTT fibre decreased 34.07%, which indicated that the dyeing was easier under ultrasonic condition.

References
[1] Y Ishigami, S Suzuki. Development of biochemicals-functionalization of biosurfactants and natural dyes.

Progress in Organic Coatings, 1997, 31:51-61.
[2] X X Feng, L L Zhang, J Y Chen, J C Zhang. New insights into solar UV-protective properties of natural dye. Journal of Cleaner Production, 2007, 15:366-372.
[3] T Bechtold, A Turcanu, E G Berger, S Geissler. Natural dyes in modern textile dyehouses — how to combine experiences of two centuries to meet the demands of the future. Journal Cleaner Production, 2003, 11:499-509.
[4] F A Nagia, R S R EL-Mohamedy. Dyeing of wool with natural anthraquinone dyes from Fusarium oxysporum. Dyes and Pigments, 2007, 75:550-555.
[5] R Shanker, P S Vankar. Dyeing cotton, wool and silk with Hibiscus mutabilis (Gulzuba). Dyes and Pigments, 2007, 74:464-469.
[6] A Moiz, M. A Ahmed, N Kausar, K Ahmed, etc. Study the effect of metal ion on wool fabric dyeing with tea as natural dye. Journal of Saudi Chemical Society, 2010, 14: 69–76.
[7] C F Wang, W D Wu, M Chen, W G Duan, etc. Emodin induces apoptosis through caspase3-dependent pathway in HK-2cells. Toxicology, 2007, 231:120–128.
[8] H Z Lee, C J Lin, W H Yang, W C Leung, etc. Aloe-emodin induced DNA damage through generation of reactive oxygen species in human lung carcinoma cells. Cancer Letters, 2006, 239: 55–63.
[9] P S Vankar, R Shanker, J Srivastava. Ultrasonic dyeing of cotton fabric with aqueous extract of Eclipta alba.Dyes and Pigments, 2007, 72:33-37.
[10] M M Kamel, M M El Zawahry, N S E Ahmed, F Abdelghaffar.Ultrasonic dyeing of cationized cotton fabric with natural dye.Part 1: Cationization of cotton using Solfix E. Ultrasonics Sonochemistry, 2009, 16:243-249.
[11] M M Kamel, R M El-Shishtawy, B M Youssef, H Mashaly. Ultrasonic assisted dyeing. IV. Dyeing of cationised cotton with lac natural dye. Dyes and Pigments 2007, 73: 279-284.
[12] M M Kamel, R M El-Shishtawy, B M Yussef, H Mashaly.Ultrasonic assisted dyeing III. Dyeing of wool with lac as a natural dye.Dyes and Pigments, 2005, 65: 103-110.
[13] M M Kamel, H M Helmy, H M Mashaly, H H Kafafy.Ultrasonic assisted dyeing: Dyeing of acrylic fabrics C.I. Astrazon Basic Red 5BL 200%.Ultrasonics Sonochemistry, 2010, 17:92-97.
[14] S Vajnhandl, A M Le. Ultrasound in textile dyeing and the decolouration/mineralization of textile dyes. Dyes and Pigments, 2005, 65:89-101.
[15] D Sun, Q J Guo, X Liu. Investigation into dyeing acceleration efficiency of ultrasound energy.Ultrasonics, 2010, 50:441-446.
[16] H X Wang, Fiber dyeing Science. Taichung, University Book Supply Agency. 1982: 100-107.
[17] Tae-Kyung Kim, Young-A. Son, Yong-Jin Lim. Thermodynamic parameters of disperses dyeing on several polyester fibers having different molecular structures. Dyes and Pigments, 2005, 67:229-234.

83 Study on the Effect of Cellulase on Cotton Properties

Yingzhe Wu, Jianzhong Shao[1]*, Lan Zhou[2]

(1. Key Laboratory of Advanced Textile Materials and Manufacturing Technology, Ministry of Education of China, Zhejiang Sci-Tech University, Hangzhou 310018, China
2. Engineering Research Center for Eco-Dyeing & Finishing of Textiles, Ministry of Education, Zhejiang Sci-Tech University, Hangzhou 310018, China)

Abstract: This work investigated the changes in structures and properties of cotton fabrics washed with cellulose-containing detergent in order to understand the mechanism of fabric soil release improvement. The results showed that the wettabiliy of cotton fabric increased along with the increase of washing cycles, but the changes in stiffness and crystal index were opposite. After washing, the fabric surface became more fragile, meanwhile generated etching and many hydrophilic groups, so that the fabric hydrophilicity and the soil release property were improved.

Keywords: Cotton; Cellulose enzyme; Soil-release property; Wettability; SEM; FTIR

1 Introduction

The textiles will be stained gradually in use, the ideal property for the stained textiles should be cleaned easily under common laundry conditions, and the removed stain should not redeposit on the fabric during washing. This performance of textile products is called soil release property. The previous experiments found that the soil release property of fabric will be improved after multi-cycles cellulase washing. This work is aimed to understand the effect of cellulase washing on textile properties and the mechanism of facilitating soil release process.

Normally, the soil release property of fabric is mainly related to its hydrophilicity, which can be enhanced by chemical and physical treatments[1-2]. The chemical treatments are set to increase the content and the hydrophilic ability of surface hydrophilic groups and the physical treatments are set to decrease the fiber crystallinity or to loose its physical structure to achieve the purpose of improving hydrophilicity. Therefore, the physical and chemical changes of flat cotton fabrics with 5,10,15 and 20 cycles washing were studied by wettabiliy, stiffness, Martindale abrasion property, SEM and FTIR analysis in this work.

2 Experimental
2.1 Materials
The flat cotton which was treated by three different pre-treatments was used in the experiments. The fabric was treated by treatment A (no cellulase washing), B (0.25ppm cellulase washing) and C (1.00ppm cellulase washing) with 5, 10, 15 and 20 washing cycles respectively.
2.2 Methods
2.2.1 Wettability
The wetting times of fabric was measured according to AATCC Test Method 79.
2.2.2 Stiffness
The bending length of fabric was measured according to GB/T18318-2001 Test Method.
2.2.3 Martindale abrasion property
The fabric was rubbed 1000 times by Martindale abrasion machine according to ASTM D 4966-1998 Test Method.
2.2.4 SEM analysis
The fabric was observed (the work distance was 3.5mm and the experiment conditon was no gilding) by Scanning electron microscope (ULTRA55-36-73; Carlzeiss Company, Germany) with magnification 100 and 2000 times.
2.2.5 FTIR analysis
The fabric was measured (the scan time was 32 times) by infrared spectroscopy (Nicolet Avator 170, US) in KBr model.

---

* Corresponding author's email: jshao@sztu.edu.cn

3 Results and discussion
3.1 Wettability

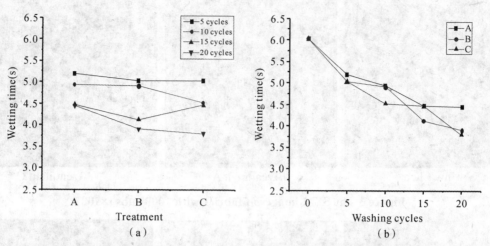

Figure 1 Average wetting time of variously washed cotton swatches
(a: Wetting time VS Treatment   b: Wetting time VS Washing cycles)

Wettability is one of the most direct indexes to reflect the hydrophilic ability of fabrics, and the fabric wettability is inversely proportional to its wetting time. The shorter the wetting time is, the better the wettability is. Figure 1 shows that wetting time of cotton fabrics decreased with the increase of cellulase concentration and washing cycles. It can be explained that the cotton fibers were affected by cellulase, detergent, mechanical force and temperature during laundry, so a part of even the whole fiber was damaged. It indicated that the water penetrates into fiber easily. Meanwhile, the various factors in the laundry process can also cause the hydrolysis of cotton fiber to increase the aldehyde groups which enhance the fiber hydrophilicity. In addition, the un-washed fabric may be treated by softener, which can be another important reason for deterioration in wettability[3].

3.2 Stiffness

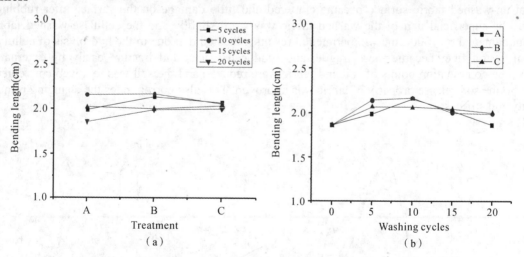

Figure 2 Average bending length of variously washed cotton swatches
(a: Bending length VS Treatment   b: Bending length VS Washing cycles)

Stiffness testing results can partially explain the change of cotton fibers during laundry. The Figure 2 shows that the bending length had no clear changes with increase in cellulase concentration, and it increased firstly then decreased with increase in washing cycles. The un-washed fabric has been treated by softener, thus the bending length became shorter. This result was consistent with previous results of wettability. The enzyme hydrolyzed fibers and made fabric structure looser, which led to a decrease in fabric stiffness after multi-washing cycles.

3.3 Martindale abrasion property

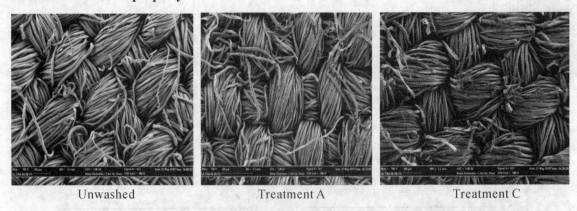

Unwashed　　　　　　　　　Treatment A　　　　　　　　　Treatment C

Figure 3 The SEM images of fabrics with 1000 rubs (×100)

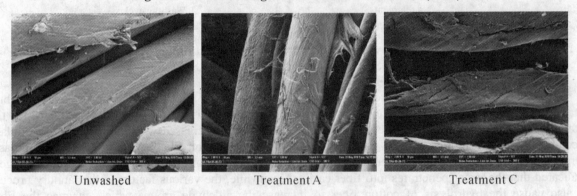

Unwashed　　　　　　　　　Treatment A　　　　　　　　　Treatment C

Figure 4 The SEM images of fabrics with 1000 rubs (×2000)

The Figure 3 and 4 were the SEM images of cotton fabric after Martindale abrasion test. Those figures show that un-washed fabric surface became cluttered and little damage on the surface after rubbing 1000 times, but the superficial part of the washed fabric was worn slightly and the cellulase-washed fabric was significant damaged by 1000 times. Experimental results indicated that due to the hydrolysis of cellulase, the surface of cotton fiber became more fragile, and produced cracks and fracture easily by external force. Therefore, the combination points of soil and fabric were reduced and the soil was removed conveniently by washing. So the soil release property of fabric was improved. This also corroborates the changing principle of wettability and stiffness.

3.4 FTIR analysis

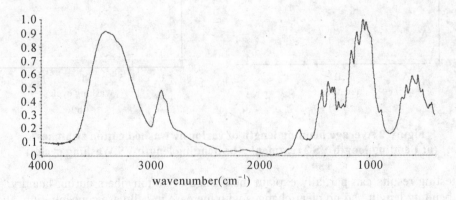

Figure 5 The FTIR spectra of untreated cotton fabrics

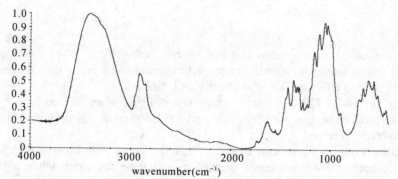

Figure 6 The FTIR spectra of cotton fabrics with treatment C

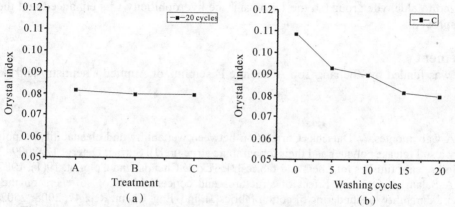

Figure 7 Average crystallization index of variously washed cotton swatches
(a: Crystal index VS Treatment   b: Crystal index VS Washing cycles)

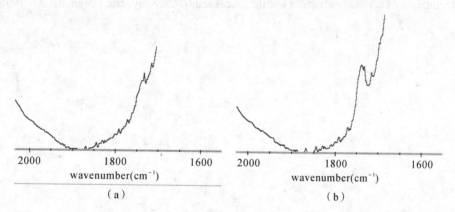

Figure 8 The FTIR spectra of untreated cotton fabrics from 2000cm^{-1} to 1600cm^{-1}
(a: Fabric unwashed   b: Fabric with treatment C)

The peak of 1372cm^{-1} was C-H stretching vibration peak of cellulose crystalline and the peak of 2900cm^{-1} was C-H stretching vibration peak of cellulose amorphous. Their ratio of peak area ($K=A_{1372}/A_{2900}$) was considered as the crystallization index of cellulose, and it was used to judge the size of cellulose crystallinity commonly[4]. The spectra shows that crystal index of cotton fiber decreased slightly with increase in cellulase concentration, and it decreased obviously with increase in washing cycles. Because of the hydrolysis of cellulose, the cellulose macromolecular chains of crystalline edge can be cut by endonuclease, and the exposed non-reducing end of cellulose molecules also may be cut by exonuclease to cellobiose[5]. So the crystal structure of the fiber was "nibbled" by the effect of endo and exonuclease and the crystalline region was further decreased. In addition, there was a carbonyl stretching vibration appeared at 1730 cm^{-1} in the Figure 8(B), but it did not be founded in unwashed fabric. It meant that fractured 1,4-β glycoside bond was oxidized to aldehyde during the laundry company with the increase of hydrophilic groups on the fiber, so the hydrophilicity of fabric was improved.

4 Conclusions

(1) The mechanical physical parameters test and analysis showed that when washing cycles increased, the wettability of flat cotton fabrics increased, but the stiffness increased firstly then decreased. The cellulase concentration showed limited effect on fabric wettabiliy and stiffness.

(2) The analysis of SEM, FTIR and copper value showed that when the washing cycle increased, the etching degree of fiber surface became more serious, and more and more 1,4-glucoside bond were fractured. The crystallinity of fabrics decreased.

(3) Cellulase hydrolyzed the superficial fibers, so that the fibers were deteriorated. During laundering, the combination force between soil and fabric washed by cellulase decreased when some damaged fibers removed. It was benefit to remove soil from fabric by external force. The cellulase etched fiber surface, so that the fiber generated some grooves and its crystallinity decreased. The fractured 1,4-glucoside bond of fiber was oxidized to aldehyde group, so the fabric surface hydrophilicity was enhanced and the soil release property was improved.

Acknowledgement

This work was funded by Zhejiang Top Academic Discipline of Applied Chemistry and Eco-Dyeing & Finishing.

References

[1] M Hasan, A Calvimontes, V Dutschk. Correlation between wettability and cleanability of polyester fabrics modified by a soil release polymer and their topographic structure[J]. Surfact Deterg. 12, 2009: 285-294
[2] T Cooke. Soil release finishes for fibers and fabrics[J]. Textile Chemist and Colorist. 19(1), 1987: 31-41
[3] N Reddy, A Salam, Y Q Yang. Effect of structures and concentrations of softeners on the performance properties and durability to laundering of cotton fabrics[J]. Ind. Eng. Chem. Res. 47, 2008: 2502-2510
[4] R O'Connor, et al. Infrared spectrophotometic Procedure for analysis of cellulose and modified collulose[J]. Anal Chem, 29, 1957: 998
[5] A Paulo. Mechanism of cellulase action in textile processes[J]. Carbohydrate Polymers. 37, 1998:273-277.

84 Double-temperature Cationic Modification Process for Uniform Reactive Dyeing of Cotton Fabrics

Min Xu[1], Jianzhong Shao[1*], Liqin Chai[1], Lan Zhou[1], Qinguo Fan[2]

(1. Engineering Research Center for Eco-Dyeing and Finishing of Textiles, Ministry of Education, Zhejiang Sci-Tech University, Hangzhou 310018, China
2. Department of Materials & Textiles, University of Massachusetts, Dartmouth, USA)

Abstract: A double-temperature cationic modification process of cotton fabrics was studied for an even and salt-free reactive dyeing. The influences of cationic modifier concentration, modification time and temperature, alkaline agent, and surfactant were investigated. The optimal double-temperature modifying process was as follows: Cotton fabric was immersed in a bath containing 4% (o.w.f) cationic Modifier H and 0.5-1g/L Peregal O at a liquor ratio 1:30. The bath temperature was firstly raised to 80-90℃ and kept for 30 min for Modifier H to diffuse into cotton fiber. Then the temperature was decreased to 30℃ at which 0.5-1g/L sodium hydroxide was added, and further 20 min was kept for fixation of Modifer H. It was found that the uniformity of cationic modification can be improved by the double-temperature modification process, resulting in an even and salt-free reactive dyeing.

Keywords: Cationization; Cotton fabrics; Reactive dyes; Level-dyeing property; Salt-free dyeing

1 Introduction

The surface of cellulosic substrates is negatively charged in alkaline dye-bath while most of the reactive dyes contain anionic groups, making the two repelling each other. In practice, sodium chloride or sodium sulfate (30-100g/L) is substantially needed to facilitate the cellulose dyeing process, which causes seriously environmental problems relating to excess salt release[1-2]. It has been demonstrated that the incorporation of cationic sites into the original molecular structure of cellulose fibers shows significant advantages of reducing the environmental impact of cotton dyeing[3]. By cationization, the Zeta potential on cotton fibers can be lowered, the electrostatic repulsion between cotton cellulose and reactive dyes is reduced, and the dye uptake rate together with the dye fixation rate are increased, so that the low-salt (no salt) reactive dyeing can be achieved. In conventional modification, cotton fabrics are placed in the bath with both cationic modifier and alkaline added simultaneously or sequentially, temperature as high as 60-80℃ is employed, so uneven modifier migration occurred during the modification, which can give rise to non-uniform dyeing[4].

In our research, a double-temperature modification process was studied to separate the additions of cationic modifier and alkali, which can significantly improve the level-dyeing of cationically modified cotton fabrics. As the first step of the process, adding cationic modifier at high temperature with no alkalis addition can improve the penetration and even distribution of the modifier. Applying alkali at low temperature in the second step can reduce the hydrolysis of the modifier and improve its utilization. In this study, Modifier H, a kind of epoxy quaternary ammonium, was used to modify cotton fibers. The modification process parameters, especially the modification time and temperature, were optimized to improve the dyeing uniformity without salt.

2 Experiments
2.1 Materials

Combed cotton double knitted fabrics were provided by Guangdong Esquel Textile Co., Ltd. Reactive Blue KN-R (C.I. Reactive Blue 19) was purchased from Runtu·Zhejiang Ruihua Chemical Co., Ltd. Reactive Red KN-3B (C.I. Reactive Red 180) and Reactive Orange KN-3G (C.I. Reactive Orange 72) were purchased from Shunlong Chemical Co., Ltd. Modifier H was synthesized in our lab. NaOH, Peregal O of AR grade and soap were purchased from Hangzhou Gaojing Fine Chemical Co., Ltd. $A(EO)_3$, $A(EO)_6$ and $A(EO)_9$ were provided by Chuanhua Co., Ltd.

2.2 Modification and dyeing process
2.2.1 Double-temperature modified process

At liquor ratio of 1:30, the modifier concentration can be applied to cotton fabric from 2% (o.w.f) to

---

* Corresponding author's email: jshao@sztu.edu.cn

16% (o.w.f). The cotton fabric was modified in the cationic modifier solution at 30℃ to 90℃ for 5 min to 50 min. An alkaline agent was then added to the modification solution in a range from 0.5 g/L to 10g/L.

The modification process scheme was shown in Figure 1.

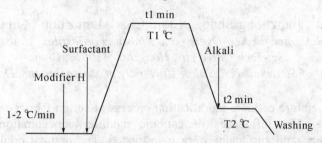

Figure 1 The process scheme of double-temperature modification

2.2.2 Salt-free reactive dyeing process

A dye solution containing 2% (o.w.f) reactive dye and 0.5 g/L Peregal O was used to dye the cationically modified cotton at liquor ratio of 1:30. The dyeing temperature was raised to 65℃ at a rate of 1-2°C/min and kept for 30min, then 10 g/L $NaCO_3$ was added. The dye was continued for another 30min, the dyed cotton fabric was taken out for washing. In soaping step, 5g/L soap solution was used at 1:50 liquor ratio at 90℃, for 10 min.

3 Results and discussions
3.1 Optimization of modification process
3.1.1 The effect of modifier concentration on K/S value and levelness

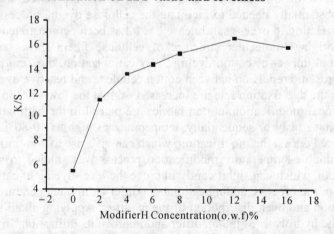

Figure 2 The effect of modifier concentration on K/S value

Figure 2 shows K/S value of dyed fabrics modified with different concentration of Modifier H. As it can be seen, the K/S value increased with the growing concentration very rapidly at the beginning, when the concentration was higher than 8% (o.w.f.), there was no significant increase in K/S value. However, when the concentration was higher than 4% (o.w.f.), the unevenness became significant. This result was attributed to the reaction between cotton fibers and the modifier. Modifier H carrying positive charge with active epoxide groups and amino groups can react with hydroxyl groups of cotton cellulose in basic solution. This reaction brings cationic groups onto cotton fibers, lowering the dye repulsion, thus increasing the dye uptake. As the modifier concentration increases, the amount of absorbed and reacted modifier increases sharply which leads to a non-uniform modification. In consideration of K/S value and levelness, 4% (o.w.f.) Modifier H was recommended for this modification process.

3.1.2 The effect of alkaline agent on K/S value and levelness

The use of alkali was studied with the results presented in Table 1.

Table 1 The effect of different alkaline agents on K/S value

Alkaline agent (5g/L)	K/S
-	10.14
$NaHCO_3$	11.34
Na_2CO_3	12.27
NaOH	14.03

The K/S value of dyed cotton fabrics modified with NaOH was higher than the other two alkaline agents, but less uniform. The results indicated that, with the increase of alkalinity, K/S value raised significantly while the levelness tended to decline sharply. Sodium bicarbonate and sodium carbonate cannot totally change –OH of cellulose fibers into –O^{2-}, resulting in reduced utilization of Modifier H. However, in view of uniformity, higher NaOH concentration might cause rapid and localized reaction that easily leads to insufficient migration of modifier giving rise to non-uniform dyeing.

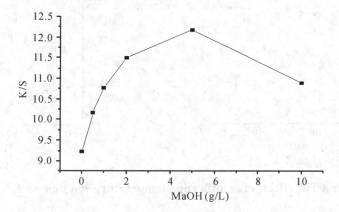

Figure 3 The effect of NaOH concentration on K/S value

Figure 3 indicated that, the K/S value increased with the rise of NaOH concentration initially, however, when the concentration was higher than 5g/L, the K/S value decreased. The excessive NaOH could accelerate the decomposition of Modifier H. On the other hand, the uniformity decreased with the increase of NaOH concentration. Therefore, 0.5-1g/L sodium hydroxide was optimal for this process.

3.1.3 The effect of modification temperature and time on K/S value and levelness

Modification temperature and modification time were both very important for the cationic process. Figure 4 illustrated the influences of temperature and time on K/S values of dyed modified cotton fabrics. In the experiments, when temperature T1 was changed, temperature T2 was fixed at 60℃, similarly, when temperature T2 was optimized, temperature T1 was fixed at 60℃. In the same way, when a series of t1 was experimented, t2 was fixed at 30min, and vice versa.

It was evident from Figure 4(a) that as the temperature T1 increased, the K/S value increased gradually, and after it was over 80℃, the K/S value started to decrease. It is well known that higher temperature is beneficial for the modifier penetration, but unfavorable for modification because of the hydrolysis of modifier which leads to lower dye uptake. Taking above factors into consideration, temperature in the range of 80-90℃ was suitable for T1.

The effect of T2, the temperature of alkaline treatment on K/S of dyed modified cotton fabrics is shown in Figure 4(b). When temperature T2 increased, K/S initially increased a little, then decreased quickly. This phenomenon demonstrated that high temperature might accelerate the decomposition of Modifier H in basic solution. As a result, 30℃ was used for T2 in the following investigation.

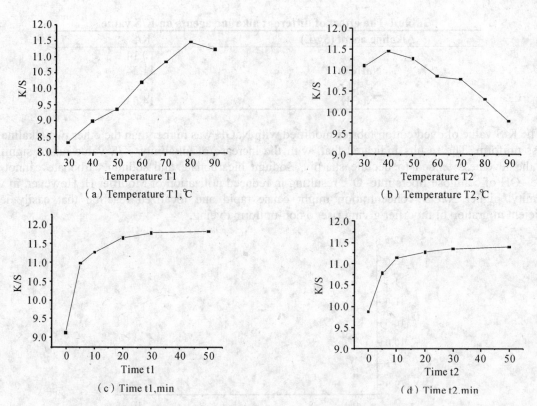

Figure 4 The effects of modification temperature and time on K/S value

As shown in Figure 4(c), when t1 was prolonged from 0 to 30min, there was an increasing in K/S value, after 30min K/S showed no significant increase. This result indicated that 30min was appropriate for modifier to migrate and penetrate sufficiently. Figure 4(d) gives the similar information as Figure 4(c), which shows the suitable time for t2 was 20min. During that time, the modifier can react with cotton fiber with low hydrolysis.

3.3 Combination with surfactant

Surfactants are used as a leveling agent to increase the rate of modifier migration on cotton fabrics. Cationic surfactants can compete with cationic modifier, which in return reduces the adsorption of Modifier H onto cotton fibers. Anionic surfactants can cancel the positive charge of the modifier. Therefore neither of them was chosen. There is little competition between non-ionic surfactants and quaternary ammonium modifier, and the non-ionic surfactants can give a good penetration and migration[5]. In order to enhance the level-dyeing, some non-ionic surfactants were chosen, which were presented in Table 2.

Table 2 Optimization of some non-ionic surfactants

Non-ionic Surfactant	Content (%, o.w.f.)	K/S Value
-	-	11.35
Peregal O	0.5	11.91
$A(EO)_3$	0.5	11.66
$A(EO)_6$	0.5	10.78
$A(EO)_9$	0.5	10.85

Note: $A(EO)_n$ was used to express the non-ionic surfactant, A means fatty alcohol, EO means polyethenoxy ether, n represents the number of polyethenoxy ether perunit.

It was concluded from Table 2 that some kinds of non-ionic surfactants may cause the reduction of K/S value, and 0.5-1g/L Peregal O was in favor of the level-dyeing.

3 Conclusions

(1) The double-temperature cationic modification of cotton fabrics can improve the uniformity of

reactive dyeing with low salt and high uptake.

(2) Adding some non-ionic surfactants in modification process can produce a good modification effects and further improve the level-dyeing of cotton fibers.

(3) To balance the levelness and modification effects, the Modifier H was applied to cotton via the following optimal formula: cotton fabric was placed in a pretreatment bath containing 4% (o.w.f.) Modifier H and 0.5-1g/L Peregal O at liquor ratio of 1:30. The temperature was raised to 80-90℃ initially and kept for 30 min for reaction. Cooling down the temperature to 30℃, 0.5-1g/L sodium hydroxide was added, then the temperature was kept for 20 min before washing.

Acknowledgement

The authors express their thanks for Zhejiang Top Academic Discipline of Applied Chemistry and Eco-Dyeing & Finishing.

References

[1] Wang L L, Ma W, Zhang S F, Teng X X, Yang J Z. Preparation of cationic cotton with two-bath pad-bake process and its application in salt-free dyeing. Carbohydrate Polymers, 2009, 78: 602-608
[2] United States Patent 6,350,872 B1, Feb, 26, 2002
[3] Hauser P J. Reducing pollution and energy requirements in cotton dyeing. Textile Chemist and Colorist & American Dyestuff Reporter, 2000, 32: 44-48
[4] Edson C, Silva F, Júlio C P, Melo, Claudio A. Preparation of ethylenediamine-anchored cellulose and determination of thermochemical data for the interaction between cations and basic centers at the solid/liquid interface. Carbohydrate Research, 2006, 341: 2842-2850
[5] Ali R. Tehrani B., Hajir B., Barahman M., Mokhtar A., Fredric M., Menger. Interactions of gemini cationic surfactants with anionic azo dyes and their inhibited effects on dyeability of cotton fabric. Dyes and Pigments, 2007, 72: 331-338

85 Modification of *Mulberry Silk* by Calcium Salt Treatment and Epoxy Crosslinking

Wei Zhang, Shizhong Cui

(College of Textiles, Zhongyuan University of Technology, Zhengzhou 450007, China)

Abstract: *Mulberry silk*s were modified by calcium salt treatment and epoxy crosslinking with glycerin triglycidyl ether (GTGE), and the silk fibers with porous structure and preferable tenacity were obtained. The effect of temperature, time and catalyst sodium carbonate on the crosslinking reaction of silk fibers had been investigated. The change in the structure and the physical properties of silk fibers after calcium salt treatment and epoxy crosslinking had been studied. The separating behavior of microfibers occurred on the surface of silk fibers after calcium salt treatment, and porous structure formed in the interior of silk. This porous structure of silk was enlarged by subsequent epoxy crosslinking, and accordingly the moisture conduction of silk fibers was improved remarkably. Breaking strength, breaking elongation and wet elastic resilience of silk fibers increased evidently after modified, and the modified silks exhibited a better flexibility.

Keywords: *Mulberry silk*; Calcium salt treatment; Epoxy crosslinking

1 Introduction

Silk is highly appreciated for its outstanding characteristics such as unique luster, comfortable hand, excellent softness, and good drape. However, it also suffers from some inferior properties, one of which is low wet resiliency.[1] The poor wet resiliency of silk fibers may be due to the lack of cystine residues and the resultant absence of chemical cross linkages between silk fibroin molecules.[2] A considerable amount of work has been carried out on the chemical modification of silk with a view to improving its low wet resiliency. The modifying processes mainly include graft copolymerization,[3] dibasic anhydrides treatment,[4] amino-formaldehyde resin finishing and epoxy crosslinking.[5]

However, these methods have drawbacks such as environmental pollution and hazard to health. This paper proposed a new modifying process of silk that silk was dissolved slightly with calcium salt solution, and subsequently crosslinked with epoxide GTGE in water solution.

2 Experimental

2.1 Materials

*Mulberry silk*s came from Henan province, China. All chemical reagents were of reagent grade and used without further purification.

2.2 Calcium salt treatment

1g degummed silk fibers were immersed in 100ml of calcium chloride solution (molar ratio of $CaCl_2:H_2O$, 1:8), and left at 70℃ for a given times (0-60min). Then the silk fibers was added into 5% EDTA solution, and heated at 100℃ for 5min to remove calcium ions remained. The fibers collected were washed extensively with deionized water, and dried in vacuum at 60℃ for 24 h. Weight loss of sample was calculated according to the difference in weight before and after treatment.

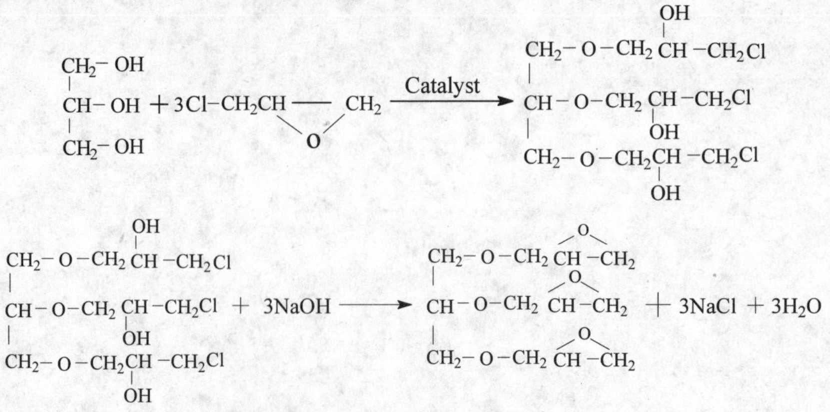

Figure 1 Reaction equation of GTGE synthesis

2.3 Epoxy crosslinking reaction

Silk fibers obtained by calcium salt treatment were crosslinked in GTGE solution with a concentration of 3-12% at a liquor ratio of 1:20, and Na_2CO_3 was used as catalyst of reaction. Epoxy crosslinking reaction was performed by three methods as follows.

Method 1	Immersed in deionized water for 1h, then immersed in crosslinking agent solution without catalyst for 1h	
Method 2	Immersed in deionized water for 1h, then immersed in crosslinking agent solution with 2% Na_2CO_3 for 1h	
Method 3	Immersed in 2% Na_2CO_3 solution for 1h, then immersed in crosslinking agent solution without catalyst for 1h	

Silk fibers collected from crosslinking agent solution were dried at 40℃ for 1h. After loosed, the dried samples were baked for 10min at a crosslinking temperature of 90-130℃. The crosslinked samples were washed extensively with deionized and dried in vacuum at 100℃ for 24 h. Weight gain of samples was calculated according to formula 1.

$$M = \frac{M_1 - M_2}{M_1} \times 100\%$$

(Formula 1)

where M_1 is the dry weight of silk fibers obtained by calcium salt treatment, and M_2 is the dry weight after epoxy crosslinking.

2.4 Measurement of physical properties

The mechanical properties of silk samples were measured on INSTRON 5582 tensile tester with a gauge length of 10mm and a stretching speed of 100 mm/min. Moisture conduction of silk fibers was determined by measuring vertical wicking height of fiber assemblies. The test has been conducted using a vertical wicking tester according to DIN 53924 method.[6]

2.5 SEM observation

Silk samples were coated with gold film in order to observe the surface morphology and microstructure. The instrument was a JEOL JSM-5600LV electron microscopic with an accelerating voltage of 15 kV.

3 Results and discussion
3.1 Calcium salt treatment

Dissolution behavior occurred on *Mulberry silks* after calcium salt treatment, and the weight loss of the silks increased linearly with the time of calcium salt treatment, as shown in Figure 2(a). Calcium ions can impregnate into the silk, and coordinate with the polar lateral groups of amino acid such as hydroxyl, carboxyl and amido to form chelate compound.[7] Therefore, the secondary bonds between silk fibroin molecules were destroyed and the intermolecular force weakened, which resulted in the dissolution of SF molecules.

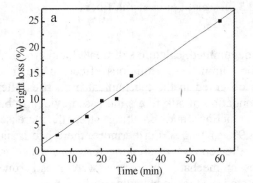 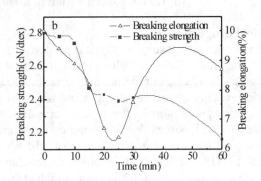

**Figure 2 The effect of the time of calcium salt treatment on
(a) weight loss and (b) tensile properties of silk fibers**

Calcium salt treatment had a major effect on the tensile properties of silk fibers, as shown in Figure 2(b). Tensile strength of the fibers decreased with the prolong of treating time, and the breaking strength appeared twice obvious reduction after calcium salt treatment for 10min and 30min, which may be due to the destroy of amorphous and crystalline regions in the fibers, respectively. The breaking elongation of silk fibers exhibited a reduced trend in calcium salt treatment, especially treated after 10min, whereas breaking elongation increased with treating time. Therefore, breaking strength and elongation of silk fibers decreased less within calcium salt treatment for 10min.

3.2 Epoxy crosslinking reaction

Silk sample treated by calcium salt solution for 10min was crosslinked by epoxide GTGE, and Figure 3(a) shows the weight gain of silk fibers using three methods under various crosslinking temperatures at 7% crosslinking agent solution. The weight gain at each temperature in method 2 and method 3 was more than in method 1, in which no catalyst was used. Using Na_2CO_3 as catalyst, epoxy groups of GTGE could be opened to form active centers by nucleophilic attack of negative ions in alkali solution on them, which would react with some groups such as amines, alcohols, phenols and carboxylic acids in the side chain of silk fibroin to form crosslinking structure.[8] Weight gain of silk fibers increased almost linearly with the elevation of crosslinking temperature, however, the weight gain decreased when the temperature was more than 120℃ in the method 2 and method 3 with Na_2CO_3 as catalyst.

It can be found by comparing the results from method 2 and method 3 that the adding mode of catalyst Na_2CO_3 also affected the weight gain of silk fibers. Immersing firstly in Na_2CO_3 solution was in favor of improving the crosslinking effect of silk fibers in the high crosslinking temperature of more than 110℃ (method 3 of Figure 3a). A possible explanation was that carbonate broke the secondary bonds between silk fibroin molecules, and the bulk effect of silk fibers resulted in the increase of the accessibility of crosslinking agent. Figure 3(b) shows the relationship between the concentration of GTGE and the weight gain of silk fibers in method 3 at the temperature of 120℃. It can be observed that the weight gain of silk fibers increased firstly with the concentration of crosslinking agent. But when the concentration was more than 7%, silk fibers demonstrated a decreasing weight gain, which was due to the increased chance of ring opening polymerization of crosslinking agent themselves in the high concentration.

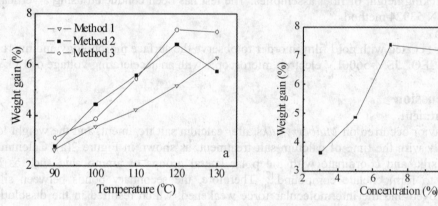

Figure 3 The effect of (a) crosslinking temperature and (b) GTGE concentration in method 3 on weight gain of silk fibers

3.3 Physical properties

The measurement results of physical properties for degummed, calcium salt treated and crosslinked silk fibers are shown in Table 1. Calcium salt treatment for 10min has no obvious effect on the mechanical properties of silk fibers, but significant variations were observed in the mechanical properties after epoxy crosslinking reaction. Breaking strength and breaking elongation of silk fibers increased evidently but initial modulus decreased slightly after epoxy crosslinking reaction, and breaking strength increased from 2.74cN/dtex to 3.31cN/dtex, breaking elongation from 8.9% to 13.6%. Furthermore, there had a significant improvement in the wet elastic resilience of silk fibers after epoxy crosslinking reaction, which increased from 64% to 81% (Table 1). Obviously these changes in mechanic properties were mostly due to the crosslinking structure formed in silk fiber, which also affected the tensile behavior of the fiber.

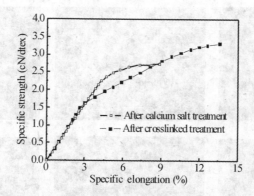

Figure 4 Comparison of tensile curves of calcium salt treated and epoxy crosslinked silk fibers

Figure 4 exhibits the tensile curves of silk fibers treated by calcium salt solution and crosslinked subsequently with epoxy GTGE. The yield point position of silk fiber shifted downward from 5% to 2.8% after epoxy crosslinking; meanwhile, a high elastic deformation region appeared in its tensile curve. The silk fiber obtained after epoxy crosslinking reaction exhibited a better flexibility. No obvious change was found in the hydroscopic property for the silk fibers obtained after calcium salt treatment and subsequent epoxy crosslinking, however, moisture conduction of silk fibers was improved markedly by the two processes, for the wicking height of silk fibers increased by 92% after calcium salt treatment, and then increased by 1.3 times further after subsequent epoxy crosslinking (Table 1).

Table 1 Physical properties of degummed, calcium salt treated and epoxy crosslinked silk fibers

Silk samples	Breaking strength (cN/dtex)	Breaking elongation (%)	Initial modulus (cN/dtex)	Breaking work (cN/dtex)	Wet elastic resilience (%)	Moisture regain (%)	Wicking height (mm)
Degummed	2.80	9.8	53.0	0.17	64	9.1	37
Calcium salt treated*	2.74	8.9	51.2	0.15	64	8.9	71
Epoxy crosslinked**	3.31	13.6	46.7	0.24	81	8.6	164

* Treated for 10 min;
** Used method 3 at the condition of a temperature of 120℃, a GPGE concentration of 7% and a time of 10 min.

3.4 Fiber morphology

Morphology changes of silk fibers obtained after calcium salt treatment and subsequent epoxy crosslinking were observed by SEM and shown in Figure 5. Degummed silk fiber showed a cross section with only few pores and a smooth surface except some residual sericin (Figure 5a, b). After calcium salt treatment a separating microfibrillar structure appeared on the surface of silk fiber, and an enlarge cross section presented apparent pore structure (Figure 5c, d).

An apparent accretion at cross sections of silk fibers could be observed after epoxy crosslinking reaction, and the pores in the interior enlarged in the size and increased in the number (Figure 5e). The secondary bonds of silk fibroin also can be destroyed by the polar epoxy groups, and the pore structure formed after calcium salt treatment enhanced the accessibility of GTGE in the interior of silk fibers, which resulted in the further swelling and dissolution of silk fibers. The remarkable improvement in the moisture conduction of silk fibers should be relation to the formation and increase of pore structure. However, the surface of epoxy crosslinked silk fibers exhibited a continuous coating layer, and the former microfibrillar structure could hardly be observed. In addition, some granules deposited onto the fiber surface could also be found (Figure 5f). The observed surface morphology was due to the epoxy crosslinking process, which caused the bonding between the silk and the epoxy.[9]

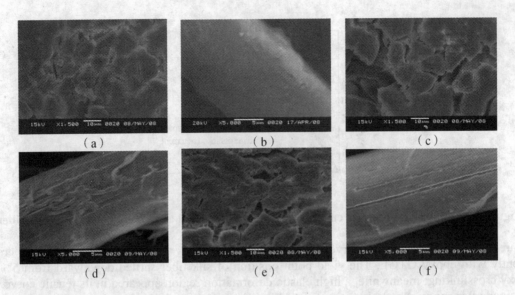

Figure 5 SEM photographs of (a), (b) degummed, (c), (d) calcium salt treated and (e), (f) epoxy crosslinked silk fibers

4 Conclusions

*Mulberry silk*s were modified by calcium salt treatment and crosslinking reaction with GTGE. The porous structure formed in the interior of silk in the calcium salt treatment. It was favorable for improving crosslinking effect to immerse 1h using 2% Na_2CO_3 solution before epoxy crosslinking reaction, and the optimum condition of reaction had been determined as a temperature of 120℃ and a GTGE concentration of 7%. The porous structure of silk fibers was enlarged further in subsequent crosslinking reaction, and the moisture conduction was improved remarkably. There was significant variation in the mechanical properties of silk fibers after epoxy crosslinking, and the breaking strength, breaking elongation and wet elastic resilience of silk fibers increased evidently. The modified silks exhibited a preferable flexibility.

References

[1] W K Lee, K M Sin. Silk Fabric Crosslinking. Textile Asia 1991, 22: 86-91.
[2] Y Yang, S Li. Silk fabric non-formaldehyde crease-resistant finishing using citric acid. J Text Inst 1993, 84: 638-644.
[3] M Tsukada. Structural characteristics of 2-hydroxyethylmethacrylate (HEMA) /methacryl-amide (MAA)-grafted silk fibres. J Appl Polym Sci 1988, 35: 2133-2140.
[4] M Tsukada, H Shiozaki. Chemical and property modification of silk with dibasic acid anhydrides. J Appl Polym Sci 1989, 37: 2637-2644.
[5] C Hu, K Sun. Easy-Care Finishing of Silk Fabrics with a Novel Multifunctional Epoxide. J Soc Dyer Col 1998, 114: 359-364.
[6] B Das, A Das, V K Kothari, R Fanguiero. Effect of fibre diameter and cross-sectional shape on moisture transmission through fabrics. Fibers and Polymers 2008, 9: 225-231.
[7] S Ha, Y H Park, S M Hudson. Dissolution of Bombyx mori silk fibroin in the calcium nitrate tetrahydrate-methanol system and aspects of wet spinning of fibroin solution. Biomacromolecules 2003, 4: 488-493.
[8] C Hu, Y Jin. Wash-and-Wear Finishing of Silk Fabrics with a Water-Soluble Polyurethane. Textile Research Journal 2002, 72(11): 1009-1012.
[9] J Prachayawarakorn, K Boonsawat. Physical, chemical, and dyeing properties of Bombyx mori silks grafted by 2-hydroxyethyl methacrylate and methyl methacrylate. J Appl Polym Sci 2007, 106: 1526-1530.

86 Prospects of Cellulose Fiber Industrial Production in Solvent Process

Qingzhang Zhao

(China Textile Academy, China)

Abstract: As people's living standard is further improved and idea for sustainable development is strengthened, people pay more and more attention to develop green process and renewable row materials for textile application. Cellulose fiber in solvent process just is a high technology with characteristic of environment friendly, renewable raw materials and without "white" pollution. It is significance to develop the technology for sustainable development of textile industry. This article briefly introduces the new developments on Lyocell at home and aboard. Industrialization feasibility and prospect of different processes were analyzed.

Global fiber output in 2009 was approximately 70 million tons, of which the natural fiber was 26 million tons, representing 37%; synthetic fiber was 41.60 million tons, representing 59% and viscose was 3 million tons, representing 4%. It is easy to understand by Analyzing raw material source that possibility to expand cotton, linen and other natural fibers are limited due to shortage of land and water and so on; oil as non-renewable raw material will be used out within certain years; viscose fiber is made from renewable cellulose, however its process causes serious pollution, greatly hindered its rapid development. As the improvement of living standards and the sustainable development concept going, people has paid more and more attention on developing green process and renewable material, cellulose fiber in solvent process is just a such environment-friendly, renewable raw materials and without "white pollution" high-tech one. It is also the way for sustainable development of textile industry.

Cellulose as the richest polymer materials in nature, with its outstanding regeneration speed, excellent performance, has been an important raw material for fiber and films. It is reported that the global yield of plants by photosynthesis reaches to 220 billion metric tons annually, providing 100-150 billion ton scale of cellulose every year. It includes wood, bamboo, straw, etc which can be used for textile. Wood has been widely used for viscose fiber. Bamboo is another rich source for fiber manufacture in China. The cultivated area of bamboo has reached to 4.2 million hectares. The annual production of bamboo is about 70 million tons. More than 1000 million tons of underutilized crop straw is available in China. This shows that the cellulose would never be used out as renewable fiber source.

Viscose has always been an important species of chemical fiber. The world's annual production capacity of viscose fiber is about 3 million Tons. Despite the fact that viscose production process continued to improve, it is hard to neglect the fact that there is at least 28 kg of carbon disulfide in exhaust emission, and 300m^3 wastewater discharge for staple fiber (1200m^3 for filament) for each tone of viscose fiber product. Wastewater in viscose process contains harmful materials such as sulfuric acid, zinc sulfide, and salt. Therefore, Environmental pollution for this process is evident. It is the reason why in recent years, capacity of viscose has been dramatically decreased in developed countries, either close down or move to developing countries. On the other hand, the demand for renewable cellulose textile in market is still booming, coupled with relatively low requirements on environmental protection in China, resulting in rapid development of viscose in our country in recent years. Annual growth rate of viscose fibers is over 12% from 2002 to 2009. It is means that China has suffered most of severe pollution bring by viscose production.

Simple molecular formula of cellulose is as $(C_6H_{10}O_5)_n$, which is composed of multiple D-glucose-based. The degree of polymerization for natural cellulose could reach to 10000. Polymerization degree of regenerated cellulose is 200 ~ 800 typically. In the cellulose chains, there are large amounts of hydrogen bonds. This kind of hydrogen and oxygen in the chain is connected each other to form a ribbon polymer chain with high stiffness. Oxygen in polymer chain is also able to bond with oxygen in neighboring polymer by hydrogen bond. It is precisely because of strong hydrogen bonding in these molecules and intermolecular that cellulose is not soluble in conventional solvents.

The so-called cellulose fibers in solvent process broadly refers to that the cellulose is directly dissolved and spinning in certain solvent system without any chemical reaction during the process, compared to the traditional route of viscose. On the other hand, because of this system are required to fully recycle the solvent, thereby reducing the pollution of the environment. Use cellulose fiber in solvent process can be traced back to a few hundred years ago. Several solvents systems which could directly dissolve cellulose have been found, unfortunately, no one had been used in industrial scale except NMMO system which have been formed large-scale production.

Akzo Nobel in Netherlands is the pioneer for manufacturing cellulose fibers in NMMO as solvent in 1976. Later Lenzing Company in Austria and Courtaulds Company in British built industrial product lines respectively basis on this patent. (International bureau for the standardization of manmade fiber has named this product as Lyocell), Lenzing now is only one who owns this high technology after Lenzing Company acquired the Courtaulds. Three factories with total capacity of the 150 thousand tons have been built in United Kingdom, the United States and Austria respectively. Other countries and regions, such as R. O. Korea, Russia, India as well as Shanghai and Taiwan of China also launched the research on Lyocell fiber.

The study on cellulose fiber in NMMO solvent was started at 1987 in China, including Chengdu University of science and technology, Yibin chemical fiber manufacture, Donghua University, China Textile Academy and so on. A lot of fundamental research on manufacturing process and solvent recovery has been done and some progresses have been made. Shanghai Textile Holdings (Group) Cooperation engaged a Lyocell project with capacity of 1000tone/year in 2001. It started test running in 2005. The project was taken the import-digest-absorption model. China Textile Academy cooperated with Xinxiang Chemical Fiber Company have built a pre-industrial production line with capacity of 1000 tone/year in 2009. The construction work has been finished. It will put into operation soon.

Lyocell processing abandon the chemical reactions used in viscose manufacture. Manufacture of fiber was directly based on physical change of NMMO/water/cellulose system. NMMO is non-toxic solvent and is reused through recycling and refining technology. Generally speaking, this process is a green processing with almost no emissions and sewage discharge. The fiber keeps all characteristics of nature fiber such as moisture absorb and breathability, comfort, luster, dye ability and biodegradable. It even has high wet tenacity like polyester. It is an ideal raw material for textile processing.

Three major technology challenges in industrialization of Solvent process for cellulose fibers is faced, that is, continue true cellulose solution preparation, spinning process and equipment, and efficient solvent recovery system. One of the biggest difficulties is the preparation of true cellulose solution. Study on ternary phase diagrams of NMMO/H_2O/Cellulose showed that cellulose can only dissolved in very narrow range of NMMO aqueous solution. The most suitable ration of NMMO:H_2O is 87:13. In order to achieve this purpose, one process is to use starting materials containing excess water. The excess water is discharged through the vacuum and higher temperature during the process. The other process is mixed pulp with NMMO/H_2O (87:13) solvent directly. Korea Institute of Science and Technology has developed twin screw extruder process using 87% of NMMO aqueous solution. In this process, pulp has to be highly smashed to avoid forming "white core" because of the higher concentration of NMMO solutions, the easy dissolved the surface of cellulose fiber. The advantage of this process is easy to implement accurate measurement and continuous operation, without vacuum systems. The drawback is that formed spinning solution has higher level of gel particles which will affect spinning ability. Unit's productivity is relatively low. LIST Kneader reactor provide another possible way to made cellulose solution in which NMMO aqueous solution with excess water mixed with additives, cellulose pulp directly in a vacuum and higher temperature, the resulting true cellulose solution. The device features excellent blending effects and uniform temperature. Its capacity per unit is much higher than twin screw process. The drawback is higher manufacturing cost due to complex mechanical structure. Small evaporation area and long operation time are also the shortage of this machine. Matured massive machine for solution preparation is the film evaporators in which pulp slurry is scraped up to a film and rapidly formed cellulose solution under certain temperature and vacuum. The process is characterized by a relatively simple device structure, lower manufacturing costs, large evaporation area and short operation time. It not only reduces the cellulose degradation, but also significantly reduces the ratio of solvent decomposition. Less held material in the system is also in favor of safety. The highly efficient thin-film evaporator also brings great unit capacity, favorable to the scale of production. The drawback is that the equipment design and control technology is highly difficult.

Unlike traditional viscose spinning process, viscosity of cellulose solution in solvent process could reach to 3 million centipoises. Therefore the special spinning equipment is needed, including transportation and distribution of spinning solution, spinning bin, and spinneret and so on. Different fiber forming mechanism is followed by the solvent process, comparing with traditional viscose process. large number of hydroxyl are esterified which the interactions among hydroxyl are eliminated in viscose process, so that wet spinning technique could be used since the macromolecules still are mobile after entering coagulation bar. Hydroxyl group in solvent process is not blocked by chemical reaction. It only depends on correct region in ternary phase of solution. As soon as primary fiber enters coagulation bar, the hydrogen bonds between molecules recovered immediately. Fiber would not allow for further draw, therefore only dry jet -wet spinning process could be used. It is also means that traditional wet spinning or melt spinning components are not available. Because drawing process must be finished in very short air gap, higher speed of quenching is needed, which is even ten times higher velocity than normal one.

Solvent recovery is the key for successful industrialization. Process should guarantee minimize of NMMO decomposition. A lower process temperature and shorter residence time are the most effective way to reduce consumption of solvent. It is the important factors for equipment selection. Industrialization of the solvent recovery also need to consider the energy-save methods since concentration of solvent need a huge amount of steam, which consists of main part of cost for solvent recovery.

The Lyocell with trademark A100 mainly is imported form UK. Lyocell with trademark LF which belong to anti-fibrillation is imported from Lenzing Austria. Annual imports of Lyocell are about 20 thousand tone in China. Chinese customers have done a lot of work in application and accumulated some experience. Excellent processing ability and higher wet tenacity has been noticed. Unfortunately all those products have to be imported from abroad. It has hindered the further development in china. We are look forward to develop the technology with own characteristics in nearer further.

87 Surface Modification of Polypropylene Nonwoven Fabric with Plasma Activation and Grafting

Ahmad Mousavi Shoushtari[1], Aminoddin Haji[2*], Azadeh Jafari[1]

(1. Textile Engineering Department, Amirkabir University of Technology, Tehran, Iran
2. Islamic Azad University, Birjand Branch, Birjand, Iran)

Abstract: Polypropylene fiber has the advantages of high tensile strength, excellent chemical and biological resistance and low production costs. But some drawbacks such as dyeing problems and low wettability, raised from highly crystalline and hydrophobic nature of polypropylene chain, has restricted the usage of this fiber in some applications.

In this study, plasma induced grafting as an effective procedure for modifying the surface characteristics of polypropylene nonwoven fabrics has been applied. Atmospheric pressure plasma was used to generate active sites on the polymeric chains of polypropylene fibers, and then the plasma activated samples were grafted with acrylic acid. The effects of plasma treatment time, monomer concentration, grafting time and temperature on graft yield and water absorption behavior of treated samples were evaluated.

To evaluate the surface morphology changes, SEM images were employed. FTIR analysis was used to examine the chemical changes made on the fabric sample surfaces. The FTIR analysis showed that carbonyl and carboxyl groups have been created on the surface of plasma treated and grafted fibers. Water absorption of untreated and plasma grafted samples were evaluated using contact angle measurements. The results show that water absorption of plasma grafted polypropylene nonwoven samples has been improved significantly.

Keywords: Plasma, Polypropylene, Graft, Acrylic acid, Surface modification

1 Introduction

Polypropylene (PP) is one of the most versatile fibers because of its chemical inertness, better tensile strength, and low cost. However, the hydrophobic nature of the polymer restricts its application in a number of technologically important areas [1]. Several chemical modification processes such as oxidation, grafting, etc. have been developed to improve the characteristics of PP fiber [2].

The functionalization of polymeric materials by plasma and high energy radiation has attracted wide attention to introduce desirable functional properties in the material surface [1]. Plasma treatment of textiles may result in desirable surface modifications, including but not limited to surface etching, surface crosslinking, chain scission, decrystallization and oxidation, without affecting bulk properties of the fibers. Compared with conventional wet finishing, plasma processes have the decisive advantage of less water, chemicals and energy usage with the possibility to obtain typical textile finishes without changing the key textile properties [3,4]. Efficacy of the treatment depends on the choice of the process gas, plasma density and energy [5]. When using special gases, a plasma-induced deposition polymerization may occur [6].

In atmospheric pressure plasmas, the surface activation typically takes place with oxygen-containing gas mixtures such as air. Surfaces with low functionality become more reactive by enhancing the concentration of oxygen containing or other polar groups at the surface. After plasma activation in air a great variety of different oxygen-containing functional groups, such as -OH, -C=O, -COOH will be introduced onto the surface of the material. Activation can also mean the increased wettability of surfaces whose surface energy is increased by the polar surface groups [7].

Many studies have already been carried out with low pressure plasma [3-4, 8-12], but atmospheric pressure plasma treatments would be ideal for the continuous processing of textile materials. Atmospheric plasma devices and treatments could be utilized for use in high-speed continuous processing operation, and when optimized, could replace or increase many current wet chemical finishing processes [13].

Several studies have been done on atmospheric pressure plasma treatment of PP fibers. McCord *et al* used He and He/O2 gases for plasma treatment of PP fabrics and showed that the surface oxygen and nitrogen of PP fabric increased significantly, along with a slight decrease in tensile strength of the fabric [5]. Kuwabara *et al* have made an atmospheric plasma jet and confirmed the etching of the material at suitable positioning of the device [14]. Cheng *et al* reported the decrease of water contact angle of PP fibers after treatment with

---

* Corresponding author's email: aahl58@yahoo.com

plasma jet using argon gas [15]. Choi *et al* have compared the effect of low pressure oxygen plasma (LPOP) and atmospheric pressure glow discharge (APGD) on the surface properties of PP fibers. After treating PP surface with both methods, they observed OH-groups, C=O groups in ester, ketone and carboxyl groups, C=O groups in unsaturated ketones and aldehydes. According to their studies, etching rates were higher in LPOP [8]. Abdou *et al* have reported antibacterial effects produced on PP fabric after Ar/O2 plasma treatment followed by chitosan grafting [16].

In this study, PP nonwoven was surface activated with N2 plasma at atmospheric pressure, then grafted with acrylic acid monomer (AA). The surface changes of the substrate were evaluated using SEM, FTIR and contact angle measurements.

2 Experimental
Materials

PP nonwoven with thickness of 109 μm and density of 20 g/m² was used in this research. Before plasma treatment, samples (5×10 cm²) were washed in ethanol for 30 minutes at ambient temperature, rinsed with distilled water and finally dried at 50°C. All chemicals used were of analytical grade made by Merck.

Methods:

Atmospheric pressure plasma was produced in a laboratory scale reactor designed in Textile Engineering Department of Amirkabir University of Technology-IRAN. Plasma treatments were done under frequency of 10 kHz and voltage of 7 kV at different times. There was 3 mm space between electrodes and industrial nitrogen was used as plasma gas. Samples were grafted after plasma treatment, with different concentrations of acrylic acid in water. Prior to enter the plasma treated samples in the grafting solution, the solution was deaerated with nitrogen bubbling through it. The grafting process was carried out at different temperatures and times. After grafting, the residual monomers and homopolymers were removed from the surface of the fibers by washing in methanol and distilled water for 30 minutes respectively.

The degree of grafting was calculated according to equation 1:

$$D.G.\% = [(W1-W2)/W1] \times 100 \quad (1)$$

where W2 is the weight of grafted, washed and dried PP non-woven sample and W1 is the weight of the non-woven sample before grafting.

The hydrophilicity of fabrics was determined by means of contact angle measurements before and after plasma polymerization, according to ASTM D724.

For contact angle measurement, five distilled water drops (10 μl volume) were uniformly dispersed on surface of each sample by a micrometer pipette. The image of each single drop was acquired by a video-camera connected to a computer. The contact angle's (θ) value was calculated by measuring the base (b) and height (h) of the drop profile and using the equation 2:

$$\text{Equation 2:} \quad \theta = \arcsin[4bh/(4h^2+b^2)]$$

The hypothesis is that the drop falls and sets on the surface with an exact spherical shape [17].

Scanning electron microscopy analyses of cotton fiber surfaces were performed with a Philips-XL30 SEM instrument. The surface chemical groups of the prepared fibers were confirmed by attenuated total reflection Fourier transform infrared spectroscopy (Nicollet Nexus 670).

3 Results and discussion

Figure 1 shows the effect of plasma treatment time on graft yield. As we see the graft efficiency increases with the increase in plasma treatment time. The reason is the creation of more free radicals and active sites as the plasma treatment increases. The grafting percent on the blank sample (grey PP nonwoven) is approximately zero which approves the positive effect of plasma treatment on grafting yield.

The decrease in graft yield after 60 sec. plasma treatment is due to over etching and non suitability of the surface for grafting.

As we see from Figure 2, grafting percent increases as the AA concentration increased from 15% to 30% after which decreased with further increase in AA concentration. This decrease can be due to more chance of AA monomers to form homo-polymer instead of copolymer with cellulose at increased concentrations.

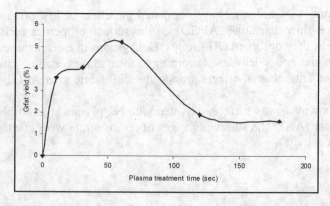

Figure 1 Effect of plasma treatment time on graft yield (25% AA, 80°C)

It is obvious from Figure 3 that graft yield increases with increase in grafting time, but after 60 minutes, more increase in grafting time has no significant effect on grafting yield. It can be because of reduced amount of AA monomer in the solution and free radicals at the fibers surface after prolonged time.

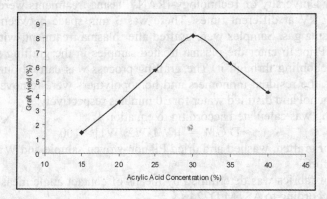

Figure 2 Effect of AA concentration on graft yield (60 sec. plasma treatment, 80°C)

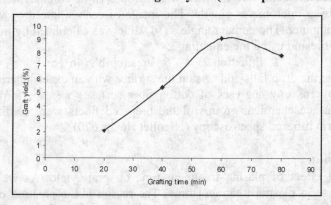

Figure 3 Effect of grafting time on graft yield (60 sec. plasma treatment, 80°C)

Table 1 shows the effect of grafting temperature on graft yield. It can be seen that graft yield increases with increase in grafting temperature because of more chemical reactivity in higher temperatures between monomers and free radicals on the fibers surfaces.

Table 1 Effect of grafting temperature on graft yield (25% AA, 60 sec plasma treatment)

Grafting temperature (°C)	Graft yield (%)
50	2.4
60	4.7
70	7.5
80	9.1

Figure 4 shows the SEM images of untreated, plasma treated (60 sec), plasma treated (60 sec) and grafted (25% AA, 60 min) samples respectively. The images show that plasma treatment has etched the surface of the PP fibers effectively; also the AA has been grafted on the surface of the fibers efficiently.

Figure 5 shows the ATR-FTIR spectra of plasma treated (60 sec) and grafted sample (25% AA for 60 min). Stretching peak at 1724 cm^{-1} indicates the existence of C=O, peakes in the range of 1100 to 1200 cm^{-1} indicates the C-O and the one at 1537 shows the COOH group on the surface of the fibers.

Figure 6 shows a remarkable decrease in contact angle of water on the surface of the PP nonwoven after plasma treatment and grafting with AA, indicating better wettability of the grafted nonwoven. This phenomenon is due to generation of polar chemical groups on the fibers surfaces which was confirmed by FTIR-ATR analysis.

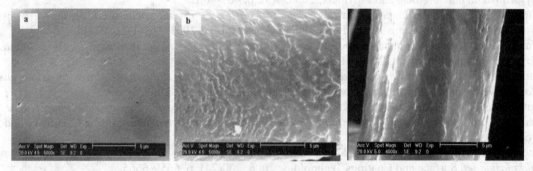

Figure 4 SEM images of untreated (a), plasma treated (60 sec) (b), plasma treated (60 sec) and grafted (25% AA, 60 min) (right) samples.

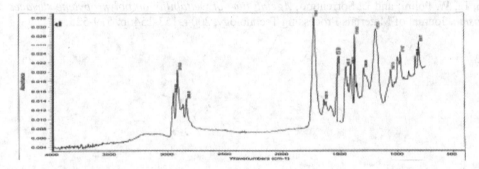

Figure 5 The ATR-FTIR spectra of plasma treated (60 sec) and grafted sample (25% AA for 60 min)

4 Conclusion

Plasma treatment as mentioned in this study is able to alter the surface of polypropylene fibers. The active species generated thereby, can facilitate the grafting of acrylic acid on the surface of the fiber. Graft yield increases with increase in plasma treatment time, grafting temperature, monomer concentration and grafting time. In the case of plasma treatment time and monomer concentration, there is a critical point, after which the graft yield decreases. SEM images and FTIR-ATR analyses confirm the grafting of AA on the fibers. Wettability of the grafted samples has been improved because of new polar groups generated on the surface of substrate.

References

[1] Gupta, B., S. Saxena, and A. Ray, *Plasma induced graft polymerization of acrylic acid onto polypropylene monofilament.* Journal of Applied Polymer Science, 2008. 107(1): p. 324-330.
[2] Kamaloddin Gharanjig, A.K., *Dyeing synthetic fibers and cellulose acetate.* 1995, Tehran: Amirkabir University Pub.
[3] López, R., et al., *Surface characterization of hydrophilic coating obtained by low-pressure plasma treatment on a polypropylene film.* Journal of Applied Polymer Science, 2009. 111(6): p. 2992-2997.
[4] Buyle, G., *Nanoscale finishing of textiles via plasma treatment.* Materials Technology, 2009. 24(1): p. 46-51.
[5] M.G. McCord, Y.J.H., P.J. Hauser, Y. Qiu, J.J. Cuomo, O.E. Hankins, M.A. Bourham and L.K. Canup, *Modifying Nylon and Polypropylene Fabrics with Atmospheric Pressure Plasmas.* Textile Research Journal, 2002. 72(6): p. 491-498.

[6] Masaeli, E., M. Morshed, and H. Tavanai, *Study of the wettability properties of polypropylene nonwoven mats by low-pressure oxygen plasma treatment.* Surface and Interface Analysis, 2007. 39(9): p. 770-774.

[7] Renáta Szabová, Ľ.Č., Magdaléna Wolfová, Mirko Černáka, *Coating of TiO2 nanoparticles on the plasma activated polypropylene fibers.* Acta Chimica Slovaca, 2009. 2(1): p. 70 - 76.

[8] Choi, H.S., et al., *Comparative actions of a low pressure oxygen plasma and an atmospheric pressure glow discharge on the surface modification of polypropylene.* Surface and Coatings Technology, 2006. 200(14-15): p. 4479-4488.

[9] Inbakumar, S., et al., *Chemical and physical analysis of cotton fabrics plasma-treated with a low pressure DC glow discharge.* Cellulose, 2010. 17(2): p. 417-426.

[10] Masaeli, E., et al., *Effect of process variables on surface properties of low-pressure plasma treated polypropylene fibers.* Fibers and Polymers, 2008. 9(4): p. 461-466.

[11] Yuranova, T., et al., *Antibacterial textiles prepared by RF-plasma and vacuum-UV mediated deposition of silver.* Journal of Photochemistry and Photobiology A: Chemistry, 2003. 161(1): p. 27-34.

[12] Wang, C. and C. Wang, *Surface pretreatment of polyester fabric for ink jet printing with radio frequency O2 plasma.* Fibers and Polymers, 2010. 11(2): p. 223-228.

[13] Yaman, N., *Improvement Surface Properties of Polypropylene and Polyester Fabrics by Glow Discharge Plasma System under Atmospheric Condition.* Tekstil ve Konfeksiyon, 2009. 1.

[14] Kuwabara, A., S.-i. Kuroda, and H. Kubota, *Development of Atmospheric Pressure Low Temperature Surface Discharge Plasma Torch and Application to Polypropylene Surface Treatment.* Plasma Chemistry and Plasma Processing, 2008. 28(2): p. 263-271.

[15] Cheng, C., Z. Liye, and R.-J. Zhan, *Surface modification of polymer fibre by the new atmospheric pressure cold plasma jet.* Surface and Coatings Technology, 2006. 200(24): p. 6659-6665.

[16] Abdou, E.S., et al., *Improved antimicrobial activity of polypropylene and cotton nonwoven fabrics by surface treatment and modification with chitosan.* Journal of Applied Polymer Science, 2008. 108(4): p. 2290-2296.

[17] Carrino, L., W. Polini, and L. Sorrentino, *Ageing time of wettability on polypropylene surfaces processed by cold plasma.* Journal of Materials Processing Technology, 2004. 153-154: p. 519-525.

88 Characterization and Mechanical Behavior of a Tri-layer Polymer Actuator

Akif Kaynak, Chunhui Yang, Yang C. Lim, Abbas Kouzani
(School of Engineering, Deakin University, Geelong 3217, Australia)

Abstract: Electrochemical synthesis of a tri-layer polypyrrole based actuator optimized for performance and stability was reported. 0.05M pyrrole and 0.05M tetrabutylammonium hexaflurophosphate in propylene carbonate (PC) yielded the optimum performance and stability. The force produced ranged from 0.2 to 0.4mN. Cyclic deflection tests on PC based actuators for a duration of 3 hours indicated that the displacement decreased by 60%. PC based actuator had a longer operating time, exceeding 3 hours, compared to acetonitrile based actuators. A triple-layer model of the polymer actuator was developed based on the classic bending beam theory by considering strain continuity between PPy and PVDF. Results predicted by the model were in good agreement the experimental data.

Keywords: Conducting polymers; Actuators; Synthesis; Mechanical behaviour; Modelling

1 Introduction

Conducting polymer actuators can convert the electrical input to mechanical output. Large bending deflections are possible at low operating voltages [1]. There has been considerable interest in improving the mechanical properties and performance of conducting polymer actuators with respect to power to force ratio and stability [2]. Most of the studies have been performed using polypyrrole and polyaniline as conducting polymers. Controllability of the electrochemical process, reasonable power to mass ratio, and controllable output make the conducting polymer actuator research attractive for potential applications [3]. Mass transport that occurs due to an applied potential difference across the width of an actuator is the basis for actuator movement [2]. Ion transport between the conducting polymer layers across the porous polymer membrane in the centre of the 3-layered actuator structure causes volumetric strain in the material due to transport of ions across the membrane [4]. The diffusion across the membrane is accompanied by movement of solvent molecules due to osmotic pressure. In this paper we report the electrochemical synthesis of a tri-layer polypyrrole based actuator optimized for synthesis and an analytical modelling of mechanical response of the actuator.

2 Design, synthesis and experimental of polymer actuator

Electrochemical polymerization can be performed either galvanostatically or potentiostatically by anodic oxidation of a solution containing the monomer and supporting electrolyte (dopant), generally in an organic solvent such as acetonitrile (ACN) or propylene carbonate (PC). Electrochemical polymerization is normally performed using a three-electrode set-up, where polymer deposits on to the working electrode. A two-electrode set-up can also be used with galvanostatic method. The electrochemical method offers a better control of the morphology, thickness and degree of doping of the conducting polymer and enable formation of polymers with better electrical and mechanical properties [5].

The distillation of pyrrole was carried out in a Kugelrohr ball–tube distillation apparatus at 160mbar at 120°C. The distilled pyrrole was stored in refrigerator. Tetrabutylammonium hexaflurophosphate ($TBAPF_6$) was used as the dopant. Propylene carbonate and acetonitrile were used as solvents. Gold coated polyvinylidene fluoride (PVDF) membranes (Sigma) were used as the working electrode. A fine stainless steel mesh (fine 250) was used as the counter electrode material. The fine mesh had higher surface area than the working electrode (PVDF) in order to ensure that the polypyrrole growth occurred only at the working electrode and not at the counter electrode. The chart software used in monitoring actuation was obtained from eDAQ company. A tri-layer actuator was fabricated [4,6], by initially sputter coating a PVDF film with approximately 100nm of gold layer, resulting in a conductive film with a surface resistance of 8-10Ω. The PVDF film was about ~145 μm thick had an approximate pore size of 45 μm. As the gold coated porous PVDF film was conductive it could be used as the working electrode on which a polypyrrole layer was deposited electrochemically. A solution containing 0.05M distilled pyrrole monomer, 0.05M ($TBAPF_6$) and 1% (w/w) distilled water in PC (propylene carbonate) solution was purged with nitrogen for 15 minutes. The deposition was carried out at a constant current density of 0.05mAcm^{-2} for 12 hours at -20°C in a freezer. The synthesis was carried out at low temperature as low temperature synthesis has been reported to yield higher conductivity and smoother surface morphology [7]. The constant current required was controlled using EG&G Princeton Applied Research Model 363 potentiostat/galvanostat. The final synthesized actuator was deep black in colour. The film was rinsed with PC solution containing 0.05M TBA$^+$ PF$_6^-$, the edges trimmed off and stored between two glass slides in the same solution before testing for actuation.

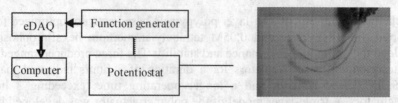

Figure 1 Instrumentation for actuation

As current density is a critical factor in the synthesis of actuators, optimization of the current density for actuation performance was investigated. Constant current density approach was chosen as opposed to constant voltage as the synthesis voltage could be maintained in the desired oxidation range of 0.7 to 1.2V. Different current densities were used while keeping the reactant concentrations at 0.05 M pyrrole and 0.05 M of dopant ion PF$_6$ in propylene carbonate (PC), which was used as the solvent. When synthesized at a current density of 0.1mA/cm^2, the film obtained had a higher surface roughness and was more rigid compared to the films obtained from 0.05mA/cm^2. These observations were confirmed in the actuator displacement tests. The actuator synthesized at lower current density of 0.025mA/cm^2 resulted in weak adhesion of the polypyrrole and some areas of the PVDF film remained uncoated. Also, the existing coating was observed to be peeling off. This could be due to the insufficient oxidation of pyrrole, resulting in poor adherence of polypyrrole to the PVDF film. Therefore, the synthesis of actuators were carried out and reported at current densities of 0.1 and 0.05mA/cm^2. The frequency was altered by using the function generator at a given voltage. The output of the function and voltage was monitored using a data acquisition system (eDAQ) interfaced with a computer, using Chart software (Figure 1). The displacement of actuator was measured at different voltages and frequencies and was recorded in millimeters (Figure 2).

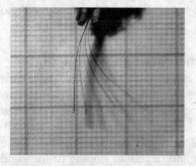

Figure 2 Superimposed sequential images to show the displacement of the actuator

Figure 3 shows the actuation displacements at different voltage and frequencies synthesized at 0.1mA and 0.05mA of PC/PPy/PF$_6$ actuators. The maximum displacement of around 5.0cm at 0.2Hz and 3V was obtained from the 4 cm actuator synthesized at a current density of 0.05mA/cm^2. The actuators synthesized with current density of 0.05mA/cm^2 had smoother surface morphology and they exhibited slightly better actuator performance. It was observed in both films that, as frequency increased the displacement decreased and eventually could not be observed. This was because the PF$_6$ ions and solvent ions did not get enough time to migrate to and fro between polypyrrole layers, through the PVDF layer at higher frequencies. Therefore, there was no observable displacement.

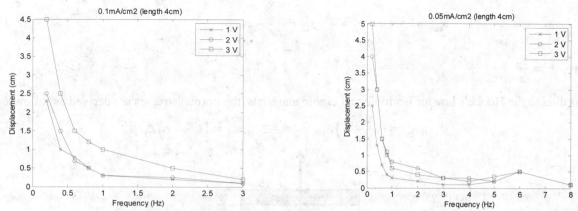

Figure 3 Actuator displacements of PF_6/PC actuator synthesized at $0.1mA/cm^2$ and $0.05\ mA/cm^2$

Figure 4 shows an SEM image for the PC/PF_6 actuator which was synthesized at $0.05mA/cm^2$. The higher magnification image shows an even deposition of PPy layer, which is about 12 microns thick (Figure 4a) and PVDF is about 152 microns thick (Figure 4b). The gold layer was not discernable as it was ~100nm and there was no morphological feature that would indentify it in the SEM image.

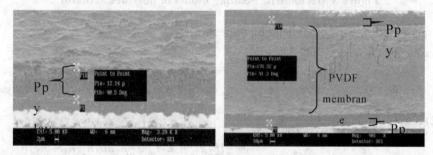

Figure 4 SEM images of the cross section of PC/PF_6 actuator synthesized at $0.05mA/cm^2$ (a) showing thickness of PPy layer (b) showing PVDF and PPy layers

Experiments were repeated with the same conditions using acetonitrile as solvent. These films had higher surface roughness compared to those with propylene carbonate. A potential difference of ±1V did not show any noticeable displacement. When ±2 and ±3 volt were applied, the displacement was similar to that of propylene carbonate based actuators. Acetonitrile based films were not as flexible and therefore were not as responsive to actuation at lower voltages.

The force and actuation life of actuator films synthesized in acetonitrile and propylene carbonate were investigated. Forces generated by the actuator were measured by tapping over a microbalance. The force produced from propylene carbonate actuators ranged from 0.2 to 0.4mN. PC based actuator had a longer operating time, exceeding 3 hours, compared to acetonitrile based actuators. Cyclic deflection tests on PC based actuators for a duration of 3 hours indicated that the displacement decreased by 60%. However, actuation could be regenerated by immersing the actuator into the electrolyte solution. In contrast, there was no significant electrical degradation after a 3 hour deflection test. Surface resistance measurements on the actuators prior to and after the continuous deflection tests did not show any significant increase in the resistance of the PPy layer within the operating range of ±1 to ±3V. Exceeding these limits may cause accelerated aging due to thermal degradation of the PPy.

3 Analytical modelling and force prediction of polymer actuator

The new triple-layer model of the polymer actuator was based on the classic bending beam theory. Different from Alici's model [8], the proposed model under pure bending was constructed by considering strain continuity between PPy and PVDF layers, as depicted in Figure 5. Therefore, the normal strains in PPy and PVDF layers are assumed to be distributed linearly along the thickness:

$$\varepsilon_{1u} = k\frac{y}{\rho} + c \quad \varepsilon_2 = \frac{y}{\rho} \quad \varepsilon_{1l} = k\frac{y}{\rho} - c, \tag{1}$$

where, $c = \frac{h_2}{\rho}$.

According to the Hooke's law for isotropic and elastic materials, the normal stresses are derived as follows:

$$\sigma_{1u} = \left(k\frac{y}{\rho} + c\right)E_1 \quad \sigma_2 = \frac{y}{\rho}E_2 \quad \sigma_{1l} = \left(k\frac{y}{\rho} - c\right)E_1. \tag{2}$$

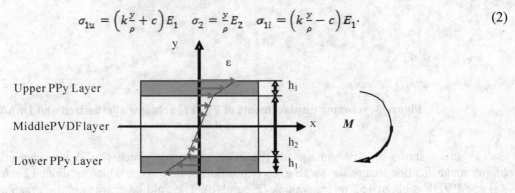

Figure 5 Mechanistic bending model of polymer actuator

Considering the quasi-static condition of bending moment,
$M = FL = \int y\sigma dA$,

$$M = \int_{-\frac{h_2}{2}-h_1}^{\frac{h_2}{2}+h_1} \sigma b y dy = \int_{\frac{h_2}{2}}^{\frac{h_2}{2}+h_1} \sigma_{1u} by dy + \int_{-\frac{h_2}{2}}^{\frac{h_2}{2}} \sigma_2 by dy + \int_{-\frac{h_2}{2}-h_1}^{-\frac{h_2}{2}} \sigma_{1l} by dy, \tag{3}$$

Substitute Eq. (2) into (3) and integrate it, we have

$$M = \frac{1}{\rho}\left[E_1 k I_1 + E_2 I_2 + \frac{1}{2}E_1 b h_1 h_2 (h_1 + h_2)\right], \tag{4}$$

where, $I_1 = I_{1u} + I_{1l} = \frac{b(h_2 + 2h_1)^3}{12} - \frac{bh_2^3}{12}$, $I_2 = \frac{bh_2^3}{12}$.

From Eq. (4), let $M = 0$,

$$k = -\left[\frac{1}{2}E_1 b h_1{h_2}(h_1 + h_2) - E_2 I_2\right]/E_1 I_1. \tag{5}$$

Further, the force at the actuator tip can be also obtained as

$$F = \frac{1}{\rho L}\left[E_1 k I_1 + E_2 I_2 + \frac{1}{2}E_1 b h_1 h_2(h_1 + h_2)\right], \tag{6}$$

As shown in Eq. (6), the force is inversely proportional to the length of the actuator and proportional to the curvature of the deformed actuator. In the present study, the thickness of PPy layers were the same, h_1 = 0.012 mm and the thickness of the core PVDF layer, h_2 = 0.145 mm. The width of the polymer film, b=5.0 mm and the total length of the actuator, L=40.0 mm. The Young's Moduli of the PPy and PVDF, E_1 and E_2, are 80 N/mm^2 and 440 N/mm^2, respectively. By using Eq. (5), k was calculated as 8.5. The prediction of the force outputs using the proposed new model are compared with those from Alici & Huynh [8] in Figure 6. According to Eq. (5), the value of k is dependent on the shapes, dimensions and Young's modulus of PPy and PVDF. Considering that the PPy layers are very thin compared to the PVDF layer, $k = 0$ can be used for constant strain condition, which led to good agreement between the experimental data and the model. From the experimental results shown in Figure 3, the minimum radius of the curved beam was found to be 0.0192 mm, which was subjected to the maximum bending moment generated by the polymer actuator. The output forces of the polymer actuator predicted by using the proposed analytic models have a good agreement with the maximum force, that was measured within the scope of 0.2-0.5 N.

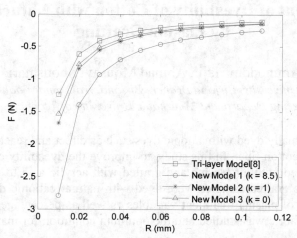

Figure 6 Predictions on force outputs of polymer actuator (unit: N)

4 Conclusion

Electrochemical synthesis of tri-layer actuators were experimentally studied to for optimum mechanical, electrical properties, actuator deflection, force output and stability. It was observed that 0.05 M pyrrole and 0.05 M tetrabutylammonium hexaflurophosphate in propylene carbonate (PC) yielded the best performance and stability. The force output ranged from 0.2 to 0.4mN. Cyclic deflection tests on PC based actuators for a duration of 3 hours indicated that the displacement decreased by 60%. However, the actuator performance could be regenerated by immersing the film back into the electrolyte solution. When compared to acetonitrile based actuator, PC based actuator had a longer operating time, exceeding 3 hours. A triple-layer model of the polymer actuator was developed based on the classic bending beam theory by considering strain continuity between PPy and PVDF. Results predicted by the model were in good agreement the experimental data. The minimum radius of curvature of the curved actuator beam was measured from sequential photographic records during actuation. The force output of the actuator was predicted by using the proposed analytic model which had a good agreement with the maximum force measured, which was within the range of 0.2-0.5 N.

Acknowledgement

The authors would like to thank Mr. Nishan Ahmed for the actuator fabrication and testing and Professor Geoff Spinks and Dr.Scott McGovern for initial discussions and training in actuator synthesis. The authors also acknowledge the support of Deakin University Grants Scheme (CRGS) for supporting the project.

References

[1] F. Vidal, Synth. Met., Vol. 156 (2006) p. 1299.
[2] R. H. Baughman, Synth. Met., vol.78 (1996) p. 339.
[3] J.D. Madden, Synth. Met., vol. 113, (2000) p. 185.
[4] Y. Fang, Mater. Sci. Eng.: C, Vol.28, (2008) p. 421.
[5] A. Kassim, et al, Synth. Met.,Vol.62 (1994) p. 41.
[6] Y. Wu, Synth. Met., Vol.156 (2006) p.1017.
[7] A. Kaynak, Mater. Res. Bull., Vol. 32 (1997) p. 271.
[8] G. Alici, N.N. Huynh, Sensor & Actuators A, 132, (2006) p. 616.

89 Improvement of Dyeability of Cotton with Natural Cationic Dye by Plasma Grafting

Aminoddin Haji[1], Ahmad Mousavi Shoushtari[2]

(1. Islamic Azad University, Birjand Branch, Birjand, Iran, email: aahl58@yahoo.com
2. Textile Engineering Department, Amirkabir University of Technology, Tehran, Iran)

Abstract: Cotton fabric is usually dyed with anionic dyes such as direct and reactive dyes. Naturally, there is no affinity for basic dyes to cotton fiber. In this study, to improve the dyeability of cotton fiber with cationic dyes, the fabric was pretreated with air plasma and grafted with acrylic acid to create acidic groups on the surface of cotton fibers. The grafted samples were dyed with natural cationic dye extracted from roots of berberis vulgaris. The effect of plasma treatment variables as well as percentage grafting of acrylic acid on the color strength of cotton fabric was studied and the optimum condition for maximum dye exhaustion was determined.

Keywords: Plasma; Grafting; Cationic dye; Dyeability; Cotton

1 Introduction

Natural dyes and pigments are found in some plants, animals, insects, bacteria, fungi and minerals. For example alizarin is a very old and common red dye found in roots of madder, rubia tinctoria. Development of synthetic dyes in the last century, reduced the use of natural dyes in modern dyeing [1,2,3]. Synthetic dyes are produced from cheap petroleum sources, and generally have easy dyeing with superior fastness properties. But there are drawbacks about synthetic dyes mainly toxicity and environmental pollution caused by waste water expelled from dye-houses.

Recently, a new tendency to natural dyes has raised mainly due to their environmentally friendly characteristics [1,2]. They are considered to give several advantages to several applications such as non toxic functions, specific medical actions and environmentally friendly finishes [2]. Natural dyes are clinically safer than their synthetic analogs in handling and use because of non carcinogenic and biodegradable nature [3,4,5]. Another factor driving interest in natural dyeing can be attributed to strict regulations and laws enacted by governments in response to consumer concerns [6].

Cotton is one of the most important natural fibers and is used extensively in textile industry alone or in blend with synthetic fibers [1]. This fiber has great characteristics such as good water and dye absorbency, comfort and stability. Synthetic dyes such as direct, vat, sulphur, azoic and reactive dyes can be used simply to dye cotton fiber [1,2]. Today due to environmental and health concerns, there is a great tendency to natural dyes [1-6]. Natural dyes can be used to dye protein fibers easily but there are several problems in using them on cotton fiber, mainly low affinity and fastness properties of natural dyes toward cotton fiber. To overcome this drawback, several studies have been carried out. P.S. Vankar etal have used mineral and biomordants to improve dyeability of cotton fiber with extract of rubia cardifia [2]. In another research P.S. Vankar etal, have used ultrasound energy to increase the absorption of rubia cardifia extract to cotton fiber [3]. Pretreatment of cotton with chitosan has been used to increase the affinity of cotton fiber for natural dyes [5]. In other researches, anionic active compounds, cross linking agents and enzymes have been used to improve the dyeability of cotton with natural dyes [4-6].

Berberis Vulgaris is a shrub which is extensively implanted in southern khorasan-Iran and many other places all over the world, for it's valuable fruit, barberry. In this study, the roots of this plant, has been used as a source of a natural colorant. There is a natural cationic yellow dye in these roots named Berberine (Figure 1) [7]. The amount of the dye is less in the woods of the plant. This natural cationic dye was extracted and applied on cotton fiber. Because of low affinity of the cationic dye to cotton fiber, the fabric was pretreated with air plasma and immediately grafted with acrylic acid to create acidic groups on the surface of cotton fibers. The grafted samples were dyed with natural cationic dye.

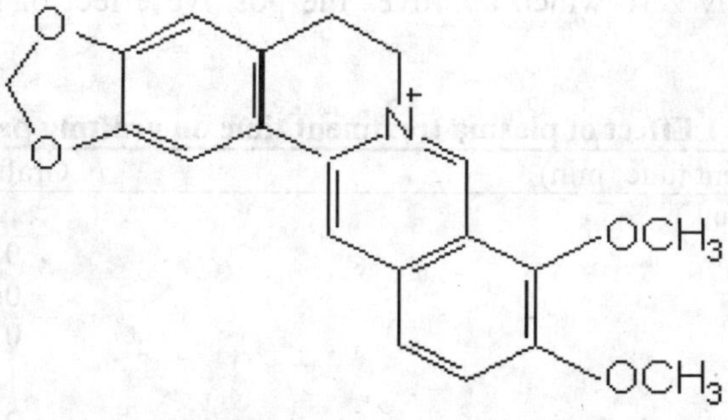

Figure 1 Chemical structure of berberine

2 Experimental
2.1 Materials
In this work scoured and bleached cotton fabric (142 g/m2) was supplied from Mazandaran textile Company, Iran. Before being used, the fabric was treated with a solution containing 1 g/L non-ionic detergent and 1g/L sodium hydroxide at 95°C for 30 minutes. Then the fabric was thoroughly washed with water and air dried at room temperature.

Berberis vulgaris roots were first washed and dried and then powdered. To prepare the original solution of the dye, each 100 gram of powder was added to 1 liter of distilled water and boiled for 2 hours and then filtered. The concentration of the resultant solution is 10% W/V. All chemicals used were analytical grade reagents from Merck.

2.2 Methods
1- Plasma treatment: the fabric samples were treated in an atmospheric pressure plasma chamber composed of two parallel electrodes with 2 mm space. The samples were placed between the electrodes. In all treatments, air was used as the processing gas with the power of 50 watts, voltage of 20 kV and frequency of 10 kHz at different time intervals.

2- Grafting: The plasma-treated cotton sample was withdrawn from the chamber and within about 90 seconds of exposure to atmospheric oxygen was placed into a reaction flask containing 200 ml of solutions of different acrylic acid (AA) concentrations in distilled water. The reaction flask was heated for different times at different temperatures. Then the fabric sample was drained and soaked first in 1/1 methanol/water (60 min. at 85°C), followed by two washings with distilled water for 15 min. to remove any non-reacted acrylic acid. The samples were then dried in an oven at 80°C for 1 h, cooled over silica-gel in a desiccator and weighed. The grafting percent is calculated according to the following equation: $G\% = [(W_1-W_2)/W_1]*100$

Where, W_1 and W_2 are the weights of the conditioned cotton fabric before and after the grafting process, respectively.

2- Dyeing: 100 cc of original dye solution was mixed with 100 cc of distilled water for each 5 gram of cotton (L:G= 40:1). The dyeing was started at 40°C and the temperature was raised to boil at the rate of 2°C per minute. Then the samples remained in that condition for appropriate 45 minutes, and then rinsed and air dried. All dyeing processes were carried out using a laboratory dyeing machine made by Rissanj co.-Iran.

3- Color measurements: the reflectance of dyed samples and color coordinates CIE L*, a*, b* values were measured on a Color-eye 7000A spectrophotometer using illuminant D65 and 10°standard observer. Color strengths (K/S) of dyed samples were calculated using kubelka-munk equation:

$$\text{Equation 1: } K/S = (1-R)^2/2R$$

where R is the observed reflectance, K is the absorption coefficient and S is the light scattering coefficient.

4- Color fastness tests: color fastness to washing, light and rubbing was measured according to: ISO 105-C01: 1989(E), ISO 105-B02: 1994(E), ISO 105-X12: 1993(E) respectively.

3 Results and discussion
3.1 Effect of plasma treatment on grafting efficiency
Cotton samples were plasma treated at different times and grafted with a 10% V/V of AA for 1 hour at 60°C. Table 1 shows the effect of plasma treatment time of graft yield of AA on cotton. As we see the graft efficiency increases with the increase in plasma treatment time. The reason is the creation of more free radicals and active sites as the plasma treatment increases. The grafting percent on the blank sample (grey

cotton fabric) is approximately zero which approves the positive effect of plasma treatment on grafting yield.

Table 1 Effect of plasma treatment time on grafting percent

Plasma treatment time (min)	Grafting %
0 (blank)	0
1	0.41
2	0.64
3	0.96
4	1.54
5	2.37

3.2 Effect of AA concentration on grafting efficiency

Plasma treated cotton samples (1 min, 20 kV, 50 W, 10 kHz) were grafted using different concentration of AA. Figure 2 shows that grafting percent increase as the AA concentration increased from 10% to 40% after which decreased with further increase in AA concentration. This decrease can be due to more chance of AA monomers to form homo-polymer instead of copolymer with cellulose at increased concentrations.

3.3 Effect of grafting time on graft yield

As we can see from Figure 3, the grafting yield increases with the increase of grafting time from 20 minute to 60 minute, after which more increase in grafting time has no significant effect on grafting yield. it can be because of reduced amount of AA monomer in the solution and free radicals at the fibers surface after prolonged time.

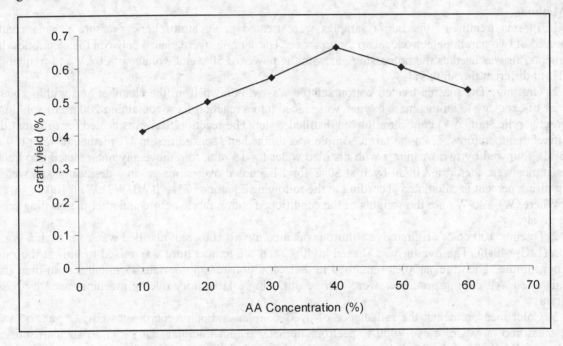

Figure 2 The effect of AA concentration on graft yield

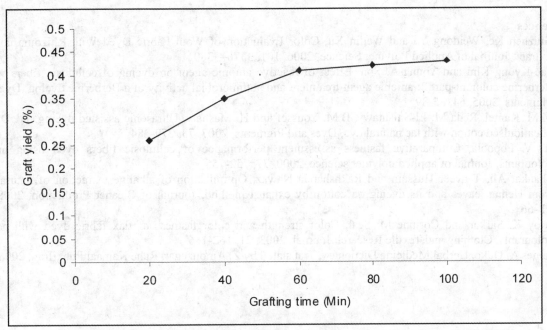

Figure 3 Effect of grafting time on graft yield (plasma treated, 10% AA, 60°C)

3.4 Effect of graft yield on color strength of dyed cotton sample

Cotton samples from different graft percents were dyed with the above mentioned process. As we see from Figure 4 the K/S of samples increases as the graft yield increases. This is due to creation of more acidic sites after grafting of AA on cotton. Cellulose has no active site to react with cationic dye, Berberine (Figure 1). After grafting of AA on cotton, the acidic COOH groups will appear on cotton surface which will promote more absorption of cationic dye to it.

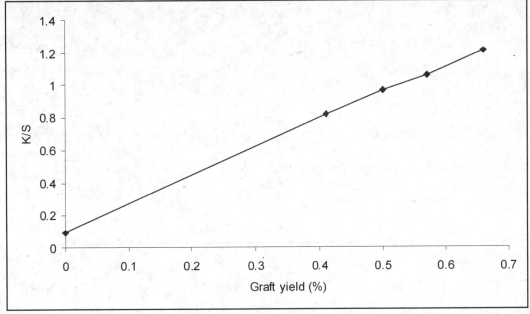

Figure 4 Effect of graft yield on color strength of plasma treated cotton

4 Conclusion

Plasma treatment of cotton fabric as described in this study improves the graft yield of acrylic acid on the fiber. Grafting of acrylic acid on cotton fiber is affected by AA concentration and time of plasma treatment and grafting time. Grafting of acrylic acid onto cotton fiber improves the absorption of Berberine natural dye on it because of ionic interactions between the dye and the modified fiber.

References

[1] Guizhen Ke, Weidong Yu and Weilin Xu, Color Evaluation of Wool Fabric Dyed With Rhizoma coptidis Extract, Journal of Applied Polymer Science, 2006, 101, 3376–3380

[2] Tae-Kyung Kim and Young-A. Son, Effect of reactive anionic agent on dyeing of cellulosic fibers with a Berberine colorant-part 2: anionic agent treatment and antimicrobial activity of a Berberine dyeing, Dyes and Pigments, 2005, 64, 85-89

[3] M.M. Kamel, Reda M. El-Shishtawy, B.M. Youssef and H. Mashaly, Ultrasonic assisted dyeing. IV. Dyeing of cationised cotton with lac natural dye, Dyes and Pigments, 2007, 73, 279-284

[4] A. V. Popoola, Comparative fastness assessment performance of cellulosic fibers dyed using natural colourants, Journal of applied polymer science, 2000, 77, 752-755

[5] Shaukat Ali, Tanveer Hussain and Rakhshanda Nawaz, Optimization of alkaline extraction of natural dye from Henna leaves and its dyeing on cotton by exhaust method, Journal of Cleaner Production, 2009, 17, 61–66

[6] Ajoy K. Sarkar and Corinne M. Seal, Color strength and color fastness of flax fabric dyed with natural colourants, Clothing and Textile Research Journal, 2003, 21, 162-166

[7] James A. Duke, Herbal Medicine Dictionary, Translated by Z. Amouzegar, Rahe Kamaal Pub., Iran, 2006

90 The Impact of Textile Fibres on the Environment

Chris Hurren, Qing Li, Xungai Wang
(Centre for Material and Fibre Innovation, Deakin University, Geelong, VIC 3217, Australia)

1 Introduction

The production of fibres and textiles consumes inputs including raw materials, energy, chemicals and water. Production generates outputs including finished articles, hard waste, liquid waste, waste heat and waste gases [1]. Each fibre type has its own environmental footprint with some fibres consuming more energy than others whereas some produce more waste streams than others. As global resources become scarcer and people target environmentally friendly products the selection of low impact textiles will become important.

The main fibres used in apparel are natural fibres (plant grown cellulose and animal fibres) and manufactured fibres (polyester, regenerated cellulose, nylon and acrylic). There is a strong debate on the benefit of natural versus manufactured fibres as natural fibres come from a renewable resource and manufactured fibres mostly come from a petroleum starting base [2]. The growing, processing and use of natural fibres can have significant environmental problems whereas the sourcing of the precursor material and processing of the fibre has the majority of the problems for manufactured fibres. Also entering into the debate are the new range of polymers produced from a naturally occurring precursor material and these include poly-lactic acid [3].

This paper compares each of the established and emerging fibres and investigates their cost on the environment. The information on dyeing within this report is derived from dyeing pattern cards produced by Ciba Specialty Chemical, Switzerland and BASF, Germany.

2 Naturally occurring cellulosic fibres

Of the naturally produced cellulosic fibres, cotton holds the market share (35.9% of 2006 total apparel market [4]). Other natural cellulosic fibres include hemp, flax, bamboo, ramie, kenaf, pineapple, jute, sisal and coir. However, for textile apparel only the first six have diameters and properties that are appropriate. These fibres are derived from the bark (bast fibres), leaf and seed husks of the plant but are still only a niche product in the fibre market. These fibres show the largest potential for growth due to their high production rates with respect to land use.

2.1 Cotton

The main advantage of cotton is that the fibre has good moisture adsorption, strength and durability. As a plant harvested fibre cotton is derived from a renewable resource with carbon bound in the fibre during growth. Cotton seed and cotton seed oil are by-products of the fibre harvest. With the assistance of irrigation cotton has moderate fibre production rates per land area. Processing methods are well known and significant research has been done in optimising energy consumption during production. The fibre is biodegradable however the fibre can be slow to degrade depending on the conditions.

The main environmental disadvantages with cotton are found during fibre growth and textile colouration. During growth cotton has high irrigation water requirements ($7m^3$/kg to $29m^3$/kg [5]). Growth requires significant inputs of chemicals (herbicides, insecticides and defoliating agents) and fertilisers. In 1999 cotton accounted for 24% of global insecticide and 7.5% of all artificial fertiliser but occupied only 2.4% of the arable land [5]. Colouration requires significant inputs of water (80-200 l/kg) and chemicals, moderate inputs of energy and produces effluent with high colour content, biological oxygen demand (BOD) and chemical oxygen demand (COD). The bleaching step before dyeing has high alkali (40-100 g/kg NaOH) and peroxide (30-80 g/kg H_2O_2) consumption which is largely still in the discharge effluent of the bleach bath and subsequent rinse baths. The dyeing step requires high to very high amounts of salt (240-16000 g/kg NaCl) and high alkali (40-100 g/kg Na_2CO_3 and 8-40 g/kg NaOH). Low fixation is seen in cotton reactive dyes with 50-85% fixed to the fibre depending on the reactive dye used. This leads to high colour content in the discharge effluent. The water holding capacity of cotton fibre makes it slow to dry after wet processing.

2.2 Bast and leaf fibres

The advantages of bast and leaf fibres are good moisture adsorption, strength, durability and some microbial attack resistance. Apart from flax the fibre is produced at high production rates per land area and requires little or no pesticide use during growth. Growth requires chemical (herbicides) and fertiliser inputs for optimum production rates. Most fibres are only part of the plant yield with seed and pulp cellulose available from bast fibres and fruit available from some leaf fibres. Like cotton the fibre is from a renewable resource and has carbon bound into it during growth. The fibre has similar biodegradability properties to that of cotton.

The disadvantages of these fibres are that most are bound together with a mixture of lignins, pectins and waxes. These need to be dissolved before yarn manufacture can be conducted by a process called retting. Of the retting systems field retting is the most environmentally friendly but ties up production land and is labour intensive. Water or chemical retting is faster but has higher effluent production with larger BOD and COD values [6]. Chemical retting is commonly used but can require large amounts of alkali (20-70 g/kg NaOH [7]) and produce high BOD and COD effluent.

If processed by the traditional processing path bast fibre yarn manufacture is inefficient and has seen very little optimisation for reduced energy consumption. The fibre can be separated to a fineness where it can be spun on cotton processing equipment but this currently requires high levels of chemical retting with its associated effluent problems. Once processed into a yarn these fibres also suffer from the same colouration problems as cotton as both use the same dyestuffs and dyeing process.

2.3 Animal fibres

There are a wide range of animal fibres available including wool, silk, alpaca, and cashmere. However the share of these fibres in the market place is relatively small. In 1996 wool accounted for only 1.7% of all textile apparel fibre sales and silk for 0.2% [4]. In this paper the environmental impact of wool fibre will be considered as it is one of the most commonly used animal fibres.

2.3.1 Wool

Wool is seen as a specialty fibre with good moisture adsorption, warmth, handle and drape. The fibre has very good durability with a high extension to break providing long lasting garments. The fibre is seen as a luxury fibre and in some countries it is worn to represent prestige of the owner. A bi-product of wool production is meat and lanolin and the fibre comes from a renewable resource. Wool fibre is biodegradable however it is slow to degrade.

The disadvantages of wool start on the farm. Sheep produce moderate emissions of methane which is a greenhouse gas and there is significant ammonia contamination of land from their manure [8]. The land production value is low when grazed by sheep and this is reflected in the purchase price of raw fibre which is almost four times the price of cotton ($8.10/kg in May 2010 [9]).

Disadvantages are seen in production during wool scouring, yarn production and dyeing. Wool scouring requires moderate inputs of energy and chemicals (detergents, builders, alkali) and produces effluent with high BOD and COD values. Yarn production is slow with top making requiring a high number of fibre processing steps. Colouration is more efficient than dyeing cellulose but still has a significant impact on the environment. One of the main dyeing systems utilises premetalised dyes in which heavy metals, including chromium and cobalt, are used to improve dye fastness. The dyeing step has a moderate input of energy and a moderate input of chemicals with salt (50-100 g/kg Na_2SO_4), acid (10-40 g/kg CH_3COOH) and detergents (10-40 g/kg) as the main components. Dye exhaustion is considerably better than cotton with 92-99% exhaustion common. Dyeing processes can be improved to yield even better dye exhaustion and improve effluent qualities. Dye effluents are better than cotton but still have a low to moderate BOD and COD demand.

2.4 Manufactured fibres

Of the manufactured fibres polyester (PET) holds around 78% of the synthetic market, followed by nylon (PA) with 9%, viscose with 7% and acrylic with 6% [10]. This discussion will centre on PET and viscose as nylon and acrylic have similar advantages and disadvantages to PET. The newly developed poly-lactic acid fibre (PLA) will also be discussed as it is derived from renewable resources.

2.4.1 Polyester

PET fibre has the advantage that it can be engineered to suit end use performance requirements. The base fibre is strong, has good elasticity and is durable. However, with the right production profile and chemical modification during synthesis and extrusion it can be given a range of different properties. These include wicking, de-lustred, coloured, flame retardant, hydrophobic, oleophobic, hydrophilic and microstructured. The processing method is well known and highly optimised for reduced energy consumption with very high production rates achievable. PET can be solution coloured by pigments during manufacture with no effluent production or significant chemical use. The product is far more durable during washing and in use than all of the natural fibres and most of the manufactured fibres and requires considerably less energy to dry after wetting or laundering. The used product can be recycled into lower grade PET items after garment end of life cycle.

The disadvantages of PET are that there is a high level of energy used in its polymer production with the precursor material derived from a non-renewable petrochemical feedstock. If dyed in fibre form with conventional aqueous dyeing methods with disperse dyes then PET has poor to very good wash fastness with very high dye fixation (90-98%) depending on the dyestuff used. The dyeing step has a high input of energy and a moderate input of chemicals with alkali (5-20 g/kg NaOH), reducing agent (10-40 g/kg Sodium hydrosulfide) in the reduction clear as the main components. Dye effluents are better than cotton but still have a low to moderate BOD and COD demand. The fibre takes a very long time to break down in land fill.

2.4.2 Regenerated cellulose fibres

There are a large range of regenerated cellulose fibres and the name is derived from the processing method. The names include Viscose, Rayon, Acetate, Triacetate, Tencel and Lyocel. Like natural cellulose fibres regenerated cellulose fibres are derived from renewable precursors and this is generally wood pulp. The fibre has good fibre properties including high moisture absorption, handle, and lustre with the fibre cross-section and diameter controlled to suit the end product. The finished product is biodegradable after garment lifecycle.

The disadvantages with regenerated cellulose fibres are the chemicals used in the production process, colouration and poor wet fibre strength. Fibre production uses chemicals that are bad for the environment. Chemicals used in production include carbon disulfide, n-Methylmorpholine n-oxide, acetic anhydride and solvents like acetone. The chemicals vary depending on the processing method but each processing method has at least one environmentally unfriendly chemical in it. Each of these chemicals is a significant risk to the environment and is released during manufacture as either liquid waste, with escaping gasses or as a residue in produced fibres. The colouration process is exactly the same as that used for cotton however wet dye fastness is generally slightly less than that seen for cotton.

2.4.3 Polylactic acid

PLA is a polymer derived from a renewable resource such as corn starch or sugar cane. The main manufacturer Cargill Dow claims that the fibre has excellent wicking properties, moisture management, low odour retention, does not support bacterial growth, is hypoallergenic and quick drying. Like PET it can be engineered to suit end use performance requirements. Colouration can be achieved by solution dyeing with pigments during manufacture with no effluent production or significant chemical use. The polymer is recyclable or biodegradable at the end of the garment lifecycle.

The main disadvantages are very high energy use in precursor growth, extraction and in polymer production. Production of the precursor from corn starch or sugar takes land away from food production. The melting point of the polymer is far lower than that of PET. The dyeing process is almost identical to that used for PET however dyeing is undertaken at lower temperatures with a resultant reduction in wash fatness.

2.5 Discussion

Within the natural fibres animal fibres will continue to be grown and used in apparel manufacture as they are a by-product of animal meat production. Cotton will remain dominant in the market until a viable and environmentally friendly processing method can be developed for bast or leaf fibres. The bast fibre hemp would be a good alternative as it has a high yield of fibre, cellulose (from the internal stem material) and seed for a given parcel of land over a growing year. It requires far less herbicides, pesticides and chemicals than cotton. Natural fibres will be used more in their natural colour if there is a significant environmental restriction placed on dyeing effluents.

Cellulose based and starch/sugar based manmade fibres will not be a viable option due to production environmental impacts. Polyester will continue to dominate the fibre market and will tend towards solution dyed to reduce effluent production during colouration. Polyester will become a far more dominant fibre if a viable and environmentally friendly production method can be developed for producing its precursor materials from waste cellulose or another waste product.

3 Conclusion

Of the fibres available PET has a relatively low impact on the environment. Cotton has significant environmental problems but has no currently viable plant derived cellulose based alternative ready to replace it. Animal fibres will continue to be used as they are produced as a by-product of meat production. Of the manufactured fibres, regenerated cellulose and PLA, are not currently an alternative to PET or cotton due to the environmental impact in their production. Aqueous dyeing will need to significantly reduce its environmental impact for it to continue to be adopted in textile colouration.

Acknowledgement

The authors would like to acknowledge the inspiration of Katie Trewhella and Tricia Pert from the Gordon TAFE for suggesting this work.

References

[1] Groff, K.A., *Textile Waste.* Water Environment Research, 1992. 64(4): p. 425-429.
[2] Kalliala, E.M. and P. Nousiainen, *Life Cycle Analysis - Environmental Profile of Cotton and Polyester-Cotton Fabrics.* AUTEX Research Journal, 1999. 1(1): p. 8-20.
[3] Ingeo. *Ingeo Apparel Product Guidelines.* 2006 [cited 2009; Available from: www.ingeofibers.com.]
[4] Rupp, J., *Ecology and economy in textile finishing.* 2009, http:www.textileworld.com.
[5] Soth, J., C. Grasser, R. Salerno, and P. Thalmann, *The Impact of Cotton on Fresh Water Resources and Ecosystems.* 1999, World Wide Fund: Zurich, Switzerland.
[6] Morrison III, W.H., D.E. Akin, G.N. Ramaswamy, and B. Baldwin, *Evaluating Chemical Retted Kenaf Using Chemical, Histochemical and Micro-spectrophotometric Analysis.* Textile Research Journal, 1996. 66(10): p. 651-656.
[7] Hurren, C.J., X. Wang, H.G.S. Dennis, and A.F.K. Clarke. *Evaluation of Bast Fibre Retting Systems on Hemp.* in *82nd Textile Institute World Conference.* 2002. Cairo, Egypt.
[8] Dahllof, L., *LCA Methodology Issues for Textile Products* in *Environmental Systems Analysis.* 2004, Chalmers University of Technology: Goteborg, Sweden.
[9] YnFx. *Pricewatch Report Highlights 24 May_2010.* 2010; Available from: YarnsandFibers.com.
[10] YnFx. *World Fiber Report.* 2008 [cited 2010; Available from: YarnsandFibers.com.

91 Thermal and Chemical Properties of Wool Powders

G. Wen, X. Liu, X.G. Wang

(Centre for Material and Fibre Innovation, Institute for Technology Research and Innovation, Deakin University, Geelong, Victoria 3217, Australia)

Abstract: Wool powders with different particle sizes were examined in terms of their crystal structures, thermal properties, surface chemical compositions and moisture regains. It was found that the crystallinity of wool powders was increased, and the moisture regains were decreased as the particle sizes of wool powders were reduced. For comparison, the properties of activated charcoal were also investigated. The higher dye uptake of activated charcoal at pH 10, compared to that of wool powder, could be due to its greater surface area and porous structure.

Keywords: Wool powders; Activated charcoal; Characterization; Dye uptake

1 Introduction

In our previous study [1], wool powders with different particle sizes have been produced using various milling techniques. The powders were characterized in particle size and surface morphology with scanning electron microscopy (SEM), in surface area and surface charges by measuring their zeta potentials, and in surface chemical composition using X-ray photoelectron spectroscopy (XPS). The rate and extent of dye uptake at pH 4.5 by finer wool powders were found to be comparable to that obtained with an activated charcoal, although the surface area of the activated charcoal was 100 times greater.

In the present study, the properties of wool powders were further investigated. The crystal structure and the thermal properties of wool powders were determined by using X-ray diffraction (XRD) and differential scanning calorimetry (DSC), respectively. The XPS spectra of C 1s, N 1s and O 1s for wool powders were reported. The moisture regains of wool powders of different particle sizes were measured using a standard method. The same measurements were also conducted on activated charcoal, which was used to compare dye sorption ability of wool powders. The comparison of dye uptake by a wool powder and activated charcoal was carried out at pH 10, at which these two sorbents have similar amount of surface charge.

2 Experimental

2.1 Materials and chemicals

Merino wool top and four wool powders, labelled as WP-A, WP-B, WP-C and WP-D, respectively, were used in this study. The production and characterization of these samples were introduced in detail elsewhere [1]. In brief, the Merino wool top was used as the starting material for wool powder production. The average diameter and the Hauteur length of the wool fibre were 20.4 μm and 72.8 mm respectively. The particle sizes on a volume basis were 61.0 μm, 51.4 μm, 6.2 μm and 4.5 μm for WP-A, WP-B, WP-C and WP-D, respectively.

Activated charcoal was supplied by Sigma® chemical company (untreated powder, No.C-5260). Albegal FFA (Ciba) was used as a wetting agent. Ammonium sulphate (Analar®) and ammonia solution (30%) were used to prepare a pH 10 buffer.

Two acid dyes, C. I. Acid Red 13 and C. I. Acid Red 18, as well as a basic dye, methylene blue (C. I. basic blue 9), were used as sorbates without further purification. Their chemical structures are given in Figure 1.

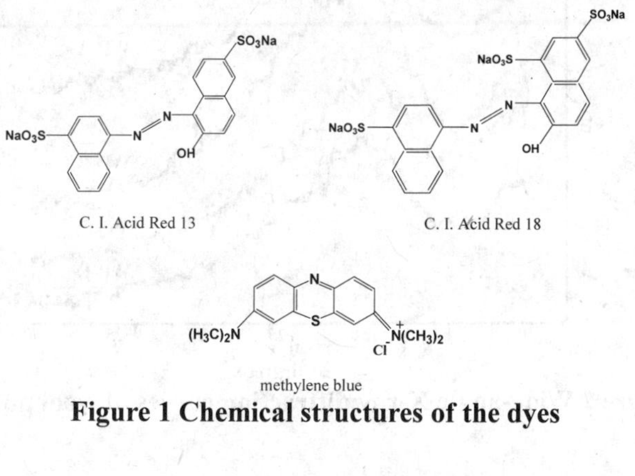

Figure 1 Chemical structures of the dyes

2.2 X-ray diffraction (XRD) analysis

The crystallinity of wool powders was determined using a Philips PW1729 diffractometer. The samples were scanned from $2\theta = 10 - 40°$ at a scan rate of 1°/min. The voltage and current of X-ray were 38 kV and 28 mA, respectively.

2.3 Thermal analysis

Differential scanning calorimetry (DSC) analysis of wool substrates was performed using a TA DSC Q200. Wool fibres or powders (~5 mg) were weighed on aluminium cells and sealed to avoid moisture evaporation. The range of temperature scanning was set from 20°C to 320°C with a heating rate of 10°C/min.

2.4 Moisture regain measurement

Triplicate samples from each wool powders or fibres (approximately 5 g for each sample) were dried at 105°C for 1 hour, and then conditioned at 65 ± 2% relative humidity and 20 ± 2°C for 24 hours and weighed (W_1), finally dried at 105°C until no further weight loss occurred (W_2). Moisture regains of wool samples were calculated from equation 1:

$$\text{Moisture regain (\%)} = \frac{\text{weight of moisture}(W_1 - W_2)(g)}{\text{weight of dry wool sample}(W_2)(g)} \times 100 \tag{1}$$

2.5 Particle size measurement for activated charcoal

The particle size of activated charcoal was measured using a Malvern Mastersizer 2000 with a hydro 2000s. The solvent used for the measurement was water. The refractive index used was 2.420 for activated charcoal, and 1.330 for water. The pump was set at 2520 RPM.

2.6 SEM of activated charcoal

The morphology of activated charcoal was studied using a LEO 1530 at a 5-kV acceleration voltage and a working distance of 2 mm.

2.7 X-ray photoelectron spectroscopy (XPS) of activated charcoal

The surface composition of activated charcoal was analysed using a Kratos AXIS Ultra DLD X-ray Photoelectron Spectrometer under the same conditions for wool samples, as described in the previous paper [1].

2.8 Dye sorption

Dye sorption by wool powder (WP-C) and activated charcoal was carried out under the conditions of 1.5 g/L dye concentration, pH 10, 25°C and a contact time of 2 hours. The sorption procedure was described elsewhere [2].

3 Results and discussion

3.1 X-ray diffraction (XRD)

The X-ray diffraction curves of wool powders in Figure 2 show that all wool powders have the typical diffraction pattern of α-keratin, with a prominent peak at $2\theta \approx 20.2°$, which corresponds to the crystalline spacing of 4.39 [3]. It was noticed that there was no significant difference in the intensities and the areas under the major peaks of WP-A, WP-C and WP-D. This suggested that the crystallinity of these three wool powders was similar, although WP-C and WP-D had undergone more mechanical milling processes. The reason for the increased intensity of WP-B, which meant more crystalline regions in this wool powder, is not clear, and needs to be further determined.

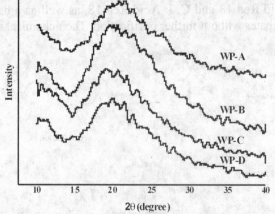

Figure 2 Wide-angle X-ray diffraction curves of wool powders

3.2 Differential scanning calorimetry (DSC)

DSC curves of wool fibre and wool powders are shown in Figure 3, and the peak temperatures and enthalpies are summarized in Table 1. The glass transition of wool (Peak 1), occurring at $T_g = 50 - 60°C$, was difficult to be observed in chopped wool fibres and wool powders because of the overlapping with the beginning of the Peak 2 and quick loss of moisture.

Peak 2 is assigned to the evaporation of water (moisture) in sample. In the present study, the water evaporation temperatures for all wool samples were found to be around 80°C, which were lower than those reported in some literatures (110 - 120°C) [4, 5]. It can be seen from Table 1 that, as the size of wool powders was reduced, the water evaporation temperatures were increased, while the evaporation enthalpies reduced. This indicated that the interactions of wool with water became stronger, but the amount of water present in wool powder was decreased [5], when wool was milled into fine wool powder.

Table 1 DSC peak temperatures and relative calculated enthalpies of transition/reaction, for the analysed wool samples

Wool sample	Water evaporation (Peak 2)		Denaturation (Peak 3)			Breakage of crosslinks (Peak 4)
	Temperature (a) (°C)	Enthalpy (J/g)	Temperature (°C) (b)	(c)	Enthalpy (J/g)	Temperature(d) (°C)
Wool fibre	78.25	268.2	232.18	240.04	13.38	-
WP-A	78.77	248.6	235.86	240.65	11.68	277.54
WP-B	80.95	278.1	237.15	-	33.79	279.12
WP-C	81.98	235.1	238.83	245.41	36.99	278.75
WP-D	84.88	186.0	234.36	245.77	40.16	281.45

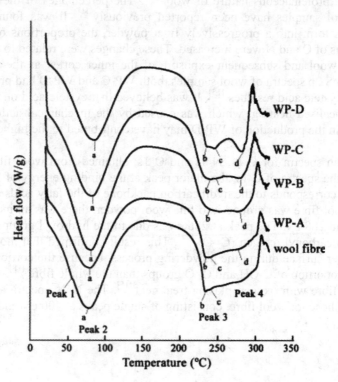

Figure 3 DSC curves of wool samples

Peak 3 is relative to the denaturation of wool, corresponding to the stages of melting and degradation of the different morphological components of wool [6]. Wool samples showed typical bimodal endotherms with two peaks at 230°C - 250°C. The bimodal endotherm of wool could be attributed to the different melting temperatures of the α-form crystallines in the ortho- and para- cortical cells, as well as the overlapping of the melting endotherm of α-form crystallites and the thermal degradation of other components (e.g. cell membrane complex (CMC)) in wool [7].

It is shown in Table 1 that, with the decrease in the wool powder size, both the denaturation temperatures and the denaturation enthalpies increased. This suggested that the finer the wool powder, the more stable and greater were its crystalline regions. A similar result for cortical cells was reported by Wortamnn and Deutz [8]. They found that the denaturation enthalpies of both ortho-and para-cortical cells, which were isolated from wool fibres with an enzyme (Subtilisin), were higher than that of whole wool fibre. They believed that the main reason for their result was the removal of the cuticles. The disrupted cuticles and the exposed cortical cells of wool powders (especially WP-C and WP-D) (as shown in SEM images of wool powders [1]) may contribute to the increased crystallinity of fine wool powders.

The breakage of crosslinks of wool is considered to be responsible for Peak 4 [9]. These crosslinks include disulfide bonds (–S–S–), hydrogen bonds and salt links. The intensity of Peak 4 increased when the wool was milled into finer powders. This indicates that mechanical milling broke some crosslinks of wool during wool powdering. The prominent Peak 4 of WP-D may be due to the treatment of wool with sodium dichloroisocyanurate (DCCA) used in the production of WP-D. DCCA released hypochlorous acid (HClO), which further released chlorine (Cl_2) [10, 11]. The chlorine oxidized cystine ($\substack{OC \\ HN}\!\!>\!\!CH\text{-}CH_2\text{-}S\text{-}S\text{-}CH_2\text{-}CH\!\!<\!\!\substack{CO \\ NH}$) into cysteic acid residues ($\substack{OC \\ HN}\!\!>\!\!CH\text{-}CH_2\text{-}SO_3^-$) [12, 13].

The crystallinity of wool powders determined using XRD varied slightly from that using DSC; the former found no remarkable difference in the crystallinity of wool powders with different sizes, while the latter reported higher crystallinity in the finer wool powders.

3.3 XPS analysis of wool powders

The major elements on the surface of both wool fibres and powders include carbon, oxygen, nitrogen and sulphur due to the proteinaceous nature of wool [14]. The percentages of these elements and the S 2p spectra of various wool samples have been reported previously [1]. It was found that, as the wool was converted from fibrous form into a progressively finer powder, the proportions of C and S were reduced, while the concentrations of O and N were increased. These changes were related to progressive breakdown of cuticle cells from the wool and subsequent exposure of the inner cortex as the fibres were mechanically milled into powders. In S 2p spectra of wool samples, both WP-C and WP-D had prominent peaks at ~168 eV, which corresponds to cysteic acid residues [15]. It was believed that cysteic acid on the wool powder surfaces could be formed by localised heating, which was caused by the mechanical action of air-jet milling. The DCCA treatment used in the production of WP-D may have contributed to the higher concentration of cysteic acid of this powder.

The high-resolution spectra for C 1s, N 1s and O 1s, obtained from wool fibres and the powders, are shown in Figure 4. In the spectra of C 1s, the major peak at the binding energy of 285 eV, which is assigned to C-C, C-H, and C-S, corresponds to the hydrocarbon backbone of the fatty acids and the side groups of the amino acids [16]. As wool fibre was ground into fine wool powder, there was an increase in the intensities of the peaks at 286 eV and 288 eV. The peak at 286 eV is due to the hydroxyl groups (C-OH), and the peak at 288 eV is attributed to carbonyl oxygen (C=O) [17]. This result confirmed that more cortical materials were exposed to wool powder surface during the powdering process, because the cortical cells are rich in cystine and contain a higher proportion of C-OH and C=O groups than the whole fibre [18]. A similar result was found for the oxidized wool fibre with oxygen plasma treatment [15]. The N 1s spectra and O 1s spectra for wool powders are similar to those for wool fibre, consisting of single peaks at 400 eV and 532 eV.

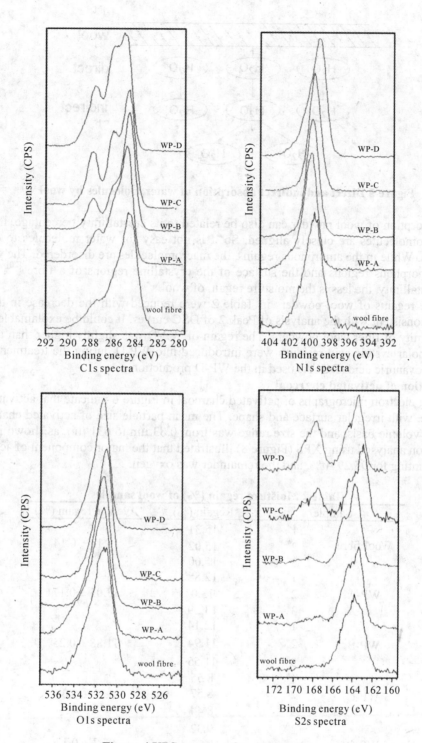

Figure 4 XPS spectra of wool samples

3.4 Moisture regain of wool powders

When moisture is absorbed into a fibre, water molecules can be directly attached to the polar groups, such as $-NH-$, $-OH$, $-NH_3^+$, $-COO^-$, $-CONH_2$ in wool. They can also be attached to the water molecules already absorbed via polar groups, as shown in Figure 5. More polar groups in fibres means more water can be absorbed [19]. The directly attached water molecules are firmly held by the polar groups via hydrogen bonds, while the indirectly attached water molecules are more loosely linked.

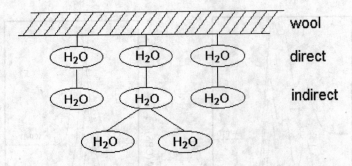

Figure 5 Direct and indirect absorption of water molecules by wool [19]

Moisture absorption of wool powder can also be related to its crystallinity percentage. In the crystalline regions, the macromolecules are closely aligned. So it is not easy for water molecules to diffuse into the crystalline regions. While in the amorphous regains, the macromolecules are disordered. The water molecules can access the amorphous regions and the surface of the crystalline regions of a fibre [20]. The higher the percentage of crystallinity, the less is the moisture regain of wool.

The moisture regains of wool powders in Table 2 were reduced with the decrease in the size of wool powders. This is consistent with the analysis of Peak 2 of DSC curves. It could be explainable in terms of the increased crystallinity of finer wool powders. The regain of WP-D was slightly higher than that of WP-C. It was because the polar cysteic acid residues were introduced into WP-D during the treatment of the sodium salt of dichloroisocyanuric acid (DCCA) used in the WP-D production.

3.5 Characterization of activated charcoal

The scanning electron micrographs of activated charcoal in Figure 6 indicate that activated charcoal has a porous structure, with irregular surface and shape. The mean particle size of activated charcoal was found to be 33 μm on a volume basis, and the size range was from 0.83 μm to 631 μm, as shown in Figure 7. The surface composition analysed using XPS (Figure 8) illustrated that the major component of activated charcoal was carbon, accounting for 95.29 At%, and the remainder was oxygen.

Table 2 Moisture regain (%) of wool samples

Sample		Regain (%)	Average regain (%)
Wool fibre	1#	15.31	
	2#	15.02	15.11 ± 0.14
	3#	15.00	
WP-A	1#	12.88	
	2#	13.0	13.05 ± 0.17
	3#	13.29	
WP-B	1#	12.14	
	2#	11.94	11.88 ± 0.25
	3#	11.55	
WP-C	1#	8.55	
	2#	8.57	8.52 ± 0.06
	3#	8.44	
WP-D	1#	9.32	
	2#	9.32	9.23 ± 0.13
	3#	9.04	

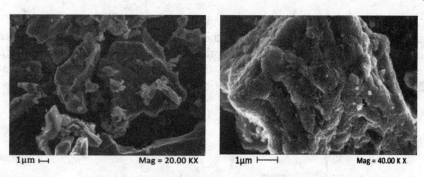

Figure 6 SEM images of activated charcoal

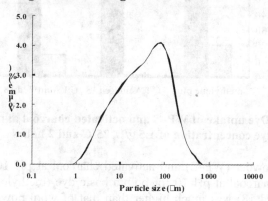

Figure 7 Particle size distribution of activated charcoal on a volume basis

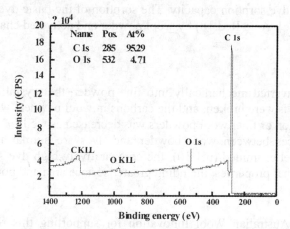

Figure 8 XPS spectrum of activated charcoal

3.6 Comparison of dye sorption by wool powder and activated charcoal at pH 10

Binding of dyes to wool involves a number of forces and interactions, such as electrostatic forces between the charged wool and the ionic dyes, hydrophobic interactions and hydrogen bonding. The sorption capacity of activated charcoal is mainly related to its surface area, degree of surface reactivity, microporous structure and the presence of acid or basic groups on the surface [21]. The dye sorption of activated charcoal has been explained in terms of electrostatic forces as well as dispersion forces [22].

In the previous study [1], uptake of dyes (C. I. Acid Red 88, C. I. Acid Red 18 and Lanasol Blue CE) by wool powders and activated charcoal was conducted at pH 4.5. Results showed that the dye sorption capacity of fine wool powders (such as WP-C) was comparable to that of activated charcoal, although the surface of activated charcoal (858 m^2/g) was much greater than that of WP-C (5.9 m^2/g), Under the condition of pH 4.5, the zeta potential of WP-C (-1.02 mV) was less negative than that of activated charcoal (-19.8 mV), which was expected to favour dye uptake of WP-C. Additionally, penetration of dye into wool particles, which was visualised using a fluorescent dye, was believed to be able to increase the dye sorption ability of wool powder as well.

At pH 10, the zeta potential of WP-C (-8.1 mV) was similar to that of activated charcoal (-9.13 mV) [2]. It could indicate that the electrostatic forces, between the negatively charged WP-C / activated charcoal and the ionic dyes, would play a similar role in the dye sorption of WP-C / activated charcoal. The dye uptake of WP-C and activated charcoal carried out at pH 10 in the present study aimed to investigate the effect of factors other than the electrostatic force on the dye sorption capacity of wool powder and activated charcoal.

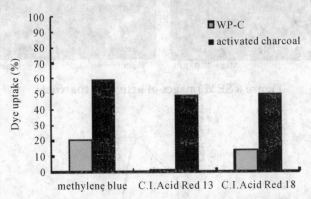

Figure 9 Dye uptake of WP-C and activated charcoal at pH 10:
dye concentration of 1.5 g/L, 25°C and 2 hours

The dye uptake of wool powder (WP-C) and activated charcoal at pH 10 is shown in Figure 9. Dye sorption capacity of activated charcoal at pH 10 for both a basic dye (methylene blue) and acid dyes (C. I. Acid Red 13 and C. I. Acid Red 18) was much higher than that of wool powder. This results suggest that other properties of the activated charcoal, such as the relatively large surface area and porous structure, would have contributed to its higher dye sorption capacity. The sorption of the basic dye (methylene blue) and acid dyes (C. I. Acid Red 13 and C. I. Acid Red 18) by wool powder and activated charcoal has been explained in detail in the previous paper [2].

4 Conclusions

As wool fibres were converted mechanically into fine powder, the crystallinity of wool powders was increased, some chemical bonds were broken, and the carbons in wool powders were oxidated by mechanical milling. The moisture absorption of finer wool powders was decreased as well.

As the electrostatic forces between wool powder and dyes are similar to those between activated charcoal and dyes the pH level examined (pH 10), the comparatively high dye sorption ability of activated charcoal could be due to its other properties, including greater surface area and porous structure.

Acknowledgement

We are grateful to the Australian Wool Innovation for supporting this research through the China Australia Wool Innovation Network (CAWIN) project. We also thank Dr. Peter Cookson for his kind assistance and valuable suggestions in the initial stage of this work.

References

[1] G Wen; J A Rippon; P R Brady; X G Wang; X Liu; P G Cookson, The Characterization and Chemical Reactivity of Powdered Wool Powder Technology 2009, 193, 200-207.
[2] G Wen; P G Cookson; X Liu; X G Wang, The Effect of pH and Temperature on the Dye Sorption of Wool Powders Journal of Applied Polymer Science 2010, 116, 2216-2226.
[3] G Freddi; M Tsukada; H Shiozaki, Chemical Modification of Wool Fibers with Acid Anhydrides Journal of Applied Polymer Science 1999, 71, 1573-1579.
[4] A M Manich; J Carilla; S Vilchez; M D de Castellar; P Oller; P Erra, Thermomechanical Analysis of Merino Wool Yarns Journal of Thermal Analysis and Calorimetry 2005, 82, 119-123.
[5] M Martí; R Ramírez; A M Manich; L Codech; J L Parra, Thermal Analysis of Merino Wool Fibres Without Internal Lipids journal of Applied Polymer Science 2007, 104, 545-551.
[6] G Freddi; R Innocenti; T Arai; H Shiozaki; M Tsukada, Physical Properties of Wool Fibers Modified with Isocyanate Compounds Journal of Applied Polymer Science 2003, 89, 1390-1396.
[7] A R Haly; J W Snaith, Differential Thermal Analysis of Wool - The Phase-Transition Endotherm Under

Various Conditions Textile Research Journal 1967, 37, 898-907.
[8] F-J Wortmann; H Deutz, Thermal Analysis of *Ortho-* and *Para*-Cortical Cells Isolated from Wool Fibers Journal of Applied Polymer Science 1998, 68, 1991-1995.
[9] W Xu; W Guo; W Li, Thermal Analysis of Ultrafine Wool Powder Journal of Applied Polymer Science 2003, 87, 2372-2376.
[10] J M Cardamone; J Yao; A Nunez, DCCA Shrikproofing of Wool. Part I: Importance of Antichlorination Textile Research Journal 2004, 74 (6), 555-560.
[11] E R Trotman, Dyeing and Chemical Technology of Textile Fibres. Third Edition ed.; Charles Griffin & Company Limited: London, 1964; p 639.
[12] R W Moncrieff, Chemical Changes in the Keratin when Wool is Chlorinated The Textile Manufacture 1967, April, 145-147.
[13] J A Rippon, in: Friction in Textile Materials, B. S. Gupta, (Ed.) 2008; pp 253-291.
[14] J A Rippon, in: Wool Dyeing, D. M. Lewis, (Ed.) Society of Dyers and Colourists: 1992; pp 1-51.
[15] R H Bradley; I L Clackson; D E Sykes, XPS of Oxidized Wool Fibre Surfaces Surface and Interface Analysis 1994, 22, 497-501.
[16] N Brack; R Lamb; D Pham; P Turner, XPS and SIMS Investigation of Covalently Bound Lipid on the Wool Fibre Surface Surface and Interface Analysis 1996, 24, 704-710.
[17] R H Bradley; I L Clackson; I Sutherland; J A Crompton; M A Rushforth, Oxygen Plasma Modification of Wool Fibre Surface Journal of Chemical Technology & Biotechnology 1992, 53, 221-226.
[18] C Popescu; H Hocker, Hair - the Most Sophisticated Biological Composite Material Chemical Society Reviews 2007, 36, 1282-1291.
[19] S B Warner, Fibre Science. Prentice-Hall, Inc.: 1995; p 316.
[20] W E Morton; J W S Hearle, Physical Properties of Textile Fibres. Second edition ed.; The Textile Institute and Heinemann Ltd.: 1975; p 660.
[21] J Díaz-Terán; D M Nevskaia; A J López-Peinado; A Jerez, Porosity and Adsorption Properties of an Activated Charcoal Colloids and Surfaces A: Physicochemical and Engineering Aspects 2001, 187-188, 167-175.
[22] M Dai, Mechanism of Adsorption for Dyes on Activated Carbon Journal of Colloid and Interface Science 1998, 198, 6-10.

92 Influence of Spacer on Spun Yarn Quality in Cotton Spinning

Sayyed Sadroddin Qavamnia, Amir Hossein Raei, Arsham Zibaee
(Islamic Azad University, Birjand Branch, Birjand, Iran)

Abstract: Hairiness is an important factor in yarn quality. There are many methods for reducing hairiness such as traveller, spindle speed and …. At this research, we study the impression of various spacers (white, yellow, and grey) on properties of some yarns. As you know in ring frame, fibers after passing through from draft zoom they reach to product roller. They must pass aprons. The distance between these aprons is very important. We used some clips such as grey, white, yellow for polyester/viscose, cotton/viscose and 100% cotton yarns with difficult yarn count (30 combed and carded, 40 combed). After we tested, it showed any yarn needs to a special clip and any clips are not good for any yarn. If we use suitable clips, it can reduce hairiness.

Keywords: Clips; Spacer; Hairiness; Short staple yarn; Thin and thick; Cv%; Nep; Yarn quality

1 Introduction

Yarn apparent shape depends highly on yarn hairiness, evenness and thick and thin places. Yarn hairiness is a complex concept, which generally cannot be completely defined by a single figure. The effect of yarn hairiness on the textile operations following spinning, especially weaving and knitting, and its influence on the characteristics of the product obtained and on some fabric faults has led to the introduction of measurement of hairiness [1]. An important yarn characteristic, which greatly influences the appearance of fabrics, is hairiness [2]. Hairiness occurs because some fibre ends protrude from the yarn body, some looped fibres arch. Pillay proved that there is a high correlation between the number of protruding ends and the number of fibres in the yarn cross-section. Torsion rigidity of the fibres is the most important single property affecting yarn hairiness. Other factors are flexural rigidity, fibre length and fibre fineness. Out from the yarn core and some wild fibres in the yarn [1]. If the length of the protruding fibre ends as well as that of the loops is considered, the mean value of the hairiness increases as the cross-sectional area increases and decreases with the length of the loops. The hairiness is affected by the yarn twist, since an increase in twist tends to shorten the fibre ends. Wild fibres are those for which hte head alone is taken by the twist while the tail is still gripped by the front drafting rollers[1]. Cotton yarns are known to be less hairy than yarns spun from man-made fibres. The possible reason for this is the prifile of the two fibres[1]. If the width of the fibre web in the drafting field is large, the contact and friction with the bottom roller reduce the ability of the fibres to concentrate themselves and hairiness occurs. This effect is found more in coarse counts with low TPI. This suggests that the collectors in the drafting field will reduce yarn hairiness[1]. According to Uster 15% of fabric, defects and quality problems stem from hairiness. Hairiness is a unique feature of staple fiber yarns that distinguishes it from filament yarns. Hairiness is generally regarded as undesirable because of the following reasons [3]. Hairiness is removed or suppressed by singeing, waxing, application of lubricant, enzyme treatment and sizing [3]. One of the other parameters that make an impression on yarn apparent is yarn evenness. Yarn evenness can be defined as the variation in weight per unit length of the yarn or as the variation in its thickness. There are a number of different ways of assessing it [6]. Yarn evenness deals with the variation in yarn fineness. This is the property, commonly measured, as the variation in mass per unit length along the yarn, is a basic and important one, since it can influence so samy other properties of the yarn and of fabric made from it. Such variations are inevitable, because they arise from the fundamental nature of textile fibres and from their resulting arrangement [7]. The spinner tries to produce a yarn with the highest possible degree of homogeneity. In this connection, the evenness of the yarn mass is of the greatest importance. In order to produce an absolutely regular yarn, all fibre characteristics would have to be uniformly distributed over the whole thread. However, that is ruled out by the in homogeneity of the fibre material and by the mechanical constraints. Accordingly, there are limits to the achievable yarn evenness [7]. Irregularity can adversely affect many of the properties of textile materials. The most obvious consequence of yarn evenness is the variation of strength along the yarn. If the average mass per unit length of two yarns is equal, but one yarn is less regular than the other, it is clear that the more even yarn will be the stronger of the two. The uneven one should have more thin regions than the even one as a result of irregularity, since the average linear density is the same. Thus, an irregular yarn will tend to break more easily during spinning, winding, weaving, knitting, or any other process where stress is applied[7]. Other fabric properties, such as abrasion or pill-resistance, soil retention, drape, absorbency, reflectance, or luster, may also be directly influenced by yarn evenness[7].

The distance between the lower edge of the top cradle and bottom apron nose bar determines the distance between the top and bottom aprons [10]. This in turn determines the intensity of pressure applied to the fibres to be under control. This distance, which is introduced by means of special device, is called Spinning Spacer. As the pressure between the aprons in the drafting zone is controlled by spinning spacer, in turn it governs the degree of control exercised on the floating fibres and as such would have influence over the drafting irregularities. Table 1 shows useable clips in spinning that suggested with SKF and Rieter[9].

Table 1 Distance clips Reiter & SKF

	Distance (Spacing) Clips Reiter & SKF		
SKF	Distance Clip olc-0964 117 Red	Rieter	Distance Clip 1.75MM Grey Color
	Distance Clip Olc-0964 118 Yellow		Distance Clip 2.5mm Brown Color
	Distance Clip Olc-0017 7065 Lilac		Distance Clip 3.0mm White Color
	Distance Clip Olc-0964 119 White		Distance Clip 3.5mm Black Color
	Distance Clipolc-00178 627grey		Distance Clip 3.75mm Beige Color
	Distance Clip Olc-0964 120 Black		Distance Clip 4.0 Mm Red Color

Evenness and total imperfection could be improved by closing down the apron spacing. SKF recommends smallest possible spacer for all the counts. It is, however, often necessary to use a wider spacer for a coarser counts. If there are undrafted places in the yarn when it leaves the front rollers, the break draft should be increased. Spacer should be increased only if the draft results remain unsatisfactory after the break draft has been increased. Under identical spinning conditions, the spacer will normally be wider for synthetic aprons than for the leather aprons. This is because with a synthetic apron the fibres will offer greater resistance to forward movement than with a leather apron using the same apron spacer. With widened apron spacing, a progressive deterioration in regularity and strength of yarn was noticed, but the effect was more noticeable only with higher break drafts. The hairiness value decreases as the spinning spacer value increases. But there is a small rise in the hairiness value at the combination of 85° shore hardness and 3.5 mm spacer. From the Figure 5, minimum hairiness is found at 85° shore hardness and 3.0 mm spacer[4]. The formula has been derived to compute the spacer number to have optimum yarn quality [5]. The formula was made up to be effected inversely by two induces called spacer factor and divider. The formula is given below:

Spacer number for Ring frames = (8.233 / (Spacer factor x Divider x vNe)) + 2.03
Spacer factor = (Micronaire value) / (Maturity coeff x 50% Span length (mm))
Divider = (Roving hank x Break draft on Ring frame) + (0.96 / vRoving frame)

The very purpose of introduction of apron drafting is to have an adequate short fibre control during drafting process. These short fibres are to be carried with full control along the surface speed of the second roller as close as possible to the nip and passed forward. Until the fibre is gripped at the front nip, it is expected to be under the full control of the apron during its forward movement. Thus a positive apron control is governed by shore hardness. Any deviation of the apron control from the optimum values will lead to a distorted drafting process, whereby revolving in proper selection of shore hardness, is also detrimental to yarn quality[4]. The spacer opening seems to have a more impact on yarn strength. From the Figure 2, it is observed that the yarn strength decreases to the minimum value 98.8 with the spacer and shore hardness combination of 3.5 mm and 70°. Therefore, 2.5 mm spacer and 85° shore hardness gives the correct combination to get the better strength as compared to the various combinations. But the minimum strength CV% is observed at 2.5 mm spacer and 70° shore hardness[4]. As spacer value increases, U% gets increased for both the shore hardness. Referring to the Figure 4, 3.0 mm spacer with shore hardness 700° shows the minimum U% of 11.71%[4]. The hairiness value decreases as the spinning spacer value increases. But there is a small rise in the hairiness value at the combination of 85° shore hardness and 3.5 mm spacer. From the Figure 5, minimum hairiness is found at 85° shore hardness and 3.0 mm spacer[4]. In general it is found that increasing the spacer size decreases the yarn quality parameters. Instead of increasing the spacer size, break draft can be slightly increased. From the tests carried out on the combed count of 30s P/C, it is found that, the spacers with 70° & 85° shore hardness are found to have more impact on the yarn quality parameters.

It is concluded that 2.5 mm spacer with shore hardness of both 85° & 70° are performing relatively well at the same rate. In particular, 2.5 mm spacer with 70° shore hardness gives more advantageous results, as compared to that of 2.5 mm spacer & 85° shore hardness.

It is found that minimum spinning spacer with minimum shore hardness will give the relatively intended quality yarn.

2 Material and method

For this research, we used various yarn counts such as 40, 30, with fibers materials such as polyester/viscose, cotton and cotton/viscose (carded and combed yarn) also roving count was 1 Hank. For experiments, we used clips white, grey and yellow.

All of these, tested on G35 ring frame. We selected four machines and selected six heads on each machine (for any yarn count), and change the clips for each test. For testing, we used Uster Tester 4 and tested thin (-40% & -50%) and thick (+35% & +50%), nep and hairiness. The results are below:

a. At first, we used combed cotton yarn with yarn count 30. Table 1 shows the results.

Table 2 the results of influence 3 kind of clips on combed cotton yarn No 30

Yarn count	Clips	Cv%	Thin -40%	Thin -50%	Thick +35%	Thick +50%	Nep |+200%	Hairiness
30 combed cotton	White	11.64	14.3	0.3	190.3	17.8	52.5	5.24
	Yellow	12.02	27.5	0.3	275.3	25.3	48.5	5.1
	Grey	11.57	14.8	0.3	175.8	15.3	41.3	5.26

b. Then, we used carded cotton yarn with yarn count 30. Table 2 shows the results.

Table 3 The results of influence 3 kind of clips on carded cotton yarn No 30

Yarn count	Clips	Cv%	Thin -40%	Thin -50%	Thick +35%	Thick +50%	Nep |+200%	Hairiness
30 carded cotton	White	14.61	162.8	6	927.3	171.3	212.8	5.2
	Yellow	14.65	188.3	3.5	913	155	217.5	5.42
	Grey	14.9	189	8	1094	171	222	5.21

c. Then, we used combed cotton/viscose yarn with yarn count 30. Table 3 shows the results.

Table 4 The results of influence 3 kind of clips on combed cotton/viscose yarn No 30

Yarn count	Clips	Cv%	Thin -40%	Thin -50%	Thick +35%	Thick +50%	Nep |+200%	Hairiness
30 combed cotton/viscose	White	11.52	23	0.0	146.0	7.5	15.8	4.81
	Yellow	11.50	21.5	0.3	155.8	9.0	14.8	4.57
	Grey	11.32	19.3	0.0	127.8	8.5	18.3	4.56

d. Then, we used carded cotton/viscose yarn with yarn count 30. Table 4 shows the results.

Table 5 The results of influence 3 kind of clips on carded cotton/viscose yarn No 30

Yarn count	Clips	Cv%	Thin -40%	Thin -50%	Thick +35%	Thick +50%	Nep |+200%	Hairiness
30 carded cotton/viscose	White	14.99	184.5	2.3	1104	189.3	198.3	4.59
	Yellow	14.61	147.3	1.5	984.3	171.8	217.5	4.74
	Grey	14.45	123.8	2.5	913	149.5	204.3	4.78

e. Then, we used combed polyester/viscose yarn with yarn count 40. Table 5 shows the results.

Table 6 The results of influence 3 kind of clips on combed polyester/viscose yarn No 40

Yarn count	Clips	Cv%	Thin -40%	Thin -50%	Thick +35%	Thick +50%	Nep |+200%	Hairiness
40 combed polyester/viscose	White	14.25	233.8	11.0	473.8	64.0	82.3	4.3
	Yellow	14.41	220.0	9.8	521.5	62.0	72.3	3.87
	Grey	13.94	192.0	7.0	397.3	53.8	73.3	4.19

Graphs

At this section, we can see the results with graphs.

Graphs 1 to 5 shows the results of Tables 2 to 6. At these graphs, we see influence of various clips on various yarns.

Graphs 1:

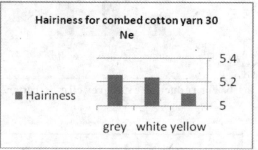
a. For reduction hairiness yellow clips is better.

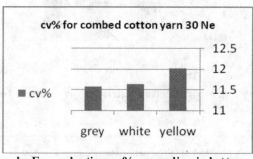
b. For reduction cv% grey clips is better.

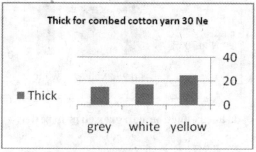
c. For reduction thick grey clips is better.

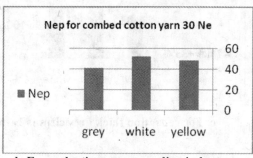
d. For reduction nep grey clips is better.

Graphs 2:

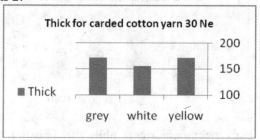
a. For reduction thick white clips is better.

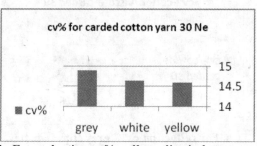
b. For reduction cv% yellow clips is better.

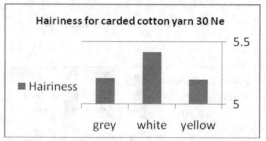
c. For reduction hairiness yellow clips is better.

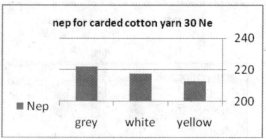
d. For reduction nep yellow clips is better.

Graphs 3:

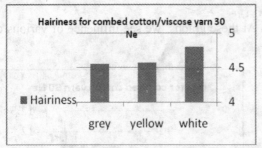
a. For reduction hairiness grey clips is better.

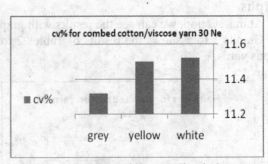
b. For reduction cv% grey clips is better.

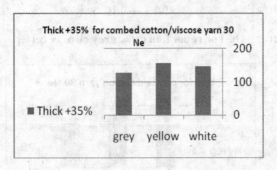
c. For reduction thick grey clips is better.

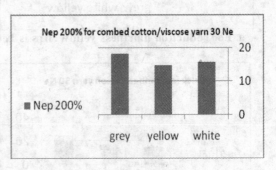
d. For reduction nep grey clips is better.

Graphs 4:

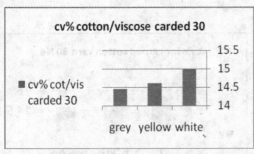
a. For reduction cv% grey clips is better.

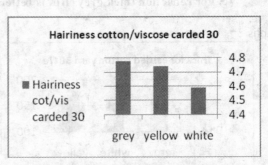
b. For reduction hairiness white clips is better.

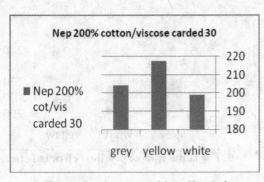
c. For reduction nep white clips is better.

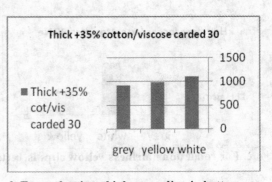
d. For reduction thick grey clips is better.

Graphs 5:

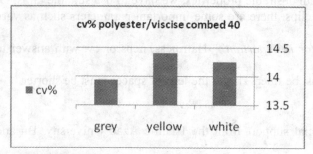

a. For reduction cv% grey clips is better.

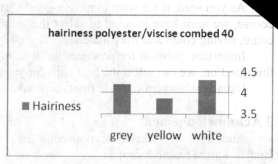

b. For reduction hairiness yellow clips is better.

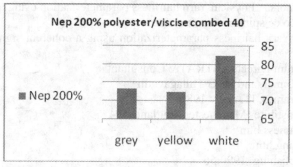

c. For reduction nep yellow clips is better.

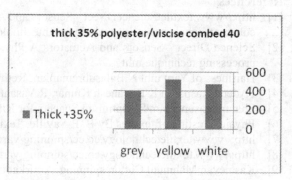

d. For reduction thick grey clips is better.

In following, we can see all of the results in Table 7.

Table 7 All of the results for 3 clips and 5 yarns

Type of yarn	Cv% clips			Thin -40% clips			Thin -50% clips			Thick +35% clips			Thick +50% clips			Nep 200% clips			Hairiness clips		
	w	y	g	w	Y	g	w	y	g	w	Y	g	w	y	g	w	y	g	w	y	g
30 combed cotton			•		•	•						•			•		•			•	
30 carded cotton	•			•			•				•	•					•			•	
30 combed cotton/ viscose		•			•							•		•	•		•	•		•	•
30 carded cotton/ viscose	•				•		•				•				•	•	•		•		
40 combed polyester/ viscose				•			•		•			•			•		•	•		•	

to have a good yarn with desirable properties, we have to use suitable clips. Of course, you must have in mind to select the suitable clips, there are some important parameters such as yarn count, ... and fibre materials.

... producer is *"Which factor is important?"* Hairiness, neps or cv. with answer to this ... select the best clips for yarn.

... yarn count is finer, clips count must be finer. That is the legs of spacer must be shorter.

Acknowledgement

Authors would like to acknowledge the financial support from the Islamic Azad University, Birjand Branch for the research project.

References

[1] http://www.textiletScienceDirect - Sensors and Actuators A Physical Yarn hairiness parameterization using a coherent signal processing technique.mhtechnology.co.cc/spinning/hairiness.htm

[2] Science Direct - Sensors and Actuators A Physical Yarn hairiness parameterization using a coherent signal processing technique.mht

[3] Hairiness of Yarn.mht N.Balasubramanian. Retired Joint Director(BTRA) & Consultant

[4] FAdetails.asp.htm K Buvanesh Kumar, R Vasantha Kumar, and Dr G Thilagavathi

[5] Shanmugasundaram S: Optimum Spacers for Ring Frames, ITJ, July 1985.

[6] Physical testing of textiles By B. P. Saville, Textile Institute (Manchester, England): 94

[7] http://www.textiletechnology.co.cc/spinning/yarnevenness.htm

[8] http://www.textiletechnology.co.cc/spinning/yarntesting.htm

[9] Kasaieyan Mahmud. Cotton spinning: Amirkabir Publishing. 1990:56

93 Thermobonding of Wool Nonwovens

R. H. Gong, K. M. Nassar, I. Porat
(Textiles and Paper, School of Materials, University of Manchester, UK)

Abstract: Due to the low surface of wool fibres it is difficult to bond wool fibres. To overcome this problem, different surface treatments were applied to the wool fibres to raise their surface energy. These were chlorination, hercosett and plasma treatment. Low melting nylon fibres were blended with the wool to provide the bonds. It was shown that wool and nylon fibres can be thermally bonded to provide a light-weight and high-loft nonwoven structure with superior thermal insulation performance. Among the three treatments, hercosett treated wool fibres yielded the strongest bonds.

Keywords: Nonwoven; Thermal bonding; Wool; Hercosett; Plasma; Chlorination

1 Introduction

Wool was arguably the first fibre to be used for producing nonwoven fabrics in the form of felts. However, the felting process produces a very dense and heavy structure. In order to exploit the thermal insulation properties of wool fibres, attempts have been made to bond wool fibres using thermal plastic fibres to produce a more voluminous structure [1, 2]. However, these have not been successful. Wool fibres are hydrophobic and have low surface energy mainly because of the epicuticle membrane, which is the very outer surface of the cuticle (scales) that consists of adhering hydrophobic covalently bound 18-methyl-eicosanoic acid lipids in addition to the highly crosslinked exocuticle. Therefore increasing the surface energy and hydrophilicity of wool fibres via surface modification is seen as a way to raise the surface energy [3, 4, 5] to increase adhesion. We report in this paper our study into the effects of three different wool fibre surface treatments on the thermal bonding properties of the nonwoven structures.

2 Experimental details

The following fibre types were used to conduct the trials:
- 20.5 micron untreated wool fibres;
- 20.5 micron plasma treated wool fibres;
- 20.5 micron chlorinated wool fibres;
- 20.5 micron hercosett wool fibres;
- 3.3 dtex and 220/115 C core/sheath melting point bi-component nylon fibres.

All wool fibres were Australian Merino wool tops. The wool fibres were blended with the nylon fibre with a 50:50 lend ratio.

The nonwovens webs were produced using an air-laying method. The webs were bonded using though-air thermal bonding. The equipment used is illustrated in Figure 1.

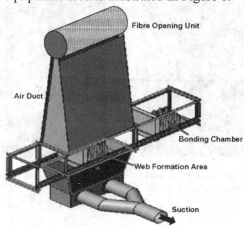

Figure 1 The nonwoven web former

The fibres were opened to individual fibres by the fibre opening unit. The fibres were then stripped off the opening unit by high velocity air. The airflow was generated by a suction fan which was connected to the area underneath the web forming moulds. The formed web was then moved to the bonding chamber. Bonding was achieved by drawing pre-heated hot air through the web using another suction fan.

Three bonding times were used, namely 30 seconds, 60 seconds and 90 seconds. The thermal bonding temperature was nominally 160 °C. This was dictated by the melting point of the nylon fibre which provided the bonds. Preliminary tests had proved that the wool fibres suffered no damage at this bonding temperature.

Tensile strength testing of the samples was performed on the Instron Series IX Automated material testing system according to ISO 9073-3:1989 [6], and under standard atmospheric conditions. The tested samples measured 100 mm in length, 40 mm in width. The grip distance was 60 mm. The samples were also examined under SEM using a variety of SEM models including the Zeiss Evo60 Environmental Scanning Electron Microscope (ESEM) which has the ability of handling nonconductive materials without coating [7].

3 Results and discussions

The tensile test results of the samples are shown in Figure 2. The SEM examples of bonding are shown in Figures 3 to 6.

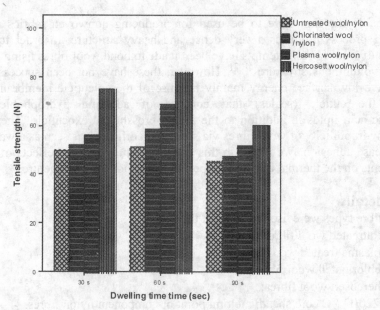

Figure 2 Tensile strength of wool/nylon nonwovens

Figure 3 Examples of failed bonds between untreated wool/nylon fibres

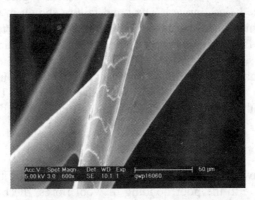

Figure 4 Examples of bonding between plasma wool/nylon fibres

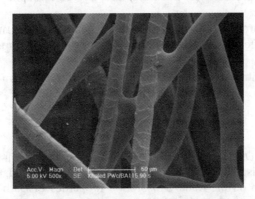

Figure 5 Examples of bonding between chlorinated wool/nylon fibres

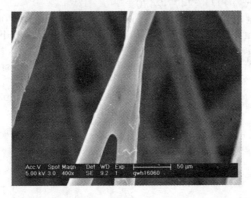

Figure 6 Examples of bonding between hercosett wool/nylon fibres

It is clear that Hercosett treated wool provides the highest bonding strength regardless of the bonding time. This is followed by plasma, chlorinated and untreated wool, in that order. For all wool fibres, the 90-second bonding time gives the lowest tensile strength.

The improved bonding performance of hercosett wool with nylon may be attributed to the cationic PAE polymer masking the outer surface of the wool fibres and hence creating a bonding surface which is both highly reactive and chemically compatible with nylon fibres. On the other hand, the inferior bonding performance of chlorinated wool with nylon as compared to hercosett and plasma treated wool may be a result of the smooth outer surface of the chlorinated wool fibres created by the degradation of the cuticle [3]. In contrast to chlorination, plasma treatment leads to a rougher outer surface due to grooves created by the bombardment of the oxygen plasma species. This is further helped by the high surface energy of the plasma treated fibres due to the polar O_2 atoms [4, 8] on the fibre surface and leads to better adhesion performance compared to chlorinated wool.

Untreated wool has the lowest bonding strength with nylon. In Figure 3, it can be seen that bonds can fail without much external force being applied. The bond failure shown in the figure existed in the bonded web without any external deformation except usual handling. Most of the tensile strength of the nonwoven from untreated wool and nylon is due to wool fibres being mechanically trapped inside a network of bonded nylon fibres.

The results also suggest that at 30 seconds bonding time, there was insufficient melting and polymer flow from the nylon fibres to form strong bonds. On the other hand, there is accelerated nylon fibre degradation with prolonged bonding time. This is the reason for the declining nonwoven tensile strength at 90 seconds bonding time.

4 Conclusions

The tensile strength of nonwovens made from wool and thermoplastic fibres may be affected by a large number of factors. However, due to the poor bonding of wool fibres, the bond strength is the most critical factor as the fibres themselves have much higher strength compared to the bonds. In through air thermal bonding, there is very limited pressure on the fibre web and consequently there is little contact pressure between the fibres.

Under these conditions, process parameters such as bonding time and binding temperature become significant and have notable effect on the bond strength, but the most important contributor is the surface treatment of the wool fibres.

All of the three surface treatments studied in this paper, hercosett, plasma and chlorination, improved the bonding of wool to nylon fibres, and the most effective surface treatment has been shown to be the hercosett treatment.

References

[1] Hoffmeyer F. J. and Watt J. D., Thermo-Bonding of Lofty Wool Batts with Low Melting Temperature Fibres. 1986, Wool Research Organisation of New Zealand Reports: New Zealand. p. 1-3.
[2] Zhu H., et al., Investigation on the Radar Absorption Properties of Carbon Fiber Containing Nonwovens. Journal of Industrial Textiles, 2007. 37(1): 94-95.
[3] Bradley R. H., Mathiesonb I., and Byrnec K. M., Spectroscopic Studies of Modified Wool Fibre Surfaces. Journal of Material Chemistry, 1997, 7:2477-2480.
[4] Höcker H., Plasma Treatment of Textile Fibers. Pure Applied Chemistry, 2002. 74(3): 244-423.
[5] Silvaa, M., et al., Treatment of Wool Fibres with Subtilisin and Subtilisin-PEG. Enzyme and Microbial Technology, 2005. 36(7): 917.
[6] ISO 9073-3:1989, Methods of Tests for Nonwovens: Determination of tensile strength and elongation. 1998.
[7] http://research.amnh.org/mif/zeiss-evo-60-ep-sem.
[8] Garg S., Hurren C., and Kaynak A., Improvement of Adhession of Conductive Polypyrole Coating on Wool and Polyester Fabrics Using Atmospheric Plasma Treatment. Synthetic Materials, 2007, 76(4): 309.

94 PEG-[Si(OEt)$_3$]$_2$ Sol Agent Applied to PP Fabrics for Moisture Management

Yi Hu[1,2], Jinghong Yuan[3], Wenjie Chen[1,2], Jinqiang Liu[1,2]

(1. Key Laboratory of Advanced Textile Materials and Manufacturing Technology, Ministry of Education of China, Zhejiang Sci-Tech University, Hangzhou 310018, China
2. Engineering Research Center for Eco-Dyeing & Finishing of Textiles, Ministry of Education, Zhejiang Sci-Tech University, Hangzhou 310018, China
3. Institute of Costume, Zhejiang Textile & Fashion College, Ningbo 315000, China)

Abstract: PEG-[Si(OEt)$_3$]$_2$ sol originated from PEG with different molecular weights were synthesized as PP (polypropylene fabrics) moisture adsorption and perspiration exhaust agent. The agents applied to PP for hydrophilicity finishing is derived through prepolymer synthesized and hydrolysis process. The result showed that the sol prepared with PEG2000 show the optimum result especially when cylodextrin was added. The perspiration exhaust effect of the PP improved greatly. It was 26.57% after finishing compared with 3.09% of the untreated one. Moreover, there was also an improvement in moisture adsorption, which increased to 8.0 cm. But the value decreased to 5.6 cm after soaping 10 times.

Keywords: PP (polypropylene fabrics); Moisture adsorption; Perspiration exhaust effect

1 Introduction

PP fabrics with hydrophobic character due to its configuration, whose molecular do not have reactive groups, such as -OH, -NH$_2$, -CONH-, -COOR, so the application of the fabric is limited, especially in family textile markets. Furthermore, higher crystallinity of the PP fabrics can also lead to water diffuse and infiltration into the fabrics being more difficult in comparison with other fibers. It would be more desirable if the moisture adsorption performance of PP fabric is significantly improved. So we decided to synthesize another agent which should have amount of hydrophilic groups[1-2], which can give PP fabric better moisture ability and should have reaction crossing groups which can produce cross linking reaction on the fabrics, thus let the PP fabric have better durable property.

2 Experimental

2.1 Materials

3-aminopropyltriethoxysilane (APTES), Polyethylene glycol (molecular weight: 2000, 4000, 6000) were dehydrated by distillation for 1 hour at 120℃ at 600 mmHg before use. Isophorone diisocyanate (IPDI) was commercial and industrial grade. 502A of Cyclofresh™ containing cylodextrin, which is express as "CD" in some of the figures afterwards, was purchased from Cognis Co.

2.2 Chemical synthesis of PEG-[Si(OEt)$_3$]$_2$ sol

Stoichiometric amounts of PEG and IPDI were added in a four-necked round-bottom flask equipped with a stirrer, an N$_2$ inlet and a cooler/heater. After the reaction temperature rose to 60℃ to form a NCO-group ended prepolymer. Then the preset dosage of 3-aminopropyltriethoxysilane (APTES) were added, keeping reaction some minutes, the NCO-group ended prepolymer was blocked by 3-aminopropyltriethoxysilane and PEG-[Si(OEt)$_3$]$_2$ was synthesized[3-8]. After that the agent achieved (Scheme 1) by adding PEG-[Si(OEt)$_3$]$_2$, water and absolute alcohol into the flask, while adjusting pH value to 5-6 with addition of HCl drop. After stirring violently under room temperature for two hours, the reactant mixture was left at room temperature and standed for 10 h.

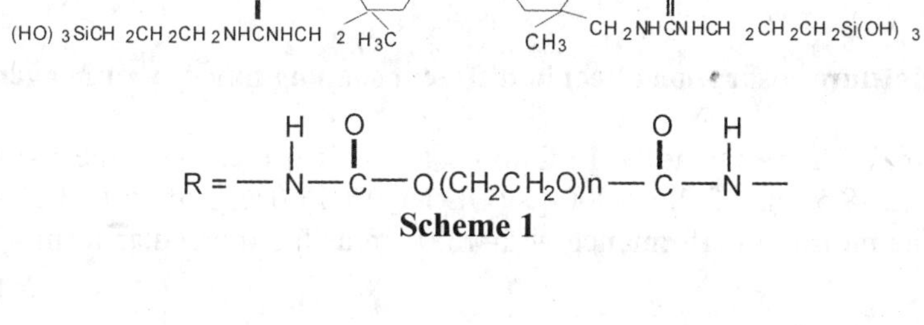

Scheme 1

2.3 Treatment of PP fabrics

PP fabrics were treated by the different agents with pad-dry-cure process (bath ratio 20: 1, pH value of 3-4). Then the samples were dried at 80℃ for 3min and cured at 120℃ temperature for 3min, respectively.

2.4 Measurements standard

The PP fabrics treated before and after were tested according to the following industrial standard: Water adsorption test (FZ/T01071 or JIS L1907 Byreck method); Washing fastness test (GB/T 3921.1-1997 eqv. ISO 105-C01: 1989); Quick drying property test (TTF 0007).

3 Results and discussion

3.1 Synthesis of the Polyurethane

To confirm the **2.2** above reactions, the prepolymer and blocked prepolymer were characterized by FT-IR. The FT-IR spectra of prepolymer (a) and blocked prepolymer (b) for PEG2000-[Si(OEt)$_3$]$_2$ as an example, are as shown in Figure 2.

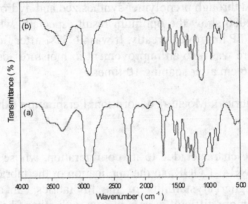

Figure 1 FT-IR spectra of (a) prepolymer and (b) blocked prepolymer

The peak at 2259 cm^{-1} in Figure 1. (a) proved an existence of –NCO groups, and the peak at 3325cm^{-1} (υ_{NH}) and 1717 cm^{-1} ($\upsilon_{C=O}$) confirms the existence of urethane groups. The FT-IR spectrum of PEG-[Si(OEt)$_3$]$_2$ is shown in Figure 1. (b). In Graph (b) it can be seen that the characteristic bands of carbamido groups at 1665 cm^{-1} appears and the peak at 2259 cm^{-1} of –NCO groups disappeared, which proves -NCO group was blocked with 3-aminopropyltriethoxysilane.

3.2 Moisture adsorption

The Moisture adsorption of PP fabrics treated by the prepared PEG-[Si(OEt)$_3$]$_2$ sol with different soaping times is compared in Figure 2. According to the agents with different PEG molecular weight and whether add CD (cylodextrin), there are different agents: PEG2000-[Si(OEt)$_3$]$_2$, PEG2000-[Si(OEt)$_3$]$_2$+CD, PEG4000-[Si(OEt)$_3$]$_2$, PEG4000-[Si(OEt)$_3$]$_2$+CD, PEG6000-[Si(OEt)$_3$]$_2$, PEG6000-[Si(OEt)$_3$]$_2$+CD. All the measurements maintain 30 min, the moisture adsorption value of the untreated fabric display null during 30 min.

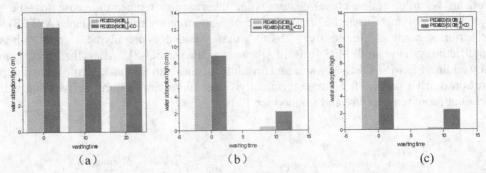

Figure 2 The Moisture adsorption effect in different soaping times by pure agent and add CD

It could be seen from Figure 2. that the PEG molecule weight increased, the moisture absorption value improved obviously from 8.5 cm of PEG2000-[Si(OEt)$_3$]$_2$ to 13.0 cm of PEG4000-[Si(OEt)$_3$]$_2$, however, there is no change of the moisture performance with 13.0 cm as the molecular form 4000 to 6000, while the value is 8.0cm, 9.0cm and 6.3cm as the agent containing CD. For the treated PP fabric washing durable observation from the above figure, it can be seen that the value decreased from 4.2 cm to 0.3 cm when the PEG molecular from 2000 to 6000 with 10 soaping times, while the value of 5.6 cm, 2.4 cm and 2.3 cm correspond with the agent containing CD with different PEG molecular. This improvement is resulted from the increased hydrophilicity with the molecular weight of PEG. It is helpful for the formation of hydrophilic coating film under curing condition, which endow the PP with improved moister property. However, the moister performance improved is limited as the PEG molecular reached a value. In addition, due to the washing durable decided by the proportion of the hydrophilic and hydrophobic groups of the agent, so higher amount of the hydrophobic groups can cause the treated sample durable dropped obviously, therefore the samples treated by the agent containing PEG 2000 can achieve better moister performance than other agent. For the reason why the treated samples durable improved by the agent with CD adding, it is may be that the CD cavity structure can protect some hydrophilic group avoiding soaping removal and after washing process completed that the protected group can regain display from CD cavity and produce function role. According to the above result, it is obviously that the PEG2000-[Si(OEt)$_3$]$_2$+CD as the optimum agent is selected using for the PP moister treatment, so it is feasible that PEG2000-[Si(OEt)$_3$]$_2$+CD as the next perspiration exhausts research object.

3.3 The perspiration exhausts of the treated PP fabric

The test results on perspiration exhausts of the PP fabrics treated by PEG2000-[Si(OEt)$_3$]$_2$ and PEG2000-[Si(OEt)$_3$]$_3$ mixed with CD are as shown in Figure 3.

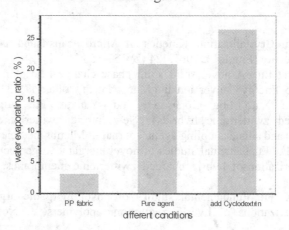

Figure 3 The perspiration exhausts effect by PEG2000-[Si(OEt)$_3$]$_3$ sol and add cyclodextrin

It is seen from Figure 3, the PEG2000-[Si(OEt)$_3$]$_2$ agent and the agent (PEG2000-[Si(OEt)$_3$]$_3$+CD) could endow perspiration exhaust effect for PP fabric. The perspiration exhaust for the PEG2000-[Si(OEt)$_3$]$_2$ sol treated fabric was 21% while the perspiration exhaust for the untreated fabric is only 3.09% in 12 min. The agent (PEG2000-[Si(OEt)$_3$]$_3$+CD) could endow a better perspiration exhaust effect with the perspiration exhaust of 26.57% in 12 min. According to above result, the samples treated by the lab synthesis agent is satisfied with that TTF 0007 standard requirements, which emphasizes that the knitting with the perspiration exhaust should higher 20% in 12 min.

3.3 SEM observation

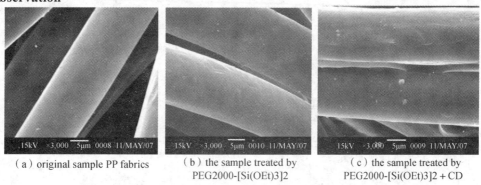

(a) original sample PP fabrics   (b) the sample treated by PEG2000-[Si(OEt)3]2   (c) the sample treated by PEG2000-[Si(OEt)3]2 + CD

Figure 4 The SEM of the samples treated before and after

From the Figure 4, the original sample display smooth cylinder shape (a) and the treated sample morphology changed a little, (b) and (c) show that the fiber surface unsmooth and some crosslinking appear at inter-fiber, which may be say that the thick hydrophilic polymer film by the agent adsorption and coating for the fiber is helped to the sample moister and durable performance improved.

4 Conclusions

PEG-[Si(OEt)$_3$]$_2$ sol originated from different PEG in molecular weights of 2000, 4000 and 6000 were synthesized as PP moisture adsorption and perspiration exhaust agent. PEG-[Si(OEt)$_3$]$_2$ sol prepared from PEG with the lowest molecule weight in 2000 has the best durable and moisture absorbing effects while the agent (PEG2000-[Si(OEt)$_3$]$_2$ added cylodextrin) could endow a better perspiration exhaust effect with the perspiration exhaust of 26.57% while the untreated fabric is only 3.09% in 12 min.

Acknowledgement

This research was funded in part through a grant by zhejiang education department research plan projects (Y200806085),Key Laboratory of Advanced Textile Materials and Manufacturing Technology of Chinese Ministry of Education (Zhejiang Sci-Tech University) subsidized project (2008QN11),and this work is supported by Science Foundation of Zhejiang Sci-Tech University (ZSTU) under Grant No. 11113332610836.

References

[1] Dong Zhi-zhi. Isothermal Crystallization Kinetics of Microencapsulated Polyethylene Glycol Particles. Journal of Donghua University (Eng.Ed.), 2007, l24 (2):157-158.
[2] Jing-Cang Su, Peng-Sheng Liu. A novel solid–solid phase change heat storage materialwith polyurethane block copolymer structure. Energy Conversion and Management, 2006, (47): 3185–3191.
[3] Junsheng Liu, Tongwen Xu, Ming Gong, Fei Yu, Yanxun Fu. Fundamental studies of novel inorganic–organic charged zwitterionic hybrids4. New hybrid zwitterionic membranes prepared from polyethyleneglycol (PEG) and silane coupling agent. Journal of Membrane Science, 2006, (283): 190–200.
[4] J.S. Liu, T.W. Xu, Y.X. Fu. Fundamental studies of novel inorganic–organiccharged zwitterionic hybrids. 2. Preparation and characterizationsof hybrid charged zwitterionic membranes. J. Membr. Sci., 2005 (252): 165–173.
[5] J.S. Liu, T.W. Xu, Y.X. Fu. Fundamental studies of novel inorganic–organic zwitterionic hybrids. 1. Preparation and characterizations of hybrid zwitterionic polymers. J. Non-cryst. Solids, 2005, (351): 3050–3059.
[6] J. S. Liu, T. W. Xu, M. Gong, Y. X. Fu. Fundamental studies of novelinorganic–organic charged zwitterionic hybrids. 3. New hybrid charged mosaic membranes prepared by modified metal alkoxide and zwitterionic process. J. Membr. Sci., 2005, (260): 26–36.
[7] C.M.Wu, T.W.Xu. Fundamental studies of a new hybrid(inorganic–organic) positively charged membrane: membrane preparation and characterizations. J. Membr. Sci., 2003, (216): 269–278.
[8] W.Chen, H. Feng, D. He, C. Ye. High resolution solid-state NMR and DSC study of poly (ethylene glycol)-silicate hybrid materials via sol–gel process. J. A. Polym. Sci., 1998, (67): 139–147.

95 Synthesis of Pyrazolone-Containing Carboxyester Dyes and Their Application to Poly(lactic acid) Fabric

Kai Liu[1], Zhihua Cui[1, 2*], Weiguo Chen[1, 2]

(1. Key Laboratory of Advanced Textile Materials and Manufacturing Technology, Ministry of Education of China, Zhejiang Sci-Tech University, Hangzhou 310018, China
2. Engineering Research Center for Eco-Dyeing & Finishing of Textiles, Ministry of Education, Hangzhou 310018, China)

Abstract: A series of hydrophobic carboxyester dyes bearing phenylazopyrazolone chromophore were prepared from carboxyl-containing acid dyes by successive chlorination and esterification. Then the molecular structures of the carboxyester dyes were characterized by FTIR, ^{1}H NMR and mass spectrometry. All of these carboxyester dyes were applied to poly(lactic acid) fabrics by exhaust dyeing at 110℃, in which dyebath pH was adjusted to 5.0. It was found that these disperse dyes exhibited high dye exhaustion and good fastnesses to washing and rubbing on poly(lactic acid) fabrics.

Keywords: Carboxyester; Synthesis; Characterization; Poly(lactic acid); Dyeing

1 Introduction

Poly(lactic acid) (PLA), an ideal biodegradable polymer, has been used in textiles, pharmaceuticals and many other fields due to its good biological compatibility and biodegradability, so it has a broad developing prospects[1]. With the development of large-scale operations for the economic production of PLA polymer in recent year, manufacture and application of PLA fiber in textile field have been realized. As a hydrophobic aliphatic polyester, the dyeability of PLA fiber is similar to the other polyester materials which could be colored with hydrophobic dyes. However, only a few disperse dyes could be applied to PLA fiber with excellent dyeing results including high sorption and good color fastness[2] after attempting lots of commercial disperse dyes. Therefore, designing and exploiting high-quality dyes suitable for PLA fiber is the key of the extension of PLA in textile field.

In view of the importance of high dye sorption on PLA, the sorption of disperse dyes onto PLA has been successfully explained according to the group contribution method[3]. According to the findings, it has been proposed that some polar groups (e.g. sulphonyl, carbonyl) introduced into dye molecule for stronger dipole-dipole forces with ester group in the macromolecular chain of PLA result in high affinity between them. To meet high dye sorption, some nonpolar end groups (e.g. alkyl, phenyl) are also supposed to be introduced into dye molecule at the same time for decreasing the improved water solubility of the dyes induced by the polar groups[4-5].

In this article, five carboxyester dyes bearing polar carbonyl group and nonpolar alkyl group were designed and synthesized. The carboxyester dyes have similar chemical structure with PLA and have been facilely prepared from carboxyl-containing acid dyes by successive chlorination and then esterification with different sorts of alcohols. Subsequently, these disperse dyes were applied to PLA, and the color fastnesses of the dyes on PLA fabric were evaluated.

2 Experiments
2.1 Materials and instrumentations

The knitted PLA fabric (250 g·m^{-2}, 18.5 tex) was supplied by Huayuan Eco-technology Co., Ltd (Dezhou, Shandong Province, China).

The reagents, ethanol, propanol, isopropanol, butanol, 3-pentanol, $SOCl_2$, DMF, K_2CO_3, p-aminobenzoic acid, 3-methyl-1-phenyl-2-pyrazoline-5-one, acetone were bought from Haochem Chemical Co., Ltd (Shanghai, China). The dispersant sodium methylenedinaphthalene disulphonate (NNO) was bought from Chuangfeng Chemical Factory (Shangyu, Zhejiang Province, China).

^{1}H NMR spectra were recorded on a Varian INOVA 400 NMR Spectrometer with TMS as internal standard in $CDCl_3$. IR spectra were measured with an FT/IR-430 spectrophotometer. Mass spectra (MS) were

---
* Corresponding author's email: zhhcui@zstu.edu.cn

determined by using a HP1100 mass spectrometer. Ultraviolet-visible (UV-vis) absorption spectra were recorded on a Lambda 900 UV/Vis spectrophotometer. Melting points were measured on a Mel-Temp capillary melting point apparatus and were uncorrected. The PLA samples were dyed by using a RY-25016 Infrared dyeing machine.

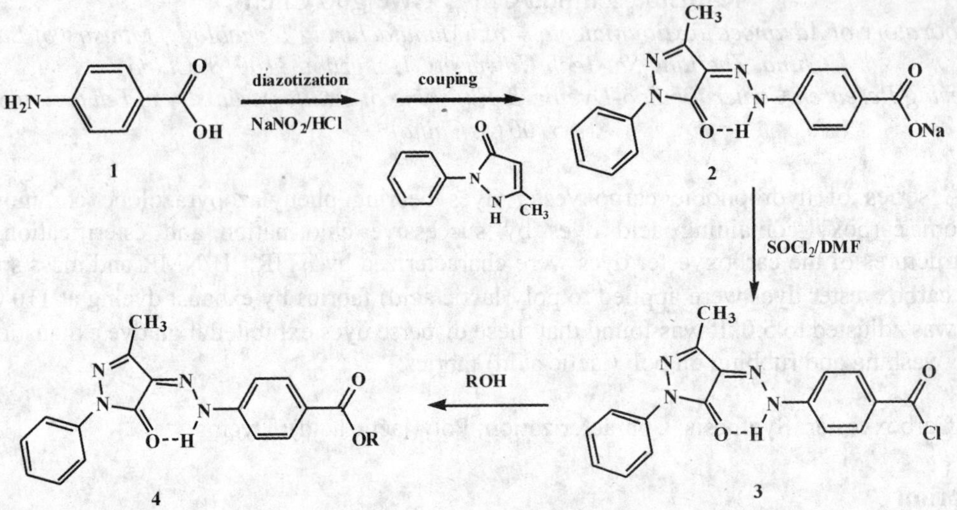

Scheme 1 Synthetic routes of pyrazolone-containing carboxyester dyes

2.2 Synthesis procedure of pyrazolone-containing carboxyester dyes
Synthesis of acid dye 2

Acid dye **2** was synthesized by conventional diazo-coupling methods: diazonium ion was prepared by diazotising p-aminobenzoic acid (10 mmol) in 5% sodium hydroxide aqueous solution (10 mmol, 10ml), by adding 12% sodium nitrite aqueous solution (11 mmol, 6ml) at a temperature of 0-5℃. The coupling reaction was carried out by adding diazo component solution to the coupling component (3-methyl-1-phenyl-2-pyrazoline-5-one) solution (10.1 mmol) at 0-5℃, pH 8.5-9.0 for 3 h. The pH value of the coupling liquor was controlled by adding sodium carbonate powder. The dye was salted out with sodium chloride, then filtered and purified by DMF-ether purification method[6].

Synthesis of acyl chloride 3

To a stirred suspended mixture of acid dye **2** (3.5 g, 10 mmol) and thionyl chloride (20 ml, 270 mmol), DMF (0.2 ml) was added at room temperature. The mixture was heated to 60℃ and stirred for 1.5 h. The solvent and excess thionyl chloride were removed by vacuum distillation, then the residue was transferred into ice water. After it was cooled to room temperature, the product was collected by filtration. The filter cake was washed with ice-cooled water till the filtrate was colorless and neutral, then dried in vacuum.

Synthesis of carboxyester dyes 4a-4e (4a as an example)

To a stirred suspension of acyl chloride **3** (1.7 g, 5 mmol) in dry ethanol (30 ml), K$_2$CO$_3$ (0.7 g, 5 mmol) and a few zeolites were added. The reaction mixture was boiled at boiling point and refluxed for 1.5 h, then the solvent was entirely distilled and the residue was poured into 10% dilute hydrochloric acid (100 ml). The product was collected by filtration, the filter cake was washed with water till the filtrate was colorless and neutral, then the carboxyester dye **4a** was air dried at room temperature.

Other carboxyester dyes were synthesized using the same process described above.

2.3 Application of pyrazolone-containing carboxyester dyes to PLA

Prior to dyeing, all fabrics were scoured at a liquor ratio of 50:1 in a solution of 1g/L sodium carbonate and 2 g/L nonionic detergent at 60℃ for 15 min. The dyes had been milled for 8 hour with pea gravel at room temperature in the presence of dispersing agent NNO (Dye:NNO = 1:1wt/wt). Then the dyes were applied to dye knitted PLA (2 g) in stainless steel, sealed dye pots in a infrared high temperature and pressure dyeing machine, using a liquor ratio 50:1, according to the recommended method shown in Figure 1[7]. Reduction clearing was carried out after dyeing using 1.5 g/L sodium carbonate and 2 g/L sodium hydrosulfite using a liquor ratio of 50:1 at 60℃ for 15 min[8]. After clearing, the samples were rinsed with cold water and air dried.

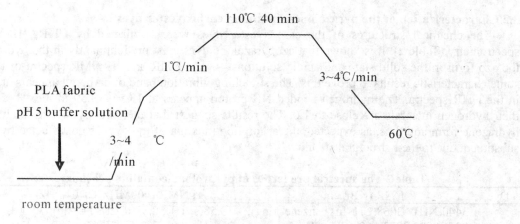

Figure 1 Dyeing procedure used for pyrazolone-containing carboxyester dyes on PLA

2.4 Measurement of dye exhaustion

The dye in the residual dye bath and staining inside the dyeing tube was dissolved into 50/50 acetone/water[9]. The optical absorbance of the solutions was determined according to Lambert-Beer's law at the wavelength of the maximum absorption of each dye by UV-visible spectrophotometer. Subsequently, the dye exhaustion ($E\%$) was calculated from Eq.(1).

$$E\% = [(A_1 - A_2) / A_1] \times 100\% \qquad (1)$$

where A_1 and A_2 are the absorbance of the dye bath dissolved into 50/50 acetone/water before and after dyeing.

2.5 Fastness testing

The dyeing samples were used to assess fastnesses to washing, sublimation and rubbing according to GB/T3921.3-1997, GB/T6152-1997, GB/T3920-1997.

3 Results and discussions

3.1 Synthesis of the pyrazolone-containing carboxyester dyes

The pyrazolone-containing carboxyester dyes were stepwisely prepared from p-aminobenzoic acid via diazotization, coupling reaction, chlorination and esterification with different alcohols according to the process similar to our previous research (Scheme 1)[10]. The yields, melting points of the dyes were measured and listed in Table 1. In order to characterize the carboxyester dyes and determine the influence of the substituents on their spectral properties[11], spectrophotometric investigation were carried out. The absorption UV-vis spectra of the dyes in acetone were recorded and the data were also presented in Table 1. It shows that carboxyester dyes exhibit the same λ_{max} in acetone (393nm), despite they have a bit of difference at the end of ester group.

Table 1 Yield, physical and chemical properties of pyrazolone-containing carboxyester dyes

Dye	Chemical structure		Molecular weight	Yield (%)	Melting point (℃)	λ_{max} in Acetone (nm)
	General structure	R				
4a		CH_2CH_3	350.1	85.1	140-142	393
4b		$(CH_2)_2CH_3$	364.2	88.7	113-116	393
4c		$(CH_2)_3CH_3$	378.2	75.4	108-110	393
4d		$CH(CH_3)_2$	364.1	80.2	190-192	393
4e		$CH(C_2H_5)_2$	392.2	70.2	196-198	393

3.2 Characterization of the pyrazolone-containing carboxyester dyes

The chemical structures of the carboxyester dyes were confirmed by FTIR, ^{1}H NMR and mass spectrometry (Table 2). It is known that azopyrazolone dyes exist predominantly in the hydrazone form over the azo form in the solid state and acidic solutions[12]. The FTIR and ^{1}H NMR spectra of the dyes provide some characteristic results to prove this. The stretching vibration band of carbonyl appears at 1658~1667cm^{-1} in the FTIR spectra. The hydrogen-bonded NH proton appears at 13.56~13.59 in the ^{1}H NMR spectra and their hydrogen integrals are close to 1.0. The results suggest that the carboxyester dyes almost exist in the hydrazone form and thus, as expected, the absorption maxima of the dyes is not affected by variation in the substituent due to their structural stability.

Table 2 The spectral properties of pyrazolone-containing carboxyester dyes

Dye	FTIR (KBr, cm^{-1})	^{1}H NMR (400MHz, CDCl$_3$, Hz)	MS m/z /%
4a	3440(NH), 3068(CH in benzene ring), 2975,2923,2872,1460(CH$_2$,CH$_3$), 1707(C=O in esteryl), 1658(C=O in pyrazolone), 1272, 1253, 1106(C-O-C)	13.58(s, 1H), 8.03(d, 2H, J = 8.0), 7.90(d, 2H, J = 8.0), 7.74(d, 2H, J = 8.0), 7.47(t, 2H, J = 8.0), 7.24(t, 1H, J = 8.0), 4.33(t, 2H, J = 8.0), 2.32(s, 3H), 1.33(t, 3H, J = 8.0)	348.8([M-H]$^-$)
4b	3317(NH), 3062(CH in benzene ring), 2968,2937,2877,1460(CH$_2$,CH$_3$), 1719(C=O in esteryl), 1666(C=O in pyrazolone), 1275, 1253, 1106(C-O-C)	13.58(s, 1H), 8.12(d, 2H, J = 8.4), 7.96(d, 2H, J = 8.0), 7.47~7.42(q, 4H), 7.23(t, 1H, J = 7.6), 4.29(t, 2H, J = 8.0), 2.39(s, 3H), 1.80(q, 2H, J = 8.0), 1.04(t, 3H, J = 8.0)	362.8([M-H]$^-$)
4c	3416(NH), 3062(CH in benzene ring), 2958,2930,2872,1461(CH$_2$,CH$_3$), 1718(C=O in esteryl), 1667(C=O in pyrazolone), 1256, 1253, 1103(C-O-C)	13.56(s, 1H), 8.10(d, 2H, J = 8.0), 7.93(d, 2H, J = 8.0), 7.45~7.41(t, 4H), 7.26(q, 1H, J = 8.0), 4.33(t, 2H, J = 7.6), 2.37(s, 3H), 1.74(q, 2H, J = 8.0), 1.48(q, 2H, J = 8.0), 0.99(t, 3H, J = 8.0)	376.8([M-H]$^-$)
4d	3327(NH), 3069(CH in benzene ring), 2976,2928,2877,1458(CH$_2$,CH$_3$), 1706(C=O in esteryl), 1661(C=O in pyrazolone), 1280, 1250, 1106(C-O-C)	13.58(s, 1H), 8.11(d, 2H, J = 8.0), 7.96(d, 2H, J = 8.0), 7.47~7.42(t, 4H), 7.21(t, 1H, J = 8.0), 2.39(s, 3H), 1.39~1.37(d, 6H, J = 8.4)	362.8([M-H]$^-$)
4e	3430(NH), 3064(CH in benzene ring), 2967,2939,2878,1460(CH$_2$,CH$_3$), 1702(C=O in esteryl), 1664(C=O in pyrazolone), 1275, 1254, 1097(C-O-C)	13.59(s, 1H), 8.12(d, 2H, J = 8.0), 7.96(d, 2H, J = 8.0), 7.48~7.43(q, 4H), 7.23(t, 1H, J = 8.0), 5.02(t, 1H), 2.39(s, 3H), 1.72(t, 4H, J = 8.0), 0.96(t, 6H, J = 8.4)	390.8([M-H]$^-$)

3.3 Dyeing characteristics and color fastnesses of the pyrazolone-containing carboxyester dyes on PLA

Dye sorption and colour fastness results were summarised in Table 3, where it could be seen that changing the R-group in the carboxyester group had great effect on dye sorption on PLA fabrics[13]. As seen in Table 3, the dyes except dye **4e** had dye sorptions greater than 80%, which is higher than those of commercial disperse dyes without polar gourps in their molecular structures. It proved that the carboxyester group introduced to the dye molecule could enhance the affinity between the dye and PLA fiber[13]. According to the difference at the end groups, the carboxyester dyes could be divided into two types: straight-chain dyes **4a**, **4b**, **4c** and branched-chain dyes **4d**, **4e**. The dye sorptions were related to the steric hindrance of the carboxyester dyes during dyeing process. The straight-chain dyes **4a**, **4b**, **4c** having little steric hindrance give over 85% dye sorption on PLA, while the branched-chain dyes **4d**, **4e** exhibit lower dye sorption than that of straight-chain dyes. It can be concluded that too big end group in dye molecule is not in favor of higher dye sorption on PLA.

The results in Table 3 also shows that washing and rubbing fastnesses of the carboxyester dyes are all good to excellent and the sublimation fastnesses are all above medium. These results demonstrate that carboxyester dyes are suitable for coloration of PLA fabric with good dyeing performances.

Table 3 Dye sorptions and color fastness properties of pyrazolone-containing carboxyester dyes on PLA

Dye	Dye sorptions (%)	Washing			Heat/sublimation(150℃)		Rubbing	
		Change	Staining		Change	Staining Cotton	Dry	Wet
			Cotton	PLA				
4a	93.8	4-5	4-5	4-5	3	3-4	5	4-5
4b	95.7	5	4-5	5	3	3-4	5	4-5
4c	85.8	4-5	4-5	4	3	3	5	4-5
4d	82.3	4-5	4	4-5	3-4	3-4	5	4
4e	37.8	4	4	4	3-4	3	5	4

PLA samples were dyed at 0.5% owf

4 Conclusions

Some carboxyester dyes designed for PLA were synthesized and their dyeability on PLA fabric was also investigated. The supposal that introducing some polar and nonpolar group into dye molecule can enhance the affinity between the dye and PLA fiber have been proven by our investigations one more time. It is found in the research that the straight-chain dyes often give higher dye sorption on PLA than that of branched-chain dyes due to their little steric hindrance during dyeing. Among these carboxyester dyes, **4b** exhibits the best dyeing performances including 95.7% dye sorption and satisfied color fastnesses.

Acknowledgement

The authors are grateful to the Zhejiang Provincial Natural Science Foundation of China (Y4090227), and Science Foundation of Zhejiang Sci-tech University (ZSTU) for financial support.

References

[1] E Lois. A C Scheyer. Application and performance of disperse dyes on polylactic acid fabric. AATCC Review, 2001, 1(2):44-48.
[2] Y Yang, S Huda. Comparison of disperse dye exhaustion, color yield, and colorfastness between polylactide and poly(ethylene terephthalate). Applied Polymer Science J, 2003, 90: 3285-3290.
[3] D Karst, Y Yang. Using the solubility parameter to explain disperse dye sorption dye sorption on polylactide. Journal of Applied Polymer Science, 2005, 96(2): 416-422.
[4] J L Liu. Synthesis of dyes and application to polypropylene. Dalian University of Technology[D], 1999.
[5] S Zhang, X Chen, J Yang. Study on synthesis and application of the azo dyes containing sulfonamido as link group. Dyestuff Industry, 1997, 34(1): 1-7.
[6] J Yang. Analysis and Anatomy of Dyes. Beijing: Chemical Industry Press, 1989, 360.
[7] O Avinc, J Bone, H Owens, D Phillips, Wilding W. Preferred alkaline reduction-clearing conditions for use with dyed Ingeo poly(lactic acid) fibres. Color Technol, 2006, 122: 157-161.
[8] E L Hu. Synthesis of carboxamide dyes and application to PLA fabric. Zhejiang Sci-Tech University[D]. 2008
[9] Y Yang, S Huda. Dyeing conditions and their effects on mechanical properties of polylactide fabric. AATCC Review, 2003, 3(8): 56-61.
[10] Z H Cui. Synthesis of sulfonamide dyes and applications to fabrics. Dalian University of Technology[D], 2007.
[11] M S Yen, I J Wang. A facile syntheses and absorption characteristics of some monoazo dyes in bis-heterocyclic aromatic systems part I: syntheses of polysubstituted-5-(2-pyride-5-yl and 5-pyrazolo-4-yl) azo-thiophene derivatives. Dyes and Pigments J, 2004, 62:173-180.
[12] S F Zhang, Y Z Tian, Z W Wu. The research on the azo-hydrazone of the azo dyes. Chemical Technology J, 1995, 46(2): 152-157.
[13] J J Lee, J H Choi. Synthesis and application of temporarily solubilised azo disperse dyes containing sulphatoethylsulphonyl. Dyes and Pigments J,2005,65(1):75-81.

96 Preparation of Baicalin-Al(Ⅲ) Complex Dye and Application in Dyeing of Silk Fabric

Zhengming Liu[1,2], Qibing Wang[1,2], Zhicheng Yu[*,1,2]

(1. Engineering Research Center for Eco-Dyeing & Finishing of Textiles, Ministry of Education, Zhejiang Sci-Tech University, Hangzhou 310018, China
2. Key Laboratory of Advanced Textile Materials and Manufacturing Technology, Ministry of Education of China, Zhejiang Sci-Tech University, Hangzhou, China)

Abstract: The baicalin-Al(Ⅲ) complex was prepared with baicalin and alum under certain conditions, the parameters of the reaction were studied, the molecular structure of the complex was characterized by FTIR, finally the prepared complex dye was applied on the silk fabric. The results show that the optimal synthesis conditions for baicalin-Al(Ⅲ) complex are as following: mole ratio of baicalin/Al(Ⅲ) is 2:1, reaction temperature is 30℃, time is 30min, pH value is 5. FTIR result shows that 5-phenolic hydroxyl group and 4-carbonyl group of baicalin are involved in the complex reaction. Compared to the dyed fabric with baicalin by the methods of meta-mordanting and direct dyeing, the fabric dyed with baicalin-Al(Ⅲ) complex has highest *K/S* value and color fastness.

Keywords: Natural dyes; Baicalin-Al(Ⅲ) complex dyes; Dyeing

0 Introduction

In recent years, the research of using the renewable, functional natural dyes in textile dyeing and finishing has got more and more achievements. Natural dyes are extracted from plants, animals or mineral resources, which can not react with fiber, but include lots of group that can react with the mordant, so most natural dyes are mordant dyes. According to the sequence of the mordant used in dyeing, the dyeing methods can be divided into pre-mordanting, meta-mordanting, after-mordanting. Pre-mordanting and after-mordanting[1] generally obtain higher color fastness, due to the dyeing condition of the two methods are more conducive to the formation of the complexes[2], but the process of the two dyeing methods are complicated, having low utilization ratio of raw material, and poor color shade reproducibility. These shortcomings hamper the process of industrialization of natural dyes. So how to improve the fastness of natural dyes and simplify the dyeing process is the focus of natural dyes.

Scutellaria baicalensis Georgi is Labiatae perennial herbaceous plant, an ancient anti-inflammatory medicine. Baicalin, the main component of Scutellaria baicalensis Georgi, belongs to flavonoid[3] derivative, has lots of functions, such as antibacterial, anti-inflammatory, anti-ultraviolet, which has been used in various fields. Baicalin is a worth studying natural dye, the fabric treated by Baicalin can not only gain bright yellow, but also obtain antibacterial and anti-ultraviolet functions[4].

In this paper, baicalin react with alum under certain conditions to produce a natural metal complex dye, the structure were characterized by FTIR, and the metal complex was directly used to dyeing silk fabric, the fabric dyed by metal complex not only improved the color depth, but also simplified the dyeing process.

1 Materials and apparatus
1.1 Materials
Silk(crepe de chine), baicalin, alum(AR), sodium carbonate(AR).
1.2 Apparatus
THZ-82 constant temperature oscillating water bath pot, DHG-9140A electrothermal constant temperature blast dry drying oven, MP502B electronic balance, DF-105S constant temperature magnetic agitator, SH2-3 circulated water multipurpose vacuum pump, QM-ISP04 planetary ball mills.

2 Experimental methods
2.1 Baicalin-Al(Ⅲ) complex preparation
1.0 g baicalin was added to 100 ml deionized water solution with shocking; the sodium carbonate

---

* Corresponding author's email: yuzhicheng8@yahoo.com.cn

solution was added dropwise until the baicalin dissolved, the pH value of the solution was in the range of 5-6, then alum solid was added, and the solution became turbid immediately. The reaction occurred under optimal temperature and time with medium-speed stirring, and then the solution was filtered. The sediment was washed by water more than three times, and then dried at constant temperature drying oven under 50℃. The final product is orange yellow powder. The yield of baicalin-Al(III) complex was calculated as the following formula:

$$\text{Yield of baicalin-Al(III) complex} = W/W_0 \times 100\%$$

Where: W: the weight of the baicalin-Al(III) complex (product);

W$_0$: the total weight of baicalin and metal ion.

2.2 Baicalin-Al(III) complex FTIR measurement

Infrared spectra was characterized with a Nicolet 5700 Fourier Transform Infrared Spectrophotometer, operated in the presser-bit mode, used to estimate the structure of Baicalin-Al(III) complex.

2.3 Color value of fabric measurements

The color values for the dyed fabric were tested by spectrophotometer (DataColour SF600- PLUS), the value of K/S, L*, a*, b* were tested under the light of source D65 and 10° observer, the values are calculated from three repetitive measurements at the different place of sample.

The dyeing process parameter is listed in the following:

   Dyes   2% (o.w.f)
   Liquor ratio   1: 30
   Temperature   90℃
   Time   60 min.

2.4 Fastness properties

Fastness of the dyed samples was tested according to GB/T3920-1997—GB/T3921-1997.

3 Results and discussion

3.1 Effect of n (baicalin): nAl (III) on the yeild of complex

The reaction temperature is 30℃, reaction time is 30 minutes, pH value is 5, the effect of molar ratio of baicalin and alum on the yield of baicalin-Al(III) complex was studied, the results are shown in Figure 1.

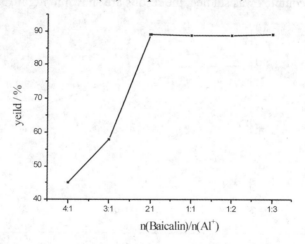

Figure 1 The effect of n(Baicalin)/ n(Al$^+$) on the yield of complex

As seen in Figure 1, when n(Baicalin)/ n(Al$^+$) is more than 2, the yeild of baicalin-Al(III) complex increases with the decrease of n(Baicalin)/ n(Al$^+$), when the molar ratio is 2:1, the yield of the complex reaches 88.9%. But with the further decrease of the molar ratio of baicalin and alum, the yield of the complex no longer increased, the main reason is that aluminum ion and the baicalin have a specific coordination number, the overdoses of alum will not react with baicalin, therefore, the optimal mole ratio of baicalin and alum is 2:1.

3.2 Effect of pH value on the yield of complex

The reaction temperature is 30℃, reaction time is 30 minutes, n(Baicalin)/ n(Al$^+$) is 2:1, the effect of pH value on the yield of baicalin-Al(III) complex was studied, the results are shown in Figure 2.

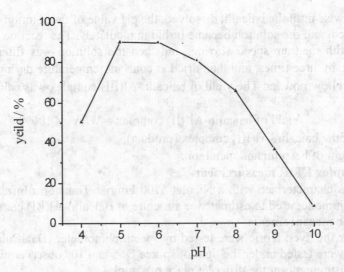

Figure 2 The effect of pH value on the yield of complex

As shown in Figure 2, when the pH value is less than 5, the yield of baicalin-Al(III) complex increased with the increase of pH value, which indicate that the baicalin is hard to react with Al as an electron donor under strong acidic conditions; the yield of baicalin-Al(III) complex reached the maximum when the pH value rise to 5, which indicate that it is beneficial to the formation of complex under weak acidic conditions; While under neutral or alkaline conditions, the yield of complex decrease, because baicalin in alkaline conditions is unstable and easily hydrolyzed, the oxidation number of baicalin increase as the pH value increase, In addition, Al ions under alkaline conditions is easily forming aluminum hydroxide deposits, leads to Al ions decreases. Therefore the pH value of the reaction solution is in the range of 5 to 6.

3.3 Effect of temperature on the yield of complex

The reaction time is 30 minutes, n(Baicalin)/ n(Al^+) is 2:1, pH value is 5, the effect of temperature on the yield of baicalin-Al(III) complex was studied, the results are shown in Figure 3.

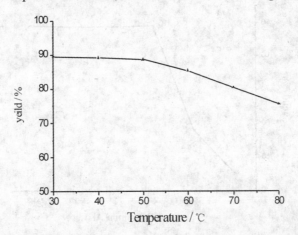

Figure 3 The effect of temperature on the yield of complex

As shown in Figure 3, the yield of baicalin-Al(III) complex is high when the reaction temperatures at 30-40℃, but, as the temperature increased, the yield of baicalin-Al(III) complex decreased, which indicate that high temperature is not conducive to complex formation, due to the baicalin in higher temperature is easily oxidized. This explains why the K/S values and fastness of the fabric in one bath dying (high temperature) significantly lower than the fabric in after-mordanting(room temperature or low temperature). Therefore, optimal reaction temperature for synthesis is 30℃.

3.4 Effect of time on the yield of complex

The reaction temperature is 30 minutes, n(Baicalin)/ n(Al^+) is 2:1, pH value is 5, the effect of reaction time on the yield of baicalin-Al(III) complex was studied, the results are shown in Figure 4.

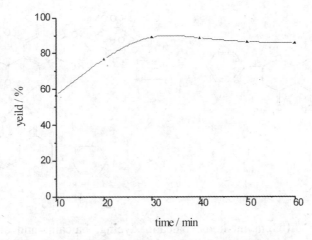

Figure 4 The effect of time on the yield of complex

It can be seen in Figure 4, when the reaction time is less than 30min, with the reaction time increases, the yield of baicalin-Al(III) complex increases; When the reaction time is more than 30min, the yield of baicalin-Al(III) complex do not further increases, so the optimal reaction time for synthesis is 30 min.

3.5 FTIR analysis

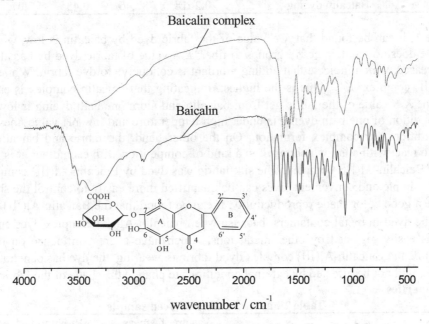

Figure 5 FTIR spectra of baicalin and baicalin complex

The spectra of baicalin and its complex are presented in Figure 5. As seen in the spectrum of baicalin, there are two absorption peaks in 3200-3700cm^{-1}, the absorption peak at 3490.08 cm^{-1} results from associating hydroxyl, which originate from the formation of intra molecular hydrogen bond (5-phenolic hydroxyl group with 4-carbonyl), while the absorption peak disappear in baicalin-Al(III) complex dyes, that results from the disappearance of intra molecular hydrogen bonds, which indicate that Al complex reaction may be located at the place of 4,5. The absorption peak at 1726.8 cm^{-1} in baicalin results from the carboxyl acid carbonyl which almost disappear in baicalin-Al(III) complex, because carboxyl acid change into carboxylic acid salt during dissolution, so baicalin-Al(III) complex in the absorption intensity becomes very weak. the absorption peak at 1660.65cm^{-1} results from 4-carbonyl in baicalin, while in baicalin-Al(III) complex the absorption peak shifted to 1634.74 cm^{-1}, redshift 26 cm^{-1}, because Al molecules have strongly electron-withdrawing effect, which leads to Make he baicalin-Al(III) complex molecular electron cloud close to Al. Final conclusion: baicalin reacts with alum to produce complex, the 5-phenolic hydroxyl group and 4-carbonyl of baicalin react with Al ion, alum and baicalin complex reaction equation is as follows[5-7]:

3.6 Dyeing

Dyeing the silk with different methods: baicalin dyeing, baicalin and alum one bath dyeing and Baicalin-Al(III) complex dyes dyeing, the results are shown in Table 1.

Table 1 Color value of dyed sample

Dyeing methods	K/S	L*	a*	b*
Baicalin-Al (III) complex dyeing	5.5811	75.58	10.93	45.61
Baicalin and alum one bath dyeing	3.3916	80.08	6.04	46.17
Baicalin dyeing	0.2618	87.45	0.45	7.07

From Table 1, it can be found that K/S value of the fabric dyed by baicalin is low, which indicate that baicalin molecule does not contain active groups to fiber; K/S value of fabric dyed by baicalin and alum one bath dyeing increased, which indicate that adding mordant is conducive to dye fabric; While the fabric dyed by baicalin-Al(III) complex, K/S value is the highest, indicating that baicalin complex is easiest for dyeing the fabric, and the K/S value of the fabric get from baicalin and alum one bath dyeing is low, partly because of the dyeing condition of one bath dyeing is under high temperature and low pH value, these conditions are not the best conditions for complex formation; On the other hand, the fabric and baicalin both have the complex group to react with alum which result in a kind of competition with each other, it is not beneficial to the formation of Baicalin-Al(III) complex. The silk fabric was dyed by Baicalin-Al(III) complex directly, the dyeing process is simple and convenient, dyes can be quantified more easily to control the shade of the dyed fabrics, and have a good color shade reproducibility. It can be found that the Baicalin-Al(III) complex dyeing process can not be dyed in metal containers, because when baicalin-Al(III) complex dyes encountered other metal ions, Al is easily replaced by other metal ions, which have a great influence on the shade of the complex dye. While the Baicalin-Al(III) complex dyed fabric in wearing, the dye has penetrated into the fiber interior, therefore, will not be affected by external conditions to change the shade of the fabric.

3.7 Fastness properties

Table 2 Fastness rating of dyed sample

Dyeing methods	Washing fastness		Rubbing fastness	
	Change in color	Staining	Dry	Wet
Baicalin-Al (III) complex dyeing	4-5	5	5	4-5
Baicalin and alum one bath dyeing	3	4	5	3

As shown in Table 2, after being dyed with baicalin-Al(III) complex dyes, the washing and rubbing fastness of silk fabric achieve above level 4 compared with baicalin and alum one bath dyeing, which is up to the wearing standard.

4 Conclusion

(1) The results show that the optimal process conditions for synthesis are as follows: mole ratio of baicalin / Al (III) is 2:1, temperature is 30℃, time is 30min, and pH value is 5;

(2) FTIR result shows that 5-phenolic hydroxyl group and 4- carbonyl group of baicalin are involved in the complex reaction;

(3) The silk fabric dyed with natural complex dyes has high color fastness.

References

[1]. Xin Sun. Research on the dyeing of silk fabrics with Scutellaria Bailalensis. Silk, 2002, 2: 11-13
[2]. Bing Zhao. Research progress in the chemistry of metallic complexes of flavanones. Chemical Regent, 2006, 28(3): 141-143
[3]. Rubens F.V. de Souza. Synthesis, spectral and electrochemical properties of Al(III) and Zn(II) complexes with flavonoids. Spectrochimica Acta, 2005, 61: 1985-1990
[4]. Le Wang. The pharmacodynamics of baicalin-metal ion chelation. Chinese archives of traditional Chinese medicine, 2007, 25(4): 548-550
[5]. S. Birjees Bukhari. Synthesis, characterization and investigation of antioxidant activity of cobalt-quercetin complex. Journal of Molecular Structure, 2008, 892: 39-46
[6]. M. Tereza Fernandez. Iron and copper chelation by flavonoids: an electrospray mass spectrometry study. Journal of Inorganic Biochemistry, 2002, 92: 105-111
[7]. Armida Torreggiani. Copper(II)-Quercetin complexes in aqueous solutions: spectroscopic and kinetic properties. Journal of Molecular Structure, 2005(744-747): 759-766

97 Influence of 1:1 Acid Metal Complex Dyes on Extractable Chromium of Wool Fabric

Baihua Wang[1, 2*]

(1. School of Materials Science and Engineering, Beijing Institute of Fashion Technology, Beijing 100029, China
2. Beijing Key Laboratory of Clothing Materials R&D and Assessment, Beijing 100029, China)

Abstract: Wool fabric was dyed with 1:1 acid metal complex dyes, Palatine or Neolan. Neolan dyes were used in the formic acid method, while Palatine dyes were utilized in the aminosulfonic acid method or the sulfuric acid method. The extractable chromium content of samples were measured by plasma emission spectrometry (ICP-AES), according to Oeko-Tex Standard 100 specifications. The results show that the extractable chromium content of wool fabric dyed with Neolan dyes in the formic acid method is less than 2ppm when dye dosage does not more than 1.5% o.w.f., however, that of fabric dyed with 1.5% o.w.f. Palatine dyes is more than 2ppm. There may be a negative correlation between the structure stability of dyes and the extractable chromium content on wool fabric dyed with 1:1 acid metal complex dyes.

Keywords: Extractable heavy metal; Acid metal complex dye; Wool fabric

1 Introduction

Acid metal complex dyes are a kind of acid dyes which contains the metal chelate structure. Those metal ions could be chromium ion, and sometimes cobalt ion. According to the difference of the proportion between the dyes and the metal chelates as ligand atoms within the dye molecular structure, there can be 1:1 metal complex dyes and 1:2 metal complex dyes [1]. The structural stability of these metal complexes results in free heavy metal ions on the fabric. According to the Oeko-Tex Standard 100, extractable heavy metals means that those free heavy metals extracted by artificial acidic sweat, and a limit value of the extractable chromium content on the fabric must be less than 2 ppm[2]. In the past study of acid metal complex dyes, more emphasis were on reducing wool injuries and improving the color fastness of these dyes, while less in extraction of heavy metals[3-6].

In this paper, we selected six 1:1 metal complex dyes involved three colors of red, yellow and blue. Wool fabric samples were dyed separately in the formic acid method, the aminosulfonic acid method and the sulfuric acid method. According to Oeko-Tex Standard 100 requirements, the extractable chromium content of each wool fabric samples were measured by ICP-AES determination. By way of theoretical analysis, the relation between the reasonable of test data and dyeing methods was determination. A reliable basis in industry of rightly use the acid metal complex dyes and effectively control the extraction of chromium was provided.

2 Experimental

2.1 Materials:
2201 pure wool gabardine

2.2 1:1 metal complex dyes and dyeing method
Dyes: Neolan P Red; Neolan P Yellow; Neolan P Blue; Palatine Fast Pink BN; Palatine Yellow GR; Palatine Blue RRN

Dyeing methods:
① Ciba-Geigy in Neolan P-type dying method (hereinafter referred to as formic acid dyeing method)
② BASF Palatine dye using amino acid (hereinafter referred to as aminosulfonic acid dyeing method)
③ The traditional Palatine dying method used in wool factories (hereinafter referred to as sulfuric acid dyeing method)

Three dyes prescription were showed in Table 1.

---

* Corresponding author's email: teacher_wang06@163.com. Project Source: Beijing Science and Technology Commission project "Ecological textile testing and early warning system establishment" number: Z0004077041411

Table 1 The wool dyeing prescription

Dyes and Auxiliaries	Formic acid method Dosage (% o.w.f)		Aminosulfonic acid method Dosage (% o.w.f)		Sulfuric acid method Dosage (% o.w.f)	
Dyes	Neolan P	X	Palatine	X	Palatine	X
Acid	Formic acid (85%)	4-5	Amino acid	6	Sulfuric acid (98%)	6
Dispersant	Albegal Plus	2	Leveler W-SX	2	Peregal O	2
	Albegal FFA	1	—		—	
Anhydrous Sodium Sulfate	—		5		5	

Note: The auxiliaries are provided by the dyes manufacturer, corresponding to the dyeing method.

The three kinds of dyeing method mentioned above, use the same dyeing temperature process curve, shown in Figure 1.

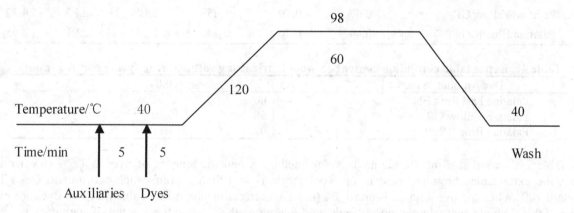

Figure 1 Dyeing curve

2.3 Measurement of extractable chromium content on fabric
2.3.1 Preparation of the artificial acidic sweat
The artificial acidic sweat was prepared according to GB/T3922-1995 (equivalent to ISO105-E04: 1994) "Textiles perspiration fastness test method".
2.3.2 Extraction
According to Oeko-Tex Standard 100 requirements for processing the samples, extraction conditions are: Weigh the chopped samples dried at 60℃ of 1g (accurate to ± 0.0001g) in 150ml flask, add 20ml these artificial acidic sweat, after fully wetting, oscillating 1h in 37 ± 2℃ in water bath, standing 1h, then filtering the extract into 50ml volumetric flask.
2.3.3 Determination of extract
Using plasma emission spectrometry (ICP-AES) (Spectro Plasma Emission Spectrometer, Germany) to determine the chromium content of the extract.

3 Results and discussion
3.1 Extractable chromium content on wool fabrics
The traditional wool fabric dyeing method by using 1:1 metal complex dye, is boiling dyeing using 5% ~ 7% (o.w.f) sulfuric acid at the pH value of 1.8 to 2.0, which is in order to reduce the dyeing rate and improve the levelness. However, it will damage the wool, rough the feel, and corrode the dyeing machine. To solve these problems, solutions were proposed:

①Adding non-ionic leveling agent polyethylene oxide ether (such as Peregal O) into the dye bath, reducing sulfuric acid dosage down to 4% ~ 6% (o.w.f.). However, problems remain.

②Replacing sulfuric acid with aminosulfonic acid, and using the appropriate leveling agent to keep the dye bath pH value between 2.5 ~ 3.5.

③Replacing sulfuric acid with formic acid and optimizing the compatibility among three primary colors, keeping dye bath pH value of 3.5 ~ 4.0.

The latter two solutions can resolve the problems mentioned and also improve the fastness of these dyes. This paper will evaluate extractable chromium content by using these three dyeing method. The results are shown in Table 2 to Table 4.

Table 2 Extractable chromium content on wool fabric using formic acid dyeing method (ppm)

Dye Amount % (o.w.f.)	0.1	0.2	0.4	0.8	1.5	3.0
Neolan P Red	0.38	0.50	0.65	0.98	1.64	2.45
Neolan P Yellow	0.29	0.27	0.22	0.36	1.10	2.32
Neolan P Blue	0.31	0.33	0.29	0.26	0.30	0.31

Table 3 Extractable chromium content on wool fabric using aminosulfonic acid dyeing method (ppm)

Dye Amount % (o.w.f.)	0.1	0.2	0.4	0.8	1.5	3.0
Palatine Fast Pink BN	0.47	0.61	0.85	1.24	2.01	3.28
Palatine Yellow GR	0.47	0.79	1.05	2.10	2.82	4.72
Palatine Blue RRN	0.44	0.50	0.24	1.35	2.01	3.68

Table 4 Extractable chromium content on wool fabric using sulfuric acid dyeing method (ppm)

Dye Amount % (o.w.f.)	0.2	0.8	1.5
Palatine Fast Pink BN	1.00	2.42	4.34
Palatine Yellow GR	1.59	5.16	15.79
Palatine Blue RRN	1.02	3.11	4.28

Table 2 shows that using formic acid dyeing method, when the amount of dyes is 1.5% (o.w.f.) and below, the extractable chromium content on wool fabric is less than 2 ppm which conform to Oeko-Tex Standard 100. When the dye amount is up to 3% (o.w.f.), extractable chromium content of two dyes are over 2 ppm. Table 3 shows that using aminosulfonic acid dyeing method, when the amount of such dyes is 1.5% (o.w.f.), extractable chromium content are over 2 ppm. The extractable chromium content is over 2 ppm of Palatine Yellow GR when in the amount of 0.8% (o.w.f.). This dye's molecular structure is the same as Neolan P Yellow's, of which C.I. number is C.I. Acid Yellow 99, molecular structure are No. C.I. 13900 [1]. Table 4 shows that using sulfuric acid dyeing method, when the dye amount is only 0.2% (o.w.f.), the extractable chromium content is up to 1.59 ppm. Considering Palatine Yellow GR. The extractable chromium content of dyed fabric using formic acid dyeing method is the smallest, and others the contents significantly increase. For industry applications, when the formic acid dyeing method is adopted and the dye amount is no more than 1.5% (o.w.f.), the the extractable chromium content on light-colored and middle-colored wool fabric is below the Oeko-Tex Standard 100 limit value, so this method is valuable to some extent. Other dyeing methods can not reach the requirement of Oeko-Tex Standard 100. So basically they are not valuable in the production of wool fabric marked Oeko-Tex Standard 100.

3.2 Influence of the structure stability of acid metal complex dyes on the extractable chromium content

Let us talk about the structure of metal complexes here. The 1:1 metal complex dye's general formula is $[Cr4G2X]^+ X^-$. The maximum coordination number of chromium is 6, G means the anion price group that bond with chromium ions. 1:1 dye's general formula shows that the structure has two anion groups to combine with chromium by ionic bond, and four neutral molecules or neutral groups to combine with chromium by coordination bond. The dye will have three empty tracks. Theoretically, they can coordinate with amino, hydroxyl or carboxyl on wool. However, under acidic conditions, the amino-groups on wool turn to cationic amino and carboxyl-groups are inhibited to turn to carboxylic acid anions. Thus water molecules will replace those groups to coordinate with chrome, increasing free chromium and extractable chromium content.

3.3 Influence of two dyeing auxiliary on the extractable chromium content

As to 1:1 metal complex dyes, there are three empty tracks of center chromium when pH value of dyeing bath is 2-4. If those tracks coordinate with water molecules, it will reduce the stability of these metal complexes and the extractable chromium content of dyed fabirc will increase, which happens in the sulfuric acid dyeing method. In order to improve the stability of these metal complexes, the sulfuric acid would be replaced by aminosulfonic acid which could take place of water and coordinate with chromium in dyes.

Albegal Plus is a organicsiloxane compounds containing coordination structure, which its molecular is bigger than aminosulfonic acid molecule and it can further increase the stabilization of these metal complexes by forming coordination bond with the chromium. Comparing fabric dyed with yellow dyes in Table 2 and Table 3, the extractable chromium content of fabric dyed in the formic acid method is 50% less than that in the aminosulfonic acid method. This can approve the analysis above.

4 Conclusions

There may be a negative correlation between the structure stability of dyes and the extractable chromium content on wool fabric dyed with 1:1 acid metal complex dyes. Aminosulfonic acid can take place of water molecular coordinate with chromium in dyes and enhance the structure stability of dyes. Albegal Plus molecular is bigger than aminosulfonic acid molecule, thus it can futher stabilize the complex. When using 1:1 acidic metal complex dyes with formic acid dyeing method, it is necessary to control the dyes amount no more than 1.5% (o.w.f.) and use the Albegal Plus in order to meet the demand of Oeko-Tex Standard 100.

References
[1] Society of dyers and colourists. Colour Index International. Fourth Edition Online.
[2] Hohenstein institute. Oeko-Tex® Standard 100, 01/2009.
[3] Jolanta Sokolowska-Gajda, Harold S. Freeman, Abraham Reife, 1994, 64, 388-396.
[4] Hough M.C., Wright D.C., Sokolowska-Gajda, Freeman HS, Reife A., Dyes and Pigments, 1996, 30,1-20.
[5] Edwards Laura C., Freeman Harold S., Coloration Technology, 2005, 121, 265-270.
[6] Edwards Laura C., Freeman Harold S., Coloration Technology, 2005, 121, 271-274.

98 Preparation of Hydrostable Fiber Crosslinking Agent Containing Glycidyl Groups[*]

Xiumei Zhang, Hualiang Wen, Jinlong Chen
(College of Materials and Textile, Zhejiang Sci-Tech University, Hangzhou 310018, China)

Abstract: Polymer containing glycidyl groups and double bonds was synthesized by copolymerization of styrene, glycidyl allyl ether and isoprene in controlled conditions. Results showed that the best fiber crosslinking efficiency was obtained at monomer molar ratio 1:1:1, when reaction temperature was 70°C and reaction time was 5 hours. The FT-IR spectra of the monomers and polymer were determined and proved that the polymer contained hydrophobic benzene, epoxy groups for fiber crosslinking and double bonds for self-polymerization. Prepared polymer was applied to make handsheet of hardwood fibers and the handsheets were cured by UV radiation. The strength of the handsheet increased with addition amount of the polymer and curing time increasing.

Keywords: Unsaturated Polymer; Fiber Crosslink Agent; Hydrostable; Glycidyl; UV curing

1 Introduction

Fiber crosslinking agent is widely used in the materials made from botanical fibers, especially in the case of moisture conditions because fiber material is very sensitive to water attack. Urea-formaldehyde and melamine-formaldehyde resins were applied to the fiber in the water-soluble, monomeric, or intermediate stage of polymerization. Polyethyleneimine was the first wet strength additive to be put on the market but was quickly superseded. The cost of manufacture and relative poor performance made it difficult to be widely used. polyaminoamide-epichlorohydrin resins possess water soluble and cationic features and gained almost immediate commercial acceptance and are by far the most widely used wet strength resins today. They were able to function at neutral/alkaline pH conditions [1] but discharging of absorbable organic halogen limit its use and make it under more pressure to be reformed [2].

In our previous studies, the polymers containing glycidyl groups were synthesized and proved to have significant crosslinking effect [3,4]. Another effort to increase the crosslinkings between fibers to restrict fiber swelling in the water, unsaturated resins were applied and initiated polymerization in fiber materials by high temperature or UV curing [5]. In this work, the combined crosslinking efficiency of glycidyl groups and double bonds were studied

2 Experimental
2.1 Materials
Styrene, glycidyl allyl ether and isoprene were purified and retention aid PG2 and photocatalyst 184 were used as received. Ethyl alcohol and chloroform were distillated. Hydrochloric acid and sodium hydroxide solutions were titrated by standard materials. Hardwood pulp was beaten to 35°SR.

2.2 Polymerization
Deioned water was added in a three-neck flask and sodium lauryl sulfonate was added as emulsifier. Then styrene, glycidyl allyl ether and isoprene were dropped in under stirring. Sodium sulfite and sodium persulfate were added in. Temperature was raised to 70°C, kept 5 hours and then stopped by cooling.

2.3 Purification of the polymer
Resulted mixture after polymerization was poured in anhydro-alcohol under shaking. The flocculated precipitation was washed by anhydro-alcohol for 3 times and then by deionized water for 3 times. The polymer was dried in vacuum oven.

2.4 Handsheet making
Handsheet was made according to ISO5269/2-2008 by handsheet maker ZQJ1-B. Diluted PAE resin was added under stirring for 5 minutes and then kept standing 5 minutes before drainage.

2.5 Determination of sheet physical properties
Tensile strength: sheet strips were tested by tensile tester DC-KZ300C according to GB/T 12914. Wet strength was tested by dipping the sheet strips into pure water for 10 minutes, taken out and then removed

---

[*] Contract grant sponsor: Zhejiang Provincial Natural Science Foundation of China; project number: Y4080359.

surface water for determination.

Burst resistance factor: Burst resistance of sheet samples were tested by DC-NPY5600 tester according to GB/T 454-1989.

3 Results and discussion

3.1 Decision of the polymerization conditions

Emulsion polymerization was selected in order to fit for wet end of handsheet making. Sodium lauryl sulfonate was used as emulsifier while redox system of Sodium sulfite and sodium persulfate was used as catalyst at molar ratio 1.05:1. Reaction temperature was selected between 60–80°C and the pH value was neutral to avoid ring open of epoxy groups. Reaction time was selected in the range of 4–6 hours according to relative references [3,6]. The reaction molar ratio of the monomers was decided by strengthening experiments and the results are listed in Table 1. Results showed that the handsheets got higher strength improvement at molar ratio of the monomers 1:1:1, especially for wet strength. Increasing proportion of styrene could produce higher burst resistance.

Table 1 Influence of molar ratio of monomers on handsheet strength (70°C, 5 hours)

Molar ratio of monomers	blank	1:1:1	1:1:2	1:2:1	2:1:1
Burst index, Kpa m^2/g	145.5	187.73	147.53	185.9	202.82
Dry tensile index, N m^2/g	31.78	41.52	33.88	35.61	41.3
Wet tensile index, N m^2/g	0.75	2.7	1.95	1.98	2.33

Molar ratio: styrene:glycidyl allyl ether:isoprene; blank is handsheet without polymer

Influence of reaction time on handsheet strength is shown in Figure 1. When molar ratio was 1:1:1 and reaction temperature was 70°C, handsheet dry strength increased by adding polymer but decreased after 5 hours reacted, while wet strength increased in the all reaction time range. Reaction time of free radical polymerization has relatively small influence to degree of polymerization but long time reaction leads to ring open of epoxy groups and loss of double bonds, which would decrease the crosslinking points therefore decrease sheet dry strength. While longer reaction time makes greater conversion degree of monomers, especially introduces more hydrophobic phenyl groups in the polymer, therefore reaction for longer time would increase wet strength of the sheets.

Theoretically, the polymerization will happen in the whole range of 60–80°C. Experiments of the reaction at different temperature were performed and results are showed in Figure 2. It indicates that paper strength increased to achieve a peak and then decreased with reaction temperature rising after 70°C. That can be interpreted that epoxy group and double bond are more sensitive to reaction temperature and the loss of the function groups for crosslinking led to the decrease of sheet strength.

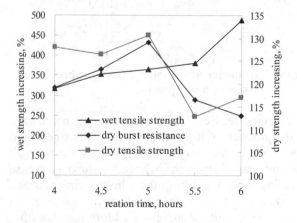

Figure 1 Influence of polymerization time on handsheet strength

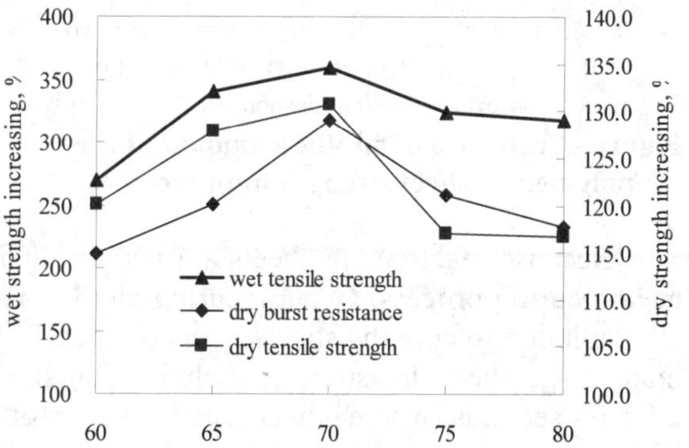

Figure 2 Influence of polymerization temperature on handsheet strength

3.2 Characterizations of the polymer

Polymer prepared at molar ratio 1:1:1 at 70°C for 5 hours was characterized. Content of epoxy groups was 0.0043mol/g determined by hydrochloric acid-acetone method. Degree of unsaturation was 0.01mol/g determined by potassium bromate–potassium bromide addition reaction. The structures of the monomers and polymer were proved by FT-IR spectroscopy as shown in Figure 3.

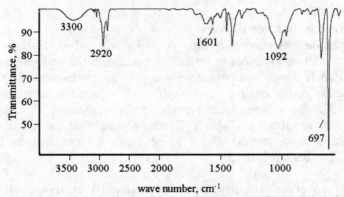

Figure 3 FT-IR spectra of the polymer

Comparing to the spectra of the monomers, absorption peaks at 1601cm^{-1} and 697 cm^{-1} are the characteristic absorbance of benzene. It proved that benzene structure remained in the polymer. Strong absorption at 1092 cm^{-1} shows epoxy response. At the same time, peaks at 3026 cm^{-1} and 2920 cm^{-1} proved that double bonds remained in the polymer. The peak at 3300 cm^{-1} shows exist of –OH groups, which illustrates that parts of epoxy rings opened during polymerization.

3.3 Applications of the polymer on paper strengthening

The polymer synthesized at molar ratio 1:1:1 at 70°C for 5 hours was applied to make hand sheets of wood fibers. Figure 4 shows the influence of addition amount of the polymer on sheet strength improvement. Sheet strength increased with addition amount, especially for wet strength, which closed to linear increasing.

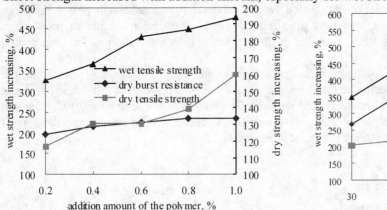
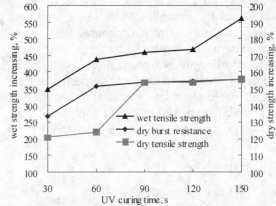

Figure 4 Influence of addition amount of the polymer on sheet strength improvement

Figure 5 Influence of UV curing time on sheet strength improvement

Because reactions in sheets are very slow at room temperature, wet fiber crosslinking agents always need a curing process. General curing method is to heat the products at high temperature. This study used UV radiation to cure the sheets with polymer. Figure 5 shows the influence of curing time on sheet strength increasing. Sheet dry strength including burst resistance and tensile strength, behaved a slowly increase after 90 seconds, whereas wet tensile strength kept increasing tendency during the whole curing process. UV curing helped to increase the polymerization of double bonds between fibers and restrict the fibers in swelling. The mechanism has relatively less effect on sheet dry strength improvement. Moreover, the fiber temperature would increase with UV radiation, that made fibers to be horny and lost swelling ability gradually.

4 Conclusions

Copolymerization reaction of styrene, glycidyl allyl ether and isoprene was performed at molar ratio

1:1:1 at 70°C for 5 hours, which were proved to be the best reaction conditions for fiber crosslinking efficiency. The polymer has glycidyl groups and double bonds, proved by determinations of epoxy content, degree of unsaturation and FT-IR spectroscopy. The polymer had significant fiber sheet strengthening efficiency. Though sheet burst resistance and dry tensile strength increased with addition amount of the polymer, the curves leveled off, whereas sheet wet tensile strength increased closing to linear pattern. UV curing time also had greater effect on sheet wet strength improvement.

References
[1] Obokata T, Isogai A. The mechanism of wet-strength development of cellulose sheets prepared with polyamideamine-epichlorohydrin (PAE) resin. chem. Eng. Aspects, 2007(302): 525-531.
[2] David I., Nancy S. Clungeon., Stephen A Fischer. Reducing organic chloride contaminants in polyaminoamide-epichlorohydrin wet-strength resins. Tappi, 1991, 74(12): 135
[3] Zhang X, Tanaka H. Copolymerization of Glycidyl Methacrylate with Styrene and Application of the Copolymer as Paper-strength Additive. Journal of Applied Polymer Science, 2001(80): 334-339
[4] Zhang X, Tanaka H. Influence of Retention System and Curing on Wet Strength Effectiveness of the Copolymer Containing Glycidyl Groups. Journal of Applied Polymer Science, 2001(81): 2791-2797
[5] Zhang X; Wu F; Zhou P. Investigation of the polymer with vinyl groups in paper strength improvement. Proceedings of International Conference on Pulping, Papermaking and Biotechnology 2008, 1, 483-486
[6] Liu Y; Li L; Zhong H. Preparation and characterization of epoxy resin with glycol glycidyl groups, Guangdang Chemical Industry, 2009:36-40

99 Silver Modified Silk Fibroin Composite Film and Its Antibacterial Property

Lan Zhou, Jianzhong Shao*, Xinxing Feng, Bin Sun

(Engineering Research Center for Eco-Dyeing & Finishing of Textiles, Key Laboratory of Advanced Textile Materials and Manufacturing Technology, Ministry of Education of China, Zhejiang Sci-Tech University, Hangzhou 310018, China)

Abstract: The Ag^+ modified silk fibroin composite films were synthesized using the different ratios of SF and Ag^+. The moisture content, water loss ratio and mechanical strength were tested. The results revealed that, compared with the pure silk fibroin films, the moisture content and the water loss ratio of modified composite films enormously decreased and the mechanical strength of modified composite films increased. The structure and dyeing property of these composite films were characterized by FTIR, XRD and correlative antimicrobial standards. The measurements indicated that two crystal structures (Silk I and Silk II) coexisted in the pure silk film, and by adding silver ions, the crystal structures of the composite silk fibroin films were transited from typical Silk I to typical Silk II. The antimicrobial activity of SF/Ag^+ composite films was excellent.

Keywords: Silk fibroin; Ag^+ ions; Crystal structure; Antimicrobial activity

1 Introduction

The development in the field of tissue engineering has accelerated the demand for biomaterials that are biodegradable, biocompatible and with considerable mechanical properties. In recent years, silk fibroin, a naturally occurring structural protein with good mechanical properties, has been considered as a good candidate for tissue engineering utilization[1-3].

However, there are some problems for silk fibroin film to be used in the tissue engineering field. The pure SF film had high water-solubility and could be dissolved nearly completely in water, which is not appropriate to repair and regenerate damaged or diseased tissue. Besides, for the implanted materials, the bacteria infection is one of the major clinical complications. Despite the strict antiseptic operative procedures, including systemic antibiotic prophylaxis and special enclosures using laminar flow, there were also lots of patients suffered the pain by the biomaterial centered infection (BCI) and numbers of bacteria trends have demonstrated an increasing resistance toward antibiotics. In addition, it has long been a problem that the regenerated SF films in the dry state show poor mechanical properties and brittleness compared to their original form. Silver ions which have been used throughout history as antimicrobial agents receive a renewed interest. Their powerful antimicrobial activity is known to be efficient against nearly 650 trends of bacteria. However, little attention has been given to silver ions modification on silk fibroin.

In the present study, we studied the antimicrobial activity of SF/Ag^+ composite films and applied FTIR and XRD techniques to probe the secondary structural transition of silk fibroin, and further investigated the corresponding inducing mechanism of the secondary structure when Ag^+ particles were formed in the composite films. We found that the joined Ag^+ ions reduced the water loss ratio of SF film, induced the secondary structural transition and endowed the silk fibroin with antimicrobial activity.

2 Experimental

2.1 Reagents and Materials

Calcium Chloride (AR) was purchased from Shanghai the Second Chemical Reagent Factory, China; Ethanol, silver nitrate and sodium hydroxide (AR) were purchased from Hangzhou Huipu Chemical Reagent Co. Ltd, China; Cocoons of B.mori silkworm silk was kindly supplied by Silk Museum, China.

2.2 Preparation of Regenerated B. mori Silk Fibroin Solutions and SF/Ag^+ composite films

To remove sericin, the cocoon was degummed using the fatty acid neutral soap (2 wt%) and sodium carbonate (0.1 wt%) in hot water (98℃) two times, each for about 1h. Degummed silk fibroin was washed thoroughly with deionized water for 30 min to remove any remaining sericin and surfactants and then it was gently dried in air.

---

* Corresponding author's email: jshao198@yahoo.com.cn

To dissolve the fibroin, the solvent system was prepared by dissolving calcium chloride in ethanol. The molar ratio of $CaCl_2$: C_2H_5OH: H_2O was 1:2:8.5 g degummed silk fibroin was dissolved in 200 ml $CaCl_2/C_2H_5OH$ solution in a round-bottom flask, and mechanical stirring was applied to provide adequate physical agitation to the system. A heater was used to keep the temperature at 75℃. It took about 2 h to dissolve the silk fibroin in the above system. The undissolved part was removed by filtration. The fibroin/salt solution was dialyzed against deionized water for 3 days, until the conductivity of the aqueous solution was less than 0.8 μS/cm. The dialyzed fibroin solution was kept in a desiccator for subsequent use. The weight-average molecular weight of dialyzed silk fibroin was measured about 40~120 kDa. Finally the product was cast on polystyrene Petri dish surfaces at room temperature in a drying oven. The thickness of these films was about 20-40 μm by a micrometer.

2g $AgNO_3$ was put into 100 ml distilled water, stirred to be dissolved, and the mixed solution was put into 1000 ml volumetric flask. 20 ml silk fibroin solution was added into 50ml beaker. The $AgNO_3$ solution was added dropwise to the prepared silk fibroin solution (Ratio of Ag^+ to silk fibroin was 0.1 (wt%)), and the mixed solution was concussed by the ultrasonification for 40 minutes. Finally the product was cast on polystyrene Petri dish surfaces at room temperature in a drying oven. The thickness of composite films was about 20-40 μm by a micrometer. For comparison, the same experiments were carried out under different SF/Ag^+ ratios.

2.3 Dissolubility in water and mechanical properties of films

0.300g SF film samples were dried at 105℃ for 24h in an oven, and the water ratio of films was determined. Meanwhile, 0.500g SF film samples were immersed in 100 ml de-ionized water. The samples were located at constant temperature cabinet at 50℃, then rinsed thoroughly with de-ionized water and dried at 105℃.

The mechanical properties of specimens (5×50×0.025 mm) were measured with stretching speed of 100 mm/min using a Shimadzu yarn tensile tester at ambient conditions.

2.4 Characterization

FT-IR spectra of samples were measured with Nicolet 5700 apparatus. The spectra of transmittance mode were recorded using ATR in the 700-1800 cm^{-1} range with 0.05 cm^{-1} resolution. A total of 128 scans were taken for each sample.

The XRD data were obtained in an ARL-X 'TRA diffractometer using $CuK\alpha$ radiation, in 2θ range of 10°~70° with a step size of 2 Deg/min and a counting time of 10s. The room-temperature measurements were performed with samples spread on a conventional glass sample holder.

2.5 Antimicrobial testing

Antimicrobial activity was investigated according to AATCC 100-2004 and JISL 1982:2002. In this work, Escherichia coli were used for test. The bacteria reduction (%) was used to quantitatively assay the antimicrobial activity of each sample in this work, which is defined as follow:

$$R (\%)=(A-B)/A\times100 \quad (2)$$

where R is bacteria reduction, A is the number of colonies on the control, B is the number of colonies on the samples.

3 Results and discussion

3.1 Dissolubility in water

The moisture content of silk fibroin is closely related with its protein conformation and crystallinity. And the moisture in silk fibroin is composed of three parts. The first part is unified with silk fibroin protein and becomes the bound moisture, which can't be removed at 105℃. The second part combines with silk fibroin dependent on capillary effect and physical chemistry adsorption, which may be removed a little at 105℃. The last free moisture existing in the cellular structure has the weakest binding force with silk fibroin, and can be removed completely at 105℃

In the compact crystalline region, silk fibroin peptide chains unify together dependent on the hydrogen bond and little moisture can be seeped in. However, in amorphous region and incomplete crystalline region, certain moisture can be absorbed. In general, silk fibroin with lower crystallinity has higher moisture sorption capcity and moisture sorption characteristics of silk fibroin film reflect its crystallization structure.

Table 1 Moisture content and water loss ratio of the silk fibroin and Ag+-modified composite films

Ag+:Silk Fibroin	0:1000	1:1000	2:1000	4:1000
Moisture Content (%)	39.603	3.473	3.247	4.046
Water Loss Ratio (%)	-	9.654	9.335	11.672

Table 1 shows the dissolubility in water of the pure SF film and the SF/Ag$^+$ composite films with different contents of Ag$^+$. It was obvious that the moisture content of the composite silk fibroin films was less than that of pure silk film. The moisture content of SF film reaches as high as 39%, but that of SF/Ag$^+$ composite films was less than 5%. The results indicated that the moisture content of the composite films tended to decrease and the samples containing 0.2% (wt) Ag$^+$ particles had the lowest moisture content.

The pure SF film had high water-solubility and could be dissolved nearly completely in water. However, the joined Ag$^+$ reduced the water loss ratio of SF film, which demonstrated that the joined Ag$^+$ disrupted the union between the macro-molecule chain of SF and the water molecule. On the one hand, Ag$^+$ formed stable complexes with carbonyl, imido-group, imidazolyl and hydroxyl in silk fibroin protein molecules, which led to the loss of hydrophilic groups. On the other hand, multi-spot cross linkage and bonding formed the spatial mesh structure, which also greatly reduced water loss ratio of SF films. The Ag$^+$ particles could promote the crystal structure of the SF and induce partial transformation from Silk I to the Silk II, a more stable silk form.

3.2 FTIR spectra of silk fibroin composite films

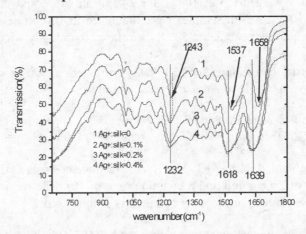

Figure 1 FTIR spectra of the silk fibroin and Ag$^+$ modified composite films

The structure of fibroin proteins in the pure film and the SF/Ag$^+$ composite films were characterized, and the conformational transition of SF composite films was determined by FTIR spectroscopy when Ag$^+$ was added. The positions of absorption bands indicated the conformation of fibroin protein.

FT-IR spectra of the pure and composite SF film samples in the spectral ranges of 1800-600 cm^{-1} are shown in Figure 1. In accordance with literature, the absorption peaks of amide I, amide II, amide III for SF were at 1658, 1540, and 1235cm^{-1}, respectively, characteristic of a Silk I structure. The SF films treated with methanol had peaks of 1625 (amide I), 1528 (amide II), and 1260 cm^{-1} (amide III), characteristic of a Silk II structure. In Figure 1, we can see that the pure SF had three characteristic vibrational peaks at 1658 (amide I), 1537 (amide II) and 1243 cm^{-1} (amide III). Obviously, the pure SF appeared to be predominantly Silk I structure. The amide I and amide II peaks of the composite films were observed around 1639 and 1518 cm^{-1}, and shifted to short-wave direction with the increase of Ag$^+$ content. The amide III peaks of the composite films were observed around 1232 cm^{-1}, and shifted to long-wave direction. The peak shifts indicated that adding Ag$^+$ was advantageous to change SF molecules conformation from Silk I to Silk II which was more regular. The peak shifts indicated that the Silk I and Silk II structure could coexist in the SF composite films, but partial conformation transition from Silk I to Silk II structure occurred when the Ag$^+$ was added in the SF film

3.3 XRD analysis of silk fibroin composite films

X-ray diffraction was used to characterize the structure of the silk fibroin. Generally, there have been two types of crystalline structures proposed for silk, Silk I and Silk II. The main diffraction peaks of silk I present at $2\theta=19.7°$ and $24.7°$, while Silk II present at $2\theta=18.9°$ and $20.7°$. Figure 2 displayed XRD spectra of the pure and composite SF films samples. In the range of $17\sim25°$, both the pure SF and the composite SF had peak package. It was the result of superimposition peak of $18.9°$, $19.7°$, $20.7°$ and $24.7°$, and we applied software Origin to divide it as shown in Figure 3 and Table 2.

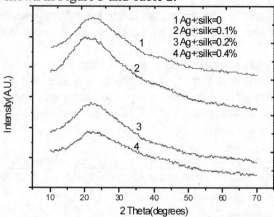

Figure 2 XRD curves of the silk fibroin and Ag+ modified composite films

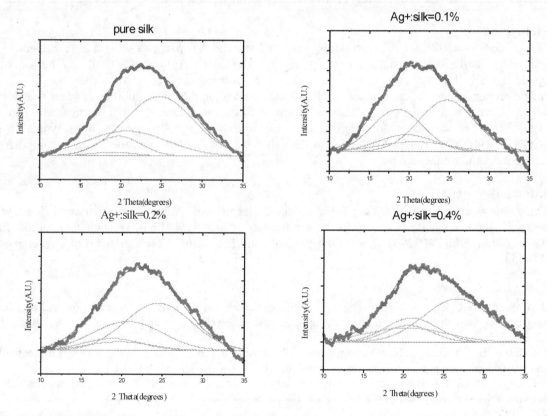

Figure 3 XRD fitting spectra of the pure silk fibroin and Ag+ modified silk fibroin composite films

Table 2 Relative percent of different configuration of the pure silk fibroin and Ag+ modified silk fibroin composite films

	Silk II		Silk I		Silk II (%)	Silk I (%)
	18.9°	20.7°	19.7°	24.7°		
Pure silk	5.6	137.5	29.6	67.7	30.1	69.9
Ag+:silk=0.1%	82.5	111.5	37.1	26.1	42.2	57.8
Ag+:silk=0.2%	22.8	104.0	18.7	66.4	42.1	57.9
Ag+:silk=0.4%	31.4	80.4	21.6	34.5	39.3	60.7

Figure 3 and Table 2 presented that Silk I and Silk II crystalline structure coexisted in both the pure SF and the SF/Ag$^+$ composite films. This result indicated that the amount of the Silk II crystalline structure in the composite film was higher than that in the pure silk fibroin film. And with the rise of Ag$^+$, the amount of the Silk II crystalline structure increased and then reduced. The crystalline structure of silk fibroin molecules could be induced to change from Silk I to Silk II by the formation of Ag$^+$ particles in the silk fibroin, but the transformation had not occurred completely. This result corresponded closely to the FT-IR spectra in Figure 1.

3.4 Antimicrobial activity

Further investigations of antimicrobial activity were performed on the various SF composite films with different ratios of SF and Ag$^+$ shown in Table 3. It can be clearly seen that the SF/Ag$^+$ composite films had high antimicrobial activity. After silver ion contacting with bacterium, the silver ions combined with -SH unions in bacterium proteinase, which caused the loss of proteinase activeness and led to the death of bacterium.

Table 3 Antimicrobial activity of SF/Ag$^+$ composite films

Sample	Number of colony forming unit (CFU)	Bacteria reduction (%)
Pure silk	9.35×10^4	—
Ag$^+$:silk=0.1%	10	>99.9
Ag$^+$:silk=0.2%	5	>99.9

4 Conclusions

SF/Ag$^+$ composite films were successfully prepared and the Ag$^+$ particles were tightly embedded in the SF films. Silk I and Silk II crystalline structure coexisted in the SF/Ag$^+$ composite films. The formation of Ag$^+$ inside the SF film may induce, in part, a Silk II conformation. Compared to the pure SF films, the mechanical properties of SF/ Ag$^+$ composite films was improved, and the dissolubility in water was decreased. This could be attributed to the partial transformation of fibroin structure to Silk II conformation in the SF composite films. The secondary structure of composite films were partially transited from typical Silk I to typical Silk II with the increase of Ag$^+$ content (less than 0.2 wt%). Moreover, the SF/ Ag$^+$ composite films had excellent antimicrobial activity.

5 Acknowledgement

This work was funded by Zhejiang Top Academic Discipline of Applied Chemistry and Eco-Dyeing & Finishing, Zhejiang education department research plan project (0901066-F), Zhejiang Natural Science Foundation (Grant No. Y407295), Changjiang Scholars and Innovative Research Team in University (Grant No. IRT0654).

References
[1] B B Mandal, S C Kundu. Non-bioengineered silk gland fibroin protein: Characterization and evaluation of matrices for potential tissue engineering applications. Biotechnology and Bioengineering, 2008, 100(6): 1237-1250.
[2] Mandal, B. B; Kundu, S.C. Non-bioengineered silk fibroin protein 3D scaffolds for potential biotechnological and tissue engineering applications. Macromolecular bioscience, 2008, 8(9): 807-818.
[3] R E Unger, M Wolf, K Peters, etc. Growth of human cells on a non-woven silk fibroin net: a potential for use in tissue engineering. Biomaterials 2004, 25:1069-1075.

100 Nondestructive Observation of Textile Reinforced Composites by Low Energy X-ray Beam

Hyungbum Kim[1], Jung H. Lim[1], You Huh[2]

(1. Dept. of Textile Eng., Graduate School, KyungHee Univ., Youngin, 449-701, R. O. Korea
2. Dept. of Mechanical Engineering, KyungHee Univ., Youngin, 449-701, R. O. Korea)

Abstract: NDT (Nondestructive Testing), NDE (Nondestructive Evaluation) technology with radiography is widely used in various fields, especially, medical science and industrial technology. Because NDT method does not destroy or change the characteristic of the sample, it is a useful technology that can inspect the sample inside, while both the form and the structural status of the specimen preserve. In taking advantage of the NDT method as an analysis tool for soft materials like textile reinforced composites the radiographic energy range plays the most important role. And we decided the low-energy X-ray of around 10-30keV suitable for our research purpose. To investigate the potentiality of the NDT method for soft materials, we have designed and constructed an experimental system, including development of the data processing software to subtract information on the structural insight of the specimens that were prepared. Based on the results achieved so far, we could confirm the feasibility of the X-ray application for the textile composites and obtain useful X-ray sample images successfully.

Keywords: NDT; NDE; Low-energy X-ray; X-ray imaging; Textile reinforced composites; Soft materials; Nondestructive observation; Structural analysis; Computed tomography; Volume rendering

1 Introduction

Nondestructive imaging technology can be useful for visualizing structures or materials containing flaws or faults which are hardly confirmed without causing sample damage. NDT methods well known can be divided into several categories in terms of energy types; i.e., ultrasonic, magnetic-particle, liquid penetrant, radiographic, and eddy-current testing, etc. But to ensure a high spatial resolution and a wide application scope the X-ray radiographic source is selected for the textile nondestructive visual imaging research.

X-ray is a sort of electromagnetic radiation and has an energy range of 300 eV to 450 keV. X-ray radiation ranging about 0.3 to 1.2 keV is classified as soft X-ray, and about 20 to 450 keV as hard X-ray, according to their penetrating abilities. Hard X-ray can penetrate objects consisting of metallic or inorganic matter, and is mainly used to take images of the inside of the objects for diagnostic radiography for medical purposes and inspection radiography for industrial applications. Because most of the textile composite materials mainly consist of organic matters which are easily penetrated by X-rays, we adopted an intermediate-to-low energy X-ray of around 10 to 30 keV in order to investigate the structural states of such soft materials as textile compound composites.

2 Principle of X-ray imaging

Entering of X-ray into material interacts with the material and results in a reduced energy intensity of X-ray with penetration depth. This reduction is described in terms of attenuation coefficient (μ). As depicted in Figure 1, transmitting intensity I_T of an incident radiation through layers of material with thickness X is related to the incident intensity Io according to the Equation (1).

$$I_T = I_o e^{-\mu x} \tag{1}$$

where x denotes the path length and μ is the attenuation coefficient.

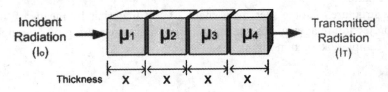

Figure 1 Attenuation coefficient (μ)

When the transmitted X-ray hits a detector cell, an electrical voltage is induced, which is proportional to the incident X-ray intensity. Every individual voltage can be treated as a pixel signal of the X-ray projection image.

3 Experimental rig

X-ray scanner is composed of an X-ray tube, an X-ray image intensifier (XRII), a zoom lens, and a CCD camera, as shown in Figure 2. The distance between the X-ray tube and the object is represented as SOD, and the distance between the X-ray tube and the image intensifier XRII is SDD. The magnification ratio (M) can be defined as SDD/SOD.

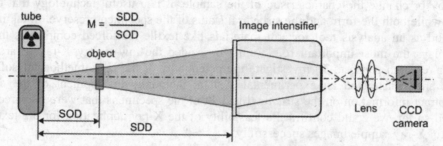

Figure 2 X-ray scanner structure and ray optics

XRII converts the X-ray to a mass of electrons and finally to a visible ray. Inside XRII a number of electrons are added, so that the intensity of the incident X-ray intensity is resultantly amplified. A CCD camera changes the visible ray to an electrical signal matrix that appears in the form of an image.

4 Detector correction

X-ray projection image includes some distortions caused by XRII and CCD device. To correct the detector image Equation (2) is used.

$$I(u,v) = \log[I_{air}(u,v) - I_{dark}(u,v)] - \log[I_{proj}(u,v) - I_{dark}(u,v)] \quad (2)$$

where $I(u,v)$ is a corrected image, and $I_{air}(u,v)$ is a flat field image that is acquired without object in the scanner. $I_{dark}(u,v)$ is a dark image that is acquired without X-ray incidence. $I_{proj}(u,v)$ represents a raw input image.

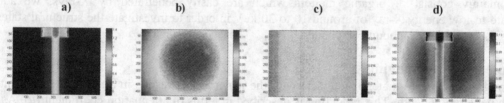

Figure 3 Image examples for detector correction.
a) $I(u,v)$: calibrated image. b) $I_{air}(u,v)$: flat field image (for XRII compensation).
c) $I_{dark}(u,v)$: dark image (for CCD compensation). d) $I_{proj}(u,v)$: raw input image.

5 Experimental results

To visualize the structure of some specimens made of soft materials by means of the NDT method, we took several X-ray projection images of the specimens and volume rendering was conducted to generate the images with selected samples, as seen in Figure 4-8.

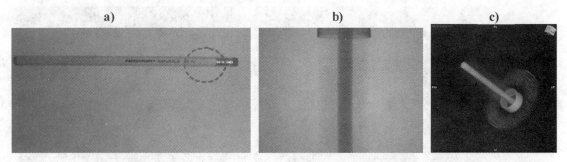

Figure 4 X-ray measurement for a pencil. (25 kV, 300 μA)
a) specimen. b) a projection image. c) a volume rendering image.

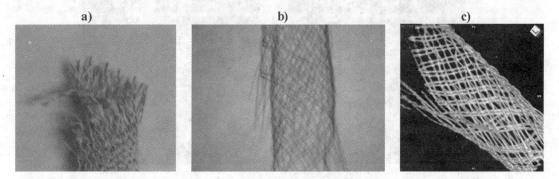

Figure 5 X-ray measurement for a basalt compound fabric. (25 kV, 300 μA)
a) specimen. b) a projection image. c) a volume rendering image.

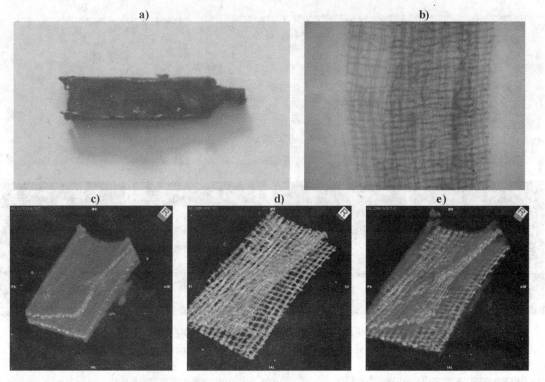

Figure 6 X-ray measurement for a basalt composite material. (Sample A, 25 kV, 300 μA)
a) specimen, b) a projection image, c, d, e) volume rendering images.

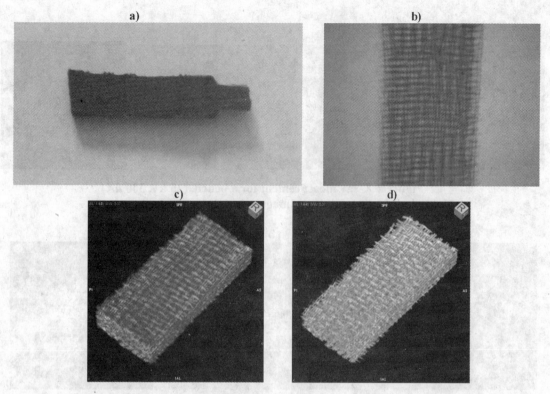

Figure 7 X-ray measurement for a basalt composite material. (Sample B, 25 kV, 300 μA)
a) specimen, b) a projection image, c, d) volume rendering images.

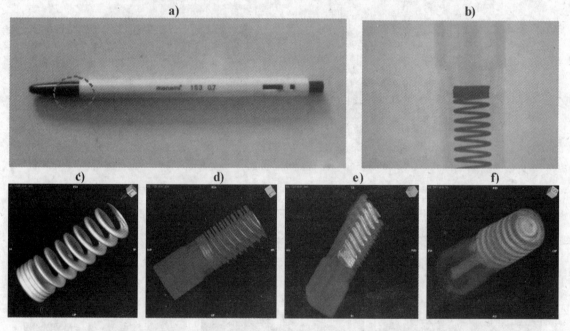

Figure 8 X-ray measurement for a ball-point pen. (35 kV, 300 μA)
a) picture. b) a projection image. c, d, e, f) volume rendering images.

As the provisory results show, the sample structures are regenerated very good.

6 Conclusions

We employed an X-ray with the intermediate-to-low energy level of around 10 to 30 keV to investigate the inside structure of textile composites. And the feasibility of the new measurement system was tested to see if the image expected could be obtained. The results showed that the experimental rig developed in this research operated so good that the images obtained regenerated the inside structure of the multilayered textile composites, which indicates that the X-ray application for the structural analysis of soft materials can provide much information on the inside states without destruction of the specimens.

References
[1] Handbook of Medical Imaging, Vol. 1 physics and Psychophysics.
[2] Willi A. Kalender, Computed Tomography: Publicis Corporate Publishing (2005).
[3] Osman R., Antoine R. and Joris H., OsiriX the pocket guide: OsiriX Imaging Software.
[4] Nondestructive Evolution and Quality Control, ASM Handbook, Vol. 17 (1997).

101 Structural Differences of Wild Silks and B. Mori Silk Characterized by FTIR Spectroscopy, XRD Diffraction and TG Analysis

Jianxin He[1,2], Yan Wang[1], Kejing Li[1], Shizhong Cui[1]

(1. College of Textiles, Zhongyuan University of Technology, Zhengzhou 450007, China
2. Key Laboratory of Textile Science & Technology, Ministry of Education, Donghua University, Shanghai 201620, China)

Abstract: Differences in secondary structure, crystalline structure and thermal property among B. mori silk and two wild silks of A. yamamai and A. pernyi are investigated by FTIR, XRD and TG analysis. The β-sheet structure is primary in three silk, and B. mori silk has the most β-sheet structure. A. yamamai silk contains more α-helix structure, whereas more β-turn and random coil structures for A. pernyi silk. B. mori silk has the aggregation structure with high crystallinity and small crystallite whereas the low crystallinity and large crystallite structure for the two wild silks. The two wild silks display a lower initial degradation temperature, but a high temperature in the most speed of thermal degradation and a longer degradation process compared with B. mori silk, and the thermal stability of A. yamamai silk is better than A. pernyi silk.

Keywords: Silk, Secondary structure, FTIR

1 Introduction

A. pernyi and A. yamamai belong to wild silk, and their silk fibroins are similar in amino acid composition. The Gly and Ala residues in the two wild silk fibroins are 77% (in mol%), which is similar to B. mori silk (73%), but the relative composition of Ala and Gly is reversed. The proportion of Gly residues is greater in B. mori silk fibroin, while the content of Ala residues (48%) is greater in the two wild silk fibroins. The complete sequence of the silk fibroin from A. pernyi and A. yamamai had been determined by Yukuhiro et al[1]. The silk fibroins mainly consist of the repeated similar sequences where there are alternative appearances of -(Ala)$_n$- region and the Gly-rich region. However, up to now there has not a detailed description about the structure of A. pernyi and A. yamamai silk, especially for the secondary structure of A. yamamai silk, whose structure should be different from A. pernyi silk because its elongation and elasticity are significantly more than A. pernyi silk. In this article, the differences in the secondary structure among the three silks were investigated by FTIR, X-ray diffraction and thermogravimetric analysis.

2 Experimental

2.1 Materials

The cocoons of B. mori, A. pernyi and A. yamamai came from Henan province of China. These cocoons were degummed three times with 0.5(w/v)% Na_2CO_3 solution at 100℃ for 30min and washed with distilled water in order to remove sericin from the surface of the fibers. The fibroin fibers were dried at 37℃ overnight prior to the next experiments.

2.2 FTIR spectroscopy

FTIR spectra of silk fibroin fibers were recorded with a Nicolet Nexus 670 FTIR spectrometer using the KBr disc technique. 100 Scans were taken with a resolution of 2 cm^{-1}.

2.3 X-ray diffractometry

X-ray diffractograms were recorded with a Rigaku-D/Max-2550PC diffractometer using Ni-filtered Cu-Kα radiation of wavelength 0.1542 nm. The X-ray unit operated at 40 kV and 30 mA. Angular scanning was conducted from 5° to 50° at 2°/min. The diffraction profile was fitted by Gaussian function, ranging from 5° to 40° of 2θ degree, and the crystallinity X_d was calculated by reference [2]:

2.4 Thermogravimetric analysis

Thermogravimetric analysis of silk fibroin fibers was carried out on Perkin Elmer TGA thermogravimetric analysis instrument. All the experiments were carried out with the same nitrogen flux and the heating-up speed was 10°C/min.

3 Results and discussion
3.1 Infrared spectral analysis

FTIR spectra of B. mori, A. yamamai and A. pernyi silks are shown in Figure 1. As reported in the literature[3-4], in the region of amide III, the signals at 1235 cm^{-1} and 1263 cm^{-1} are associated with random coil and β-sheet conformations. A. yamamai and A. pernyi silks exhibit a single peak at 1238 cm^{-1}, whereas B. mori silk shows a doublet at 1230 cm^{-1} and 1263 cm^{-1} (Figure 1), which suggest that B. mori silk contains most β-sheet structure compared with the two wild silks. Amide I bands centered between 1610-1640 cm^{-1} are generally considered to be characteristic of β-sheet structure, and the random and α-helix conformations are usually associated with the bands at 1640-1650cm^{-1} and 1650-1660cm^{-1}, respectively[5-7]. Infrared bands between 1660-1700 cm^{-1} are usually assigned to β-turn structure, moreover, the investigation of Opella etc also indicated that there presents an absorption peak of anti-parallel β-sheet structure at 1699 cm^{-1}[8]. As shown in Figure 1, a triplet at 1699, 1637 and 1620cm^{-1}, assigned to β-sheet structure, can be observed in IR spectra of B. mori and A. yamamai silk but A. pernyi silk shows a single peak at 1650cm^{-1}, which is attributed to random and/or α-helix conformations. This indicates more β-sheet structure in B. mori and A. yamamai silks than in A. pernyi silk. In addition, it is worth notice that B. mori silk has an apparent signal of anti-parallel β-sheet structure at around 1699 cm^{-1} while A. yamamai silk is an inconspicuous shoulder peak, suggesting B. mori silk has more β-sheet structure than A. yamamai silk and has the highest β-sheet structure content among the three silks, which is consistent with the result in the analysis of amide III bands.

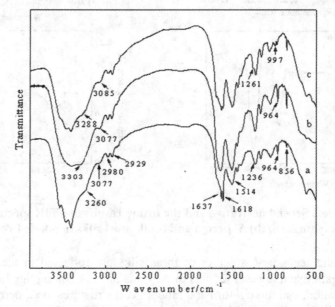

Figure 1 FTIR spectra of (a) A. yamamai, (b) A. pernyi and (c) B. mori silks

Figure 2 shows the second derivative functions and the fitting results of the spectra examined in the amide I region The amount of different secondary structures in silk fibers can be calculated by correlating the percentage values of the amide I components, and the result is shown in Table 1. As mentioned above the β-sheet structures in three silk are dominant, most β-sheet structure can be found in B. mori silk, and A. pernyi silk has the lowest content of β-sheet structure. In spite of considerable similarity in amino acid composition for A. yamamai and A. pernyi silks, there are rather differences in the secondary structures of two wild silks. A. yamamai silk contains more α-helix conformation (37.8%, as shown in Table 1), of which small random coil is also possibly included because the amide I component corresponding to α-helix conformation covers a wider wavenumber range, however, A. pernyi silk contains more β-turn structure (33.7%), in addition, considerable random coil structures (9.2%) are also observed in A. pernyi silk whereas random coil structure is absent in B. mori silk.

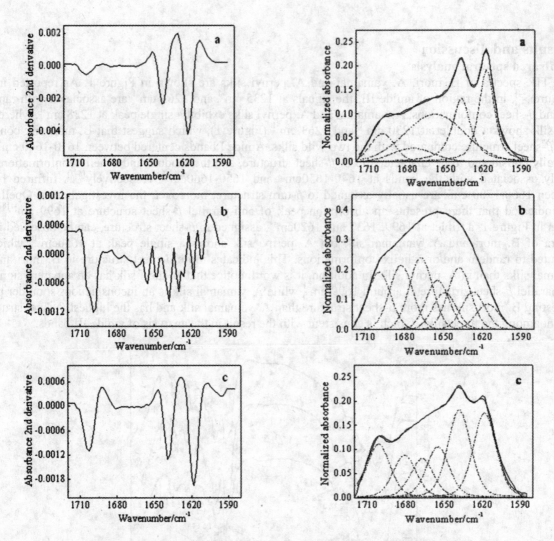

Figure 2 Second derivative and the fitting results of FTIR spectra of
(a) A. yamamai, (b) A. pernyi and (c) B. mori silks in amide I region

The differences in the secondary structure of three silks are reflected in their mechanical properties (Table 2). Maximal elongation and lowest modulus of A. yamamai silk among the three silks are correlated to its more α-helix conformation, and more β-turn and random coil structures in A. pernyi silk also contribute to its larger elongation. It is obvious that the highest modulus and rigidity of B. mori silk are induced by its most β-sheet structure, which is as high as 63.1%.

Table 1 Secondary structure contents of A. yamamai, A. pernyi and B. mori silks

Silk fibers	β-sheet (%)	α-helix (%)	β-turn (%)	Random coil (%)
A. yamamai	54.77	37.76	7.47	—
A. pernyi	45.04	12.13	33.67	9.16
B. mori	63.09	14.19	22.72	—

Table 2 Mechanical properties of A. yamamai, A. pernyi and B. mori silks

Silk fibers	Breaking strength		Breaking elongation		Initial modulus		Work of rupture	
	Mean (cN/dtex)	CV (%)	Mean (%)	CV (%)	Mean (cN/dtex)	CV (%)	Mean (cN/dtex)	CV (%)
A. yamamai	2.37	16.7	31.38	19.8	22.60	30.6	0.44	28.2
A. pernyi	2.84	11.6	24.42	22.4	23.17	24.9	0.38	27.8
B. mori	2.80	19.2	9.40	27.5	43.05	20.2	0.17	46.6

3.2 X-ray diffraction analysis

X-ray diffractograms of A. yamamai, A. pernyi and B. mori silks are fitted by Gauss function and the results are shown in Figure 3. Crystallinity, crystallite size, and crystallite orientation for the three silks are calculated and shown in Table 3. Both wild silks are congenetic and the positions of their crystalline peaks are consistent, a couple of main diffraction peaks appear at 2θ angle of 16.7° and 20.4°, in addition, there has also a diffraction peak at 24.3°, while B. mori silk shows a prominent diffraction peak at 20.4°, the peak at 16.7° disappears and a diffraction peak at 9.3° occurs.

Wild silks contain both parallel and antiparallel β-sheets whereas the latter is dominant in B. mori silk, therefore this difference in diffraction peaks between wild silks and B. mori silk is associated with the difference of their β-sheet structures. Obviously the diffraction peaks of B. mori silk at 9.3°, 20.4° and 24.3° can but be attributed to the anti-parallel β-sheet, especially the peak at 20.4° should be the main crystalline peak of anti-parallel β-sheet, however, for both wild silks another diffraction peak at 2θ angle of 16.7° should be associated with parallel β-sheet besides the peak of the anti-parallel β-sheet at 20.4°.

It is found that the values of crystallinity are the same as those of β-sheet contents for three silks in sequence, however, the crystallinity difference between A. yamamai and A. pernyi silks is not as large as the difference of their β-sheet contents. As shown in Table 3, B. mori silk has the aggregation structure with high crystallinity, high orientation and small crystallite, while wild silks with low crystallinity, low orientation and large crystallite. The crystallite sizes of A. yamamai silk are greater than those of A. pernyi silk but A. yamamai silk has the lower crystallite orientation, which may be also a cause of its lower tensile strength.

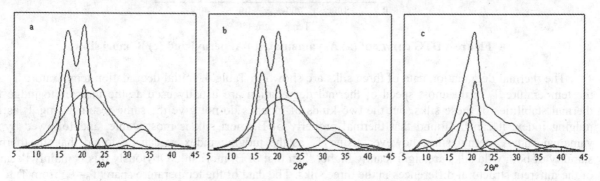

Figure 3 X-ray diffractogram curve and fitting results of (a) A. yamamai, (b) A. pernyi and (c) B. mori silks

Table 3 Crystallinity, crystallite size and crystallite orientation of A. yamamai, A. pernyi and B. mori silks

Silk fibers	X_d (%)	Crystallite size (nm)			OG (%)
		$L/2\theta$	$L/2\theta$	$L/2\theta$	
A. yamamai	46.63	2.93/16.7	4.26/20.4	5.23/24.3	80.3
A. pernyi	45.68	2.91/16.7	4.03/20.4	4.83/24.3	80.6
B. mori	57.83	3.03/9.3	3.63/20.6	2.72/24.1	82.5

3.3 Thermogravimetric analysis

DTG curves of A. yamamai, A. pernyi and B. mori silks are shown in Figure 4. Obviously, at the first peak between 40 and 100°C in DTG curves, the weight loss is due to water vaporization (drying). After this peak, the DTG curves of these silks display differently. B. mori silk only exhibits an endothermic peak at 320℃, which corresponds to silk fibroin decomposition of β-sheet crystalline conformation, while the maximal decomposition peaks of two wild silks assigned to silk fibroin in crystalline area appear around 360℃ [9], and the degradation temperature of A. yamamai silk is slightly more than of A. pernyi silk. It is apparent that the larger crystallite sizes in both wild silks contribute to their higher degradation temperature of β-sheet crystalline structure.

Shoulder peaks appear before the decomposition peak of β-sheet crystalline structure in the DTG curves of A. yamamai and A. pernyi silks, which can not be found in the DTG curve of B. mori silk (Figure 4). A. yamamai silk has an apparent shoulder peak at 300℃, while there is also a small shoulder peak about 250℃ in the DTG curve of A. pernyi silk besides the decomposition peak at 300℃. This difference in thermal property indicates that the secondary structure and aggregation structure of wild silks are complicated than of B. mori silk. As our previous analysis, B. mori silk assumes the dominant β-sheet structure with high crystallinity, high orientation and small crystallite, while there are also covered with α-helix, β-turn and random coil structures except for β-sheet conformation in both wild silks, which assumes the structure of low crystallinity, low orientation and large crystallite. Therefore, the decomposition peak of both wild silks at 300℃ should be attributed to α-helix or β-turn, while the small peak at 250℃ in the DTG curve of A. pernyi silk may be associated with the decomposition of random coil structure because this peak appeared in a lower temperature and is absent from A. yamamai silk. It can be seen from FTIR that random coil conformation is almost lacking in A. yamamai silk.

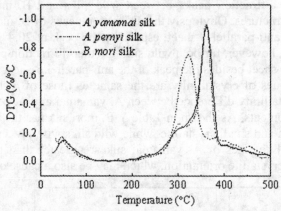

Figure 4 DTG curves of (a) A. yamamai, (b) A. pernyi and (c) B. mori silks

The thermal degradation data of three silks are shown in Table 4. Initial degradation temperature T_i and the temperature T_p in the most speed of thermal degradation are usually used as the indexes to judge the thermal stabilities of these silks, but the two kinds of indexes do not give the same result. Using T_i as the judging index, it can be found that thermal property of B. mori silk is most stable, and followed by A. yamamai and A. pernyi silks, however, using T_p as the judging index, it can be found that the thermal stability of both wild silks are significantly higher than that of B. mori silk. Obviously this is mainly because of the inherent structural differences in the three silks. The data of the temperature span ($T_p - T_i$) from T_i to T_p and the whole temperature interval of decomposition ($T_e - T_i$) indicate that although initial degradation temperature of B. mori silk is the highest, it reaches the highest decomposition speed and make decomposition finish in a very short time once its thermal decomposition begins while both wild silks ware just the reverse.

Table 4 Thermal degradation data of A. yamamai, A. pernyi and B. mori silks

Silk fibers	T_i(℃)	T_p(℃)	T_e*(℃)	$(\frac{d\alpha}{dt})_p$**(%/min)	$T_p - T_i$	$T_e - T_i$
A. yamamai	253.0	361.5	378.5	9.5	108.5	125.5
A. pernyi	249.6	360.4	378.1	9.6	110.8	128.5
B. mori	292.5	321.4	351.6	7.5	28.9	59.1

* The end temperature of decomposition, ** The highest velocity of decomposition

4 Conclusion

B. mori silk has the most β-sheet structure, mainly anti-parallel β-sheet structure, while the two wild silks contain both anti-parallel and parallel β-sheet structure, and the latter is advantage. There have distinct differences for the secondary structure of both wild silks. The content of β-sheet structure in A. yamamai is more than in A. pernyi silk, the former contains more α-helix structure whereas more β-turn and random coil structures for the latter. Despite of the lower β-sheet content and crystallinity in wild silks compared with B. mori silk, they possess more perfect crystallites with greater sizes. The crystallite sizes of A. yamamai silk are greater than those of A. pernyi silk but A. yamamai silk has the lower crystallite orientation.

Acknowledgement

This research was financially supported by China Postdoctoral Science Foundation funded project (20080430079), and we also acknowledged the support from the Programme of Introducing Talents of Discipline to Universities, B07024, China.

References

[1] K Yukuhiro, T Kanda, T Tamura. Preferential codon usage and two types of repetitive motifs in the fibroin gene of the Chinese oak silkworm, Antheraea pernyi. Insect Mol. Biol., 1997, 6: 89–95.
[2] JX He, Y Tang, SY Wang. Differences in morphological characteristics of bamboo fibres and other natural cellulose fibres. Iran. Polym. J., 2007, 16: 807-818.
[3] X Chen, W Li, T Yu. Conformation transition of silk fibroin induced by blending chitosan. J. Polym. Sci., Part B, Polym. Phys., 1997, 35: 2293-2296.
[4] Z Arp, D Autrey, J Laane, GJ Thomas. Tyrosine Raman Signatures of the Filamentous Virus Ff Are Diagnostic of Non-Hydrogen-Bonded Phenoxyls. Biochemistry, 2001, 40: 2522-2529.
[5] YN Chirgadze, OV Fedorov, NP Trushina. Estimation of amino acid residue side-chain absorption in the infrared spectra of protein solutions in heavy water. Biopolymers, 1975, 14: 679-685.
[6] NV Bhat, GS Nadiger. Crystallinity in silk fibers: partial acid hydrolysis and related studies. J. Appl. Polym. Sci., 1980, 25: 921-932.
[7] PW Holloway, HH Mantsch. Crystal and solution structures of the B-DNA dodecamer d(CGCAAATTTGCG) probed by Raman spectroscopy. Biochemistry, 1989, 28: 931-938.
[8] SJ Opella, MH Frey. Selection of nonprotonated carbon resonances in solid-state nuclear magnetic resonance. J. Am. Chem. Soc., 1979, 101: 5854-5861.
[9] J Magoshi, Y Magoshi, H Kakudo. Physical properties and structure of silk-V. Thermal behaviour of silk fibroin in the randomcoil conformation. J. Polym. Sci. Polym. Phys. Ed., 1977, 15:1675-1682.

102 The Tearing Strength of Plain Woven Fabric with the High Strength

Xinling Li, Zhiyu Zheng, Xiaohong Zhou*

(Key Laboratory of Advanced Textile Materials and Manufacturing Technology, Ministry of Education of China, Zhejiang Sci-Tech University, Hangzhou 310018, China)

Abstract: The plain-woven fabric with the equal density in warp and weft is used in skeleton material for composite products. Some fabrics with equal warp and weft density made of Aramid 1313 were designed and woven on the ASL2000 rapier loom. Tearing test of single rip method was carried out. Experimental results show that the tearing strength of fabric with high strength is inversely proportional to warp and weft density and yarns are slippaged, gathered and pulled out throughout the tearing process. It will help to choose the skeleton material for composite products and optimize composite process.

Keywords: Aramid 1313; Tearing strength; Woven fabric with the equal density in warp and weft; Tearing test of single rip method

1 Introduction

The phenomenon that part yarns are teared for the collective load is called tearing. Not only is the property of tearing the important project of quality testing, but it is the main research content both in cloth textile and industrial textile as well. Zhang Haixia researched the tearing strength of abraded denim. In this paper, relationship between the tearing strength of the abraded denim and yarn strength as well as yarn breaking elongation and fabric density has been found. The material of sample is denim, and most structure is 3/1 twill. Jiang Qigang compared the tearing strength with different tearing method. In this paper, four pieces of polysulfonamide woven fabrics with different weaves and the same specifications were measured by using four different tearing methods. The relationships among different tearing methods were obtained. The experimental results showed that the 2/2 basket fabric have both good tearing strength and stretch strength. The warp and weft density of sample is 280×160 yarns/10cm.

There are many influence factors of tearing strength, such as the tensile strength of yarns, warp and weft density, woven fabric structure and so on. The property of matrix and interfacial condition between cloth and matrix are also the influence factors of tearing strength while the cloth is compound with matrix. It is know that plain-woven fabrics made of Aramid 1313 are often used in skeleton material for composite products. In the paper, two kinds of fabrics made of Aramid 1313 are designed and woven on the ASL2000 rapier loom. Tearing test of single rip method is carried out to research the relationship between tearing strength and weft and warp density.

2 The design and preparation of sample

2.1 The properties of yarn

Two kinds of Aramid 1313 yarns are used. 24tex yarn has 585cN of breaking strength and 14.95% of breaking elongation ratio, 22tex×2 yarns has 1095cN of breaking strength and 17.72% of breaking elongation ratio.

2.2 Maximum warp and weft density of fabric

The fabric which is used in foundation for Composite Material is always plain woven fabric with equal density in warp and weft. According to empirical formula of Bnierley, the maximum warp and weft density of fabric is calculated:

$$P_{max} = 10 \cdot K \cdot \sqrt{\frac{10}{N_t}} \cdot f^m \qquad (1)$$

Where, P_{max} is the maximum warp and weft density of fabric (yarns/10cm); K is constant of yarn category, $k = 14.0$; f is the average float length of weave structure, $f = 1$; m is constant of weave, $m = 1$; N_t is the fineness of yarn. Put data in formula (1) and get:

while N_t equals 24tex, $P_{max} = 260$ yarns/10cm

---

* Corresponding author's email: zhouxh314@163.com

while N_t equals 44tex, $P_{max} = 200$ yarns/10cm

By using the empirical formula of Bnierley, the warp and weft density of fabric whose linear density is 24tex are designed as follows: 260×260, 240×240, 220×220, 200×200, 180×180, 140×140, 120×120 (yarns/10cm); and the density in warp and weft of fabric whose linear density is 24tex×2 are designed as follows: 200×200, 180×180, 160×160, 130×130, 120×120, 100×100 (yarns/10cm). All fabrics were weaved by ASL2000 rapier loom and finished by heat setting machine in temperature 150℃, time of 50s. Each fabric was weaved about 250×1200mm, and 5 samples about 50×200 mm were cut off from each fabric.

3 Experimental

3.1 Experimental apparatus and experimental method

Experimental apparatus of tearing test is YG028-3000 electronic strength tester. This machine whose designed standards include GB/T 3917.2-1997, GB/T15788-1995, GB/T13763-1992, GB/T14800-1993, GB/T16989-1997 can be widely used in draw, tearing, bursting test of textile materials.

The thesis refers to single rip method of GB/T3917.2-1997. It is need to cut a 100mm incision in the middle paralleled to the length and mark out the tearing end in the middle of the sample 25mm away from the uncut end. The principle of single rip method is infibulating two tongue pieces of the sample in the tearing test machine, initializing the distance between two collets as 100mm and setting the sample cut line as beeline. After starting the machine, force in the cut direction will make the uncut transverse yarn broken to the tearing end. Then the auto graph plotter draws out the tearing curve, the displacement of the collet as abscissa and the force endured by the fabric in the tearing process as y-axis.

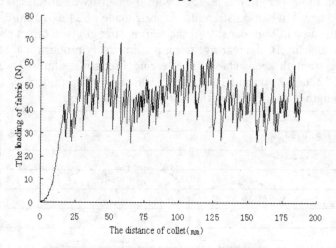

Figure 1 The curve charts of tearing

Table 1 Tearing strength of samples

number	Fineness / (tex)	Weave structure	Density(warp×weft) / (yarns/10cm)	Weight / (g/m^2)	Tearing Strength / (N)
1#	24	Plain structure	120×120	58.60	-
	24	Plain structure	140×140	69.20	57.47
	24	Plain structure	180×180	88.40	50.96
	24	Plain structure	200×200	97.80	43.96
	24	Plain structure	220×220	107.60	43.78
	24	Plain structure	240×240	118.20	34.08
	24	Plain structure	260×260	130.21	50
2#	22×2	Plain structure	100×100	88.67	-
	22×2	Plain structure	120×120	105.63	73.93
	22×2	Plain structure	130×130	114.52	72.7
	22×2	Plain structure	160×160	140.90	69.77
	22×2	Plain structure	180×180	158.49	61.37
	22×2	Plain structure	200×200	178.32	62.46

3.2 Experimental result

The loading of fabric throughout the tearing process is described in Figure 1. Based on the way of GB/T 3917.2-1997, the stress-strain curve is divided into four areas. The tearing strength of fabric is the average of 12 extrema which find out two minimum and two maximum from the second, third and forth area. The tearing strength of samples are calculated in Table 1, several classic patterns of cloth after being teared are given in Table 2.

Table 2 Several classic patterns of cloth after being teared

Charact-eristic	Yarns are broke completely	Parts of yarns are pulled out	Most of yarns are pulled out	All of yarns are pulled
1#				
2#				

Based on the data of Table 1, Figure 2 is made with the density as abscissa and the tearing strength as ordinate. The relationship between tearing strength of fabric made of 24tex Aramid 1313 yarn and warp/weft density is as follows: in the area of high density A, the tearing strength is in direct proportion to the density of fabric; in the zone of low density B, the tearing strength is inversely proportional to the density of fabric; in the area of C, the tearing strength keeps stable. The tearing strength is almost inversely proportional to the density of fabric made of 22tex×2 Aramid 1313 yarn. In the area of high density D and in the zone of low density E, the tearing strength is unaffected by the warp and weft density.

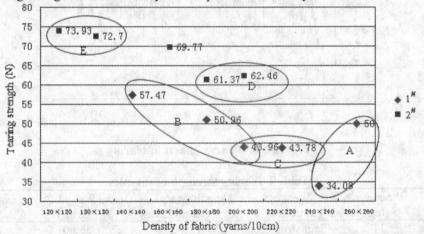

Figure 2 The relationship between tearing strength and density of fabric

2.3 Mechanism of tearing

Figure 3 is the force diagram of fabric during tearing. Vertical yarn pick up the load F given by the movement of collet. It will elongate the parallel yarns and make them have the tendency of being pulled out. It is clear that the displacement of collet is based on the elongation of parallel yarns and the yarn length being pulled out. Collet move at the beginning of tearing and the first parallel yarn become elongate. As the displacement of collet is enlargement, the transformation of the first parallel yarn are increase, and the load is increase too. This load will transmit to every yarns of tearing zone. Parallel yarn will be pulled out if the load of yarn is larger than weaving resistance which is in direct proportion to warp and weft density.; while it will be broken if the breaking strength of yarn is smaller than weaving resistance. Tearing strength is not in direct proportion to the warp and weft density when the breaking strength of yarn is very large. For example, the tearing strength of fabric whose linear density is 22tex×2 is almost inversely proportional to warp and weft density because yarns are slippaged, gathered and pulled throughout the process of being teared. In same occasions, tearing strength is not influenced by warp and weft density.

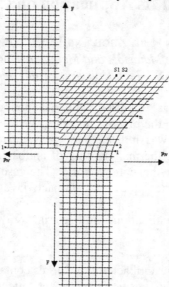

Figure 3 Conceptual sketch of tearing

3 Conclusions and prospect

Plain cloth is usually used for the foundation of Composite Material. It has excellent tearing property while the density in warp and weft is maximal. However, there is no doubt that it will increase the weight of fabric and cost. It often choose light fabric as foundation. For example, the plain weave fabric of which density is 180×180 yarns/10cm and fineness is 24tex has high tearing strength, but many yarns are pulled because the weaving resistance of yarns is lower than yarn tension. It is necessary to increase the adhesive power between foundation and fabric, and make sure that yarns will not be pulled out.

References

[1] Zhu Weimin, Study on shearing property of woven fabric
[2] Zhang Haixia, Zhang Xichang. Study Oil tear strength of the abraded denim. Shanghai Textile Science & Technology. Vol.32, No.2, 2004
[3] Jiang Qigang, Li Yuling, Cheng Xuwei and Shen Mingfia. Study the strength of polysulfonamide on teanng ot polysultonamide ftabric. Industrial Textiles, No.2, 2007
[4] Yu Weidong. Textile Materials[M]. Beijing: China Textile Press, 2006.5:95-296

103 The Preparation of Mesoporous Silica Inorganic Antibacterial Material and Its Application on Cotton Fabrics

Bin Sun, Xinxing Feng*, Lan Zhou, Na Liu, Zhangwei Wu, Jianyong Chen

(Key Laboratory of Advanced Textile Materials and Manufacturing Technology, Ministry of Education of China, Zhejiang Sci-Tech University, Hangzhou 310018, China)

Abstract: The mesoporous SiO_2 was modified with 3-aminopropyl-triethoxysilane(APTES) via post-synthesis silylation methods. Then dip the powder into the $AgNO_3$ solution and the antibacterial agent was produced. The mesoporous structure was indicated by x-ray diffraction, and the amino group inside the channel was identified by IR, and the thermal stability of antibacterial agent was shown by TG. Then the antibacterial material was applied on the cotton fabric. The result of antibacterial test showed that the antibacterial property of the cotton fabric treated with antibacterial agent was great.

Keywords: Mesoporous materials; Modification; Antibacterial

1 Introduction

Recently, more and more inorganic antibacterial materials containing silver have been developed and the research on the inorganic antibacterial has attracted much attention. Silver exhibits good antibacterial properties and in recent years has been used in a variety of medical applications ranging from wound dressings to urinary catheters. They have the advantages of long lasting biocide with high temperature stability and low volatility. The antimicrobial activity of silver ions has been well established. Silver is well known as a significant resource for topical antimicrobial for its great antimicrobial properties.

At present, there are many methods for preparing silver-loaded antimicrobials, most of them are based on physical adsorption or ion exchange, silver ions are fixed in the zeolite, phosphate or other porous materials. Wang Hongshui[1] et al developed a kind of silver-loaded zeolite antimicrobial. The zeolite powder was dipped into $AgNO_3$ solution, then Na^+, Ca^{2+} were exchanged for silver ions. The antibacterial agent was prepared after filtration, washing, drying and sintering. Wu Yuehui[2] et al used self-made amorphous aluminum silicate as a carrier, mixed with $AgNO_3$ solution for ion exchange. After filtration, washed with deionized water and dried, the aluminum silicate antibacterial agent was obtained. Li Bowen[3] et al selected bentonite for the carrier, silver-loaded bentonite antibacterial agent were prepared, as based on the same principle.

Ordered mesoporous silica has the characterizations of narrow pore size distribution, ordered pore structure. Moreover, the mesoporous structure is controllable and the preparation technology of mesoporous silica is mature. It has indicated many incomparable properties compared with other porous materials. The appearance of mesoporous materials is multifarious, the composition of hole wall is controllable[4], the thermal stability is great and the hydrothermal stability can be obtained by optimizing the reaction condition. Ordered mesoporous silica material is used as the carrier, the controllable release property is better than other drug release system [5].

The most common method used to prepare Ag/SiO_2 mesoporous composites is the post-synthesis impregnation of pure or organic-functionalized mesoporous silica host matrices with silver nitrate solutions. In the present study, mesoporous silica was used as the carrier of antibacterial agent. The surface of mesoporous channel was modified with APTES, and the silver was loaded in the channel. In order to verify the antibacterial properties, the antibacterial agent was successfully applied to cotton fabrics.

2 Experiment

2.1 Reagents and equipment

P123 ($EO_{20}PO_{20}EO_{20}$,Ma), TEOS(AR), sodium chloride, nitrate, ethanol, $AgNO_3$(AR), toluene, glucose, ammonia, 3-aminopropyl-triethoxysilane(APTES), PBS buffer solution, beef extract(BR), peptone(BR), agar(BR).

2.2 Preparation of antibacterial agent

The preparation of SBA-15 was following the reference literature [6]. The SBA-15 was modified with silane coupling agent APTES at room temperature. The modified SBA-15 was poured in $AgNO_3$ solutions (50 ml) of different concentrations and the solutions were stirred vigorously for 6 h at room temperature.

After being washed, the samples were filtered and dried for one day.

2.3 Characterization of antibacterial agent

The XRD data were obtained in an ARL-X 'TRA diffractometer using CuK$_\alpha$ radiation, in 2θ range of 0°~10°, λ=0.15406nm.

Infrared spectra of samples were obtained by FT/ IR-610.

TG analysis was performed in N_2 and a heating rate of 10℃/min (Pyris Diamond TGA).

2.4 Antimicrobial functional finishing

The antimicrobial (24g/L) was dispersed in the deionized water, some dispersant (10g/L) was added. Then the solution was mixed with ultrasound for 2 h, the adhesives(40g/L) was dropped in the solution slowly, stirred for 30 min. Antibacterial finishing on cotton fabric was carried out for 2 times. Baking temperature: 150℃, 3 min, the remaining content: 80%, liquor ratio1:5.

2.5 Assessment of antibacterial finishes on cotton fabric

Swatches of test and control textile materials were tested qualitatively for antibacterial activity by AATCC Method 147.Those showing activity were evaluated quantitatively. Test and control swatches were inoculated with the test organisms. After incubation, the bacteria were eluted from the swatches by shaking in known amounts of neutralizing solution. The number of bacteria present in this liquid was determined, and the percentage reduction by the treated specimen was calculated using the formula: $R = 100 \times (B-A)/B$.

Where:

R=%reduction

A=the number of bacteria recovered from the inoculated treated test specimen swatches in the jar incubated over the desired contact period

B=the number of bacteria recovered from the inoculated treated test specimen swatches in the jar immediately after inoculation (at "0"contact time)

According to the test, part of E. coli colonies were inoculated to 100 ml nutrient broth, shaken at 37℃,150r/min and cultured for 24 h. Preparation of the bacterial suspension with normal saline diluted to 3.0×10^5 cfu/ml.

Cotton fabrics were treated with different antimicrobial concentrations (0.50%, 1.00%, 1.50%, 2.00%), cut into pieces and immersed in the buffer solution, 1ml bacterial suspension was added to the sample solution, the sample was in touch with bacteria at 37℃, shaken for 1h. Took 0.1ml and 10 times dilution, and finally 1ml of the solution was taken and painted on nutrient agar surface, placed in an incubator at 37℃ for 24h.Then bacterial flora was observed and the amount was counted, antibacterial efficacy was calculated by the reduction percent of bacteria.

2.6 Laundering

After antibacterial application, all fabric samples were washed different times. The laundering process was carried out according to BSEN ISO 26330 standard (5A program). A laundering machine was used with 4 g/L of soap.

3 Results and discussion

3.1 XRD analysis

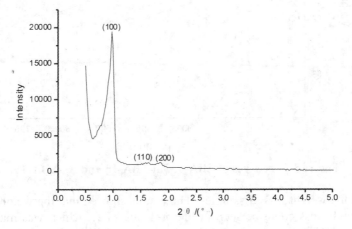

Figure 1 Low angle X-ray diffraction patterns of SBA-15

Low angle powder X-ray diffraction pattern of the mesoporous materials was shown in Fig1. From the low angle XRD pattern, three diffraction peaks appeared, which corresponding to the (100), (110) and (200). One very intense diffraction peak indexed as (100) and another two weak peaks indexed as (110), (200). The XRD patterns indicated that the mesoporous material prepared had a 2D hexagonal (p6mm) structure.

3.2 IR analysis

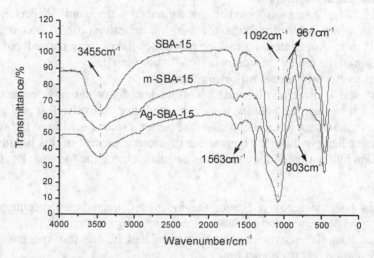

Figure 2 FTIR spectra of SBA-15, m-SBA-15 and Ag-SBA-15

In Figure 2, it could be seen that a strong absorption peak appeared at 1092 cm^{-1}, which attributed to Si-O-Si peaks. Absorption bands at 803 cm^{-1} were observed. From the infrared spectrum of m-SBA-15, it showed that the absorption peak of Si-OH at 967cm^{-1} disappeared after modification, for Si-OH on the surface of SBA-15 were instead of -NH$_2$ in this reaction, which resulted the number of Si-OH on the surface reducing. At 1563 cm^{-1} the -NH$_2$ absorption peakoccurred[7], indicated that the amino was grafted on the surface of mesoporous successfully. In addition, -NH$_2$ absorption band was covered by hydroxyl[8], due to the strong absorption of hydroxyl at 3300~3500 cm^{-1}.

3.3 TG analysis

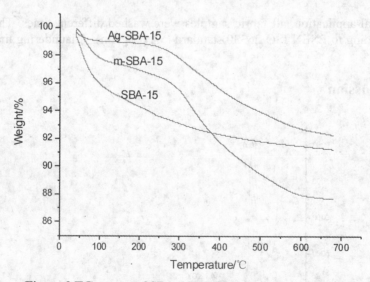

Figure 3 TG curves of SBA-15, m-SBA-15 and Ag-SBA-15

Figure 3 indicated the weight loss of samples. The results showed that the weight change of SBA-15 in the heating process, weight loss could be seen in the previous stage, which was mainly due to the loss of water molecule that adsorbed in the sample; the weight loss of m-SBA-15 could be divided into two steps, the first stage was from room temperature to 100℃, the adsorbed water of the sample gradually reduced in this stage, the corresponding weight loss was approximately 2%. When the temperature reached about 300℃, the organic groups began to decompose, this was the second stage; when the temperature reached 700℃, the organic groups were burned totally and the stability of the sample became stronger. From the weight loss of Ag-SBA-15 at high temperature, it showed that the process was similar to the m-SBA-15.when the temperature got close to 300℃, the amino groups which grafted in the channel began decomposing gradually, a obvious weight loss stage emerged, which showed that the thermal stability of antibacterial agent was great. The amino groups grafted on the surface of the SBA-15 decomposed gradually, so a obvious weight loss peak could be seen. It indicated that the thermal stability of antimicrobial was very well.

3.4 Antimicrobial activity

Table 1 The effect of different dosage of antibacterial agents

Antimicrobial content	Antibacterial efficacy
0.50%	98.76%
1.00%	99.99%
1.50%	100%
2.00%	100%

Further investigations of antimicrobial activity were performed on the antibacterial cotton fabrics. It could be seen from Table 1, when the dosage of antimicrobial agent was up to 1.00%, the antibacterial efficacy of samples could reach 99.99%, E.coli was killed completely.

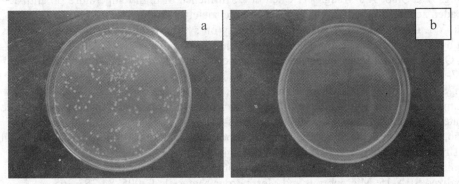

Figure 4 colonies on blank sample (A), colonies on antibacterial sample (B)

Figure 4 showed that the bacterial reduction of the treated samples compared to control fabrics. Sample B exhibited strong bactericidal activity after treatment. Results showed that antibacterial agent had a strong antibacterial activity even at a low concentration.

Table 2 The effect of different washing times

Laundering times	Antibacterial efficacy
0	100%
10	99.06%
20	98.60%

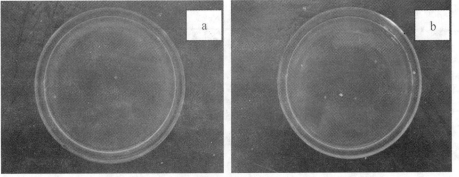

Figure 5 The antibacterial activity of sample after washed several times against Escherichia coli
(a) fabric washed 10 times; (b) fabric washed 20 times

Sample bacterial reduction was also examined after 0, 10, 20 laundry cycles. From the Figure 5, samples treated with the antibacterial agents preserved their strong antibacterial activities even after 20 launderings. The antibacterial activity of the samples treated with agent slightly decreased after laundering. Although the test method was adapted to test for this agent's activity, testing of non-leaching type agents with dynamic test methods (e.g., ASTM 2149) could provide better results.

These antibacterial test results showed that the bonding capacity of antibacterial agent on the fabric was very high and that long washing durability could be obtained by these chemicals.

4 Conclusions

(1) Modified SBA-15 is used as the carrier of inorganic antimicrobial agent, the antibacterial properties and the thermal stability is excellent.

(2) The inorganic antimicrobial agent applied to the cotton fabric can get great antibacterial effect, the fabric antibacterial efficacy remains on 98% after 20 times launderings.

Acknowledgement

This research was funded by Zhejiang education department research plan project (0901066-F).

References

[1] H S Wang, X L Qiao, X J Wang, J G Chen et al. Influence of high temperature on antibacterial property of silver-loaded zeolite. Journal of the Chinese ceramic society J. 2006, 34(2): 171-174.
[2] Y H Wu, Y L Liu, A M Deng, J H Wu et al. Study on synthesis of antibacterial amorphous aluminosilicate material by ion-exchange reaction. Journal of the Chinese ceramic society J. 2004, 32(5): 564-569.
[3] B W Li, Q H Xiao et al. Study on antibacterial properties of Ag-carried bentonite. Non-Metallic Mines J.2001, 24(5): 17-18.
[4] A Sayari, S Hamoudi, Y Yang. Chem Mater J. 2005, 17: 212-216.
[5] H D Zhang. Research progress in controlled delivery of drug assembled in nano-pores of ordered mesoporous silica materials. Materials Review J.2009, 23(8): 88-91.
[6] D Y Zhao, J L Huo, G D Stucky et al. Tri-block copolymer syntheses of mesoporous silica with periodic 50 to 300 angstrom pores. Science J. 1998, 279 (5350): 548-552.
[7] H Norihito, Y Katsunori, Y Tatsuaki. Adsorption characteristics of carbon dioxide on organically functionalized SBA-15. Microporous and Mesoporous Materials J.2005, 84:357-365.
[8] X Fu, J P Li, N Zhao et al. CO_2 adsorption on mesoporous molecular sieves modified by aminosiliane. Petrochemical Technology J.2008, 37(10): 1021-1025.

104 Observation and Modeling of the Microstructure of Heavyweight Hydroentangling Nonwoven Fabrics

Hong Wang, Xiangqin Wang, Xiangyu Jin, Haibo Wu, Baopu Yin
(Key Laboratory of Textile Science & Technology, Ministry of Education, Donghua University, 201620, China)

Abstract: The structure of the hydroentangling nonwoven fabrics is a key characteristic parameter which determines other properties of the fabrics. While there exists many available studies on the structure of different hydroentangling nonwoven fabrics, a universal model has not been studied well. In this work, we present a novel model of the heavyweight hydroentangling nonwoven fabrics based on the microstructure observation of the fabrics by using SEM and X-ray tomography techniques. This new model is useful to study the entangling mechanism of fibers during the hydroentangling process and the mechanical properties of the hydroentangling nonwoven fabrics.

Keywords: Hydroentangling; Nonwoven; 3D; Model

1 Introduction

Hydroentangling is a mechanical bonding process that exploits the energy transfer capabilities of high-pressure water to entangle loose assemblies of fibers or fabrics carried and supported by a support conveyor [1]. The water jets push the fibers off the crests of the knuckles into the open inter weave spaces of the support conveyor and then a sheet of randomly oriented fibers is transformed into a kind of hydroentangling nonwoven fabric with a specific microstructure.

The structure of the hydroentangling nonwoven fabrics is crucial to the properties of the final products, whilst there is only fewer detailed technical information of these studies. We can find lots of images or micrographs of hydroentangling nonwoven fabrics acquired by optical microscopy, digital camera, SEM, DVI, etc, whilst there is no detailed technical paper to summarize all the information and set up a universal model about the microstructure of the hydroentangling nonwoven fabrics. Pourdeyhimi etc summarized the structure of lightweight hydroentangling nonwoven fabrics as shown in Figure 1 and a model of the entangling point is developed as shown in Figure 2 [2].

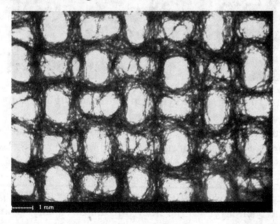

Figure 1 Structure of lightweight hydroentangling nonwoven fabrics

They find that the lightweight hydroentangling nonwoven fabrics show a definite cellular structure with the open cells formed at the knuckes, as shown in Figure 1. Based on the structure of the lightweight hydroentangling nonwoven fabrics, a model of the entangling point is depicted schematically in Figure 2.

The author tried to analyze the energy absorption during the hydroentangling process by using this model. However, only 1% of the energy of the high speed water jets is absorbed by the fiber webs, which is far from the real situation [3]. Therefore, a new model of the hydroentangling nonwoven fabrics should be set up.

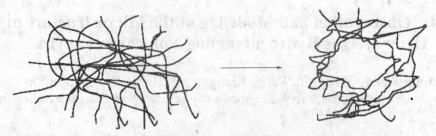

Figure 2 Model of the entangling point of the lightweight hydroentangling nonwoven fabrics

2 Experimental
2.1 SEM observations
The surface and cross-section of the heavyweight hydroentangling nonwoven fabrics (the mass per unit area is in the range of 120-200g/m^2)were observed by using a JSM-5600LV (GEOL Co., Japan) Scanning Electron Microscope (SEM).

2.2 X-ray tomograph observations
3D images of the heavyweight hydroentangling nonwoven fabrics were tested and reconstructed in 3D by using BL13W1 X-ray imaging facility.

3 Results and discussion
3.1 SEM micrographs of heavyweight hydroentangling nonwoven fabrics
In order to set up a model of the heavyweight hydroentangling nonwoven fabrics, SEM micrographs of the surface and cross-section of the fabrics were tested and shown in Figure 3 and Figure 4, respectively.

Figure 3 SEM micrograph of heavyweight hydroentangling nonwoven fabrics

It can be seen from Figure 3 that there are visible jet marks on the surface of the heavyweight hydroentangling nonwoven fabric, which were formed by the impacting of the water jets with the fiber web. Some fibers in the jet mark area were deflected downwards the support surface while fibers in the ridges among the jet marks were reoriented along the process direction.

It can be seen from Figure 4 that two jet marks (pointed by vectors) with fibers pushed down by the water jets from the surface to the back layers of the fiber web are obviously visible on the cross-section of the hydroentangling nonwoven fabric. At the same time, a great amount of cross-sections of single fibers appear among the jet marks, which is consistent with the result of the SEM micrograph of the surface of the hydroentangling nonwoven fabric.

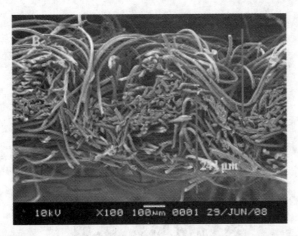

Figure 4 SEM micrograph of the cross-section of heavyweight hydroentangling nonwoven fabrics

3.2 X-ray micrographs of heavyweight hydroentangling nonwoven fabrics

3D images of the heavyweight hydroentangling nonwoven fabrics were tested by using X-ray imaging and biomedical application facility and the micrographs of the two cross-sections of the fabrics were shown in Figure 5

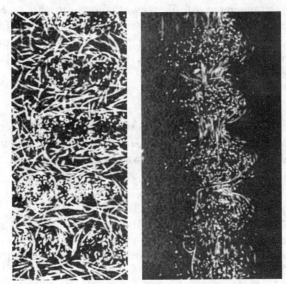

Figure 5 X-ray micrographs of heavyweight hydroentangling nonwoven fabrics

It can be found from Figure 5 that some fiber segments were orientated while some fiber segments were penetrated from the surface down to the back of the fabrics, which is consistent with the result of the SEM micrograph.

3.3 Model of the heavyweight hydroentangling nonwoven fabrics

The structure of the heavy weight hydroentangling nonwoven fabrics in three directions, i.e, surface, cross-section in the machine direction and cross direction were observed by SEM and X-ray micrographs. Based on the micrographs obtained, a model is set up by using the image software as shown in Figure 6.

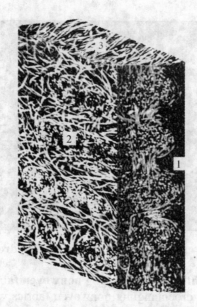

Figure 6 Computer reconstructed 3D image of heavyweight hydroentangling nonwoven fabrics

In Figure 6, a patterned arrangement of fibers was shown in three sections of the heavyweight hydroentangling nonwoven fabrics. Jet marks were shown on section 1 of the fabric, with fibers penetrated from the jet marks down to section 2 of the fabric. We can see a great amount of cross-sections of fibers on section 1 and reoriented fiber segments on section 2, which means that fibers among jet marks were reoriented during the hydroentangling process.

4 Conclusions

In this paper, a 3D model of heavyweight hydroentangling nonwoven fabrics were set up by reconstructing the micrographs acquired by SEM and X-ray tomography. This model can represent the real microstructure of the heavyweight hydroentangling nonwoven fabrics and will be used further to study the hydroentangling mechanism in the future study.

References
[1] N Anantharamaiah, K Ro¨mpert, H V Tafreshi, B Pourdeyhimi. A novel nozzle design for producing hydroentangled nonwoven materials with minimum jet-mark defects. J Mater Sci, 2007, 42: 6161–6170.
[2] A M Seyam, D A Shiffler, H Zheng. An Examination of the Hydroentangling Process Variables. International Nonwovens Journal, 2005, 14(1): 25-33.
[3] E Ghassemieh, M Acar, H K Versteeg. Improvement of the efficiency of energy transfer in the hydro-entanglement process. Composites Science and Technology, 2001, 61: 1681–1694.

105 Frequency Features of Acoustic Emission on Failure Mechanisms in PE Self-reinforced Laminates

Xu Wang, Huipin Zhang, Xiong Yan[*]

(Key Lab of Textile Science and Technology, Ministry of Education, College of Textiles, Donghua University, Shanghai 201620, China)

Abstract: The purpose of this study is to investigate frequency features of acoustic emission (AE) on different failure mechanisms in PE self reinforced laminates. Three kinds of specimen pure LDPE resin, single fiber composites (SFC) and UHMWPE/LDPE unidirectional laminate were prepared to obtain desired damage modes. AE signals of matrix damage, interface debonding and fiber breakage can be activated by quasi-state tension of LDPE resin, [90] laminate and SFC respectively. Fast Fourier transform (FFT) was used to obtain frequency information of four different failure mechanisms including matrix plastic deformation (F1), matrix fracture (F2), interface debonding (F3) and fiber breakage (F4). The result revealed, frequency features were different on four failure mechanisms. Frequency distribution of F1 and F2 are below 100 kHz with peak frequency at about 40 kHz. Frequency distribution of F3 has two main regions 0-100 kHz and 200-300 kHz with peak frequency at about 40 kHz and 240 kHz respectively. Frequency of F4 also has two main regions 100-200 kHz and 400-500 kHz with peak frequency at about 150 kHz and 450 kHz respectively. The result can be applied to identify failure mechanism in PE self-reinforced composites.

Keywords: PE self-reinforced laminates; Acoustic emission; Frequency features; Failure mechanisms

1 Introduction

Acoustic emission (AE) can be defined as a transient elastic wave generated by the rapid release of energy within a material. Because the AE feature of a material depends on different failure mechanisms, AE inspection is a powerful aid to materials testing and the study of deformation and failure mechanisms. Typical failure mechanisms of composites such as matrix cracking, fiber breakage, interface debonding and interlayer delamination have been confirmed as AE sources by many researchers [1-6]. Many researchers have already studied frequency features of AE signals on thermoset matrix, metal matrix and ceramic matrix composites [7-11]. A large number of experiments showed frequency features of AE signals can be used to identify different failure mechanisms [7, 9-11]. The purpose of present study is to investigate frequency features of AE signals on different failure mechanisms in PE self reinforced laminates. During the failure process of different model specimens, AE signals of desired failure mechanisms were recorded. The relationship between failure mechanisms and AE signals can be established by frequency feature. The result revealed frequency feature can be applied to identify damage mechanism in PE self reinforced composites.

2 Experimental

2.1 Materials and specimens

The density of LDPE and UHMWPE fiber is 0.92g/cm3 and 0.97 g/cm3 respectively. The melt temperature of LDPE and UHMWPE is about 112℃ and 147℃ respectively. The tensile strength of UHMWPE fiber is about 3GPa with fracture strain at 3.3%. Three kinds of specimen were fabricated to activated desired failure mechanisms. LDPE resin and unidirectional laminate were prepared by compression molding at 120℃, 1.0MPa for 5 minutes and 120℃, 1.5MPa for 10 minutes respectively. Single fiber composite (SFC) were fabricated by a single filament of UHMWPE embedded in LDPE resin. All tensile specimens were cut into coupon with dimension size $120\times20\times0.8mm^3$.

2.2 Tensile testing monitored by AE system

All tensile testing were tensioned till fracture by universal testing machine (WDW-20, Hualong Shanghai, China).Tensile velocity of LDPE resin and SFC were 20mm/min, [90] laminate was 1mm/min. Failure process was monitored by AE system (PCI-2, Physical Acoustic Corporation, USA). A wide band sensor was attached to the middle position of the specimen by vacuum grease and fixed by elastic tape. According to testing environment and pencil lead break procedure, the parameters of AE system were as follows: peak definite time (PDT), 50us; hit definite time (HDT),100us; hit lock time (HLT), 300us; system

---

[*] Corresponding author's email: yaxi@dhu.edu.cn

threshold, 40dB; preamplifier, 40 dB. Sampling rate and length of AE system was set at 1MHz and 1k bytes.

3 Results and discussion
3.1 Frequency features of AE on matrix damage

AE signals of matrix damage can be activated by quasi-state tension of LDPE resin. According to experimental observation, there are two kinds of failure mechanisms during matrix failure process. The first is plastic deformation (F1) occurring after the yield stress. The second is fracture (F2) occurring at the end of loading process. Fast Fourier transform (FFT) which can change AE waveforms from time domain to frequency domain was used to obtain frequency information of matrix damage modes mentioned above. Figure 1 shows failure mechanisms of LDPE. Frequency distribution of F1 and F2 are all below 100 kHz with peak frequency at about 40 kHz. However, it is obvious that the duration time of F2 is longer than F1 and the amplitude of F2 is higher than F1.

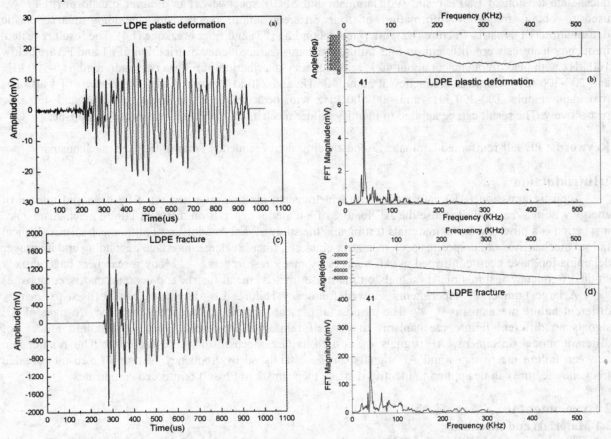

Figure 1 Frequency feature of matrix damage (a) time domain signal of plastic deformation, (b) amplitude-frequency diagram of plastic deformation, (c) time domain signal of fracture, (d) amplitude-frequency diagram of fracture

3.2 Frequency features of AE on interface debonding

AE signals of interface debonding (F3) can be activated by quasi-state tension of [90] laminate. Due to natural inert fiber surface, the bonding property between fiber and matrix is certain weak. According to transverse tensile strength of unidirectional laminate is smaller than pure LDPE and fracture exist in joint surface between fiber and matrix, the interface debonding can be regarded as dominant damage mode in [90] laminate. Figure 2 shows the frequency features of F3 mainly include three types. Type1 and type2 has only one peak frequency at about 40 kHz and 240 kHz respectively. However, type3 has two peak frequency mentioned above. Frequency distribution of F3 has two main regions 0-100 kHz and 200-300 kHz.

3.3 Frequency features of AE on fiber breakage

AE signals of fiber breakage can be activated by quasi-state tension of unidirectional SFC. Comparing with Figure 1 and Figure 2, Figure 3 shows the frequency feature of fiber breakage (F4) are obviously different with matrix damage and interface damage. F4 mainly include two types. Type1 has only one peak frequency at about 140 kHz and type 2 has two peak frequency at about 140 kHz and 450 kHz respectively. Frequency distribution of F4 has two main regions 100-200 kHz and 400-500 kHz.

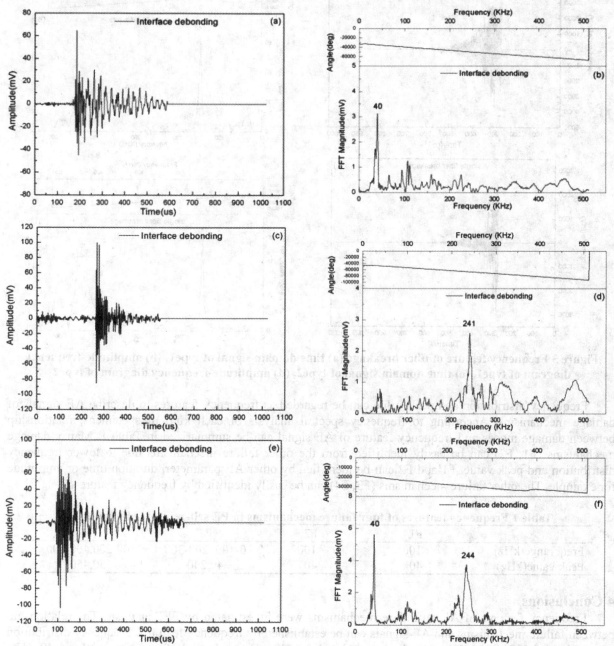

Figure 2 Frequency feature of interface debonding (a) time domain signal of type1, (b) amplitude-frequency diagram of type1, (c) time domain signal of type2, (d) amplitude-frequency diagram of type2, (e) time domain signal of type3, (f) amplitude-frequency diagram of type3

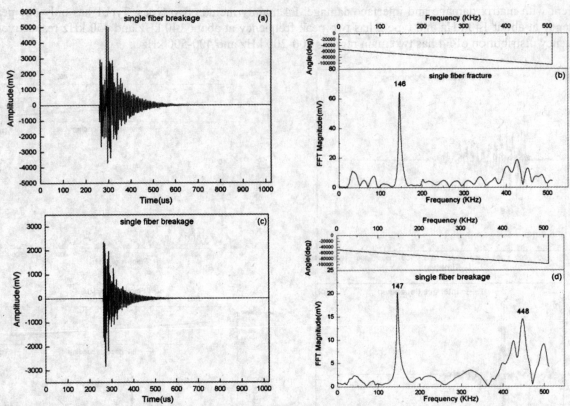

Figure 3 Frequency feature of fiber breakage (a) time domain signal of type1, (b) amplitude-frequency diagram of type1, (c) time domain signal of type2, (d) amplitude-frequency diagram of type2

Frequency distribution and peak value can be regarded as frequency features to describe AE signals of damage mechanisms. According to frequency spectral analysis of each kind of specimens, relationship between damage modes and frequency feature of AE signal can be summarized in Table 1. Matrix damage mechanisms (F1, F2) can be easily identified from the other failure mechanisms due to lower frequency distribution and peak value. F1 and F2 can be identified by other AE parameter, duration time or amplitude for example. The other failure mechanisms (F3, F4) can be easily identified by frequency features.

Table 1 Frequency features of four failure mechanisms in PE self-reinforced laminates

	F1	F2	F3	F4
Freq. range(kHz)	<100	<100	0-100, 200-300	100-200, 400-500
Peak value(kHz)	40	40	40, 240	150, 450

4 Conclusions

Frequency features of AE on failure mechanisms were investigated by FFT analysis. The relationship between failure mechanisms and AE signals can be established by frequency features. Frequency distribution of matrix deformation and matrix fracture are below 100 kHz with peak frequency at about 40 kHz. Frequency distribution of interface debonding has two main regions 0-100 kHz and 200-300 kHz with peak frequency at about 40 kHz and 240 kHz respectively. Frequency distribution of fiber breakage also has two main regions 100-200 kHz and 400-500 kHz with peak frequency at about 150 kHz and 450 kHz respectively. The result revealed frequency features of AE signals can be further applied to identify failure mechanism in PE self-reinforced composites.

References

[1] S. Huguet, N. Godin. Use of acoustic emission to identify damage modes in glass fibre reinforced polyester, Comp. Sci. and Tech., 2002, 62:1 433-1444.
[2] V. Kostopoulos, T.H. Loutas, K. Dassios. Fracture behavior and damage mechanisms identification of SiC/glass ceramic composites using AE monitoring, Comp. Sci. and Tech., 2007, 67:1740-1746.

[3] N. Godin, S. Huguet.Integration of the Kohonen's self-organising map and k-means algorithm for the segmentation of the AE data collected during tensile tests on cross-ply composite, NDT&E International, 2005, 38: 299-309.

[4] C. Bhat, M.R. Bhat, C.R.L. Murthy. Acoustic emission characterization of failure modes in composites with ANN, Composite Structures, 2003, 61: 213-220.

[5] X.M. Zhuang, X.Yan.Investigation of damage mechanisms in self-reinforced polyethylene composites by acoustic emission, Comp. Sci. and Tech., 2006, 66: 444-449.

[6] T.H. Zhang, X. Wang, X. Yan, H.P. Zhang.An investigation into the propagation characteristics of AE signals in PE/PE composite laminates, Insight, 2007, 49: 665-668.

[7] Y.H. Yu, J.H. Choi, J.H. Kweon, D.H. Kim. A study on the failure detection of composite materials using an acoustic emission, Composite Structures, 2006, 75: 163-169.

[8] J. Mikael, G. Peter. Broad-band transient recording and characterization of acoustic emission events in composite laminates, Comp. Sci. and Tech., 2000, 60: 2803-2818.

[9] S.C. Woo, N.S. Goo. Analysis of the bending fracture process for piezoelectric composite actuators using dominant frequency bands by acoustic emission, Comp. Sci. and Tech., 2007, 67: 1499-1508.

[10] J.K. Lee. AE characteristic of the damage behavior of TiNi/Al6061 SMA composite, Composite Structures, 2003,60:255-263.

[11] M. Gioradano, A. Calabro, C. Esposito, A.D. Amore, L. Nicolais. An acoustic emission characterization of the failure modes in polymer composite materials, Comp. Sci. and Tech., 1998, 58: 1923-1928.

106　Research on Performance of Filaments/Short Fibers Composite Yarn

Yu Xie, Ruicai Jing, Min Guo, Shiqin Song, Bin Jiang, Tonghua Zhang*
(College of Textile & Garments, Southwest University, Chongqing 400715, China)

Abstract: This paper studies the structure and performance of a kind of novel composite yarn made of polyester filaments and cotton fibers. The relationships between the performance and the structure of the yarn are discussed, and comparative analysis of characteristics is conducted between the composite yarn and single yarn. Research results show that the novel composite yarn exhibits predominant performance, such as extensional strength and abrasion resistance, etc.

Keywords: Composite yarn; Performance of yarn; Structure of yarn

1 Introduction

At present, the studies on composite yarn made of filaments and short fibers has reach to a high level and this kind of yarn was produced for a long time. A clear conclusion has been drawn that this kind of composite yarn has improved its performance compared with single yarn in many studies. Whereas, most of those analyses and studies mainly focus on core-spun yarn or wrapped yarn, and etc [1]. This paper presents a kind of novel yarn, which is different from those composite yarns on aspect of structure and some of yarn's performance. And this paper also analyzes the advantages of the yarn's performance in way of comparing with single yarn. The relationship between the advantages and yarn's structure is discussed.

2 Analysis of components of composite yarn

Filaments and short fibers composite yarn is one kind of multi-component yarns, thus, there must be a consequence that this kind of yarn not only make use of each component of fiber's predominance but also offset their own limitation [2]. In this paper, filaments and cotton fibers are used to generate the novel composite yarn, and the two materials take possession of their own special performance.

Cotton fiber can be made into a low-density yarn because of its lower linear density. It is endowed with upper intensity, upper anti-crease, upper heat durability and well affinity to dye. In addition, the fabric made of cotton yarns has a remarkable performance of washing, without phenomena of static electricity and pilling, possessing well draping quality and pleasing handle. However, cotton fiber has disadvantages of lower acid resistant and lower fungus resistance which lead it easily to be damaged by microbe [3], otherwise, it exhibits poor tensile behavior, worse elasticity and rebound elasticity and easy to generate pillars on the surface fabric [4].

Polyester fiber is one of synthetic fibers with a great deal of favorable performance. Say, it has upper intensity, and better thermal stability, acid resistance, antimicrobial property, and micro-organism resistance, and etc. Whereas, its poor moisture absorption leads to polyester fabric's poor qualities of absorbing humidity and helping relievesweat, even influences on its wearability. What is worse, fabric's dyeing behavior is also affected because of the fiber's low wet pickup, neither the fiber nor its fabric is hard to be colored. Furthermore, polyester fabric can cause electric spark easily because of its superior electric resistance and static electricity caused by virtue of friction with other objects [5].

In comparison with polyester yarn or cotton yarn, the composite yarn not only exhibit higher intensity and abrasion resistance, but also enhances dyeing behavior.

3 Materials and the structure of the composite yarn
3.1 Materials and spinning procedure

In the experiment, three polyester filaments (12D/1F) and cotton fibers silver are used to form composite yarn. The composite yarn is 40s+36d. During the process of spinning, three filaments are fed into the front nip at a certain distance by a special input feed system and commix with cotton fibers. When the filaments and cotton fibers are exported through the front roller nip, they are twisted into a composite yarn by spindle. Inside of the composite yarn, there is a uniquely three-dimensional structure formed between filaments and cotton fibers.

---

* Corresponding author's email: zhtonghua@yahoo.com.cn

3.2 Longitudinal structure of composite yarn

Figure1 shows the optical microscope image of longitudinal structure of composite yarn. Obviously, the three separate polyester filaments' make geometric helix distribution and conformation in the composite yarn. In the yarn, polyester filaments are transparent and biggish in diameter.

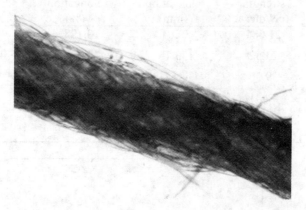

Figure 1 Longitudinal structure of composite yarn

Fed into cotton roving by special input feed system, the three polyester filaments neither converge into compound filament, nor uprightly disperse in the yarn during the process of twisting. Fig1 also shows that each filament is twisted with cotton fibers simultaneously, and presents spiral structure with a lower twist. What is more, the yarn does not form a structure like core-spun yarn or wrapped yarn, but a special structure in the three-dimensional space.

3.3 Cross section of composite yarn

Figure2 shows the cross section structure optical microscope image of the composite yarn. In the yarn, the polyester filaments have biggish diameter.

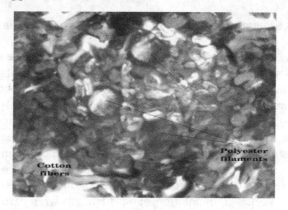

Figure 2 Cross section structure of composite yarn

It is clear to observe the cross section conformation of composite yarn from image: three polyester filaments equably distribute in composite yarn, and the position presents an approximate triangle. This kind of structure is related to the method of feeding and different transfer principle of the two kinds of fibers during the process of spinning.

4 Experiments and analyses

In order to make contrastive analyses, the novel composite yarn and the identical counts of single yarn are studied. During the experiment, strength and abrasion resistance of yarns are tested. All parameters, set at two times of spinning, are completely same. The single yarn is made in 40 counts.

4.1 Strength of yarn

This test is carried out on YG061FQ electronic single-strand strength machine with 500 mm experimental yarns' length and 500mm/min drawing speed. Tests are processed according to GB/T3916—1997. And the two kinds of yarns both need to be tested for 60 times. The results are shown in Table 1 and the typical strength-extension curves are exhibited in Figure 3.

Table 1 Summary of tensile test on composite yarn

Merit index	Tensile strength (cN)	Breaking strength (cN/dtex)	Rupture elongation (mm)	Rupture elongation (%)	Breaking work (mJ)	Initial modulus (cN/dtex)
Mean	527.1	21.084	50.64	10.128	143.9	2.757
Max	588	23.52	133.16	26.632	410.555	3.88
Min	484	19.36	34.671	6.934	85.878	1.36
CV%	4.441	4.441	33.848	33.848	39.953	22.034
Corrected strength	527.1	21.084	50.64	10.128	143.9	2.757

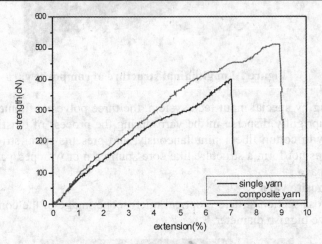

Figure 3 Strength-extension plot of composite yarn and single yarn

It can be seen from Table1 and Figure3, the breaking strength and rupture elongation of composite yarn are obviously higher than those of single yarn. Because of higher strength of polyester than that of cotton fiber, composite yarn is enhanced in strength in comparison with the same counts single yarn [6]. This result also has to do with the structure of yarn: three filaments distribute in yarn evenly and symmetrically observed from cross section of composite yarn. Therefore, the sections of yarn shrink during the process of extension and outer cotton fibers force almost equal pressure to each filament [5]. Because of cotton fiber's lower extensibility than polyester, an extra load is averagely assigned on each filament when outer cotton fibers rupture one by one [2], which lead to get an even force on every cross section of yarn. During the process of extension, the filaments with spiral structure inside of yarn are not undertaken an external force in a longitude direction until they are enough stretched. In addition, spiral filaments wrap parts of cotton fibers which result in reducing the possibility of fibers breaking. These causations all potentially lead to increase rupture elongation of yarn. From analysis of testing data, the composite yarn has more 55 percent of strength than same-count single yarn at maximum degree.

4.2 Test of abrasion resistance

Abrasion resistance test is undertaken on apparatus Y731, a cohesion meter [4]. Both of the composite yarn and single yarn are tested 40 times. Testing results are shown in Table 2.

Table 2 Abrasion resistance of composite yarn and single yarn

Test yarn	Max (times)	Min (times)	Mean (times)	CV (%)
single yarn	420	100	210	42.46
composite yarn	3480	403	989	70.97

From test data shown in Table2, the abrasion resistance of composite yarn is obviously better than single yarn's. The fibers in yarn are abraded by abradant to be attenuating, breaking or rupture, and some fibers even have a possibility to be cut off or pulled out [5]. In terms of quality of yarn, compared the same linear density of chemical fiber with cotton fiber, the former owns better abrasion, that is to say, yarn with filaments will present higher strength. In the other hand, there is a close relationship between abrasion resistance and structure of yarn. The fibers transform their spiral forms into axial direction gradually when the yarn endures repeating abrasion, so the yarn is easy to be abraded to breaking. When there are filaments in yarn, some cotton fibers are encircled and twisted because of filaments' spiral. This structure makes cotton fibers steadier and tends to keep fibers' original locations, even though fibers on the surface are easy to be cut off or pulled out. Furthermore, the entanglement crosslinks of fibers' are reinforced due to the filaments in yarn, that is to say, many double- entanglement crosslinks come into being among fibers. Under this circumstance, fibers gain minimum energy during the process of abrasion [2] and own better stability to avert to slip away. The abrasion resistance of composite yarn is 4-35 times higher than that of single yarn.

5 Conclusions

Filaments and short fibers composite yarn is confirmed with many better performances based on experiments, such as high strength and well abrasion resistance. There is a high correlation between performances and structure of composite yarn. Furthermore, performances of both polyester filaments and cotton fibers in this yarn overcome the breaking strength, elongation and abrasion resistance limitations of single fibers. Therefore, this novel composite yarn has promising application prospect.

Acknowledgement

This work is supported by Fundamental Research Funds for the Central Universities (XDJK2009A006) and Science and Technology Project of Chongqing (CSTC, 2009AC4208).

References

[1] D G Li. Textile Material. Beijing: China Textile Press, 2006:309-311.
[2] B C Goswami, J. G. Martindale, F. L. Scardino. Textile Yarns: Technology, Structure, and Applications. New York: Wiley, 1977.
[3] S B Zhang. Development of multiple components jet-vortex spun yarn and research of its quality. Shandong Textile Technology, 2008, 6: 11-14.
[4] S J Zhao. Experiment Tutorial of Textile Material. Beijing: China Textile Press, 1989:7-11.
[5] M Yao, et al. Textile Material. Beijing: China Textile Press, 1990:377,429.
[6] Z H Song, et al. The strength and extensibility test of yarn and its application. Beijing: China Textile Press, 1995.

107 Research on the Constant Tension Yarn Feeder of Weft Knitting Machine

Xiaochuan Bian, Yuan Fang

(School of Materials and Textiles, Zhejiang Sci-Tech University, Hangzhou 310018, China)

Abstract: This paper introduce a method that how the spandex are fed into knitting program with constant tension. In the feeding process, the tension fluctuation mainly comes from two reasons: The first reason is the spandex filament it self's longitudinal vibration; the second reason is the spandex filament formed an air balloon models which generates the tension fluctuations. Constant tension feeder adopts the principle of PID closed-loop system; it can perfectly balance the tension fluctuation of spandex filament and meet the requirement of knitting program.

Keywords: Constant tension feeder; Spandex filament; PID; Closed-loop system

1 Introduction

With the improvement of living standard, people have a higher standard demand of knitting products. In weft knitting, spandex filament can effectively increase the flexibility of the knitted fabrics, especially in swimwear, sportswear, knitting underwear and other products are widely used. However, because of the high tensile stretch(>400%) of spandex in different packages and elastic recovery rate of these physical properties, feeding process will produce a greater tension fluctuation which leads to a serious of the retraction of spandex fabric shrinkage and affects the product's stability of appearance and quality. Therefore, weft knitting machine must be installed the constant tension feeder, it can make spandex to maintain a certain speed and allowable tension variations to make the process of knitting well done.

2 The modeling analysis of spandex filament feeding tension fluctuation
2.1 The influence of spandex filament surface adhesion and friction

According to the literature, Static coefficient of friction depends on the surface of the yarn spool winding shape, but can ignore this part of the adhesion factor because during the unwinding process the tension fluctuation is several times the adhesive force level. The resistance between spandex filament is not strictly tend to vertical pressure[1]. Classic Amontons Law does not apply to the characterization of fiber friction, a formula based on actual production experience shows: $F = aN + bN^c$, simplified form can be $F = bN^c$, b, c is constant. Friction coefficient $\mu = \dfrac{F}{N}$ is not constant; it is a function of vertical pressure.

2.2 Fiber vibration model and Tension fluctuation during feeding process

In the knitting process such as knit swimwear, underwear and other special required product, the main raw material to feed the way of spandex filament, package may be one of many types, a cone, a tube, a dye tube or a spool, depending upon the next operation the yarn must encounter. Conical bobbin unwinding condition makes better production efficiency, one of the three truncated cone-shaped cheese, it's yarn layer shorten the length of sequence from both ends, therefore, in addition to the middle of conical outer ends of cheese was also conical, spool's flaring angle is equal to the bobbin flaring angle[2]. The half of cone-apex angle is 3°30'.

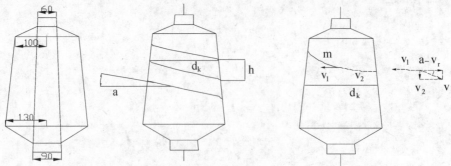

Figure 1 Winding form and take-off speed of cones

α - Angle of spiral;
h - The distance between two corresponding points on a helix;
v_1 - Take-up speed;
v_2 - Liner velocity (reciprocating traverse of the yarn);
From the vector triangle can obtains:

$$v = \sqrt{v_1^2 + v_2^2} \qquad \alpha = \arctan \frac{v_2}{v_1} \qquad (1)$$

To reduce tension fluctuation, v_2 should be changing constantly while the winding diameter has changed. Any layer of the cone near the large end of the take-up speed should large than the other end. In order to stabilize the take-up speed, the v_2 is required lower than the small side of the cone's take-up speed. Fibers under tension can pass along their wave, take a Cartesian coordinate system XYZ, =with Z axis is the static fiber axis, while Z-axis has become the wave of the transmission direction. When the vibration wave propagate direction is as same as the direction of particle vibration, it's called a longitudinal wave. When the vibration wave propagates direction is parallel with XY plane and perpendicular to the Z axis, it's called a transverse wave. Before the feeding, spandex filament also makes axial expansion, that is, the direction of vibration along the longitudinal wave. It as special flexible material, elastic force and friction in the vibration system is a nonlinear change and process as a nonlinear model.

As Figure 2 shows[3], as infinitesimal δz, mass: $\delta m = \rho A \delta z$. is density of spandex filament, A is area of section, z is the original position. The infinitesimal shift away from its bottom place is $\xi(t)$, from its top place is $\xi + \partial \xi$, elastic modulus is E. Infinitesimal section of the initial length under no tension:

$$\delta z_0 = \partial z_0 / (1 + \frac{\sigma_0}{E}) \qquad (2)$$

Tension stress: $\sigma - \sigma_0 = E(E - E_0) = (E + \sigma_0)\frac{\delta \xi}{\delta z} = E_0 \frac{\delta \xi}{\delta z} \qquad (3)$

Equation of motion: $\rho \frac{\partial^2 \xi}{\partial t^2} = E_0 \frac{\partial}{\partial z}(\frac{\partial \xi}{\partial z}) = E_\sigma \frac{\partial^2 \xi}{\partial z^2} \qquad (4)$

Acceleration of Infinitesimal section: $\frac{\partial^2}{\partial t^2}(z + \xi) = \frac{\partial^2 \xi}{\partial t^2} \qquad (5)$

The tension difference value of both ends: $\delta T = A \delta \sigma \qquad (6)$

The general solution is the Z axis direction wave: $\xi = F(z \pm \sqrt{\frac{E_\sigma}{\rho}} t) = F(z \pm v t) \qquad (7)$

The F is decided by initial conditions[3].

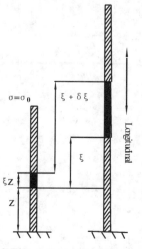

Figure 2 Fiber portrait vibration model

When the filament separates from the cone and feeds to the knitting machine, this trip can be treated as a balloon modeling. This modeling ignores air friction and inertia force. In this process, filament tension fluctuations come from its vertical vibration force and balloon centrifugal force[4].

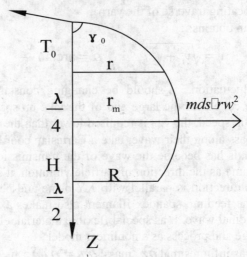

Figure 3 Yarn balloon model

Take infinitesimal $dm = mds$, ignore Coriolis force, air friction and own gravity, so the filament only affords centrifugal force $mds\Box rw^2$ and T_0. The r direction of the force is:

$$mds\Box rw^2 = -d(T_z \frac{dr}{dz}) \tag{8}$$

Set semicircle's apex as the feeding point, according to the yarn mechanics problem. Based on the tested balloon, the vertex angle γ_0 and quarter wavelength can be measured and the tension formula is[5]:

$$T = m(f\lambda_0)^2 = m(f\Box\lambda_r \Box \frac{\pi/2k}{(1-V^2/f^2\lambda_0'^2)\cos\gamma_0 \sqrt{1+k^2}})^2 \tag{9}$$

m -Tex of filament

f -Balloon rotational speed

λ_r -Wavelength of balloon

$\lambda_0 = \sqrt{T_0/m}/f$ -wavelength (T_0 =balloon vertex angle)

$k = \tan(\gamma_0)$

$K = \int_0^{\frac{\pi}{2}} (1-\tan^2(\gamma_0)\sin^2\psi)^{\frac{1}{2}} d\psi$

$R = \frac{d_k}{2}$ -Let-off radius

$\lambda_0' = \lambda_r \frac{\sqrt{1+k^2}}{1-k^2} \frac{\pi}{2k}$

V = Knitting speed

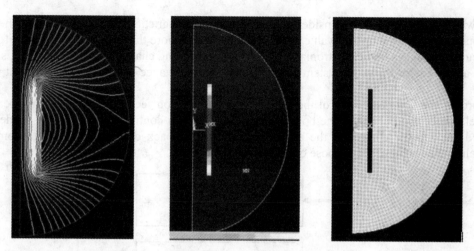

Figure 4 Yarn balloon model in ANSYS

The yarn balloon may suppose to be simulated and modeled in ANSYS. The filament balloon is treaded as a semi-circle model; the filament itself is treated as a rectangle model. The model is divided into tens of thousands of grids to apply loads. From its modeling results, it can plot simulated figures as follows. The filament main deformation is from length direction. In feeding process, the balloon shape and stress is always changing, we should to use rotating coordinate system and set the rotating speed is same as take-up speed. In this model, the balloon shape is steady. With the unwinding, the balloon shape is elongated, the length of friction segment is increased, and the unwinding tension is becoming higher from the last lap of yarn balloon to the first lap. From above modeling and analysis, it can indicates that the in order to control yarn tension fluctuation, it must balance tension fluctuation comes from yarn balloon. Balloon shape, length of friction segment, balloon vertex angle and balloon wavelength also affects yarn tension.

3 Analyses of constant tension feeder and tension force testing
3.1 Analyses of constant tension feeder and automatic control theory model

In order to balance spandex filament tension fluctuations, this installment is needed. It is not only measure the spandex tensions changing range accurately but also speed; tension can be passed the digital-analog conversion so that fluctuations in the value of stability in a small range of permissible to meet the needs of the process.

The main design ideas is putting a pair of piezoelectric sensors in the front of the device, Transmit the electronic signal comes from physical deformation, then guided by the servo motor-driven wheels and yarn storage device, transport filaments to the extremity of the feeder, yarn tension sensor's voltage change will be transmitted to the controller to measure the fluctuation of filaments constantly. Tension damper is must needed, it uses electromagnetic field changes. The changes come from a voltage signal created by piezoelectricity sensors.

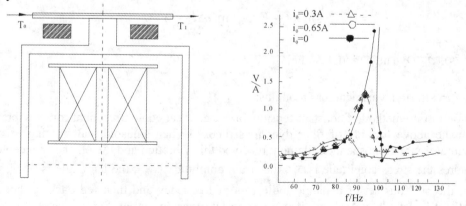

Figure 4 Tension damping and damp-rotor amplitude-frequency characteristic

The wheel has a magnet coil inside; it generates electromagnetic damping applied to the wheel. When the rotor speed exceeds the critical value, amplitude decreased more than 1 times, damper's effect is obvious. With the current changes in the electromagnetic coil corresponding changes have taken place, simultaneously, resonance amplitude of the rotor system changes, and, the resistance applied to the filament changes at the same time.

According the automatic control theory, the constant tension feeder institutions can create PID closed loop model, select and put to use the boundary method. System controlled object is a spandex feed tension will be fed back to the controller, the controller use the feedback data and settings to compare with the original value, then achieve the purpose of tension control[6].

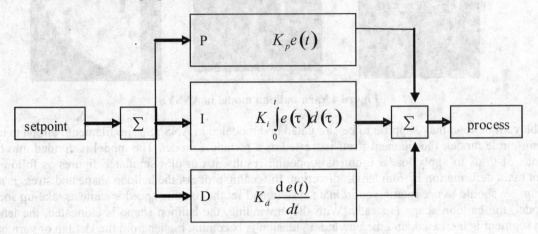

Figure 5 Block diagram of PID control model

3.2 Mechanical properties testing of constant tension feeder

Modal analysis in the actual process only considers the tension in the air balloon and the fiber longitudinal tensile change; it ignores the adhesion between the fiber, the damping force and inertial force affect. Fiber polymer in the dynamic stress and strain, strain always fall behind stress a phase angle as δ. In order to facilitate the theoretical study, it often considered as sinusoidal oscillation, so dynamic mechanical behavior of fiber has the following basic relations:

$$\begin{cases} \sigma = \sigma_0 \sin(\omega t + \delta) \\ \varepsilon = \varepsilon_0 \sin \omega t \end{cases} \quad (9)$$

$$\sigma = \sigma_0 \cos\delta \sin\omega t + \sigma_0 \sin\delta \cos\omega t$$

$$= \varepsilon_0 \frac{\sigma^0}{\varepsilon_0} \cos\delta \sin\omega t + \varepsilon_0 \frac{\sigma^0}{\varepsilon_0} \sin\delta \sin(\omega t + \frac{\pi}{2})$$

$$= E'\varepsilon_0 \sin\omega t + E''\varepsilon_0 \sin(\omega t + \frac{\pi}{2}) \quad (10)$$

$E' = \dfrac{\sigma_0}{\varepsilon_0}\cos\delta$ - Dynamic modulus of elasticity;

$E'' = \sigma_0/\varepsilon_0 \sin\delta$ - Dynamical loss modulus;

The tension testing method of constant tension yarn feeder called non-resonant forced vibration method. It is based on the above formula of fiber dynamics theory. When sinusoidal alternating driving force to promote the vibrator vibration, spandex filament received this vibration and the other end received the stress. When it measures the stress amplitude is σ_0, the strain amplitude is ε_0, the lag angle is δ [7], value can be obtained from the previous set of dynamic indicators of frequency and then fed back to the target system control, adjust to the default. In the end, it makes constant tension output. Sensor passes the signals to the capture card, transport to the computer and the print the result.

In order to verify the validity of the modeling and the effects of the constant tension yarn feeder, it should to test the tension of the feeding yarn. The experimental method is: spandex filament exports from the feeder, pass through a tension sensor chip and goes into the knitting area. Fluctuations induced tension sensor signal, through information collection card, transfer to computer, print results.

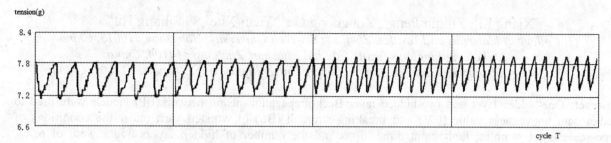

Figure 6 Spandex yarn tension fluctuations range in one cycle

The testing is the tension fluctuations of spandex filaments which fed from constant tension feeder in one cycle. The results show that the yarn controlled by the device, fluctuations range from 7.2g to 7.8g. Through the constant tension feeder, the spandex filament can output with a steady yarn tension, it fully meets the need of knit feeding.

4 conclusions

In knitting process, spandex is always feeding in the knit area as filament. The three truncated conical form of cone is more suitable for feeding. The main reason for the tension fluctuation is the air balloon and spandex longitudinal vibration, adhesion force, air resistance and its own inertia can be ignored. Constant Tension feeder device uses PID closed-loop system to achieve automatic control. The filament tension can be fed back to the controller and achieve the purpose.

If PID control system uses servo technology, it will have better effects. Because servo system can be able to track to the input command signal and obtain accurate position, velocity and power. It may be control the tension fluctuations accurately, such as 0.1g.

References
[1] W D Li, C Y Chu. Textile Physics.Shanghai: China Textile University Publishing Company, 2002:252-254
[2] L S Xu, H R Long. Knitting Technology and Equipment. Shanghai: China Textile University Publishing Company,1998:28-29
[3] R Z Chen. Yarn mechanics. Beijing: Textile Industry Publishing Company, 1989:158
[4] D.G.Padfield.A note on friction between yarn and package. Journal of the Textile Institute, 1955(46):71-77
[5] C J Du, J H Li. The research of tension fluctuation factor of take-off in section warping. Textile Machine. 2009(1):23-26
[6] L B Song, J S li, Y Q Fei. Motion Control System Theory, structure and design.Shanghai:Shanghai Science and Technology Reference Publish Company, 2009:211-215
[7] W D Li, C Y Chu. Textile Physics.Shanghai: China Textile University Publishing Company, 2002:90-98

108 Optimization of Prediction Performance for Worsted Yarn Based on Neural Network

Xiang Li[1], Zhiqin Peng[1], Zongdong Gu[2], Yuan Xue[3], Guoliang Hu[1*]

(1. College of Materials and Textiles, Zhejiang Sci-Tech University, Hangzhou 310018, China
2. Zhejiang Linglong Textile Co., Ltd., Jiashan, Zhejiang 314104, China;
3. College of Garment and Art Design, Jiaxing University, Jiaxing, Zhejiang 314001, China)

Abstract: One-hidden layer and two-hidden layer Back-propagation neural network(BP) models were used to predict both unevenness value (CV) and breaking strength (BS) of worsted yarn under the conditions of large-scale input samples, high input dimensions and the number of hidden layers nodes were of 6,7,8, respectively. And the optimum model in predicting CV or BS was selected to make a comparison with Radial basis function network (RBF) model. The results showed that one-hidden layer BP neural network with 8 hidden layer nodes was the most suitable in the prediction of CV, while two-hidden layer BP neural network with 7 hidden layer nodes was better than others in predicting BS. In addition, both BP neural network models were superior to RBF neural network models in predicting performances.

Keywords: Worsted yarn; One-hidden layer; Two-hidden layer; Back-Propagation; Radial Basis Function

1 Introduction

Mechanism of artificial neural network technology in information processing is similar to that of human beings neural networks for external information processing. Hecht-Nielsen[1] proved that a closed-interval continuously function could be approximated by using a one-hidden layer BP neural network. A BP network with one-hidden layer structure could be used to accomplish mapping from N dimensions to M dimensions. Therefore, high prediction accuracy could be reached by using models with one-hidden layer to predict yarn-related performance with BP neural network technology[2,3]. However, BP neural network models that used by most researchers were of low dimensions in input layers and too simple to use usually. Mustafa E. Üreyen[4,5] used artificial neural network technology and multiple linear formula to model and predict the hairiness, unevenness and tensile of yarn. Owing to the input layer dimensions were up to 14 and two-hidden layer was used, high prediction accuracy was obtained in that study. However, when the prediction is carried under the conditions of large-scale input samples, high input dimensions, it is still not clear whether the one-hidden layer or two-hidden layer BP neural networks has higher prediction accuracy. Few related researches were reported according to our knowledge.

In this paper, both one-hidden layer and two-hidden layer BP neural network models were used to predict performances of worsted yarn in the cases of large-scale input samples, high input dimensions. Optimal models were chosen to compare with RBF neural network models. Correlation coefficients(R) were chosen to evaluate prediction performance of each model.

2 The design of neural networks
2.1 BP and RBF neural networks

Back-Propagation (BP) neural network is a kind of feed-forward neural networks. A BP neural network normally has a multilayer structure including input layer, hidden layer and output layer. A basic working principle of BP neural network is that the signals are forward-propagation while the error is back-propagation. Specific formulae are not further derived after the error is below a pre-determined value or steps stop training, by amending the network weights and threshold values to render the error function and decrease it along the gradient direction.

The following Figure 1 and Figure 2 are one-hidden layer and two-hidden layer BP neural network models with 7 nodes in hidden layer, respectively. BP neural network models with 6 and 8 nodes in hidden layer can be analogized.

---

* Corresponding author's tel:13326139637, email: zisthugl@sina.com

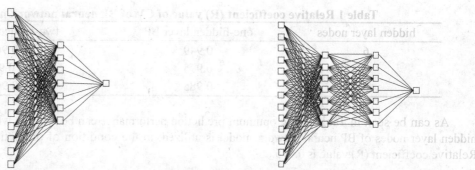

Figure 1 One-hidden layer BP neural network model   Figure 2 Two-hidden layer BP neural network model

Radial basis function network can be called RBF network for short. Its structure and calculation are similar to that of one-hidden layer BP neural network. The distinctive between them is that activation function in the hidden layer is selected Gaussian function. Newrb function in matlab software is used in this paper.

2.2 Experimental materials

Input nodes were indicators got from actual detection in factory and reported in documents[6,7], including fiber diameter, diameter coefficient of variance, mean fiber length, length coefficient of variance, short fiber content, nep, top moisture regain, fore-spinning total draft, roving draft, roving weight, yarn count, yarn twist and traveler number. Output nodes were based on several worsted yarn quality indicators which were concerned by the factory. CV and BS indicators were selected to be predicting in this study.

A total of 85 yarn samples were used, which were all worsted yarns. 60 yarn samples were selected at random as a set for training used artificial neural networks, and 10 yarn samples were chosen as a tested set. Each model should be trained about 15 times so as to obtain one set as predictive values (10 samples), which were closest to achieved values. Then a comparison was made between the two kinds of models. 15 samples were selected as a validated set for models.

2.3 Number determination of hidden layer nodes of BP neural network

When input layer nodes and output layer nodes were determined, more and more scholars were attempting to use an empirical formula provided by Gao daqi [8]:

$$S = \sqrt{0.43mn + 0.12n^2 + 2.54m + 0.77n + 0.35} + 0.51$$

where m represents the number of input layer nodes. In this trial, m is 13; n represents the number of output layer nodes. One output has been used in this trial, and n is 1; S is approximate 7 through the calculation, but 6 or 8 hidden nodes were still need to use in the practical operation. One-hidden layer and two-hidden layer BP neural networks have been used to establish properties prediction models of worsted yarn in the conditions of hidden layer nodes are of 6,7,8, respectively.

2.4 Training parameter setting of models

Training parameters of BP neural network models were set as the follows. Learning rate was 0.02. Accuracy was 0.01. Training steps were 50,000. The length of training step was 50. RBF neural networks' shape parameters were set to be 1.1 (CV) and 1.4(BS) ultimately.

2.5 Establishment methods of models

Matlab7.0 software was used for mathematical modeling. Relative coefficient (R) was gradually applied to reflect the close relationship between predicted and achieved values, so relative coefficient (R) value was utilized to assess prediction performance of two models.

3 Results and discussion

3.1 Establishment of BP neural network models

As can be seen in Table 1, the best prediction performance can be achieved when 1 hidden layer and 8 hidden layer nodes of BP neural network model is used, in the case of predicting CV of worsted yarn. Relative coefficient (R) value is 0.985.

Table 1 Relative coefficient (R) value of CV of BP neural network models

hidden layer nodes	one-hidden layer BP	two-hidden layer BP
6	0.949	0.967
7	0.975	0.97
8	0.985	0.964

As can be seen in Table 2, the optimum prediction performance can be attained when 2 hidden layer and 7 hidden layer nodes of BP neural network model is utilized, in the condition of predicting BS of worsted yarn. Relative coefficient (R) value is 0.975.

Table 2 Relative coefficient (R) value of BS of BP neural network models

hidden layer nodes	one-hidden layer BP	two-hidden layer BP
6	0.951	0.96
7	0.934	0.975
8	0.968	0.935

3.2 Characterization of relative coefficient (R) for performances of neural networks

Relative coefficient (R) is a statistical indicator that reflects close correlation between predicted value and achieved value. And relative coefficient (R) is used to analyze the level of prediction performances of models, which is a common method. As shown in Figure 3, in the process of predicting performances for worsted yarn, one-hidden layer BP neural network models with 8 hidden layer nodes reach the best prediction performances. While two-hidden layer BP neural network models with 7 hidden layer nodes achieve the optimal prediction performances. Analysis from models shows that when information of input layer is processed by a one-hidden layer BP neural network model, more hidden layer nodes are needed to globally analyze and process owing to less training layers. While there are more training layers existing in two-hidden layer BP neural network, so fewer nodes are needed to analyze and process information comprehensively in each hidden layer.

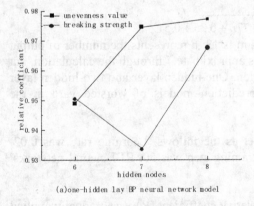
(a) one-hidden lay BP neural network model

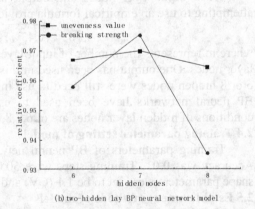
(b) two-hidden lay BP neural network model

Figure 3 Relative coefficient (R) values of two kinds of BP neural network models

3.3 Comparison between RBF and BP neural network

Our previous work[9] proved that BP neural network model had strong fault-tolerance ability under the conditions of large-scale input samples and high input dimensions. Rational prediction results could be inferred in BP neural network and false results would not be attained in the influence of partial interference, which were according to large inputted information. Samples should be screened or abnormal samples should be excluded so as not to reduce prediction performances of models in establishing mathematical model of RBF neural network. Therefore, after abnormal samples were excluded, BP and RBF neural network models were established for both CV and BS of worsted yarn under the conditions of large-scale input samples, high input dimensions, and same accuracy. BP neural network models were found to be the best ones from each kind of model studied in the previous experiments. After screening, number of verified samples was reduced from the original 15 to 14. Relative coefficient (R) value of each model is shown in Figure 4.

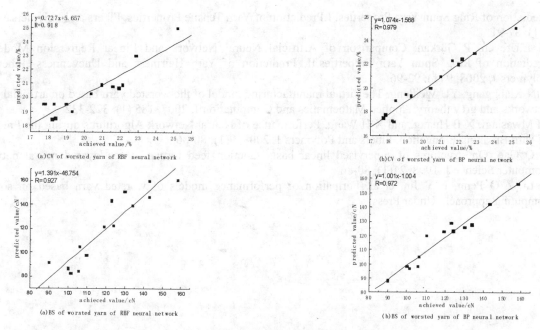

Figure 4 Relative coefficient (R) values of RBF and BP models for CV and BS of worsted yarn after eliminating abnormal samples

It can be seen from Figure 4 that prediction performances of BP neural network model are significantly higher than that of RBF neural network model. Resulting in this phenomenon might be due to activation function in hidden layer of RBF neural network model is partial response for inputted signals. These networks have partial approximation ability and weights are partial adjusted. Error increases between predicted values and achieved values due to the partial adjustment of weights. As for BP neural network model, global approximation approach is used, which means slower convergence rate but overall adjustment of weights. Global approximation approach is one reason for the increase of accuracy. Global approximation approach is one of unique advantages for BP neural network model by which large-scale and high input dimensions are possessed.

4 Conclusions

(1) One-hidden layer and two-hidden layer BP neural network models had been used to predict both the performances of worsted yarn in the conditions of large-scale input samples, high input dimensions and the number of hidden layers nodes were of 6,7,8, respectively. The results showed that one-hidden layer BP neural network with 8 hidden layer nodes was the most suitable in the prediction of CV. Two-hidden layer BP neural network with 7 hidden layer nodes was better than others in predicting BS.

(2) When one-hidden layer and two-hidden layer BP neural network model had been applied to predict the performances of worsted yarn, the former met the best prediction performances with 8 hidden layer nodes, while the latter achieved the optimal prediction performances with 7 hidden layer nodes.

(3) Prediction performances of BP models were significantly better than that of RBF neural network models in predicting CV and BS of worsted yarn.

Ackonwledgement

This work was supported by Major Science and Technology of Zhejiang (subject number 2008C01069-3) and Zhejiang Linglong Textile Co., Ltd.

References

[1] R Hecht-Nielsen. Counterpropagation Networks. Applied OpticsJ, 1987, (26): 4979-4984.
[2] A Majumdar, M Ciocoiu, M Blaga. Modelling of Ring Yarn Unevenness by Soft Computing Approach.Fibers and Polymers J, 2008, 9 (2): 210-216.
[3] A Babay, M Cheikhrouhou, B Vermeulen. Journal of the Textile Institute J, 2005, 96(3): 185-192.
[4] M E.Üreyen, P Gürkan. Comparison of Artificial Neural Network and Linear Regression Models for

Prediction of Ring Spun Yarn Properties. I.Prediction of Yarn Tensile Properties. Fibers and Polymers J, 2008, 9(1): 87-91.
[5] M E.Üreyen, P Gürkan. Comparison of Artificial Neural Network and Linear Regression Models for Prediction of Ring Spun Yarn Properties.II. Prediction of Yarn Hairiness and Unevenness. Fibers and Polymers J, 2008, 9 (1): 92-96.
[6] YIN Xianggang, YU Weidong. The virtual manufacturing model of the worsted yarn based on artificial neural networks and grey theory. Applied Mathematics and Computation J, 2007, 185 (1): 322-332.
[7] J I Mwasiagi, X B Huang, and X H Wang. Performance of Neural Network Algorithms during the Prediction of Yarn Breaking Elongation. Fibers and Polymers J, 2008, 9(1): 80-86.
[8] D Q Gao. On structures of supervised linear basis function feed forward three-layered neural networks. Computer Science J, 1998, 21(1): 80-86.
[9] X Li, Z Q Peng, F Y Jin, et al. Formation of performance models of worsted yarn based on software computing approach. Under Press.

109 Effect of Bending Stiffness on Drape of Soft Weft-Knitted Fabric

Chengxia Liu[1*], Caiqian Zhang[2], Weilai Chen[1]

(1. Zhejiang Sci-Tech University, Hangzhou 310018, China)
(2. Shaoxing University, Shaoxing 312000, China)

Abstract: Drape is one of the most important properties that play essential part in the appearance of fabrics. It has very close relationship with bending stiffness. But very little literature is about the effect of bending stiffness on drape of knitted fabric. In this paper, eighteen soft weft-knitted fabrics were chosen. After testing their drapeability and bending stiffness, correlation and regression were introduced to explain the interrelation of them. From our research, it had been found that the course wise B had much higher correlation coefficient with fabric drape than wale wise B did, except L_0, i.e. course wise B had played more important part in the drape of knitted fabric, such as F, Ac and T. A soft weft-knitted fabric with larger course wise B would have larger F, Ac and T.

Keywords: Fbric drape; Kitted fabric; Bnding stiffness; Crrelation and regression

1 Introduction

Drape is considered to be one of the most important fabric properties, because it is the main contributory factor in designing and retailing garments. The subject of fabric drape has been studied by a number of researchers (Peirce,1930[1]; Chu et al.,1950[2]; Cusick,1968[3]; Vangheluwe and kickens,1993). Hearle et al.(1969[4]) had reported the early work on fabric drape. The most widely accepted method of measuring fabric drape is by using the so-called 'Drapemeter' (Booth, 1968[5]), developed by the Fabric Research Laboratories in Dedham, MA. USA. A circular specimen about 10 in. (254 mm) in diameter is supported on a circular disk of 5-in. (127-mm) diameter.

The drape of a fabric is defined as a description of the deformation of the fabric produced by gravity when only part of it is directly supported. Drape is a unique property that allows a fabric to be bent in more than one direction. Cusick studied the dependence of fabric drape on bending and shearing stiffness[6]. Sufficient literature [7-10] is found on study of drape on fabrics, but very few or scanty information on drape of knitted fabric, especially the effect of bending stiffness on drape of knitted fabric. This paper presented the results of an examination of the effects of the bending stiffness of a series of weft-knitted fabrics on their drape characteristics.

2 Experimental
2.1 Choice of fabrics

Eighteen weft-knitted fabrics were chosen for the experiment, which were made of cotton, viscose, bamboo, modal and tencel. They were all of cellulosic fiber to preclude the influence of fiber content. Since each of the specifications of knitted fabric, such as thickness, weight, yarn density and stitch density, etc, would influence bending stiffness of fabric and these fabrics were mostly used for summer wear, especially suitable for women's dress as it was required of beautiful draping contour. This is the reason why we selected the weft-knitted fabrics with different specifications possessing thin and light structure generally.

Their specifications were as following Table 1.

---

* Corresponding author's email: 907953074@qq.com

Table 1 Fabrics' specification parameters

No.	Content	Density of yarn	Weight of square meter(g/m^2)	Thickness (mm)	Course wise density (number of loops/5cm)	Wale wise density (number of loops/5cm)
1#	M/SP 95/5	32S+30D	215	0.609	110	85
2#	R/C/SP 95/5	30S+30D	280	0.748	130	90
3#	R/SP 95/5	40S+20D	190	0.623	160	90
4#	T/R 65/35	32S	137	0.460	90	75
5#	100%R	32S	200	0.637	100	95
6#	B/SP 95/5	32S+20D	160	0.678	140	85
7#	M/C 50/50	32S	140	0.459	80	80
8#	V/C 50/50	32S	154	0.501	90	85
9#	V/SP 95/5	40S+30D	200	0.577	120	85
10#	C/SP 95/5	30S	230	0.713	110	85
11#	VSP 95/5	32S+20D	210	0.655	110	85
12#	B/SP 95/5	40S+20D	180	0.590	120	85
13#	100%T	40S	150	0.427	90	75
14#	100%M	80S	75	0.273	90	75
15#	100%T	80S	75	0.317	90	90
16#	100%M	32S	65	0.292	55	60
17#	100%T	32S	95	0.425	55	60
18#	100%C	40S	145	0.480	90	85

2.2 Drape testing of the knitted fabrics

Testing instrument: YG (L) 811—DN fabric Static and Dynamic drapemeter
Size of fabrics: circular specimen with the diameter of 24cm
Testing conditions: standard atmosphere
Drape indexes and their formula:

Drape coefficient: $F = \dfrac{(S_t - S_x)}{(S - S_x)} \times 100\%$

S_t — The actual projected area of the specimen; S_x — tbe area of the supporting disk;
S — The area of the specimen.

Lively rate: $L_0 = \dfrac{(F_1 - F_0)}{(1 - F_0)} \times 100\%$

F_0 — static drape coefficient F_1 — dynamic drape coefficient

Beautiful coefficient: $Ac = F \times \dfrac{6}{R_m} \times \left[1 - \dfrac{1}{(1+N)^2}\right] \times 100\%$

R_m — the average radius of projected contour N — Number of ripples

Rigid coefficient: $T = \dfrac{R_m - R_0}{R_0}$

R_m — the average radius of projected contour R_0 — Radius of the supporting disk (6cm)
 Every fabric was tested three times for the drape according to GB-T 23329-2009.

2.3 Testing bending stiffness

Testing instrument: FAST-2 bending instrument
$B = W \times C^3 \times 9.807 \times 10^{-6}$ (bending stiffness B: uN.m; bending length C: mm, weight of fabric W: g/m^2)
Testing conditions: standard atmosphere
Course wise B and wale wise B were both tested.
The testing results of fabric drape and bending stiffness were as in following Table 2.

Table 2 Fabrics' drape and bending stiffness

No.	F	L_0	Ac	T	Course wise B	Wale wise B
01	20.61%	0.61%	16.01%	27.21%	2.005	1.373
02	25.80%	0.67%	19.06%	33.19%	4.745	1.5
03	19.96%	1.08%	15.61%	26.44%	1.954	1.515
04	22.24%	1.35%	18.02%	29.12%	1.277	0.688
05	30.29%	1.27%	21.55%	38.15%	5.007	0.943
06	33.81%	1.37%	23.44%	41.94%	4.305	1.144
07	29.89%	1.02%	21.33%	37.71%	2.79	0.471
08	28.81%	0.96%	20.74%	36.57%	2.95	0.681
09	21.57%	1.62%	16.61%	28.33%	2.731	0.722
10	45.79%	1.78%	28.95%	54.03%	7.869	0.89
11	29.49%	0.38%	21.11%	37.28%	4.701	1.586
12	19.34%	1.27%	15.22%	25.71%	1.287	1.287
13	20.04%	0.83%	15.66%	26.53%	1.703	1.623
14	16.20%	1.19%	13.18%	21.80%	0.252	0.083
15	19.49%	1.28%	15.29%	25.81%	0.401	0.146
16	18.17%	0.96%	14.67%	24.66%	0.512	0.152
17	27.01%	1.45%	19.70%	34.52%	0.539	0.154
18	47.88%	1.12%	29.73%	56.10%	4.486	0.926

3 Results and discussions

Correlation analyzing is a statistical method that is often used to study the relation between parameters. The results of correlation analyzing are correlation coefficient. The bigger the correlation coefficient between the two parameters is, the closer relationship the two parameters have. This correlation method was used to process and analyze the data in Table 2. The correlation coefficients between B (course wise bending stiffness and wale wise bending stiffness) and drape performances were as in Figure 1.

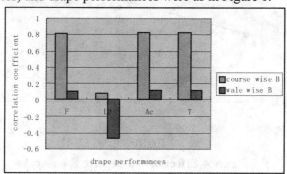

Figure 1 Correlation coefficient between drape performances and B

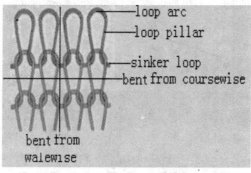

Figure 2 Structure of plain knit

From Figure 1, it could be clearly shown that course wise B had much higher correlation coefficient with fabric drape than wale wise B did, except L_0. That is to say, course wise B had played more important part in the drape of soft weft-knitted fabric, such as F, Ac and T. The correlation coefficients between them were all more than 0.8, while wale wise B did not have so close relation with fabric drape. This result could be explained from the structure of weft-knitted fabric (Figure 2). Loop is the fundamental component of knitted fabric and a loop consists of two loop pillars, a loop arc and a sinker loop. As a result, if the fabric is bent from course wise, the number of loop pillars doubles that of loop arc or sinker loop when the fabric was bent from wale wise. Besides, from the specifications in Table 1, we could see that course wise density was bigger than wale wise density. Therefore, the fabric bent easily in the wale wise direction than in the course wise direction. Probably because of this, the change of B in course wise would have relatively obvious effect on the fabric drape.

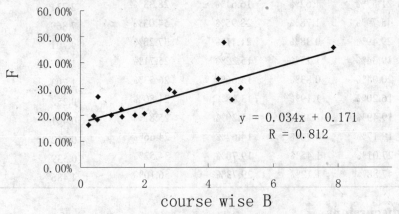

Figure 3 Effect of course wise B

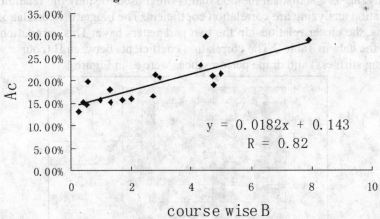

Figure 4 Effect of course wise B on Ac

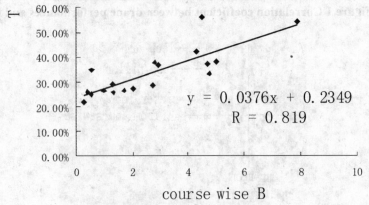

Figure 5 Effect of course wise B on T

To determine the effect of bending stiffness on fabric drape, linear regression was performed as predictors of fabric drape. Equations (1)-(3) showed the regression results with the coefficient of determination (R):

$$F(\text{drape coefficient}) = 0.034 * \text{course wise B} + 0.171 \ (R = 0.812) \quad (1)$$
$$Ac(\text{beautiful coefficient}) = 0.0182 * \text{course wise B} + 0.143 \ (R = 0.82) \quad (2)$$
$$T(\text{rigid coefficient}) = 0.0376 * B + 0.2349 \ (R = 0.819) \quad (3)$$

where B was the bending stiffness.

From the above equations, it could be drawn that with the increase of course wise B, the drape index of F, Ac and T all increase. The more rigid a fabric was, the larger the projected area would be, which resulted in the larger F and T. As for Ac, the possible reason would be that a very thin and soft knitted fabric was bent too easily and had too small F. The projected contour of it would not be so beautiful as the other knitted fabric with moderate bending rigidity. The fabrics chosen for our experiment were all thin and soft (as shown from data in Figure 1). So with the increase of course wise B, Ac also showed the tendency of increasing. As a result, the soft weft-knitted fabric with relatively larger bending stiffness would have more beautiful drape.

4 Conclusions

The literature concerning fabric drape rarely treats the relationship between the bending stiffness and drape of weft-knitted fabric. In our research, correlation and regression were introduced to investigate the effect of bending stiffness on drape of soft knitted fabric. The following conclusions had been drawn: The course wise B had much higher correlation coefficient with fabric drape than wale wise B did, except L0. i.e. course wise B had played more important part in the drape of knitted fabric, such as F, Ac and T. A soft knitted fabric with larger course wise B would have larger F, Ac and T. The relation of course wise B and drape of soft knitted fabric was as following: F (drape coefficient) = 0.034 * course wise B $+0.171$ (R = 0.812); Ac (beautiful coefficient) = 0.0182 * course wise B $+0.143$ (R = 0.82); T (rigid coefficient) = 0.0376 * B $+0.2349$ (R = 0.819) (3).

Acknowledgement

The authors are grateful to the anonymous reviewers for their reading of the manuscript, and for their suggestions and critical comments.

References

[1] Peiree. F.T. 1930. J Text.Inst. 21. 377.
[2] Chu. C.C., Ciimmings. C.L. and Texeira. N.A., 1950. Text. Res. J.20. 539.
[3] Cusick. G.E. 1968.J Text. Inst. 59. 253.
[4] Hearle. J.W.S. Grosberg, P, and Backer. S., 1969. Structural Mechanics of Fibers, Yattis and Fabrics. Wiley-Inlerscience. New York. NY. USA. Chapter 12.
[5] Booth. J.E. 1968. Principles of Textile Testing. Heywood, London, 3rd edition.
[6] Cusick. G.E. 1965.J Text. Inst. 56.596.
[7] Lin H. J Text.Inst. 2009,100. 35-43
[8] Kenkare, Narahar, Lamar, Traci A.M. J Text.Inst.2008, 99. 211-218
[9] Panduranga, Pradeep.J Text.Inst.2008,99. 219-216
[10] Jeong,.Y.J. J Text.Inst. 1998,89.59-69

110 Regression Analysis of the Influences of Cylinder Speed on Drawn Sliver in the Irregularity

Kui Mu, Chongqi Ma
(Tianjin Polytechnic University, Tianjin 300160, China)

Abstract: The cylinder speed had effect on the quality of card web under given weight of feeding; furthermore, it influenced the uniformity of drawn sliver. In order to know the specific relation between cylinder speed and the evenness of drawn sliver, the impact of cylinder velocity on the drawn sliver was analyzed by the single factor analysis of variance. In addition, the relation between cylinder speed and Sashi irregularity of the drawn sliver was also investigated by means of origin75 software. The empirical equation had been established after the analysis on correlation and regression. The result demonstrated that cylinder rate had the conspicuous influence on Sashi irregularity of the drawn sliver and did have the strong linear relevance to Sashi irregularity of the drawn sliver.

Keywords: Cylinder speed; Drawn sliver irregularity; Variance analysis; Regression analysis; Empirical equation

1 Introduction

The irregularity of the drawn sliver, which has a direct relation to the evenness of roving and spun yarn, is one key factor of deciding yarn quality. The yarn irregularity, as an important index of yarn quality, has significant effect on yarn strength and strength unevenness, twist and twist unevenness as well as fabric appearance and products quality [1]. A basic task in the process of spinning is to control drawn sliver uniformity. Additionally, improving the evenness of the drawn sliver is crucial for raising the quality and grade of products.

Measures to improve drawn sliver quality are taken from drawing process at present. Moreover, attentions had been greatly paid to the process of eliminating impurities but hardly in improving the quality of drawn sliver in the carding process. The higher speed is advantageous to enhance carding effect, but increasing cylinder speed blindly will bring about hidden dangers. Doffer output and the quality of card web have strong dependence on cylinder rate under given cotton feeding, thus the uniformity of drawn sliver was also influenced by cylinder rate [2]. Theoretical instructions were given to the improvement of drawn sliver uniformity by using the statistic knowledge in this paper. Impact of cylinder velocity on the drawn sliver was not only researched by the single factor analysis of variance, but also the empirical equation had been established after the analysis of correlation and regression.

2 The reason of the cylinder speed causing sliver unevenness

The cylinder speed affects the quality of card web: the higher cylinder speed is, the better the effect of removing impurity is. It is advantageous for carding ability to increase cylinder speed in a certain scope; the carding frame is removing cotton neps and increasing them at the same time if cylinder speed is too high. These cotton neps in the card web can not be eliminated in the drawing process and finally, they retain in them and further deteriorate the drawn sliver. If the cylinder speed is slow, the time of single fiber carded is longer, so it is beneficial to carding cotton neps and writhen fibers; the cylinder runs slowly and fewer fibers are transferred into the doffer per rotation under given weight of feeding, therefore the fibers brought back by the cylinder is increased, and fibers following per cylinder revolution through carding area are also decreased. Fibers from former rotation left and latter rotation added were uniformly mixed together. The fewer the doffer outputs fibers per rotation, the thinner the card web is, and the more the quantity of card web layer is, so web layers are formed uniformly between the former layer and the latter. So the uniformity of drawn sliver can be improved by the even card web.

3 Experiments

3.1 Experiment equipment

The experiment was carried on by DSCa-01 digital miniature carding frame and DSDr-01 drawing frame, which were developed by Tianjin Jiacheng Machinery Electronic Co., Ltd. The primary controller of the

DSCa-01 digital miniature carding frame was PLC; the touch screen was its man-machine interface; the convertor and servo controller was its motion controller. The process parameters of the DSCa-01 digital miniature carding frame included cylinder speed, the linear velocity ratios between the lickerin and feed roller (note R_{lf}), between cylinder and licherin (note R_{cl}), between cylinder and doffer (R_{cd}), respectively. The process parameters could be set up through the touch screen. The linear velocity ratios were shown in Table 1.

Table 1 The parameters of the carding frame

Parameters	R_{lf}	R_{cl}	R_{cd}
Ratio	3000	3	100

3.2 Experiment raw material and index

The import American cotton was used in this experiment, and the specification of the American cotton was shown in Table 2. The test condition was temperature 20-23℃ and humidity 40%-45%. The cylinder speed range of DSCa-01 digital miniature carding frame was 500r/min to 850r/min. In this experiment, the initial speed of cylinder was 500r/min, and then the cylinder speed increased 50 rotations per experiment. Every 25 grams of cotton were fed into the carding frame, which was repeated 4 times, and finally 4 carding webs were formed. In the breaker drawing, the card web was drawn into the sliver after 5.24 times drawing; then 4 slivers were drawn into one sliver in the finisher drawing through drawing of 7.85 times.

Table 2 The performance of American cotton

Grade	Modal length(mm)	Breaking strength(N/tex)	Maturity Index	Process length(mm)	Micronaire value	Moisture regain percentage (%)
2.85	29	29.5	0.86	30.5	3.87	10.07

The irregularity was tested by Y311 evenness tester for sliver and roving. Each 10 meters of drawn sliver as unit was tested at 20-23℃ temperature and 40%-45% relative humidity.

3.3 Experiment data

The result was listed in Table 3.

Table 3 Experiment data

Experiment numbers	Cylinder speed (r/min)	Irregularity 1 (%)	Irregularity 2 (%)	Irregularity 3 (%)	Average (%)
1	500	16.63	16.06	16.31	16.33
2	550	16.79	17.06	16.26	16.70
3	600	17.34	17.13	17.00	17.16
4	650	16.76	16.93	17.48	17.06
5	700	16.61	17.34	17.84	17.26
6	750	18.22	17.25	17.09	17.52
7	800	17.50	17.28	18.57	17.78
8	850	18.99	17.64	19.00	18.54

4 Data analysis

4.1 Variance analysis

The assumptions ($H_0: \mu_1 = \mu_2 = ... = \mu_r$) were made to the matrix for T test. The data in Table 3 was analyzed by the single factor analysis of variance, and its analysis result was in Table 4. Seeing from analysis result, $F > F_{0.5}(7,16)$, the initial assumptions were rejected, which showed the cylinder velocity had a significant influence on the irregularity of drawn sliver. Keeping the conditions same, the conspicuous change of drawn sliver was caused by cylinder velocity. It was known that cylinder speed had significant impact on the irregularity of drawn sliver by the analysis of variance, but it could not explain the mathematical relationship between cylinder speed and the irregularity of drawn sliver. Therefore, the mathematical relationship between cylinder speed and the irregularity of drawn sliver was researched by regression analysis.

Table 4 The result of variance analysis

Source	SS	df	MS	F	P-value	F crit
Model	9.597733	7	1.371105	4.846749	0.004288	2.657197
Error	4.526267	16	0.282892			
Total	14.124	23				

4.2 Regression analysis
4.2.1 Establish regression model

X (cylinder speed) was a controllably variable factor, and Y (the irregularity of drawn sliver) depending on X was a random factor. The mathematical relationship was $Y = \alpha + \beta x + \varepsilon$ (α, β were constant numbers; ε obeyed normal distribution.) According to experimental data, α, β were calculated by least square regression analysis, and sum of square Q ($Q = \sum_{i=1}^{n}(y_i - \alpha - \beta x_i)^2$) was calculated least. The value of α, β could be computed when first-order partial derivatives of α and β respectively were zero [5].

4.2.2 Hypothesis testing

Hypothesis was $H_0 : \beta_1 = 0$. As a significant level $\alpha = 0.05$, the linear regression could be tested by T statistic. Besides, sample correlation coefficient R could illustrate the effect of the linear regression.

4.2.3 Data analyzing by origin75 software

Correlation coefficient of linear model could be achieved by origin75 software, and the analysis result was shown in Table 5. From the analysis result, statistic $T>T_{0.025}(6)$, so the linear effect was significant. The linear model of cylinder velocity and the drawn sliver irregularity could be expressed by $y = \alpha + \beta x$, where y represented the irregularity of drawn sliver (%); x represented cylinder speed (r/min). Sample correlation coefficient R(R=0.95594) was near to 1, which explained the linear regression was effective. The P value was smaller, and the linear regression model was more accurate. The linear relation between cylinder speed and drawn sliver irregularity was Y=13.73393+0.00527*X, and the regression coefficient was achieved on the same experiment conditions. The fluctuation of the regression coefficients α and β was smaller on the same experiment conditions, but the regression coefficients α and β should be calculated again if experiment conditions were changed [6].

Table 5 The result of regression analysis

Intercept	X coefficient	R	SD	T	$T_{0.025}$	P-value
13.73393	0.00527	0.95594	0.21425	36.922	2.4469	2.06844×10^{-4}

The scatter diagram of cylinder velocity and the irregularity of drawn sliver were shown in Figure 1. Seeing from Figure 1, the relation between cylinder speed and drawn sliver irregularity was linear; the scatter points scattered around and near to the line, and their fluctuation range was smaller. The trend of scatter points was consistent with the regression line, which showed that the regression model was significant. Scatter points conversion would change the trend of the regression line, which illustrated that cylinder velocity had a conspicuous impact on the irregularity of drawn sliver.

It was known from $Y = 13.73393 + 0.00527 * X$ that the cylinder speed was slower, and the regularity of drawn sliver was better within limits of cylinder speed. Under conditions of the low cylinder velocity, the time that single fiber was carded was longer, and fibers could be enough to be carded, so more cotton neps could be eliminated. These were advantageous to improve the quality of card web and the uniformity of drawn sliver.

5 Conclusions

The cylinder velocity was one of important parameters in the carding process. The slow cylinder velocity would make contributions to elimination of impurities in fibers and enforce the carding ability, however, the low cylinder speed would result in the delaying of time as the single fiber was carded, which was beneficial to scattering fibers. More fibers were transferred into the doffer with quick cylinder velocity under the same feeding of cotton, which lead to the worse quality of carding web. In order to illustrate the relations between cylinder velocity and the irregularity of drawn sliver, the mathematical relation was built by statistic knowledge in this paper.

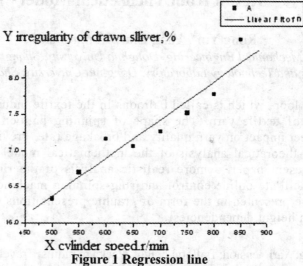

Figure 1 Regression line

(1) The impact of cylinder speed on the irregularity of drawn sliver was studied by the single factor analysis of variance.

(2) Empirical equation between cylinder velocity and the irregularity of drawn sliver was established and the mathematical relation was explained.

(3) Experiment data was obtained and empirical equation was established on the same experiment conditions. Empirical regression coefficients were calculated again if the experiment conditions were changed.

(4) Empirical equation was only used under the same cotton feeding.

References
[1] J L Shao, C J Liu. Processing optimization configuration of enhancing spinning evenness. Cotton textile technology J, 2007, 35(1): 9-12.
[2] P Z Sun. Discussion of cylinder speed of carding machine, Cotton textile technology J, 2006, 34(8):15-19.
[3] X Z Yu, P Z Sun. Test and discussion of relationship between cylinder take-in speed ratio and card sliver quality, Cotton textile technology J, 2009, 37(3): 18-20.
[4] P Zhao. Regression analysis on effect of cashmere properties on pilling rates of its knitted fabric. Wool textile journal, 2009, 37(9): 47-49.
[5] S S Mao, J X Zhou, Y C. Experiment Design. Beijing: China Statistical Publishing Set, 2004, 8-25.
[6] X L Xu, P L Li, Z Z, H Z. Analysis of woolen cashmere yarn evenness, Textile research J, 2009, 30(6): 29-33.

111 The Effect of Ring-spinning Parameters: Result from Theoretical Model

Rong Yin[1], Yang Liu[2], Hongbo Gu[1]*

(1. College of Mechanical Engineering, Donghua University, Shanghai 201620, China
2. Textile Materials and Technology Laboratory, Donghua University, Shanghai 201620, China)

Abstract: Rotating yarn loop, which is called balloons in the textile industry, is the last phase of the manufacturing process of textile yarn. The shape of spinning balloon is closely related to yarn tension which has a direct impact on yarn quality and breakage rate. Previous studies in the literature paid much attention to theoretical analysis of the mathematical model with regard to bifurcation phenomenon. In the present paper, a more realistic analysis of the ring-spinning parameters are discussed in order to facilitate online control and ring-spinning machine design. The results of the numerical simulation are presented in the form of graphic presentations of the balloon shape and of the yarn tension-balloon height dependencies.

Keywords: Ring-spinning; Yarn tension; Balloon height; Package radius; Traveler mass

1 Introduction

Ring-spinning is the principle method for production of high quality yarn from spun fibers. The principle of ring-spinning is shown in Figure 1. After output from the front drafting rollers, the yarn threads through the guide-eye and then forms a spinning balloon between the guide-eye and the traveler. Finally, the yarn is wound onto a bobbin mounted on a driven spindle. When the bobbin and spindle rotates, the yarn drags the traveler gyring along the ring which adds a twist to the yarn.

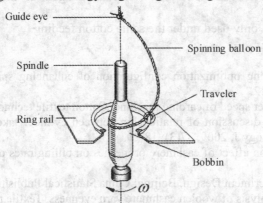

Figure 1 Ring-spinning frame

Research on ring-spinning balloon could be traced back to nineteenth century. A modern ring-spinning frame could be found in Mack[1] and DeBarr[2]. Batra[3] introduced numerical solutions for steady state balloon shapes. Fraser[4,5,6] did much work to theoretical and experimental investigation of spinning balloon. He studied the effect of yarn non-uniformity on the stability of the ring-spinning balloon; the effect of balloon-control rings and the transient solution of the ring-spinning balloon equations for inextensible yarn. Stump[7] have investigated the stability of the quasi-stationary solutions to small velocity perturbations of the yarn loop shape. Zhu[8,9] studied the steady state response and stability of ballooning strings under the influence of air drag and used the Hopf-like bifurcation to predict the limit cycles of ballooning strings. Fan[10] calculated the natural frequencies and mode shapes of a single loop balloon. Skenderi[11] presented a theoretical analysis of average spinning tension based on measurements of two forces at the yarn guide. More recently, Tang et al[12,13,14] estimated the air drag coefficients on ballooning cotton and wool yarns. Yarn tension and balloon shape were obtained by experimental data. Yin[15] derived linear elastic yarn model and investigated the effect of yarn elasticity.

∗ Corresponding author's email: zjh@dhu.edu.cn

However, previous studies in the literature paid much attention to theoretical analysis of the mathematical model with regard to yarn snarling and balloon flutter instability or bifurcation phenomenon which should be prevented from happening for industry production. As a result, it is possible to offer some results useful to help optimization the ring-spinning parameters. The most important parameters for ring-spinning system are: m_T (traveler mass), h (balloon height), b (package radius), T_0 (yarn tension at the guide-eye) and R (balloon radius). So this paper is focused on the above parameters which can be easily used in online control and its machine design.

2 Theoretical model

We adopted previous spinning balloon model[16] in the cylindrical coordinate system. The system has the dimensionless form (a bar over the parameter) as follows:

$$\bar{r}'' = [\bar{r}(\bar{V}_s\bar{\theta}'+1)^2 - \bar{T}\bar{r}\bar{\theta}'^2 + \bar{T}'\bar{r}' - \frac{1}{16}p_{\tau 0}(\bar{V}_s+\bar{r}^2\bar{\theta}')^2\bar{r}' + \frac{1}{16}p_{n0}\bar{r}^3\bar{r}'\bar{\theta}'\sqrt{\bar{r}'^2+\bar{z}'^2}]/(\bar{V}_s^2-\bar{T})$$

$$\bar{\theta}'' = [-2\bar{r}'\bar{V}_s(\bar{V}_s\bar{\theta}'+1) + 2\bar{T}\bar{r}'\bar{\theta}' + \bar{r}\bar{\theta}'\bar{T}' - \frac{1}{16}p_{\tau 0}(\bar{V}_s+\bar{r}^2\bar{\theta}')^2\bar{r}\bar{\theta}' - \frac{1}{16}p_{n0}\bar{r}^2(\bar{r}'^2+\bar{z}'^2)^{\frac{3}{2}}]/(\bar{r}(\bar{V}_s^2-\bar{T}))$$

$$\bar{z}'' = [\bar{T}'\bar{z}' - \frac{1}{16}p_{\tau 0}(\bar{V}_s+\bar{r}^2\bar{\theta}')^2\bar{z}' + \frac{1}{16}p_{n0}\bar{r}^3\bar{\theta}'\bar{z}'\sqrt{\bar{r}'^2+\bar{z}'^2} + \bar{g}]/(\bar{V}_s^2-\bar{T})$$

$$\bar{T}' = -\bar{r}\bar{r}' + \frac{1}{16}p_{\tau 0}(\bar{r}^2\bar{\theta}'+\bar{V}_s)^2 - \bar{g}\bar{z}'$$

with boundary conditions

$\bar{r}(0) = 0$, $\bar{\theta}(0) = 0$, $\bar{z}(0) = 0$, $\bar{\theta}'(0) = 0$, $\bar{z}'(0) = \sqrt{1-\bar{r}'(0)}$, $\bar{r}(\bar{s}_l) = 1$, $\bar{z}(\bar{s}_l) = \bar{h}$.

where $()' = d()/ds$, $\bar{T}$ is the tension in the yarn, $\bar{s}_l$ is the yarn length in balloon, $\bar{h}$ is the balloon height, $\bar{V}_s$ is the yarn delivery speed winding onto the bobbin, $\bar{g}$ is acceleration of gravity, p_τ and p_n are tangential normal air drag coefficients at a material point $\mathbf{R}(r,\theta,z)$, respectively.

The dimensionless equation for the motion of traveler is given by:

$$\bar{T}_1[f\sin\phi - \bar{\theta}'(\bar{s}_l)] = \mu\sqrt{\{\bar{T}_1[\bar{r}'(\bar{s}_l)+f\cos\phi]-M\}^2 + [\bar{T}_1\bar{z}'(\bar{s}_l)-Mg]^2}$$

where $M = m_T/ma$ is the dimensionless traveler mass parameter, $\bar{T}_1$ is the yarn tension on the balloon side of the traveler, $\phi = \sin^{-1}(b/a)$ is the angle between the traveler and lay point makes with the radial direction, μ is the coefficient of friction between the yarn and the traveler and $f = e^{\mu\alpha}$, where α is the angle of wrap.

3 Result and discussion

In the ring-spinning process, the traveler mass is a constant value with regard to different spinning fibers. The balloon height is changed by up and down motion of ring rail and the package radius is increased with more and more yarn winding onto the bobbin, which leads to the variations of yarn tension and balloon radius. In order to illustrate the effect of these parameters more clearly, results have been given for a range of values of traveler mass (M=50-150), package radius (b=0.5-0.85a) and balloon height ($\bar{h}$=4-18) according to Manual of Cotton Spinning[17]. As discussed elsewhere[4-14], this system of equations and boundary conditions have Hopf-like bifurcation phenomenon. For ease our study, figures below give the maximum values of the multiple solutions.

The relationship between yarn tension and balloon height of various values of traveler masses for a fixed package radius are shown in Figure 2(a). For a certain traveler mass (M=100) for example, the yarn tension is increased with increasing balloon height from 4 to 12 and then yarn tension is decreased sharply from 19.78 to 12.68 because of balloon fluttering between single loop and double loop. The Hopf-like bifurcation occurs in this area that the ring-spinning balloon is unstable which should be prevented from happening in the manufacture process expect for ring-spinning having a control ring. After this balloon height, the yarn tension has a small value $\bar{T}_0 = 11.78$ and is still increased with the balloon height increasing. For a designated balloon height, the heavier the traveler mass is, the higher yarn tension is. Various values of traveler mass from 150 to 50, the guide-eye tensions are 16.16, 13.74, 11.36, 9.22 and 7.64 for $\bar{h}=6$, respectively. And the Hopf-like bifurcation is likely to happen in higher balloon height for heavier traveler mass. With regard to control ring, the designers took advantage of a double loop balloon having a lower tension that greatly reduced yarn breakage rate for high-speed spindle machine.

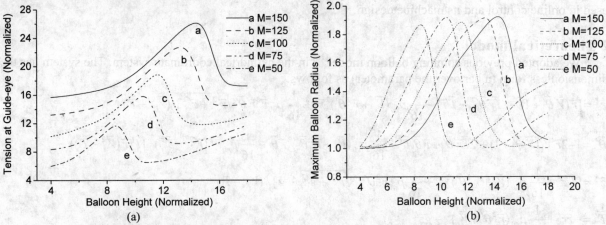

Figure 2 (a) relationship between yarn tension and balloon height (b=0.5a);
(b) relationship between balloon radius and balloon height (b=0.5a)

Balloon radius versus balloon height for the same package radius above is displayed in Figure 2(b). Maximum balloon radiuses of various traveler masses are almost the same that can reach about 2 at different balloon height. The Hopf-like bifurcation points can also be seen that after reaching the maximum balloon radius, all of the curves are decreased rapidly from 2 to 1 with balloon height further increasing. Compared to Figure 2(a), yarn tension and balloon radius have the same trend which can be applied to on-line and real time control. The operators can adjust the traveler mass to ensure high quality production by predicting yarn tension according to various balloon radiuses.

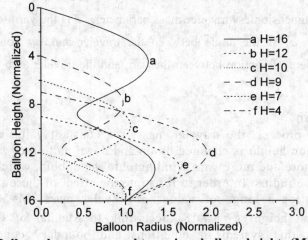

Figure 3 Balloon shapes compared to various balloon heights (M=50, b=0.5a)

Balloon profiles from single loop to double loop are displayed in Figure 3. In fact, some of these situations should lead to yarn breakage in the manufacture process. For double loop balloons, the minimum balloon radiuses are somewhat less than 0.5a which cause yarn balloon touch on the yarn package. On the other hand, if balloon radius exceeds to 1.5, it is likely to interfere other spinning balloon nearby.

Yarn tension is also sensitive to package radius, shown in Figure 4(a). With more yarn winding onto the bobbin, the bobbin radiuses are increased from 0.5a to 0.8a for a fixed balloon height 8 for instance; yarn tensions at guide-eye are decreased by 7.69%, 12.38% and 15.69%, respectively.

Balloon radiuses of various package radiuses are not decreased like yarn tension, displayed in Figure 4(b). Yarn tensions and balloon radiuses have the same trend of different package radiuses which are familiar with the trend of various traveler masses.

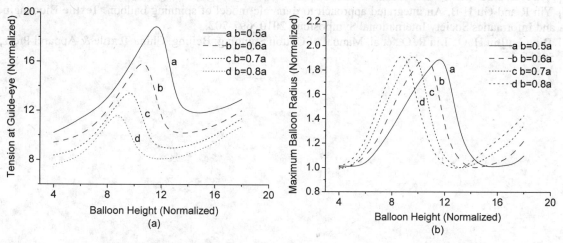

Figure 4 (a) relationship between yarn tension and balloon height (M=100);
(b) relationship between balloon radius and balloon height (M=100)

4 Conclusion

This paper investigates the ring-spinning parameters from theoretical model to present some useful results for online control and ring-spinning mechanism design. Theoretical analysis of the result shows that yarn tension is significantly influenced by traveler weight, bobbin radius, and balloon height. Moreover, it is observed that traveler weight, bobbin radius, and balloon height do have a significant effect on balloon radius. As the balloon height increases, the tension of ballooning yarn experiences a non-linear relationship with various traveler masses and package radiuses. These results also demonstrate that yarn tension and balloon radius are closely related which can be used to predict yarn tensions according to various balloon radiuses. These conclusions provide new insight into the problems of yarn breakage in yarn spinning and facilitate further study of the ring-spinning process.

References

[1]   Mack C. Theoretical study of ring and cap spinning balloon curves. J Text Inst, 1953, 44: 483-498.
[2]   Barr A E D. The role of air drag in spinning. J Text Inst, 1961, 52: 126-139.
[3]   Batra S K, Ghosh T K, and Zeidman M I. An integrated approach to dynamic analysis of the ring spinning process—part II: with air drag. Text Res J, 1989, 59(7): 416-424.
[4]   Fraser W B. On the theory of ring spinning. Phil Trans R Soc Lond A, 1993, 342: 439-468.
[5]   Fraser W B, Farnell L, and Stump D M. Effect of yarn non-uniformity on the stability of the ring-spinning balloon. Phil Trans R Soc Lond A, 1995, 449(1937): 597-621.
[6]   Freser W B, Clark J D, Ghosh T K, and Zeng Q. The effect of a control ring on the stability of the ring-spinning balloon. Phil Trans R Soc Lond A, 1996, 452(1944): 47-62.
[7]   Stump D M, and Freser W B. Dynamic bifurcations of the ring-spinning balloon. Math Engng Ind, 1996, 5(2): 161-186.
[8]   Zhu F, Hall K, and Rahn C D. Steady state response and stability of ballooning strings in air. Int J Non-linear Mech, 1998, 33(1): 33-46.
[9]   Zhu F, and Rahn C D. Limit cycle prediction for ballooning strings. Int J Non-linear Mech, 2000, 35(3): 373-383.
[10] Fan R, Singh S K, and Rahn C D. Modal analysis of ballooning strings with small curvature. J Appl Mech, 2001, 8: 332-338.
[11] Skenderi Z, Oreskovic V, Peric P. Determining yarn tension in ring spinnning. J Text Inst, 2001, 71(4): 343-350.
[12] Tang Z X, Fraser W B, and Wang X. Modelling yarn balloon motion in ring spinning. Appl Math Model, 2007, 31(3): 1397-1410.
[13] Tang Z X, Wang X, Fraser W B. Simulation and experimental validation of a ring spinning process. Simul Model, 2006, 14(7): 809-816.

[14] Tang Z X, Wang X, and Fraser W B. An investigation of yarn snarling and balloon flutter instabilities in ring spinning. J Text Inst, 2006, 97(5): 441-448.
[15] Yin R and Gu H B. Numerical simulation of quasi-stationary ring spinning process, Text Res J. (accepted)
[16] Yin R and Gu H B. An integrated approach to dynamic model of spinning balloon. Textile Biooengineering and Informatics Society International Symposium, 2010, 697-702.
[17] Ding L., Liu H. Q., Liu R. Q., et al. Manual of Cotton Spinning: Beijing, China Textile & Apparel Press, 2004: 592.

112　Repeatability Research of New Fabric Anti-crease Evaluation Method

Caiqian Zhang[1], Chengxia Liu[2]

(1. Shaoxing University, Shaoxing, Zhejiang, China)
(2. Zhejiang Sci-Tech University, Hangzhou, China)

Abstract: A new fabric anti-crease evaluation method was used to test elastic recovery performances of fabric, which can imitate the cloths' actual wear course of people. The four sides' high peaks of five pieces of fabrics under the smooth and creased fabrics were compared for three times, and the elastic recovery rate indicators were calculated. The result shows that the repeatability and stability of the new fabric anti-crease evaluation method can meet the testing requirements.

Keywords: Anti-crease; Elastic recovery rates; Repeatability

1 Introduction

The anti-crease evaluation methods of fabric mainly fall into two categories: subjective methods and objective methods.

The popular subjective evaluation of fabric wrinkle resistant properties is the AATCC Appearance Law [1], which makes use of visual rating method. The reference samples were classified five levels according to AATCC 128-2003[2], and the tested fabrics are compared with the reference samples to determine the anti-wrinkle fabric grades. The test results express as DP. From 1 to 5 levels. In general, if DP value of fabric is more than 3.5, it has a good anti-crease performance. The AATCC Appearance Law is closer use state status to clothing, and it has been widely used. While the method has a clearly lack: DP value is larger, the errors is ever larger, especially the DP value is more than 3.5, because of the inconspicuous differences among the reference samples[3]-[5].

Objective evaluation methods use varieties of instruments and equipments in accordance with recognized standards to evaluate the fabric crease situation. And the most widely used anti-crease method is crease-recovery-angle method, such as GB/T 3819-1997 and AATCC66-2003. The measuring principle of two standard methods are same, and the only difference of the two methods is that the load of AATCC is adjustable, while GB/T 3819-1997 is un-adjustable. KESF measurement method is another method, which refers to Fabric Style Instrument, and can test the fabric anti-crease properties indirectly. But KESF misrepresents the situation of the fabric crease recovery properties because of the shorter time interval during the course of testing. In addition, the results of KESF measurement are the bending stiffness, which is not a direct index for the fabrics' anti-crease performances [6]. As well, the objective evaluation methods have two problems. Firstly, it is difficult to guarantee the accuracy of measurement because of the apparatus and reading errors, Secondly, the measuring methods and actual usage condition of fabric has a certain gap[7]. Actually, the fabric crease situation are usually three-dimensional[8], while the fabric crease-recovery- angle is two-dimensional, so the test results may not be able to accurately reflect the fabric anti-crease performances [9].

2 New Anti-crease evaluation methods

This article introduced a simple and high-precision testing method for fabric anti-crease performances, as following:

Fabric was cut into 20cm diameter circular specimens, then was ironed and spread out on a flat slab. The pictures from the four sides of fabric were taken with cameras.

The circular fabric was put into opening top and canopied bottom (the cylindrical pipe was fixed above the flat slab). A light plastic piston and load (100cN) was placed on the fabric in turn from cylindrical pipe opening for 16 hours which approached one day of dressing on human.

The load and light plastic piston were removed in turn. The lid of cylindrical pipe was taken off, and the fabric fell into the flat slab in the gravity. The pictures were taken from the four sides of fabric with cameras in 30 minutes. The fabric was imaged under pressure and no pressure were contrasted, and the crease recovery performance parameters of fabrics were obtained.

Fabric anti-crease performance testing device is shown in Figure 1. The diameter of cylindrical pipe 1 is 5cm and the height is 8cm. The pipe 1 was fixed on the flat slab through the trestle 9. The lid 2 has a good fastness to the pipe 1 through buckle. The light cylindrical plastic piston 3 has a smooth surface with 4.95cm diameter and 1cm height. The load 4 is 3cm diameter cylinder.

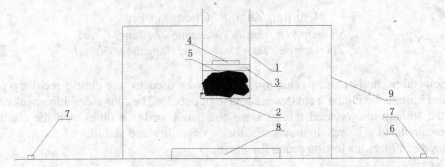

Figure 1 Structure diagram of fabric anti-crease performance testing device
1-cylindrical pipe 2- lid 3- light cylindrical plastic piston 4- load 5- fabric
6- flat slab 7- cameras base 8- test flat slab 9- trestle

Fabric anti- crease performance testing flat slab is shown in Figure 2. The Flat slab diameter is 110cm, and four Canon A710IS digital cameras are on cameras bases 7, which are 50cm away from the center of flat slab. The test flat slab 8 is placed in the center of flat slab with a diameter of 20cm. The trestles 9 set up in the Northeast edge and south-westerly direction of the flat slab in order to the work of photography.

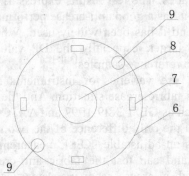

Figure 2 Fabric anti-crease performances testing flat slab

As shown in Figures 3 and 4, 10 represents the five peaks thickness W_0 of the fabrics' images under smooth state (Figure 3) and 10' represents the five peaks thickness W_1 of the fabrics' images under creased state (Figure 4). The images from fabrics' four directions were processed with CAD software and 20 peaks thickness values were obtained. If $\overline{w_0}$ represents the smoothed fabrics average value for 20 peaks thickness values, $\overline{w_1}$ represents the creased fabric(3min) average value for 20 peaks thickness values, then the fabric acute elastic recovery rate represents $\delta = (\overline{w_1} - \overline{w_0})/\overline{w_0}$.

Figure 3 Five peaks thickness for smoothed fabrics   Figure 4 Five peaks thickness for creased fabrics

3 Results

The repeatability and stability of the new testing method for the fabric anti-crease performance was evaluated: Five pieces of fabrics were selected randomly, the same fabric was measured for three times with fabric anti-crease performance testing method, and the average value of 20 peaks thickness was computed, such as in Table 1.

Tables 1 20 peaks thickness average value for five pieces of fabrics (Fabric number as a, b, c, d, c)

Fabric number		a	b	c	d	e
First test	Smooth	0.33mm	0.25mm	0.27mm	0.34mm	0.19mm
	Creased	0.52mm	0.45mm	0.52mm	0.51mm	0.27mm
	δ	57.6%	80.0%	92.6%	50.0%	42.1%
Second test	Smooth	0.34mm	0.25mm	0.27mm	0.36mm	0.18mm
	Creased	0.53mm	0.46mm	0.52mm	0.52mm	0.25mm
	δ	55.9%	84.0%	92.6%	44.4%	38.9%
Third test	Smooth	0.33mm	0.23mm	0.26mm	0.36mm	0.20mm
	Creased	0.51mm	0.42mm	0.51mm	0.54mm	0.28mm
	δ	54.5%	82.6%	96.2%	50.0%	40.0%
Standard deviation of δ (%)		1.27	1.66	1.70	2.64	1.33

As shown in Table 1, the standard deviation of δ for five pieces of fabrics are all less than 2.64%, so the repeatability and stability of the new fabric anti-crease evaluation method can meet the testing requirements.

Then, twenty pieces of fabrics were selected (from 1# to 20#) randomly. And the anti-crease performances were tested with the GB/T 3819-1997 crease-recovery-angle method and the new fabric anti-crease performance testing method, respectively. The results were shown in Table 2 and Table 3:

Table 2 Crease-recovery angles

No.	Warp direction (°)					Latitude direction (°)					Recovery-angle (°)
1[#]	128	135	144	155	140	135	145	130	140	137	278.9
2[#]	160	155	158	145	150	150	143	155	150	148	301.9
3[#]	160	160	155	158	165	177	171	168	169	175	332.2
4[#]	84	90	115	110	95	160	155	140	148	150	253.5
5[#]	160	165	162	163	160	158	153	155	160	160	319.2
6[#]	165	167	163	165	165	167	172	170	170	175	336.1
7[#]	170	172	167	170	171	163	164	163	165	164	333.5
8[#]	160	161	158	160	162	168	165	166	167	165	326.7
9[#]	162	165	165	165	165	138	140	138	136	140	301.7
10[#]	160	165	155	170	160	155	160	170	165	160	324.2
11[#]	135	156	154	153	155	160	165	160	165	163	315.4
12[#]	156	160	160	157	160	151	155	160	160	155	314.9
13[#]	155	160	155	165	160	158	160	163	160	155	318.6
14[#]	158	158	160	158	160	155	158	160	155	158	316.0
15[#]	168	155	168	172	163	165	168	172	170	159	331.8
16[#]	170	172	171	170	171	165	162	160	165	162	333.3
17[#]	150	155	153	151	152	155	150	150	155	152	304.8
18[#]	158	150	164	155	162	150	155	162	164	158	315.6
19[#]	160	162	160	163	160	155	155	158	152	153	315.4
20[#]	152	150	145	150	148	113	127	128	120	115	267.9

Table 3 Fabric elastic recovery rates

No.	Smooth	Creased	δ	No.	Smooth	Creased	δ
1[#]	0.32mm	0.45mm	40.6%	11[#]	0.32mm	0.37mm	15.6%
2[#]	0.28mm	0.35mm	25.0%	12[#]	0.26mm	0.31mm	19.2%
3[#]	0.18mm	0.22mm	22.2%	13[#]	0.50mm	0.62mm	24.0%
4[#]	0.11mm	0.17mm	54.5%	14[#]	0.16mm	0.20mm	25.0%
5[#]	0.28mm	0.34mm	21.4%	15[#]	0.18mm	0.21mm	16.7%
6[#]	0.17mm	0.19mm	11.8%	16[#]	0.41mm	0.47mm	14.6%
7[#]	0.16mm	0.18mm	12.5%	17[#]	0.35mm	0.43mm	22.9%
8[#]	0.20mm	0.24mm	20.0%	18[#]	0.29mm	0.35mm	20.7%
9[#]	0.31mm	0.40mm	29.0%	19[#]	0.70mm	0.86mm	22.9%
10[#]	0.36mm	0.41mm	13.9%	20[#]	0.34mm	0.52mm	52.9%

A coordinate axis was built up with crease-recovery angles as x-axis and the elastic recovery rates as the y-axis, and the linear relation equation is y = ax + b, as shown in Figure 5.

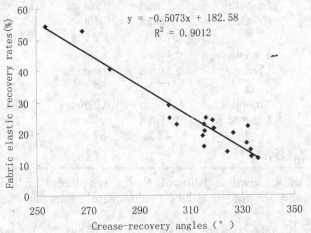

Figure 5 The linear relation between crease-recovery angles and elastic recovery rates

As is shown in Figure 5, the fabric crease recovery angles are larger, the elastic recovery rates are smaller and the anti-crease properties are better. And the correlation coefficient between the two variables has reached 0.9012, which shows that the new anti-crease performance testing device is an accurate measurement to fabrics anti-crease performance.

4 Conclusions

In this paper, the repeatability, exactness, and stability of the new fabric anti-crease performance testing method was analyzed based on image analysis. The result shows that the new method can meet the testing requirements, and has an accurate measurement to fabrics' anti-crease properties.

While this method still needs further improvement, such as anti-crease performance test system software and the load selected to meet the different thickness of fabric. And all these issues need to be resolved through further study.

Acknowledgement

This support are granted from the resources of Analyzing and Testing Foundation of Zhejiang Province (2009F70022), The authors are also grateful to the anonymous reviewers for their reading of the manuscript, and for their suggestions and critical comments.

References
[1] AATCC 128 Wrinkle Recovery of Fabrics: Appearance Method
[2] AATCC 66 Wrinkle Recovery of Woven Fabrics: Recovery Angle
[3] GB / T3819-1997. Wrinkle Recovery of Textiles: Recovery Angle
[4] George K. Stylish. New measurement technologies for textiles and clothing. International Journal of Clothing Science and Technology, 2005,(17): 135- 149
[5] Aroon Shenoy. A dynamic oscillatory test that fulfills the objective of the elastic recovery test for asphalt binders. Materials and Structures. 2008(7): 1039-1049
[6] Mohamed Hashem, Nabil A. Ibrahim, Wfaa A. El-Sayed, etc. Enhancing antimicrobial properties of dyed and finished cotton fabrics. Carbohydrate Polymers. 2009(10):502-510
[7] Moustafa M.G. Fouda, A. El Shafei, S. Sharaf and A. Hebeish. Microwave curing for producing cotton fabrics with easy care and antibacterial properties. Carbohydrate Polymers. 2009(9):651-655
[8] Ma Chongqi, Zhao Shulin, Huang Gu. Anti-static charge character of the plasma treated polyester filter fabric. Journal of Electrostatics, 2010(2):111-115
[9] Amira El Shafei, S. Shaarawy, A. Hebeish. Application of reactive cyclodextrin poly butyl acrylate preformed polymers containing nano-ZnO to cotton fabrics and their impact on fabric performance. Carbohydrate Polymers, 2010(5): 852-857

113 Embedded Microcontroller Based Flat Knitting Machine Controller Design

Zhang Hua*, Xudong Hu, Yanhong Yuan, Xianmei Wang, Laihu Peng, Jianyi Zhang

(Faculty of Mechanical Engineering & Automation, Zhejiang Sci-Tech University, Hangzhou 310018, China)

Abstract: There are kinds of flat knitting machines; they combine basic functions like jacquard, yarn color selection, knit loop traverse, etc. These machines can be used to knit sweaters or flat pieces of clothing. In this paper, authors design a distributed control based controller. In the controller, individual functional modules are designed to perform special functions. Every individual functional module is a CAN node of the controller. CAN are used to transfer signals and commands between nodes. The control system, based on distributed controls and modularizing designs, are easy to design and can meet the needs of reliability and real-time control with low cost.

Keywords: Flat knitting machine; CAN; Distributed control

0 Introduction

The flat knitting machine is a kind of weft knitting machine, which can be used to knit sweaters, flat pieces of clothing, collars etc. The fundamental configuration of knit fabric is knitting texture. It is manufactured by a series of flat cam systems called triangles. Triangles are mounted on the carriages of the flat machine. Fabric texture changing has to replace triangles. So it is time consuming when changing knit fabrics. With the application of electronic technology in the knitting industry, it raises the automation of the knitting technology. Many knitting processes can be executed automatically. Computer aided fabric design, electronic yarn selection, yarn tension, knitting tension controls, electronic patterns, and jacquard devices. All of these electronic device applications extend knit fabric from decorate to industry fields[1].

There are many types of controllers for flat knitting machines. Some of them are designed based on the 8-bit Microcontroller with LED display and simple keyboards. They are limited in function and cannot satisfy the incremental requirements of modern knitting machines. There are also some of types that work with PC and Windows operating systems. Those systems could be programmed human-machine interface friendly and beautifully. They can also realize all functions of knitting. But they are costly and could not support system well at real-time. Recently the embedded microprocessor and the embedded digital signal processor have been widely applied in many areas. They are proved to be good at machine controller. Thus for the knitting machine there are more selections for its control systems.

1 The design scheme of the control system of the flat knitting machine

Different flat knitting machines have different functions to some extent, though the main functions are similar. The SE808 flat knitting machine control system, which we designed, needs following configuration. There are: 30 channels of opto-isolated digital output, 10 channels of opto-isolated digital input, 5 sets of step motors driven output, 1 set of AC motor driven output, 1 set of DC motor driven output, a USB port for pattern files transfer, Internet enabled, large pattern memory, perfect human-machine interface with LCD display and 24-key board.

Considering of the requirements mentioned before, it looks hard to design the control system based on single 8-bit Microcontroller or DSP. Meanwhile, PC plus MCU architecture will not only raise real-time problems, but also tend to raise cost. In our design, authors adopt the concept of distributed control system. Individual MCUs are used to perform individual tasks. A 32-bit MCU with embedded operating system is used to interface with people and the Internet. All MCUs are connected with star network LAN. The main embedded module is based on 32-bit ARM microprocessor. It communicated with subordinates via CAN bus, and configured Internet interface to reach remote date access. It's also configured with a USB port on it. The subordinate 8-bit MCUs were used both in the jacquard control module and the motor control module. Figure 1 is the whole scheme of the SE808 flat knitting machine control system.

---

* Corresponding author's email: zh0121@yahoo.com.cn

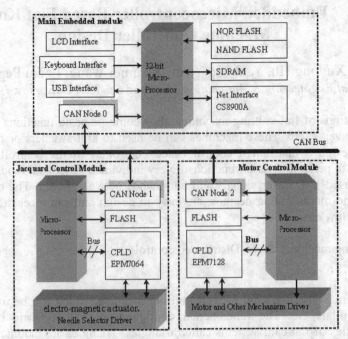

Figure 1 Scheme of SE808 flat knitting machine control system

In this scheme, the main embedded module is responsible for storing and interpreting pattern files and sending distributing control commands to the related MCU. It also performs LCD display and keyboard input. During knitting, the Internet access is turned off. Thus the main embedded microprocessor is only used for pattern file reading, interpreting, and sending to subordinate modules. A Linux OS is embedded to the microprocessor, which improves system's adaptive, reliable, and real time performance.

In this control system, digital I/Os are not only used for controlling electro-magnetic actuators, to realize needle selection and yarn selection, but also for sensing carriage position and others protections. The digital I/Os in this system are much more important, designers take many ways to improve its reliability and efficiency such as opto-isolate in each channel. The critical improvement is to use CPLD to pre-process the signals.

The speed control of individual motors, especially the step motor is also important. Software PWM and accelerate/decelerate techniques will consume a lot of computer time. In this design, again the CPLD is used to finish these functions. It will of course improve the performance of motor controlling.

Figure 2 is the information flow of a flat knitting machine. The CAD system prepares the pattern image and transfers to the pattern data files. According to this data, it interprets individual commands that can be recognized by machine controllers. The pattern data files can be passed to embedded microprocessor modules directly via Internet or downloaded from disks and USB devices. The data file includes machine control information and fabric information. The fabric information involves the number of knitting systems, the number of electro-magnetic actuators, the number of yarn carries, the initial positions of yarn carries, the fabric width, length and the number of total needles and rows. The machine control information refers to the control of every row. It includes the all control commands relating to needle selection, yarn selection, carriage movement, needle bed traverse, step motors running for stitch controlling, jacquard and any special technique requirements. All kinds of information are stored in a definite data form.

The embedded microprocessor can display pattern images and states on LCD according the pattern file. Users can edit the pattern information through the keyboard. After the user confirms the states of all preparation, he can turn on the knitting machine. During knitting, the embedded microprocessor module communicates with the jacquard control module and the motor control module via CAN Bus[2]. Subordinate MCU modules receive the commands or data from embedded microprocessors and control actuators via CPLD. Meanwhile it detects the states of actuators and sends the testing data back to the main controller. If any malfunction takes place, the main controller will react immediately and display the error message on the LCD.

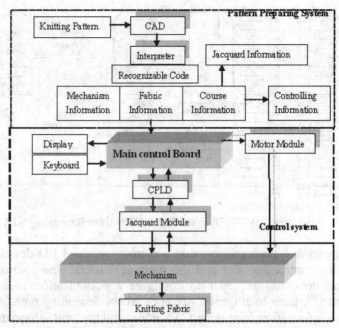

Figure 2 The information flow of flat knitting machine

2 CAN interface design and its realization

Controller Area Network (CAN) protocol was firstly developed in 1980s' by the German company, Bosch[3]. At that time, as a kind of serial communication protocol, it was used for data exchanging among numerous control and measurement instruments of modern automobiles. CAN protocol is based on ISO/OSI models. To achieve design transparency and implementation flexibility, CAN has been subdivided into three layers. They are: physical layer, transfer layer and object layer [4]. It transmits signals through twisted-pairs. Its transfer speed can reach 1Mbps, and the communication distance can range to 10km/5kbps. Now the application areas of CAN range from high speed networks to low cost multiplex wire [5]. Most modern microprocessor families now have embedded on-chip support for CAN protocol.

CAN bus has an arbitration mechanism to resolve the bus access conflicts. In order to achieve the utmost safety of data transfer, powerful measures for error detection, signaling and self-checking are implemented in every CAN node. CAN nodes are able to distinguish short disturbances from permanent failures. Defective nodes are switched off [4]. Thus the communication based on CAN achieve utmost anti-jamming. Meanwhile in CAN, system Nodes can be added to the CAN network without requiring any change in the software or hardware, thus making the system flexible. It is important in flat knitting machine control systems because the system can be reconstructed by adding or taking off the control modules depending on the type of flat knitting machines.

In this design, we adopt the SJA1000 flat knitting machine controller. The SJA1000 is a stand-alone controller for CAN used in general industrial environments. It is the successor of the PCA82C200 CAN controller (BasicCAN) from Philips Semiconductors. The chip can be interfaced with a variety of microprocessors. In this control system, the jacquard control module and the motor control module are two CAN nodes. The two modules are designed based on MCU W77E58. MCU W77E58 is extended a CAN interface using SJA1000. Figure 3 illustrate the design of CAN interface.

541

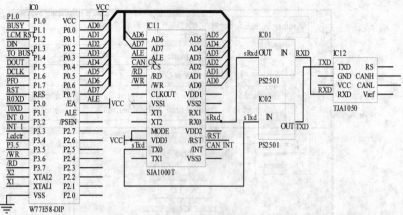

Figure 3 W77E58 extended a CAN interface using SJA1000

The SJA1000 appears to a microprocessor as a memory-mapped I/O device. The address area of the SJA1000 consists of the control segment and the message buffers. The control segment is programmed during an initialization download in order to configure communication parameters, such as its work frequency and Baud rate [6]. Each intelligent CAN node works depending on SJA1000 in an interrupt way. Each module can obtain information from other CAN nodes and transmit information to others. The program relating the CAN node consists of three sects: the SJAT1000 initialization, the program sect about receiving messages and the program sect about transmitting messages. The part of program with C language is written as follows:

```
    #include "CAN51.h";                    //Defin1tion of CAN Communication register
    uchar can_init(void)
    {
    ……                                     //Character define
        for(i=0;i<13;i++)     { *p++=0;  } //Buffer reset
        XBYTE[iCan_CR]=byte;              //reset mode
         XBYTE[iCan_CDR]=0xC8;            //PeliCAN mode
            XBYTE[iCan_ACR0]=0x88;        //set to single filter   mode
         XBYTE[iCan_BTR0]=0x00;           //16M crystal frequency
         XBYTE[iCan_BTR1]=0x14;           //1M transfer speed
         XBYTE[iCan_OCR]=0x1A;            //normal output mode
    XBYTE[iCan_IER]=0x1;                  //receive interrupt admit
    ……
    }
    uchar can_send(void)    interrupt 1   //send data
    {
    ……
    }
    void can_rec(void)      interrupt 2   //receive data
    {
    ……
    }
```

Following the standard of CAN communication, designers use three communication modes for command data, knitting course data and motor control data transfer individually. The command data protocol is as "command head flag + command + parameters + command end flag". The command data includes file information, pattern information, mechanism configuration; yarn carries information, knitting density data etc. There are a mass of knitting course data and the motor control data to be transferred. The same protocol, which is as "the number of data bytes + data segment", is used to transfer those two kinds of data. This protocol means the length of data can be different. It is useful for real-time control.

3 System achievement and conclusion

When the system has been constructed, the CAN protocol and the communication ability of CAN nodes were tested in the lab using ZLGCAN Test. Only correct initialization of CAN nodes can pass this test. After testing and making some design improvements, the control system was embedded in a flat knitting machine to run in a real circumstance. The flat knitting machine runs at knitting speed of 1.2m/s. The whole control system has been running successfully for three months. Figure 4 is a picture taken from the workshop.

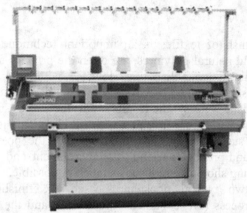

Figure 4 A picture of flat knit machine using SE808 controller

This study has investigated the use of distributed microprocessors to develop CAN-based systems. This kind of control system is easy to design and can meet the needs of reliability and real-time control.

Acknowledgement

This project was supported by The Nature Research Fund of Zhejiang Province (No. D1080780).

References

[1] Zhang Hua, Hu Xudong, The Study of the Computerized Flat Knitting Machine Control System Based on Embedded Linux, Journal of Zhejiang Sci-tech University, 2006.23.03:268-270.
[2] Rao Yuntao, Zhou Jijun, Zheng Yong yun, Field Bus CAN Principle and application, Beihang University Press, 2003.6:137-166.
[3] Li Jinggang, Liu Yonghong, CAN Intelligent Node design based on AT89C51, Foreign Electronic Elements, 2006, 8: 26-29.
[4] R.G. Bosch, CAN specification version 2.0: Robert Bosch GmbH, Postfach 50, D-7000 Stuttgart 1, Germany, 1991.
[5] J.P. Thomesse, A review of the fieldbuses, Annual Reviews in Control 1998, 22: 35–45.
[6] Philips, SJA1000 stand-alone CAN controller, 2004.

114　A Durable Flame-retardant Finish for Cotton

Bin Fei, Zongyue Yang, John H. Xin[*]

(Institute of Textiles & Clothing, The Hong Kong Polytechnic University, Hong Kong, China)

1 Introduction

Flame-retardant (FR) finish of textiles is an important technique to prevent fire disaster. Textiles, including various synthetic and natural polymers, are often the primary source of ignition and contribute to rapid fire spread due to their highly ignitable nature. However, without textiles, it will be impossible for daily human activities and for the surroundings, it will not have the desired aesthetics. Currently, it is relatively easy to impart flame retardancy to synthetic fibres in the fibre spinning process via the blending incorporation techniques. However, a post-finishing with FR chemicals is necessary for natural fibres. With the increasing awareness of sustainability and diminishing of the potential pollution, it is necessary to develop finishing chemicals and processes friendly to environment and people. The carbon foot-print and chemical toxicity in FR finishing should be reduced as much as possible.

Nowadays, there are two typical phosphorus-based FR finishing systems in the market: the "pre-condensate"/ammonia process (brand named "Proban") [1] and the reactive phosphorus process (e.g., N-methylol dimethyl phosphonopropionamide (MDPPA), brand name "Pyrovatex") [2], which are suitable and wash-durable to cellulosic substrates. However, both systems have problems of formaldehyde release and other shortcomings such as stiffness for the treated fabrics. A new generation of green and durable FR finishing system for cellulosic substrate has great demand by users and manufacturers.

In this work, we developed a novel FR finishing product named Neo-FR, composed of two components, which can provide highly efficient FR protection and durable against multiple-washing. The process includes the separate and consecutive application of chemically reactive crosslinker and phosphorus-based FR compound to the fabric. The first step is dip-pad water-soluble crosslinker (small molecules, oligomers, or polymers with three or more reactive groups), followed by drying. In the second step, phosphorus-based compound with multi-functional groups is padded onto the fabrics which can react and condensate with the pre-applied crosslinker forming insoluble networks within the fiber matrix (Figure 1). This two-pass process fixes sufficient phosphorus to the cellulosic fabric to impart a FR function which is durable to multiple launderings. Application of the phosphorus-based compound by increased amounts leads to a higher FR performance at the expense of hand feel. The properties of finished fabric were compared with that of Pyrovatex CP New.

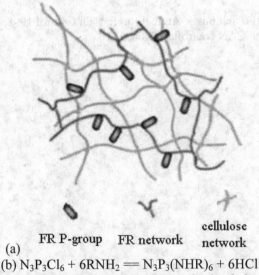

(a) FR P-group　FR network　cellulose network

(b) $N_3P_3Cl_6 + 6RNH_2 == N_3P_3(NHR)_6 + 6HCl$

Figure 1 (a) FR network within fiber matrix; (b) synthesis of phosphorus compound

---

[*] Corresponding author's email: tcxinjh@inet.polyu.edu.hk

2 Experiments

Twill cotton fabrics (from China Dye Ltd., Hong Kong) were finished in a two-step way: the crosslinker solution was applied onto cotton fabric by padding (2.75 kg/cm^2, with a wet pick-up 70 wt%) and drying (80°C, 5 min); then the fabric was padded again with the FR compound solution (synthesized as described in Figure 1), and left to dry at room temperature. The same fabric is also treated by Pyrovatex CP New composition (according to the guideline from supplier) for comparison.

The FR cotton fabric after finishing was measured by a thermo-gravimetric equipment (TG, NETZSCH STA 449C, air atmosphere, 10°C/min). Its tensile strength was tested using a tension instrument Instron 4411 according to ISO 13934.1: 1999.

Flammability assay of cotton fabrics was performed according to the ASTM D6413-99 standard (vertical test). Before the vertical flammability test, all samples are dried, and conditioned overnight at 20°C, relative humidity 70%. The FR cotton fabrics after finishing and burning were also observed under a scanning electron microscope (FE-SEM, JEOL, JSM-6335F, at 3 kV). The washing of FR fabrics is processed according to the standard AATCC 61-2003, using an AATCC standard wash machine (Atlas Launder-Ometer) and the AATCC Standard Detergent WOB (9 minutes = 1 cycle).

3 Results and discussion

After FR finishing, the fabric by our FR composition showed a higher tensile strength and extension than that by commercial Pyrovatex CP New in both warp and weft directions, although there is also a visible reduction in comparison to the control fabric (Table 1). The better retain of mechanical properties by Neo-FR finishing is benefited from the low temperature finishing process and mild aqueous reaction.

Table 1 Tensile measurement results of fabrics: control, that treated by Pyrovatex CP New, and that treated by our FR composition

Samples	Warp Broken Strength	Weft Broken Strength	Warp Tensile Extension	Weft Tensile Extension
Control Fabric	941.47N	483.08N	32.72mm	30.19mm
Pyrovatex CP New	816.36N	417.49N	24.80mm	27.14mm
Neo-FR	878.04N	475.92N	29.82mm	29.14mm

The FR finishing significantly increased the char residue of cotton fabric during thermal degradation, as confirmed by the TG results (Figure 1). Pure cotton fabric rapidly loses its 80% mass around 350°C, and completely loses its mass at 500°C. The thermal decomposition of cotton around 350°C produces volatiles including both combustible and non-combustible species. This drastic process in air generates a flame combustion phenomenon. The pyrolysis at higher temperature generates a smoldering phenomenon, leading a slow mass loss [3]. After finishing by our Neo-FR composition, the temperature, at which the weight loss reaches 5%, is clearly lowered by nearly 100 degree. While a significant increase in char formation above 500°C compared with the control cotton is also noticed. This enhancement occurs in a region where char oxidation is normally occurring and previous research shows that enhanced char formation over this temperature region is synonymous with improved FR performance [4].

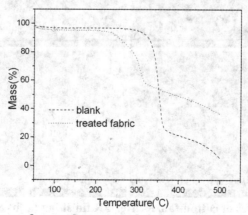

Figure 1 TG curves of control cotton and that treated by our FR composition

The flammability of finished fabrics was tested by the vertical test. The apparatus and test residue were showed in Figure 2. The FR property of treated fabrics is quite well without after-flame and after-glow, which has a char length of only 76 mm (Table 2). And the fine FR property is durable against machinery washing even after 30 cycles, where is only a little increment in the char length. This durable FR performance is comparable to the data from Parovatex CP New. These results confirm the highly FR efficiency of the Neo-FR composition, which are in good agreement with the TG results.

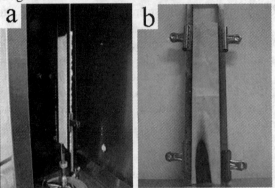

Figure 2 Vertical flammability test equipment with attached frames:
(a) the fixed sample frame and lighter; (b) treated fabric after test

Table 2 Vertical flammability test results of cotton fabrics treated by our FR composition and washed for different periods

Washing cycles	After-flame time (s)	After-glow time (s)	Char length (mm)
0	0	0	76
20	0	0	103
30	0	0	110

The FR cotton fabrics were further observed under SEM. As showed in Figure 3 (a, b), there are not thick coatings over cotton fibers, indicating that most FR chemicals have penetrated and resided in the fiber matrix. The wearer-friendly surface property of cotton fibers are thusly well retained after the FR finishing. While after burning, swollen membranes and blown balloons appear between the fibers and over the fiber surfaces (Figure 3 - c, d). This morphology strongly supports an intumescent FR mechanism [5], and well explains the high efficiency of the Neo-FR product.

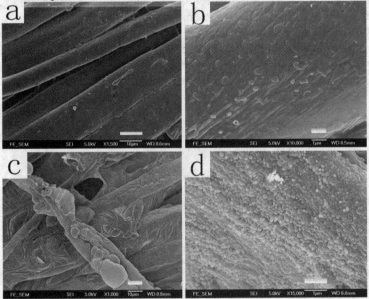

Figure 3 SEM images of cotton fabric after FR finishing (a, b) and burning (c, d);
scale bars are 10 ìm in (a, c) and 1 ìm in (b, d)

4 Conclusions

In summary, our Neo-FR composition can provide an efficient and durable FR property to cotton fabrics by a low temperature finishing process. In addition, this finishing well retains the mechanical property of cotton fabrics. Therefore, the Neo-FR composition is quite superior to the commercial product Parovatex CP New in the view of sustainability.

Acknowledgement

The ITF Funding ZR06 and Research Grants Council GRF project (PolyU5309/07E) of the Hong Kong SAR government are acknowledged.

References

[1] U.S. Pat. 4,494,951; 4,078,101; 5,238,464.
[2] Germany patent DOS 2,136,407.
[3] Chen Y, Frendi A, Tewari SS, Sibulkin M. Combustion properties of pure and fire-retarded cellulose. Combust Flame 1991, 84, 121-40.
[4] Kandola BK, Horrocks S, Horrocks AR. Evidence of interaction in flame-retardant fibre-intumescent combinations by thermal analytical techniques. Thermochim Acta 1997, 294, 113-25.
[5] Zhang S, Horrocks R. Substantive intumescent flame retardants for functional fibrous polymers. J Mater Sci 2003, 38, 2195-8.

115 Influence of Extension on the Side-glowing Properties of POF

Rui Wu, Jinchun Wang, Bin Yang*

(Key Laboratory of Advanced Textile Materials and Manufacturing Technology, Ministry of Education of China, Zhejiang Sci-Tech University, Hangzhou 310018, China)

Abstract: In this paper two kinds of polymer optical fiber (POF) were stretched with different elongation at different temperature to investigate the effects of axial tension on the side-glowing properties of POF. Luminance colorimeter BM-5A was used to measure the side-glowing luminance of POF after stretching, and the surface morphology of POF was characterized by SEM. The results showed that light emitting fabric with good brightness and uniformity could be prepared by virtue of the extension during weaving process.

Keywords: The side-glowing properties; POF; SEM; BM-5A; Axial tension

1 Introduction

Smart textiles with photoelectric came into being due to the traditional textiles can no longer meet people's needs with the development of technology. Among which a new type light emitting fabric with various color and decorative features begin to attract much attention. There are several ways to prepare light emitting fabric such as combined polymer optical fiber (POF) with common textile yarn by paste, implant or weaving and then through hot melting, Spotting, cutting, Sandblasting, lapping or laser to broken the surface of POF in the light emitting fabric [1-6]. There are some disadvantages of these methods as they are complex and difficult to control the brightness of light emitting fabric. Our group had already prepared the light emitting fabric using optic fiber as weft weaving with common yarn through reagent coating the optic fiber on the fabric surface [7-10]. There are some defects in this processing due to the interference of other lines in the operation and potentially damage to other yarns so that light damage to the fabric pattern. We discovered that part of the transmitted light to be leaked out from the crack which appeared in the surface of POF due to the tensile skin-core interface damage, and this phenomenon can be used to prepare light-emitting fabric. In this paper POF was stretched at different elongation and temperature, and the result shows the side-glowing luminance of POF increases with the elongation. So the light-emitting fabric could be prepared by virtue of the extension during weaving process.

2 Experiment

2.1 Material

The performance indexes of the two POF used in the experiments with the same fiber diameter are shown in Table 1.

Table 1 The parameter of POF used in the experiment

Fiber Parameter	POF ①	POF ②
core/cladding	PMMA/fluorouracil	PMMA/fluorouracil
Index of refraction ($n_{core}/n_{cladding}$)	1.404/1.490	1.404/1.490
Breaking elongation (%)	88.38	22.03
Fracture stress(MPa)	32.01	48.86

2.2 Stretching

1. POF ① and ② were stretched with YG061-1500 Yarn flexibility instrument in constant speed to a certain elongation (5%, 15%, 20%, 25% and 30%) at room temperature, and then return to the loading zero. Take 10 samples stretched at the same elongation as one bunch to measure the side-glowing luminance.

2. POF ① was stretched with elongation of 10% and 15% at certain temperature (95℃, 100℃, 105℃ and 110℃).

2.3 Measuring luminance

The side-glowing luminance of POF after stretched was measured with the luminance meter (BM-5A,

---

* Corresponding author's email: Yangbin5959@yahoo.com.cn

TOPCON). White LED lamps were connected to the two ends of bunch as light source. The test angle of BM-5A was set as 1°, and the diameter of measuring spot is 0.97-1.30mm. In order to see the measurement area clearly, we adjusted the distance between luminance meter and the measurement area to 251-258mm. The schematic of the luminance measuring is shown in Figure 1.

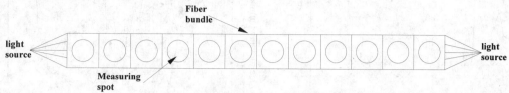

Figure 1 Schematic of the luminance measuring

3 Results and discussions
3.1 Influence of elongation to the side-glowing luminance of POF

Two kinds of POF were stretched at different elongation, and the side-glowing luminance was measured with the luminance meter. As the results showed in Figure 2, the side-glowing luminance of the two kinds of POF decreases exponentially with the distance from light source increases, and increases with elongation. The surface morphology of POF after stretched in different elongation was characterized by SEM as shown in Figure 4. There are more cracks on the surface of POF when the elongation increased and the transmitted light leaked out from the crack, so the side-glowing luminance increased with the elongation.

In Figure 2 (a), the side-glowing luminance of POF ① was maximum when the elongation is 30% due to the several continuous cracks on the surface shown in Figure 4 (e), and part of the transmitted light leaked out from the crack.

The relation curves between the side-glowing luminance and the distance from the light source of the two kinds of POF stretched at 15% and 20% are shown in Figure 3 (a) and (b) separately. The results show the side-glowing luminance of POF ② is higher than that of ① at the same elongation. There are two reasons: firstly, as shown in Figure 4 (a) and (a'), there are several small cracks on the surface of original POF ②, while the POF ① is uniform and smooth; secondly, compared with (b) and (c'), (c) and (d') in Figure 4, the cracks on POF ① are less than that of ② at the same elongation which was caused by the broken elongation of POF ② is much higher than ①. Although the side-glowing luminance of sample ② is higher than ① at the same elongation, we still chose POF ① to make light emitting fabric due to the sample ② fracture easily during weaving.

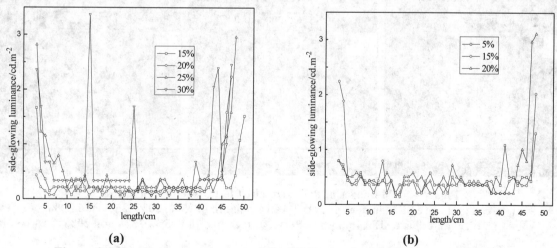

Figure 2 Side-glowing luminance of the POF stretched in different elongation
(a) POF①; (b) POF②

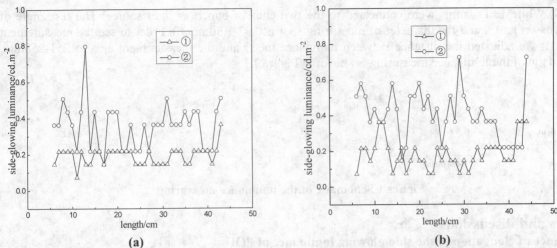

Figure 3 Side-glowing luminance of the POF stretched in same elongation
(a) Elongation of 15%; (b) Elongation of 20%

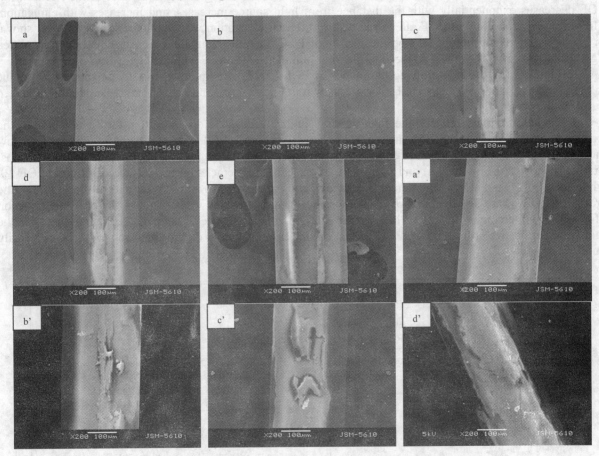

Figure 4 SEM images of POF stretched in different elongation
a, b, c, d and e are SEM images of POF ① stretched in 0, 15%, 20%, 25% and 30%;
a', b', c' and d' are SEM images of POF ② stretched in 0, 5%, 15% and 20%

3.2 The influence of stretch to the side-glowing luminance of POF at certain temperature

The conclusion obtained above shows that the side-glowing luminance increased with the elongation but not uniformity. So we consider increasing the side-glowing luminance of POF① by lower elongation and higher temperature. The elongation is set to 10% and 15%, and the stretched temperature is 95℃, 100℃, 105℃ and 110℃ respectively. Compared the side-glowing luminance of POF① at different temperature at the same elongation as shown in Figure 5 (a) and (b). In Figure 5 (a), the side-glowing luminance of POF① stretched at different temperature is higher than that of the original one, and increases with the temperature. In Figure 5 (b), only the side-glowing luminance of POF① heated by 95℃ is higher than that of the original one, whereas the minimal luminance appears at 110℃. As the influence of temperature to the axis deformation of POF ①, the damage on the surface of POF increased with the temperature at the same elongation, so the side-glowing luminance of POF increased with the temperature stretched at elongation of 10%, but when the elongation is above 15% and temperature above 95℃, the side-glowing luminance decreased with the temperature increase as there are excessive damages appeared and the transmitted light in POF reduced.

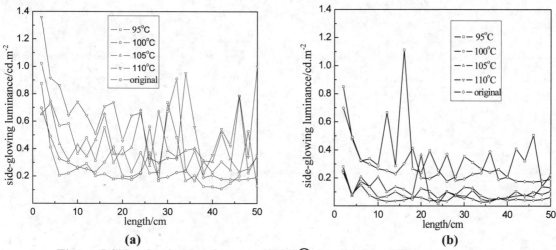

Figure 5 Side-glowing luminance of POF①elongated in different temperature
(a) Elongation of 10%; (b) Elongation of 15%

4 Conclusion

After stretching two kinds of end-lighting POF with different elongation at different temperature and the analysis of the extension effect to the side-glowing luminance of POF, the following conclusions are as drawn:

The side-glowing luminance of the two kinds POF has different sensitivity to stretch due to their tensile capacity. The POF ① has higher broken elongation and toughness, so the luminance decays slowly and uniformly after stretching, and suitable to make light emitting fabric.

A certain degree side-glowing luminance of the POF can be achieved by smaller elongation at a given temperature, such as the POF ① after 10% stretching at 110℃ having the same luminance as that of the POF ① after 25% stretching at room temperature.

Acknowledgement

This work was supported by a program for Changjiang scholars and innovative research teams in university (No. IRT 0654).

References

[1] FENG Qing-rong. A type of luminance fabric: CHINA Parent, CN02250063.4[P].2004-10-20.
[2] JIANG Ji-chun, LI Ji-yang. New type luminance fabric: CHINA Parent, CN00219365.5[P].2001-01-24.
[3] HUANG Bo-xiong, HUANG Hong-wen, et al. The method of using optical fiber to make luminous fabrics: CHINA Parent, 200410048136.6[P].2005-12-21.
[4] Maurice Daniel. LIGHT EMITTING FABRIC: US Patent, 4, 234, 907[P]. 1980-11-18.
[5] Maurice Daniel. LIGHT EMITTING OPTICAL FABRIC ASSEMBLIES AND METHOD FOR FORMING THE SAME: US Patent, 4, 519, 017[P]. 1985-05-21.
[6] PENG Qizong, LIU Shengde, WANG Qingshan. The equipment of light emitting fabric: CHINA Parent, CN1783169A[P].2006-06-07.
[7] YANG Bin, CHEN Yuan-yuan, GAI Guo-ping. Preparation and Characterization of PMMA Side-glowing Optical Fiber[J]. Journal of Materials Engineering, 2007, S1, 36-41.

[8] CHEN Yuan-yuan, YANG Bin, JIN Zi-min. Manufacture of controllable luminous fabric and characterization of the luminance[J]. Journal of Textile Research, 2008, 29(8): 38-41.
[9] CHEN Yuan-yuan, YU Ling-ling, YANG Bin, et al. Research on a New Type of Fabric a Luminous Fabric[J]. Silk Monthly, 2008, (5): 18-21.
[10] YANG Bin, CHEN Yuan-yuan, JIN Zi-min, et al. A type of controllable luminous fabric: CHINA Parent, CN1948578A[P].2007-04-18.

116 A Rechargeable Antibacterial Poly (ethylene-co-methacrylic acid) (PE-co-MAA) Nanofibrous Membrane: Fabrication and Evaluation

Jing Zhu, Dong Wang, Gang Sun
(Polymer and Fiber Science, University of California, Davis, California, USA)

1 Introduction

Nano-fibrous membranes have attracted much attention for various unique applications including sensors, filtration media, biomedical materials and protective clothing.[1-4] Their extraordinary high surface area to volume ratios and open porous structures also make the membranes an excellent candidate for antibacterial materials since they provide maximum surface contact to bacteria cells. A number of antimicrobial agents can be incorporated to the nanofibers matrix, i.e. heavy metals, quaternary ammonium compounds or antibiotics[5-8]. Among these biocides, N-halamine structures have features of high efficiency, durability and rechargeability to kill bacteria, making them ideal functions for air and water filtering materials. Some antibacterial nanofibrous membranes with N-halamine precursors have been developed by using electrospinning method.[11, 12] In these cases, N-halamine moieties were physically incorporated into the nanofibrous membrane through simple mixing two or more components in electrospinning solutions. However, the embedded agents could be released and become completely exhausted during usage, affecting durability and rechargeability of the products. An alternative strategy is to covalently incorporate N-halamine precursors onto the surface of nanofibers, which may introduce antimicrobial function permanently.

As a demonstration of covalently immobilizing N-halamine onto nanofibers, poly (ethylene-co-methacrylic acid) (PE-co-MAA) nanofibrous membrane were prepared following a melt extrusion process developed in this group; an N-halamine precursor, hydantoin derivative, was immobilized onto the surface via solid phase synthesis. Consequently, these modified membranes are chlorinated and then challenged against both gram-negative (*E. coli*) and gram-positive bacteria (*S. aureus*) to test their antibacterial efficiency and rechargeability.

2 Experimental

2.1 Materials

Poly (ethylene-co-methacrylic acid) (PE-co-MAA) with 15 wt% methacrylic acid, phosphorous pentachloride, 1,3-diaminopropane, 1,8-diamine-3,6-dioxaoctane, 4,9-dioxa-1,12-dodecanediamne and N,N-Diisopropylethylamine were purchased from Sigma-Aldrich Co. Cellulose acetate butyrate (CAB; butyryl content 44–48%), *N*, *N'*-diisopropylcarbodiimide (DIC), 1-Hydroxybenzotriazole (HOBt) and all organic solvents were purchased from Acros Chemical.

2.2 Fabrication of PE-co-MAA nanofibrous membrane

PE-co-MMA nanofibers were prepared according to a previously published procedure[13]. In brief, mixtures of CAB/PE-co-MMA with the blend ratio of 85/15 were fed into a co-rotating twin-screw extruder, and the blends were extruded through a strand (2 mm in diameter) rod die and air cooled to room temperature. Then the PE-co-MMA nanofibers in the form of continuous bundles were obtained by soxhlet extraction of CAB/PE-co-GMA composite fiber for 24hours with acetone to remove the sacrificial matrix CAB. Subsequently, the nanofibers were dispersed in ethyl ether and poured into a glass plate. After dried in vacuum, the nanofibrous membrane formed with the average thickness at 0.8 mm.

2.3 Surface modification of PE-co-MAA nanofibrous membrane

Figure 1 showed a three-step modification route to immobilize N-halamine precursor on the surface. Firstly, the carboxylic acid groups on the surface of PE-co-MAA membrane were activated into acid chloride in phosphorous pentachloride saturated DCM/MeCN (2:1, v/v) mixture solution at room temperature for 4 hours, and then washed with DCM/MeCN and MeCN thoroughly. Secondly, the activated membranes were mixed with excess amount of diamines and DIPEA in DCM/MeCN (2:1, v/v) mixture with gentle shaking at room temperature for 8 hours, and rinsed via the same method as the first step. Thirdly, DIC and HOBt were used as coupling agents to activate hydantoin-acetic acid in DMF for 30 min followed by addition of the aminated membranes. After shaking overnight, the membranes were washed by DMF and MeCN three times, respectively.

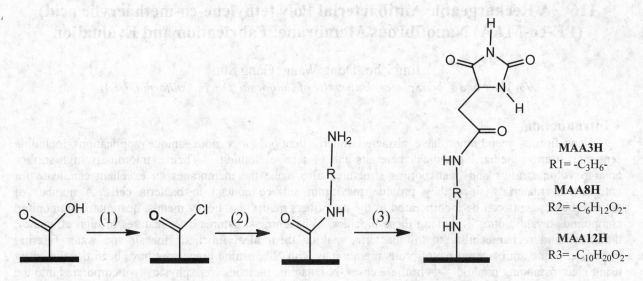

Figure 1 Three-step reaction route: (1) PCl$_5$ activation of PE-co-MAA surface; (2) addition of diamines linker moieties; (3) surface fixation of hydantoin via coupling reaction

2.4 Characterization

The morphologies of pristine and functionalized PE-co-MAA nanofibrous membranes were observed by a FEI XL-30 SFEG Scanning Electron Microscope (SEM). The fiber diameter distribution was measured using Image J analysis software based on five randomly selected SEM images. Fourier transform infrared spectroscopy was used to characterize the structural change in the frequency range of 4000-400 cm^{-1}. The amount of diamines and hydantoin grafted onto the membrane surfaces were measured by nitrogen element analysis with the original PE-co-MAA membrane as a control.

2.5 Chlorination of hydantoin-immobilized PE-co-MAA nanofibrous membrane

The functionalized PE-co-MAA NFMs were immersed in diluted chlorine bleach solutions containing 1500 ppm of active chlorine and 0.05w% nonionic surfactant, Triton TX-100 at room temperature overnight. Then the membranes were washed thoroughly with a large amount of distilled water to remove any free chlorine and dried in vacuum. To test the rechargeability, the chlorinated membrances were quenched with a 0.001N sodium thiosulfate solution for 2 hours at room temperature and then rechlorinated under the same condition as first time.

The iodometric titration method[14] was used to quantitatively test the available active chlorine on the functionalized membranes. Herein, about 0.2g membrane samples were immersed in 20 ml 0.001N sodium thiosulfate solution. Then excess amount of sodium thiosulfate was titrated with 0.001N iodine standard solution. The active chlorine content of the samples was calculated according to the following Equation:

$$Active\ Chlorine\ Content\ (ppm) = \frac{35.45 \times (V_c - V_s) \times 1000}{2 \times W}$$

where V_c and V_s are the volume (ml) of the iodine solution consumed in the titration of the control and samples, respectively; W is the weight of the chlorinated membrane.

2.6 Antibacterial assessment

Antibacterial activities of the PE-co-MAA membranes were evaluated using a modified American Association of Textile Chemist and Colorists (AATCC) Test Method 100 against Escherichia *coli* (*E. coli*, K-12, UCD microbiology lab) and *Staphylococcus aureus* (*S. aureus*, ATCC 12600). 0.2 ml of diluted bacterial suspensions containing about 10^5 colony forming units per milliliter (CFU/ml) bacteria were loaded onto the surfaces of the modified membranes in a sterilized glass jar. After varied period of contact, 20 ml 0.001N sodium thiosulfate solution was added in to quench the active chlorine. The mixture was vigorously shaken for 2 min. An aliquot of the solution was serially diluted and 0.1 ml of each dilution was placed on an agar plate. After incubation at 37 for 18 h, viable bacterial colonies on the agar plates were counted. The same method was also applied to unchlorinated pristine membrane as a control. The percentage reduction of bacteria was calculated according to the following equation:

$$R = \frac{(B - A) \times 100}{B}$$

where A and B are the number of bacteria colonies on the agar plates from the control and modified membranes, respectively.

3 Results and discussion
3.1 Morphology of PE-co-MAA nanofibrous membranes

The morphologies of these membranes before and after modification were observed via scanning electron microscope. Figure 2(a) and (b) are the surface of unmodified membranes with different magnification. As the image shown, the nanofibers are homogeneous with the average diameter at 148nm (Figure 2(d)), which exhibited really large specific surface area and high density of reactive sits for further chemical functionalization. The modified nanofiber membrane is shown in Figure 2(c). Comparing Figures 2(b) and 2(c), only slight changes of the nanofibers were observed before and after surface modification, indicating the surface morphology basically unchanged after these chemical reactions (Figure 2).

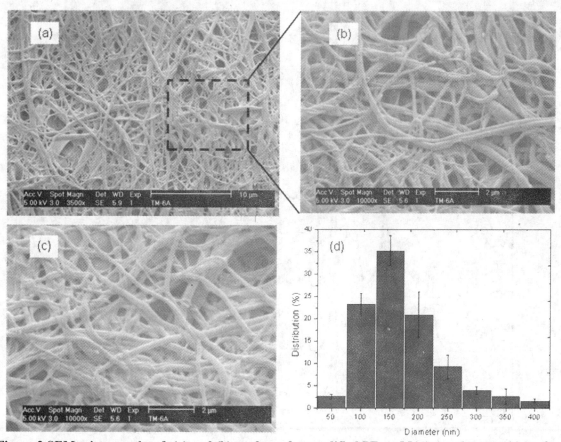

Figure 2 SEM micrographs of: (a) and (b) surface of unmodified PE-co-MAA membranes; (c) surface of modified PE-co-MAA membranes; (d) fiber size distribution of the unmodified PE-co-MAA membranes

3.2 Characterization of the modified PE-co-MAA nanofibrous membrane

The strategy for covalently bonding N-halamine precursor involved activation the carboxylic acid group on the nanofiber surface followed by incorporating different diamine linkers. In solid phase synthesis, linkers play an important role since they could dramatically reduce steric hindrance from bulky surface and enhance the reactivity of solid supports. In addition, the bounded functional agents with long and flexible spacers will act much similar to their free status in solution.[15] Thus, three diamine linkers, 1,3-diaminopropane, 1,8-diamine-3,6-dioxaoctane and 4,9-dioxa-1,12-dodecanediamne were introduced onto the nanofibers to study their influence on the modification process and future antibacterial activities of nanofibrous membrane. The hydantoin moiety was selected as a biocidal site since it is one of the most effective halamine precursors.[16,17]

To covalently link this substrate, a typical coupling reaction with the presence of DIC and HOBt was used to provide three antibacterial nanofibrous membranes: MAA3H, MAA8H and MAA12H.

The successfully grafting process was confirmed by FTIR analysis. As shown in Figure 3, the virgin PE-co-MAA membrane processes a characteristic peak at 1700 cm^{-1} attributing to the carboxylic acid group on the surface. After activation in PCl$_5$ saturated DCM for 4 hours under room temperature, all the acid groups were converted into acetyl chloride structure, which was identified by a new strong peak at 1800 cm^{-1}. In the case of amination reaction, two new peaks at 1650 cm^{-1} assigned to C=O stretching (Amide I) and 1560 cm^{-1} assigned to N-H bending (Amide II) demonstrated three diamine linkers were attached to the membrane surface. After coupling reactions, all three membranes show two twin peaks at 1720 and 1760 cm^{-1} which are from the amide and imide bonds of hydantoin structure, respectively.[18]

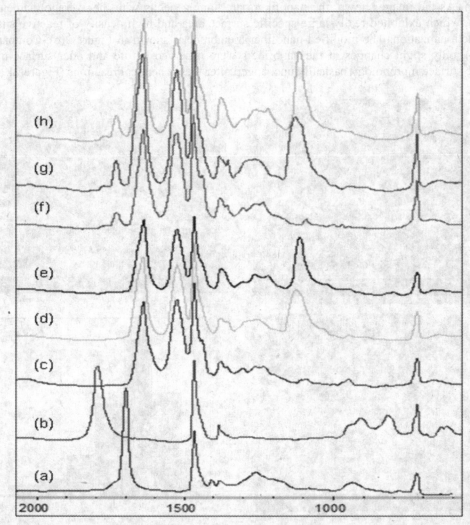

Figure 3 FTIR spectra of: (a) pristine PE-co-MAA membrane; (b) activated PE-co-MAA membrane; (c) aminated membrane MAA3H with 1,3-propanediamine; (d) aminated membrane MAA3H with 1,8-diamine-3,6-dioxaoctane; (e) aminated membrane MAA3H with 4,9-dioxa-1,12-dodecanediamne; (f) hydantoin immobilized membrane MAA3H; (g) hydantoin immobilized membrane MAA8H; (h) hydantoin immobilized membrane MAA12H.

3.3 Active chlorine content

Figure 4 shows the chlorination of the modified membranes. The chlorination reaction can take place on the amide groups and the NH groups on hydantoin rings. After the iodometric titration test, we found all modified membranes exhibited sufficient active chlorine even after three recharges (Figure 4), indicating hydantoin had been successfully immobilized onto PE-co-MAA membranes. Compared with the hydantoin-grafted microfibers[16] reported previously, the distinct more active chlorine content is attributed to the high specific surface area of nanofibers, providing higher N-halamine density on the surface and easy access to chlorination solution. It is interestingly to note that three different membranes did not show much difference in the contents of active chlorine. To further evaluate the rechargeability, the chlorinated samples were treated with 0.01M sodium thiosulfate solution to quench the active chlorine, and rinsed thoroughly. As we expected, there is no essentially change of active chlorine content for all three modified membranes after repeating tests 3 times.

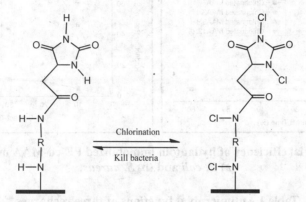

Figure 4 Chlorination of hydantoin immobilized PE-co-MAA membranes

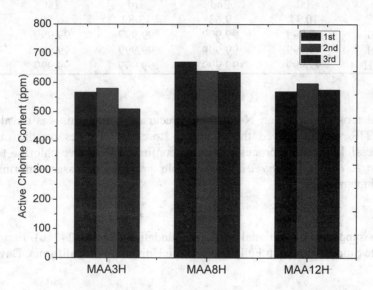

Figure 5 Active chorine content of modified PE-co-MAA nanofibrous membranes

3.4 Antibacterial evaluation

The biocidal activities of hydantoin-immobilized PE-co-MAA nanofibrous membranes were examined against both gram-negative (E. coli) and gram-positive (S. aureus) bacteria according to a modified AATCC 100 method, and the results are shown in Figure 6. All of the modified membranes after chlorination showed powerful antimicrobial functions against both bacteria at 30 min contact time. Meanwhile, the chlorinated pristine membrane showed slight reduction against bacteria, possibly due to the activation of acid groups on the fiber surface. The MAA3H membrane with the shortest linker presented slower reduction rate than the other two samples, which could be attributed to the hydrophobicity difference of the membrane surfaces. Since the longer linkers contain oxygen, the modified membranes, MAA8H and MAA12H, are more hydrophilic than MAA3H, attaining a better contact between bacteria and nanofibrous membrane. Another interesting fact is that the reduction rate of E. coli is slower than that S. aureus; even though the complete kill of both bacteria was obtained after 30 min contact. This is due to the structural difference of cell walls of both bacteria, longer time is usually required to penetrate the cell wall of E. coli than that of S. aureus[12]. In addition, Table 1 reveals rechargeability of the biocidal functions on the three modified membranes even after reused for three times.

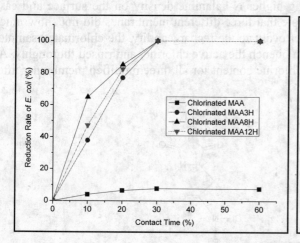 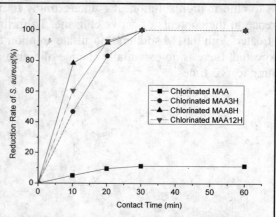

Figure 6 Antibacterial efficiency of hydantoin immobilized PE-co-MAA membranes against (a) *E. coli* and (b) *S. aureus*.

Table 1 Antimicrobial functions of three recharges

Bacterial	*E. coli*			*S. aureus*		
Recycle Times	1st	2nd	3rd	1st	2nd	3rd
Chlorinated MAA	10.34	9.63	13.82	6.50	8.37	7.77
Chlorinated MAA3H	99.999	99.999	99.999	99.999	99.999	99.72
Chlorinated MAA8H	99.999	99.999	99.999	99.999	99.999	99.88
Chlorinated MAA12H	99.999	99.999	99.999	99.999	99.999	99.82

4 Conclusion

To achieve antibacterial function, an N-halamine precursor, hydantoin, was covalently fixed onto the membrane surface. FTIR spectra proved the chemical structural changes on the modified membranes. Furthermore, the successfully grafting process was also confirmed by active chlorine test and rechargeable biocidal effects against *E. coli* and *S. aureus*. These results provided necessary information for developing nanofibrous functional membranes.

Acknowledgement

This research was supported by National Science Foundation (CTS 0424716). Jing Zhu is grateful for a Jastro-Shields Graduate Student Research Fellowship at the University of California, Davis.

References

[1] J. Venugopal, S. Ramakrishna, *Applied Biochemistry and Biotechnology*, Vol. 125, January 2005, pp147-157
[2] S. Agarwal, J. H. Wendorff, A. Greiner, *Polymer*, Vol.49, September 2008, pp 5603–5621
[3] W. E. Teo, S. Ramakrishna, *Nanotechnology*, Vol.17, June 2006, pp89-106
[4] B. Ding, M. Wang, J. Yu, G. Sun, *Sensors*, Vol. 9, March 2009, pp1609-1624
[5] Y. Wu, W. Jia, Q. An, Y. Liu, J. Chen, G. Li, *Nanotechnology*, Vol. 20, March 2009, pp245101-245109
[6] G.Amitai, J. Andersen, S. Wargo, G. Asche, J. Chir, R. Koepsel, A. J. Russell, *Biomaterials*, Vol. 30, September 2009, pp6522–6529
[7] R. Chena, N. Coleb, M. D.P. Willcox, J. Parka, R. Rasul, E. Cartere, N. Kumar, *Biofouling*, Vol. 25, No. 6, August 2009, pp517–524
[8] C. Yao, X. Li, K.G. Neoh, Z. Shi, E.T. Kang, Applied Surface Science, Vol. 255, October 2009, pp3854–3858
[9] X. Ren, A. Akdag, H. B. Kocer, S.D. Worley, R.M. Broughton, T.S. Huang, *Carbohydrate Polymers*, Vol. 78, March 2009, pp220–226
[10] X.g Ren, H. B. Kocer, S.D. Worley, R.M. Broughton, T.S. Huang, *Carbohydrate Polymers*, Vol. 75, September 2009, pp683–687
[11] X. Ren, A. Akdag, C. Zhu, L. Kou, S. D. Worley, T. S. Huang, *Journal of Biomedical Materials Research Part A*, Vol. 91, November 2008, pp385 – 390
[12] K. Tan S,. K. Obendorf, *Journal of Membrane Science*, Vol. 305, November 2007, pp287-298

[13] D. Wang, G. Sun, B-S Chiou, J. P. Hinestroza, *Polymer Engineering and Science*, Vol. 47, pp1865 – 1872
[14] Z. Chen, J. Luo, Y. Sun, *Biomaterials*, Vol. 28, December 2006, pp1597–1609
[15] S. M. Ribeiro, R. Gonsalves, *Tetrahedron*, Vol. 63, May 2007, pp7885–7891
[16] M. R. Badrossamay, G. Sun, *Macromolecules*, Vol. 42, February 2009, pp1948-1954
[17] M. R. Badrossamay, G. Sun, *Journal of Biomedical Materials Research Part B: Applied Biomaterials*, Vol. 89, August 2008, pp93 – 101
[18] Y. Sun, G. Sun, *J App Polym Sci*, Vol. 80, March 2001, pp2460-2467

117 Recent Developments in Electrospinning of Nanofibers and Nanofiber Yarns

Tong Lin, Xungai Wang
(Centre for Material and Fibre Innovation, Deakin University, Geelong, VIC 3217, Australia)

Abstract: Electrospinning is a simple, but efficient and versatile, technology to produce polymeric nanofibers for diverse applications in both textile and non-textile areas. In this paper, recent research developments in electrospinning and electrospun nanofibers, especially those from the Centre for Material and Fiber Innovation, Deakin University, are introduced. Important findings on needleless mass-electrospinning and direct electrospinning of highly-twisted continuous nanofiber yarns are presented.

Keywords: Electrospinning; Nanofibers; Yarns

Electrospinning is a very useful technique to produce nanofibers for diverse applications in areas such as tissue engineering scaffolds, control release, filtration, protective clothing, catalyst carriers, chemical sensors and many others [1].

In our previous studies, we have developed effective techniques to improve the fiber uniformity by adding a small quantity of surfactants [2] or polymeric thickener to a polymer solution for electrospinning, and used a microfluidic device as the spinneret to electrospin side-by-side bicomponent nanofibers [3]. We have also developed techniques to improve the mechanical properties of nanofiber membranes [4-6], to understand the fiber morphology evolution during electrospinning [7-8], and to produce nanofiber webs with a superhydrophobic surface [9-10] and the ability to recover trace precious metals from aqueous solutions [11]. In this report, our recent research developments on needleless electrospinning and direct electrospinning of nanofiber yarns are presented.

1 Needleless electrospinning for increased nanofiber production

In an effort to increase the productivity of electrospinning, we have recently developed two new needleless electrospinning systems, disc and coil electrospinning (Figure 1). A disc spinneret was initially chosen and compared with the cylinder electrospinning setup [12]. The disc spinneret needed a relatively low applied voltage to initiate the fiber formation, and the fibers were mainly initiated from the disc edges. For the cylinder spinneret, the fibers were initiated from the cylinder ends first, and then from the entire cylinder surface only if the applied voltage was increased to a certain level. Nanofibers electrospun from the cylinder surface showed a higher dependence on the applied voltage. More recently, a spiral coil was used as the spinneret for needleless electrospinning (Figure 1) [13]. Nanofibers can be easily electrospun from the coil surface, with a critical applied voltage similar to that of the disc system.

All three needleless electrospinning systems could produce uniform nanofibers, but the fibers produced from the coil spinneret were the finest with the narrowest diameter distribution when the same voltage was applied to the electrospinning systems. Disc electrospinning produced finer nanofibers than the cylinder one. A thin disc (diameter 8 cm and thickness 2 mm) could produce nanofibers at a similar rate to a cylinder of the same diameter but 100 times wider (i.e. 20 cm long). However, a spiral coil spinneret that has the same length and diameter as the cylinder can double the fiber production rate.

Compared with the conventional needle electrospinning, the cylinder or disc spinneret produced coarser nanofibers, but the dependency of fiber diameter on the polymer concentration showed a similar trend. The coil spinneret, however, produced even finer nanofibers with a narrower diameter distribution than the conventional needle electrospinning system.

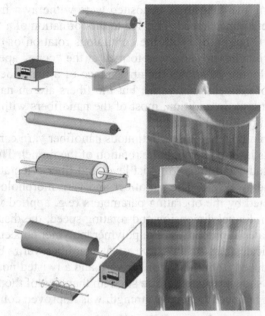

Figure 1 Schematic of disc, cylinder and coil electrospinning systems and the electrospinning processes

2 Direct electrospinning of nanofiber yarns

Although nanofiber nonwovens have been shown enormous application potential, nanofiber bundles that have long continuous length and interlocked fibrous structure, i.e. nanofiber yarns, are expected to create new opportunities for the nanofiber materials. While progress has been made in nanofiber yarn production, considerable challenges remain in this area.

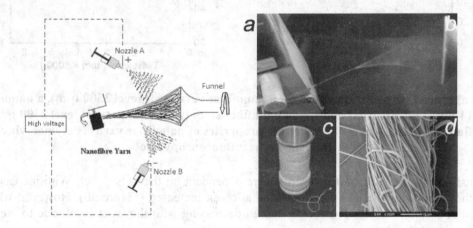

Figure 2 a) Schematic illustration of the basic setup for electrospinning of nanofiber yarns, b) Cone-shaped fiber deposition due to the rotation of the funnel collector and continuous drawing of nanofibers from the funnel, c) As-produced nanofiber yarns, d) SEM image of a nanofiber yarn section

Recently, we have developed a novel and simple process to directly electrospin continuous nanofiber yarns with well-controlled twist level using a dual-nozzle electrospinning setup and a yarn winding system[14]. Figure 2a schematically illustrates the basic setup for electrospinning of nanofiber yarns, which consists of two needle nozzles, a rotating funnel collector, a cross winder, and a high-voltage DC power supply. Two nozzles were placed in front of either side of the funnel collector, and a yarn-winding system was set between the nozzles with a distance further than that between nozzle and funnel collector. During electrospinning, two needle nozzles were connected separately with the positive and negative polarities of the DC power supply. Nanofibers electrospun from the oppositely charged nozzles were deposited onto the rotating funnel collector to from a fibrous membrane covering the funnel surface.

To form a nanofiber yarn, the nanofibers deposited were withdrawn from the funnel surface, while the funnel was rotating at a predetermined speed. This led to the formation of a "cone" shaped fibrous membrane attaching to the funnel edge (Figure 2b). With the continuous rotation of the funnel and withdrawal of the fibrous cone, a twisted fiber bundle, or yarn, was formed at the "cone" apex, and the yarn was then wound onto a package (Figure 2c). Figure 2d shows the appearance of a nanofiber yarn produced by this process, which looks similar to that of conventional yarns, but the fibers are on nanometer scales (with an average diameter of 486 nm in this study). In addition, most of the nanofibers within the yarn are aligned at certain angles along the yarn length direction (Figure 2d).

With this novel electrospinning system, a continuous nanofiber yarn can be directly produced, with up to 7400 turns-per-meter of twists inserted through the rotation of the funnel. The maximum yarn production rate was 5 m/min in this initial study, and the yarn twist level and fiber orientation was controlled by the funnel rotating speed and yarn withdrawal rate. The nanofiber and yarn morphologies, yarn dimension, twist level and production rate are affected by the operating parameters (e.g. applied voltage, electrospinning distance, flow-rate of polymer solution, funnel dimension and rotating speed, the distance between funnel and winder, and winding speed) and/or material properties (e.g. polymer, polymer concentration).

Figure 3a shows a typical stress-strain curve of a nanofiber yarn. For comparison, the strain-stress curves of a braided yarn produced from the nanofiber yarn and a twisted nonwoven nanofiber strip were also included. The nanofiber yarn showed higher tensile strength than that of the twisted nanofiber nonwoven strip. When the nanofiber yarn was braided, the tensile strength was improved considerably.

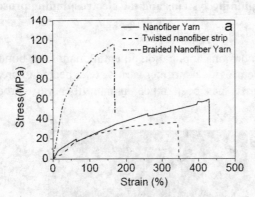 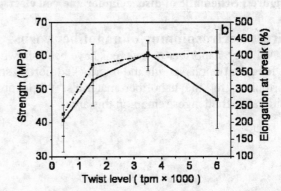

Figure 3 a) typical stress~strain curves of a nanofiber yarn (twis level 3500 tpm), a nanofiber yarn braid (4-ply), and a twisted nonwoven nanofiber strip (width 2mm, twist level 3500 tpm), and b) influence of twist level on the tensile properties of nanofiber yarns (Polymer, vinylidene fluoride-co-hexafluoropropylene)

The tensile properties of nanofiber yarns were dependent on the twist level. With increasing the twist level, both the tensile strength and the elongation at break increased (Figure 3b). However, when the twist level reached 3500 tpm, the tensile strength started decreasing with the twist level, due to the well known obliquity effect in staple fiber yarns.

References
[1] Fang J, Niu HT, Lin T and Wang XG 2008 Applications of electrospun nanofibers *Chinese Science Bulletin* 53 2265-86
[2] Lin T, Wang H, Wang H and Wang X 2004 The charge effect of cationic surfactants on the elimination of fibre beads in the electrospinning of polystyrene *Nanotechnology* 15 1375
[3] Lin T, Wang H and Wang X 2005 Self-crimping bicomponent nanofibre electrospun from polyacrylonitrile and elastomeric polyurethane *Advanced Materials (Weinheim, Germany)* 17 2699-703
[4] Fang J, Lin T, Tina W and Wang X 2007 Toughened electrospun nanofibres from crosslinked elastomer-thermoplastic blends *Journal of Applied Polymer Science* 105 2321-6
[5] Naebe M, Lin T, Tian W, Dai L and Wang X 2007 Effects of MWNT Nanofillers on Structures and Properties of PVA Electrospun Nanofibers *Nanotechnology* 18 225605
[6] Naebe M, Lin T, Staiger M, Dai L and Wang X 2008 Electrospun single-walled carbon nanotube/polyvinyl alcohol composite nanofibers: structure-property relationships *Nanotechnology* 19 305702

[7] Fang J, Wang H, Niu H, Lin T and Wang X 2010 Evolution of fiber morphology during electrospinning *Journal of Applied Polymer Science* DOI:10.1002/app.32569
[8] Fang J, Wang H, Niu H, Lin T and Wang X 2010 Evolution of Fiber Morphologies during Poly(acrylonitrile) Electrospinning *Macromolecular Symposia* 287 155-61
[9] Wang H, Fang J, Cheng T, Ding J, Qu L, Dai L, Wang X and Lin T 2008 One-step coating of fluoro-containing silica nanoparticles for universal generation of surface superhydrophobicity *Chemical Communications* 877-9
[10] Xue Y, Wang H, Yu D, Feng L, Dai L, Wang X and Lin T 2009 Superhydrophobic Electrospun POSS-PMMA Copolymer Fibres with Highly Ordered Nanofibrillar and Surface Structures *Chemical Communications* 6418-20
[11] Wang H, Ding J, Lee B, Wang X and Lin T 2007 Polypyrrole-Coated Electrospun Nanofibre Membranes for Recovery of Au(III) from Aqueous Solution *Journal of Membrane Science* 303 119-25
[12] Niu H, Lin T and Wang X 2009 Needleless Electrospinning. I. A Comparison of Cylinder and Disk Nozzles *Journal of Applied Polymer Science* 114 3524-30
[13] Lin T, Wang X, Wang X and Niu H 2010 Electrostatic spinning assembly.
[14] Ali U, Zhou Y, Wang X and Lin T 2010 Direct Electrospinning of Highly Twisted, Continuous Nanofiber Yarns. In: *Fiber Society Spring Conference,* May 2010, Bursa, Turkey.

118 Fabric Wrinkle Characterization and Classification Using Modified Wavelet Coefficients and Support-Vector-Machine Classifiers

Jingjing Sun[1], Ming Yao[1], Patricia Bel[2], Bugao Xu[1]

(1. School of Human Ecology, University of Texas at Austin, Austin, TX 78712, USA
2. USDA Southern Regional Research Center, New Orleans, LA70124, USA)

Abstract: This paper presents a novel wrinkle evaluation method that uses modified wavelet coefficients and optimized support-vector-machine (SVM) classifications to characterize and classify wrinkling appearance of fabric. Fabric images were decomposed with the wavelet transform, and five parameters were defined based on the modified wavelet coefficients to describe wrinkling features, such as directionality, hardness, density, and contrast. These parameters were also used as the inputs of optimized SVM classifiers to obtain overall wrinkle grading in accordance with the standard AATCC smoothness appearance (SA) replicas. The SVM classifiers based on a linear kernel and a radial-basis-function (RBF) kernel were used in the study. The effectiveness of this evaluation method was tested by 300 images of five selected fabrics that had different fiber contents, weave structures, colors and laundering cycles. The cross-validation tests on the SA classifications indicated that the SA grades of more than 75% of these diversified samples could be recognized correctly. The extracted wrinkle parameters provided useful information for textile, appliance, and detergent manufactures to inspect wrinkling behaviors of fabrics.

Keywords: Fabric wrinkling; Objective evaluation; Wavelet transform; SVM classification

1 Introduction

Wrinkling caused in wear and care procedures is a vital performance characteristic of fabric. Conventionally, fabric wrinkling is evaluated by visual examination of fabric samples performed by trained experts in accordance to the wrinkling standards (e.g., the AATCC Smoothness Appearance (SA) replicas)[1]. Recently, many instrumental methods have been developed and introduced to the industry to meet the needs for more objective and reliable measurements of fabric wrinkling appearance[2-14]. 3D imaging technology, including laser structure light[2-4] and stereovision[5-8], has been used to generate 3D surfaces or profiles from which wrinkles can be detected. Various digital image analysis methods have been adopted for automatic wrinkle characterization and grading based on 2D texture information [9-14]. Of these image processing methods, the wavelet transform (WT) appears to be a robust algorithm for locating and evaluating wrinkles and other surface features[11-14]. Fabric wrinkles are localized (transient) and disordered disturbance from a regular background, and are particularly suitable for the analysis using the WT, because the WT can break up an image into different frequency domains, and allows each domain to be studied with a resolution matching wrinkles' scale. Therefore, the WT is a powerful tool to remove unwanted information associated with weave structures, colors, and illumination variations which are separable at different decomposing levels or frequency domains[13,14].

The AATCC Test Method 124 defined five wrinkling (or smoothness) grades, ranging from SA-1 (severely wrinkled) to SA-5 (very smooth)[1], for visual assessment of fabric appearance. Another grade, SA-3.5, was added to the standard to describe a fairly smooth, non-pressed appearance. Although replica SA-3.5 follows the general trend of smoothness change over the complete replica set, its instrumental measurement of roughness disrupts the incremental differences seen in the replicas[3]. Since the six SA grades are not equally spaced, it is difficult to discern fabrics with relatively smooth surfaces. How to classify these grades reliably remains challenging in developing new wrinkle evaluation methods since the overall evaluation accuracy largely depends on the solution of this problem.

Fabrics with different fiber contents or weave structures can result in different styles and severities of wrinkles, even if they undergo the same laundering treatment. It is perplexing to compare a cotton sample to a polyester sample, even if both are rated at the same SA grade with the AATCC replicas. The replicas cannot represent all wrinkling styles. A visual grader often has to infer a SA grade based on personal experience, when the sample does not fit into the wrinkling style of the replicas. An automatic evaluation system must be able to rate samples that differ in fiber contents, weaves or other parameters in a way consistent to the severity levels defined in the standard.   In other words, an effective system must be able to be trained separately for major categories of fabrics.

In this study, we will explore a reliable way for wrinkling evaluation by using off-the-shelf hardware and effective WT techniques. Firstly, the frequency domains of wavelet decompositions in which fabric wrinkles of various sizes are retained are determined. The wavelet coefficients are then modified to curb the noise associated with non-wrinkle features, such as weave structures. Wrinkle directionality, hardness, density and other characteristics are defined and measured directly from the modified wavelet coefficients. The overall SA rating of a sample is performed by employing SVM classifiers, whose parameters are optimized by the "grid searching" method during the cross-validation process. 300 images of five selected fabrics with different fiber contents, weave structures, colors and laundering cycles are used to train and test the linear-kernel and RBF kernel SVM classifiers [15].

2 Methods

The designed wrinkle evaluation system mainly consists of an infrared digital camera, four infrared LED light fixtures each containing 40 LED lights, a copy-stand, a computer and the software that controls the camera, processes the images, and measures wrinkles (see Figure 1). To intensify wrinkle contrasts in the image, the LED lights are placed in oblique positions to case directional lighting, generating high intensity gradients across the image field. The four lights can be turned on in sequence to highlight wrinkles from different angles. Therefore, the same sample can be imaged four times without being moved, and wrinkles in different directions can be illuminated equivalently in separate images. It is important to use this multi-lighting scheme for an unbiased evaluation because one directional light highlights wrinkles more in that direction than in other directions.

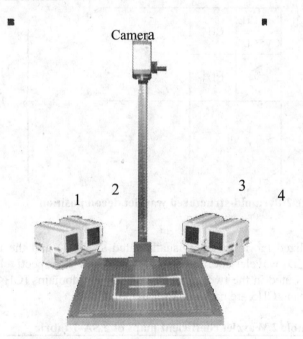

Figure 1 Fabric imaging system with infrared digital camera and four LED light fixtures

3 Materials

Five fabrics with different fiber contents and weave structures were chosen by an appliance company to validate the evaluation results of the wrinkle evaluation system. 25 swatches of 380x380 mm (15x15 in) were cut for each fabric and laundered in various cycles to create diversified wrinkling appearances. Each sample was graded with the AATCC SA replicas by the trained personnel, and then imaged four times by the system. Ten sample images were selected at each SA grade to construct the datasets shown in Table 1.

Table 1 Fabric characteristics and smoothness appearance

Fabric	Fiber content	Weave	Color	SA					
				1	2	3	3.5	4	5
F1	100%C[1]	Plain	White	10	10	10	10	10	10
F2	100%P[2]	Plain	White	10	10	10	10	10	10
F3	60C/40P	Plain	White	10	10	10	10	10	10
F4	100%C khaki	Twill	Khaki	10	10	10	10	10	10
F5	100%C jean	Twill	Blue	10	10	10	10	10	10

1: C — cotton;   2: P — polyester

4 Image analysis

The grayscale image of a treated sample was firstly normalized by dividing its pixels with the maximum value. The pyramid-structured 2-D Haar WT was performed using the Matlab® Wavelet Toolbox for image decomposition (see Figure 2)[16]. This commonly used wavelet algorithm can break down image signals into shifted and scaled versions of the Haar (original) wavelets. Wrinkles normally appear in lower frequency domains than the fabric texture (weave patterns, yarns, etc.), and in higher frequency domains than the oblique lighting [14]. After the trial tests with original fabric images, it was found that only the wavelet coefficients at decomposing levels 4 and 5 were needed for wrinkle characterization. Components in other domains were either insignificant or irrelevant in representing wrinkles.

Figure 2 Pyramid-structured wavelet decomposition

Table 2 shows a normalized fabric image (visually rated at SA-1) and the maps of horizontal (CH), vertical (CV) and diagonal (CD) wavelet coefficients at levels 4 and 5, respectively.   For this SA-1 image, predominant powers are distributed in the two horizontal coefficient domains (CH4 and CH5). As displayed in Table 2, the wrinkling scales in CH5 are much coarser than in CH4.

Table 2 Wavelet coefficient maps of a SA-1 fabric

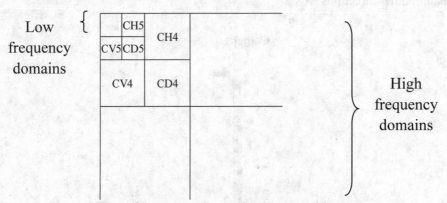

The image of the SA-1 fabric can be reconstructed through the inverse wavelet transform (IWT) using the coefficients at decomposing levels 4 and 5, as displayed in Figure 3. The normalized image (Figure 3a) has apparent non-uniform illumination—a bright stripe along one diagonal and dark regions at the two other corners. In the reconstructed image (Figure 3b), the background variation has been filtered out because its corresponding coefficients are not contained in the domains at levels 4 and 5. However, most of the wrinkles are preserved with a smoothed background whose value is set to zero.

(a)　　　　　　　　　(b)

Figure 3 Normalized image (a) and reconstructed image (b) of the SA-1 fabric

Table 3 displays the normalized images and the maps of wavelet decomposing coefficients at level 4 (higher) and level 5 (lower) for fabrics rated at the six AATCC SA grades, respectively. It is noted that the six coefficient maps (CH4, CV4, CD4, CH5, CV5 and CD5) reveal the wrinkling differences in intensity, scale and direction among the six grades.

Table 3 Wavelet coefficient maps of fabrics rated at the six SA grades

Fabric image	Horizontal	Vertical	Diagonal

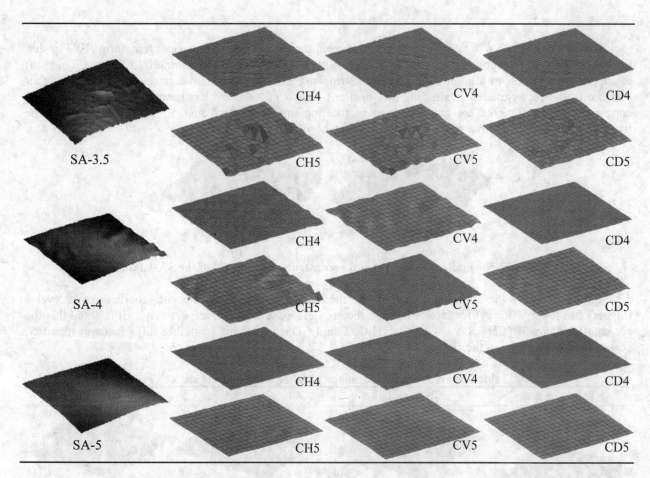

However, these selected coefficient maps still contain some small peaks, which correspond to short variations on fabric surfaces that may not be wrinkles. These peaks should be removed prior to wrinkle characterization. A global threshold was applied to remove noisy peaks in the coefficient maps. In order to obtain an appropriate threshold, 30 SA-5 images of F1, F2 and F3 were used for the threshold calculation because these images should be wrinkle-free. The mean value of the coefficients in CH4, CH5, CV4, CV5, CD4 and CD5 of these images appears to be a good universal threshold for modifying all the map images. The coefficients under the threshold are set to zero and the rest remain unchanged. The modified coefficient maps and the reconstructed image of the SA-1 fabric in Table 2 are shown in Table 4. By using the modified coefficient maps at levels 4 and 5, the reconstructed image further eliminates the influence of noise arising from undesired textures (e.g., weave patterns).

Table 4 Modified coefficient maps and the reconstructed image

5 Wrinkle characterization

Useful wrinkling information, including wrinkle orientation, height, size, shape and density, can be extracted from the modified wavelet decomposing coefficients (i.e. CH4, CH5, CV4, CV5, CD4 and CD5). In this study, three types of wrinkling parameters were defined from different perspectives: texture energy, geometrical measurements and contrast.

1. The total energy (ENERGY) of the selected coefficients: a quantitative measure to reflect the texture content information. The average energies, E_{4H}, E_{5H}, E_{4V}, E_{5V}, E_{4D} and E_{5D} of the coefficients CH4, CH5, CV4, CV5, CD4 and CD5 are calculated as follows:

$$E_{ld} = \frac{1}{M \times N} \sum_{i=1}^{M} \sum_{j=1}^{N} C_{ld}(i,j)^2 \quad (l=4, 5; d= \text{horizontal, vertical and diagonal}).$$

Where $M \times N$ is the size of the wavelet coefficient, $C_{ld}(i,j)$ are the values of the wavelet coefficient at level l and in direction d.

The total energy is a simple summation of these individual energies, that is,

$$\text{ENERGY} = E_{4h} + E_{5h} + E_{4v} + E_{5v} + E_{4d} + E_{5d}$$

Figure 4 exhibits three images with various SA grades. From a highly wrinkled sample (Figure 4a) to a smooth sample (Figure 4c), the wrinkling severity diminishes quickly, and the ENERGY values decrease dramatically, as demonstrated in Table 5.

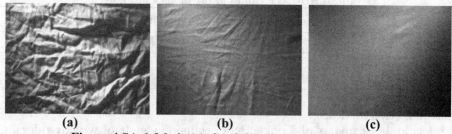

(a)　　　　　　　　　　(b)　　　　　　　　　　(c)

Figure 4 SA-1 fabric (a), SA-3 fabric (b), and SA-5 fabric (c)

Table 5 ENERGY values of the images in Figure 4

Image	Figure 4a	Figure 4b	Figure 4c
SA Grade	SA-1	SA-3	SA-5
ENERGY	3.35×10^3	2.80×10^2	4.30

2. The wrinkle geometrical measurements: the directionality (orientation), hardness (shape), and density of wrinkle are used to describe whether wrinkles are oriented in a dominant direction, how sharply wrinkles are formed, and how densely winkles are distributed.

5.1 Wrinkle directionality (DIRECT)

The WT decomposes an image into horizontal, vertical, diagonal detail components at various levels. Therefore, the decomposing coefficients in different domains can be used to reveal the overall directionality of wrinkles. In this study, a parameter, DIRECT was defined based on the energy content information: and are the total horizontal energy (E_H), total vertical energy (E_V) and the total diagonal energy (E_D) at levels 4 and 5, that is,

$$E_H = E_{4H} + E_{5H},$$
$$E_V = E_{4V} + E_{5V},$$
$$E_D = E_{4D} + E_{5D},$$

$$\text{DIRECT} = \text{MAX}(E_H, E_V, E_D) / \text{ENERGY} \times 100\%.$$

Based on the observations with the selected images, DIRECT can be used to discern wrinkle appearance into three categories:
1) Disordered and directionless if DIRECT < 50%;
2) Slightly ordered wrinkles if 50% ≤ DIRECT < 70%;
3) Highly ordered (oriented) wrinkles if DIRECT ≥ 70%.

Three SA-1 images with different wrinkling orientations and their corresponding DIRECT values are shown in Figure 5 and Table 6. Although the appearance of the three images was rated as SA-1 by the experts, wrinkles in Figure 5a appear to be highly oriented while wrinkles in Figure 5c are almost random. The difference in directionality of these wrinkled appearances is correctly ordered by their DIRECT values (Table 6). It is necessary to point out that DIRECT is not associated with wrinkling severity, and therefore it will not be included in the parameter set for the classification of wrinkling appearance.

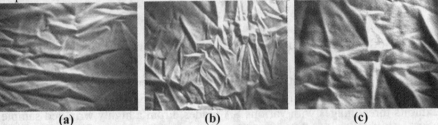

(a) (b) (c)

Figure 5 SA-1 fabric images, (a) highly oriented, (b) slightly oriented and (c) randomly oriented

Table 6 DIRECT values of the images in Figure 5

Image	Figure 5a	Figure 5b	Figure 5c
Wrinkle Orientation	Highly ordered	Slightly ordered	Disordered
DIRECT	87.4%	50.2%	43.2%

5.2 Wrinkle hardness (HARDNESS)

Hardness refers to the sharpness of wrinkle, representing the angle of wrinkle ridges. Compared with round and soft wrinkles in SA grades of 3.5, 4 and 5, severe wrinkles in SA grades of 1, 2 and 3 have sharp tip edges, which means that they are hard-pressed and more difficult to recover or remove.

In the WT, the decomposing coefficients at the fourth level (CD4, CV4 and CH4) contain higher frequency components (e.g., sharper wrinkles) than those at the fifth level (CD5, CV5 and CH5). Thus, wrinkles HARDNESS can be characterized by the proportion of the energy of the coefficients at the fourth level to the ENERGY (total energy) as follows:

$$HARDNESS = (E_{4H} + E_{4V} + E_{4D}) / ENERGY$$

A higher value of HARDNESS corresponds to more severe and hard-pressed wrinkles; a lower value of HARDNESS means that majority of the wrinkles on the fabric are soft, rounded lumps. HARDNESS helps to differentiate wrinkle appearances from SA-3 to SA-3.5 as seen in Figure 6.

(a) (b) (c)

Figure 6 Fabric SA-3 (a), fabric at SA-3 (b) and fabric at SA- 3.5 (c)

Table 7 Wrinkle HARDNESS measurements of hard and soft wrinkles in Figure 6

Image	Figure 6a	Figure 6b	Figure 6c
SA Grade	SA-3	SA-3	SA-3.5
Wrinkles	Hard	Hard and soft	Soft
HARDNESS	0.71	0.57	0.26

A HARDNESS value can be interpreted as the estimated percentage of hard-pressed wrinkles in the image. For example, the HARDNESS of Figure 6a is 0.71, which implies that 71% of the wrinkles on this fabric are pressed wrinkles with sharp angles. Although both Figure 6a and Figure 6b were rated visually as SA-3, The HARDNESS of Figure 6b is noticeably lower, suggesting that the wrinkling appearance grade should be rated between SA-3 and SA-3.5. The imaging system seems to be more sensitive to changes in wrinkling style than human eyes. Figure 6c is a fabric rated at SA-3.5 and its HARDNESS value is 0.26, which indicates that the majority of the wrinkles are round lumps. Therefore, HARDNESS is an effective parameter to indicate the wrinkle shape on the fabric surface.

5.3 Wrinkle density (DENSITY)

Winkle density is a parameter to indicate the ratio of wrinkled areas to the image area. This can also be calculated using the selected wavelet coefficients. The mean value of each coefficient domain is firstly computed as a threshold to choose wrinkle points whose coefficients are higher than the mean value, and then the DENSITY is calculated by using the ratio of the total wrinkle points to the total coefficient point:

DENSITY = $(Num_{CH4}+ Num_{CV4}+ Num_{CD4})/3 \times M_4 \times N_4 + (Num_{CH5}+ Num_{CV5}+ Num_{CD5})/3 \times M_5 \times N_5$,

where Num_{CH4}, Num_{CV4}, Num_{CD4}, Num_{CH5}, Num_{CV5} and Num_{CD5} are the numbers of wrinkle points whose values are higher than the mean coefficient value in the different frequency domains. $M_4 \times N_4$ is the total coefficient point at level 4 and $M_5 \times N_5$ is the total coefficient point at level 5. According to the Haar WT definition, $M_4 \times N_4 = M_5 \times N_5 \times 2$. The DENSITY values of the three fabric images shown in Figure 4 are listed in Table 8, which are highly consistent with the SA grades assigned by the experts. DENSITY is also an important factor to be used for wrinkling classification.

Table 8 DENSITY values of the images in figure 4

Image	Figure 4a	Figure 4b	Figure 4c
SA grade	SA-1	SA-3	SA-5
DENSITY	0.53	0.12	6.10×10^{-3}

3. The contrast parameter (CONTRAST): a supplementary reference to help predicting the SA grade of a fabric. CONTRAST can be calculated from the modified coefficients as follows:

CONTRAST=,l--,d--,$1-M \times N$,,$i=1$-M-,$j=1$-N-,,C-ld.(i,j)-,MI-

$$\text{CONTRAST} = \sum_l \sum_d \frac{1}{M \times N} \sum_{i=1}^{M} \sum_{j=1}^{N} |C_{ld}(i,j) - MI_{ld}|,$$

where, MI-where MI_{ld} is the mean intensity of the coefficients at level l and in direction d. Table 9 displays the CONTRAST values of the three images in Figure 4. The SA-1 image (Figure 4a) has the highest CONTRAST value among the three images, while the SA-5 image (Figure 4c) has no contrast due to a wrinkle-free surface.

Table 9 CONTRAST values of the images in Figure 4

Image	Figure 4a	Figure 4b	Figure 4c
SA grade	SA-1	SA-3	SA-5
CONTRAST	0.81	0.16	5.20×10^{-4}

6 SVM for wrinkle classification

Prior to the SA classification using a SVM scheme, the numeric attributes (i.e. classification features) were scaled to [0, 1] in order to prevent the dominance of attributes in large ranges (i.e. ENERGY) over attributes in small ranges [17]. This scaling can also improve the efficiency of the SA classifications by curbing large kernel computations caused by the large numeric range, and reduce the data variability arising from different weave structures and fiber contents. The classification takes wrinkle features (ENERGY, HARDNESS, DENSITY and CONTRACT) as inputs and SA grades as outputs that target the visual SA ratings. The SVM classification scheme was used with two different kernels for comparison [17, 18]:

Linear kernel: $K(x_i, x_j) = x_i^T x_j$.

Radial basis function (RBF) kernel: $K(x_i, x_j) = exp\left(-\gamma \|x_i - x_j\|^2\right), \gamma > 0$.

The two kernel functions describe the relationships between the training data set (x_i) and the testing data set (x_j). γ is the kernel parameter of the RBF kernel function.

The RBF non-linear kernel can deal with the case when the relationship between groups and features is nonlinear by mapping the original data sets, which may not be separable in a linear space, into a higher dimensional space. Moreover, the RBF kernel has computational superiority compared to other non-linear functions (e.g., sigmoid kernel) [19].

In order to prevent the over-fitting problem, a 5-fold cross-validation was used. In the 5-fold cross-validation, the data sets were randomly separated into 5 subsets, called folds. Each time, the SVM classifier was first trained with four subsets and then tested by the last subset. This procedure was repeated five times to generate the average classification accuracy, which is the percentage of the samples that were classified in the same SA grades as visually rated by the experts.

There are two classifier parameters for the RBF kernel: penalty factor C and kernel parameter γ, while only penalty factor C for linear kernel. Since the effectiveness of the SVM classification scheme largely depends on the selection of the classifier parameters, a "grid searching" procedure was used to determine the best selection of the parameters [20]. During the cross-validation process, various C or pairs of (C, γ) values exponentially grow for the linear kernel or the RBF kernel, respectively. In this study, the range of parameter sequence was preset as: $C = 2^{-10}, 2^{-8}, ..., 2^{10}$; $\gamma = 2^{-10}, 2^{-8},..., 2^{10}$. Therefore, eleven C values were attempted for the linear kernel SVM, and $11\times11=121$ combinations of (C, γ) for the RBF kernel SVM. The one giving the best rating accuracy was chosen as the optimal C or (C, γ). The rating accuracy based on such optimized classifier was discussed in the following section.

7 Results and discussion

Table 10 shows the box plots of the four wrinkle feature measurements (ENERGY, HARDNESS, DENSITY and CONTRAST) of the three plain weave fabrics: F1 (100% cotton), F2 (100% polyester) and F3 (60c/40p). A box plot is an effective descriptive tool to indicate the distribution of a data set. In the plot, the central line in the box signifies the median of the data, and the borders of the box indicate upper quartile and lower quartile. The data range, excluding outliers which are specified by "+", is shown by the whiskers. From the plots in Table 10, it is clear that all of the five features show similar declining trends and similar numeric ranges as the SA grade ascends. F2 fabrics (100% polyester) have lower ENERGY values than to F1 and F3 because they have smoother surfaces and smaller wrinkles than fabrics containing cotton fibers. This difference is also noticed in the CONTRAST and DENSITY plots, especially at SA-1, meaning polyester fabrics contain shallower (flatter) and fewer wrinkles than cotton fabrics even if they are rated at the same SA grades. The HARDNESS values of the three fabrics at the first SA stage (SA-1, SA-2 and SA-3) appear to be in the same level, but start to exhibit bigger differences after SA-3.5. The difference of the wrinkle hardness between first SA stage and second SA stage (SA-3.5, SA-4 and SA-5) is clear for F1 (100% Cotton fiber). But this trend for F2 and F3 is not as clear as that of F1 due to the changes of fiber content. After a small increase at SA-3, the HARDNESS of F2 remains the same at SA-3.5 and then decreases. This implies that polyester fabrics are more likely to have hard-pressed wrinkles than fabrics containing cotton fibers.

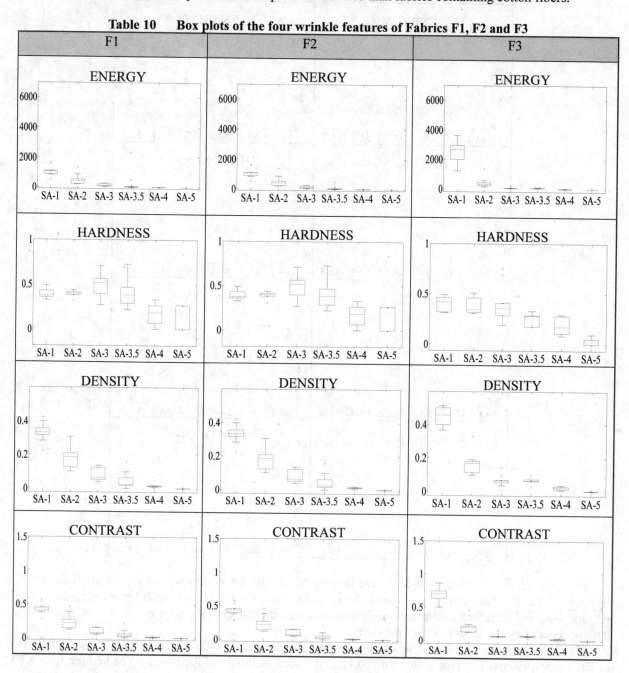

Table 10 Box plots of the four wrinkle features of Fabrics F1, F2 and F3

The box plots of the wrinkle features of the two twill-weave fabrics, F4 (khaki) and F5 (blue jean), were illustrated in Table 11. Both F4 and F5 are thicker and rougher than the plain weave fabrics, and therefore they are supposed to have higher ENERGY and CONTRAST values. However, only F4 meets this expectation. The ENERGY and CONTRAST values of F5 are sharing the same numeric ranges with that of F1, F2 and F3 because the darker color (blue) of F5 helps concealing wrinkles. F4 and F5 are thick fabrics, which are difficult to generate hard or sharp wrinkles. Therefore, the HARDNESS values of F4 and F5 are lower than that of F1, F2 and F3, except those at SA-5. This is because the fabrics of F4 and F5 at SA-5 are not completely smooth. Permanent lumps and small-scale wrinkles always exist on the surface. Nonetheless,

HARDNESS is still an important supplemental feature for the SA classification, because it reveals important wrinkling information.

Table 11 Box plots of the four wrinkle features of Fabrics F4 and F5

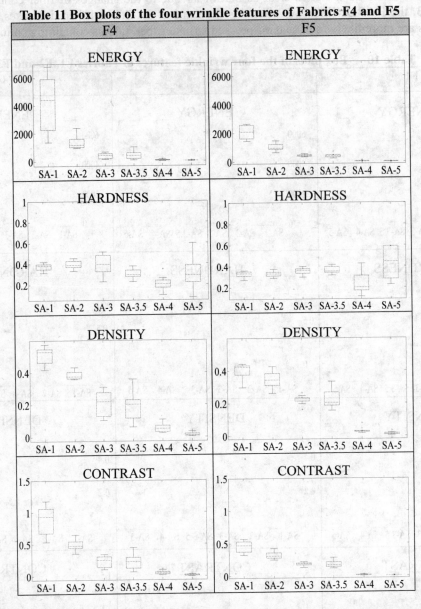

From the plots in Tables 10 and 11, overlapping across the SA grades was observed in all the wrinkle features. Of the five fabric samples, F3 and F5 had more scattered distributions with less overlapping than F1, F2, and F4., and therefore could have higher accuracy in the SA classification because data classification largely depends on the separability of data sets.

The measurements of the four wrinkling features, ENERGY, HARDNESS, DENSITY and CONRAST, of F1, F2, F3, F4 and F5 were scaled into [0,1], and used as inputs of the linear SVM and the RBF SVM classifiers. The five-fold cross-validation (four training sets and one testing set) was used to implement a "grid searching" procedure that can help to achieve unbiased SA classifications. To illustrate the grid searching process, a contour map was plotted for the RBF kernel SVM used to classify F3 images (see Figure 7). Different isolines represent different numeric regions of ($\log_2 C$, $\log_2 \gamma$) associated with different levels of classification accuracy. The highest accuracy (90%) was firstly archived when $(C, \gamma) = (16, 16)$. Later, six more pairs of (C, γ) on the inner isoline also archived 90% accuracy as indicated in Figure 7.

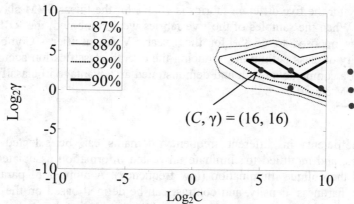

Figure 7 Contour map of grid searching

An one-against-one scheme was employed to construct the multi-class classifiers [17]. There were 15 (6×5/2) combinations for the six SA classes/grades. The scheme constructed binary classifiers to differentiate SA grades. Figure 8 displays an example of using the linear kernel and the RBF kernel SVMs to classify the SA-1 and SA-2 images of F4. For the sake of visualization, only two wrinkle features, HARDNESS and ENERGY, were used for the classification. Eight of ten SA-1 samples and eight of ten SA-2 samples, which are marked as hollow circles and triangles in the figure respectively, were randomly selected to train the classifier. The rest two SA-1 and two SA-2 samples, marked as solid circles and triangles respectively, were used to test the classifier. As shown in the Figure 8a, the linear kernel SVM classifier generated a linear decision line that separates the SA-1 and SA-2 classes with one training sample and one test sample being misclassified. The RBF kernel SVM can form a hyperplane which correctly divides all the training data with only one test sample being misclassified. Therefore, the RBF classifier is more robust in separating overlapped classes, although the linear classifier can provide satisfactory results.

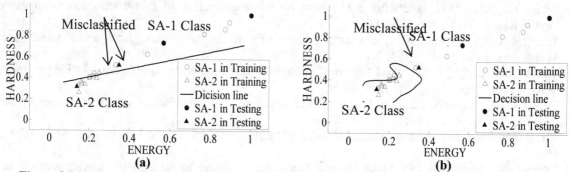

Figure 8 Linear kernel(a) and RBF kernel(b) SVM classifications of SA-1 and SA-2 samples of F4

Table 12 SA classification accuracy of two SVM classifiers

Fabric	F1	F2	F3	F4	F5	Combined
Number of samples	60	60	60	60	60	300
RBF kernel SVM	71.7%	73.3%	91.7%	71.7%	81.7%	73.3%
Linear kernel SVM	71.7%	73.3%	85.0%	70.0%	76.7%	66.3%

Table 12 provides the SA classification accuracy (the percentage of the samples classified in the same SA grades as the experts' visual rating) of the RBF and the linear SVM classifiers for the five selected fabrics. The classifications were performed separately for each fabric to avoid the influence of the weave structures and the fiber contents. Various levels of the classification accuracy ranging from 70.0% to 91.7% were achieved. In addition to the difference in fabric structure, the quality of the visual rating was another important factor that had an impact on the classification accuracy. For F1, F2 and F5, the accuracies of both the linear and RBF classifiers are at comparable levels, but are lower than those of F3 and F5. This is because the wrinkle feature data of F3 and F5 were more separated as mentioned previously. Overall, the average classification accuracy for the five different fabrics is 75.3% for the linear SVM classifier and 78.0% for the RBF SVM classifier. When the samples of the five fabrics were combined, the RBF SVM classifier had an accuracy of 73.3% as opposed to 66.3% for the linear SVM classifier. Combining the data together increased the ambiguity of class definitions, reducing the overall classification accuracy. However, the RBF SVM classifier output a comparable result and demonstrated an advantage in classifying overlapped wrinkle data over the linear SVM classifier.

8 Conclusion

The wavelet coefficients in different frequency domains can be selected to represent wrinkling appearances of fabrics, and modified to eliminate unwanted information associated with weave structures (high frequency) and the oblique illumination (low frequency). A number of parameters that characterize wrinkle directionality, hardness, density, and contrast can be defined based on the modified coefficients as well. It is demonstrated that these parameters are useful in revealing basic wrinkling features (such as oriented or random, hard or soft), and are effective in discriminating fabric images in terms of the AATCC SA grades when used as inputs for the SVM classifiers. The five-fold cross-validation scheme can prevent biased classifications, yielding an accuracy of 78% and 75%, respectively, for the RBF SVM classifier and the linear SVM classifier. The new method provides an efficient way for both the detailed characterization of wrinkling features and the overall grading of smoothness appearance.

References

[1] Technical Manual of Standard Testing Methods, American Association of Textile Chemists and Colorists, Research Triangle Park, NC (1991).
[2] Xu, B., Cuminato, D.f. and Keyes, N.M., Evaluation of Fabric Smoothness Appearance Using a Laser Profilometer, Textile Res. J., 68(12), 900-906,1998.
[3] Su, J. and Xu, B., Fabric wrinkle evaluation using laser triangulation and neural network classifier, Opt. Eng., 38(10), 1688-1693, 1999.
[4] Abidi, N., Hequet, E., Turner, C., and Sari-Sarraf, H., Objective evaluation of durable press treatments and fabric smoothness ratings, Textile Res. J., 75(1), 19-29, 2005.
[5] Kang, T.J., Cho, D.H., and Kim, S.M., New objective evaluation of fabric smoothness appearance, Textile Res. J., 71(5), 446-453, 2001.
[6] Yang, X.B., and Huang, X.B., Evaluating fabric wrinkle degree with a photometric stereo method, Textile Res. J., 73(5), 451–454, 2003.
[7] Yu, W. and Xu, B., A Sub-pixel Stereo Matching Algorithm and Its Applications in Fabric Imaging, Mach Vision Appl., 20(4), 261–270, 2009.
[8] Yu, W., Yao, M. and Xu, B., 3-D Surface Reconstruction and Evaluation of Wrinkled Fabrics by Stereo Imaging, Textile Res. J., 79(1), 36–46, 2009.
[9] Xu, B. and Reed, J.A., Instrumental evaluation of fabric wrinkle recovery, J. Text. Inst., 86(1), 129-135, 1995.
[10] Zaouali, R., Msahli, S., El Abed, B. and Sakli, F., Objective evaluation of multidirectional fabric wrinkling using image analysis, J. Text. Inst., 98(5), 443-451, 2007.
[11] Kang, T.J., Kim, S.C., Sul, I.H., Youn, H.R., and Chuang, K., Fabric surface contrast evaluation using wavelet fractal method: part I: wrinkle smoothness and seam puckers, Textile Res. J., 75(11), 751-760, 2005.
[12] Lin, S. and Xu, B., Evaluating fabric fuzziness using wavelet transforms, Opt. Eng., 39(9), 2387-2391, 2000.
[13] Han, J., Yang, M. and Matsudaira, M., Analysis of wrinkle properties of fabric using wavelet transform, J. Text. Eng., 49(3-4), 54-59, 2003.
[14] Chang, T. and Kuo, C., Texture analysis and classification with tree-structured wavelet transform, IEEE Trans. Image Proc., 2(4), 429–441, 1993.
[15] Chang, C. and Lin, C., LIBSVM: a library for support vector machines, 2001. Software is available at http://www.csie.ntu.edu.tw/~cjlin/libsvm.
[16] MATLAB Wavelet Toolbox, The MathWorks Inc., 2004.
[17] Hsu, C., Chang, C. and Lin, C., A practical guide to support vector classification, http://www.csie.ntu.edu.tw/~cjlin/papers/guide/guide.pdf.

[18] Fan, R., Chen, P. and Lin, C., Working set selection using second order information for training support vector machines, J. Mach. Learn. Res., 6, 1889-1918, 2005.
[19] Vapnik, V., The Nature of Statistical Learning Theory. Springer-Verlag, New York, NY, 1995.
[20] Hsu, C and Lin, C., A comparison of methods for multi-class support vector machines, IEEE Tran. Neural Network, 13(2): 415–425, 2002.

图书在版编目(CIP)数据

第二届先进纺织材料及加工技术国际会议论文集:英文 / 先进纺织材料与制备技术教育部重点实验室等主编. —杭州:浙江大学出版社,2010.9
ISBN 978-7-308-07958-7

Ⅰ.①第… Ⅱ.①先… Ⅲ.①纺织纤维—国际学术会议—文集—英文 Ⅳ.①TS102-53

中国版本图书馆 CIP 数据核字(2010)第 177818 号

第二届先进纺织材料及加工技术国际会议论文集
先进纺织材料与制备技术教育部重点实验室等　主编

责任编辑	诸葛勤
封面设计	雷建军
出版发行	浙江大学出版社 (杭州市天目山路 148 号　邮政编码 310007) (网址:http://www.zjupress.com)
排　　版	杭州中大图文设计有限公司
印　　刷	杭州杭新印务有限公司
开　　本	889mm×1194mm　1/16
印　　张	37.5
字　　数	1888 千
版印次	2010 年 9 月第 1 版　2010 年 9 月第 1 次印刷
书　　号	ISBN 978-7-308-07958-7
定　　价	78.00 元(含光盘)

版权所有　翻印必究　　印装差错　负责调换
浙江大学出版社发行部邮购电话　(0571)88925591